Lecture Notes in Computer Science 5617

Michael J. Smith Gavriel Salvendy (Eds.)

Human Interface and the Management of Information

Designing Information Environments

Symposium on Human Interface 2009
Held as Part of HCI International 2009
San Diego, CA, USA, July 19-24, 2009
Procceedings, Part I

Springer

Volume Editors

Michael J. Smith
University of Wisconsin
Department of Industrial and Systems Engineering
459 Mechanical Engineering Building
1513 University Avenue, Madison, WI 53706, USA
E-mail: mjsmith@engr.wisc.edu

Gavriel Salvendy
Purdue University, Grissom Hall, Room 263
315 North Grant Street, West Lafayette, IN., 47907-2023, USA
E-mail: salvendy@purdue.edu

Library of Congress Control Number: Applied for

CR Subject Classification (1998): H.5, H.3, H.4, K.4.3, D.2

LNCS Sublibrary: SL 3 – Information Systems and Application, incl. Internet/Web and HCI

ISSN 0302-9743
ISBN-10 3-642-02555-2 Springer Berlin Heidelberg New York
ISBN-13 978-3-642-02555-6 Springer Berlin Heidelberg New York

springer.com

Printed in Germany

Typesetting: Camera-ready by author, data conversion by Scientific Publishing Services, Chennai, India
Printed on acid-free paper SPIN: 12704187 06/3180 5 4 3 2 1 0

Foreword

The 13th International Conference on Human–Computer Interaction, HCI International 2009, was held in San Diego, California, USA, July 19–24, 2009, jointly with the Symposium on Human Interface (Japan) 2009, the 8th International Conference on Engineering Psychology and Cognitive Ergonomics, the 5th International Conference on Universal Access in Human-Computer Interaction, the Third International Conference on Virtual and Mixed Reality, the Third International Conference on Internationalization, Design and Global Development, the Third International Conference on Online Communities and Social Computing, the 5th International Conference on Augmented Cognition, the Second International Conference on Digital Human Modeling, and the First International Conference on Human Centered Design.

A total of 4,348 individuals from academia, research institutes, industry and governmental agencies from 73 countries submitted contributions, and 1,425 papers that were judged to be of high scientific quality were included in the program. These papers address the latest research and development efforts and highlight the human aspects of the design and use of computing systems. The papers accepted for presentation thoroughly cover the entire field of human–computer interaction, addressing major advances in knowledge and effective use of computers in a variety of application areas.

This volume, edited by Michael J. Smith and Gavriel Salvendy, contains papers in the thematic area of Human Interface and the Management of Information, addressing the following major topics:

- Supporting Work, Collaboration and Business
- Product Design and Development
- Interacting with Information, Documents and Knowledge
- Novel Devices, Interfaces and Interaction Environments
- User-Centered Design and User Experience in Information Systems

The remaining volumes of the HCI International 2009 proceedings are:

- Volume 1, LNCS 5610, Human–Computer Interaction—New Trends (Part I), edited by Julie A. Jacko
- Volume 2, LNCS 5611, Human–Computer Interaction—Novel Interaction Methods and Techniques (Part II), edited by Julie A. Jacko
- Volume 3, LNCS 5612, Human–Computer Interaction—Ambient, Ubiquitous and Intelligent Interaction (Part III), edited by Julie A. Jacko
- Volume 4, LNCS 5613, Human–Computer Interaction—Interacting in Various Application Domains (Part IV), edited by Julie A. Jacko
- Volume 5, LNCS 5614, Universal Access in Human–Computer Interaction—Addressing Diversity (Part I), edited by Constantine Stephanidis

- Volume 6, LNCS 5615, Universal Access in Human–Computer Interaction—Intelligent and Ubiquitous Interaction Environments (Part II), edited by Constantine Stephanidis
- Volume 7, LNCS 5616, Universal Access in Human–Computer Interaction—Applications and Services (Part III), edited by Constantine Stephanidis
- Volume 9, LNCS 5618, Human Interface and the Management of Information—Information and Interaction (Part II), edited by Gavriel Salvendy and Michael J. Smith
- Volume 10, LNCS 5619, Human–Centered Design, edited by Masaaki Kurosu
- Volume 11, LNCS 5620, Digital Human Modeling, edited by Vincent G. Duffy
- Volume 12, LNCS 5621, Online Communities and Social Computing, edited by A. Ant Ozok and Panayiotis Zaphiris
- Volume 13, LNCS 5622, Virtual and Mixed Reality, edited by Randall Shumaker
- Volume 14, LNCS 5623, Internationalization, Design and Global Development, edited by Nuray Aykin
- Volume 15, LNCS 5624, Ergonomics and Health Aspects of Work with Computers, edited by Ben-Tzion Karsh
- Volume 16, LNAI 5638, The Foundations of Augmented Cognition: Neuroergonomics and Operational Neuroscience, edited by Dylan Schmorrow, Ivy Estabrooke and Marc Grootjen
- Volume 17, LNAI 5639, Engineering Psychology and Cognitive Ergonomics, edited by Don Harris

I would like to thank the Program Chairs and the members of the Program Boards of all thematic areas, listed below, for their contribution to the highest scientific quality and the overall success of HCI International 2009.

Ergonomics and Health Aspects of Work with Computers

Program Chair: Ben-Tzion Karsh

Arne Aarås, Norway
Pascale Carayon, USA
Barbara G.F. Cohen, USA
Wolfgang Friesdorf, Germany
John Gosbee, USA
Martin Helander, Singapore
Ed Israelski, USA
Waldemar Karwowski, USA
Peter Kern, Germany
Danuta Koradecka, Poland
Kari Lindström, Finland
Holger Luczak, Germany
Aura C. Matias, Philippines
Kyung (Ken) Park, Korea
Michelle M. Robertson, USA
Michelle L. Rogers, USA
Steven L. Sauter, USA
Dominique L. Scapin, France
Naomi Swanson, USA
Peter Vink, The Netherlands
John Wilson, UK
Teresa Zayas-Cabán, USA

Human Interface and the Management of Information

Program Chair: Michael J. Smith

Gunilla Bradley, Sweden
Hans-Jörg Bullinger, Germany
Alan Chan, Hong Kong
Klaus-Peter Fähnrich, Germany
Michitaka Hirose, Japan
Jhilmil Jain, USA
Yasufumi Kume, Japan
Mark Lehto, USA
Fiona Fui-Hoon Nah, USA
Shogo Nishida, Japan
Robert Proctor, USA
Youngho Rhee, Korea
Anxo Cereijo Roibás, UK
Katsunori Shimohara, Japan
Dieter Spath, Germany
Tsutomu Tabe, Japan
Alvaro D. Taveira, USA
Kim-Phuong L. Vu, USA
Tomio Watanabe, Japan
Sakae Yamamoto, Japan
Hidekazu Yoshikawa, Japan
Li Zheng, P.R. China
Bernhard Zimolong, Germany

Human–Computer Interaction

Program Chair: Julie A. Jacko

Sebastiano Bagnara, Italy
Sherry Y. Chen, UK
Marvin J. Dainoff, USA
Jianming Dong, USA
John Eklund, Australia
Xiaowen Fang, USA
Ayse Gurses, USA
Vicki L. Hanson, UK
Sheue-Ling Hwang, Taiwan
Wonil Hwang, Korea
Yong Gu Ji, Korea
Steven Landry, USA
Gitte Lindgaard, Canada
Chen Ling, USA
Yan Liu, USA
Chang S. Nam, USA
Celestine A. Ntuen, USA
Philippe Palanque, France
P.L. Patrick Rau, P.R. China
Ling Rothrock, USA
Guangfeng Song, USA
Steffen Staab, Germany
Wan Chul Yoon, Korea
Wenli Zhu, P.R. China

Engineering Psychology and Cognitive Ergonomics

Program Chair: Don Harris

Guy A. Boy, USA
John Huddlestone, UK
Kenji Itoh, Japan
Hung-Sying Jing, Taiwan
Ron Laughery, USA
Wen-Chin Li, Taiwan
James T. Luxhøj, USA
Nicolas Marmaras, Greece
Sundaram Narayanan, USA
Mark A. Neerincx, The Netherlands
Jan M. Noyes, UK
Kjell Ohlsson, Sweden
Axel Schulte, Germany
Sarah C. Sharples, UK

Neville A. Stanton, UK
Xianghong Sun, P.R. China
Andrew Thatcher, South Africa
Matthew J.W. Thomas, Australia
Mark Young, UK

Universal Access in Human–Computer Interaction

Program Chair: Constantine Stephanidis

Julio Abascal, Spain
Ray Adams, UK
Elisabeth André, Germany
Margherita Antona, Greece
Chieko Asakawa, Japan
Christian Bühler, Germany
Noelle Carbonell, France
Jerzy Charytonowicz, Poland
Pier Luigi Emiliani, Italy
Michael Fairhurst, UK
Dimitris Grammenos, Greece
Andreas Holzinger, Austria
Arthur I. Karshmer, USA
Simeon Keates, Denmark
Georgios Kouroupetroglou, Greece
Sri Kurniawan, USA
Patrick M. Langdon, UK
Seongil Lee, Korea
Zhengjie Liu, P.R. China
Klaus Miesenberger, Austria
Helen Petrie, UK
Michael Pieper, Germany
Anthony Savidis, Greece
Andrew Sears, USA
Christian Stary, Austria
Hirotada Ueda, Japan
Jean Vanderdonckt, Belgium
Gregg C. Vanderheiden, USA
Gerhard Weber, Germany
Harald Weber, Germany
Toshiki Yamaoka, Japan
Panayiotis Zaphiris, UK

Virtual and Mixed Reality

Program Chair: Randall Shumaker

Pat Banerjee, USA
Mark Billinghurst, New Zealand
Charles E. Hughes, USA
David Kaber, USA
Hirokazu Kato, Japan
Robert S. Kennedy, USA
Young J. Kim, Korea
Ben Lawson, USA
Gordon M. Mair, UK
Miguel A. Otaduy, Switzerland
David Pratt, UK
Albert "Skip" Rizzo, USA
Lawrence Rosenblum, USA
Dieter Schmalstieg, Austria
Dylan Schmorrow, USA
Mark Wiederhold, USA

Internationalization, Design and Global Development

Program Chair: Nuray Aykin

Michael L. Best, USA
Ram Bishu, USA
Alan Chan, Hong Kong
Andy M. Dearden, UK
Susan M. Dray, USA
Vanessa Evers, The Netherlands
Paul Fu, USA
Emilie Gould, USA

Sung H. Han, Korea
Veikko Ikonen, Finland
Esin Kiris, USA
Masaaki Kurosu, Japan
Apala Lahiri Chavan, USA
James R. Lewis, USA
Ann Light, UK
James J.W. Lin, USA
Rungtai Lin, Taiwan
Zhengjie Liu, P.R. China
Aaron Marcus, USA
Allen E. Milewski, USA
Elizabeth D. Mynatt, USA
Oguzhan Ozcan, Turkey
Girish Prabhu, India
Kerstin Röse, Germany
Eunice Ratna Sari, Indonesia
Supriya Singh, Australia
Christian Sturm, Spain
Adi Tedjasaputra, Singapore
Kentaro Toyama, India
Alvin W. Yeo, Malaysia
Chen Zhao, P.R. China
Wei Zhou, P.R. China

Online Communities and Social Computing

Program Chairs: A. Ant Ozok, Panayiotis Zaphiris

Chadia N. Abras, USA
Chee Siang Ang, UK
Amy Bruckman, USA
Peter Day, UK
Fiorella De Cindio, Italy
Michael Gurstein, Canada
Tom Horan, USA
Anita Komlodi, USA
Piet A.M. Kommers, The Netherlands
Jonathan Lazar, USA
Stefanie Lindstaedt, Austria
Gabriele Meiselwitz, USA
Hideyuki Nakanishi, Japan
Anthony F. Norcio, USA
Jennifer Preece, USA
Elaine M. Raybourn, USA
Douglas Schuler, USA
Gilson Schwartz, Brazil
Sergei Stafeev, Russia
Charalambos Vrasidas, Cyprus
Cheng-Yen Wang, Taiwan

Augmented Cognition

Program Chair: Dylan D. Schmorrow

Andy Bellenkes, USA
Andrew Belyavin, UK
Joseph Cohn, USA
Martha E. Crosby, USA
Tjerk de Greef, The Netherlands
Blair Dickson, UK
Traci Downs, USA
Julie Drexler, USA
Ivy Estabrooke, USA
Cali Fidopiastis, USA
Chris Forsythe, USA
Wai Tat Fu, USA
Henry Girolamo, USA
Marc Grootjen, The Netherlands
Taro Kanno, Japan
Wilhelm E. Kincses, Germany
David Kobus, USA
Santosh Mathan, USA
Rob Matthews, Australia
Dennis McBride, USA
Robert McCann, USA
Jeff Morrison, USA
Eric Muth, USA
Mark A. Neerincx, The Netherlands
Denise Nicholson, USA
Glenn Osga, USA

Dennis Proffitt, USA
Leah Reeves, USA
Mike Russo, USA
Kay Stanney, USA
Roy Stripling, USA
Mike Swetnam, USA
Rob Taylor, UK
Maria L.Thomas, USA
Peter-Paul van Maanen, The Netherlands
Karl van Orden, USA
Roman Vilimek, Germany
Glenn Wilson, USA
Thorsten Zander, Germany

Digital Human Modeling

Program Chair: Vincent G. Duffy

Karim Abdel-Malek, USA
Thomas J. Armstrong, USA
Norm Badler, USA
Kathryn Cormican, Ireland
Afzal Godil, USA
Ravindra Goonetilleke, Hong Kong
Anand Gramopadhye, USA
Sung H. Han, Korea
Lars Hanson, Sweden
Pheng Ann Heng, Hong Kong
Tianzi Jiang, P.R. China
Kang Li, USA
Zhizhong Li, P.R. China
Timo J. Määttä, Finland
Woojin Park, USA
Matthew Parkinson, USA
Jim Potvin, Canada
Rajesh Subramanian, USA
Xuguang Wang, France
John F. Wiechel, USA
Jingzhou (James) Yang, USA
Xiu-gan Yuan, P.R. China

Human–Centered Design

Program Chair: Masaaki Kurosu

Gerhard Fischer, USA
Tom Gross, Germany
Naotake Hirasawa, Japan
Yasuhiro Horibe, Japan
Minna Isomursu, Finland
Mitsuhiko Karashima, Japan
Tadashi Kobayashi, Japan
Kun-Pyo Lee, Korea
Loïc Martínez-Normand, Spain
Dominique L. Scapin, France
Haruhiko Urokohara, Japan
Gerrit C. van der Veer, The Netherlands
Kazuhiko Yamazaki, Japan

In addition to the members of the Program Boards above, I also wish to thank the following volunteer external reviewers: Gavin Lew from the USA, Daniel Su from the UK, and Ilia Adami, Ioannis Basdekis, Yannis Georgalis, Panagiotis Karampelas, Iosif Klironomos, Alexandros Mourouzis, and Stavroula Ntoa from Greece.

This conference could not have been possible without the continuous support and advice of the Conference Scientific Advisor, Gavriel Salvendy, as well as the dedicated work and outstanding efforts of the Communications Chair and Editor of HCI International News, Abbas Moallem.

I would also like to thank for their contribution toward the organization of the HCI International 2009 conference the members of the Human–Computer Interaction Laboratory of ICS-FORTH, and in particular Margherita Antona, George Paparoulis, Maria Pitsoulaki, Stavroula Ntoa, and Maria Bouhli.

Constantine Stephanidis

HCI International 2011

The 14th International Conference on Human–Computer Interaction, HCI International 2011, will be held jointly with the affiliated conferences in the summer of 2011. It will cover a broad spectrum of themes related to human–computer interaction, including theoretical issues, methods, tools, processes and case studies in HCI design, as well as novel interaction techniques, interfaces and applications. The proceedings will be published by Springer. More information about the topics, as well as the venue and dates of the conference, will be announced through the HCI International Conference series website: http://www.hci-international.org/

General Chair
Professor Constantine Stephanidis
University of Crete and ICS-FORTH
Heraklion, Crete, Greece
Email: cs@ics.forth.gr

Table of Contents

Part I: Supporting Work, Collaboration and Business

Crafting Contemporary Enterprise Application User Experiences 3
Jeremy Ashley and Misha W. Vaughan

A Management System for Visual Communication Design at Film Studio 13
Hong-Sheng Chen

A Mobile User Interface for Business Knowledge Management (BKM) 22
Danco Davcev, Marjan Arsic, and Dalibor Ilievski

Designing a Peer Reviewing Tool on Lecture Video with Handwritten Annotation 31
Hironori Egi, Shotaro Houri, Yukari Kato, Tatsuya Terada, Wataru Tsukahara, and Masaki Nakagawa

Usability Analyses of CRM Systems in Call Centers: The Data Mining Approach 40
Ádám Horváth, László Laufer, and András Kangyal

Urgent Collaboration Service for Inclusive Use 49
Naotsune Hosono, Hiroyuki Miki, Michio Suzuki, and Yukata Tomita

Improvement of Member's Concentration during Discussion 59
Junko Ichino, Kazuhiro Takeuchi, and Hitoshi Isahara

Representation Method for Engineering Perspective 69
Gaku Ishii and Masatake Sakuma

A Skill Transfer Method for Manual Machine Tool Operation Utilizing Cutting Sound 77
Takashi Kawashimo, Noriaki Sato, Daisuke Doyo, Michiko Anse, and Tsutomu Tabe

A Business Performance Measurement Model for Mobile User Interface 87
Min-Jeong Kim and Jonghun Park

Web Service Systems for Cooperative Work Support in Knowledge Creation Processes 94
Hiroyuki Kojima, Kentaro Funaki, and Tsuyoshi Inoue

Designing International Enterprise Software 104
Janaki Kumar, Dan Rosenberg, Paul Hofmann, and Michael Arent

Proactive Fault Detection Schema for Enterprise Information System Using Statistical Process Control 113
ChiHoon Lee, Doohyung Lee, Jahwan Koo, and Jinwook Chung

A Human Factors Model for Enterprise Resources Planning System Implementation 123
Chiuhsiang Joe Lin, Chih-Wei Yang, Shi-Bin Lin, and Shiau-Feng Lin

A Study on Effective Methods of Visual Inspection for Reused-Parts by Inspectors 131
Toshiyuki Matsumoto, Hideki Shirai, and Keisuke Shida

Practical Use of Task Models for Building and Modeling Operations Chart in the Industrial Production 140
Tomasz Mistrzyk and Alexander Redenius

A Model for Measurement and Analysis of the Workflow Processes 149
Pavel Ocenasek and Miroslav Sveda

Tactical Access to Complex Technology through Interactive Communication (TACTIC) 154
Alexei Samoylov, Christopher Franklin, Susan Harkness Regli, Patrice Tremoulet, Kathleen Stibler, and Peter Gerken

A Method for Evaluating the Relationship among Four Perspectives of the Balanced Scorecard 163
Yumiko Taguchi, Takumi Kaneko, and Tsutomu Tabe

Exploring Employee Perspectives on Information Privacy and Security in the Mobile Environment 171
Peter Tarasewich, Benjamin Ngugi, and Mansi Sanghavi

Organization Diagnosis Tools Based on Social Network Analysis 181
Takanori Ugai and Kouji Aoyama

Designing for the Next Generation: Generation-Y Expectations 190
Patañjali Venkatacharya, Sean Rice, and Lulit Bezuayehu

A Study on a Method of Support for Improving the Motivation of Employees 197
Daisuke Yatsuzuka, Yumiko Taguchi, and Tsutomu Tabe

Research on the Supplier Promise Management Based on the Lean Six Sigma 205
Wang Ying-chun, Wang Shi-jie, Liu Yong-xian, and Zhang Xinmin

Part II: Product Design and Development

Supporting Scenario-Based Product Design and Its Adapters: An Informal Framework for Scenario Creation and Use ... 217
Irene Anggreeni and Mascha van der Voort

How to Learn from Intelligent Products; The Structuring of Incoherent Field Feedback Data in Two Case Studies ... 227
Renate de Bruin, Yuan Lu, and Aarnout Brombacher

Successful User Experience in an Agile Enterprise Environment ... 233
Melissa Federoff and Catherine Courage

From Research to Product: Integrating Treemaps into Enterprise Software ... 243
Joseph H. Goldberg, Jonathan I. Helfman, and John Beresniewicz

How Product Differentiation Affects Online Shopping ... 253
A. Walkyria Goode

Design and Evaluation of the Customized Product Color Combination Interface Based on Scenario Experience ... 263
Ying-Jye Lee, Cheih-Ying Chen, and Fong-Gong Wu

Distribution of Human-Machine Interfaces in System-of-Systems Engineering ... 271
Sandro Leuchter and Dirk Mühlenberg

Determination of Inspection Threshold Using Perceptive Sensitivities of Experienced Panel ... 279
Masao Nakagawa, Hidetoshi Nakayasu, and Tetsuya Miyoshi

Reversing the Simon Effect with Prior Practice of Noncorresponding Location Words ... 287
Andrea Rottermann and Kim-Phuong L. Vu

Object-Oriented Interactive Processes in Decentralized Production Systems ... 296
Thomas Schlegel

A Retrospective and Prospective View of Information Technology Professionals' Use of Tools: Maturing the User Experience ... 306
Candace Soderston

Customer Boards as Vehicles of Change in Enterprise Software User Experience ... 316
Anna M. Wichansky

Fashion Support from Clothes with Characteristics ... 323
Yuri Yonezawa and Yoshio Nakatani

Part III: Interacting with Information, Documents and Knowledge

Exploring History with Narrative Timelines 333
Robert B. Allen and Sumanth Nalluru

Facilitating Search through Visualized Results 339
Takahisa Ando, Satoko Shiga, Tomoya Iwakura, and Seishi Okamoto

The 7 Basic Functions of a Digital Library - Analysis of Focus Groups about the Usefulness of a Thematic Digital Library on the History of European Integration 345
Eric Brangier, Jérôme Dinet, and Laurent Eilrich

Exploiting Browsing History for Exploratory Search 355
Wei-Lin Chen and Wei-Guang Teng

A Proposal of Awareness Services for the Construction of Quality Community Knowledge Supported by the Knowledge Management System KnowCat 365
Ruth Cobos, Ivan Dario Claros Gómez, and Jaime Moreno-Llorena

Electronic Document Tracking System (EDTS): A Prototype 375
Rochin Demong, Lailatul Faizah Abu Hassan, Tuan Badrol Hisham Tuan Besar, and Zulhaimi Zulkifli

Adaptive Visual Clustering for Mixed-Initiative Information Structuring 384
Hakan Duman, Alex Healing, and Robert Ghanea-Hercock

An Empirical Analysis of Personal Digital Document Structures 394
Sarah Henderson and Ananth Srinivasan

Search Mathematical Formulas by Mathematical Formulas 404
Yoshinori Hijikata, Hideki Hashimoto, and Shogo Nishida

Metadata-Based Reminder Classification in Reminiscence Engineering 412
Masashi Ishibashi and Yoshio Nakatani

InfoScape: A Browser for User Behavior-Based Information Retrieval System 419
Masaaki Kawata and Katsuhiko Ogawa

The Roles of Profession and Gender in Some PIM Tasks 429
Guangfeng Song and Chen Ling

Efficient Text Classification Using Best Feature Selection and Combination of Methods 437
Mettu Srinivas, K. Pujari Supreethi, E.V. Prasad, and S. Anitha Kumari

Designing Sticky Knowledge-Network SNS for Japanese Science Teachers 447
Yoshihisa Wada, Masayuki Sakoda, Hiroshi Tsuji, Yuuki Aoki, and Kazuhisa Seta

Part IV: Novel Devices, Interfaces and Interaction Environments

A Proposal of EMG-Based Training Support System for Basketball Dribbling 459
Seimei Abe, Takayuki Nozawa, and Toshiyuki Kondo

UbiSOA Dashboard: Integrating the Physical and Digital Domains through Mashups 466
Edgardo Avilés-López and J. Antonio García-Macías

Internal Aspects of the Relationship between Pressing Force and Training Difficulty 475
Daisuke Doyo

A Tangible Mixed Reality Interface for the AMI Automated Meeting Assistant 485
Jochen Ehnes

Brain Activities Supporting Finger Operations, Analyzed by Neuro-NIRS 495
Miki Fuchigami, Akira Okada, and Hiroshi Tamura

Effects of Practice with Foot- and Hand-Operated Secondary Input Devices on Performance of a Word-Processing Task 505
Fredrick P. Garcia and Kim-Phuong L. Vu

Manipulation with Fingers in a 3-D Physical Space 515
Yuki Kanai, Makoto Oka, and Hirohiko Mori

Design of Wearable Interface Considering Touch Communications 524
Mariko Kato and Naoki Saiwaki

f-MRI Study of Brain Activation in Tactile Feeling 534
Yuka Koda, Maki Taniguchi, Yukiyasu Kamitani, and Naoki Saiwaki

Using 3D Touch Interaction for a Multimodal Zoomable User Interface 543
Florian Laquai, Markus Ablassmeier, Tony Poitschke, and Gerhard Rigoll

A Framework for Fairness Guaranteed Multi-streaming Transport Protocol 553
Wonhyuk Lee, Hyuncheol Kim, Doowon Seo, and Jinwook Chung

A Study on Computing Resource Partition for Increasing Efficiency of Grid System 563
Changsun Lim, Seongjin Ahn, and Jinwook Chung

Richbiff: E-Mail Message Notification with Richer Clues 570
Mitsuru Minakuchi and Hisashi Miyamori

An Approach for the Design of Secure Communication in Embedded Systems 579
Pavel Ocenasek

Towards Security Issues in ZigBee Architecture 587
Pavel Ocenasek

A Study on Color Conversion for Color Deficient People to Identify Color 594
Makoto Oka, Naoki Ozawa, Hirohiko Mori, and Akito Sakurai

Widgets for Faceted Browsing 601
Jan Polowinski

Applications of Visible Light Path Laser Projector 611
Nobuchika Sakata, Shu Okamoto, and Shogo Nishida

Virtual Convex Polygon Based Hole Boundary Detection and Time Delay Based Hole Detour Scheme in WSNs 619
Inyoung Shin, Ngoc Duy Pham, and Hyunseung Choo

iFeel_IM! Emotion Enhancing Garment for Communication in Affect Sensitive Instant Messenger 628
Dzmitry Tsetserukou, Alena Neviarouskaya, Helmut Prendinger, Naoki Kawakami, Mitsuru Ishizuka, and Susumu Tachi

An Interactive System Based on Semantic Graphs 638
Johann Vandromme, Samuel Degrande, Patricia Plénacoste, and Christophe Chaillou

Evaluation for Adjustment Method of Vehicle's Location by Recognizing Crosswalks 648
Yoshihisa Yamaguchi, Takashi Nakagawa, Hirokazu Kato, and Shogo Nishida

Part V: User-Centered Design and User Experience in Information Systems

Improving the User Experience of Our Own UX Ideas 659
Joel Eden

Construction of the Data Save System of the Cusp Surface Analysis Using Web Application 665
Yasufumi Kume and Jee Seob Seo

Data Pattern for Allocating User Experience Meta-Data to User Experience Research Data 675
Li Li, Hong Ji, Xuejiao Chen, and Xiaowei Yuan

Transcending Human-Centered Design by Service Sciences 685
Hiroyuki Miki, Naotsune Hosono, and Sakae Yamamoto

Effects of a Mnemonic Technique on Subsequent Recall of Assigned and Self-generated Passwords 693
Deborah L. Nelson and Kim-Phuong L. Vu

Control of Personal Tempo to Support Individual Action and Cognition 702
Naoya Nobutani and Yoshio Nakatani

The Effects of Practice and Speed Stress with Different Stimulus-Response Mappings 709
Kim-Phuong L. Vu, Audrey Rabas, and Richard Roberson

Application of Population Stereotypes to Computerized Tasks 718
Jeff Wiebe and Kim-Phuong L. Vu

Designing Transportation Services Based on HCD 726
Kiko Yamada-Kawai, Naotake Hirasawa, Shinya Ogata, and Shou Ohtsu

Author Index 737

Part I

Supporting Work, Collaboration and Business

Crafting Contemporary Enterprise Application User Experiences

Jeremy Ashley and Misha W. Vaughan

Oracle USA
500 Oracle Parkway, MS 2op10
Redwood Shores, CA 94065
USA
jeremy.ashley@oracle.com, mvaughan@acm.org

Abstract. This paper outlines the field research program defined by an enterprise application user experience organization designed to collect data from globally distributed corporations. The goal of the research program was to drive the next generation user experience of enterprise applications. This paper discusses the top five lessons learned and their design solutions.

Keywords: enterprise applications, user experience, site visits, field studies.

1 Historical Perspective

Ten to 15 years ago, enterprise software applications were designed as siloed, transactional, front ends for databases. These applications were used by experts who received weeks of training and who conducted their work on big, expensive business machines. At the end of the day, what mattered most was the ability to complete a particular business task.

These applications were products of the development view of the time. Because not all end users were database programmers, applications were written to make working with the data a little easier. For the most part, these applications were direct expressions of the database, only with a graphical user interface rather than a command-line user interface. Each database—whether it was sales, service, marketing, payables, inventory, or receivables—had its own user interface.

Around 2000, the Internet began pushing those same applications to Web-based experiences. Developers began crafting HTML interfaces for those same applications [1]. Web-based businesses designed for consumers, such as Amazon.com, were also emerging. These Web-based businesses were changing the competitive landscape. As executives and end users alike saw how straightforward the user experience was, they placed the same expectation on these emerging applications [2].

Today, users' expectations have made another quantum leap. Now, when users go home at night, they procure on Ebay and invoice on Citibank. These experiences are straightforward and transparent. In addition, new technologies have emerged, such as Asynchronous JavaScript and XML (AJAX), which enable designers to build richer and more interactive experiences. Finally, new methods to document business processes have been developed, such as business process modeling (BPM), a method

M.J. Smith and G. Salvendy (Eds.): Human Interface, Part I, HCII 2009, LNCS 5617, pp. 3–12, 2009.

that enables us to document and design business best practices. The convergence of these elements means that we can deliver a user experience that enables users to concentrate on their jobs, not the software.

This paper describes a plan of user research for the field embarked upon by an enterprise application user experience organization to meet these emerging needs. Some of the challenges this plan had to address included: coordinating this at a group level, creating consistency across teams, and conducting the research globally. The paper discusses methodologies, deliverables, and the top five lessons learned and their related design solutions.

2 User Research

Starting in 2004, the Oracle Applications User Experience organization embarked on a substantial field study program to inform the design of the next generation of user experiences for enterprise applications. The field study program spanned multiple years (2004-2008), domains, and countries. The goal of the field research was to understand:

- The current state of end user experiences with a wide variety of enterprise applications
- What was successful about end user experiences that should be carried forward
- The pain points of existing enterprise software
- The workarounds that end users created when software did not meet their needs

2.1 Scope

To better understand the broad range of problems, the research was divided into several domains: human capital management, customer relationship management, financials, supply chain management, procurement, corporate performance monitoring, enabling technologies, mobile, and business intelligence. In each of these domains, companies were selected for observation based on their mix of enterprise applications on site and on their strategic relationship with our organization. This relationship proved to be necessary in order to negotiate the complex legal agreements required for each site visit. In total, approximately 1,300 hours of site visit data were collected from over 600 end users across 8 countries.

2.2 Methodology

Because we had such a large organization (approximately 150 people), we chose to create a series of best practice guidelines for designing, conducting, analyzing, and reporting on our field studies. This section covers the general recommendations for conducting the research.

Preparation. To prepare for the site visits, each Oracle user experience team was responsible for: identifying the target corporation profile, understanding prior research, obtaining legal access, scoping the functional areas within a domain, and defining the target user profiles. We will detail a specific example of these steps for the Financials domain.

The lead usability engineer (UE) for a given domain, working with an internal strategy organization, first developed a corporation profile for each domain: corporation size, location, industry, geographies, and installed applications. In financials, the team identified the following criteria for the corporation profile:

- Revenue: annual revenue of one billion or more
- Industry: public sector, financial services, and communications
- Geographic area: North America, Europe, India, and China
- Installed enterprise applications: JD Edwards, Oracle E-Business Suite, and PeopleSoft

While completing the corporation profile, the lead UE also reviewed any publicly available or internally created user and market research data about the domain. This analysis of existing data enabled the UE to work with other members of Oracle– particularly in the strategy organization– to define the functional areas within a given domain and the user profiles of interest. For example, again in Financials, the team narrowed its interest to the following functional areas: expenses, general ledger, accounts payable, accounts receivable, and asset lifecycle management.

Table 1. Sample user role description: accounts payable manager

Job Titles	Details	Education	Computer Skills
• Accounts payable manager • Disbursements director or manager • Manager of disbursements	In charge of the whole accounts payable process: • Handling regular and strategic reporting, roll up data, and volume reporting • Dealing with special accounting issues, such as checks and chart of accounts values • Enforcing fiscal discipline	• Four-year college accounting degree • Sometimes a CPA, depending on the size of the corporation	• Well-versed in Excel • Functional user • Well-versed in reporting tools • Not necessarily an accounts payable expert end user—more focused on reports and metrics

The team then developed several user profiles for each functional area. For example, for accounts payable, the team identified the following user roles: accounts payable manager, expenses supervisor, supplier invoices supervisor, payments supervisor, and processor. For each user role, the team created a detailed description of expected characteristics. For example, the above table details a description of an accounts payable manager (Table 1).

Starting Discussions with the Customer. The UE then embarked on the complex and lengthy task of contacting corporations with which our organization had a pre-existing relationship and fit the target profile. Once a corporation expressed initial interest, the UE delivered a presentation to educate the customer about what would be involved. If the customer agreed to participate, it identified its own internal stakeholders and the UE would then present a specially created legal agreement that protected both parties, yet still allowed our organization to leverage the findings in the development of software user interfaces.

Next, the UE scheduled a follow up conference call and presentation for all of the stakeholders – in which the UE would once again describe for the customer what was involved in a site visit. During this same call, the UE would determine the following things about the corporation: additional stakeholders, the information technology environment, the relevant job titles, and the top user-experience issues. After several follow-up phone calls, the UE identified an executive sponsor and a logistical point of contact for the actual visits.

The UE would then begin recruiting the actual participants. The UE typically contacted managers who oversaw the needed individual contributors and gave the same presentation. The potential observation candidates were then screened to ensure that they met our user profiles.

For each site visit, which lasted approximately a week, the UE then created a detailed hour-by-hour agenda. Maps of each location and a contact list of individuals at the customer locations were all prepared in advance.

On Site. For each site visit, a team of three to seven individuals from our organization were deployed, including a UE. On day one, the UE presented a basic "Who we are and why we are here" message. This presentation was often attended by a wide variety of individuals from the corporation who were curious about the site-visit process. At the end of this meeting, a focus-group-style conversation was often held in order for customers to do the one thing most on their minds: vent their frustrations about enterprise software.

The Oracle team would then begin the one-on-one observations. Each observation included three basic components:

- Administering a brief questionnaire to understand whether our assumptions about demographics and job experience were correct.
- Observing the end users for 1 to 4 hours, depending on the kind of work performed, seeking to discover exceptions and unexpected findings
- Clarifying observations, asking to view artifacts, recording particpants photographically (if within scope of the legal agreement), and pursuing discoveries.
- Over the next 2 to 5 days, the following events and artifacts would emerge:
- The Oracle user experience team would observe each user profile for 1 to 4 hours.
- The Oracle user experience team would develop a detailed business process diagram and information technology map based on these observations and on the discussions with information technology managers at the corporation.

After the last day of observation, the Oracle team would present their preliminary findings on site to the users. This presentation was illuminating to the customer, and motivating to the Oracle team to begin the data analysis.

Data Analysis and Deliverables. Because site visit data is rich and produces a large set of artifacts, e.g., diagrams, interview notes, photographs, and transcripts, we chose to use affinity diagramming as a method to distill down all of this data to a manageable set. Affinity diagramming is an analytic technique used particularly in businesses to synthesize trends and patterns. For more information, please see [3].

After the affinity diagramming analysis, we asked each team to generate a particular set of deliverables that further refined our results. We asked each user experience team to construct:

- A "day-in-the-life" scenario of their primary user roles
- A presentation comparing the Oracle team's assumptions with the observed realities about the user roles and the work environment
- The key lessons learned about each user role: what the user roles wanted and needed from their enterprise applications
- A presentation documenting the top five features that our organization could develop as a means of improving enterprise application user experience
- A detailed presentation documenting the results of the questionnaire
- A chart illustrating the corporation's organizational structure
- Task flow diagrams for users based on the one-on-one observations

3 Top Five Lessons Learned and the Design Solutions

Based on our field research, the number one problem facing customers of enterprise applications today is end-user productivity. By conducting site visits and observational interviews with customers, we identified five problem areas impacting productivity and then came up with design solutions for how we could improve the user experience in these areas.

3.1 End Users Spend Too Much Time Navigating between Applications, Which Translates into a Loss in Productivity

When it comes to Oracle applications (e.g., E-Business Suite, PeopleSoft, JD Edwards, and Siebel), end users of these applications have been exposed to a variety of user interface paradigms. Navigating among these various interfaces takes time and results in lost productivity. In our next generation of designs, users are exposed to only one user interface paradigm: the user interface shell (Fig 1).

Heavily informed by PeopleSoft's user research and strategic design work, the user interface shell provides a single point of access to all of a user's applications, analytics, collaboration, search, and even help. Consistent placement, presentation, and behavior of common user interface elements in the user interface shell results in a cohesive user experience that is easily learned and used by a diverse range of workers, thus saving time and money spent on training and support.

The navigation area spanning the top of the user interface shell provides the tools that are available across the suite of applications: menu navigation, search, recent items, and favorites. The region on the left contains quick access to tasks, a targeted search, and reports. The large transaction area in the center is the focus of the user's work and contains all the information and actions required to accomplish his or her tasks. The region on the right contains tools (such as analytics and help) that are driven by the work being performed in the center area.

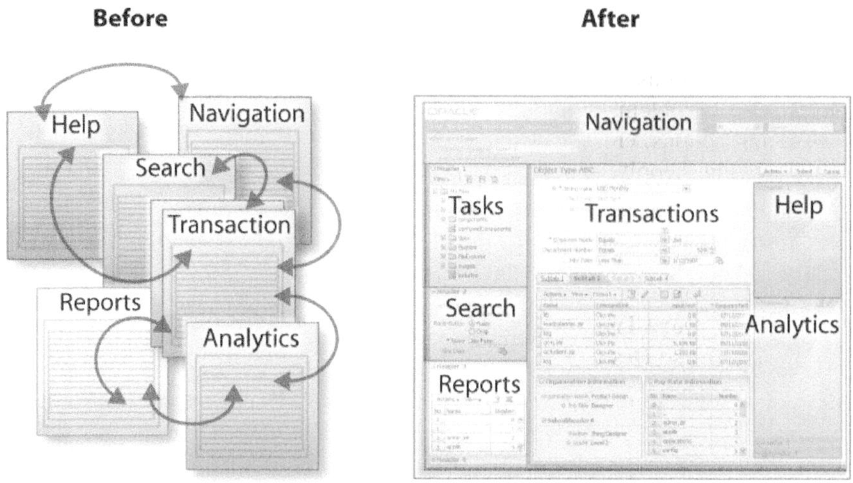

Fig. 1. How the user interface shell unifies the user experience

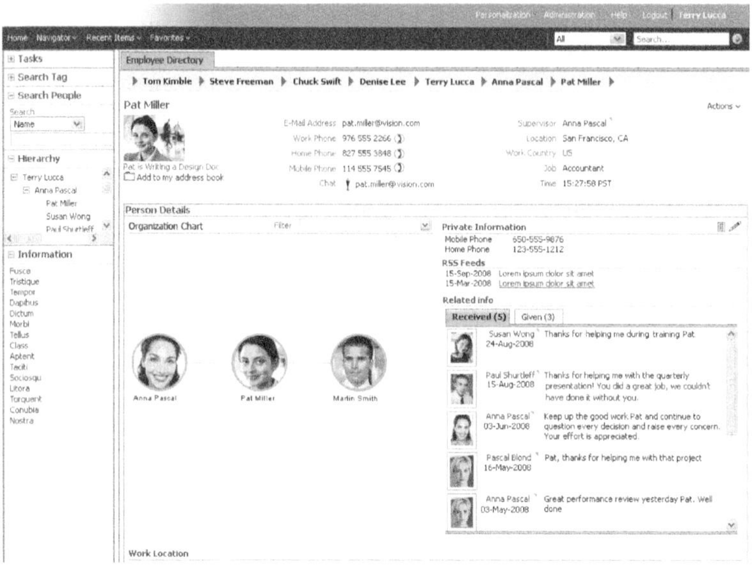

Fig. 2. Example Work Area (with right side pane collapsed) for a manager providing commonly needed information and actions together (e.g., Tasks, Search Tags, Hierarchy, Information, Employee Directory, and Person Details)

3.2 End Users Have Clear Ideas of Business Best Practices, and They Create Work Arounds, Short Cuts, and Job Aids to Capture the Information That Is Not Reflected in Their Applications

A deeper level of analysis of our customer data revealed that end users lack software experiences that map efficiently to business processes. End users used sticky notes, calendar reminders, job aids, and other documents outside the system to help them remember procedures, perform "best practices," monitor tasks, and find other information and actions in the system.

To tackle this problem, Oracle created work areas (Fig 2), where users can perform their core transactions: creating and reviewing invoices, closing books, entering time cards, managing supply chains, and so on. Work areas are context-specific assemblies of tasks, reports, analytics, and other information that help users complete core enterprise transactions. Work areas are designed from top to bottom to increase end user productivity.

3.3 End Users' Core Transactions Happen in a Larger Context—A Context Not Reflected in the Design of Current Applications

One of the chief complaints of end users from bioscience, manufacturing, test and measurement, high tech, public sector, and consumer goods was the need to communicate with peers on an as–needed basis, or rather the importance of supporting *ad hoc* collaboration. Who better for us to learn from than sales and service representatives—the masters of constant communication. By talking with sales representatives, sales managers, help desk personnel, technical support representatives in high tech, telecommunications, pharmaceuticals, manufacturing, and retail, we guided the new user experience to a core principle of keeping users in the context of the work that they are performing.

With this design goal in mind, we designed activities, notes, interactions, and contextual actions (Fig 3). These elements provide end users with the ability to work collaboratively on projects, invoices, and other such tasks and to leverage elements of social networking — all in the context of their existing transactions and using an interface that is familiar and easy to use.

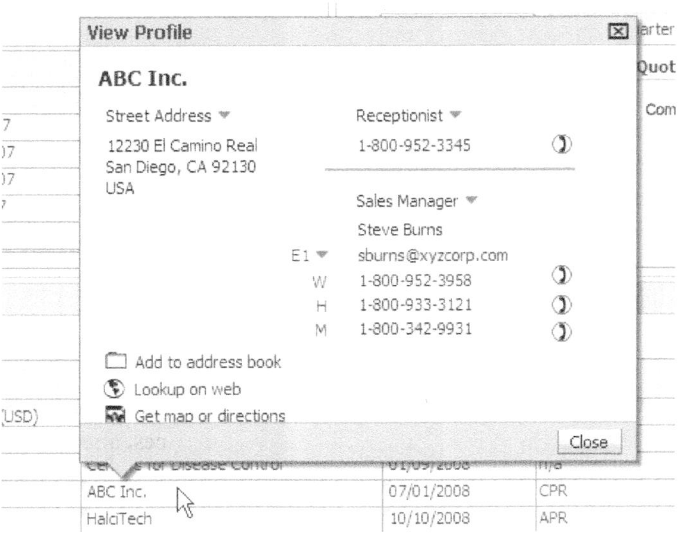

Fig. 3. Contextual actions related to a company contact

3.4 End Users Want to Get Work Done Faster with Rich, Interactive, Desktop-Like User Experiences

End users often needed access to information in an enterprise application but wanted to avoid the hassle and lost time of logging in to the full enterprise application. Enter Oracle's Desktop widgets (Fig 4), beginning with search and the worklist. These desktop widgets are simple (each widget focuses on one primary task), small (unobtrusive on the desktop), and speedy (provide quick access to key information).

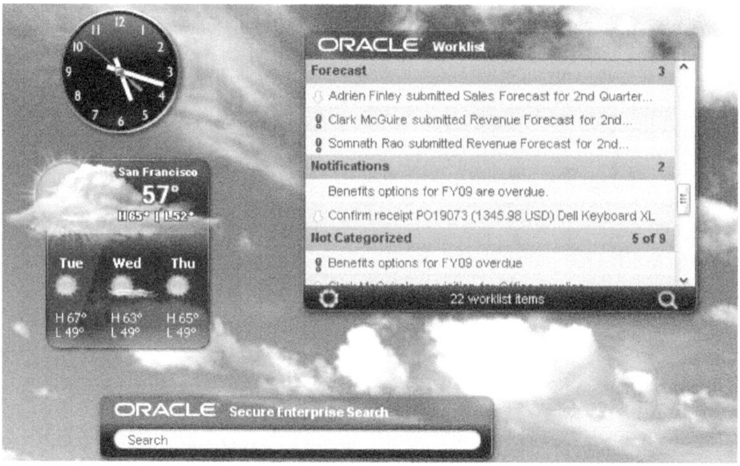

Fig. 4. Worklist widget and search widget on the desktop

3.5 End Users Want a User Experience That Is as Easy to Use as the Internet

In the business world, users often need to get back to a previously visited business object. For example, users would ask , "How do I go back to the last expense report that I submitted?" The Recent Items feature (Fig. 5), always presented as part of the menu in the navigation area, enables users to quickly navigate back to recently visited business objects (for example, to expense reports, purchase orders, or appraisals). Internet users take advantage of this same functionality when browsing the Web, going back to recently visited Web sites with a simple click of the mouse. Oracle's Recent Items functionality provides an instant productivity gain by enabling users to quickly and efficiently access needed information without drilling down into an application.

Along with getting back to recently used business objects, users just wanted an easy way to find information in general. "According to research firms IDC and Delphi Group, the average knowledge worker spends about a quarter of his or her day looking for information" [4]. Good search tools help them find this information (and then use it) more quickly. Users' expectations about the ease of use for search are high due to their experiences with Internet searching. At the same time, search in an enterprise context has to balance the need for security and access control. Oracle's

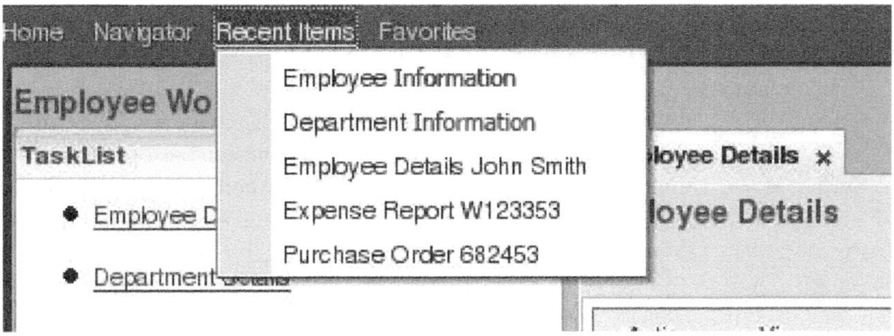

Fig. 5. Recent Items menu

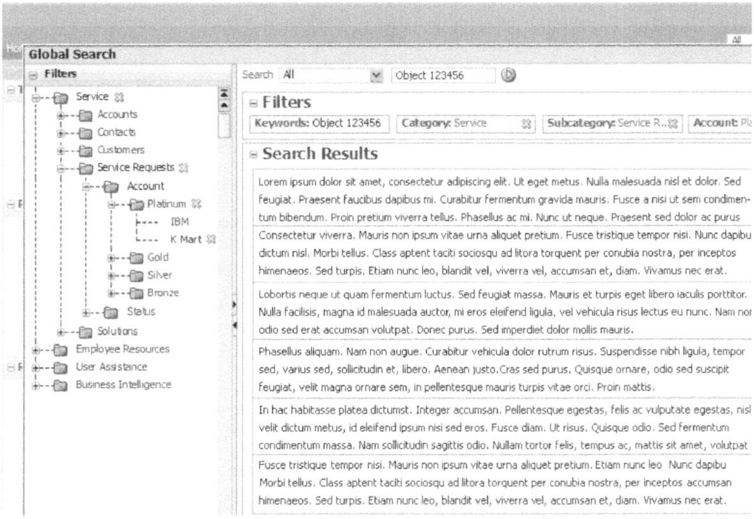

Fig. 6. Search and search results

search balances the simplicity of an Internet-like search (Fig 6) against the specificity that users of enterprise applications demand.

4 Conclusion

Our one key conclusion from all of our research findings is that context is important to the next generation of applications. Staying in the context of a user's work—be it sales orders, purchase contracts, or shop floor processes—radically improves productivity. This conclusion translated into a top-to-bottom design approach that provides users with the contextual information that they need to complete their tasks when they need it.

References

1. Beier, B., Vaughan, M.W.: The Bull's-Eye: A Framework for Web Application User Interface Design Guidelines. In: Proceedings of CHI 2003: Human Factors in Computing Systems, pp. 489–496. Association for Computing Machinery, New York (2003)
2. Vaughan, M.W., Dumas, J.D.: Web-Based Programs and Applications for Business. In: Proctor, R.W., Vu, K.L. (eds.) Handbook of Human Factors in Web Design, pp. 495–513. Lawrence Erlbaum, Mahwah (2005)
3. Holtzblatt, K., Beyer, H.: Making Customer-Centered Design Work for Teams. Interactions 36, 92–103 (1993)
4. Google.: Simplicity and Enterprise Search: A New Model for Managing Your Enterprise Information (2003), http://www.google.com/enterprise/pdf/google_simplicity_enterprise_wp.pdf#search=%22Knowledge%20Workers%20Delphi%20Group%20Google%22

A Management System for Visual Communication Design at Film Studio

Hong-Sheng Chen

No.1, Ling-Tung Rd., Ling Tung University, Taichung City, Taiwan
chsh@ntu.edu.tw

Abstract. This research aims to propose a design management system for helping designer to make arrangement of the furniture in a film shooting studio. The main issue is that furniture has its own characters, and it matches to other furniture in different degree. Therefore, designer need help to calculate the appropriateness. The appropriateness comes from items' color, texture and its image to audience. This management system help designer to visualize the result of combination by simulation.

Keywords: visual design, film studio.

1 Introduction

The last two decades have seen growing importance placed on research in design decision support system [8-10] and decision maker [2]. Recently there has been a shift in attention from a focus on design rules [8] to an emphasis on space layout problem [3,6]. Researchers of design decision support system (DDSS) have often suggested the application for management [1], teaching [7] and housing refurbishment [5]. However, research which has documented the application for film studio is scant. Therefore, my objective in this paper is to study a DDSS application for space layout at film studio.

2 Purposes

This research is a cooperation project for my university and Sanlih E-Television Company. They need an assistant system for environmental visual design at TV film studios. It should provide art workers to create visual environment according to the story by their equipments. We propose an academic research project for developing a design assisting and management system for film studios. Controlling visual communicating factors for story telling, the system should, at a same tine, assist designers for making visual environment at least 5 medium studios. And, it should solve the uncertainty of visual image problem.

This research presents three questions at a film studio.

1. Which furniture are better choices that can describe the characters of the roles in the drama?

M.J. Smith and G. Salvendy (Eds.): Human Interface, Part I, HCII 2009, LNCS 5617, pp. 13–21, 2009.

2. Which other items can match those chosen furniture?
3. Are all goods in this setting coherent with each other in terms of visual conception?

For answering above questions, researchers observed a real film studio and proposed a method as a solution. According to our observation, the basic elements in a film studio are sofa, table, lamp, wall, chair, door, curtain and so on. Only those elements facing to cameras need to be designed with visual collaboration. Those no need to be shown could be roughly constructed. Figure 1 is an example for explaining the construction of a living room at a film studio. It looks like a "half-completed" living room. The concepts of this study are explained as following sections.

Fig. 1. A film studio at Sanlih E-Television shows an environmental visual design plays an important part of film shooting

3 Concepts

For developing the system, I follow the concepts to make the rules.

3.1 Images Concept

Every item in a space has its image, which conveys a visual property to audiences. Images are important factors contributing to visual communication. When audiences watch an item, they associate with their daily experience and give the item a corresponding image. The image conveys properties of the users of this place in the drama or story. Even without speaking in words, an audience knows those qualities of the scene and understands the meaning of those items.

Through visual communication, a good environmental design speaks itself by conveying correct images.

3.2 Collocation Concept

The collocations among furniture are different. In some circumstances, designers probably select furniture that is not collocated. A room needs a set of coordinate furniture. In terms of material and color, they have to show a harmony perception.

Therefore, how to assist designers to build a proper image and a harmony perception of an environment? In this paper, I propose a theoretical framework to analysis it and use a computational system to implement it.

4 Images of Furniture

According to the story, two parts are divided and needed to be collaborated. Part 1, Image Attributes: Functions, Characters and Quality. Part 2, Furniture Items: Sofa, Table, Lamp, Wall, Chair, Door, Curtain.

Besides, the part 1 is noted in detail as follows:

1. Function attributes: Living room, Office, Bedroom, Kitchen, Pavilion, Hospital, Station, and Jail.
2. Character attributes: Gender, Income level, Personality.
3. Quality attributes: Color system, Texture, Luxury.

Figure 2 is an example for explaining how a chair is denoted for its characters and its proper user.

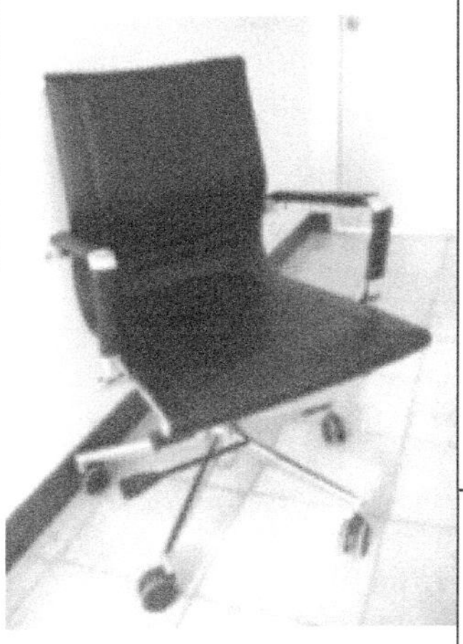

Function: Chair @ office	
Characters of the user of this chair.	Gender: for male
	Income level: 0.8 (high income)
	Personality: 0.3 (bad temper)
Image of this chair.	Color system: 6 (black)
	Texture: 0.7 (coarse)
	Luxury: 0.6 (expensive)

Fig. 2. Analysis for an office chair shows how the images are denoted. Any furniture can express some information of the story.

Then, this study explains its concepts in a logic form, which is interpreted as a formula for analyzing the collaboration between these two parts. The purpose of presenting in a formal logic form is to express the concepts. That is to say, according to these concepts, we have no need to use a specific computer language. Users can use other program language to make this system, if they are good at that program language.

5 Formal Language for Expressing Concepts

5.1 Declaration

F = { x | x is function attribute of the place} ⊃ { Living room, Office, Bedroom, Kitchen, Pavilion, Hospital, Station, Jail }
C = { x | x is character attribute of the owner} ⊃ { Gender, Income level, Personality }
Q = { x | x is quality attribute of the space} ⊃ { Color system, Texture, Luxury }
X = { x | x is a furniture} ⊃ { Sofa, Table, Lamp, Wall, Chair, Door, Curtain }

5.2 Predicates

Scene-A (F, C, Q): Scene A is a setting. In the story, owner of this place have characters of C. This place is used as F. This space is designed in a tone of Q.

C = { x | x is the attributes of the role} ⊃ { C_{Gender}, C_{Income} , $C_{Personality}$ }

Q = { x | x is the attributes of the place} ⊃ { $Q_{Texture}$, Q_{Luxury} , Q_{Color} }

C_{Gender} is a number, 1 for male, 0 for female.

C_{Income} is a number describing person's income. Range from 0 to 1, 1 for rich, 0 for poor.

$C_{Personality}$ is a number describing person's personality. Range from 0 to 1, 1 for bright, 0 for dark.

$Q_{Texture}$ is a number for describing the texture of the place. Range from 0 to 1, 1 for coarse, 0 for smooth.

Q_{Luxury} is a number for describing the quality of the place. Range from 0 to 1, 1 for high, 0 for low.

Q_{Color} is a number describing the color tone of the place, 0 for white, 1 for red, 2 for orange, 3 for yellow, 4 for green, 5 for blue, 6 for magenta, 7 for purple, 8 for black.

α : α is furniture. It has three attributes F, C and S, denote as α (Fi, Ci, Qi).

Examples are shown as follows.

Sofa-I (Fi, Ci, Qi): Sofa I has attributes Fi for Function, Ci for Character, and Qi for Space.

Table-J (Fj, Cj, Qj): Table J has attributes Fj for Function, Cj for Character, and Qj for Space.

Lamp-K (Fk, Ck, Qk): Lamp K has attributes Fk for Function, Ck for Character, and Qk for Space.

Wall-L (Fl, Cl, Ql): Wall L has attributes Fl for Function, Cl for Character, and Ql for Space.

5.3 Computation

$$d_c = ((C - C_{Gender})^2 + (C - C_{Income})^2 + (C - C_{Personality})^2)^{1/2} \quad (1)$$

$$d_q = ((Q - Q_{Color})^2 + (Q - Q_{Texture})^2 + (Q - Q_{Luxury})^2)^{1/2} \quad (2)$$

$$d = \triangle d_c + \triangle d_q \quad (3)$$

5.4 Decision Rules

R1: $(x \in F) \wedge F \rightarrow$ x matches F
R2: Scene-A (F, C, Q) $\wedge$ x matches F $\wedge$ mim($\triangle$d)
$\rightarrow$ Apply x in Scene-A (F, C, Q)

6 System Implementation

This study use AutoCAD and AutoLisp for representing and programming the concepts of above sections. A system can management the furniture items, attributes and decision rules. Some examples of the space layout are shown as figure3.

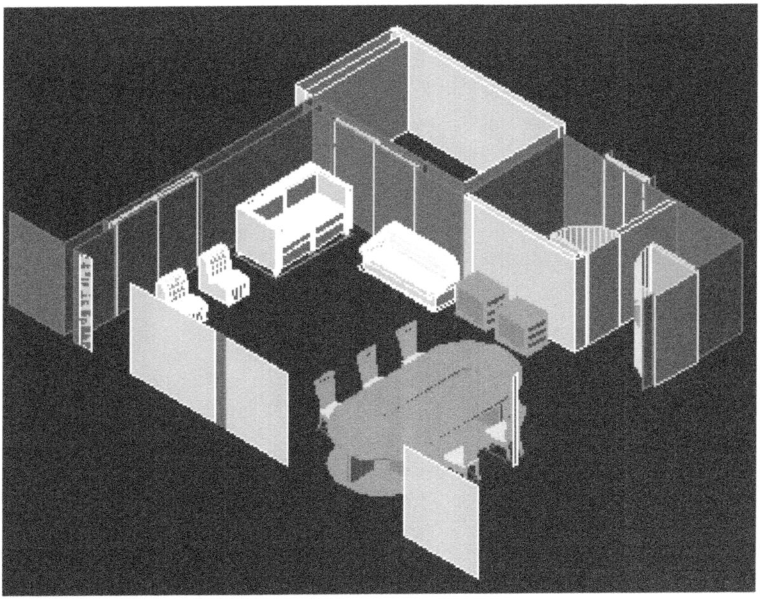

Fig. 3. A layout of furniture shows the placement in AutoCAD

Several outputs could be made by this system. For example, according the story, a director needs a layout of a living room, which owner is a female office-lady; a luxury living room and a dinning room, which owner is an old president. All articles of furniture in these film studios have coherent relation in tune with materials and color. They also have to display the wealthy of the owner in terms of social and financial levels.

A director then can see the layout design provided by this system. If he doesn't satisfy the result, he can change other values of the attribute, make difference and improve the visual quality. After the quality is fine enough, workers bring the items to their locations in studio. Then, the setting is ready for film shooting.

7 Examples on Screen

As costume design, interior decoration shows lots of information about the characters. The actors need proper hair and costume, as well as the space need proper decoration and layout. They are both important elements in terms of visual perceptions. Figure 4 - 7 are examples of space layout by which the visual images can display the quality of the room.

Figure 4 shows a female manager's reception room in an office building. Displaying the smart female quality, film studio uses bright white and red as main colors. Red color represents the outgoing and sunny personality of the owner. The texture of the furniture is smooth flannelette.

Figure 5 shows an old rich man's living room. The tone is mature. Brown color dominates the space. Leather sofa occupies the center place. This setting displays a successful and friendly man with a big family. Figure 6 is the old rich man's dining room. The colors are perfectly tuned to each other. The seat also shows a hierarchy of those people. Figure 7 is a modern office with expensive table, chairs and decoration. The drawing on the wall is a colorful background in this setting. It is easy to perceive the color contrast between furniture and wall. Small stuffs around the wall also represent owner's hobby and personality.

Fig. 4. A completed layout – a layout shows the quality of the office. The right-hand side lady is the user of this office. How we can know the information of the lady by this picture? We can get some clues of the sofa, table, china and so on. (http://www.sanlih.com.tw)

Fig. 5. A completed layout – a living room. (http://www.sanlih.com.tw)

Fig. 6. A completed layout – a dinning room. (http://www.sanlih.com.tw)

Fig. 7. A completed layout – a modern office. (http://www.sanlih.com.tw)

8 Conclusion

This research proposed computational concepts for constructing a design decision supporting system. With the concept that every item in the film studio has its collations to other goods, this system aims to help design to choose a better collocation for visual effects.

I followed the concepts and construct a DDSS by using AutoCAD and AutoLISP. It displays the layout of environment items by visualizing effects, and lists the decision results for working sheet. The visual designers input the system the qualities of figures and places. Through the rules in professional design database, searching the fine attributes in storages. Finally, the system suggests resolutions. Designers modify these simulations, then give it to workers and make the scenery.

The findings of this study highlight the need for research to investigate DDSS for film studio, and in particular, methods for improving coordinating design, as well as in providing appropriate layout for film shooting with a better understanding on visual communication.

Acknowledgements. I thank the assistant come from the members of Sanlih E-Television for their abundant assistance.

References

1. Barton, J., Parolin, B., Weiley, V.: A Spatial Decision Support System for the Management of Public Housing. In: Van Leeuwen, J.P., Timmermans, H.J.P. (eds.) Recent Advances in Design & Decision Support Systems in Architecture and Urban Planning, pp. 69–84. Kluwer Academic Publishers, Dordrecht (2004)
2. Chitchian, D., Henco, B.: A Decision Support Tool for Decision Makers, Digital Design: The Quest for New Paradigms. In: 23nd eCAADe Conference Proceedings, Lisbon, Portugal, September 21-24, pp. 715–721 (2005)
3. Gero, J.S., Kazakov, V.: Learning and reusing information in space layout problems using genetic engineering. Artificial Intelligence in Engineering 11(3), 329–334 (1997)
4. Hu, Y., Ding, W., Yuan, G.: The Adjacency Topological Model for Housing Layout and Its Generating Algorithm, Based on the Case Study of Chinese Apartment Layouts Since 90s. In: CAADRIA 2007, Proceedings of the 12th International Conference on Computer Aided Architectural Design Research in Asia, Nanjing, China, pp. 19–21 (2007)
5. Huang, D., Shih, S.: A Case-Based Decision Support System for Housing Refurbishment. In: CAADRIA 2005, Proceedings of the 10th International Conference on Computer Aided Architectural Design Research in Asia, New Delhi, India, April 28-30, vol. 2, pp. 288–299 (2005)
6. Jagielski, R., Gero, J.S.: A Genetic Programming Approach to the Space Layout Planning Problem. In: CAAD Futures 1997 Conference Proceedings, München, Germany, August 4-6, pp. 875–884 (1997)
7. Lin, C.-J.: Space layout game: An Interactive Game of Space Layout for Teaching and Representing Design Knowledge. In: CAADRIA 2005, Proceedings of the 10th International Conference on Computer Aided Architectural Design Research in Asia, New Delhi, India, April 28-30, vol. 1, pp. 130–141 (2005)

8. Rosenman, M.A., Gero, J.S., Oxman, R.E.: An Expert System For Design Codes and Design Rules, pp. 745–758. Springer, Berlin (1986a)
9. Rosenman, M.A., Gero, J.S., Hutchinson, P.J.: Expert Systems Applications in Computer-aided Design, pp. 218–225. Butterworth, Guildford (1986); reprinted in Computer Aided Design 18(10), 546–551 (1986b)
10. Rosenman, M.A., Gero, J.S.: Design Codes as Expert Systems. Computer Aided Design 17, 399–409 (1985)

A Mobile User Interface for Business Knowledge Management (BKM)

Danco Davcev, Marjan Arsic, and Dalibor Ilievski

University "Ss Cyril and Methodius", Skopje, Macedonia,
Faculty of Electrical Engineering and Information Technologies
etfdav@feit.ukim.edu.mk, a_marjan@t-home.mk, i_dalibor@yahoo.com

Abstract. In this paper, we introduce an adaptable mobile interface for efficient business knowledge management (BKM). The user (client) via the Helpdesk operator can post a request for consultation to the business Experts and/or web services. The goal is to make business knowledge available on demand at any time and any place. This approach gives faster business problem solutions, more productive expert and/or web services, and high availability of the business knowledge.

Keywords: Adaptable mobile interface, Web services, XML Agents, Fuzzy logic, Pocket PC.

1 Introduction

Exchange of knowledge and consultation process among customers and available expert authority via helpdesk operators are very important aspects of quick problem solution using the mobile devices.

The goal of the mobile technology is to make information instantly available to customers. The new business processes can benefit a lot by using mobile connection between customers and experts. The exchange of the various content types of media data (text, drawing, and sound) will enhance the efficiency of consultations between the customer and the expert advisor. It gives benefits in reducing of costs, avoiding future problems or possible claims, saving time, increasing profit and customer satisfaction.

There are many desktop applications that support multimedia communication among participants. Communication among several participants that includes multimedia transfer is at the beginning of the development and utilization in the sphere of mobile handheld devices. Hence, new demands are imposed to these devices: greater processing power to support real time multimedia transfer (video stream, voice stream, and file exchange), greater memory space. Modern feature rich applications demand space on the screen (interface) for displaying all available features, which could be an issue when handheld devices are used.

A great consideration should be given to the readability and utilization of small mobile user interfaces. In this paper, we present our approach in creating an adaptable mobile interface for demanding and exchanging of various content types and knowledge on mobile devices. The interface adapts the dimensions and visibility of the user

M.J. Smith and G. Salvendy (Eds.): Human Interface, Part I, HCII 2009, LNCS 5617, pp. 22–30, 2009.

controls according to the user's preferences and utilization of the controls and also, the current usage of the specific media contents within the business process. Our adaptable mobile interface is managed by fuzzy logic and many XML based agents.

The related work in the second section gives some comments to similar work. The architecture of our business knowledge management (BKM) system is elaborated in the third section, which is the main contribution of this paper. The fourth section describes the design of the system. Implementation and Evaluation of system are presented in fifth section. Finally, in the sixth section we conclude the paper.

2 Related Work

The agent-based approach that uses fuzzy logic to determine importance of certain information is elaborated in [1]. The agents described in this paper are based on access to the Web. We use similar approach to determine importance of user interface features in regards to user preferences and media contents used in current session. However, in our approach, wireless communication between handheld devices is used.

In [2] an agent based adaptive user interface control for desktop applications is presented. Although we share similar ideas, our approach is multi-agent based adaptive interface which uses fuzzy logic for handheld devices.

In [3], collaborative applications can be built using two types of agent collaborations, agent communications through XACL and agent visiting. Agents are represented as XML entities, not programming language entities. Secondly, agent hosting services are implemented as web services with published WSDL, not programs in certain programming languages with published APIs. Thirdly, although XML agents' behaviors have to be coded in some supported programming languages, the interactions between agents and hosts in agents' behavior code are through invoking local hosts' web services, which is neutral to programming languages and host operating systems.

Although we also use XML entities as agents and we share similar ideas of using web services (when appropriate), in our approach we use more flexible protocol for communication among agents which is based on fuzzy-based knowledge.

In [4], an XML-based multiagent recommender system for supporting online recruitment services is proposed. Although the main purpose of this system is to provide Online Recruitment Services, it is also agent and XML based and as a consequence, it can easily cooperate with company information systems. The so called ontology of their user agent stores the profile of a given user concerning the job search. It is different from our system because we plan to build a general purpose knowledge management system (not only for recruitment services). For the time being, our objective is to have a self - adaptive system from the point of view of an efficient communication between user and the advisor according to the user's preferences and the current possibilities of the communication system.

In [5] a feasible framework that combines agent mobility and intelligence for consumer-oriented e-business applications is proposed. This framework complements the current Web-based systems by adding the wireless channel of mobile agents. In our work the mobile agents use an adaptive communication protocol based on fuzzy logic. In addition, our framework includes web services.

3 BKM System Architecture

There are three kinds of users like customers, helpdesk operator and experts. Helpdesk operator is the link between experts and customers.

The goal of the helpdesk operator is to response any kind of requests on demand by the customers in appropriate formats (text, draw, VoIP) and to send the complex requests to the experts and/or web services which should be resolved by them. Client applications are implemented on Pocket PCs.

There is only one helpdesk operator and many experts in one consultation session.

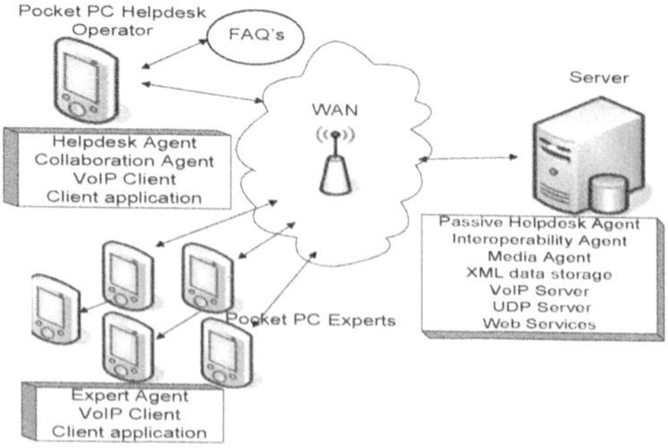

Fig. 1. Architecture of BKM system

The network communication is realized in a way that clients send UDP datagrams to the server, and the server resends the UDP datagrams, according to the contents of the received message.

The communication protocol between the agents and the information flow rely on fuzzy logic. In this case, the fuzzy logic is used for enabling or disabling (adaptation) some controls depending on the quality of the service (QoS), i.e. the signal strength. So, if the signal strength is good, the features of chat, draw and VoIP will be enabled. However, for medium signal the VoIP feature will be disabled. For bad signal, beside the VoIP, the draw feature should be also disabled, and the only active communication feature will be chat.

Streams are used for the voice transfer. Desktop based clients (for both helpdesk operator and experts) can be also connected to the server.

The Web services are used to find the solution of the problem using the wireless Internet access. If the Web services are unable to find the solution, the Experts are activated. Some types of Web services are used for the responses. These Web services support and convert various types of media data.

The experts interface has controls for chat, draw, VoIP, file upload/download, authority control (VoIP channel).

The helpdesk operator can give the control over the interface (or by giving the speaking possibility) and reclaim the control later. Helpdesk operator has the role of a moderator in the consultation process.

Figure 1 shows the architecture of the BKM system. Client applications reside on pocket PCs and they have three main parts: Helpdesk agent, which manages operator interaction within the application and communicates with interoperability agent. VoIP client serves for the voice transfer, and the client application represents all the features mentioned above (chat, draw, voice, file up/download, communication among clients). Expert Agent receives and sends the resolved requests using various file types (like text, image, and voice). Collaboration Agent calculates the cost of the corresponding expert's service. The communication is realized through wireless area networks connected to the Internet.

The server contains Interoperability Agent which redirects the complex request to the Experts and/or Web services and exchanges reformatted media types; Passive helpdesk Agent calculates the helpdesk interface; Media Agent monitors the media contents (text, image or voice); Experts and helpdesk preferences are stored in XML data storage; VoIP Server is responsible for voice transfer and UDP server application manages the login, chat, draw and control. For the purpose of saving the memory and processing power of the clients handheld devices, Passive Helpdesk Agents and Media Agent reside on the server side. The agents are XML based.

4 BKM System Design

In this paper, we describe our approach in development of a BKM system by using multi XML agents and fuzzy logic. Helpdesk agent resides on the client side, and monitors the helpdesk operator's interaction by using the features of the application and communicates with interoperability agent (e.g. sends the request to the available competent experts and/or web services, receives the response in corresponding format and regulates the customers billing). The triggers that demand helpdesk agent's actions are: drawing, sending chat messages, and communicating using VoIP. This agent is named Helpdesk because it helps the customers to simply resolve their requests. At the start of the communication session, helpdesk agent communicates with the Passive helpdesk agent and interoperability agent that reside on the server. The Passive agent demands the information from the Media agent about the type of the files to be used for the particular communication session. This agent is named passive helpdesk since it waits to be invoked; it recalculates the interface structure and sends the data to the helpdesk agent which finally draws the interface. The Expert Agent receive the requests in some format (text, file, voice) from the interoperability agent, concerns about the solution of the requests, sends the solutions to interoperability agent, modifies the expert's profile, checks the QoS and billing status.

The main tasks of the interoperability agent are: receiving the requests from helpdesk agent and forwarding them to the expert agents and/or web services. Also it exchanges reformatted data files with helpdesk agent. Media agent monitors the folder with the data files for the communication session, and sends this information to the Passive helpdesk agent. The Passive helpdesk agent recalculates the helpdesk interface according to the information from the Media Agent, the history of interface affinities (chat_aff, draw_aff, sound_aff, QoS) for the particular user and the helpdesk agent's message about the use of the features. Then, it sends the corresponding information (for the type of the helpdesk interface) to the Helpdesk agent which adapts the helpdesk interface according to this information. The collaboration Agent calculates the time session for realization of the expert task and the cost of the service.

Helpdesk and Expert profiles with personal information about the qualification and the knowledge for all of them are stored in the XML database. Additionally a history of affinities is stored for the helpdesk operator and the experts. According to the history, Passive helpdesk agent adapts the helpdesk interface on the beginning of the communication session.

There are currently three predefined user interfaces, which can be preloaded on the client's device, according to his preferences and the media contents for the current communication session. The dimensions and the positions of the controls would be stored on the helpdesk's affinity table on the server. One kind of an interface displays chat on the larger part of the screen, the second interface has larger drawing view on the screen, and the third one has equal space for the chat and the drawing space. VoIP feature doesn't occupy much space on the screen, so the buttons for voice are always displayed and they will be active if the Signal Strength is satisfied, i.e. QoS is strong.

Since there is no universal conclusion for making decisions of which user interface should be preloaded, fuzzy logic approach is used to model that kind of the imprecise information.

The interaction made by the helpdesk operator using the features of the application (chat, draw and voice feature), is represented as a vector (chat_aff, draw_aff, sound_aff). A linguistic variable named *RESULT_PREFERENCE* is introduced, and it accepts values from the set of terms {increase_chat, increase_draw, the_same}. This variable represents the visualization of the helpdesk interface, produced by (1) the helpdesk operator interaction with the interface; (2) the contents of media folder and (3) the quality of service of the signal strength that decides more accurately which interface to be preloaded.

The class diagram for WiFi signal strength is shown on Figure 2. Two classes are developed, WiFiPeek and SignalStrength. The WifiPeek class implements all the Wi-Fi query related elements. The class uses the NDIS User mode I/O driver (NDISUIO) to perform Access Point (AP) scanning.

The GetAdapters function can be used to query names of network adapters. It calls the built-in NDIS (not NDISUIO) driver. The function fills a buffer with adapter names separated by commas. The RefreshBSSIDs function requests that the driver initiate an AP survey. It takes one parameter: an adapter name. The GetBBSIDs function returns the list of available stations, i.e. peers and Access Points. The function

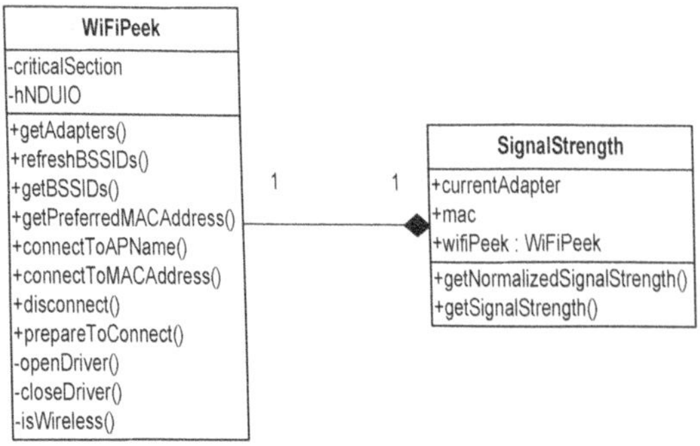

Fig. 2. WiFi signal strength – class diagram

getPreferredMACAddress returns the MAC address of the connected (associated) Access Point.

In order to calculate the user interface visual features, the following fuzzy variables are defined for this expert system: U_A (user affinity) which presents the most used feature by the user, i.e. it has the greatest affinity for the feature; M_P (media profile) presents the profile of the interface that should be used according to the media contents; QoS (Quality of Service) presents the strength of the wireless signal. Output variable named I (action for interface adaptation) presents the necessary interface to be preloaded onto the user's device screen, according to the input variables.

The linguistic variable Quality of Service (QoS) accepts values from the set of terms {Strong, Medium, Weak}.

The linguistic variable U_A accepts values from the set of terms {Chat, Draw, VoIP} and it has normalized values of utilization between 0 and 1 (0% to 100%). It is equivalent to the number of times of usage of the features chat, draw or VoIP respectively.

The linguistic variable M_P accepts values from the set of terms {Text, Drawing, Sound}. It has normalized values of utilization between 0 and 1 (0% to 100%), which is equivalent to the number of text, drawing and sound files in the media profile respectively.

In the table 1 is presented the simulation for fuzzy linguistic values for U_A, M_P and QoS and the value of I is calculated, based on the fuzzy rules. The action needed to adapt the interface is described with the following XML code like the example 1 of the Table 1.

```
<IF U_A="DRAW" M_P="DRAWING" QoS="WEAK">
<ASSIGN I="THE_SAME" />
</IF>
```

Table 1. Some examples for adapting interface using fuzzy linguistic variables

Examples	1	2	3	4	5	6
User Affinity (U_A)	Utilization %					
CHAT	38,00%	57,69%	57,69%	11,90%	31,58%	41,86%
DRAW	46,00%	23,08%	23,08%	28,57%	31,58%	23,26%
VOIP	16,00%	19,23%	19,23%	59,52%	36,84%	34,88%
Media Profile (M_P)	Utilization %					
TEXT	26,67%	30,30%	30,30%	39,47%	29,09%	76,19%
DRAWING	50,00%	54,55%	54,55%	34,21%	30,91%	14,29%
SOUND	23,33%	15,15%	15,15%	26,32%	40,00%	9,52%
QoS	Signal Strength %					
STRONG	0,00%	0,00%	10,00%	0,00%	100,00%	100,00%
MEDIUM	10,00%	25,00%	90,00%	100,00%	0,00%	0,00%
WEAK	90,00%	75,00%	0,00%	0,00%	0,00%	0,00%
Action for Interface Adaptation (I)	Fuzzy Value %					
INCREASE_CHAT	38,00%	54,55%	30,30%	39,47%	31,58%	41,86%
INCREASE_DRAW	10,00%	23,08%	23,08%	34,21%	31,58%	14,29%
THE_SAME	48,00%	25,00%	54,55%	28,57%	36,84%	23,26%

In the first example from the Table 1, the draw affinity is more utilized than chat and VoIP and the media profile has more drawing objects than text or sound. It is expected to be increased the drawing interface, but because the signal is weak and the drawing features are not supported on such signal, the interface stay unchanged (the_same). The total numbers of fuzzy rules are 27, the combination of the values of all linguistic variables. The next examples are similar like the first one.

The interaction among agents is shown on the Figure 3.

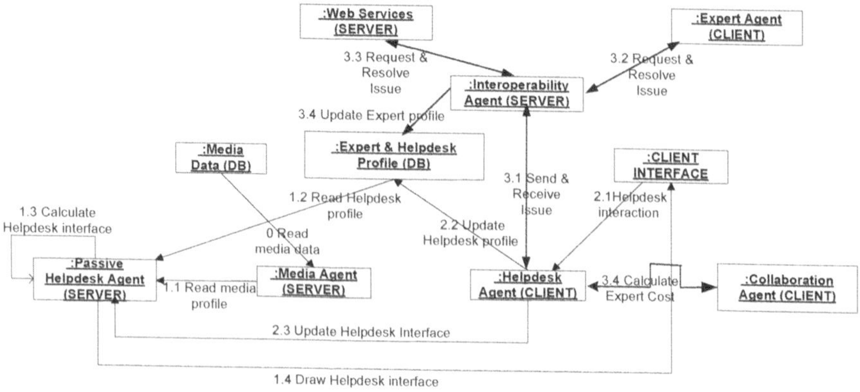

Fig. 3. Agents' collaboration diagram

5 BKM System Implementation and Evaluation

BKM system is developed in C++ Visual Studio .NET 2008 development framework with using of the MFC (Microsoft Foundation Classes) library. Operating systems used on the pocket PCs are Windows CE, Windows Mobile. For BKM system testing is used WLAN Internet provided from T-Home, Macedonia. The clients which reside on desktop PCs and the server use the standard .NET Framework. Operating system for desktop machines on which the application is practically deployed, is Windows XP SP2. The interface adaptation is realized on different screen resolutions.

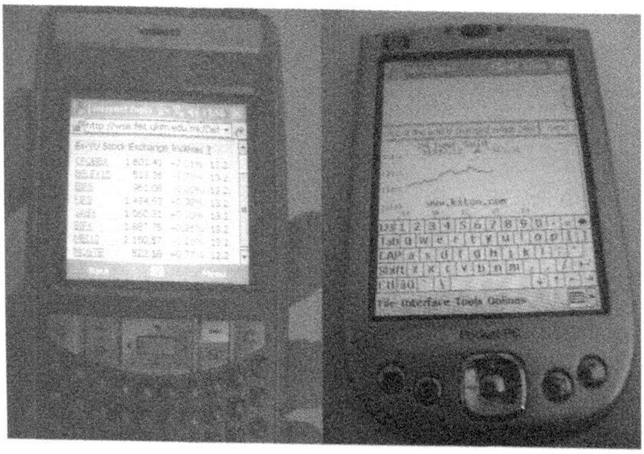

Fig. 4. Two instances of the BKM system

The two instances of BKM interface are shown on Figure 4 for different screen sizes of Pocket PC-s.

The initial feedback of BKM system was given by 30 colleagues in ICT. They were asked to use the system for stock exchange analysis in last week. General opinion among participants for the interface usability, functionality and visibility is average (71,3%). Interface is functional and suggestions for interface rearrangment are mostly done according to the user needs.

The questions that examine interface usability, functionality, visibility and provide information for future upgrades are:

1. Are you satisfied with the way of adaptive presentation?
2. Can you easily select the parameters of BKM system?
3. Were the results of the BKM system solutions are clearly displayed?
4. How much the BKM system helped in the business process?
5. Does the BKM system satisfy the quality of the service?

The questions are answered with "Yes", "No" or "I Don't Know". The answers are presented in the chart on Figure 5.

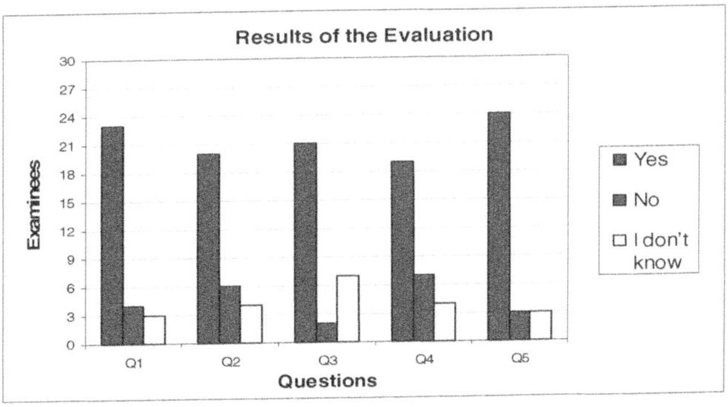

Fig. 5. Results of the examinees' opinion about the BKM system

6 Conclusion

This paper presents our approach that uses XML agents and fuzzy logic in order to achieve adaptive interface for a BKM system. We have developed such mobile system which adapts the features of the interface according to the contents for the specific session. Fuzzy logic approach is used for the communication protocol between XML agents and for interface adaptation. It can be also used to dynamically reconfigure the interface according to the preferences and the type of request. In this way, we proposed an adaptable interface for our BKM system to all mobile device users. General opinion among participants for the BKM system usability is positive.

References

1. Camacho, D., Hernandez, C., Molina, J.M.: Information Classification Using Fuzzy Knowledge Based Agents. In: IEEE International Conference on Systems, Man, and Cybernetics, pp. 2575–2580 (2001)
2. Lu, L., Kitagata, G., Suganuma, T., Kinoshita, T.: Adaptive user interface for multimedia communication system based on multiagent. In: Proc. of the 17th International Conference on Advanced Information Networking and Applications, pp. 53–58. IEEE Computer Society, Los Alamitos (2003)
3. Du, W., Li, H.: XML Agents Technology for Building Collaborative Applications. In: Collaborative Technologies and Systems, CTS International Symposium, pp. 289–297 (2006)
4. Meo, P.D., Quattrone, G., Terracina, G., Ursino, D.: An XML-Based Multiagent System for Supporting Online Recruitment Services. IEEE Transactions on Systems, Man and Cybernetics, Part A 37(4), 464–480 (2007)
5. Weng, Z., Tran, T.: An Intelligent Agent-Based Framework for Mobile Business. In: International Conference on Management of Mobile Business, ICMB 2007, p. 30 (2007)

Designing a Peer Reviewing Tool on Lecture Video with Handwritten Annotation

Hironori Egi[1], Shotaro Houri[2], Yukari Kato[3],
Tatsuya Terada[2], Wataru Tsukahara[2], and Masaki Nakagawa[2]

[1] Information Media Center
[2] Department of Computer and Information Studies
[3] Center for Higher Education Development,
Tokyo University of Agriculture and Technology
2-24-16, Naka-cho, Koganei City, Tokyo, 184-8588, Japan
egichan@cc.tuat.ac.jp, 50007258508@st.tuat.ac.jp,
{kathy,t-terada,w-tsuka,nakagawa}@cc.tuat.ac.jp

Abstract. To improve lessons in accordance with students' comprehension levels, evaluation of teaching performance has been carried out by questionnaire and lesson study. However, lesson study in the context of faculty development is difficult for participants to spend long time for classroom observation and reflective discussion. Even if they have time for discussion, unfocused discussion often causes less fruitful and mutual understanding. We are developing an annotation tool not only to solve the above-mentioned time and space constraint but provide benefits owing to archiving lessons and their reviews. We design a pen-based annotation tool by which reviewers can write comments and draw marks onto on-line and off-line movie of class observation. The details of the annotation tool and user feedback from preliminary experiment in actual classrooms are described in this paper.

Keywords: Annotation, lecture video, faculty development, peer reviewing, pen-based interface.

1 Introduction

With the spread of digital devices and networks, it has recently become possible to capture lectures and share the lectures. Course material is digitalized, provided to students or open to the public. These trends brought to fruition as the evolution of e-learning or Open Course Ware (OCW). In order to promote these activities, it is often that devices like electric whiteboards, projectors, PC with tangible interfaces and video cameras are equipped for recording lectures. Mechanisms of sharing created contents are also invented and packaged as Content Management Systems (CMS) or Learning Management Systems (LMS).

On the other hand, to improve lesson in primary and secondary education, it is common practice that a few evaluators observe a lecture, and then have their observations discussed in meetings [1]. While we can expect similar practice to improve lectures in higher education, it is difficult to schedule for all participants to come together, due to various reasons, for both lectures and meetings.

M.J. Smith and G. Salvendy (Eds.): Human Interface, Part I, HCII 2009, LNCS 5617, pp. 31–39, 2009.

In this paper, we propose a peer-reviewing system, which allows the users to add handwritten annotations on video images, both during recording or at a later time. In this way, peer review of lectures can be done flexibly. Evaluation studies, conducted in undergraduate courses, indicated that recorded movies with annotations promoted meaningful discussion and collaboration between teachers and reviewers for faculty development. We are planning to construct rooms that suit to make full use of this annotation tool, which has multiple cameras to shoot both lectures and reactions of students. Facilities required to use the annotation tool are very similar to those required to archive lectures for e-learning education. The architecture of the designed rooms is mentioned at the end of this paper.

2 Scenario of Peer Review

The goal of this research is to provide teachers with tools that are useful for reflecting and improving their lectures. Roughly teaching skills and materials are important elements that determine a lecture is good or not. Materials like handouts, slides and other digital resources are generally organized and easy to be revised. However, a teacher may be unaware of problems in teaching skills including speaking, gesture, usage of the blackboard, interaction with students and so on. Peer review by other teachers is helpful in point of objective analysis. Teachers in the same department or whose specialty is related to faculty development or instructional design are supposed to be reviewers.

We assume the system is used under different situations such as following three cases.

2.1 Real-Time Annotation at Class

Reviewers attend a lecture and review activities in the classroom. In order to grasp the activities during class, video cameras are settled and connected to tablet computers for reviewing in the classroom. They observe the lecture and write comments with the tablet computers. All of the handwritten comments and the movies from the video cameras are stored to the tablet computers with timestamps for synchronizing the comments with the movies.

2.2 Annotation to Recorded Class

Reviewers observe a lecture with a recorded movie and write comments with the tablet computers the same as shown in 2.1. The movie of the class must be recorded in advance. They do not have to attend the lecture in the classroom, but they may meet difficulty in grasping the atmosphere only with the movie.

2.3 Annotation at Review Meeting

The teacher calls a meeting with reviewers after a lecture. Then they watch the movie of the lecture that is recorded the same as the above two cases. It may be already reviewed by the reviewers as shown in 2.1 or 2.2 and comments are written. During the discussion at the meeting, reviewers or the teacher can add further comments at any point of the movie.

By using the annotation system in the above asynchronous situations, reviewers are free from time and space constraint. Reviewers annotate comments directly onto movie played on the tablet computers, which allow them to write comments on exact targets in the movie. As a result of these features, discussion will be more focused and condensed referring annotation during the lesson study procedure. Figure 1 shows the outline of the peer-reviewing process proposed in this paper.

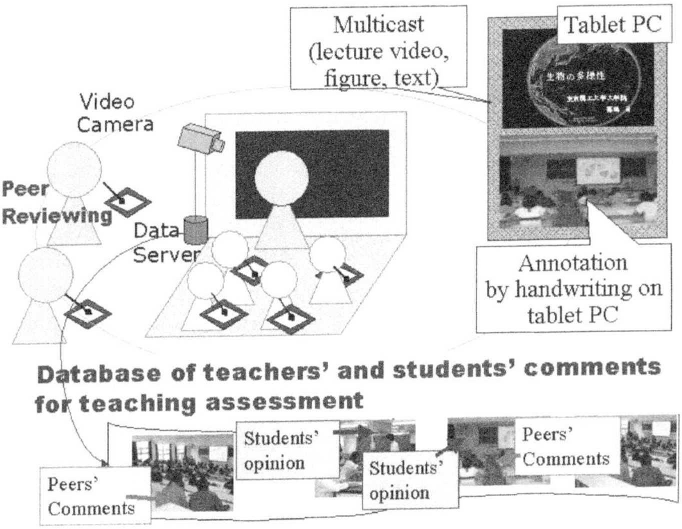

Fig. 1. The outline of the annotation procedure. The cameras shoot both the lecture of the teacher and the reactions of the students. The movie is sent to the tablet computers and reviewers write comments directly on it with the stylus pen. All of the handwritten comments and the movies from the video cameras are stored.

3 Related Work

Movie files became much easier to handle with the progress of computers, cameras, storage devices and their interfaces. It is also easy to release and share those movie files on the networks. Now video sharing websites like Youtube and Yahoo!Video attract many users all over the world. The users post comment or rate each movie through BBS form or chat, and the feedback appears on those websites. Text-based annotation is the major technique to enrich the contents. Other annotation media are also considered like handwritten messages, voice messages, and movie files as a response to original one.

Annotation to the movie files is carried out for various purposes. Tagging metadata for indexing is important to perform useful operations such as content search and content recommendation [2]. TAGGER [3] captures meeting sessions and allows individuals to manually structure these sessions by tagging with laptop computers. M4Note [4] is a Java based annotation tool with pen-based electronic ink and voice operation. It is possible to play the annotation synchronously with video and, during the playback, to add new annotations or to edit existing ones. It is also extended to the editing of interactive TV multimedia documents by end-users[5]. However, M4Note converts all the annotations into text with handwriting or voice recognition. In addition, the user has to stop the video capture and select a video frame to be annotated. It is not directed to moving pictures themselves.

LEAN [6] is a video editing software that enable casual annotation with the fluid control interfaces. Its primary interface is a digitizer tablet with a pressure-sensitive pen. LEAN allows for the manipulation of a video stream by using a small set of gestures that let users start, stop, and move to any arbitrary point in time in the stream. Annotation can be marked with a size and position directly related to both the number of annotations and the moment a particular annotation was made. Users also have the ability to 'pin' a note into the workspace, making it visible at all times, regardless of the current frame being displayed. Annotation can be written directly over the video frame. This research mainly focuses on the intuitive interfaces and it is free from restriction that aims for a specific purpose. An intuitive interface for video teleconference including annotation that can be written directly over the video frame is also proposed [7].

Several annotation tools that focus on practical use in lectures are designed [8]. Livenotes [9] is intended to be used for cooperative note-taking by students. Slides and blank sheets are prepared as handwriting area. They are also enabled to be written directly over slides, but not over video streams.

4 Design of Annotation Software for Peer Reviewing

The main objectives of our project are to support peer-reviewing process. We come to the conclusion that software specially made for the process is needed. According as the general process of peer-reviewing, prototypes are developed and the functions of the prototypes are revised with the comments from the trial of teachers. The features of developed software are described in this section.

4.1 Features of Annotation Software: FD Commons

We named the developed software "FD Commons" that focuses on the peer-reviewing process toward the activities of FD(Faculty Development)[10]. FD Commons is an annotation tool that runs on tablet PC and makes use of the movies of lectures.

By developing handwriting interface with easy operability, it aims to provide teachers outside class with on-line peer reviewing opportunity that is necessary to and relevant to their teaching/learning improvement. Moreover, the database of reviewer annotation has capability of reusing collected comments in order to suggest weak and strong points of class lectures and to design the rubric to evaluate lectures. This system development realizes ubiquitous peer reviewing and reuse of comments of reviewers for assessment of teaching/learning in higher education.

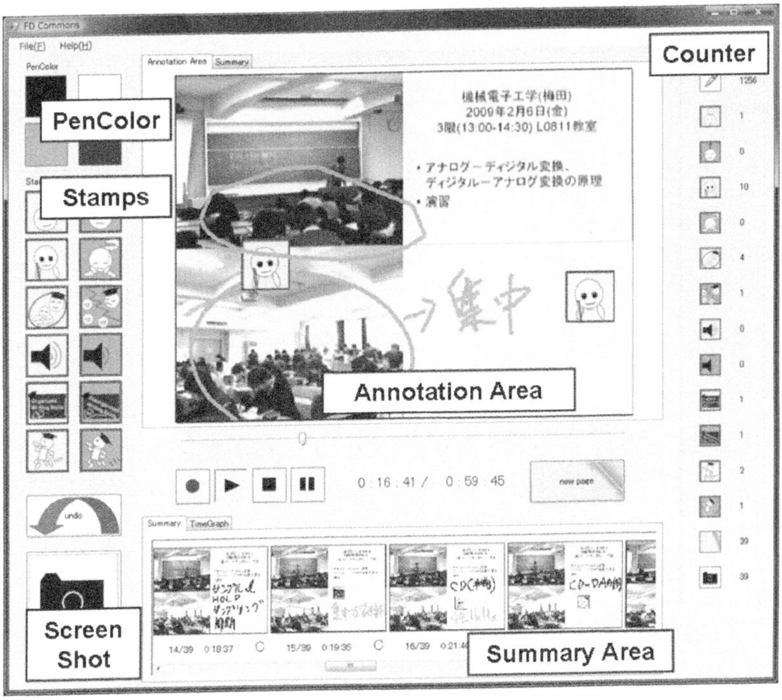

Fig. 2. The screen shot of the annotation software "FD Commons". Mainly it consists of four panes: Toolkits (Left), Annotation Area (Upper middle), Summary Area (Lower middle) and Counter (Right).

Figure 2 shows a screenshot of the annotation tool. Functions of the implemented software are explained as follows. Mainly it consists of four panes. Palettes of pen colors, Stamps stand for typical comments to the lecture, Undo and ScreenShot buttons are settled in the left pane. The lecture being recorded with multiple cameras are shown in the Annotation Area, in the upper middle pane. Screenshots of the Annotation Area are added to the Summary Area in the lower middle pane. The number of each action by reviewers is shown in the right pane.

4.2 Annotation Area

The reviewers can mark, point and write comments onto this area using a digital pen. Several stamps are also available for evaluation of the lecture. Movies are shown inside of the area from the cameras connected to the tablet PC as live recording. Instead of recording, pre-recorded movies can be replayed and annotated. The reviewers annotate comments both directly onto the movies and to the blank area next to movies, which allows them to write comments on exact targets in the movies.

4.3 Handwriting

Comments and signs that are written by the digital pen are displayed in the Annotation Area with a color selected by reviewers in the palettes. Black, green, red and white with grey contour (in order to be seen on the blank area) are prepared as pen colors. Undo button is used when the reviewers want to cancel the previous action.

4.4 Stamps

Stamps are prepared as substitutes for frequently used comments so that reviewers do not have to write the same comments repeatedly. The purpose of stamps is not only to save the time for annotation but also to evaluate the lesson quantitatively by counting the number and types of stamps used. The iconized stamps are bitmap files and easy to be exchanged. The meanings of current stamps are shown in Table 1.

Table 1. The stamps and their meaning of current version of FD Commons

Stamp Image (Positive / Negative)	Meanings
 / 	well organized for understanding / not organized for understanding
 / 	attractive and motivative for studying / not attractive and motivative for studying
 / 	interactive and aware of students' reaction / not interactive and aware of students' reaction
 / 	volume of voice is appropriate / volume of voice is too loud or low
 / 	blackboard writing is well organized and tidy / blackboard writing is not organized and poor
 / 	lecture speed is appropriate / lecture speed is too slow or fast

4.5 ScreenShot Button and Summary Area

In order to point out a certain comment, the reviewers tap "ScreenShot" button and a snapshot of Annotation Area including handwritings, stamps and movie is taken and added to the right end of Summary Area. When the reviewers tap a certain snapshot while the movie is being played, the movie player jumps to the point that the snapshot is taken. By tapping "New Page" button, all the handwritings and stamps are erased and a snapshot is added to Summary Area the same as by ScreenShot button. After finishing peer-reviewing, all the comments are gathered as a list of snapshots and saved as a bitmap file in order to handle easily. The list of snapshots is shown with tapping the "Summary" tab (Figure 3) and enables users to see the outline of the reviewing.

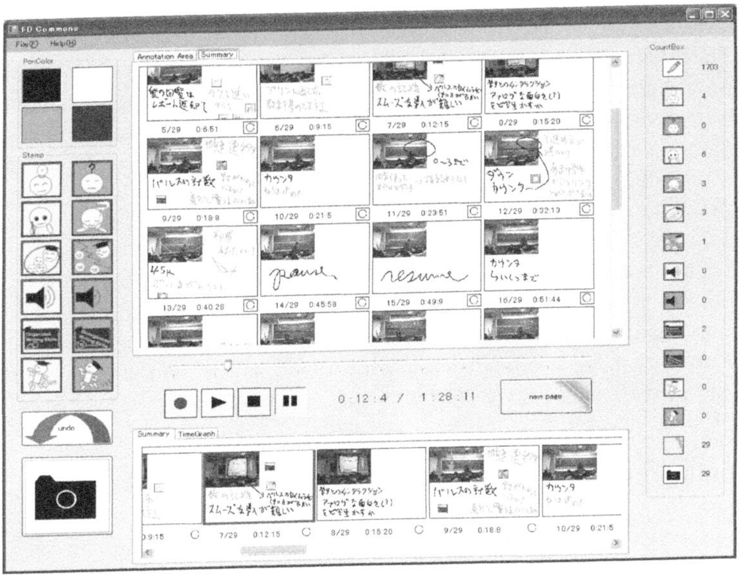

Fig. 3. The screen shot of the list of snapshots (Summary). It appears on the Annotation Area by tapping the "Summary" tab and enables users to see the outline of the reviewing.

5 User Feedback and Discussion

In the pilot studies, we investigated the effects and operability of online peer reviewing system on five reviewers those were all faculty members of our university.

One was instructional designer who belonged to center of educational development (faculty developer). The others were academic staffs majoring computer science and mechanical engineering. Totally, nine trials were conducted. All reviewers who used "FD Commons" had no problem with system operability and their views of accessibility. Figure 4 shows the preliminary experiment in a classroom.

Fig. 4. The preliminary experiment in a lecture. Cameras are equipped in the front and the rear of the classroom and connected to the tablet computer. The reviewer is sitting in the rear of the classroom and writing comments (Left) of the lecture in the classroom (Right).

The reviewers attend and review the class with writing annotation from the beginning of the class. Reflective session was held immediately after the class. In this evaluation, roughly two feedbacks are investigated. First, by using the annotation tool, the focus of discussion was cleared and retained. Previously discussion had tended to be divergent in reflective session. Secondly, annotation differed depending on the locations of video cameras. The video cameras shooting students provided teacher's viewpoint while others shooting the teacher could provide the students' viewpoint. The amounts of comments were different between these conditions and we observed from the comments that the video camera shooting students was better for evaluation.

From these observations, the annotation tool facilitates the reflective discussion in lesson study, and it requires shorter time for the reflection. Furthermore, we obtained an implication that the use of different views with several video cameras in a classroom could provide better reviewing with rich information of the classroom.

6 Conclusion and Future Work

In this paper, we propose a peer-reviewing system, which allows the users to add handwritten annotations on video images, both during recording or at a later time. In this way, peer review of lectures can be done flexibly in order to improve lessons in accordance with students' comprehension levels. Evaluation studies, conducted in undergraduate courses, indicated that recorded movies with annotations promoted meaningful discussion and collaboration between teachers and staffs for faculty development.

In the next step, we will add functions of handwriting recognition and search so that reviewers and teachers can look up specific comments. Also we are planning to construct rooms that suit to make full use of this annotation tool, which has multiple built-in cameras to shoot both lectures and reactions of students.

Acknowledgements. This work has been partly supported by Microsoft Research Asia (MSRA) under Mobile Computing in Education Theme "Innovation Based on Recognition Research Platform" and by the grant of CASIO Science Promotion Foundation.

References

1. Wiburg, K., Brown, S.: Lesson Study Communities. Crown Press (2007)
2. Rahn, N.C., Lim, Y., Groth, D.P.: Redesigning Video Analysis: An Interactive Ink Annotation Tool. In: ACM Proc. of CHI 2008, pp. 3339–3344 (2008)
3. Richer, H., Miller, C., Abowd, G.D., Hsi, I.: An Empirical Investigation of Capture and Access for Software Requirements Activities. In: ACM Proc. of Graphics Interface 2005, pp. 121–128 (2005)
4. Goularte, R., Cattelan, R., Camacho-Guerrero, J., Inacio Jr., V., Pimentel, M.: Interactive Multimedia Annotations: Enriching and Extending Content. In: Proc. of ACM DocEng. 2004, pp. 84–86 (2004)

5. Cattelan, R.G., Teixeira, C., Goularte, R., Pimentel, M.: Watch-and-Comment as a Paradigm toward Ubiquitous Interactive Video Editing. ACM Trans. on Multimedia Computing, Communications and Applications 4(4), Article 28, 1–34 (2008)
6. Ramos, G., Balakrishnan, R.: Fluid Interaction Techniques for the Control and Annotation of Digital Video. In: Proc. of ACM UIST 2003, pp. 105–114 (2003)
7. Liao, C., Liu, Q., Kimber, D., Chiu, P., Foote, J., Wilcox, L.: Shared Interactive Video for Teleconferencing. In: ACM Proc. of Multimedia 2003, pp. 546–554 (2003)
8. Cuthbert, A., Kubinec, M., Tanis, D.O., Ieong, F., Wei, L., Schlossberg, D.: Advanced technology for streamlining the creation of ePortfolio resources and dynamically-indexing digital library assets: a case study from the digital chemistry project. In: ACM Proc. of CHI 2005, pp. 972–987 (2005)
9. Kam, M., Wang, J., Iles, A., Tse, E., Chiu, J., Glaser, D., Tarshish, O., Canny, J.: Livenotes: A System for Cooperative and Augmented Note-Taking in Lectures. In: ACM Proc. of CHI 2005, pp. 531–540 (2005)
10. Kato, Y., Egi, H., Tsukahara, W., Houri, S., Nakagawa, M.: Development of e-Teaching Portfolio for Ubiquitous Peer Reviewing Process. In: Proc. of Society of Information Technology & Teachers Education (SITE) (2009)

Usability Analyses of CRM Systems in Call Centers: The Data Mining Approach

Ádám Horváth[1], László Laufer[1], and András Kangyal[2]

[1] Budapest University of Technology and Economics, Egry J.u.1., 1111 Budapest, Hungary
{Adam,Laufer}@erg.bme.hu
[2] Guanxi, Garas u. 9, 1052 Budapest, Hungary
kangyal@guanxi.hu

Abstract. We carried out a usability research based on the methodology of Software Usage Mining of the CRM system used in the Call Center of a major Hungarian telecommunication company. This paper describes the specific usability aspects of CRM systems used in Call Centers. By analyzing the communication model of the customer-company interaction, we identify the key areas of usability and data mining in the process. Our research has revealed how Web Usage Mining provides invaluable insights for identifying usability issues on a quantitative basis. Our results showed how a wizard-like interactional model would suit the process of customer care better, than the currently applied menu driven architecture. In the specific context of a Call Center, Web usage Mining is proposed to used as an evaluation tool to provide fairer assessment for Call Center operators.

Keywords: usability, data mining, web usage mining, CRM, call center.

1 Introduction

Customer Relation Management software (CRM) has an important role in handling day to day operations for companies. It can be used in the front office to handle direct interactions with customers and in the back office to keep track of operations related to the customer as well. [2] It also provides a source for analyzing the success business strategies, possibilities of churn and other customer related information. Probably the majority of the interactions with the customers take palace through a Call Center, therefore it is major application area of CRM systems.

The CRM software used in a Call Center is a bit different from HCI point of view than the majority of usability research areas. It is not obvious for example who the actual user is, the customer or the Call Center operator who executes the customer's requests? And whose needs have priority when designing such a system? Is it the operator, the customer or maybe the company? These questions raised our interest towards CRM systems used in Call Centers. A further interesting aspect was to realize that the operators are not only reacting to customer requests but they have to actively serve the company's business interests and act as salespersons for instance. Hence they are being evaluated in what extent they were up to this task.

M.J. Smith and G. Salvendy (Eds.): Human Interface, Part I, HCII 2009, LNCS 5617, pp. 40–48, 2009.

The latter issue clearly points out the necessity of usability analysis of CRM systems for two reasons. First, this is a fairly unique area where the usability of the system can be directly translated into monetary value through the time spent with each customer. Second, the operators' salary is partly influenced by their evaluation which takes into account their communication style, professionalism and the time they needed to resolves the customer's request. It means that they could be penalized for something that stems from the software and not from their skills. This clearly creates an unfair situation for the operator. If we also take into account that the average operator utilization can average between 90 and 95% we can imagine the amount of workload they are facing with. [3, 7]

In this paper we will take you through the communication model of a CRM system to identify what is the role of usability and data mining in the customer – company communication process of Call Centers. Based on this we will introduce a novel approach for usability analysis via web usage mining that can have twofold advantages. It not only identifies possible usability problems but can serve as a tool to provide fairer evaluation for Call Center operators. We also suggest changing the interaction model of CRM systems used in Call Centers from a menu driven to a wizard-like interaction. We will illustrate these findings via a research that we carried out at a major Hungarian telecommunication company.

2 Call Centers as Computer Mediated Communication Tools

In order to better understand this very special usability area let's take a look at first how the communication works between the related parties.

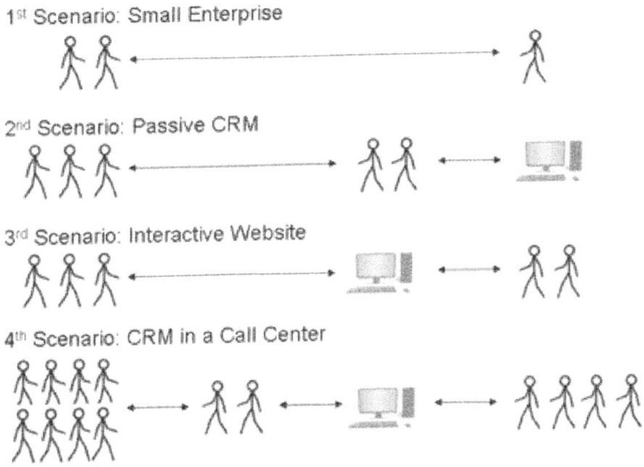

Fig. 1. The model of Customer – Company communication in Call Centers

On Figure 1. we depicted four different prototypes of this process. In the *1st scenario*, typically occurring in the case of family, or even smaller enterprises, there is no need for a computer system to mediate between the "one man" company and its clients. The person running the company is keeping in mind all his resources, the feedback from the clients, and also communicates with the clients during day to day business.

If the company is bigger, depicted in the 2nd *scenario*, there is no such person who would be the center of all information. Usually there is a need to set up a CRM keeping track of the interactions with the clients, making it possible for the company to interact with the customers as a single entity. In a typical medium-sized enterprise there are a few people communicating with one customer, recording the interactions in the CRM. In this case the CRM is a tool for mediating between the employees of the company, but it has a passive role in the communications between the company and the clients.

In the 3rd *scenario* an interactive website of the company can serve as a mediator between the two sides. The customers are providing information about themselves and their intentions either directly, by entering data in the system, or indirectly by their browsing patterns on the site. If we compare the second and the third scenario, it is clear that in the case of a website the customers interact directly with the computer system, which in some cases is not only a passive marketing tool, but also plays an important role in the information ecosystem of the company. In the case of some Web 2.0 companies, if all business processes are taking place on the website, the website serves as the CRM tool as well, and the customers are keeping track of their interactions with the company in the CRM either directly on indirectly.

As the number of customers grows and the complexity of services and business processes increases there is a need for setting up call centers to help customers solving their company related problems. As shown in the 4th *scenario* the call center personnel are the mediator between the CRM and the customers.

But is it really necessary to have mediators in this communication? Yes it is, because the customers know neither the business terms, nor the business processes of the company, and they obviously can not use the CRM system. The CC operators are translating the requests of the customers into the terminology the company implemented in the CRM system. The colleagues in the CC are also interpreting for the customer the "responses" of the CRM (e.g. choices offered by the system). This Call Center mediated communication between the CRM and the Customer is going back and forth till the original goals of the customer are being achieved. As a further argument that is also a common phenomenon that customers prefer to call a human operator rather using an automated voice menu or a website. [3]

Ultimately the interaction is between the Customer and the Company, but the information is stored in the CRM system and being mediated by the Call Center operator.

3 Assessing Interactions in Call Centers

As we mentioned, the data stored in the CRM system is the main source of information for the Company about its customers. Unfortunately there is no opportunity to sit down with every customer to discuss their personal needs therefore from the business perspective it is vital to correctly understand the needs once the customer connected them through a Call Center. The CC provides two different types of data to analyze to what extent was the interaction successful. First, CCs routinely record every verbal communication along with the footage of the CC operators screen. Second, the CRM system's log files stores every action that the operator has done in the system.

(At this point we should mention that we didn't take into consideration in our research churn and other customer assessment that is carried out on the data stored in CRM. This process is represented on Fig.1. in the 4th scenario with the arrow on the right.)

In order to assess whether the CC operators has acted in favor of the business needs they are usually being evaluated by the recorded phone calls they carried out with the customers. This is represented with the arrow on the left in the 4th scenario on Fig.1. The evaluation is done by monitoring experts who give percentages for different aspects of the operators' work. Due to the large number of recordings only about 1% of all interactions are being evaluated.

However companies often miss the opportunity to analyze what happens between the CRM and the CC operator. This data is invaluable in two different ways. On one hand it can be analyzed on a much larger scale than voice recordings, on the other hand it is a telltale sign how good the usability of the system is. Although all communication between the customer and the company goes through this bottleneck, usually very little attention is being paid on the usability of the CRM interface. It has a special importance from HCI perspective, as the usability of the software not just implicitly, but directly affects expenses of the company, as it is a key factor in how quickly operators deal with customer requests. The more difficult it is to learn and use the system, the longer the interaction takes, and the more mistakes are being made by the operators, which affects the operators' salary as well.

These interactions are available for us in the form of database log files. As a properly designed system stores the details of every transaction, we have the opportunity to take every interaction into account when performing an analysis. With data mining methods we are not only able to list the usage frequencies of different functions, but we can examine the series of actions ("clickstreams") of each unique user which enables us to see what they used and how they used it. [4]

3.1 Data Mining of CRM Usage

As we mentioned earlier the key area for data mining in CRM was so far to calculate different metrics about the customers, such as monetary value for the company or the possibility of churn. Data mining in the area of usability (Usage Mining or Web Usage Mining) is gaining more and more ground in several different areas such as e-learning. [5, 6]

Web Usage Mining can help us to better understand specific areas in the usability of CRM systems:

- On one hand it reveals how the CC colleagues use the software interface. With the data mining analyses of their click streams, we can find out what are their habits, which functions do they use the most, which functions they rarely click on.
- On the other hand it also provides information on the aims of the customers, what problems they wanted to solve by making phone calls to the CC.
- Another aspect of the analysis is assessing the usability of the CRM interface by comparing the optimal click streams corresponding to the original aims of the customer and the click stream actually have been carried out by the CC operator.
- The performance of the CC operators can also be assessed by analyzing his/her interactions with the customers as they are represented by his/her recorded click

streams on the CRM interface. This methodology of performance evaluation is on a more fair base than the one being commonly used analyzing only around 1% of the CC operators' work.
- At last the reactions of the customers on the business strategy of the company can also be analyzed from the log files by calculating different metrics of what kind of direct offers they accept or refuse during their phone calls, and what influences their decision.

4 Our Research and Results

To better understand the above mentioned issues we have carried out a research at a major Hungarian ICT service provider company that has implemented a web-based integrated CRM system. The software gathered and displayed the data from different sources such as general customer data, previous interactions, churn warnings and offers. These data are displayed on different pages in separate box-like entities, called portlets. At the time of the research the system was running for about 6 months.

We had two different datasets to analyze. The most important ones were the access logs from the web servers in Common Log Format. This was the main source of information we analyzed to identify the important functionalities being used in the system and to reconstruct the clickstreams of the users'. The second type of logs was from the database servers. We used these logs to define which functions were activated in each portlet, and how the operators have defined in the end of the call the aim of the customers calling in (it was a piece of information they had to enter into a specific portlet). As both type of logs contained the username we were able to segment the users according to their roles such as CC operator, customer service point employee or back office employee. As the time resolution was different in these logs, they couldn't be combined.

Our results are based on the usage statistics of October 2007. In this period we analyzed about 6000 unique users' activity. We did the analysis in SPSS Clementine with the help of Web Mining for Clementine add-on in the case of access logs. Unfortunately, due to its sensitive nature we are not authorized to tell the exact number of customers calling in as it is one of the most confidential piece of information related to the CC, but the magnitude of the callers was tens of thousands per day.

In order to confirm our hypotheses based on usage patterns and to get better understanding of the reasons behind specific patterns, a series of interviews has been carried out with users on the functions of the system. To have an overview on how the operators work and handle the customers while using the CRM, we carried out a series of in depth interviews with a monitoring expert.

4.1 Usage Patterns in the System

Although the software supports a wide variety of actions, there are some common steps that every usage has to include. Every session of customer interaction starts with a search for the customer's profile which is usually done by searching on his or her phone number. After displaying the profile, the customer has to be identified before most of his/her requests could be fulfilled. Before finishing the session the operator has to communicate at least one of the sale offers that the system lists for the specific customer. After the offers are either accepted or rejected the session can be finished with pressing an "End interaction" button which prompts the operator to define what the call was about, in which category did the request fall. (E.g. giving information on a product, modifying a detail of a service, handling complaints, etc.)

An important discovery was how the users started with the "Search" function. The search box has two buttons: "Search and "Reset". Interestingly enough, the data analyses showed that one third of the clicks in this box were "Reset". To interpret this data, we asked a monitoring expert who revealed the reason for this phenomenon. When the operator receives a call, an additional piece of software automatically fills in the search field with the caller's number. However, a lot of operator doesn't trust this function as they remember having incorrect search results in some occasions. In order to make sure it doesn't happen, they started the session by clicking the "Reset" button first which they believed "cleared out" the memory of the system. This is followed by the manual entry of the customer's number and then pressing "Search". The real issue with this action is that it reloads the whole page which causes an unnecessary load to the system and it results in slower response times. As it was mentioned in the user interviews, the slow response time of the system was on of the major usage problems with the system. Although the monitoring group knew about this misuse, they did not realize the extent of the problem. In 90% of all customer calls, the CC operators only used 3 functions, one of them being phone number search, a third of which being totally redundant.

4.2 Business Oriented Functions: Intentions and Real Usage

When the new CRM system was implemented, the company integrated couple of functions in it that were meant to increase the business opportunities between using different services of different branches of the telecommunication company. As in the business strategy of the management these functions were of great importance, the way they wanted to communicate it to the customers was to put these portlets on the vast majority of pages in order to encourage operators using them. This is good example of how the company communicates through the CC (see Fig.1 4th scenario). However our analysis revealed that there were huge discrepancies in the usage frequencies of different software functions and these portlets were not used to an extent that they were aimed at. On average every forth time a portlet was displayed the operator used it (by clicking in it). In one of the cases of these strategic portlets this ration was 1 usage to 2210 displays. For a similarly implemented function, the ratio between downloads and actual usages was significantly better, but it still was only about 200. This data shows that there are functions in the software that are rarely used, however they occupy screen space and they cause unnecessary load to the system which results in slower response times.

But why were these functions so underutilized? The simple answer is that users didn't know the purpose of these functions because for them they had a meaningless name. (E.g "Context" or "Flexible attributes") The log analyses showed that these functions were ignored and not used at all by specific usergroups.

These examples show not only the poor design of the system but also the lack of appropriate training that could have given the opportunity to the users to familiarize themselves with these functions. As these functions are still considered strategically important, the operators should be motivated to start using them. Data mining can help here to identify those user groups that are still having challenges using novel functions in the system. As some of these analyses could be applied regularly, the tailor-made training needs of the CC operators could be constantly developed and applied.

4.3 Clickstream Analyses

The main purpose of our research was to define what routes have the operators used to deal with the requests of the customers, and how can we help optimize the interface to help the operators to carry out their jobs better. As we had hundreds of thousands of sequences, each corresponding to a customer call the sample was large enough to use data mining methods successfully.

Menu system. The interface of the analyzed CRM system resembled a website, with a menu bar consisting of several items, and numerous submenu items under each of them. As we mentioned before on each page of these submenus there were boxes with different functionalities, and pieces of information called portlets. This type of interface empowered the operators to jump from any page (submenu) to almost all other submenus, thus enabling them to use all functions of the CRM in any combination. When we analyzed the sequences of the CRM functions being used by the operators, we examined the pairs of CRM portlets being used after each other. We called these "function switches", each of them representing a click on another portlet or on another submenu. The interface of the software made it possible to carry out over 1000 different function switches altogether. We found that during the analyzed on month period, there were only 150 of these switches present in the usage of the system. 32 of these covered 99% of all switches, and just 11 of the possible function switches were used in 90% of the cases.

Wizard. These figures show that a CRM system cannot be treated as "just another" complex software with many functions. After the task analyses of customer inquiries, which we carried out together with the monitoring experts, it became clear that the vast majority of functions at a certain phase of a specific task are irrelevant, and operators never switch to them. The small number of switches being actually used clearly suggests that a wizard type of interaction design would serve the operators better in the case of this CRM. For the most customer inquires, the wizard would minimize cognitive load on the operator, as he/she would have less choices, and thus could focus more on the conversation. On the other hand the system could suggest the different functions to continue the interaction with, and preload those pages (it is possible in the majority of cases) most likely the operator is going to use, shortening the response times of the system in the interaction, and as a result shortening the customer calls radically.

We examined the switches, and categorized them according to the customer requests (as the operators identified them at the end of the call). If the operators would have chosen the aim of the customer in the beginning of the interaction, the number of switches could be decreased even further. The sequence of functions being used in the interaction is also well determined by the purpose of the call.

All these findings suggest that a wizard type of interaction, where the operator sets the aim of the call, and the system supports him/her in his navigation in the function helps the operator in carrying out his/her task better, making less mistakes, decrease system load, and response time. As a wizard can radically cut down the length of the calls, eventually it can significantly increase customer satisfaction and the efficiency of the Call Center.

4.4 Monitoring

As we mentioned earlier, the salary of the CC operators depended on how their customer interactions (screen capture + voice recording) were assessed by the monitoring experts. Because of the big number of customers calling in, only around 1% of all calls were actually evaluated which raises the question of how fair it is to draw conclusions based on such a small sample, if all calls are recorded anyway.

There are numerous CC operator activities that the company either incentivizes or penalize if they are found in the evaluated sample, such as making offers shown in the system or failing to check whether the customer has any unpaid balance towards the company. Such actions can be checked throughout the whole activity of the operator from the log files, and could be taken into consideration in his/her evaluation. This approach in our opinion would provide more accurate information for the assessment.

However, based on our existing results we are not in the position yet to propose such a change in the remuneration system. The main reason is that it should be avoided to draw false conclusion based on the log analysis. First, such a system must be validated to find out, what are the real causes of a user's behavior. It means that an event in the log file must be unambiguously related to a specific action. Second, the reliability of the logging system must be assessed as well, as today it is not designed for such functions.

The operators` usage of the CRM is a kind of reflection of their customer care. We should emphasize though, that the analysis of this reflection only serves as a good base for the evaluation of the CRM users if the usability of the CRM software is satisfying. If the interaction with the software uses up a lot of cognitive resources from the operators, and thus stets back their work, rather then supports their interaction with the customers, the different metrics of their software activity sequences won't provide reliable bases for their evaluation. However we firmly believe that introducing data mining of software usage as an evaluation method of CC operators is a universal and fair assessment supplement to the ones currently being applied.

5 Conclusions

CRM software in a Call Centers require special considerations from the usability standpoint as it has a direct financial impact both on the company and on the Call Center operators` salary. We have introduced the method of Web Usage Mining in this area, which helped us to get insights on how the operators used software interface and what problems did they encounter. We followed up on these results with user interviews and expert analyses. The data mining approach has proved that it is a powerful tool not only to identify usability issues but with its help it is possible to quantify and identify the user groups affected by these problems. In this respect it is invaluable for businesses to set up the priorities for possible system updates and to create tailor made employee trainings.

Our analyses showed that a wizard type of interaction suits these types of CRM systems better, making it possible to adapt the choices of functionalities offered to the operators restricted to the ones which have been previously used at a similar phase of a similar interaction.

The better sampling of the customer interaction makes it possible to develop the evaluation of the operators, as almost all of their activities with the customers are reflected in the log files of the CRM system.

References

1. Åberg, J., Shahmehri, N.: The role of human Web assistants in e-commerce: an analysis and a usability study. Internet Research: Electronic Networking Applications and Policy 10(2), 114–125 (2000)
2. Aksin, Z., Armony, M., Mehrotra, V.: The Modern Call Center: A Multi-Disciplinary Perspective on Operations Management Research 16(6), 665–688 (2007)
3. Gans, N., Koole, G., Mandelbaum, A.: Telephone Call Centers: Tutorial, Review, and Research Prospects. Manufacturing & Service Operations Management 5(2), 79–141 (2003)
4. González, M.P., Lorés, J., Granollers, A.: Enhancing usability testing through datamining techniques: A novel approach to detecting usability problem patterns for a context of use. Information and Software Technology 50(6), 547–568 (2008)
5. Liu, B.: Web Data Mining: Exploring Hyperlinks, Contents, and Usage Data. Springer, Heidelberg (2007)
6. Scime, A.: Web Mining: Application and Techniques, Idea (2005)
7. Zapf, D., Isic, A., Bechtoldt, M., Blau, P.: What is Typical for Call Centre Jobs? Job Characteristics, and Service Interactions in Different Call Centres. Call Centre Work: Smile by Wire Special Issue of the European Journal of Work and Organisational Psychology 12(4), 305–310 (2003)

Urgent Collaboration Service for Inclusive Use

Naotsune Hosono[1], Hiroyuki Miki[2], Michio Suzuki[3], and Yukata Tomita[4]

[1] Oki Consulting Solutions Co., Ltd.
4-11-15 Shibaura, Minato-ku, Tokyo 108-8551 Japan
[2] Oki Electric Ind. Co., Ltd.
[3] Architectural Association of Japanese DEAF
[4] Keio University

Abstract. At the modern universal design society, this paper discusses the usefulness of the urgent collaboration service that is originally proposed by deaf people. They will be in trouble at such the occasion of disasters or accidents. At the very first of the survey many deaf people are asked their difficulties. From the point of context of use, it is best useful, effective and particularly efficient at the deaf patient complaint of pain/ache/grief among the selected ten items. Those requirements are drawn visually by pictograms or icons that are easy to understand even such emergency situation. Proposed urgent collaboration service placard is then evaluated by deaf people in the manner of the usability test. The results with this placard are that the time to communicate is shorter for about 30%. The placard is also used by foreign people and is translated into English, Spanish, Korean and Chinese for them. This placard can be also downloaded by specified URL in personal. It is also being proposed to Japanese government to have the available both in the ambulance and hospitals or public spaces.

Keywords: collaboration service, inclusive use, human interface, usability, accessibility, universal design.

1 Introduction

At the modern universal design society, this paper discusses the usefulness of the urgent collaboration service. Their appearances are not different in the daily life. However at the unexpected situation they will be in trouble at such the occasion of disasters or accidents. For instance a deaf patient is carried by an ambulance, the urgent collaboration service will one of the solution to make interface between the patient and lifesavers. Since urgent collaboration service placard is simple menu like picture cards and the patient is just to point the pain portion or severe level of the picture on the card by the finger.

This contributes the effectiveness and efficiency of the communication between them since the selections of the picture are summarized by disabled people requirements and hospitals. This urgent collaboration service can be then used not only by the deaf people but by universal users such as foreign people or hard of speaking people.

M.J. Smith and G. Salvendy (Eds.): Human Interface, Part I, HCII 2009, LNCS 5617, pp. 49–58, 2009.

2 Demand for Necessitate Support by Hearing Disabilities

Normally deaf people look same as non-disabled ordinal people. Hence there are little barrier or hardships in their daily life. However they are in problem in such a situation of emergency. Then the deaf passenger is suddenly discriminated and isolated [1].

One of the authors of this paper is a deaf person, urgent collaboration service placard is originally proposed by him. He was in problem in the communication in the ambulance. He tried to tell where the severe pain was but the lifesaver asked his name and address. He was suffered not only physical pain but mental frustration. There are similar requirements from emergency rescue personnel in the ambulance, nurses and doctors in the hospital for communication tools with deaf people.

Urgent collaboration service placard was created such an environment to communicate between deaf people in problem and their helpers. The placard is used to tell the pain portion in the picture of the body figure just menu like. The deaf patient simply points the portion of body by the finger to tell how deep and severe. It is also possible to tell the history of the disease. Therefore the placard is to improve the communication effectiveness, efficiency and then satisfaction.

3 Requirement Survey

The survey began by asking people with hearing disabilities about their difficulties and experiences concerning sudden illnesses and problems that occurred in daily life. To start with those deaf people are interviewed about their hardship by the way of inquiry. 12 deaf people were participated in the inquiry. Eleven experienced the communication problems in the hospital and five did in the ambulance. Some of them

Table 1. List of complaints by Tokyo Fire Dept. and Keio Univ. Hospital

	Tokyo Fire Dept.	Keio Univ.Hospital
	290,471	2,421
Pain problems	30.10%	38.30% Contents of Keio Univ. Pain Data: - Stomachache; 22.7% - Headache: 19.7% - Lumbago, backache: 11.9%
uncousciousness	11.20%	15.90%
breath trouble	8.40%	5.10%
fever	8.10%	2.30%
vertigo	6.60%	6.50%
convulsion	5.40%	3.00%
vomit	4.80%	2.50%
hard of standing	4.60%	2.40%
malfunction in cardiac and lung	N/A	2.30%
injury	N/A	21.60%
others	20.80%	0.10%

pointed that in the hospital they are in problem at the registration, consultation with doctor, Magnetic Resonance Imaging (MRI), Computed Tomography (CT), echo checkups, and operation. Particularly doctors and nurses usually worn mask and this situation is hampering to read the shape of their lips.

In parallel the data of patient complaints collected by Tokyo Fire Department (TFD, 290,471 patients) and Keio University hospital (2,421 patients) were analyzed (Table 1). The collected data in this case are not only deaf but all patients. From the collected both data more than 30% complaints cover pain problems.

The followings are the survey summaries.

- Ten selected patient complaint items were pain/ache/grief, unconsciousness, difficulty breathing, fever, faintness, convulsions, vomiting, difficulty standing and walking, cardiopulmonary problems and external injury.
- The urgent collaboration service placard is focused to use at the time of pain problem by the deaf patient.
- The eight pain portions are head, face, chest, back, waist, berry, hands and legs/feet.
- The three pain depths are surface skin, visceral organs and bones.
- They are drawn with pictograms, animations, metaphors and icons.
- The histories of the past illness are literary written since deaf people may recognize though they may be their second language. Through this categorization basic communication requirements between hearing disabled patients, emergency rescue personnel, and nurses and doctors in both the ambulance and at the hospital were clarified.

4 Functional Items in the Urgent Collaboration Service Placard

The former section introduced the listed requirements or needs from deaf people in urgent conditions. Reviewing TFD and university underneath the context of use [2], not all requirements are necessary to be drawn on the placard particularly at the situation of use in the ambulance. Those patients will simply require pointing their pain portions to appeal. For instance it is difficult or impossible to communicate with patients who are losing consciousness, experiencing difficulty breathing, or having severe cardiopulmonary problems. However, emergency rescue and medical personnel are accustomed to recognize the symptoms of external injury, vomiting, and fever without direct communication. From the perspective of context of use, the placard will not be necessary to cover all situations.

It was concluded that among the ten selected items, the placard was most useful, effective and particularly efficient in determining pain/ache/grief complaints for hearing disabled patients. Aches and pains were isolated in the head, face, chest, back, stomach, waist, hands, and leg/foot. The depth of the ache was in the skin surface, viscerally or at the bone level. During emergencies, patients view easy-to-understand pictograms, metaphors or icons depicting pain/ache/grief and external injury [3,4,5,6]. A minimum of carefully selected keywords augments placard icons. Ache areas and ache depth and severity are presented horizontally (Figure 1).

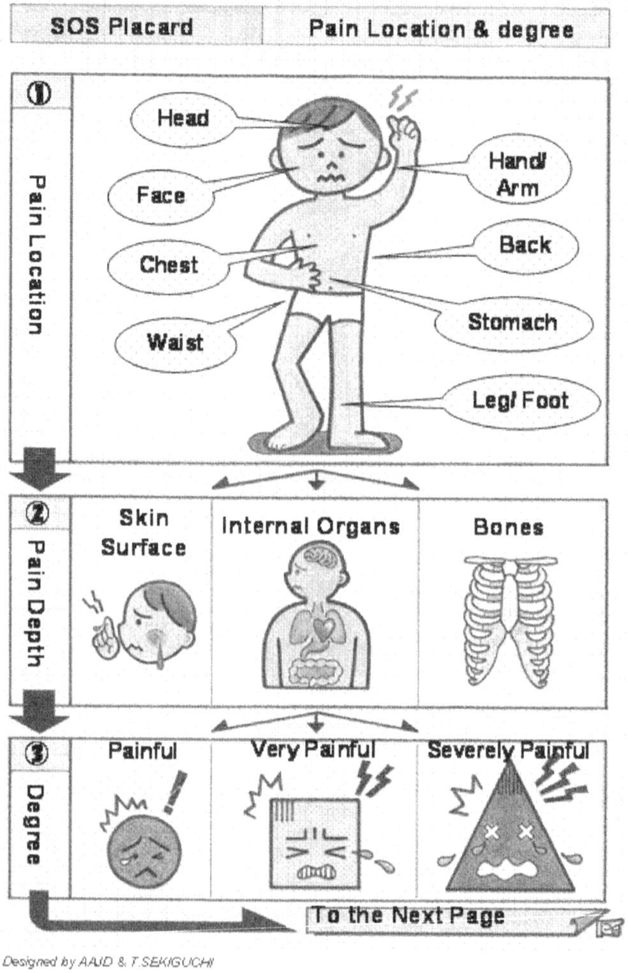

Fig. 1. The urgent collaboration service placard

Considering context of use, they will be four similar but different in detail types of placards. The main differences are the places to be used. Firstly disabled persons may download from the particular Internet home page and carry a printed small urgent collaboration service placard or put on mobile phone with them as a safety measure. It can be used the following public space, in the ambulance or in the hospital. Secondly, it is always prepared just like Automatic External Defibrillator (AED) in public places such as railway station, airport or department store for the emergency situation of sudden illness by the disabled person. In this case the patient mainly requires calling the ambulance. Thirdly it is installed in the ambulance to communicate between the person and emergency rescue personnel. In this case the person will point by the finger that their ache points and its severity (Figure 2).

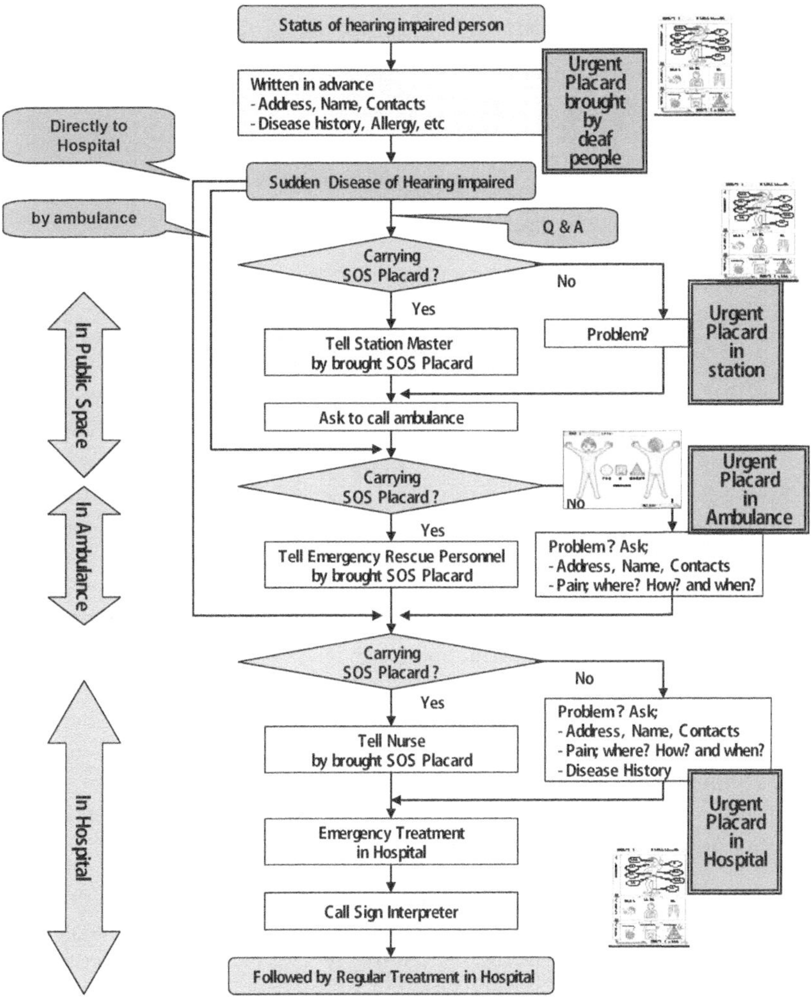

Fig. 2. The urgent collaboration service placard usage flow chart

The placard is also useful when the emergency rescue personnel will ask the patients name address and somebody relative to contact. This case is the first priority to prepare the placard for its effectiveness and efficiency nature. Finally it is in the hospital to communicate between the person and nurses, therapists or doctors. This placard will be useful when the disabled patient visit the hospital directly at the first stage of the registration. In the hospital matter the situation is normally not so much urgent since the first aid will be completed in the ambulance and sign interpreter may be with the patient [7,8].

The placard is to be used not only by the deaf people but also foreign people without speaking native language, elderly people with hearing problem and those of throat disease.

5 Usability Rvaluation

International Standard Organization (ISO) established ISO13407 as Human Centred Design (HCD) in 1998 [9]. It consists of following five steps;

1. Plan the human centred process,
2. Specify the context of use,
3. Specify user and organizational requirements,
4. Produce design solutions,
5. Evaluate designs against user requirements.

Until now the first HCD design cycle with the requirement survey and evaluation were experienced with paper prototyping over the urgent collaboration service placard. Usability is measured by effectiveness, efficiency and satisfaction [2]. In this case, the effectiveness means to inform the patient pain portion to the emergency rescue personnel. The efficiency is to tell in quicker time comparing with situation without the placard. The satisfaction is rather mental relief by carrying the placard. Considering the nature of the placard in such urgent situation to use, efficiency is the most important factor. Then there were prepared three simulated situations at the time of evaluations [10].

In usability evaluation testing, hearing disabled persons evaluated the placards. Three tasks were evaluated using and not using the placard. The first task is that the disabled person asks the conductor to call the ambulance in the railway station. The second is to tell the patient pain portion to the emergency rescue personnel in the ambulance. In this case the patient is necessary to inform the emergency rescue personnel about the name, home address, birth date or age, somebody to contact general practice and pain portion and its severity. The final one is to tell the complaint to the nurses in the hospital. In this case the general inquiry items are the same as the ambulance but there are extra are the patient's disease history and allergic reaction. These three simulated situations are compared with and without the placard. Four subjects with deaf disabled are participated. They are one for the railway task, two for ambulance task and the other one for hospital task. At the rail way station task, it is recognized that without placard and with was 51 seconds and 22 seconds respectably. At the ambulance task, it is found that without placard and with was 234 seconds and 152 seconds respectably for the first subject. The second subject was 321 seconds and 234 seconds respectably. In average it is about 30% swiftness improvement in the communication. At the hospital task, it is found that without placard and with was 269 seconds and 201 seconds respectably. The total results showed about 35% improvements of the communication time through all situations (Table 2).

Looking at hearing impaired recent activities they carry and use mobiles mail of mobile phones. Until now phones are out of reach for them. However the introduction of mobile mail becomes essential tool for them. Then the Emergency placard is downloaded into not only Personal Computer (PC), but mobile phones such as third generation of mobile phone standards and technology (G3) and iPhone by Apple Inc.

Table 2. Evaluation results

Places	Tasks (Communication)	Subjects initials	Person to correspond	Without the card	With card
Railway station	Call the ambulance	SU	Conductor (HI)	51 sec.	22 sec.
Ambulance	Inform the emergency personnel - Name - Address - Age, Birth date - Someboy to contact - General Practice - Pain portion, severeness	IS	Lifesaver (YA)	3 min. 54 sec.	2 min. 32sec.
		YO		5 min. 21 sec.	3 min. 54 sec.
Hospital	Inform the emergency personnel - Name - Address - Age, Birth date - Someboy to contact - General Practice - Pain portion, severeness - Disease history - Allegic reaction	FU	Nurse (SA)	4 min. 29 sec.	3 min. 21 sec.

Then the urgent collaboration service placard is appeared on WEB site, (http://www1s.wisnet.ne.jp/~aajd/) in order to be downloaded by anybody who are necessary. In this case the main users are hearing impaired with personally carrying it. There are also some inquiries from fire brigade to carry it in their ambulance.

At the second HCD design cycle, these two models of Emergency placard on the mobile phones are evaluated by 34 subjects who are between 20 to 60 years olds and 22 are hearing impaired among them. The questions are about its usefulness and measure of relief applied Semantic Differential (SD) Method between from 5 (Max.) to 1 (Min.). The usefulness result was 4.52 favorable with 0.51 of standard deviation (SD). Measure of relief result was 4.55 favorable with 0.62 SD.

6 Extended Applications

Basically the concept of universal design must be readily achievable [11] and be undue burden [12] for both disable people and manufactures. The urgent placard is one of the solutions to help the difficulties for deaf people without requiring many expenses. Currently the placard is reviewed by the deaf people, local fire brigades and foreign people from U.S.A. or Spain by interviews. Experimental trial to use this placard has been done in Kasuga Fire Department in Kyushu, Japan by carrying it in the ambulance (Figure 3). Observing at the actual situation of ambulance, it was found that the placard is much easier to use not portrait mode but landscape mode since the carried patient can only move the face right and left on the stretcher bed (Figure 4).

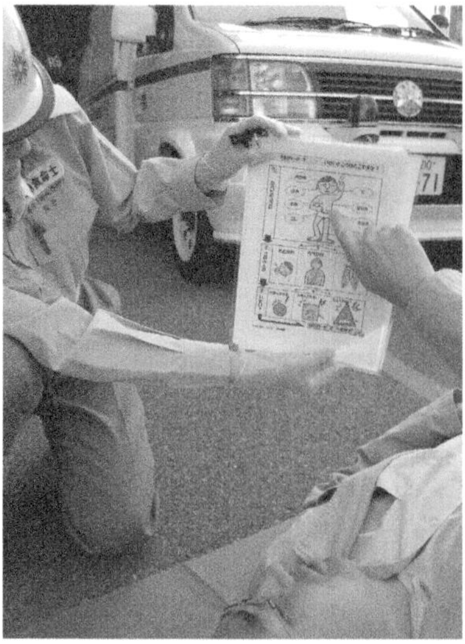

Fig. 3. The urgent collaboration service placard usage

From the stand point of universal and inclusive design, the urgent placard is extensively applicable for foreign people who can not speak local languages. It helps people such as hard of speaking and the patients of aphasia. Table 3 shows the foreign patient percentage in Keio University hospital during the years from 2000 to 2004. The figure is less than two-percent which must be not negligible small numbers since they are excluding Chinese and Korean people since their names seem to be similar to Japanese and hard to choose. Considering such background, several variations are prepared for not only Japanese but English, Spanish, Korean and Chinese languages (Figure 4).

Table 3. Foreign patient percentage in Keio University Hospital

Year	Japanese Male	Japanesse Female	Total Japanese	Foreign Male	Foreign Female	Total Foreign	Total Whole	Foreign Percentage
2000	2246	1660	3906	39	21	60	3966	1.5%
2001	2124	1627	3751	37	17	54	3805	1.4%
2002	1908	1501	3409	39	20	59	3468	1.7%
2003	1790	1487	3277	34	22	56	3333	1.7%
2004	1753	1434	3187	28	21	49	3236	1.5%

Average age: Japanese 46±23 years old
Average age: Foreign 36±18 years old

There are unexpectedly proposal from blind people by putting brails on the urgent placard. From their experience this will be helpful at the time of suddenly hard of speaking by the shock.

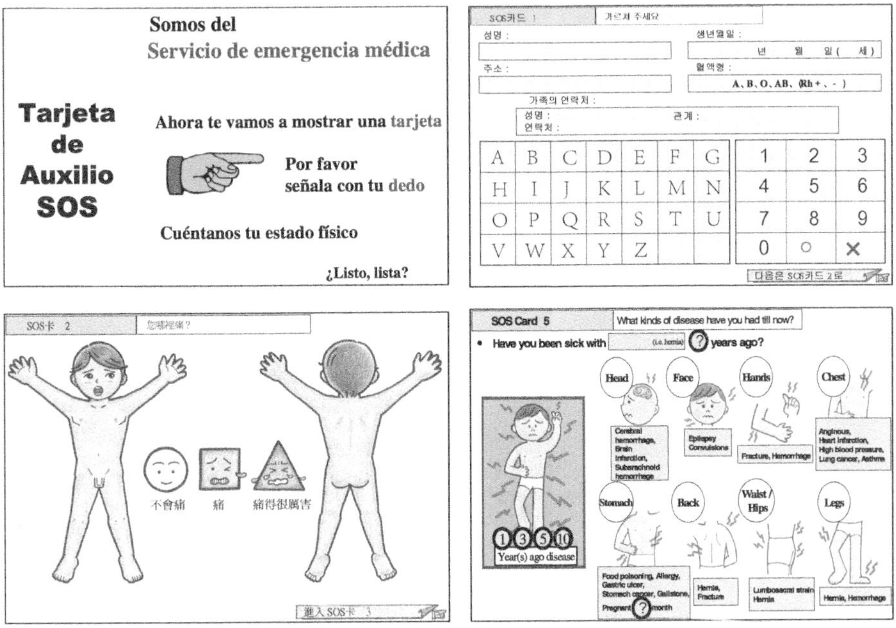

Fig. 4. The urgent collaboration service placards in ambulancefor Spanish, Chinese, Korean and English

7 Conclusion

This paper discusses the usefulness of the urgent collaboration service that is originally proposed by deaf people at such the occasion of disasters or accidents. At the very first of the survey many deaf people are asked their difficulties. From the point of context of use, the deaf patient complaint of pain/ache/grief ten patient complaint items are selected. Those requirements are drawn for visualization by pictograms or icons. It is then evaluated in the manner of the usability test. The results are that the time to communicate is shorter for about 30%. During the process it is found that the credibility to carry everyday brings much mental comfortableness and relief for the deaf people. The urgent collaboration service placard is also used by foreign people and is translated into English, Spanish, Korean and Chinese for the foreign people.

It is also being proposed to Japanese government to have the available both in the ambulance and hospitals or public spaces. This activity was received an official commendation by Japanese Fire Defense Agency in spring 2008.

Acknowledgements

Collected requirement data was provided by the Keio University Hospital Department of Emergency Medicine and the Tokyo Fire Department. The evaluations were done by Kasuga Fire Department in Kyushu/Japan and Suginami-ku Deaf Society in Tokyo. This research was supported by the Oki Volunteer Fund.

References

1. Miki, H., Hosono, N.: IT Universal Design Maruzen, Tokyo (2005)
2. ISO. Ergonomic requirements for office work with visual display terminals (VDTs) - Part 11: Guidance on usability 9241-11 (1998)
3. Buxton, B.: Sketching User Experience. Morgan Kaufmann, San Francisco (2007)
4. Ware, C.: Visual Thinking for Design. Morgan Kaufmann, San Francisco (2008)
5. Gallagher, M., Savard, L.: 1000 icons, symbols + pictograms, Rockport (2006)
6. Abdullah, R., Hubner, R.: Pictograms icons & Signs, Thames & Hudson (2006)
7. Hosono, N., Tomita, Y.: Emergency placard for deaf people. Journal of JES 41(4), 254–255 (2005)
8. Suzuki, M., Hosono, N., et al.: Proceedings of JES (2005)
9. Kurosu, M., Hosono, N., et al.: Usability Testing, Kyoritu, Tokyo (2003)
10. ISO. Human-centred design processes for interactive systems Human-centred design processes for interactive systems, 13497 (1999)
11. FCC. Section 255 Telecommunication access for people with disabilities (1996)
12. DOJ. Section 508 Accessibility Standards for E&IT (2000)

Improvement of Member's Concentration during Discussion

Junko Ichino[1], Kazuhiro Takeuchi[2], and Hitoshi Isahara[3]

[1] University of Electro-Communications
1-5-1 Chofugaoka, Chofu-shi, Tokyo 182-8585, Japan
[2] Osaka Electro-Communication University
18-8 Hatsu-cho, Neyagawa, Osaka 572-8530, Japan
[3] National Institute of Information and Communications Technology
3-5 Hikaridai, Seika-cho, Soraku-gun, Kyoto 619-0289, Japan
ichino@is.uec.ac.jp, takeuchi@isc.osakac.ac.jp,
isahara@nict.go.jp

Abstract. We developed face-to-face single display groupware (SDG) that has a novel circular frame and enhanced multiple pointers. The aim of the study is to encourage groups to concentrate on their discussions and thereby improve their cooperative decision-making. We analyzed the information input methods and information that should be shared in a face-to-face SDG by surveying related work. The interface was designed based on this analysis. Three experiments were carried out, with a general parallel input system tested for comparison. The results proved that group task accuracy, participants' memory, and their subjective evaluation improved when our SDG was used.

Keywords: face-to-face meeting support, concentration on discussion, single display groupware (SDG), pen-based, awareness.

1 Introduction

Various groups make critical decisions by coming together for sharing of information and to make choices on a set of options. We are interested in building interfaces that assist groups in improving their cooperative decision-making [3]. We treat the case of face-to-face and synchronous meetings in which a few people (about 4-5 participants) gather in front of a single display. Our goal is to encourage groups to concentrate on their discussions, in order to promote higher quality group decision-making.

For this goal, we examined multiple input methods and information-sharing methods as an interface to support a visual interaction. It can be difficult for many people to verbally communicate because only one person may talk at a time, and coordination among speakers is necessary. However, it is possible for many people to communicate visually. The visual interface has a lot of flexibility. It may be able to expand and narrow the range of behavior of each member of the group. We discuss the impact of changes in the visual interaction. We study how changes in the visual interaction affect the groups' whole communication including the verbal interaction.

M.J. Smith and G. Salvendy (Eds.): Human Interface, Part I, HCII 2009, LNCS 5617, pp. 59–68, 2009.

2 Related Work

This section surveys of some of the research on information input methods and information-sharing in groupware.

2.1 Information Input Methods for Single Display Groupware

Single display groupware is for users who share one display between them and write or edit on it. Input methods in studies on SDG are roughly categorized as parallel input and serial input. Parallel input methods allow multiple users to input at the same time, whereas serial input methods allow only one user who acquires the operational authority to input. Colab [10] pioneered research on the parallel input method. In their system, opinions were written on the single display and all users were able to focus on the input action. The display's context was reportedly difficult to understand as the users found it difficult to understand what all the others were writing at the same time. Inkpen revealed that children playing on a computer with multiple input devices tended to be more active [4]. CaptureLab [8] is a meeting support using the serial input method. The authors of the report on CaptureLab claimed that serial input makes participants focus on the content written on the display, and gives time to throw in verbal explanations during the input. Some studies (e.g., [9, 12]) have compared both input methods. Prante et al. [9] pointed out that although the serial input results in fewer opinions, these are still difficult to structure. The developers of Roomware [12] analyzed participant's glances. They reported that participants using the parallel input method kept gazing at the individual input terminals. On the other hand, participants using the serial input method mostly kept their gazes on the shared display rather than their own terminals. We compared the parallel and serial input methods of using shared displays (Table 1).

Table 1. Features of general input method in SDG

	General parallel input method	General serial input method
Direction of user's attention	Individual input terminal	Shared display
User's input frequency	High	Low
Consistency of display's context	Difficult to maintain	Easy to maintain
Focusing on the most important points	Difficult	Easy

2.2 Information Sharing in Groupware

There is obviously presence in a face-to-face environment. Such presence is missing from groupware operating in a distributed environment. There have been many studies [7, 11] on how to supplement for the lack of presence in a distributed environment. Dourish et al. [2] defined them by using the word "awareness".

On the other hand, participants train their eyes on input or output devices rather than on the participants when groupware is used in a face-to-face environment. In fact, users tend to busy themselves with the operation of the input device and do not look at others' faces [5]. Users fix their eyes on another's input or operations for passing operational authority on the shared display [12]. The above observations suggest the possibility that in a face-to-face environment, the use of a shared display may actually decreases awareness about surrounding persons and may disturb rather than enhance communications. There are only a few studies that consider support in a face-to-face environment from such an angle. DiMicco et al. [1] and Iqbal et al. [6] support face-to-face meetings by visualizing information about participant's activities on the screen. However, the user's burden is still large because s/he should take special care about the information presented during the discussion.

We consider information that should be shared in a meeting group in order to support concentration during discussion in a face-to-face environment, because new groupware media will become part of that and other meeting environments.

3 Face-to-Face Single Display Groupware Encouraging Member's Concentration during Discussion

Here, we gain insight into the user interface requirements for an input method and information sharing to support concentration during discussion in face-to-face meetings using SDG based on the previous section.

3.1 SDG User Interface Requirements

Input method to encourage independent interaction. To establish a consensus based on a mutual understanding between participants, a discussion should include various perspectives. Therefore, various opinions need to be presented on a shared display. New ideas synergistically develop from viewing the others' ideas presented on the display. To write positively on an SDG, the user must easily be able to adjust the position, range, and timing of her or his writing. During a meeting, a user has to plan not only "What do I say?" but also "When do I start writing?", "Where do I write?", and so on. In a word, input operations in SDG require ease of coordination of when or where each user writes on a shared display. The requirements for the input method to support positive participation in a face-to-face SDG are therefore that a user should be able to interact with others through an easy operation for transferring the right to write.

Information sharing to help user's concentrate on discussion. Firstly, to make continuous involvement that leads to consensus building, participants should always be able to understand a presented idea; there be shared attention among participants. Secondly, a user should be able to promptly perceive the represented information, to check its meaning, and to give responses to make the conversation go smoothly. Therefore, it is necessary to present information that encourages spontaneous participation, starting with noticing changes in others' behavior. Thirdly, enabling participants to reflect upon their behavior is important for activating a discussion. If a user becomes aware of someone with a negative attitude, s/he can encourage that person to write or speak. Moreover, a participant who over participates may find it easier to refrain from speaking. With such behavioral control, we believe that a group will have an animated discussion expressing various viewpoints. The requirements for information sharing to support positive participation in a face-to-face SDG are therefore that a user should be able to grasp the focus of an argument easily, that a user should be able to easily perceive change in another's behavior, and that a user should be able to easily assess how s/he and others have been participating.

Summary of user interface requirements. Table 2 summarizes the user interface requirements for a face-to-face SDG to support positive participation.

Table 2. Interface requirements

Requirement 1	**Positive attitude:** a user should be able to interact with others positively.
Requirement 2	**Concentration:** a user should be able to grasp of the point of discussion easily.
Requirement 3	**Awareness of others:** a user should be able to be easily aware of the change in the behavior of others.
Requirement 4	**Historical awareness:** a user should be able to easily assess how s/he and others have been participating.

3.2 User Interface Design

We designed a user interface of the meeting support tool that addresses the four requirements summarized in Table 2. The features are a novel circular frame and enhanced multiple pointers.

Interface that meets requirements 1 and 2. With requirement 2 in mind, we designed a circular frame whose center is the center of balance of all users' pen pointer locations on the shared display (Figure 1). Its appearance calls the user's attention. The frame interlocks with the movements of each pointer so that all participants always maintain a common recognition of the discussion on the shared display. The frame size is fixed according to the result of the preliminary test. To satisfy requirement 1, we designed a parallel input interface that limits the input range within the frame. A user's pointer might be located outside of the frame because the frame is located at the center of balance of all users' pointer locations. Users whose pointers are inside the frame can write in the frame of the display at the same time. That makes it unnecessary for users to take turns writing, and it helps to activate writing interactions.

Interface that meets requirement 3. Our research deals with the activities of writing and speaking as observable user behaviors in SDG. In this case, a change in the behavior indicates events such as beginning writing, stopping writing, beginning speaking, and stopping speaking. With requirement 3 in mind, we visually represented the user's writing behavior by making multiple pointers transparent on the display (Figure 2). Multiple pointers are spontaneously attracts users' gazes. User pointers while writing are displayed with a translucent color, while other pointers are displayed with an opaque color. Translucent colors are used because everyone should be able to see the writing under the pointer as the user writes it. Pointing pointers are displayed with a shadow.

Interface that meets requirement 4. With requirement 4 in mind, we made it so that the size of the user's pointer represents the relative proportion of characters a user has written from the beginning of the meeting to the present moment. The size of each pointer shows the relative proportion of each user's writing participation (Figure 3). The size of the pointer changes in real time to show everyone how much a user has been writing during the meeting.

With requirement 4 in mind, we visually represented the proportion of time a user has spent speaking from the beginning of the meeting to the present moment, by using the angle of the user's arc in the circular frame. The frame consists of different colored arcs for each user. The arcs show the relative proportion of each user's speaking participation as in a pie chart (Figure 4).

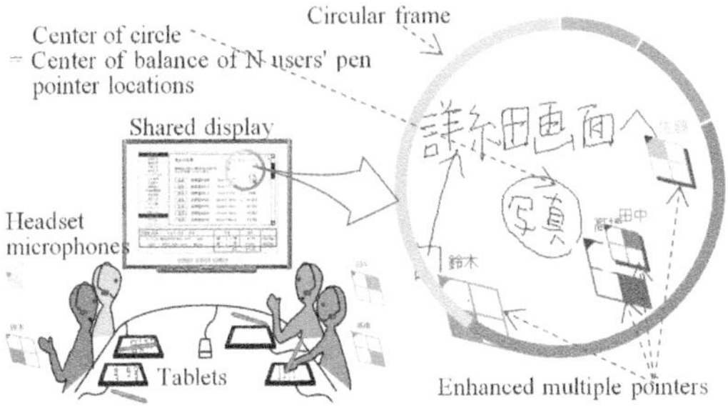

Fig. 1. Circular frame and enhanced multiple pointers

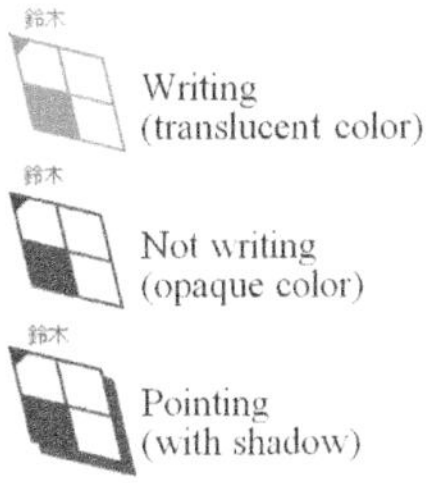

Fig. 2. Representation of writing states using multiple pointers

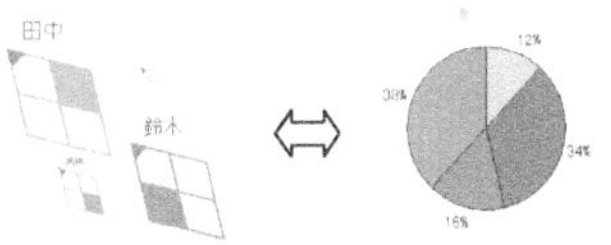

Fig. 3. Representation of proportion of characters written using multiple pointers

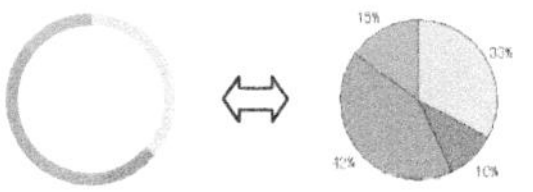

Fig. 4. Representation of proportion of time spent speaking using frame arcs

4 Basic Experiments

We developed a prototype and carried out two experiments. The experiments compared the prototypes with a general parallel input system. The reason we chose parallel input method as the object of comparison is that it is more similar to our system than the serial method. The experiments assessed the basic functions to meet requirements 1, 2, and 3 in Table 2. The interfaces corresponding to requirements 1, 2, and 3 are closely related to the user's interactions. On the other hand, the scale of the interaction involved in requirement 4 is larger.

4.1 Viewpoints

Group task accuracy. As for the first aspect, we measured the accuracy of the task that the group completed.

Participants' memory accuracy. As for the second aspect, we measured the accuracy of each participant's memory about the results of the completed task. Each participant was given a questionnaire after a fixed period of time had elapsed from the task's completion.

4.2 Methods

Design. Our system (called "frame mode") was compared with a general parallel input system (called "parallel mode"). For the frame mode, the systems were prepared without the function to satisfy requirement 4 (Figure 3 and Figure 4). Therefore, a colorless frame was prepared because the function to indicate user's speaking percentages was not included. The two experimental systems are shown in Figure 5.

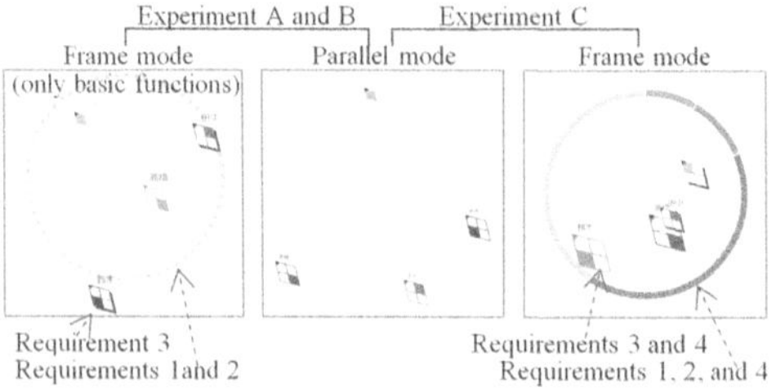

Fig. 5. Experimental conditions in the experiments

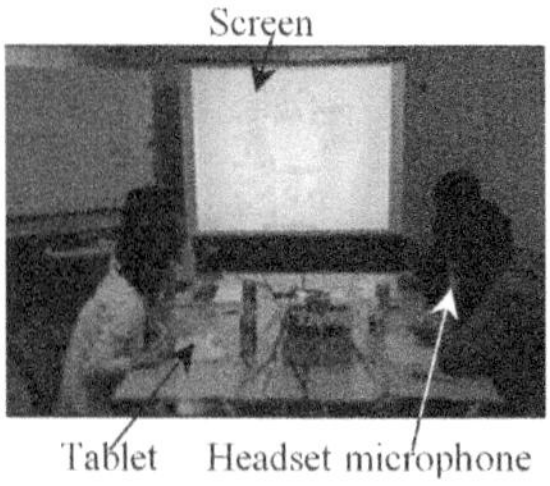

Fig. 6. A scene during the experiment

Participants. Twenty-four men and women in their 20s and 30s participated in experiment A. They were randomly assigned to six four-person groups. Three groups performed the tasks in frame mode, and the other three groups performed the tasks in parallel mode. Thirty-two people participated in experiment B. They were assigned to eight four-person groups. Four groups performed the tasks in frame mode and the other four groups performed the tasks in parallel mode.

Environment. Four users of a group sat face to face at a rectangular table. Each user had a tablet device (WACOM FAVO CTE-640) and wore a microphone (Logicool Internet Chat Headset A450). The screen was placed in front of them as a shared display. The input data of the four people were displayed in this screen.

4.3 Task for Experiment A: Artificial Task

We prepared a simple task to test the interface's usability. In this experiment, there were no differences in knowledge and memory among the participants. The group put down the locations, names, and symbols of twelve stores on a blank map on information sheets given to each subject. A total of twenty-four different information cards composed of twelve cards with the locations and names of stores and twelve cards with symbols of the stores were prepared. The cards were equally divided among the participants in a group. All participants were able to check these cards during the task. Figure 7 shows information cards distributed to participants and an example of a completed store map. They were told to answer as quickly as possible.

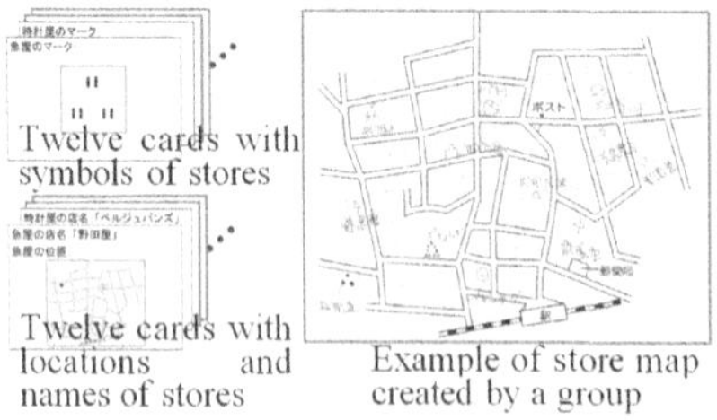

Fig. 7. Task for experiment A

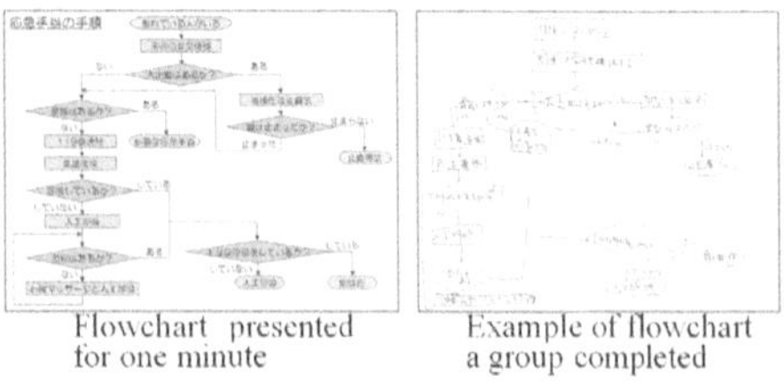

Fig. 8. Task for experiment B

4.4 Task for Experiment B: General Task

Generally, there are differences in participants' knowledge and memory. We assumed that this task would simulate a realistic situation wherein participants would each have vague and non-symmetric knowledge and would need to solve a problem cooperatively. First, a group was shown a flowchart that showed a procedure of emergency operations, for one minute. The flowchart was hidden afterwards. They then reproduced the flowchart on a shared display while discussing what they remembered. The word "reproduce" means not a copy of the chart's appearances but a copy of the substance of the flowchart. The time allowed for a group to reproduce the flowchart was 20 minutes. Figure 8 shows the flowchart that was presented for one minute and a flowchart reproduced by a group.

4.5 Results and Considerations

Group task accuracy. We evaluated how consistent the figure that each group wrote was with the original one. In experiment A, we determined whether the store's map that the participant group completed corresponded to the information on the cards distributed to the participants. In experiment B, we determined how well the flowchart completed by the group completed corresponded to the flowchart presented at the beginning. Table 3 lists the group task accuracies on a 100-point scale and the times taken to complete the task. In experiment A, the groups for both modes completed the map perfectly. We asked the participants to answer as quickly as possible. There was no significant difference in time taken between modes, although the parallel mode allowed participants to fill in the map faster than the frame mode did. In experiment B, the frame mode groups reproduced flowcharts with a significantly higher degree of accuracy than the parallel mode groups did.

Participants' memory accuracy. The participants answered the questions on the results of the task that their group had completed. In experiment A, they were asked if they recognized the locations, names, and symbols of the twelve stores written by the group. In experiment B, they were asked to recall the parts of the flowchart written by the group. Table 3 lists the participants' memory accuracy on a 100-point scale. In both experiments, the frame mode enabled participants to memorize the results of the task the group completed with a significantly higher degree of accuracy than the parallel mode. Frame mode's accuracy was especially superior in experiment A, although the time to complete the task was also a little longer.

Table 3. Results of experiments

		Frame		Parallel		One-tailed t-test	
		Mean	Std	Mean	Std		
Group's task accuracy	Exp. A	100.00	0.00	100.00	0.00	–	–
	Exp. B	93.33	8.16	78.00	17.89	t=1.891	**p<0.05**
Group's task time (minutes)	Exp. A	7.75	1.56	6.29	0.78	t=1.444	p=0.111
	Exp. B	fixed time (twenty minutes)					
Participants' memory accuracy	Exp. A	26.62	17.98	12.04	7.97	t=2.568	**p<0.01**
	Exp. B	67.15	29.64	44.17	22.34	t=2.246	**p<0.05**
	Exp. C	100.00	0.00	62.50	45.21	t=2.346	**p<0.05**

5 Application Experiment

Next, we devised a realistic task in which a group discusses graphical user interface for a new web site and determines functions to be included. This experiment determined whether the functions could meet requirement 4.

5.2 Methods

Design. For the frame mode, the function to meet requirement 4 was added to the system used in experiments A and B. For the parallel mode, the same system as in experiments A and B was used. The two experimental systems are shown in Figure 5.

Participants. Sixteen men and women in their 20s and 30s participated. They were randomly assigned to four four-person groups. Two groups performed the task in frame mode, and the other two groups performed the task in parallel mode.

5.3 Task for Experiment C: Realistic Task

We devised a realistic task for discussing online stores that a machine tool company had just established. The group discussed how it was necessary to improve the draft web site so that customers could order items without error. They refined the draft version (Figure 9) on the shared display. We divided the group discussion into a first half and latter half. The each member of the group presented their opinions for 20 minutes in the first half. Afterwards all their opinions had been stated or written down, they corrected or added to the draft version on the shared display. They discussed their opinions and selected the best three.

5.4 Results and Considerations

Participants' memory accuracy. Participants were asked to recall the three opinions that the group finally chose, and described them in the order that the group chose. Moreover, they were asked to recall the details of each opinion, the reasons about why the site should be improved, and the reason for their choice. Table 3 lists the results on a 100-point scale. This result show that the frame mode participants could memorize the results of the task with a higher degree of accuracy than the parallel mode participants could memorize.

Fig. 9. Task for experiment C

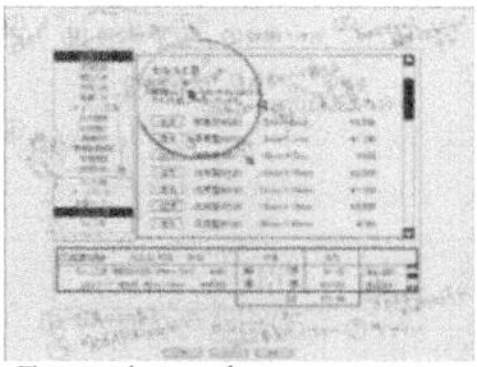

Examples that groups completed in frame mode

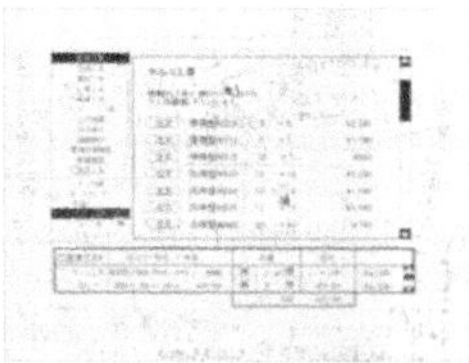

Examples that groups completed in parallel mode

Fig. 10. Screenshots of shared displays in experiment C

Variation in amount of written and spoken opinions. The proportions that each participant wrote and spoke in experiments A an B were compared with those in experiment C. The presentation functions of these proportions were embedded in experiment A and B, but not in experiment C. Figure 11 plots the proportion of distances that each participant drew with their pen. Figure 12 plots of proportion of words that each participant spoke.

The task of experiment A is artificial task that would not cause a difference in knowledge or memory among the participants. In fact, the comparison of these variations in the frame mode and these variations in parallel mode in experiment A is the simplest without presentation functions showing the proportion of expressed opinions written and spoken by the participants. The variation in the amount of written opinions was small in both modes. It seems that this is because the cards containing the same amounts of information were written on the shared display, since we asked the participants to answer as quickly as possible. The variation in the amount of spoken opinions was wider than the variation in amount of written opinions in both modes.

The task of experiment B supposed that there were the differences in knowledge and memory among the participants. The frame mode resulted in wider variations in the amount of expressed opinions, both written and spoken. Moreover, the variation in experiment B was wider than in experiment A. We think that this difference reflected the difference between experiment A (which controlled knowledge and memory) and experiment B (which did not control them).

The task of experiment C was a realistic one supposing that there were differences in the knowledge and memory among the participants. The frame mode resulted in wider variations in amount of expressed opinions, both written and spoken, for the participants of experiment B, but resulted in narrower variations in experiment C. This difference suggests the possibility that the presentation functions for showing the proportion of written and spoken expressed opinions had some effect on the participants. Instead of exploring this possibility directly, we attempted to analyze the following qualitative changes. For instance, which kind of written or spoken opinions influenced the change? Or, did the content of the written or spoken opinions change?

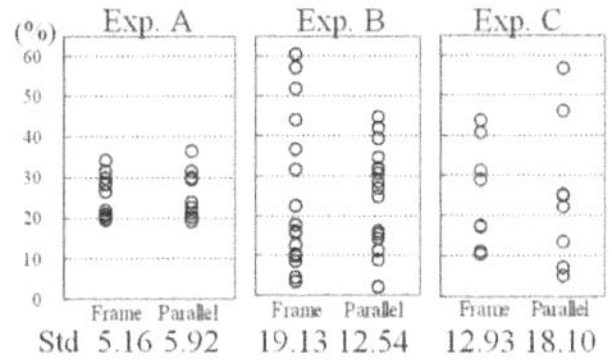

Fig. 11. Plot of individual writing rates

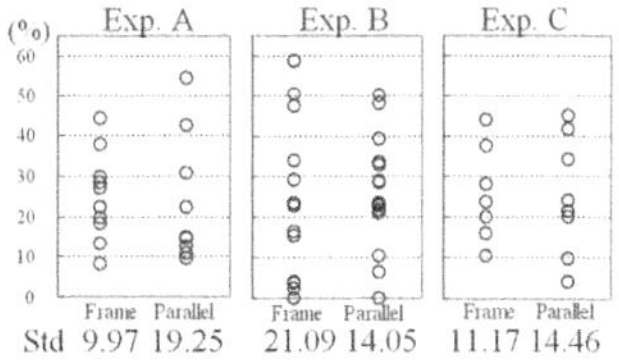

Fig. 12. Plot of individual of speaking rates

6 Conclusion

We described face-to-face single display groupware that has a novel circular frame and enhanced multiple pointers. We carried out three experiments (two basic, one application). The groupware's effectiveness was confirmed by comparing it with a general parallel input system. The experimental results indicated that the group task accuracy and the participants' memory improved when our system was used.

References

1. DiMicco, J.M., Pandolfo, A., Bender, W.: Influencing group participation with a shared display. In: Proc. CSCW 2004, pp. 614–623. ACM Press, New York (2004)
2. Dourish, P., Bellotti, V.: Awareness and coordination in shared workspaces. In: Proc. CSCW 1992, pp. 107–114. ACM Press, New York (1992)
3. Ichino, J., Takeuchi, K., Isahara, H.: Face-to-face Single Display Groupware Encouraging Positive Participation. In: Proc. UIST 2006, Poster and two-page paper (2006)
4. Inkpen, K.M., Ho-Ching, W., Kuederle, O., Scott, S.D., Shoemaker, G.B.D.: This is fun! We're all best friends and we're all playing: Supporting children's synchronous collaboration. In: Proc. CSCL 1999, pp. 252–259 (1999)
5. Ishii, H.: Design of Realtime Groupware. Journal of IPSJ 34(8), 1017–1027 (1993)
6. Iqbal, R., Sturm, J., Kulyk, O., Wang, J., Terken, J.: User-centred design and evaluation of ubiquitous services. In: Proc. SIGDOC 2005, pp. 138–145 (2005)
7. Kam, M., Wang, J., Iles, A., Tse, E., Chiu, J., Glaser, D., Tarshish, O., Canny, J.: Livenotes: a system for cooperative and augmented note-taking in lectures. In: Proc. CHI 2005, pp. 531–540 (2005)
8. Mantei, M.: Capturing the Capture Lab Concepts: A Case Study in the Design of Computer Supported Meeting Environments. In: Proc. CSCW 1988, pp. 257–270 (1988)
9. Prante, T., Magerkurth, C., Streitz, N.: Developing CSCW tools for idea finding - empirical results and implications for design. In: Proc. CSCW 2002, pp. 106–115 (2002)
10. Stefik, M., Foster, G., Bobrow, D.G., Kahn, K., Lanning, S., Suchman, L.: Beyond the chalkboard, Computer support for collaboration and problem solving in meetings. Comm. of the ACM 30(1), 32–47 (1987)
11. Tang, A., Boyle, M., Greenberg, S.: Display and presence disparity in Mixed Presence Groupware. In: Proc. AUIC 2004, pp. 73–82 (2004)
12. Watanabe, S., Obata, A., Matsukura, R., Sasaki, K.: A Comparative Study of Information Input Method to an Electronic White Board in face-to-face Meetings. IPSJ SIG Notes 97(046), 31–36 (1997)

Representation Method for Engineering Perspective

Gaku Ishii and Masatake Sakuma

Toshiba Corporation
Komukai-Toshiba-cho, Saiwai-ku, Kawasaki 212-8582, Japan
{gaku.ishii,masatake.sakuma}@toshiba.co.jp

Abstract. For most Japanese Companies, knowledge succession is important for retaining technical competitiveness in the context of the rapid change of human resources. While efficient and effective knowledge succession requires the support of an IT system, the requirements for the supporting system are severe. That is, the system not only must offer the knowledge users need, but also must represent how to apprehend and utilize the knowledge.This paper introduces a method of representing technical knowledge considered in the development process of a knowledge succession support system (IVYs). It focuses on the user interface representing the standard or situation-dependent perspective of technical knowledge that experts define.

Keywords: KM (knowledge management), knowledge succession, engineering-support system, web-based application, user interface.

1 Introduction

Japan is confronted by "the year 2007 problem," which was identified and named by Teiichi Ariga, a former director of CSK Holdings Corp. The year 2007 problem refers to the concern that the retirement of Japan's baby boomers, i.e., those born in the period 1947-1952, will cause serious problems for Japanese companies and industry. The primary concern is that the accumulated experience and know-how of the baby boomers will be lost. In particular, there is concern that there will be insufficient engineers to operate computer programs of corporate mission-critical systems, because those programs were developed mostly by baby boomers [1]. Fig.1 shows the demographics of Japan [2].

Subsequently, there was a growing realization in Japan that the year 2007 problem might have a much larger impact on various technical fields than originally anticipated [3]. Moreover, the current global recession is compelling many companies to execute structural reforms and the related downsizing of workforces, in addition to natural attrition, in exacerbating companies' concerns about the risk of losing knowledge.

In the circumstances, knowledge succession to maintain technical competitiveness has become a pressing issue. Increasingly, Japanese companies are constructing systems to support knowledge succession [4] [5]. Toshiba is one such company and this paper introduces its activities to ensure knowledge succession.

M.J. Smith and G. Salvendy (Eds.): Human Interface, Part I, HCII 2009, LNCS 5617, pp. 69–76, 2009.

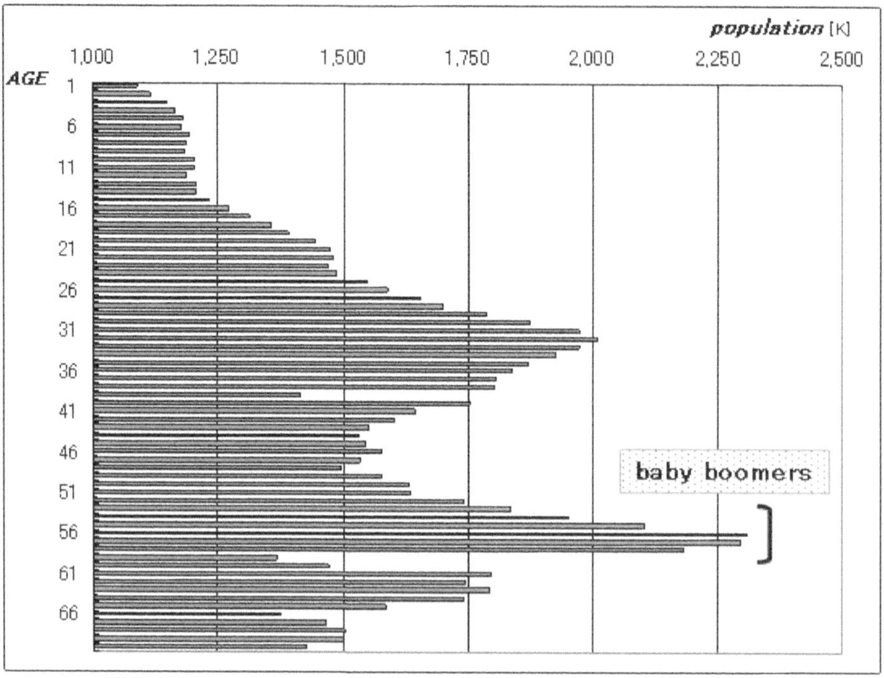

Fig. 1. Demographics of Japan, 2005

2 Knowledge Succession Support System: IVYs

IVYs (Intellectual Voyage System) is a web-based application designed by Toshiba that supports knowledge succession. IVYs has been used since April 2008 as an engineering-support system by a mechatronics engineering department consisting of about 60 engineers.

IVYs-knowledge, i.e., the knowledge registered in IVYs is mainly technical knowledge, such as technical standards and exemplary cases applied in the process of product design and development, and various functions making technical knowledge easy for engineers to use are implemented. IVYs-knowledge is mainly applied in engineering subprocesses, such as the design of size, form, and mechanism, or the selection of parts or materials.

We confirmed by interview that everyone in the department was aware of IVYs and used it for these subprocesses. On the other hand, science these subprocesses are not routine or part of daily work, engineers don't require or use IVYs-knowledge daily. Therefore, access frequency is not a reliable indication of the usefulness of IVYs. Although it has already been confirmed that IVYs has reduced the searching time for required technical knowledge by 80 to 90 %, it is important to establish a new method of evaluating the true value of IVYs.

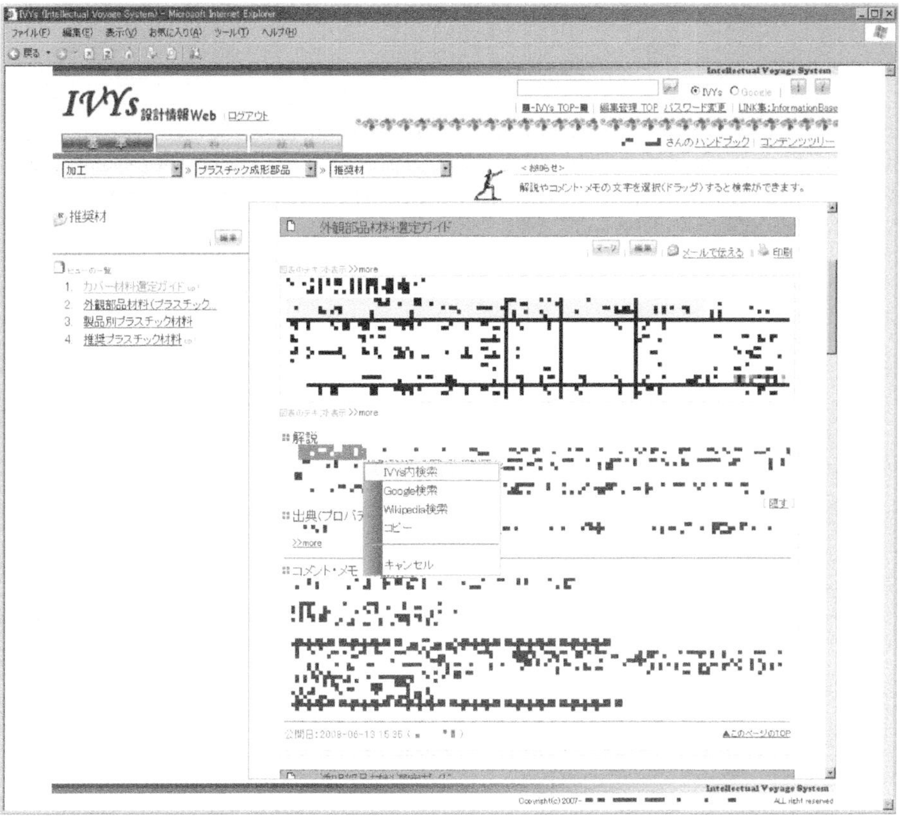

Fig. 2. IVYs (Intellectual Voyage System)

3 Requirements for Knowledge Succession in Engineering Fields

In order to practice knowledge succession in engineering fields, the support system not only must offer the knowledge engineers need, but must also represent the context of knowledge so that engineers can deepen their understanding of the knowledge. Since technology advances rapidly, engineers have to find the technical knowledge they require by themselves, i.e., engineers must have the capability to discover the field of required knowledge and its usage. An engineer with this capability can understand what query should be put to which system or whom, i.e., the engineer can generate the optimal inquiry for acquiring knowledge.

Therefore, for an engineer to strengthen this capability, the support system must represent technical knowledge, including its classification and usage scenarios, i.e., "the perspective of technical knowledge." Through the experience of knowledge succession in the mechatronics engineering department, functions for representing and sharing the perspective are implemented in IVYs.

4 Functions of Representing and Sharing the Perspective

4.1 Representing the Standard Classification of Technical Knowledge

In order to represent the standard classification of technical knowledge, IVYs-knowledge is arranged in a classification tree according to the characteristics of the network structure that allows a child node to have two or more parent nodes. The classification tree in IVYs is expressed using three forms that support engineers understanding of the classification of technical knowledge. Fig.3 shows these two forms for representing the classification tree.

(1) The list box that is a standard component of HTML

(2) The contents tree illustrating the classification tree (IVYs original UI)

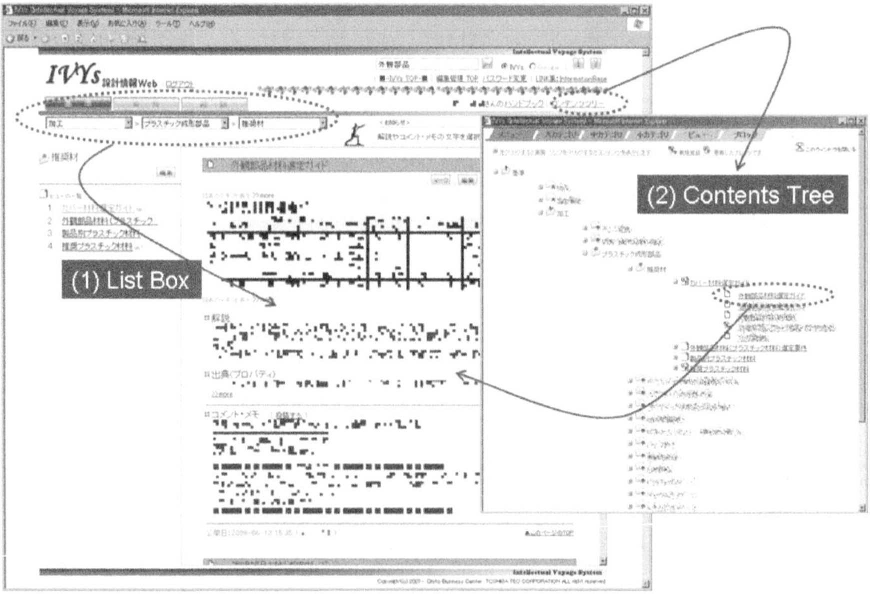

Fig. 3. HTML UI for representing the classification tree

Since most users are experienced in operation of the list box (1), it is an efficient interface for the user who understands the classification of technical knowledge. However, since the user cannot look at all choices of all classes at once, it is difficult for the user to trace a quick path by selecting the category to browse the knowledge.

Next, (2) has an advantage in that it is easier to look through the classification classes and categories of technical knowledge compared with (1). On the other hand, since the user may have to learn the operation of the application, beginner-oriented help and an intuitive user interface become more important.

In any event, since both (1) and (2) are premised on understanding the classification of technical knowledge, the situation for a user who does not understand the classification is akin to being lost in a maze. Therefore, IVYs' search function displays the "category chain" representing the classification to which the searched technical knowledge belongs. The user can access searched technical knowledge from the hyperlink in the "category chain" (Fig.4), and this function servers as a refinement that makes the classification tree more familiar for the user.

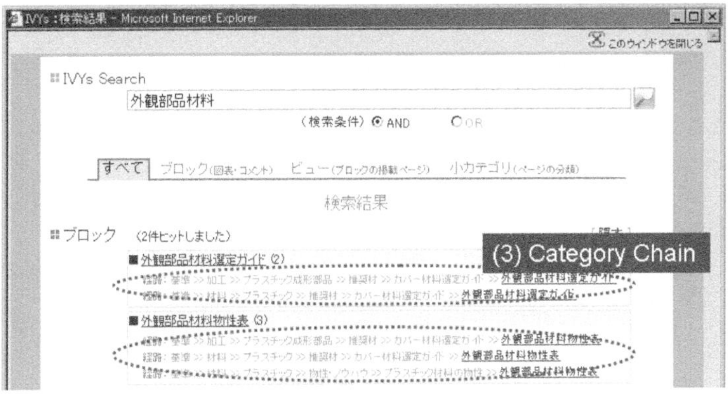

Fig. 4. Category Chain for representing the classification tree

4.2 Representing the Usage Scene of Technical Knowledge

IVYs' knowledge manager disassembles original engineering documentation into the minimum unit, called "BLOCK", which engineers can utilize, and registers them in IVYs as technical knowledge. Furthermore, the knowledge manager defines the required BLOCKs for every usage scenario as "VIEW" (Fig.5). VIEW equivalent to a page of a website is linked to the classification tree, and each VIEW represents a usage scenario of technical knowledge (Fig.6).

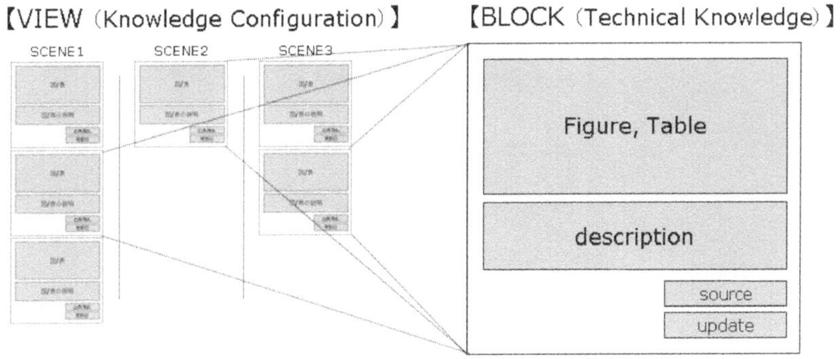

Fig. 5. IVYs' BLOCK & VIEW

A usage scenario is a certain scene in the engineering process such as "selecting a plastic component." Technical knowledge is needed for the consideration process of such selection. Therefore, in the case of the scenario "selection of a plastic component", technical knowledge required for the selection is constituted in accordance with the sequence of consideration, and shown to the user.

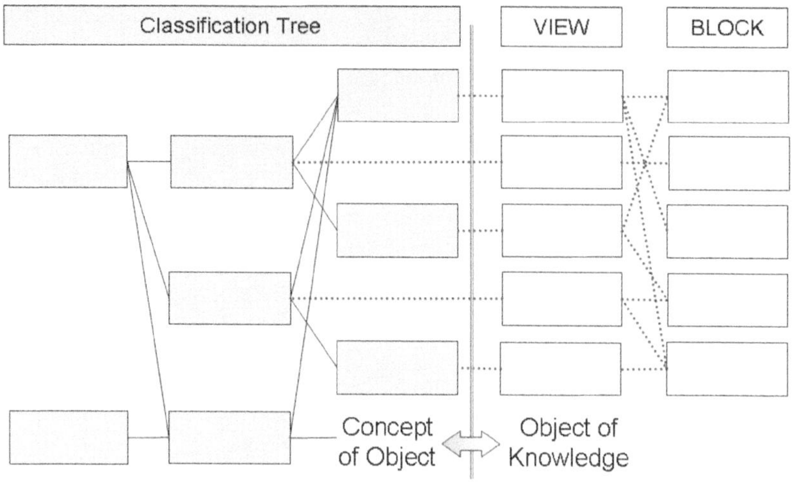

Fig. 6. Data Structure of IVYs

4.3 Representing and Sharing Situation-Dependent Perspective: Social Bookmarking

The usage scenarios of technical knowledge are defined in advance by the expert or the knowledge manager, and they can be considered to be the perspective represented by the expert that is valuable for inexperienced engineers. However, this is the standard

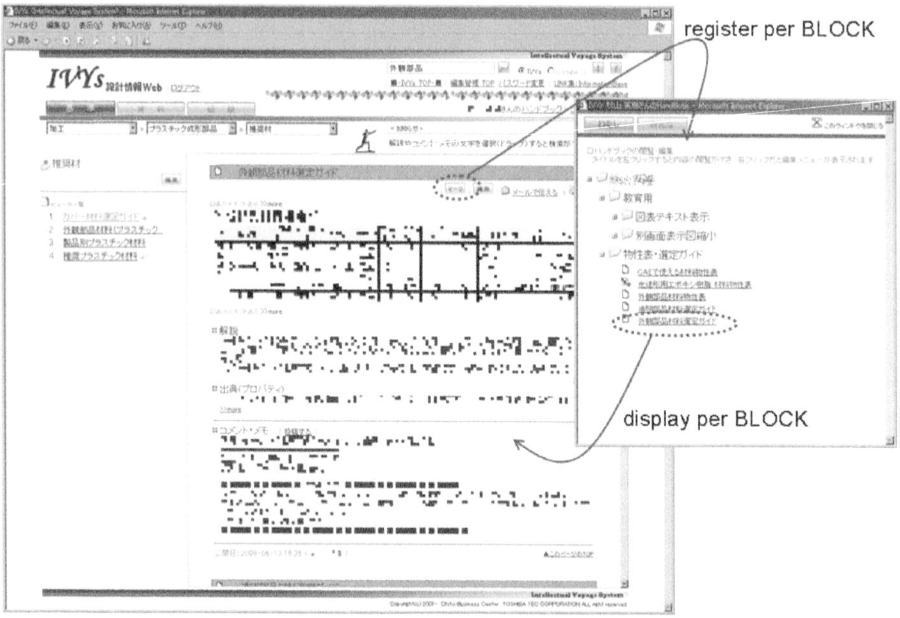

Fig. 7. IVYs HANDBOOK: Situation-dependent Classification Tree

perspective defined by the expert and it may not be applicable to actual engineering. Therefore, a function allowing each engineer to classify technical knowledge is also needed. Furthermore, in the actual engineering process, it is necessary to ensure that members of a project team have a good understanding of the technical knowledge that the project leader considers especially important, and, for that purpose, the members of the project team need a function enabling them to share the original classification.

This function named "HANDBOOK," which a form of Social Bookmarking, is intended to support succession of practical perspectives (Fig.7). The interface, which is the same as that of the contents tree explained above, can perform edit of an addition, deletion , display order change, etc. on a browser. In Addition, when maintaining social bookmarking on a company intranet, it is important to minimize the maintenance const. Therefore, the mechanism whereby the classification tree of each project is not neglected is implemented in HANDBOOK by setting up the display term.

5 Conclusion

At Toshiba, we are developing IVYs, a support system of knowledge succession, and are studying the method of representing "the perspective of technical knowledge" through the practice of IVYs.

It has already been confirmed that IVYs has reduced the searching time for required technical knowledge by 80 to 90 %. On the other hand, the impact of IVYs on knowledge succession will be evaluated from now on.

The ideal situation of knowledge succession in the engineering process is for the technical knowledge to be applied in engineering correctly. Therefore, the impact of knowledge succession should be evaluated on the basis of the observance rate of IVYs-knowledge in the engineering process. In this case, it is insufficient to evaluate the user's activity concerning the inspection of the knowledge on IVYs. It is necessary to evaluate the engineer's activity that applies IVYs-knowledge to engineering. For that purpose, a function enabling technical knowledge to be used in CAD (Computer Aided Design system) is required, and a mechanism connecting IVYs and CAD is currently under development. Analysis of the utilization of IVYs-knowledge on the extended mechanism is a subject for future work.

References

1. Yajima, N.: The 2007 problem of information system caused by the retirement of experts (April 2003) (in Japanese), `http://www.nikkeibp.co.jp/style/bizinno/conduct/article20030404.shtml`
2. Ministry of Internal Affairs and Communications, Population Census (in Japanese) (October 2005) World Wide Web, `http://www.stat.go.jp/data/kokusei/2005/kekkagai.htm` (retrieved 2/27/09)
3. Cabinet Office of Japan, The policy research report about participation in society by the elderly (in Japanese) (July 2005) World Wide Web, `http://www8.cao.go.jp/kourei/kenkyu/kenkyu.html` (retrieved 2/27/09)

4. Ishii, G., Nakayama, Y.: Method of utilizing the valuable knowledge of a company - integration of information lifecycle management and knowledge management. In: Papers presented at HCI International Conference 2005 [CD-ROM] (2005)
5. Ishii, G.: A Study of Knowledge Succession in Engineering Process Management. In: Papers presented at PICMET 2007 Conference [CD-ROM] (2007)
6. von Hippel, E.: "Sticky Information" and the locus of problem solving: Implications for innovation. Management Science 40(4), 429–439 (1994)
7. Szulanski, G.: Exploring Internal Stickiness: Impediments to the transfer of best practice within the firm. Strategic Management Journal 17, 27–43 (1996)
8. von Krogh, G., Ichijo, K., Nonaka, I.: Enabling Knowledge Creation: How to Unlock the Mystery of Tacit Knowledge and Release the Power of Innovation. Oxford University Press, New York (2000)

A Skill Transfer Method for Manual Machine Tool Operation Utilizing Cutting Sound

Takashi Kawashimo[1], Noriaki Sato[2], Daisuke Doyo[1], Michiko Anse[1], and Tsutomu Tabe[1]

[1] Aoyama Gakuin University 5-10-1 Fuchinobe, Sagamihara-shi, Kanagawa, 229-8558, Japan
a5705030@cc.aoyama.ac.jp
{doyo,anse,tabe}@ise.aoyama.ac.jp
[2] Murata Manufacturing Co., Ltd 1-10-1 Higashikoutari, Nagaokakyo-shi, Kyoto, 617-8555, Japan
sato_noriaki@murata.co.jp

Abstract. This study focuses on the inheritance of skills that has faced difficulties through the years in Japan. Sensorimotor knowledge, which is hard to be described by words alone, is often seen in technical skills in a manufacturing environment in Japan. Interpreting sensorimotor knowledge as formal knowledge, attempts have been made to impart technical skills but have faced barriers particularly in lathe processing due to no visual check of finished quality being allowed. Thus, this study suggested training on the inheritance of skills utilizing the level of cutting sound generated in process. Novice workers served as experimental subjects for an experiment to verify the effectiveness of proposed training.

1 Introduction

Japan is rapidly graying in recent years, which has wielded an influence on manufacturing industry. In 2007, baby boomers including skilled workers who anchored a high-growth period in Japan have reached the retirement zone, and this phenomenon has raised the issues of "the inheritance of skills" that many of manufacturers are groping for an avenue to hand down proficient skills to the generations. Upon enforcement of the amended Law concerning Stabilization of Employment of Older Persons, the manufacturers have taken preventive measures including the postponement of the employment period and the acquisition of skilled retirees from other companies to stem the outflow of technical skills temporarily. The measures, however, provide no fundamental solution, and it is still in need of instituting essential measures for skill inheritance.

Manufacturing Research [1] indicates that "machining and assembling" are processes in the course of manufacturing where skill inheritance issues are likely to manifest themselves. This study examines the inheritance of skills in lathe processing, Manual Machine Tool Operation, which is a major process in "machining and assembling."

Intended for novice and skilled workers, a field hearing as a pilot study was conducted in a medical precision equipment manufacturing plant on the subject of issues

M.J. Smith and G. Salvendy (Eds.): Human Interface, Part I, HCII 2009, LNCS 5617, pp. 77–86, 2009.

in the inheritance of skills in lathe processing. Preliminary findings showed that visual observation of a contact between a cutting tool and an object during precise and quick lathe processing was difficult and that a question as to what could be information for making a decision had arisen among novice workers. In their answers, skilled workers relied on their sensations in the hand and cutting sound for judgment. Their judgment criteria such as sensations in the hand and cutting sound are classified as sensorimotor knowledge that is indescribable. In particular, hand sensory information that skilled workers have will be acquired only after work is conducted on a level with them, which is difficult to get the difference in work level between novice and skilled workers across to novices. There have been earlier studies [2] and [3] on cutting sound, but the analysis of cutting sound has not taken shape and not been applied to the inheritance of skills. Focusing on auditory information, cutting sound generated in lathe processing, this study is to not only analyze cutting sound but devise a method for skill inheritance based on operation analysis so as to facilitate the inheritance. It is also intended to evaluate the possibility of improvement in work through the adoption of the proposed method and cutting sound-based training.

2 Lathe Processing and Cutting Sound

2.1 Sound Structure and Analysis Method

Lathe processing is defined as a machining operation that a rotating cylindrical object is turned and slotted with a cutting tool.

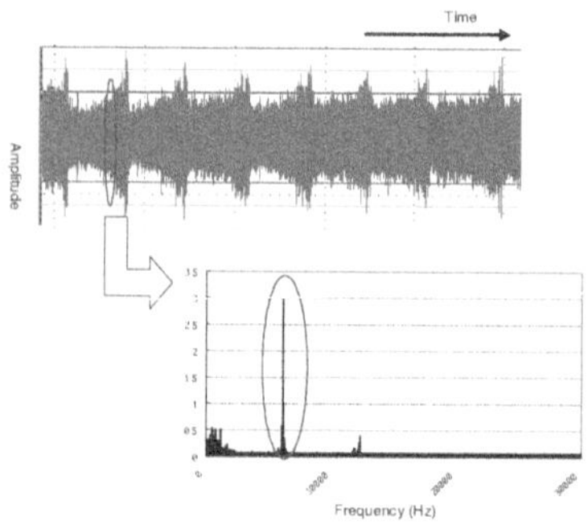

Fig. 1. FFT Separation of Frequency Components

The cutting tool operates interlocked with the lever and handle (this movement is hereinafter referred to as feed motion, the rate of feed motion as a feed speed, and the movement of the cutting tool as a feed rate), and cutting sound is produced by turning the object. Feed speed increases and decreases proportionately with a pushing force on the lever or rotating force on the handle, and variations in feed speed achieve a change in pitch of cutting sound.

Target operation in this study is slotting an 80-mm-dia SUS304 stainless steel bar 0.1mm.

Sound falls into two types, "pure sound" and "complex sound", and sound that is usually heard is classified as complex sound. The frequency analysis is a test to examine what type of pure sound is contained in complex sound producing multifrequency and sound pressure. As with the dispersion of sunlight into a seven-color light pattern by a prism, FFT (Fast Fourier Transform) analysis allows sound to be split into frequency elements (Figure 1).

In this study, FFT analysis of data on cutting sound that is split every 0.1 second is carried out to detect the frequency of the maximum amplitude spectrum (except that of a motor) and create time-series data on the frequency (hereinafter referred to as a cutting sound chart) for analyzing cutting sound.

2.2 Graduated Lathe

A graduated lathe offers easy check of a feed rate. Slotting is defined as a combination of making a zero adjustment to a contact between the cutting tool and object and moving the cutting tool until the graduation mark indicates 0.1mm. An experiment was conducted with the utilization of a lathe with a stopper that curbed excessive slotting more than a specified amount, assuming that no error is observed in the depth of slot. The experiment was to analyze subject's feed motion based on time-series data (hereinafter referred to as a feed rate chart) that was derived and created from a video of graduation marks taped during the experiment.

2.3 Cutting Sound Chart and Feed Rate Chart in Target Operations

The examples of the cutting sound and feed rate charts of FFT-analyzed slotting are presented in graphical form in Figure 2. Frequencies rise at the instant when the cutting tool comes into contact with the object (I: hereinafter referred to as a rise segment) and remain relatively stable until the object is slotted for the specified amount and the cutting tool starts being moved back (II: hereinafter referred to as a stable segment). Also, frequencies fall upon restoring the cutting tool to its original position (III: hereinafter referred to as a fall segment).

Chatter vibrations are created if too low or too high feed speed is observed in the rise segment or stable segment. Chatter vibration describes actual motion of the cutting tool associated with the resistance produced by the tool cutting the object. A cutting tool vibration develops a type of wave-patterned flaw in the surface of the object, which causes deterioration in quality. The cutting tool is to be used while it is vibrating in a wide range if low feed speed is obtained, which causes large chatter vibrations and forms a large wave-patterned flaw in the cut surface. In contrast, the tool is to be used while it is vibrating in a narrow range if high feed speed is obtained, which causes small chatter vibrations. If feed speed is excessively high, however, the cutting tool vibrates concomitantly with the scattering of an impulsive force developed on contact between the cutting tool and the object in other directions, which creates chatter vibrations. Thus, these findings suggest that the cutting tool be operated at appropriate speed. "Appropriate feed speed" varies with the type of machine, cutting tool, and object, and it is what only skilled workers are capable of acquiring from practical experience.

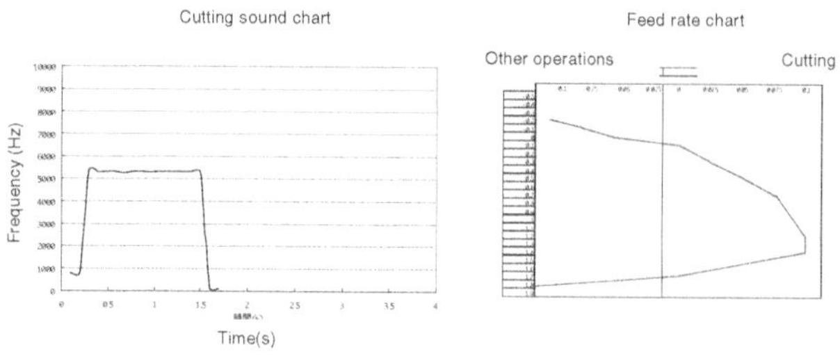

Fig. 2. Cutting Sound Chart and Feed Rate Chart in Slotting

The feed rate chart in Figure 2 shows a segment in which no fluctuations were generated in graduation marks at the tail end of the stable segment. This is a phenomenon which occurs at the instant when the handle is stopped once a slot reaches a specified depth. A momentary stoppage of the handle permits the cutting tool to be secured that keeps the shape (depth) of a slot on a spirally-cut surface even. A long stoppage of the handle, however, causes the cutting tool to develop chatter vibrations involving a high-pitched sound.

2.4 Characteristic Values of Cutting Sound during Slotting

Characteristic values for the analysis of data on cutting sound produced during target operation are comprised of the followings: "rise time", "stable time", "fall time", "maximum frequency", and "instability index." The characteristic values are described below.

- Rise time .This is defined as time in the rise segment. Long rise time denotes that the cutting tool is touching the object slowly that may cause chatter vibrations.
- Stable time .This is defined as time in the stable segment. Long stable time denotes that the cutting tool is producing slow feed motion that may cause chatter vibrations.
- Fall time. This is defined as time in the fall segment. The cutting tool may develop deflections if it is returned slowly to its original position in the fall segment. Quick return is desirable.
- Maximum frequency. This is defined as the maximum frequency detected in the progression of cutting sound. The maximum frequency increases with increase in feed speed, and it decreases with decrease in feed speed. The handle causes an increase in the maximum frequency if it is rotated with undue force in the stable segment. Operation becomes unstable if performed with undue force, which may lead to inconsistent quality.
- Instability index Lathe processing requires the constant-speed turning of the handle as a basis. Variations in frequencies in the stable segment will be minimized if the handle is turned with a uniform speed. The standard deviation of frequencies in the stable segment for each trial run is defined as an instability index so as to evaluate the stability of the operating handle movement performed by subjects.

The standard deviation of the above-listed characteristic values is used as a means of judging that constant and stable lathe processing is gained each time in several attempts.

3 Experiment

3.1 Experimental Procedure

The verification experiment was carried out on one skilled worker and two novice workers as experimental subjects, according to the following procedure.

1. As a present data analysis, each subject performed slotting five times consecutively. Cutting sound was recorded, and handle graduation marks were videotaped.
2. The novices received explanations on that different feed motion would result in a difference in the waveform of cutting sound as an outcome of cutting, according to the cutting sound chart and feed rate chart. Given an explanation for differences in feed motion and cutting sound between the novices and a skilled worker, the novices received 30-minutes of training by comparing cutting sound produced by them to that produced by the skilled worker, which allows them to establish criteria to judge cutting sound during work.
3. The two novice workers performed slotting five times consecutively after training to evaluate achievements of training. As with the present data analysis, cutting sound was recorded, and handle graduation marks were videotaped.

Roland EDIROL R-09 IC recorder was used for recording cutting sound, and Dell Precision470 XEON3.6GHz computer was used for conducting FFT analysis.

3.2 Experimental Results and Discussion

Figure 3 shows the characteristic values of cutting sound created by the subjects, and Figure 4 gives the examples of the cutting sound chart and feed rate chart.

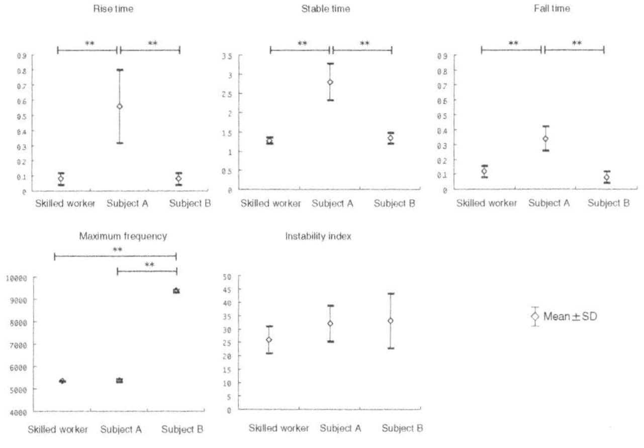

Fig. 3. Characteristic Values of Subjects

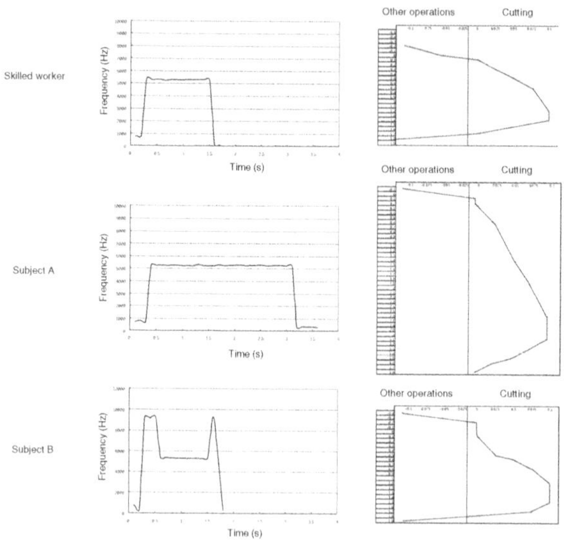

Fig. 4. Examples of Individual Cutting Sound Charts and Feed Rate Charts

The skilled worker's work was characterized by the fact that he moved the cutting tool off the object and then started cutting at initial rate. Frequencies reached the maximum (approx. 5300Hz) at the beginning of the stable segment and made a smooth progression over a stable frequency band ranging between 5200 and 5300Hz. The rise time and fall time was recorded the average of 0.1 second each, and the stable time was recorded that of 1.3 seconds. The feed rate chart testifies that the skilled worker commenced a slow adjustment once the graduation mark reached 0.075 to bring the rate close to a desired value.

Subject A's work showed that frequencies in the stable segment made a gradual progression over a stable frequency band ranging between 5250 and 5230Hz, which was an inconsistent shift, and a high frequency was obtained at the end of the stable segment. The average rise time was 0.5 seconds, the average fall time was 0.3 seconds, and the average stable time was 2.8 seconds, which demonstrated longer overall operation time than the skilled worker.

In Subject B's work, a 9000-Hz high-pitched sound was produced at the end of the rise segment and remained in a frequency band ranging between 5250 and 5300Hz in the stable segment. The same high-pitched sound was again produced at the beginning of the fall segment. The rise time and fall time was recorded the average of 0.1 second each, and the stable time was recorded that of 1.3 seconds.

Differences between skilled worker and Subject A

As to the rise segment, Subject A started gradual feed motion with respect to 0 on the graduation mark, while the skilled worker rotated the handle backward by generating an initial rate. The cutting tool that Subject A was holding was slightly touching the bar while he was making sure of 0 on the graduation mark, which resulted in longer rise time as compared with the skilled worker.

In the stable segment, Subject A performed feed motion a little at a time at approx. a third of the speed that the skilled worker took, which caused the stable time to be longer than that the skilled worker obtained. These findings inferred that Subject A exercised considerable care in work operations. In Subject A's work, low feed speed was obtained that caused large chatter vibrations. A large wave-patterned flaw in the cut surface of the object was observed under a microscope.

As with the skilled worker, it is believed that a worker is capable of quick rotating of the handle after moving the cutting tool off the object and commencing cutting at initial rate, which deliver stable frequencies in the stable segment.

- While the skilled worker rotated the handle upon loosening his hold of it in the fall segment, Subject A was trying to restore the handle once before actually rotating it. Also, his gradual restoring of the handle developed rattles of the cutting tool and rasp (wide-ranging frequency components). These findings showed that the handle was required to be jerked back at the moment when it was loosened.
- A large standard deviation subdivided into the rise time, stable time, fall time, and maximum frequency was observed in Subject A, which could be interpreted as inconsistent feed motion every time attributed to slow operation.

Differences between skilled worker and Subject B

Although the findings revealed that Subject B's and the skilled worker's work was similar in operation time and motion, a 9000-Hz high-pitched sound was developed at the end of the rise segment in Subject B's work. This is a phenomenon which is caused by undue force on feed motion. As with Subject A, Subject B started feed motion with respect to 0 on the graduation mark but rotational speed of the handle was much faster. A probable cause of handle rotation with undue force may be that Subject B was bearing down the cutting tool against high resistance generated on contact between the tool and the object. Also, Subject B's quick work allowed a small standard deviation in the rise time, stable time, and fall time. However, inconsistent exerting of a force every time caused a large standard deviation of maximum frequency.

These findings inferred that Subject B was unable to control the handle due to undue force. Therefore, for Subject B to start cutting at initial rate as with the skilled worker will be a solution to the problems of bearing down and high-pitched sound.

As with Subject A, Subject B was also trying to loosen and restore the handle once before rotating it. Then, Subject B rotated the handle with force rattling the cutting tool, which developed a 9000-Hz high-pitched sound at the beginning of the fall segment.

3.3 Instructions

The following instructions were provided to Subject A and Subject B using the cutting sound chart and feed rate chart based on the results of the above analysis, and 30-minute training followed.

Subject A

Commence cutting at initial rate.

- To ensure stable cutting with constant feed motion
- To reduce the rise segment

Gain a threefold speedup to enhance feed motion

- To improve wiggly feed motion
- To accelerate operation time
- To maintain a frequency band in the stable segment constant
  Loosen the hand and handle at the same time.
- To reduce the fall segment and have the cutting tool free of vibrations

Subject B

Commence cutting at initial rate.

- To remove undue force
- To maintain a frequency band in the stable segment constant by adjusting the force of feed motion

When securing the handle, restore the handle immediately when a high-pitched sound arises.

- To have the cutting tool free of vibrations

3.4 Results of Training

Figure 5 shows the characteristic values of cutting sound created by the subjects, and Figure 6 gives the examples of the cutting sound chart and feed rate chart.

Subject A accelerated operation time from 3.8 seconds to 1.4 seconds after training, which shortened the overall operation time. Also, reductions in the rise segment, stable segment, and fall segment were accomplished and the cutting sound chart showed signs of improvement. The result proves that Subject A moved closer to the skilled worker.

Subject B eliminated a 9000-Hz high-pitched sound that had been observed at the end of rise segment and at the beginning of the fall segment before training.

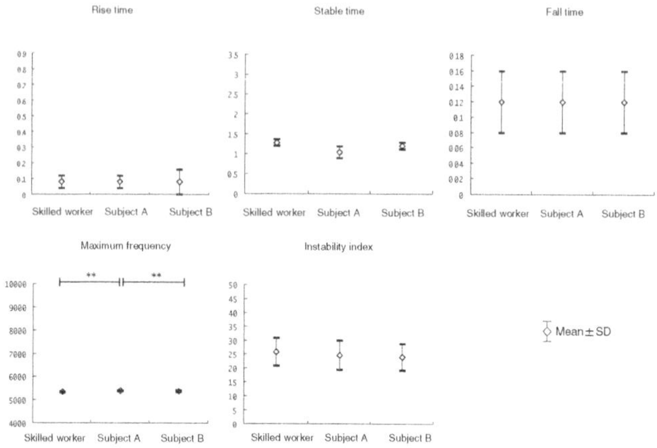

Fig. 5. Characteristic Values of Subjects after Training

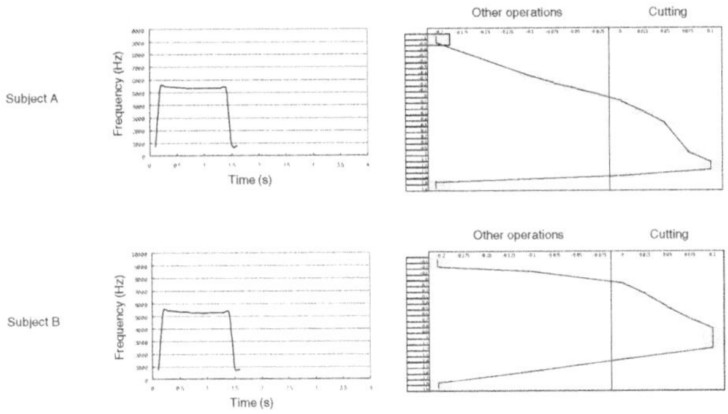

Fig. 6. Examples of Individual Cutting Sound Charts and Feed Rate Charts after Training

Subject A and Subject B attained smaller instability index as well and ensured stable operations while paying attention to cutting sound.

In the fall segment, both subjects became capable of restoring the handles to the 0 position on the graduation mark in 0.1 second upon loosening their hold of them, as with the skilled worker.

The skilled worker checked the cut surface of the subject's trial samples under a microscope, and the check found their finished quality was much the same as the skilled worker's.

4 Conclusions and Future Development

This study examined the inheritance of skills with the utilization of cutting sound produced in lathe processing, Manual Machine Tool Operation.

The pilot study ascertained that the extraction of skills was practicable with the use of cutting sound. By analyzing differences in operations according to the cutting sound chart and feed rate chart, this study suggested training on the inheritance of skills utilizing cutting sound developed in skilled worker's work operation. With two novice workers as subjects, the analysis of cutting sound generated in their slotting process under the same operating conditions was carried out along with the analysis of a feed rate, in accordance with the proposed training, to clarify the differences between the skilled worker and novice workers. The findings of cutting sound data-based training revealed the two novice workers refined their skills that allowed themselves to get close to the level of skilled worker's work.

This study adopted a simple operation, slotting, to conduct an experiment, but it poses a potential problem that it is conceivable that complicated shapes could complexity cutting sound change in process. The potential problem prods verification of the effectiveness of cutting sound data-based training in operations requiring complex changes in cutting sound.

This study still leaves room for future studies on sound type based on discriminability, given the fact that cutting sound varies with the type of materials.

References

1. Monozukuri, N.: Manufacturing Research 2005(10) (2005)
2. Shimizu, R., Shimosaka, H., Ehara, N.: Estimation of Machining Skill Using Cutting Sound. Proceedings of the Japan Society of Mechanical Engineers 1997(2), 159–160 (1997)
3. Suzuki, T., Shimosaka, H., Ehara, N.: Propriety Distinction at Byte Sending Speed Using Cutting Sound. In: Proceedings of the Japan Society of Mechanical Engineers Tohoku Branch, vol. 41, pp. 171–172 (2006)

A Business Performance Measurement Model for Mobile User Interface

Min-Jeong Kim and Jonghun Park

Terminal Laboratory, KTF
Dept. of Industrial Engineering, Seoul National University
kimi@ktf.com, jonghun@snu.ac.kr

Abstract. In this paper, we present a new framework for measuring business performance of mobile user interfaces (UIs). Until now, the performance measurement of mobile UI was mainly to assess usability performance. Yet, as mobile phones are increasingly replacing wired devices, stakeholders in mobile phones began to invest a lot of time and money to mobile UI development. As a result, it is desirable that the business performance of mobile UI has to be measured qualitatively as well as quantitatively. This paper develops a measurement model for business performance of mobile UI based on BSC (Balanced Score Card). The proposed model contains "Financial", "Customer", "Internal Business Processes", and "Learning & Growth" perspectives. We applied the presented measurement model in a real world mobile UI design project. Finally, we demonstrate the benefit of applying the proposed model to quantitatively measure the business performance.

Keywords: User Interface, Mobile Phone, Business Performance, BSC (Balanced Score Card).

1 Introduction

User Interface (UI) plays an important role in traditional software development processes since UI can improves software usability performance and reduce potential maintenance work. In order to model UI's desired business benefits, there have been many quantitative and qualitative approaches proposed in software engineering field. However, due to the relatively short lifetime of a mobile phone, there have not been many attempts for measuring the business performance in the context of mobile UI. Furthermore, as mobile phones are being adopted as direct replacements of wired devices, they have suddenly become platforms for entertainment and commerce as well as tools for information management and media consumption [8]. As a result, stakeholders in mobile phone industry are investing a lot of time and money in mobile UI. Before and after development stage of mobile phone and service, it is necessary to estimate ROI (Return on Investment) for mobile UI.

This paper develops a measurement model for business performance of mobile UI based on BSC (Balanced Score Card) [4]. The primary reason to employ BSC is that the business performance of mobile UI has to be assessed both qualitatively and quantitatively. By using measurement model, we can utilize them as data for verification

M.J. Smith and G. Salvendy (Eds.): Human Interface, Part I, HCII 2009, LNCS 5617, pp. 87–93, 2009.

before a development stage, and in case of multiple projects, we can prioritize the projects according to the measurement value to see if there's anything we need to develop now, or delay later.

In this paper, we analyze some relevant work in software engineering and related areas, and develop a framework for measuring the business performance of mobile UI. In order to demonstrate the advantages of the proposed measurement model, we applied the model to a real world project that estimates monetary value of mobile UI in a commercial service stage.

The organization of this paper is as follows. Section 2 describes some related work. Section 3 defines details of the proposed framework. In section 4, we estimate monetary value when we apply the proposed measurement model to a real mobile service project. Finally the conclusions and some remarks are given in section 5.

2 Research Reviews

In 1988, Mantei and Teorey [9] described cost-benefit analysis when human factors methods are applied in software development and classified tangible and intangible factors for costs and benefits. They calculated costs and benefits of general examples for tangible factors and listed conceptions for intangible factors. Since then, many quantitative and qualitative methods for demonstrating UI's desired business benefits have been proposed by calculating costs and benefits of UI in the actual development projects. In particular, Cost-Justifying Usability [2] of Randolph G. Bias and Deborah J. Mayhew, Presentation material [12] of Deborah J. Mayhew and others [5, 10, 11] showed that UI played an important role in traditional software development process and website development process through costs and benefits analysis. That is, UI contributed to better usability, increased productivity, and reduced cost of development and service.

Unfortunately, most existing research results in mobile UI were mainly to develop new UI techniques or evaluate usability performance of a new technique in the laboratory environment [6]. In addition, since mobile device has a small screen size compared to that of a computer, there have been many research results for improving user experience in order to overcome the hardware limits of mobile phones.

One of widely used methods for measuring business performance is cost-benefit analysis which measures financial benefit of projects quantitatively. In the 1990s, Robert S. Kaplan and David P. Norton publicized BSC concept for performance planning and measurement framework. By focusing not only on financial outcomes but also on the human issues, BSC helps provide a more comprehensive view of a business so that managers focus on performance metrics while balancing financial objectives with customer, internal process, and employee perspectives [1].

3 Measurements Model

In this section, we propose a measurement model for business performance based on BSC. BSC consists of four perspectives that are labeled "Financial", "Customer", "Internal Business Processes", and "Learning & Growth". Evaluating business performance based on BSC requires a few good measures for each perspective. For financial perspective, increased sales volume, ROI, payback period and EVA are generally used. Main measure of customer perspective is customer satisfaction. The measures selected for internal process perspective include process for operations management and customer management. Finally, learning & growth perspectives contain internal skills and capabilities that are required to support the internal process.

In this paper, we use sales volumes for financial perspective, increased ARPU (Average Revenue per User) for customer perspective, experience curve effects [3, 14] for internal process and learning and growth perspectives since experience curve effects can be applied in most situations [3]. In order to calculate sales volume of a mobile UI project, we multiply the number of original service subscribers right before starting mobile UI project for the original service by the increased ARPU during payback period in the service. The reason to use this number is because it is difficult to justify that mobile UI project efforts actually have resulted in the increased number of subscribers via positive word-of-mouth. Upgraded mobile UI can contribute to the increased number of subscribers and we propose the conservative measurement model. Therefore, instead of using the increased number of subscribers in the service with upgraded mobile UI, we use the increased ARPU since improved mobile UI leads to increased perceived quality, customer satisfaction as well as increased ARPU [13]. More formally, we define the following notations.

N : payback period of project
t : order number of project
α : coefficient of decreased cost, $1 : \alpha : 2$
Z_t : total revenue of tth project, $t = 1, 2, \cdots$
a_t : number of original service subscriber right before t^{th} project, $t = 1, 2, \cdots$
x_t^i : increased ARPU of tth project at time i, $t = 1, 2, \cdots$, $i = 1, \cdots, N$
y_t : cost of tth project
y_t^1 : cost of tth project when we produce it as 1st project by experience curve effects

The following assumption is made in this paper:

(A) Only the mobile UI projects for UI revision of existing mobile services are considered.

That is, we don't consider new mobile service project with new UI.

By using the above notations, we obtain a measurement model for the business performance of mobile UI as follows.

$$\textbf{(M)}\ Z_t = \sum_{i=1}^{N} \left(a_t \times x_t^i\right) + \alpha \times \left(y_t^1 - y_t\right) \tag{1}$$

(M), total revenue of mobile UI project consists of original sales volume and decreased costs by experience curve effects. **(M)** provides business performance after payback period of mobile UI project and evaluates real cost-benefit analysis. We utilize **(M)** as data for verification before development stage begins. In order to use **(M)** before the development stage, we formulate the measurement model with statistics. Since we already have done several mobile UI projects which need costs and lead to benefits such as revenue, we were able to utilize several numerical data consisting of a dependent variable such as revenue and an independent variable such as cost. The revenue in the regression equation can then be modeled as a function of the cost, corresponding parameters and an error term. Using regression analysis, we can obtain the statistical measurements model **(M_s)** of **(M)**.

$$(\mathbf{M_s})\ Z_t = \beta_0 + \beta_1 y_t + \alpha \times \left(y_t^1 - y_t\right),\ \beta_0, \beta_1 : \text{parameters} \tag{2}$$

where $w_t = \beta_0 + \beta_1 y_t$

(M_s) corresponds the case of linear regression. We now obtain a solution for the model **(M)** solving **(M_s)** before the development stage of a project. After payback period of the project, we can compare **(M)** with **(M_s)** for difference between estimation and actual value.

4 Results

We apply the proposed measurement model to estimate business performance of mobile UI in a real mobile service. For this purpose, we define the following procedure.

<u>Step 1</u> : Calculate **(M_s)**.
<u>Step 2</u> : Calculate the mediated **(M_s)** considering only UI.
<u>Step 3</u> : Compare **(M)** with the mediated **(M_s)** using actual value and recalculate the mediated **(M_s)**.
<u>Step 4</u> : Analyze BCR (Benefit-Cost Ratio) using **(M_s)**.

For step 1, we first construct a regression model using data which consider revenue as a dependent variable and cost as an independent variable, and obtain parameters for regression equation of **(M_s)**. That is, we fix β_0, β_1 parameters of **(M_s)** using regression analysis. In this study, 31 service cases in the mobile service provider were analyzed. Each service case is defined as the previous service with renewal UI. The regression equation about cost of renewal UI and revenue in 31 service cases is as follows :

$$w_t = 12.31\, y_t - 186{,}411{,}568,\ R^2 = 0.8595,\ \ Adj.R^2 = 0.8545 \tag{3}$$

Table 1. Regression Diagnostics(Coefficients)

	B	Std. Error	T	Sig.
Constant	-186,411,568	120,685,939.0394	-1.5446	0.1337
X	12.3062	0.9401	13.0899	0.0000

Table 2. Regression Diagnostics(ANOVA)

	df	Sum of Squares	Mean Squares	F	Sig.
Regression	1	3.04229E+19	3.04E+19	171.3465	0.0000
Residual	28	4.97146E+18	1.78E+17		
Total	29	3.53944E+19			

For regression diagnostics, we can confirm the goodness of fit of the model since R^2=0.8595. Statistical significance can be confirmed by an F-test of the overall fit, followed by t-tests of individual parameters. Then, we estimate y_t^1 with y_t by experience curve effects [3]. We use NASA's learning curve calculator [7]. The learning percent is usually determined by statistical analysis of actual cost data for similar products. That information is not available, we use the following industry guideline [7]: 1. Aerospace 85%, 2. Shipbuilding 80~85%, 3. Complex machine tools for new models 75~85%, 4. Repetitive electronics manufacturing 90~95%, 5. Repetitive machining or punch-press operations 90~95%, 6. Repetitive electrical operations 75~85%, 7. Repetitive welding operations 90%, 8. Raw material 93~96%, 9. Purchased parts 85~88%. Among them we use 96% learning percent which is guideline for raw material because UI design and development is not repetitive operations but creative operations. Because the decreasing costs by experience curve effects between **(M)** and **(M_s)** are the same, we apply the decreasing costs when we calculate the cost-benefit analysis in step 4.

For step 2, we calculate the mediated by considering only UI. We can classify several effects for increasing the service revenue. The effects for increasing the service revenue are as follows: service rate, UI, word of mouth, advertising, etc. In general, most effects for increasing the mobile service revenue come from marketing promotion such as advertising, service rate. But we assume that mobile UI project is defined as the previous mobile service with renewal UI in this paper and each effect has the same effect. And then, we calculate the mediated **(M_s)** by multiplying w_t of **(M_s)** by UI effect value (25%).

In step 3, we compare **(M)** with the mediated **(M_s)** using actual value. During the time horizon of one year, we collect actual values for a specific project for calculating the first term of **(M)**. And we apply the cost of specific project to (3) for calculating the first term of **(M_s)**. Because the second term of **(M)** and **(M_s)** is the same, we don't consider the second term for comparing **(M)** with **(M_s)**. The number of original service subscriber right before 32nd UI project is 11,506 and the following (4) includes the increasing ARPU of 32nd project for 1 year. The UI expenditure of 32nd project is 100,000,000 won.

The first term of **(M)** :

$$11{,}506 * (403 + 450 + 487 + 642 + 688 + 865 + 983 + 1{,}128 + 1{,}724 + 1{,}711 + 2{,}629 + 3{,}438) = 174{,}281{,}382 \text{ won} \quad (4)$$

The first term of $(\mathbf{M_s})$:

$$(12.31 * 100{,}000{,}000 - 186{,}411{,}568) * 0.25 = 261{,}147{,}108 \text{ won} \quad (5)$$

Through comparing $(\mathbf{M})$ with the mediated $(\mathbf{M_s})$ using actual value, we find that the first term value of $(\mathbf{M_s})$ is greater than the first term value of $(\mathbf{M})$ since all the effects for increasing the mobile service revenue are included. UI effect value has to be decreased for using the statistical measurements model $(\mathbf{M_s})$ of $(\mathbf{M})$. We can get 16~17% as UI effect value in order to revise the mediated $(\mathbf{M_s})$ in comparison to $(\mathbf{M})$. We recommend that the UI effect value is less than 20%.

In step 4, we calculate the projected BCR. Assuming the time horizon of one year, the UI expenditure of the 32nd project is 100,000,000 won (Korean money unit). We estimate $(\mathbf{M_s})$ of the 32nd project using 16% as UI effect value and 0.3 as α and estimating y_t^1 with y_t by experience curve effects with 96% learning percent.

$$(\mathbf{M_s})\ 0.16 * (12.31 * 100{,}000{,}000 - 186{,}411{,}568) + 0.3 * (130{,}197{,}071 - 100{,}000{,}000) = 176{,}193{,}270 \quad (6)$$

Therefore, the projected BCR for this project is

$$\text{BCR} = 176{,}193{,}270/100{,}000{,}000 = 1.762 \quad (7)$$

A 1.762 is a very healthy return. This makes it possible for the investment in mobile UI to be considered alongside other investments the firm might make, rather than merely being considered a cost.

5 Conclusions

In this paper, we presented a new framework for measuring the business performance of mobile UI. Until recently, performance measurement in mobile UI was mainly carried out to assess the usability performance, and the investment to mobile UI was considered a cost. However, as mobile phones are increasingly replacing wired devices, stakeholders in mobile industry began to invest a lot of time and money to mobile UI development. Therefore, it becomes necessary that the business performance of mobile user interface is considered qualitatively as well as quantitatively. This paper developed a measurement model for business performance based on BSC (Balanced Score Card). We proposed two models, $(\mathbf{M})$ total revenue model of mobile UI project, and $(\mathbf{M_s})$ statistical measurements model of $(\mathbf{M})$. In the proposed model, we considered "Financial", "Customer", "Internal Business Processes", and "Learning & Growth" perspectives. We obtained a solution for the model $(\mathbf{M})$ solving $(\mathbf{M_s})$ using real mobile UI project data. Through comparing $(\mathbf{M})$ with $(\mathbf{M_s})$, we were able to compute UI's effect quantitatively among all the effects for increased mobile service revenue. In contrast to the existing approaches, the presented method considers mobile UI development efforts as investment and it also emphasizes that these efforts must be financially accountable.

Acknowledgements. This work was supported in part by the Korea Science and Engineering Foundation (KOSEF) grant funded by the Korea government (MOST) (No. R01-2007-000-11167-0).

References

1. Balanced Scorecard, wikipedia, `http://en.wikipedia.org/wiki/Balanced_scorecard`
2. Bias, R.G., Mayhew, D.J.: Cost-Justifying Usability: An Update for the Internet Age. Morgan Kaufmann, San Francisco (2005)
3. Experience curve effects, wikipedia, `http://en.wikipedia.org/wiki/Experience_curve_effects`
4. Kaplan, R.S., Norton, D.P.: The Balanced Scorecard: Translating strategy into action. Harvard Business School Press (1996)
5. Karat, C.M.: Usability Engineering in Dollars and Cents. IEEE Software (1989)
6. Kjeldskov, J., Graham, C.: A Review of Mobile HCI Research Methods. In: Mobile HCI 2003 (2003)
7. Learning Curve Calculator, `http://cost.jsc.nasa.gov/learn.html`
8. Lindholm, C., Keinonen, T.: Mobile Usability: How Nokia Changed the Face of the Mobile Phone. McGraw Hill, New York (2003)
9. Mantei, M.M., Teorey, T.J.: Cost/Benefit Analysis for Incorporating Human Factors in the Software Lifecycle. Communications of the ACM 31, 4 (1988)
10. Marcus, A.: Return on Investment for Usable User-Interface Design: Examples and Statistics, AN+A (2002)
11. Mauro, C.L.: Professional usability testing and return on investment as it applies to user interface design for web-based products and services, White paper ver 2.0, MauroNewMedia, Inc. (2002)
12. Mayhew, D.J.: The Business Case for Usability Engineering, HFES Potomac Chapter (2007)
13. Rust, R.T., Zahorik, A.J., Keiningham, T.L.: Return on Quality(ROQ): Making Service Quality Financially Accountable
14. Wright, T.P.: Learning Curve. Journal of the Aeronautical Sciences (1936)

Web Service Systems for Cooperative Work Support in Knowledge Creation Processes

Hiroyuki Kojima[1], Kentaro Funaki[2], and Tsuyoshi Inoue[1]

[1] Department of Information Systems and Management
Hiroshima Institute of Technology
2-1-1 Miyake, Saeki-ku, Hiroshima, 731-5193, Japan
{kojima,b205003}@cc.it-hiroshima.ac.jp
[2] Hi-Elecom Co., Ltd.
funaki@tkym.hek.co.jp

Abstract. The emergence of Web service results from the business needs as mentioned below. Web service as the evolution of software engineering is also required to explore the possibility of adapting to cooperative work support with knowledge management. Web service technology is becoming the focus of interest as an open standard for web software, because of its service-oriented architecture. Web service facilitates Internet commerce, or e-commerce, because it enables the execution of inter-organization in a typical business application or logic area such as dynamic supply chain composition. Web service technology is successful in the above business with a fixed procedure. It is therefore desirable to adopt an atypical task requiring a wide variety of task procedures. This research focuses on a method of applying Web service to atypical tasks with a non-fixed procedure, such as a knowledge creation task, to improve the task performance in a cooperative work. The results presented in this paper are expected to validate the performance of Web service in carrying out a dynamic search for a document from various information resources.

Keywords: Web Services, Knowledge Sharing, Document Retrieval, Cooperative Work Support, SOAP, WSDL.

1 Introduction

The rapid progress of information and communication technology has recently produced many software application systems, such as automatic seat reservation systems, Internet shopping systems, and so on, that contain several business processes in which business objects interact with one another. For example, in the automatic seat reservation system, the web-based self-service solutions include quick booking for go-show passengers and on-line payment functions. The practical use of these systems is enabled by using Web service technology, which is defined by the W3C (World Wide Web Consortium) as "a software system designed to support interoperable machine-to-machine interaction over a network" [1]. Web services can also implement application software components according to service-oriented architecture (SOA) concepts, where the basic unit of communication is a message rather than an operation. Therefore, Web

M.J. Smith and G. Salvendy (Eds.): Human Interface, Part I, HCII 2009, LNCS 5617, pp. 94–103, 2009.

service is attracting much attention because it allows access over a network such as the Internet and is executed on the remote system hosting the requested services.

Generally, business processes are carried out through a business life cycle such as the Plan-Do-See cycle. The Do phase is typical work with a fixed process, such as a business transaction. Web service is well suited to business processes with a fixed task pattern. The above-mentioned service systems of quick booking for go-show passengers and online payment functions have primarily been adapted to a fixed procedure.

However, knowledge is one of the most important assets in an organization. Real concern exists about the knowledge and know-how drain as well as the lack of technology transfer following the retirement of the current experts. For a business organization to enhance its competitiveness, it must acquire collective knowledge in the context of their business, and effectively put this knowledge and information into practical use for various task performances [2].

Both information and knowledge in the business community depend on particular individuals or organizations that are identified as senders or receivers in a collaboration. The challenge is to transform this information into knowledge that can be reused and shared in knowledge-intensive work, such as in design offices and business communities. In addition, because this knowledge and information exist in many types of resources, such as web sites and local site documents, varied but effective application software tools are needed to support cooperative work. Web service systems can provide a retrieval method that enables individuals and organizations to manage contents and application software by using an XML-based protocol for exchanging information over an HTTP protocol.

These knowledge workers in cooperative work environment typically are part of a planning or analyzing task in the Plan-See phase. These tasks are called atypical work tasks with a non-fixed pattern. Since information system problems are possible in this area, SOA will be needed to support atypical work performance.

This research focuses on a method of applying Web service to atypical tasks with a non-fixed procedure, such as a knowledge creation task, to improve task performance in cooperative work. As an example, this research uses a "literature survey in research work" as a target application for knowledge creation and reuse work. This paper describes the Web service implementation procedure applied to the atypical cooperative task, in which knowledge and information in a community is shared through dynamic retrieval of various contents from several location sites; here, the Web Serve procedure includes an information organizing system that is based on ontology to represent task models.

2 Knowledge Sharing in Cooperative Creation Work as a Target of Web Service

Due to the number of documents shared in physical environments and database systems, conventional knowledge management or groupware system development in the business field needs improved electronic document retrieval and management functions. Knowledge systems should be able to enable knowledge workers to create knowledge and reuse the knowledge of others.

Knowledge can be categorized into tacit knowledge and explicit knowledge. Tacit knowledge is knowledge that is normally difficult to formalize and articulate. It often exists only in a person's mind. Explicit knowledge is knowledge that has been codified and formalized into procedures and rules. Transforming tacit knowledge to explicit knowledge is very important to an organization, as it enables people to derive knowledge instead of only information. However, it is not easy to share or learn tacit knowledge that is more instinctive than rules and procedures. Personal tacit knowledge can be shared and sanctioned in a group or an organization, and the knowledge becomes larger and twice the original size through four transformation processes: socialization, externalization, combination and internalization (SECI model) [3].

Though the SECI model is only conceptual, all progressive company practice this model mechanism. Therefore, the computer implementation of the SECI model is very significant. This model is called the "knowledge spiral" because the interaction process between tacit and explicit knowledge continually spirals upward as newly created knowledge continually evolves. There are two issues in the implementation of the SECI model. One is dialoguing system to support the externalization of tacit knowledge. and another is network system to support the combination of knowledge, knowing what anyone is knowing best among peers of an organization. This paper pays attention the former system. In other words, organizational knowledge creation and sharing is a spiral process, starting at the individual level and moving up through an organization. In addition to the people working in an organization, an information-organizing tool should also drive this spiral. Codification of this knowledge puts the organizational knowledge into a language that is accessible to all members of the organization [4], [5].

This SECI spiral process is accompanied by information structuring of knowledge using any document retrieval tool and a comprehensive framework of the domain task procedure. Therefore, Web service has the potential to provide effective implementation support of information systems based on the SECI model.

3 Web Service Systems for Atypical Business Task Processes

This research investigates the system requirements and implementation feasibility of the interoperating between different software applications supporting knowledge management and atypical cooperative work. In accordance with the task characteristics, knowledge workers select and combine the most suitable application tools from application service sites. A Web service method can be applied to this information processing as a standard means of interoperating between different software applications running on a variety of platforms. Figure 1 shows a schematic diagram of the system architecture of the Web service.

A Web service method is conventionally applied to interoperating between typical business software applications with a fixed logic pattern. It does not adopt an atypical task requiring task knowledge. This research considers a "literature survey in research work" as an example of knowledge creation and reuse work. The task procedure of literature retrieval is strategic and needs to use various search means because of different contents media or sites. So it is used to evaluate capability of the Web service method to improve task performance in cooperative office work.

Web services provide a standard means of interoperating between different software applications running on a various platforms. Web services are characterized by their great interoperability and extensibility, as well as their machine-processable descriptions using XML. The descriptions can be combined in a loosely coupled way to achieve complex operations. Programs providing simple services can interact with each other to deliver sophisticated added-value services.

A W3C Web Services Activity is to design the infrastructure by defining the architecture and creating the core technologies for Web services. Web service components are themselves equivalent to general web application software and communicate with each other through the SOAP protocol and XML data exchange. With a Web service, the service requester uses the service from a service provider.

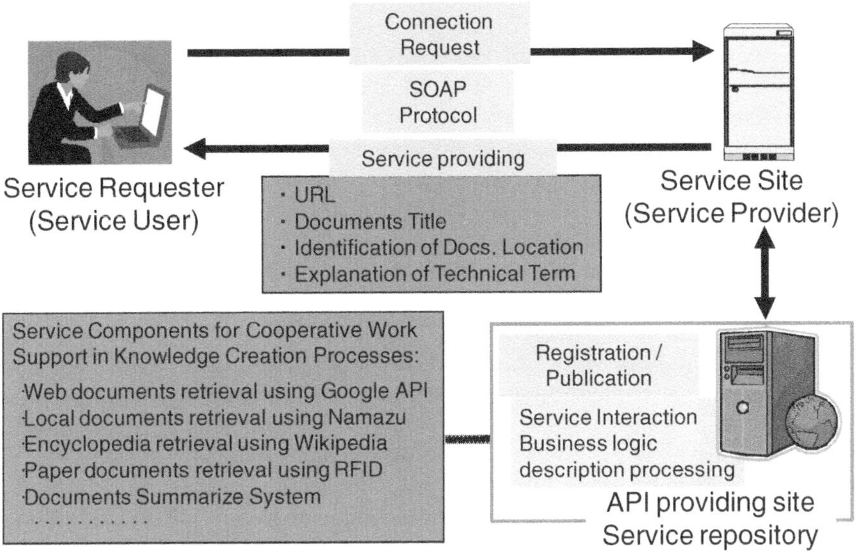

Fig. 1. Service systems for task knowledge sharing support

4 Web Service Systems for Supporting Cooperative Creation Work by Retrieval Application Interoperation

This paper uses search tasks of reference documents as a cooperative work example with a non-fixed task process. The Web service implementation and design procedure is a means of interoperating between different software applications, including web document retrieval using Google API, local document retrieval shared in a group or work organization, and encyclopedia retrieval using Wikipedia API. The information organizing and retrieval system [5] is a core application software program for Web service trial and is used as follows.

4.1 Web Service Systems Development Environment and Procedure

Knowledge and document retrieval based on information organizing is an example task for a Web service of business logic. The retrieval objects here are publication documents on the web and local site documents. Google and Namazu [6] are used for the document retrieval engine.

Figure 1 shows the Web service systems for task knowledge support implemented using Java platform technology. This figure also includes JSP, Servlet, JavaBeans, a service requester and a service provider site. Apache Tomcat is the server side software and Apache Axis is used as the Web service framework. This figure also shows an interoperating scheme between different software applications, running on a variety of platforms, that is, web documents retrieval and local documents retrieval software to share the task knowledge and information.

(1) Information organizing and retrieval software with Web service. Using a former research paper for reference [4], [5], the information organizing and sharing system is explained as follows. Such a system is desirable in a knowledge community of creative workers, such as in the product research, design and development departments. This paper expresses task knowledge as a task model consisting of a work procedure and documents of explanatory contents. Here is taken up a creative task example as literature search in research work, and a task model is expressed as the table form structure. A description of task model, information organizing and document retrieval are as follows.

(2) Description of Task Model, information Organizing and Document Retrieval. The information retrieval process attempts to sort out very complex, important issues as precisely as possible by selecting the appropriate information resources and planning the retrieval strategy. This process of sorting out issues indicates the task model is a problem-solving method that forms task knowledge. Ontology is defined as the specification of concepts to be used for expressing knowledge. Our system task model is based on the task ontology concept. The most typical kind of ontology for the web has a taxonomy and a set of inference rules. Task ontology consists of the terminology, vocabulary and ontology of a task. The task model consists of the task procedure and task ontology.

In this paper, a table form is typically considered as the expression of a task model for document collection [4], [5]. The table form is suitable for expressing a relationship and presents an organizing structure and at-a-glance understanding of the relationship of items classified in the table. One example given in this paper is mentioned about a literature searching problem for assist of one's research. Figure 2 shows the task knowledge spiral cycle using a task model of the table form expression. A good researcher can compose the task model in a table using task items from his own knowledge and register it in a database. The retrieval engine would collect documents using these items and organize them in the table. This information of organizing results with a table form of documents could be referenced, changed and updated with the new ideas of other workers in the case of task progression or other expert participation.

For example, the task case in figure 2 is a survey of literature documentation about the "method of installation and setting of computer server". The researcher selects "WWW server" and "mail server" for the server service software and "Windows Server 2003" and "Fedora Core Linux" for the related server OSs. These selected keywords are deployed in table rows and columns. Document retrieval is carried out and produces an information organizing result with the combination of keywords. The XML sample of this task model is also shown in figure 2. The data can be represented as follows: <name>: the task name, <author>: the creator of the task model, <keyword>: combinations of the selected keywords, <row>: the arranged keywords in a row of the table, <column>: the arranged keywords in a column of the table, <task-model>: the container for the prescribed element of task model in the task model list <task-model-list>.

Information retrieval is carried out based on the combination of items in this table. Radio buttons are used for the selection of Web site documents, local site documents in an office, paper documents at a local site, or a mixture of paper and electronic documents, in accordance with the classification of the information source. Retrieval processing is carried out by using an adequate search engine, such as Google for web documents and Namazu [6] for electronic documents, paper documents, and also task model definition data in a local site.

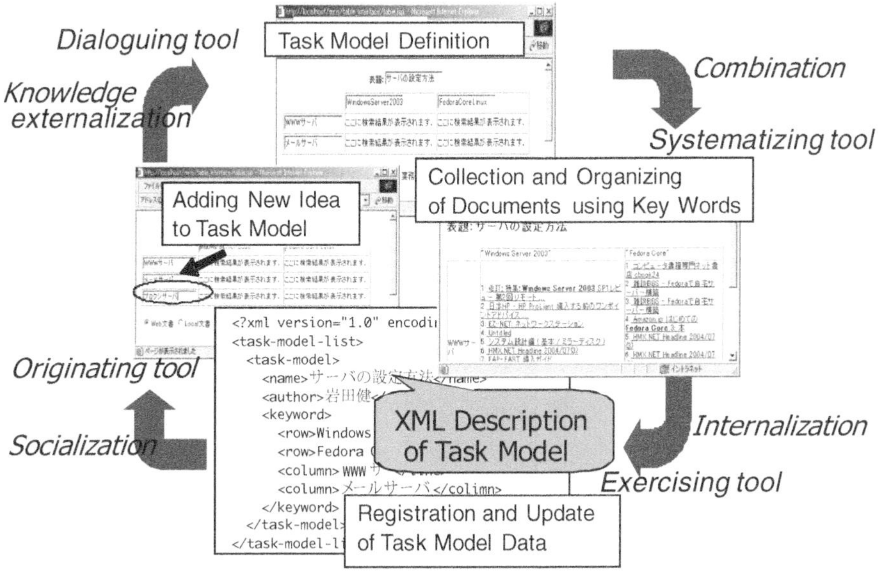

Fig. 2. Task model sharing using an example of a table form, and information collection and organizing

(3) Web Document Retrieval Component using Google API. In this paper, a table form is typically adopted to represent the task model. Terms in the rows and columns of the table are given as retrieval query words for the Google engine. The row data and column data are viewed by the user with task knowledge. The number of rows and columns is variable and depends on the task.

Google Web API as a Web service component provides a programming interface to query web pages through Google from our own computer programs. This library is a processing as a resource in our applications for Google Web API. Google Web API has great potential for creating table organizing information, especially since a huge amount of data is available through Google. Therefore, web document retrieval in this study uses the Google API because it helps to decrease system development by providing a set of scripts for a wide range of platforms that can be used to launch almost any Java Application controlled by our library.

(4) Local Documents Retrieval Component using NAMZU. A pre-developed, ready-made web application is reused as the Web service. In this paper, a software application for local content retrieval is used to search a local file repository site at a laboratory or organization by using the table form of the information organizing and sharing system [5]. The local content retrieval components are customized for JavaBeans. JavaBeans components are reusable software programs that can be developed and assembled easily to create Web service applications for the task knowledge support system. The Namazu component is executed as the command line form in a JSP program source.

The Web service deployment procedure [7] on the server side of local document retrieval is summarized as follows, using the abovementioned JavaBeans. Figure 3 illustrates the Web service arrangement.

(1) Creation of application program (JavaBeans, in this case) for service deployment
(2) Application program is added in the Apache Axis engine
(3) Execution of deployment program using Deployment Descriptor for Web service
(4) Creation of WSDL (Web Services Description Language) and publishing the application service program on the web server.

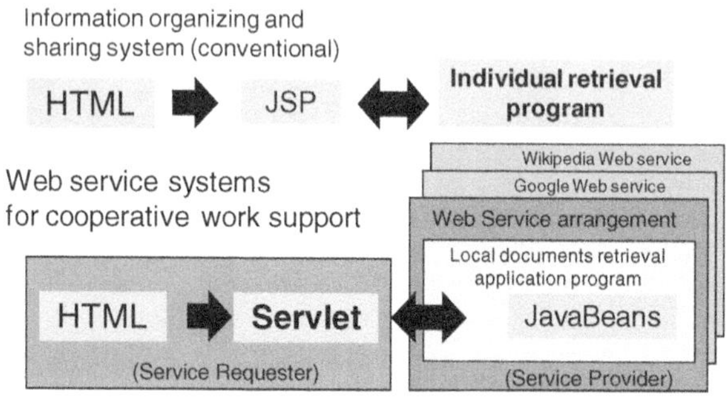

Fig. 3. Web service arrangement using JavaBeans for usual web application

(5) Encyclopedia Retrieval using Wikipedia API. Wikipedia is a multilingual, Web-based, free-content encyclopedia project. Because it is written collaboratively by volunteers from all over the world, anyone can edit it. Encyclopedia retrieval is useful for creative collaborative work. Therefore, this application software is used for the implementation trial of Web service. Implementation is carried out in the same way as the abovementioned web document retrieval application. The JSP source for Wikipedia API implementation is as follows.

```
<%@ page contentType="text/html;charset=UTF-8"
      import="javax.xml.transform.*,javax.xml.transform.stream.*" %>
<%
      request.setCharacterEncoding("UTF-8");
      String keyword=request.getParameter("keyword");
      String xml=("http://wikipedia.simpleapi.net/api?keyword=" + keyword);
      String xsl=application.getRealPath("table.xsl");
      StreamSource xmlSS=new StreamSource(xml);
      StreamSource xslSS=new StreamSource(xsl);
      StreamResult outSR=new StreamResult(out);
      TransformerFactory fac=TransformerFactory.newInstance();
      Transformer tran=fac.newTransformer(xslSS);
      tran.transform(xmlSS,outSR);
%>
```

4.2 Web Service System Prototype for Knowledge Sharing in Cooperative Creation Work and Discussion

As stated above, this research focuses on an application method of Web service to an atypical task such as a knowledge creation task for improving task performance in cooperative office work. A Web service interoperating system is produced as a prototype using the Google Web service, local document retrieval Web service, and the Wikipedia API.

Conventional system construction for interoperating of web application software is costly and time consuming. Web service technology provides the interoperability of the server's call server instead of the user's services call operation. Web services are expected to carry out creative work effectively, because the contents are retrieved from the appropriate site by using tools suitable for sharing creative work with an atypical type procedure.

For the interoperating system of document retrieval, Google Web service is provided as the existing web retrieval engine. This service performs web document retrieval in an accompanying WSDL file that can be imported into our Web services programming environment. The local content retrieval components are customized by JavaBeans. JavaBeans components are reusable software programs that can be developed and easily assembled to create Web service applications for the task knowledge support system. Encyclopedia retrieval is useful for the identification of unfamiliar technical terms in any creative collaborative work. The implementation of Wikipedia API service provides encyclopedia articles. Consequently, these Web service systems support cooperative work in the knowledge creation process. Figure 4 shows the Web service construction of the user interface for creative work support. Service selection and providing are based on the creative work process.

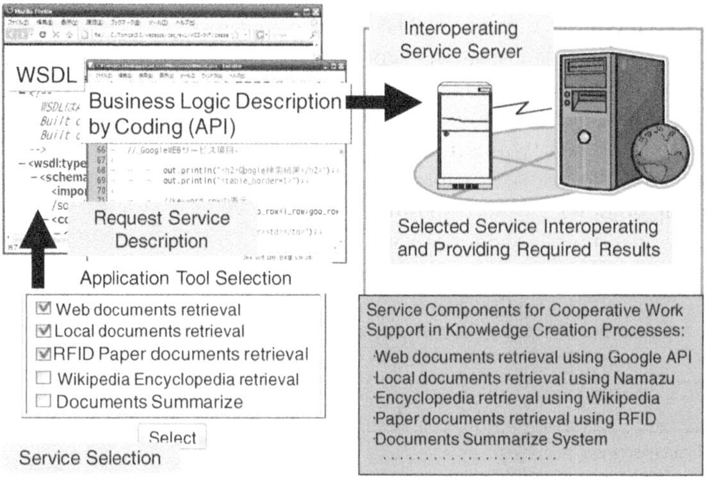

Fig. 4. Service selecting and providing based on the creative work process

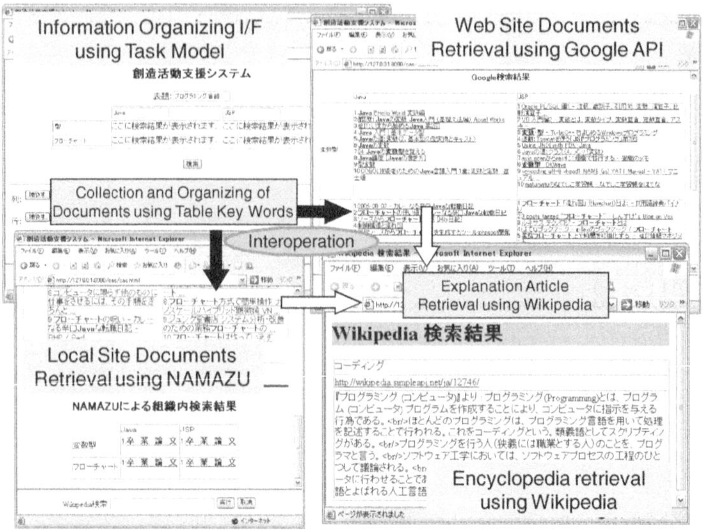

Fig. 5. Interoperating between information organizing and documents retrieval application

Figure 5 shows an overview of the interoperating document retrieval management system. A task model is used as an input key of the retrieval process. A piece of meta-knowledge is made by retrieving the keywords included in the task model, and by embedding links that direct the retrieval results into the task model. These retrieval processes are automatically interoperated according to the location of the document resource. Wikipedia service is allowed to run on this system if the need for an encyclopedia article arises.

5 Conclusion

For effective knowledge creation and sharing in a cooperative environment, the applicability of Web service technology to perform an atypical task is studied. A pre-developed integrated method of information organizing is used, in which a task procedure is structured and shared in tabular form, and document retrieval systems are managed by an open Web service API. The following results were obtained.

1. By using a "literature survey in research work" as an example of a target application of knowledge creation and reuse work, Web service architecture for the atypical work of knowledge creation in an organization is provided. The conceptual design requires interoperating between different software applications running on a variety of platforms.
2. Web service specification is provided, in which web documents and local site documents are automatically retrieved and organized by the table form task model, thus enabling the sharing of task knowledge.
3. A Web service prototype environment for document retrieval and task knowledge support is implemented by using Google for web documents, Namazu for electronic documents and paper documents, and encyclopedia functions by Wikipedia API.

References

1. W3C (World Wide Web Consortium), `http://www.w3.org/2002/ws/`
2. Von Krough, G., Ichijo, K., Nonaka, I.: Enabling Knowledge Creation. Oxford University Press, Oxford (2000)
3. Nonaka, I., Takeuchi, H.: The Knowledge Creating Company: how Japanese companies create the dynamic innovation. Oxford University Press, Oxford (1995)
4. Kojima, H., Yamaguchi, Y., Woodworth, C.: Cooperative Creation Support Interface for Collective Knowledge. In: Proc. of the 8th World Multi-Conference on Systemics, Cybernetics and Informatics (SCI 2004), IIIS, vol. X, pp. 35–40 (2004)
5. Kojima, H., Iwata, K., Nishimura, N.: Document Management and Information Organizing Method Using RFID Tags. In: Jacko, J. (ed.) Human-Computer Interaction Part IV. LNCS, vol. 4533, pp. 601–610. Springer, Heidelberg (2007)
6. Baba, H.: Development and practical use of Namazu system. Soft Bank Publishing (1998)
7. Makino, T.: Web service, `http://www.xmlconsortium.org/websv/kaisetsu/B1/main.html`

Designing International Enterprise Software

Janaki Kumar, Dan Rosenberg, Paul Hofmann, and Michael Arent

3410 Hillview Ave., Palo Alto, CA, 94304, USA
{janaki.kumar,dan.rosenberg,paul.hofmann,
michael.arent}@sap.com

Abstract. This paper provides a framework to understand the various aspects of creating international enterprise software. These aspects could be used to evaluate and prioritize a software product's investment in internationalization. The intended audiences for this paper are user experience designers, product managers and other members of the software development product team, interested in creating world ready enterprise software.

Keywords: internationalization, enterprise software, business software, user experience design, product design, world ready software, global software, SAP.

1 Introduction

In today's economy, enterprise software vendors consider globalization a key part of their strategy to reach the world market. Over the decades, the criteria for "world-ready" software have gone beyond mere translation of texts and labels. It takes a cross-functional team to create software that is truly international. Developers need to make sure the underlying code is Unicode compliant, and that the software is architected in a flexible manner to enable localization. The technical writers and translators need to provide meaningful screen texts and documentation in the languages supported. And user experience designers along with the product managers need to define the product functionality with an international user base in mind.

2 Difference between Enterprise Software and Consumer Software

Enterprise software refers to a software suite comprised of common business applications such as Human Resources, Financials, Supply Chain Management, Customer Relationship Management etc., along with tools for modeling how the entire organization works (e.g. task flows and business process flows). Enterprise software is typically purchased by companies and made available to its employees. Consumer software is typically purchased by an individual for his or her personal use, such as email, word processing, instant messenger, web browsing.

This paper provides user experience designers a framework to understand the various aspects of creating international enterprise software. This framework is based on our years of experience working for SAP, a global enterprise software provider,

M.J. Smith and G. Salvendy (Eds.): Human Interface, Part I, HCII 2009, LNCS 5617, pp. 104–112, 2009.

headquartered in Germany with offices around the world. Roughly 74% of SAP revenues come from outside the US, with customers in 120 countries. Almost all of SAP's customers have a multinational presence requiring SAP software to adapt to their business needs.

3 Framework for Internationalization

We examine internationalization of enterprise software along two dimensions: the *Deployment Market* and *Global Differences*. This framework is designed as an aid to software companies in determining their investment in internationalization. All cells in this framework are not of equal importance to all companies. For example, the Legal and Compliance differences need to be addressed only by software handling international trade that is governed by such laws and regulations.

Fig. 1. Framework for Internationalization of Enterprise Software

3.1 Deployment Market: National vs. Multi National

Deployment market refers to the market where the software will be used. Business software makers need to understand where their software will be deployed and utilized to make a rational investment in its internationalization capabilities.

Product teams need to ask early in the design process, "Will the software be deployed by national enterprises in one country or region, (e.g. Pacific Gas and Electric (PG&E) and other utility companies) or multinational companies with a presence in multiple countries (e.g. Coca-Cola)?" Software targeted at business entities based in a single country has different design needs than software designed for multinational companies. Multinational companies have a global business presence with processes that cross national boundaries, while a company that operates within one country may have a strong national identity and will need their software to reflect it. The degree of centralization of business processes by the headquarters has an influence on the level of internationalization required in a multinational company. For example, if the headquarters establishes common business practices pertaining to workflow and approvals for all its subsidiaries, the level of country-specific internationalization with regard to this aspect of enterprise software will be limited.

These differences impact the timing of localization efforts within the software development life cycle. Typically, multinational companies require software that has internationalization built into the underlying framework early in the development cycle, to enable processes that span multiple locales. Software for national companies, on the other hand, could be built specifically for one location and localized for other national companies at a later stage of the software development process. The disadvantage is that the software needs to be localized individually for each new national market, if it is not built on an appropriately flexible and dynamic framework right from the start.

It is important to consider future deployment markets as well as the most immediate ones. If the software being created is for a national market, but there is a possibility that the software could be sold to multinational corporations, then the upfront investment of considering internationalization early in the design process will pay dividends by allowing the software maker to tap into new markets easily in the future.

3.2 Global Differences

The second dimension is the spectrum of **global differences** that enterprise software needs to address. These can be divided into three categories: *Functional differences*, *Legal differences* and *Business Practice* differences around the world.

3.3 Functional Differences

Languages. Although the actual translation will be done by a team of professional translators, user interface designers need to consider this during screen design. Text can be longer in some languages, so there needs to be adequate space on the user interface to display it, or the software must be able to dynamically adjust the screen layout. Bi-directional languages such as Arabic and Hebrew need special consideration since text is presented to the user from right to left, but numbers and Latin alphabetic strings are presented from left to right.

The words chosen in the base language need to be simple, clear and domain-relevant, since ambiguous terms may be translated incorrectly, leading to customer issues. In most cases the language is determined at log in, either explicitly by the user or implicitly based on user preferences or settings. In some cases, the language may need to be changed dynamically. For example, agents in call centers serving multiple locales may need to switch languages based on the caller's preference. These design considerations need to be made early in the process to provide an optimal user experience.

It is not recommended that developers hard code any text within the user interface, since this could be missed during translation. It is important to keep error messages in a central location and translate them thoughtfully to avoid displaying messages in a language unknown to the user - particularly when they are experiencing an error.

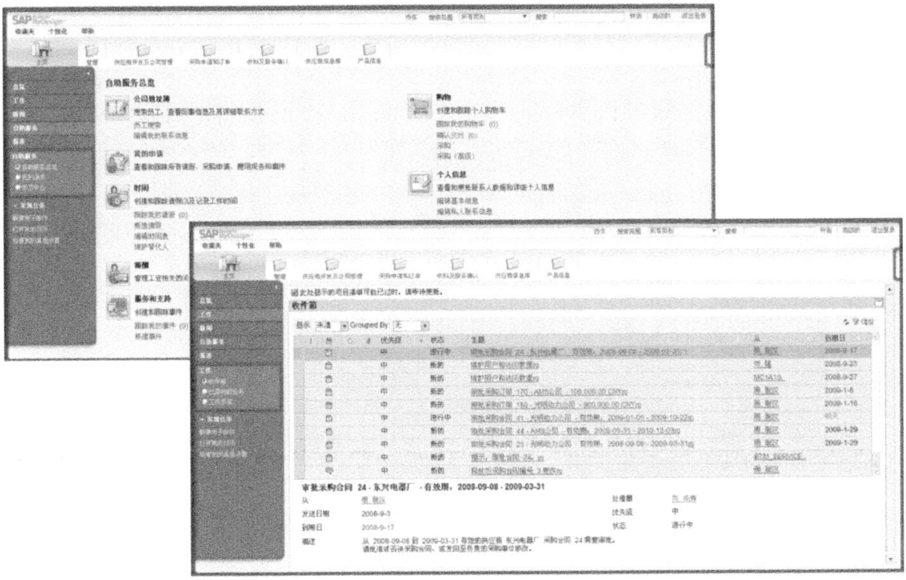

Fig. 2. Example SAP screens in Chinese

Date and Time formats. Date and time formats are critical parts of internationalization. 12/9/2008 can be interpreted as 9th December in the United States or 12th September elsewhere in the world. Typically, the display format is determined via a preferences dialog box in the software UI. However, it may not be sufficient to store date and time in user's preference alone when there are multiple time zones involved in a business transaction. With global businesses requiring that documents be sent across nations, it is important to store the date and time in Coordinated Universal Time (UTC) format, as per the International Standards Organization (ISO), and explicitly convert the format when displaying to the user according to their location and preference [1].

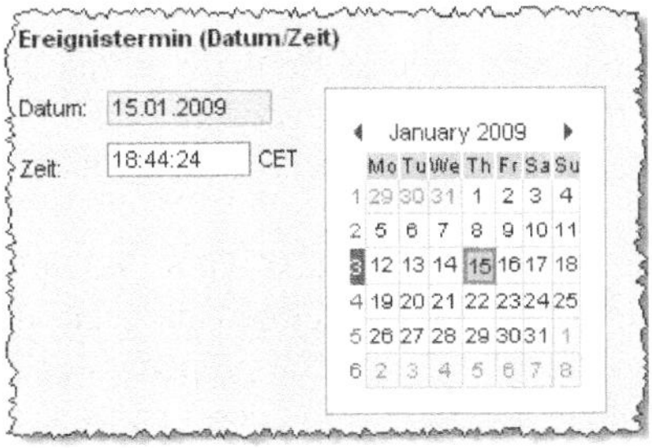

Fig. 3. Date and Time in Central European format

Furthermore, in multinational companies, servers may be located in a country different from its users. In such cases, the timestamp should be the system time in UTC, converted to the local time when displayed to users.

Currencies. In addition to the currency symbol (e.g. $, €, ¥) there are regional variations in display formats specifically in relation to the use of commas and periods. For example, 1,000.02 in the US would be 1.000,02 in Europe. These display formats need to determined from system preferences and not hard coded into the user interface.

As business transactions cross national borders in many cases, there may be multiple currencies represented in one business document. Exchange rate conversions need to be done dynamically to ensure accurate calculations. These product design considerations need to be made early in the process to ensure an optimal user experience.

In some cases, business documents such as Bill of Materials may have to be stored in multiple versions, depending on the exchange rate conversion agreements. The currency could either be converted on a fixed date, or the rate could float for a period before it is locked. This impacts the valuation of the business document, and should be thoughtfully considered at design time.

Fig. 4. Currency selection on business document

Measuring Systems. When designing business software, the unit of measure (the English system customary units, such as pounds, gallons, miles, or the metric system units, such as kilograms, liters, kilometers.) cannot be hard coded in the user interface. The user interface needs to be prepared to accept data in user preferred format and convert if needed. Surprisingly, this is a controversial issue and has legal consequences. According to the Units of Measure directive[1] in the European Union, companies engaging in trade globally have to label their products using the metric system starting in 2010, with dual labeling requirements until then. This has a direct impact on physical product design as well as the software. Business software needs to be flexible to handle multiple units of measure, whether it is displaying product information or printing product labels to be affixed on store shelves.

Similar to currency, unit of measure conversion may also cause businesses to maintain multiple versions of a business document or decide on a strategy for conversion based on their customer's needs. In manufacturing scenarios, companies may measure raw materials by weight or by volume, depending on the unit used in the Bill of Material (BOM). When using the raw material for a production order, its unit (e.g. liter) has to be converted to the unit represented in the final product (e.g. pound). Using milk as ingredient for milk powder or cake mixture would be an example where one has to convert the raw material measured in liters to kilos and then to pounds in order to use it in the right unit for a British end product. This is typically handled by storing the raw materials density (kg/liter, for the example above) in an additional field to ensure accurate conversion.

[1] Unit of Measure directive: http://en.wikipedia.org/wiki/Directive_80/181/EEC

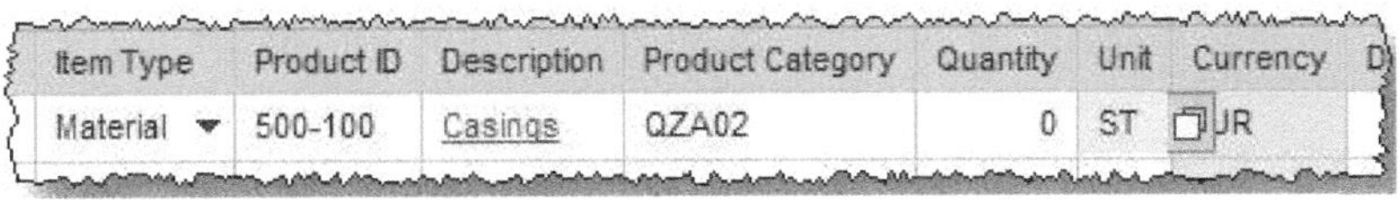

Fig. 5. Unit of Measure dynamically selected

Address Format. Address formats vary throughout the world. In the US, the format is street number, street name, city, state and zip code. In Japan, the format is: postal code, prefecture, city, ward (ku), block(s) (chome), lot (ban), building (go), addressee name (family name first). Given these wide variations, software architects and UX designers need to work together to gracefully switch to the appropriate format based on the user preference and their information needs.

Phone Number. The length and format of phone numbers vary throughout the world as well. When presenting forms to international users, the user interface needs to be flexible to accommodate variations by country.

3.4 Legal Differences

It is preferable to consider international laws and regulations at design time rather than retrofitting them as an after-thought. Here are a few examples of laws and regulations that have a direct impact on the user experience design.

Sales and Value Added Tax laws. Enterprise software products intended for international use need to consider various tax regulations during design time. For example, national differences in Value Added Tax (VAT) calculations have a direct impact on the value of goods transacted. VAT is considered on a line item level, and it applies only under certain conditions. Therefore, not only do the rates vary from country to country, but the conditions on which the rates are calculated vary as well. For example, VAT is impacted by specific ingredients contained in products such as alcohol, which may be found in common household products such as shaving cream. Enterprise software makers need to put in place a flexible framework that accommodates different rates and conditions to enable accurate VAT calculations.

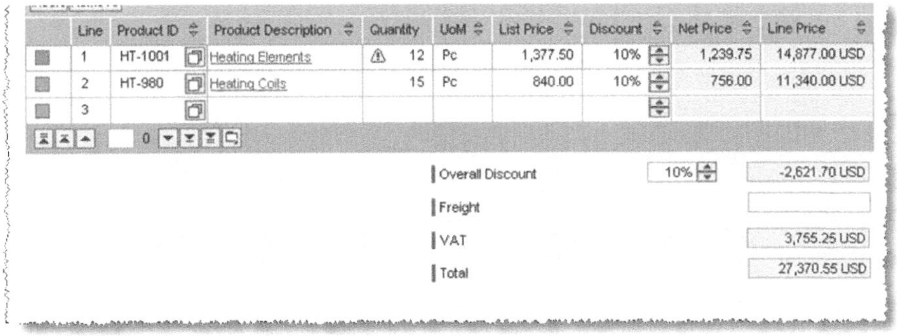

Fig. 6. VAT calculation on a Sales Order document

Privacy Laws. Businesses today are digitizing all their information, and enterprise software enables them to share this information easily across their organization. However, in some countries the data is protected and should not be made available due to government regulations.

For example, customers in the "do-not-call" list should be removed from the list of call center agents in some countries. In the US, customers opt-in to this list, whereas, in Europe, customers are automatically added to this list, and they explicitly opt-out of it to receive calls from companies.

Similarly, in HR systems, managers in different locations have different access privileges to their employees' data based on their countries' privacy laws. In the US, managers cannot view the year of their employees' birthday, whereas in Europe they can. In cases of multinational companies, where a manager oversees teams in multiple locations, the privacy laws of the employee's country apply, and the manager's view of the data needs to display the information appropriately.

Financial and Other Regulations. Companies are required to comply with a steady flow of mandates that impact the way businesses gather, store, manage and report information. The Sarbanes-Oxley Act (SOX), US Health Information Portability and Accounting Act (HIPAA), Food and Drug Administration (FDA) regulations, Restriction of Hazardous Substances (RoHS), and Waste from Electrical and Electronic Equipment (WEEE), are a few examples.

These laws and regulations vary by industry and region. For example, Japan has enacted JSOX that is based on, but is different from the US version. Regulations change over time and new regulations come about, either periodically (e.g. annual updates to tax laws) or based on world events (e.g. Food and Drug Administration updates due to mad cow disease, avian flu and the melamine-influenced product recall). Companies are expected to stay up-to-date with these changing laws and regulations, or face stiff penalties.

Sometimes businesses have to maintain a dual set of books, for example, when transitioning from the US accounting rules GAAP to international accounting rules IFRS.

Financial laws and regulations have a significant impact on enterprise software, especially due to the serious consequences of non-compliance. To effectively keep up with the changing laws and regulations, and minimize risk of non-compliance, enterprise software makers need to incorporate a flexible, reliable and scalable framework, based on a rule-based compliance approach, instead of hard coding specific laws into the software.

Trade Regulations. The World Trade Organization (WTO) has issued a set of directives to harmonize global trade practices. However, not all countries subscribe to the WTO and even among the countries that do, there are enough regional variations that add complexity to global trade.

Many of these variations fall into three categories: process differences, valuation differences, and format differences. The customs declaration process varies across the globe. For example, companies importing goods into the US have to file customs declarations before the goods reach US; whereas, EU countries allow customs declarations to be filed when goods reach the warehouse, which is after the goods reach EU soil.

Valuation of goods is affected by the process variations. For the US, the costs of transit have to be estimated at the time of custom declaration. For the EU, the customs declaration document contains actual transportation costs. These variations can be significant for companies that are involved in large volumes of global trade.

Finally there are both content and format differences in which physical and electronic documents are required for companies to clear customs.

Enterprise software makers have to consider these trade regulation differences in order to effectively internationalize their product offerings.

3.5 Business Practice Differences

Workflow and Approvals. One of the key areas in enterprise software where the interaction of business practices and cultural differences manifests itself is in how decisions, such as purchase requisitions, supplier agreements or even simple expense requests, are approved. Enterprise software systems typically contain embedded workflow engines that route approvals along the management chain, based on a set of rules. For example, an order under $10,000 might require only a Vice President level approval while an order over this amount would require a Chief Financial Officer's approval.

Research sponsored by SAP in both China and South Africa found that the typical user interface design for such workflow systems was not a good match for either local culture. While in Western culture, it is acceptable for the request with relevant data to appear in a workflow inbox screen unannounced, in these two highly differentiated emerging economies, the approval was always sought verbally up the entire management chain before utilizing the automated system. This limited the utility of the automation capabilities of the system.

Another process difference is that the users understood the concept of "Approve" and "Reject" buttons, but considered pressing the "Reject" button rude and a loss of face for the sender. In Western cultures, it is acceptable to reject a workflow request with a comment (e.g., to negotiate a better price with a supplier). In Chinese and African cultures, it is considered unfriendly to use the "Reject" button in business contexts.

Standardization of business process. One of the few international standards, accepted by more than a 100 countries, is the United Nations ISO 9000 quality management system standards. Its purpose is to standardize procedures for key business processes, thereby facilitating international trade. These standards have multiple benefits such as ensuring auditability and trackability of the quality of products and services.

Multinational companies that comply with these standards are certified as such. These standards benefit enterprise software makers by focusing internationalization efforts and thereby reducing costs.

Cultural differences impacting business practices. Aaron Marcus and Emile Gould, document the application of Hofstede's cross-cultural dimensions to Web design [2]. Cultural differences impact enterprise software as well. In some cases these differences may be so severe that the product does not apply. For example, software time recording was not successful in Korea, because it is based on the physical metaphor of "time card" machines which are generally not used there. In other cases, the prevalence of the World Wide Web has erased some of the differences. For example, it is no longer an issue to show an error message with a red icon in China, where red is regarded as a lucky color. Again, multinational companies play an important role in defining the global business culture of the next generation information worker.

4 Conclusion

Investing upfront in internationalization, based on this framework, will allow enterprise software makers to design scalable, world-ready products. It enables them to more easily enter new markets. Furthermore, the lack of a systematic approach to internationalization will prove to be a severe impediment to the product's success and growth since designing for one country and retrofitting to others one at a time can quickly become an expensive proposition. Ultimately, the level of investment made in the internationalization of enterprise software is a business decision, and the purpose of this paper is to provide a framework to analyze the requisites, impacts and trade-offs from a user experience perspective.

References

1. W3C Date and Time Formats, http://www.w3.org/TR/NOTE-datetime
2. Cultural Dimensions and Global Web User-Interface Design, http://www.amanda.com/resources/hfweb2000/hfweb00.marcus.html

Proactive Fault Detection Schema for Enterprise Information System Using Statistical Process Control

ChiHoon Lee[1], Doohyung Lee[1], Jahwan Koo[2], and Jinwook Chung[1]

[1] School of Information and Communication Engineering, Sungkyunkwan University, Chunchun-dong 300, Jangan-gu, Suwon 440-746, South Korea
chlee10@naver.com, drlee07@skku.edu, jwchung@songgang.skku.ac.kr
[2] Computer Sciences Department, University of Wisconsin-Madison, 1210 W. Dayton Street, Madison, WI 53706, USA
jhkoo@cs.wisc.edu

Abstract. This paper proposes a proactive fault detection schema using adaptive statistical approaches in order to enhance system availability and reliability in the heterogeneous & complicated information system environment. The proposed system applies Six Sigma SPC (Statistical Process Control) techniques already validated in industries in order to monitor the application system in the information system. This makes it possible to reduce false alarm rates for system faults and accurately detect faults by creating a control chart based on past performance data and controlling the distribution of performance based on the chart. The early detection of faults is also enabled through a fault prediction model. Therefore, the aforementioned system not only detect unknown or unseen faults but also resolve potential problems for system administrator by detecting abnormal behaviors before faults occur. In the experiment we show the superiority of our proposed model and the possibility to early detect system faults.

Keywords: System management, Proactive Fault detection, Statistical Process Control, Early Detection, EWMA.

1 Introduction

An explosive rise in Internet penetration has caused a massive traffic overload and required computer performance and capacity to be upgraded. Under these circumstances, the key challenge in service continuity is to ensure the availability and reliability of the information system. Such an issue can be resolved to a certain degree by preventing single-point-failures through system dualization and clustering, but which is accompanied by heavy costs. Therefore, the accurate and early detection of system faults must be guaranteed in order to achieve such a goal with limited physical and financial resources and to avoid potential serious issues.[1][2][3]

Fault management, a key area in system administration, is designed to early identify and detect faults, which has continuously attracted more attention. The information system, which most organizations heavily depend on, has become core infrastructure in an enterprise environment. The downtime caused from the system

M.J. Smith and G. Salvendy (Eds.): Human Interface, Part I, HCII 2009, LNCS 5617, pp. 113–122, 2009.

faults force companies to pay significant costs. Furthermore, a diversity of challenges should be handled in order to overcome such fault issues. First of all, the hardware or software of various vendors should be organically linked and operated in an open environment. At the same time it should meet user needs and up-to-date technology requirements and then it makes increasing fault risks [4]. Furthermore, continuous application changes and improvements to satisfy user requirements make it harder to control faults.

The biggest issue in fault detection is the fact that a fixed threshold method is used, based on the experiences of system administrators and experts. This makes it difficult to flexibly respond to changes and the extension of the information system. If a threshold value is set too high, it is not easy to detect faults. If it is too low, the system may suffer from frequent false alarms. Even though the threshold is properly set, lack of prior information makes it hard to early detect the symptoms of faults. [1].

The existing research on fault management focused on a rule-based expert system [5], finite state machines [6], a statistical model [7], and a data mining model [8]. The rule-based expert system requires the specifications of potential faults [1]. This is exposed to performance restrictions because all possible faults can not be handled, which is also vulnerable to unknown or novel types of faults and changes in the system environment. The statistical model can resolve part of the aforementioned issues but hardly forecast faults in advance. Lastly, the data-mining model can detect faults in advance to a certain degree but can not be easily applied in real time. This paper proposes a new proactive fault model that can accurately detect and forecast information system faults. This model is designed to 1) minimize FAR, 2) detect changes in distribution/average to forecast information system faults, 3) and detect in real time under the online environment by applying an adaptive threshold method based on the SPC validated in manufacturing processes. The purpose of this paper is to substantially apply the model to the management of enterprise information system faults.

In this paper Section 2 deals with system management overview and the background of SPC and EWMA. Section 3 describes proposed schema. Section 4 gives experimental results. Finally section 5 mentions conclusion.

2 Background

2.1 System Management

These days the reliance of the information system in the enterprises has increased and also been reinforced to ensure business continuity. The information system can stably support business activities only when it is harmonically activated, based on servers, a network, a database, and applications. However, the sudden deactivation of all or parts of such components can do great damage to the enterprise, with direct effects on the survival thereof. Up to now, staffs by area have supported stable operation by monitoring their partial field from time to time. However, in the current information system environment characterized by a rapidly rising number of servers & database and more complicated networks & applications, such monitoring by area has to be exposed to its own limitations. The System Management System was introduced to overcome such an obstacle. It is designed to ensure service continuity by immediately notifying the system administrator of the occurrence of faults in real-time through a sophisticated agent and helping them quickly take corrective actions. A general SMS consists of managers and agents. An agent installed on a monitoring target gathers data (faults or performance items) in real-time, transferring them to Manager. Then, Manager gathers and analyzes data before reporting faults to system, network, DB, and application administrators.

The detected faults represent the state of real faults, the notified operator should promptly take corrective actions to fix faults. However, because such signals are detected after the fault occurs, the system is already in the state of fault at the point when the operator recognizes it, causing the business to stop. Therefore, the system operator is in dire need of a system that can monitor abnormal symptoms and prevent faults in advance.

2.2 Statistical Method-Based Fault Detection

SPC (Statistical Process Control). The SPC is designed to identify, interpret, and resolve problems based on accumulated statistic data, rather than intuition or guessing. In other words, it scientifically analyzes basic data to manufacture that meet quality requirements. The key to the SPC is to produce uniform quality goods featuring little quality dispersion. The main purpose of the SPC is to reduce process variability and enhance processes in order to manufacture higher-quality products. The SPC is currently applied to a wide range of industries including manufacturing, to control manufacturing facilities, curtail logistics costs, and improve software quality. [9][10] The control chart devised by Dr. Walter Shewhart [13] is used to detect process changes from special causes through the graph that can monitor data variability over time. It can be classified into variable control charts for continuous data and attribute control charts for discrete data. Specific examples can be illustrated as follows [15]:

Variable Control Chart: $\overline{X}$ - R, $\overline{X}$ - S, I&MR chart

Attribute Control Chart: P, NP, C, and U chart

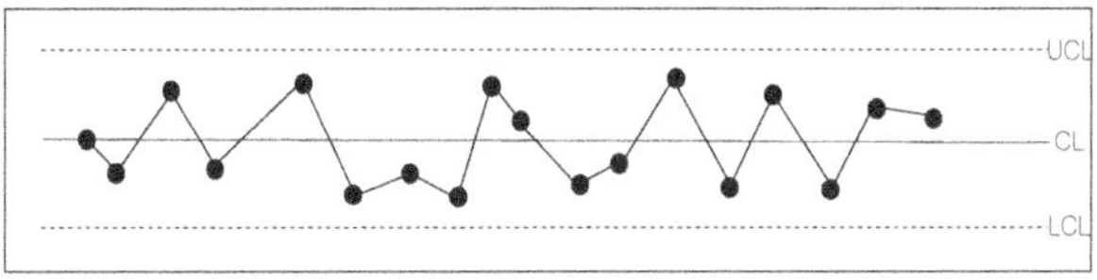

Fig. 1. General form of the control chart

Fig. 1 represents the average of quality characteristics around the Center Line. The UCL (Upper Control Limit) and the LCL (Lower Control Limit) are calculated by adding or deducting "each quality characteristic SD(Standard Deviation) multiplied by k" to or from the mean (average) of the quality characteristics. If the values calculated through data are scattered on the graph, all the points are placed between the control limits. Unless special factors are identified, the process is judged to be in a control or steady state. This paper applied the SPC in order to resolve fixed threshold issues.

EWMA (Exponentially Weighted Moving Average). The Exponentially Weighted Moving Average (EWMA) [12] is a statistic for monitoring the process that averages the data in a way that gives less and less weight to data as they are further removed in time. For the Shewhart chart control technique [13], the decision regarding the state of control of the process at any time, t, depends solely on the most recent measurement from the process and, of course, the degree of 'trueness' of the estimates of the control limits from historical data. For the EWMA control technique, the decision depends on the EWMA statistic, which is an exponentially weighted average of all prior data, including the most recent measurement. By the choice of weighting factor, λ , the EWMA control procedure can be made sensitive to a small or gradual drift in the process, whereas the Shewhart control procedure can only react when the last data point is outside a control limit.

As with all control procedures, the EWMA procedure depends on a database of measurements that are truly representative of the process. Once the mean value and standard deviation have been calculated from this database, the process can enter the monitoring stage, provided the process was in control when the data were collected. This paper used the EWMA to forecast faults in advance.

3 Proposed Scheme

3.1 Limitations of the Existing SMS

The biggest issue facing the SMS is to properly set the threshold. This requires expert knowledge of system operation, accompanied by trials and errors. For example, if a threshold value is set not to exceed in any environment, faults can not be detected. In case it is set too low, system administrators will be flooded with false alarms. It will also be impossible to flexibly respond to changes in the system environment such as system introduction & enlargement and application addition. Even if a performance value exists within the threshold range, abrupt variances in the range can be thought of as the symptoms of the faults. However, the current SMS regards them as normal.

3.2 Proposed Fault Detection System Model

This paper proposes two types of fault detection models in order to overcome the limitations and errors of the existing fault detection method described in section 3.1. One is an adaptive threshold identification model to which the SPC applies. The other is a fault forecasting model to which the EWMA applies.

SPC-based Adaptive Threshold Identification. After analyzing performance data for a certain period of time by applying the SPC methodology, an optimized threshold is automatically identified, based on which the state of performance data gathered in real-time can be assessed. To this end, the following procedures are implemented:

(1) Select a proper control chart after choosing performance items to monitor. Most of the performance data gathered in management are considered to be variable ones. A proper control chart (Xbar-R chart, Xbar-S chart, I&MR chart, etc.) should be selected, depending on the quantity of collected data and the collection interval. In case the collection cycle is short and a large quantity of data is gathered, the following choice of the control chart is made by subgroup size. If the subgroup size is set to be 10 or larger, the Xbar-S chart is applied. If the size is smaller than 10, the Xbar-R

chart is used. If the data collection cycle is long and a small quantity of data are gathered, the I&MR chart will be a proper choice.

(2) In case a control chart and a subgroup size are chosen, an automatic control limit renewal cycle is set. The cycle is set, considering changes in system capacity and the quantity of data.

(3) Based on the selected control chart and the automatic control limit renewal cycle, a Daemon program which automatically generates a control limit at a specific time every day is implemented in order to generate the control chart, which was set after reading gathered performance data by item. As a result, an optimized threshold (control limit value) for the performance item can be obtained.

(4) The SMS transmits performance date collected in real-time to a SPC processing module in real time. The SPC processing module conducts calculations in accordance with the formula of the control chart designated in (1), after the size of real-time data reaches the level that was set in (1). In case the result is placed outside the control limit identified in (3), an event is generated. The resulting event is not influenced by performance data from instant rapid variances (too large or small value), reducing FAR and enabling accurate & reliable fault detection. In addition, the value is automatically calculated and renewed, depending on the control limit setting cycle, making it possible to respond to changes in the system environment and produce an adaptive threshold.

Early Fault Detection by EWMA. The EWMA model is applied in order to analyze data that are not in control or show big variances over time. Precisely this model can forecast subtle variances in the control limit to detect faults in advance. At the current point (t), variances (r_t) in performance data are calculated as follows:

$$r_t = \ln(p_t / p_{t-1}) \tag{1}$$

where P_t is the performance data at the point of t and ln() is the natural logarithm. Variability in the change rate can be calculated by using the sample standard deviation. The standard deviation (δ) for the sample size 'n' is calculated as follows:

$$\delta = \sqrt{\sum_{t=1}^{n} (r_t - \bar{r})^2 / n} \tag{2}$$

Above formula shows that the sum of the square of deviations from the mean, (r), of performance data change rates is divided by the sample size. It can also be regarded as averaging 'the square of deviations from the mean' multiplied by a certain weight (1/n). The variability can also be estimated by using the EWMA. This is is represented as follows:

$$\delta = \sqrt{(1-\lambda)\sum_{t=1}^{n} \lambda^{t-1} (r_t - \bar{r})^2} \tag{3}$$

Where λ is a decay factor. Because the variability estimated by the EWMA has the following recursive characteristic, it is easier to predict variability:

Let $\delta_{t+1|t}^2$ be $(1-\lambda)\sum_{i=0}^{\infty} \lambda^i r_{t-i}^2$

Then we get equation (4)

$$
\begin{aligned}
\delta^2_{t+1|t} &= (1-\lambda)\sum_{i=0}^{\infty} \lambda^i r^2_{t-i} \\
&= (1-\lambda)(r_t^2 + \lambda r_{t-1}^2 + \lambda^2 r_{t-2}^2 + \cdots) \\
&= (1-\lambda) r_t^2 + \lambda(1-\lambda)(r_{t-1}^2 + \lambda r_{t-2}^2 + \lambda^2 r_{t-3}^2 + \cdots) \\
&= \lambda \delta^2_{t|t-1} + (1-\lambda) r_t^2
\end{aligned}
\tag{4}
$$

Hence $t+1|t$ means predicting $t+1$ data by using t-point performance data.

In the above equation, the $t+1$ variance can be represented through the linear equation of the pervious change rate. Therefore, if the t point variance is identified, the $t+1$ point variability can also be repeatedly estimated.

If the sigma level (s; s= 1, ... , 6) is applied to the above formula, the UCL and LCL at the point t can be calculated. When t point data are obtained, we can check whether they are outside the t point UCL and LCL or not in order to measure the reliability of performance data.

When t-1 point performance data are obtained, we will make a judgment on whether they are within the range of UCL and LCL, based on obtained t point performance data(P_t). If they are outside the range, possible faults can be predicted in advance by raising an alarm. The concept of fault tolerance can be additionally applied in order to reduce FAR.

3.3 Implementation

The proposed system is mainly composed of an SPC server, an SPC generator, an SPC console, and API modules.

The SPC generator reads monitoring item once a day (the minimum cycle for automatic control limit renewal is one day). Based on this, both UCL and LCL are calculated and saved in the SPC DB. A user selects monitoring targets by using an SPC console before entering details for monitoring. He or she should make a decision on whether to use the SPC or EWMA. If the SPC model is selected, a control chart type, a data collection cycle, an automatic control limit renewal cycle and a minimum subgroup size are set and then saved in the SPC DB. If the EWMA is chosen, the SPC generator is not activated. The reason is that the EWMA calculates the next performance data estimate in real-time but the SPC needs calculated UCL and LCL to analyze current data.

The SPC server conducts a real-time analysis in accordance with the method (SPC or EWMA) set by monitoring target, immediately after receiving performance data through SPC API. In case of SPC analysis, the mean and standard deviation of the subgroup is calculated by referring to UCL and LCL pre-calculated by the SPC generator at the point when real-time performance data corresponding to the size of the subgroup are gathered. Then, a judgment is made on whether the value is outside UCL and LCL by control chart before generating fault events on a need-to-do basis. The subgroup is calculated in accordance with the sliding window method. For example, if the subgroup size is set to 7, the first subgroup is (p1, p2, p3, p4, p5, p6, p7). The subgroup at the point of collecting the 8th data is shifted to (p2, p3, p4, p5, p6, p7, p8,). As the subgroup is sequentially shifted in this way, the mean and standard deviation thereof are calculated in real-time in order to check whether they are within the range of UCL and LCL. As for EWMA analysis, future point (Pt+1) performance data are forecast at the point when current point(Pt) performance data are collected. Therefore, UCL(t+1), LCL(t+1) values are calculated by using the change rate (rt) in order to check whether Pt+1 performance data are within the range of UCL and LCL. Only if they are outside the range, a fault event is generated.

A user selects monitoring targets and an analysis method (SPC or EWMA) through the SPC console. Also the SPC console makes it possible to query events generated from the SPC server's real-time calculation and analysis.

4 Experimental Results

The experiment was conducted for the following two purposes: One is to validate the effects of the fixed threshold and the SPC-based adaptive threshold. The other is to check whether to forecast faults in advance. A separate experimental environment was set up to generate faults, through which data were collected. However, the reliability of the deliverables may not be high enough, considering the fact that it is not easy to gather enough real world fault data. The web page response time was used for SPC validation while server CPU utilization data were applied to the EWMA analysis.

4.1 Results from SPC-Based Adaptive Threshold Model

The SPC model was applied to the web page response time. The Xbar-R chart was selected as a control chart, setting the subgroup size to 5 and designating the auto control limit renewal interval as one day. As a result, the control limit is automatically renewed on a daily basis. The Anderson-Darling test [14] was used as target data normality testing. The normality of the data was validated, as the p-value is larger than 0.05.

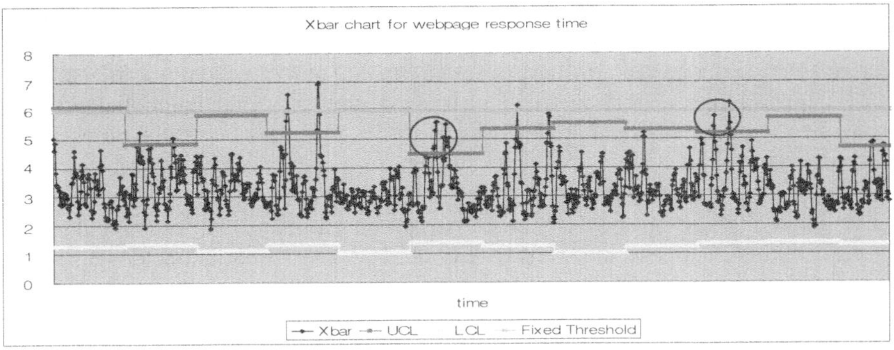

Fig. 2. Xbar chart for webpage response time

Fig. 2 shows that the control limit is renewed on a daily basis, resulting in the application of the adaptive threshold. The proposed model identified the points in the ellipse not detected by the fixed threshold as abnormal behaviors. This implies the feasibility of monitoring optimized to the system environment. Namely, in case environmental changes occur such as set based on past data, flexibly responding to endless changes in the system environment.

4.2 Results from EWMA-Based Early Fault Detection

CPU utilization data for three days including the day faults occurred were collected by using experimental devices. During the first and third days, no fault was detected but on the second day, real faults were sensed. The data were analyzed by using the EWMA model. The λ (decay factor) and s (Sigma Level) were set 0.85 and 3, respectively.

Fig. 3(a) shows the results from EWMA analysis of the first day performance data. Around 08:30, a false alarm was raised, even though a real fault did not occur. Steep changes in performance data at the point of the false alarm clearly suggest server problems, even if it has yet to reach the level of a fault. Furthermore, taking into account that the false alarm was not frequently given, the system administrator needs to be pre-cautions against the situation.

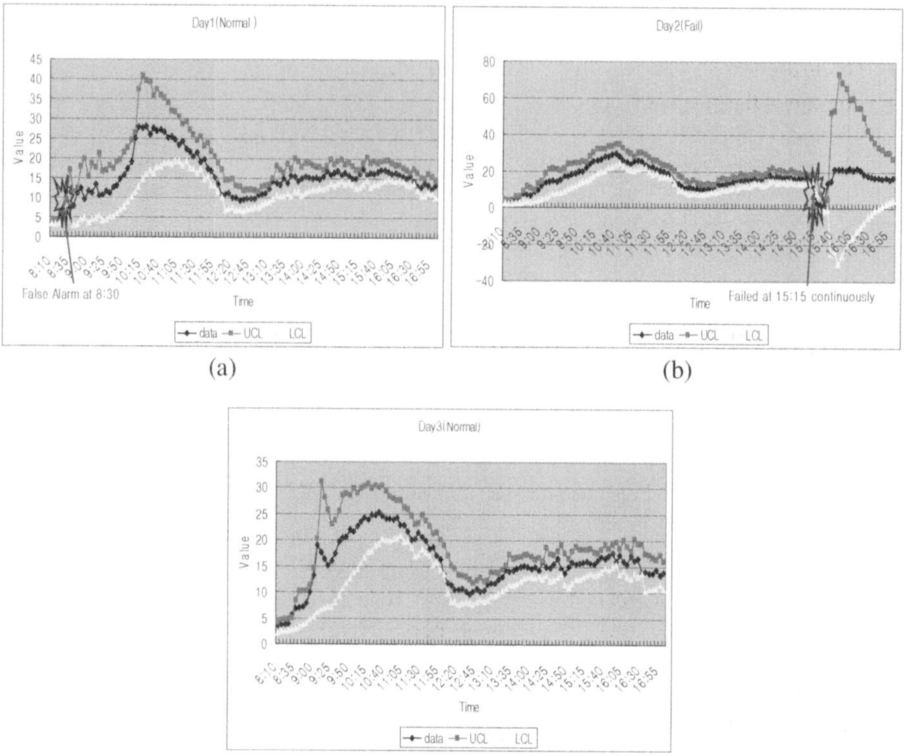

Fig. 3. CPU Utilization's performance data of specific server

As shown in Fig. 3(b) the Online Daemon Process hanged up around 15:15 on the second day. Therefore, if the current state had continued, system-down could not have been avoided. We reasoned the Daemon Process Hang Up based on variability in CPU utilization value, rather than monitoring the process. As a result, the system administrator was successfully alerted to abnormal behaviors, solving serious potential issues by restarting the Daemon Process. As displayed in Fig. 3(c), it was normally processed without any false alarm on Day 3. This shows that the proposed model is stably activated.

5 Conclusion

Fault management plays a pivotal role in ensuring service continuity in the information system environment. This research presented two types of fault detection models. One is the SPC-based adaptive threshold model designed to resolve fixed threshold issues. This was very effective in accurately detecting potential faults and reducing FAR. The other is the EWMA-based early fault detection model. This enables system administrators to avoid serious potential issues by detecting abnormal behaviors before real faults occur.

This research can apply to the information system in a manufacturing or financial environment. The SPC model is suited to manufacturing processes featuring less variability in performance data while the EWMA paradigm is more effectively applied to the financial sector characterized by larger variability in performance data. The proposed models can accurately detect and estimate system/application faults and minimize downtime, enhancing the yield in manufacturing processes and system reliability in the financial arena. The follow-up research will focus on not only detecting faults but also statistically inferring their root causes in order to make it possible to both detect and fix faults

References

1. Hood, C.S., Ji, C.: Proactive Network-Fault Detection. IEEE Transactions on reliability 46(3), 333–341 (1997)
2. Hellerstein, J.L., Zhang, F., Shahabuddin, P.: An approach to predictive detection for service management. In: Sloman, M., Mazumdar, S., Lupu, E. (eds.) Proc. 6th IFIP/IEEE Int. Symp. Integrated Network Management (IM 1999), p. 309. IEEE Publishing, New York (1999)
3. Thottan, M., Ji, C.: Fault prediction at the network layer using intelligent agents. In: Sloman, M., Mazumdar, S., Lupu, E. (eds.) Proc. 6th IFIP/IEEE Int. Symp. Integrated Network Management (IM 1999), p. 745. IEEE Press, New York (1999)
4. Yemini, Y.: A critical survey of network management protocol standards. In: Aidarous, S., Plevyak, T. (eds.) Telecommunications Network Management into the 21st Century (1994)
5. Jakobson, G., Weissman, M.D.: Alarm correlation. IEEE Network 7, 52–59 (1993)
6. Rouvellou, I.: Graph identification techniques applied to network management faults, Ph. D Dissertation. Columbia University (1993)
7. Deng, R.H., Lazar, A.A., Wang, W.: A probabilistic approach to fault diagnosis in linear lightwave networks. IEEE J. Selected Areas in Communications 11, 1438–1448 (1993)

8. Garofalakis, M., Rastogi, R.: Data Mining Meets Network Management: The Nemesis Project. In: ACM SIGMOD Int'l Workshop on Research Issues in Data Mining and Knowledge Discovery (May 2001)
9. Florence, A.W.: The MITRE Corporation. CMM Level 4 Quantitative Analysis and Defect Prevention with Project Examples, 2000 Technical Papers (September 2000)
10. Radice, R.: Statistical Process Control for Software Projects (November 1997)
11. ERETEC INC., MINITAB Release 14 (November 2005)
12. NIST/SEMATECH e-Handbook of Statistical Methods: EWMA Control Charts
13. Shewhart, W.A.: Statistical Method from the Viewpoint of Quality Control (1939)
14. Anderson, T.W., Darling, D.A.: Asymptotic theory of certain "goodness-of-fit" criteria based on stochastic processes. Annals of Mathematical Statistics 23, 193–212 (1952)
15. Oakland, J.: Statistical Process Control (2002)

A Human Factors Model for Enterprise Resources Planning System Implementation

Chiuhsiang Joe Lin[1], Chih-Wei Yang[2], Shi-Bin Lin[1], and Shiau-Feng Lin[1]

[1] Department of Industrial and Systems Engineering, Chung-Yuan Christian University, 200, Chung Pei Rd., Chung Li, Taiwan 32023, R.O.C.
[2] Nuclear Instrumentation Division, Institute of Nuclear Energy Research, 1000, Wenhua Rd., Jiaan Village, Longtan Township, Taoyuan, Taiwan 32546, R.O.C.
hsiang@cycu.edu.tw

Abstract. The implementation and adoption of enterprise resource planning (ERP) system is currently a key issue for companies. The critical problem of implementing and adopting ERP system is that the system architectures are quite complex. This study focuses on the integration of the human factors issues in the ERP implementation process. The mental or cognitive differences between information systems concerning people and end users are compared in this study based on information related theories. This paper proposes a human-centered system implementation and adoption model. The main purpose of this model is to assist the ERP system implementation and adoption and creates maximum customer value. The paper shows how to carry out the model step by step with the final goal of increasing the efficiency and acceptability of the system to be implemented and adopted.

Keywords: Enterprise resource planning, human factors, human-centered, information system.

1 Introduction

Business information systems are currently implemented through enterprise resource planning (ERP) systems, which are supposed to comprehensively deal with all concerned functions of a company. Nevertheless, the implementation and adoption of such enormous systems is still considered as a complex problem. One of the key points of the implementation and adoption is the appropriation of the system by its users [6], which is often difficult. The origin of this problem is often considered as related to business process definition, in which many human factors (HF) issues are implicated in the process.

Humans will become less willing to use difficult interfaces of ERP systems and search for alternatives when information systems development fails to meets users' requirements [10]. Numerous researches have suggested that the user is the most critical factor leading to information systems success ([1], [24]). The usability of user-interfaces can be seen as one of the critical factors that influences end-user satisfaction [23].

M.J. Smith and G. Salvendy (Eds.): Human Interface, Part I, HCII 2009, LNCS 5617, pp. 123–130, 2009.

Following the development of information technologies, the organization structure and production type have changed. Many companies tried to enhance its performance by adopting ERP systems. In fact, there are many failed cases in the past [8]. Generally, the communication between system developers and end-users is a key factor to the system success or not. So, using appropriate implementation approaches and consulting resources to ensure the system success is a challenge to many companies.

Although ERP provided standard business object and process, the design principle of ERP system is not completely suited to the target corporation ([5], [17]). Two points of view are demonstrated in this study. First, only few researches discussed the key causes and processes successfully implementing and adopting ERP to companies. ERP system is very important to companies. It is useful to discuss the utilization of ERP software. By improving the utility of ERP software, the failure rate of ERP will be decreased. Second, information systems implemented with human-centered consideration is important. When the ERP information systems are implemented in companies, the managers of these companies usually face much end user resistance. It is important to recognize that the end users are not supervisors or managers. They are the employers or operators of the systems. Therefore, an effective information system development will focus on how human users accept and use the system. So, ERP studies must also emphasize the human thinking or cognition problems.

ERP system researchers and developers believe that their system is developed based on practical business processes. The professional knowledge about how to use information technology enhances the management efficiency and should be included in ERP systems [7]. Many companies tried to adopt and implement ERP systems, but failed. There are usually several problems in ERP systems, including: (1) companies ignored ERP system and without enough investment; (2) companies excessively relied on consultant organizations; (3) consultants do not join improvement and maintenance activities.

This study discusses several causes to make ERP system implementation and adoption successfully, including: (1) to develop a formal planning system; (2) to approach consensus to a plan by the internal part of corporation; (3) to accurately acquire corporation data; (4) to enhance management planning changes; (5) to plan and manage excellent production ability; (6) to make sure that clients trust the ERP system; (7) to execute the system efficiently; (8) to consider the role of users while planning ERP system; (9) to make sure ERP systems were operated and maintained by entire company; (10) to use measurement tool to make sure and solve corporate problems.

2 Human-Centered ERP System

Compared to information systems research, theories of ERP system implementation and adoption have gained little attention and lack of theoretical support. This study reviews information systems literature to identify the theories which can be utilized to develop a human-centered ERP system implementation and adoption model.

2.1 Clients Technologies Accepted Level and Perception

An ERP system attempts to satisfy clients' information requirement. Thus, information and clients become the key factors of ERP systems. The theory of reasoned action (TRA) is rooted in social psychology, usually used to discuss human behavior intent and predict human behavior ([11], [12]). Attitude, subjective thinking, and behavior intent composed the theory. Generally, attitudes and subjective variables are influenced by others variables. By adding perceived behavioral control to TRA theory, Ajzen and Madden [2][3][4] developed a theory of planned behavior (TPB). Davis [9] discussed perception and technology which use the TRA and developed technology acceptance model (TAM), which is usually utilized to understand how external factors influence clients' internal perceptions, attitudes, intents, and information uses.

2.2 External Factors on Technology Configuration and Acceptance

Social cognitive theory (SCT) explained human behavior based on interrelations between people, behaviors, and environments. Kwon and Zmud [18] believed that the principal efficiencies in information system execution are user characteristics, organizational characteristics, information technology characteristics, and environment characteristics. Thus, Diffusion of Innovations Theory (DIT) was generated based on the characteristics identified by Kwon and Zmud [18]. When the information provided by users of information systems meets user requirements, the degree of user satisfaction will be improved. Consequently, researchers have attempted to identify the relationship between ERP systems and user satisfaction based on a theoretical model of working—technique fitness [13][14][15]. In this model, different people may make different decisions based on the same information.

2.3 User Process Mode

Goodhue [15] proposed that the central features of a successful information system are user satisfaction and information system conditions (e.g., information identification, information gathered, information integrated, and information explanation). Leitheeiser and Wetherbe [19] discussed the central causes of a successful information system based on user perspectives. Summer [26] identified several factors of a successful information system based on a case study. Magal et al. [20] proposed 22 key factors to successful information system. Computer system acceptability is a combination of social acceptability and practical acceptability. There are several levels and types of consideration in practical acceptability-- cost, compatibility, usability, and reliability. Norman and Draper [22] proposed eight user-interface design principles: consistency; short-cuts used by users familiar with the interface; feedback (informing users on how to use the interface); efficient use of dialog windows, fool-proof, cancel button, reducing memory loading.

2.4 End-Users Classification

The classifications of users can be identified as: novice or first-time users, knowledgeable intermittent users, and expert frequent users [25]. Novice or first-time users are those users who know the information about ERP system roughly. When the novice or first-time users first use the system, they will feel anxious and nervous. Sometimes, they may resist the system. To solve the above described problem, the related information should emphasize the principles, including: (a) to provide necessary dialog windows design; (b) to provide on-line guide; (c) to provide system feedback design; (d) to provide mistake messages while users make a mistake; (e) to provide on-line assistance explanation.

Knowledgeable intermittent users have necessary knowledge to their task in the information system. However, knowledgeable intermittent users did not usually operate and use the information system. Therefore, the users do not easily remember the relative architectures, positions, and characteristics of menus. To solve the problem, the information should emphasize the principles, including: (a) menus should be well organized; (b) the jargon should be consistent; (c) message dialog should be meaningful; (d) the system movement should be consistency; (e) meaningful dialogs should be provided; (f) system organization should be emphasized.

Expert frequent users are familiar with system contents and task items. Therefore, the information should emphasize the principles, including: (a) to provide quick response system; (b) to avoid distraction feedback; (c) to provide hot-key mechanism; (d) to decrease necessary steps; (e) to provide short-cut design. These principles are related to users' satisfaction while using the ERP system based on the classifications of users. Based on these principles, this study proposed a human factors model for ERP system implementation, which was described detailed in the next section.

2.5 The Conceptual Framework Development

Five major stages are proposed in the process, including: (1) Planning; (2) Business process analysis; (3) System development; (4) On-line preparation; (5) Information system implementation. Based on the five implementation stages, this proposed design can help program team shorten ERP implemented schedule and help the ERP system solve users' problem (See Figure 1).

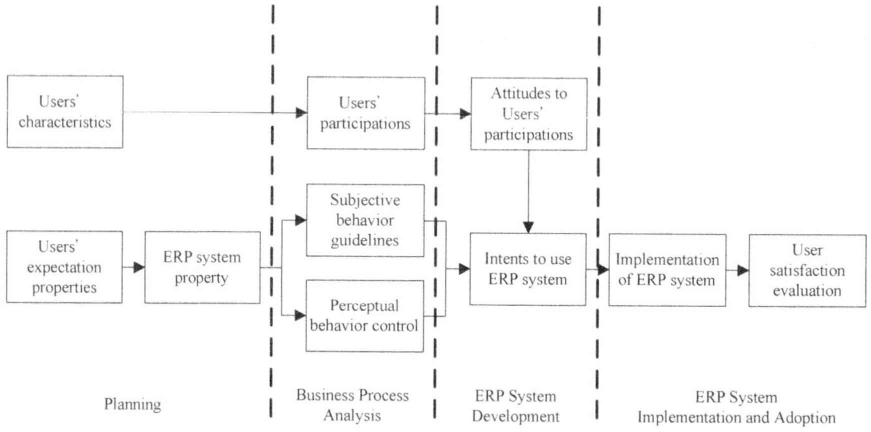

Fig. 1. The conceptual framework of a HF model for ERP system implementation

1) *Planning*: There are four steps in the planning stage. First, standards should be identified for planning management and development. Second, the planning team should consist of three categories of members. Third, education and training strategies

should be developed. Finally, the ERP system can satisfy system users. Generally, system users should be trained in software use.

2) *Business process analysis*: There are four elements in this phase. The first is the task analysis. In redesigning the operation process, customer requirements should be identified and met. The second is users' participation. User participation influences directly attitudes toward the ERP system. The third is the system integration. The ERP modules should be suited to business processes. The last but not the least element is the training program members. Absence of information and lack of understanding for a system are the major causes resulting in user resistance. Additionally, users usually have different perceptions of the system. Thus, an excellent ERP system should provide training that is tailored to user education.

3) *System development*: There are six elements in this phase. The first step is to understand the context of ERP system, to verify the human-machine interface, to make sure system hardware specification and data process of the system. The second step is to design the interface of ERP system by emphasizing the usability of user interface. Then, system designers should know that system acceptance and organization characteristics are the major causes resulted in information quality. The information quality was composed by several components. The fourth element is to change the business operation process usually result in users' resistance and complaint. Then, the commissioner of project management should be trained. Finally, sufficient managers' support can reduce the fear, worry, and animosity.

4) *On-line preparation*: There are five elements in this phase. The first element is training of users. Generally, training courses focus on software learning. Second, users will do a test to the ERP system in this phase. The third is to develop the system support team. The fourth is the on-line interchange formally. Finally, sufficient managers' support is important here.

5) *Information system implementation*: There are three elements in this phase [16]. The first is user support. Second, continued training should be provided. Third, sufficient managers' support in this phase again can reduce the fear, worry, and animosity of users.

3 Suggestions for System Implementation and Adoption

3.1 Suggestions to Information Department

Users' cooperation is the necessary conditions when developing or implementing a new information system. There are several causes in the information department. (1) Identifying the objective and scope of information system. (2) Service attitude should be kind and friendly. (3) Treat the important requirement from the users. (4) Let users' expectation updated with the system. (5) Let users take participation in the system. (6) To explain for the users that the system can help them finish more tasks. (7) To help users communicate with each other. (8) Understand different users' characteristics.

3.2 Suggestions to System Developer

System development should be user-centered. In this study, we proposed several causes which induced human errors: (1) Users cannot find appropriate object while executing their tasks; (2) Icons or labels confused users, so users usually cannot find the required object correctly; (3) Users do not know how to describe or execute the required object clearly; (4) Users usually receive incorrect information or wrong message of system feedback. The proposed methodology provided effective assistance to programmers. The below descriptions will propose four levels top-down process to assist system designers [25], shown in Table 1.

Additionally, Norman and Nobel [21] clearly defined the differences between users' intention and activities. He proposed the seven differences of the human machine interaction, including: object identification, intention, output identification, execution, system situation acceptance, explanation and description of system situation, and results assessment [25]. Norman and Nobel [21] also proposed four items to assist ERP system designers while designing well-designed system. First, situation and activity choice must be obvious. Second, system should be consistent in concept. Third, interface should have good mapping to demonstrate the mapping between different levels. Fourth, users should continue to accept the system feedback [25].

3.3 Suggestions to Education and Training

Following system improvements, business implementing ERP systems increase their competition. However, each consulting team has its own educational and training course. We offer the following suggestions to ensure that courses meet business requirements. System developers should meet the requirement of different industries to meet requirement of different operation styles. Thus, developers should classify users into the following four classifications according to their familiarization with and perceptions of the system and task content [25].

Table 1. Font sizes of headings. Table captions should always be positioned *above* the tables.

Level	
Conceptual	In interactive system, the first layer is to make an architecture based on users' mind model.
Semantic	Transferring users' mind into linguistic to send meanings of system based on the input and output of computer system.
Syntactic	How to combine the linguistic architecture to the syntactic level task items, transmission to computer and execution.
Lexical	According to classification of vocabulary, let the users easily to complete their task items according to the lexical level.

3.4 Suggestions to Education and Training

The system designers should understand the users, including: age, sex, characteristics, educational background, culture, training, motivation, objects, and attitude. Computer use is important to most current competitive environments. ERP system is beneficial to corporations and typically improves task performance and effectiveness. Additionally, users should enhance their computer-based skills and understanding of system objects and functions. Most importantly, users should have confidence.

4 Conclusions

This study aims to improve understanding of critical human factors that affect ERP implementation and adoption success and develop a human-centered model to practice this concept. Qualitative methodology is used in this study. It is already clear that consideration of the interaction between human and human systems interface is an important step in system implementation. This study hopes that the proposed model can contribute to both the academic field and business industry. Therefore, this study proposed human factors suggestions to successfully implement and adopt ERP systems. The conceptual model could be used in business reengineering process (BRP) to test its applicability. Moreover, researchers could focus on more specific areas such as human resource management or organizational viewpoint impact within one corporation so that more detailed and in-depth information could be identified.

Acknowledgements. This paper is financially supported by a project from the National Science Council of Taiwan under contract No. NSC- 97-2221-E-033 -033 -MY2.

References

1. Al-Khaldi, M.A., Wallace, R.S.O.: The influence of attitudes on personal computer utilization among knowledge workers: the case of Saudi Arabia. Information & Management 31, 185–204 (1999)
2. Ajzen, I.: From Intentions to Actions: A Theory of Planned Behavior. In: Kuhl, J., Beckmann, J. (eds.) Action Control: From Cognition to Behavior, pp. 11–39. Springer, New York (1985)
3. Ajzen, I.: Attitudes, Personality and Behavior. The Dorsey Press, Chicago (1988)
4. Ajzen, I., Madden, T.: Prediction of Goal-Directed Behavior: Attitude, Intentions, and Perceived Behavioral Control. Journal of Experimental Social Psychology 22, 453–474 (1986)
5. Bancroft, N.H., Seip, H., Andrea, S.: Implementing SAP R/3. Manning, Greenwich (1998)
6. Calisir, F., Calisir, F.: The relation of interface usability characteristics, perceived usefulness, and perceived ease of use to end-user satisfaction with enterprise resource planning (ERP) systems. Computers in Human Behavior 20(4), 505–515 (2004)
7. Curran, T., Keller, G., Ladd, A.: SAP R/3 Business Blueprint: Understanding the Business Process Reference Model. Prentice Hall, NJ (1998)
8. Davenport, T.: Putting the Enterprise into the Enterprise System. Harvard Business Review 76(4), 121–131 (1998)
9. Davis, F.D.: A Technology Acceptance Model for Empirically Testing New End-User Information Systems: Theory and Results, Doctoral Dissertation. MIT Sloan School of Management, Cambridge (1986)
10. Eason, K.: Information technology and organization change. Taylor and Francis, London (1998)
11. Fishbein, M., Ajzen, I.: Belief, Attitude, Intention and Behavior: An Introduction to Theory and Research. Addison-Wesley, MA (1975)
12. Fishbein, M., Ajzen, I.: Understanding Attitudes and Predicting Social Behavior. Prentice Hall, NJ (1980)

13. Goodhue, D.L., Thompson, R.L.: Task-Technology Fit and Individual Performance. MIS Quarterly 3(2), 213–236 (1995)
14. Goodhue, D.L.: Understanding User Evaluations of Information Systems. Management Science 41(12), 1827–1845 (1995)
15. Goodhue, D.L.: Development and Measurement Validity of a Task-Technology Fit Instrument for User Evaluation of Information Systems. Decision Science 29(1), 138 (1998)
16. Henry, C., Lucas, J.R.: Implementation: The Key To Successful Information System. Columbia University Press, NY (1981)
17. Holland, C., Light, B.: A Critical Success Factors Model for ERP Implementation. IEEE Software 16(3), 30–36 (1999)
18. Kwon, T.H., Zmud, R.W.: Unifying the Fragmented Models of Information Systems Implementation. In: Boland, R.J., Hirschheim, R.A. (eds.) Critical Issues in Information Systems Research, pp. 227–251. John Wiley and Sons Ltd., Chichester (1987)
19. Leitheiser, R.L., Wetherbe, J.C.: The Successful Information Center: What Does It Take. In: Proceedings of the 21st Annual Conference, Minneapolis, MN, pp. 56–65 (1985)
20. Magal, D.R., Houston, H.C., Watson, H.J.: Critical Success Factors for Information Center Managers. MIS Quarterly 12(2), 420–421 (1988)
21. Norman, M., Nobel, F.: User Involvement as an Interaction process: A Case Study. Information Systems Research 1(1), 89–113 (1990)
22. Norman, D.A., Draper, S.W. (eds.): User centered system design: New perspectives on human-computer interaction. Lawrence Erlbaum Associates, Hillsdale (1986)
23. Park, K.S., Lim, C.H.: A structured methodology for comparative evaluation of user interface design using usability criteria and measures. International Journal of Industrial Ergonomics 23, 379–389 (1999)
24. Szajna, B., Scamell, R.W.: The effects of information system expectations on their performance and perceptions. MIS Quarterly 17(4), 493–516 (1993)
25. Shneiderman, B.: Designing the interface: strategies for effective human computer interaction. Addison Wesley, UK (1998)
26. Summer, M.: Organization and Management of the Information Center. Journal of Systems Management 36(1), 10–15 (1985)

A Study on Effective Methods of Visual Inspection for Reused-Parts by Inspectors

Toshiyuki Matsumoto[1], Hideki Shirai[1], and Keisuke Shida[2]

[1] Aoyama Gakuin University 5-10-1 Fuchinobe, Sagamihara-shi, Kanagawa, 229-8558, Japan
matsumoto@ise.aoyama.ac.jp
[2] Nagaoka University of Technology 1603-1 Kamitomioka, Nagaoka-shi, Niigata 940-2188, Japan
shida@kjs.nagaokaut.ac.jp

Abstract. In order to promote recycling in the recycling industry, it is essential that suppliers perform a strict inspection before shipping the products using reused or recycled parts. The industry is facing pressures to further improve and standardize inspection quality and efficiency on recycled parts or products. This study thus developed an effective visual inspection method for products using reused parts by inspectors at Recycling Plant N, for assuring product quality. First, a survey was conducted to investigate the current situation at the subject plant. Secondly, another set of surveys was conducted for the 13 inspectors. The survey included a video analysis, a questionnaire survey analysis, and an analysis of eye movements using an eye-tracker, and after the analyses, the differences were specified. Thirdly, by considering the inspection method and inspection orders used by the experienced inspectors, appropriate inspection items, inspection order, and inspection method could be devised.

1 Introduction

Businesses today are facing two significant issues. One is the globalization of the economy, and the other is the growth of ever-increasing global environmental problems. Responding to such trends, businesses are forced to alter their management policies to cope with environmental issues as well as the strengthening of economic growth.

To build a sustainable recycle-oriented society, businesses are working achieve a broad adoption of the 3Rs: Reduce, Reuse, Recycle. In order to promote recycling in the recycling industry, it is essential that suppliers perform a strict inspection before shipping the products using reused or recycled parts. It is also important for them to assure customers of the quality. The industry is thus facing pressures to further improve and standardize inspection quality and efficiency on recycled parts or products.

Two types of inspection systems are currently being implemented: visual inspection which is conducted by quality inspectors, and automatic inspection, which is conducted using instruments. This study deals with the former, that is to say, visual inspection. With conventional visual inspection, quality, efficiency and inspection

M.J. Smith and G. Salvendy (Eds.): Human Interface, Part I, HCII 2009, LNCS 5617, pp. 131–139, 2009.

method vary with each inspector. As a result, inspectors may fail to reject defective products. Furthermore, inspection time may vary with each individual. In order to resolve such issues, methods for analyzing visual inspection from the structural viewpoint of the human visual system taking a cognitive scientific approach has been proposed [1,2,3,4,7], and ways to determine efficient visual inspection methods utilizing an industrial engineering approach have also been presented [5,6] However, for structurally-complicated reused parts, it has been often difficult to apply the proposed visual inspection methods.

This study thus aims to develop visual inspection methods for structurally-complicated reused parts at Recycling Plant N of an actual supplier. The objectives of this study are to determine appropriate visual inspection orders for covering the necessary inspection items, to create recommendable inspection methods, and to validate the effectiveness of the developed orders and methods upon implementing them at actual inspections.

2 Subject of Research

2.1 Target Parts and Visual Inspection

This study deals with flash lamps shown in Figure 1, disassembled at the stage of the disassembling process, and the visual inspections for testing the quality of the parts in question.

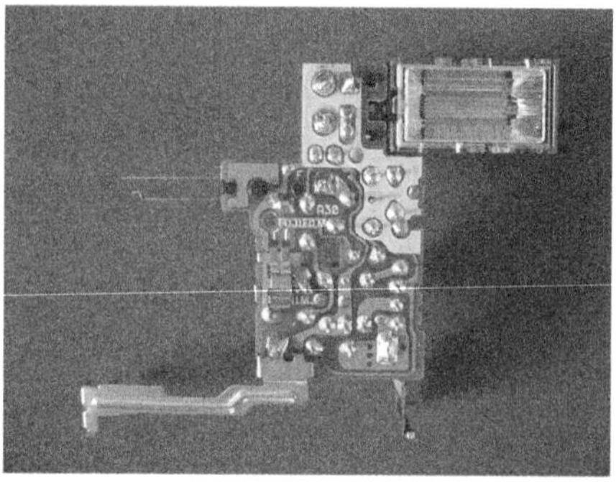

Fig. 1. Target parts

At Plant N, two types of inspections are being conducted to test the quality of flash lamps: performance inspection and appearance inspection. "Performance inspection" tests the flash function of lamps as well as the power distribution of the connecting sections of battery units by sending an electric current. Furthermore, it checks whether products function correctly upon use. Another type of inspection is "appearance inspection", and this checks if there is any change in the products' appearance, for example, if the product is rusted or deformed. Conventionally, both of these inspections were conducted by inspectors. However, as the automation of the inspection process became more popular, performance inspections came to be conducted by automatic testers, and also the part of the appearance inspection has come to be conducted automatically by the automated testing system. Currently, visual inspections handle items to be inspected, or inspection items, for subtle deformations or discolorations in appearance, which are hard to detect when using the automatic testing apparatus, and therefore, such visual inspections are conducted by humans.

Table 1. Current inspection and defective items

Item	Defect	Part picture
Protector	Flaw Stain Erased logo	
Reflector	Flaw Stain	
Synchro armature	Deformation Change of color Rust	
Battery armature(-)	Deformation Flaw Change of color Rust	
Battery armature(+)	Deformation Change of color Rust	
SM armature	Deformation Change of color Rust	
Transister	Deformation	
LED		
Xe armature	Change of color Rust	
Condenser	Deformation	
Trans	Attachet sand	
Xe terminal	Crimp Deformation	
Board	Flaw Lack of shape Stain	

Inspection items and defective items that are currently being examined in the inspections are shown in Table 1. Inspection items range over 10 different items, such as protectors, reflectors, synchronizer sections or connecting sections of battery units.

Defective items require 25 different types of defects to be detected, such as rust, deformations, scratches or stains.

2.2 Current Visual Inspection and Its Problems

Visual inspections of the targeted flash lamps are conducted using a tray which can hold up to 48 flash lamps, as shown in Figure 2. Inspectors hold the trays with both hands and move them, changing the angles at which they are held to check the position of each flash lamp, as well as to inspect the inspection items. At the same time, inspectors move their eyes in certain directions and check if the parts have any defective items.

First, to investigate the current visual inspections, visual inspection processes were video-recorded for 13 inspectors (whom were comprised of 12 experienced inspectors and 1 novice inspector). While the experienced inspectors spent 117 seconds on average to finish inspection of one tray, the novice inspector spent 149 seconds on average; hence it was found that experienced inspectors were more efficient in terms of inspection time.

Fig. 2. Picture of inspection process

Second, since each inspector used different methods and had different ideas on defective items, ideal visual inspection methods needed to be established. Basically, inspectors check parts arranged on a tray, looking at the position of each inspection item in order. However, it was found that all inspectors did not take the same approach. Therefore it was found necessary to establish ideal inspection methods by comparing and analyzing characteristics of visual inspections of experienced inspectors who identify defective items at a high rate, as well as the methods of novice inspectors.

3 Analysis of the Questionnaire Survey on Visual Inspection Items

To investigate how inspectors conduct visual inspections, first, a questionnaire survey was conducted targeting the 13 inspectors. The questionnaire covered questions in 2 categories: questions regarding inspection orders with the inspection items and questions regarding inspection methods.

After analyzing the result from the questionnaires, as for inspection orders, it was confirmed that the inspection orders were inconsistent, and that there was no standardized order. Regarding inspection orders considering positional relationship of parts, it was commonly accepted to go over the inspection items from the same direction consecutively, since that can reduce the number of times the parts are handled, as well as improving and increasing productivity of operations. Also, by checking the important items first and by reexamining the items as time allows, it was hypothesized that inspection quality could be improved.

As for inspection methods, basically the inspectors' eyes moved in a vertical S-shaped path, although according to some inspection items, some moved their eyes in vertical or horizontal directions. Thus it was found that inspection angles or directions were not consistent either. For this, the authors considered it was necessary to obtain knowledge and views from inspectors who are experienced and can identify defective items at a high rate, analyzing the result of the eye-tracking test which will be discussed later.

4 Analysis of Visual Inspection Items through Analyzing Inspectors' Eye Movements

After analyzing the video-recorded operations and questionnaire, it was found that experienced inspectors and novice inspectors detected defective items in a different manner. To verify this finding, Eye Mark Recorder, or an eye-tracking apparatus (called "Eye Camera", produced by NAC Image Technology Inc.) was used, and the operators' eye movement was analyzed.

To conduct this test, 3 examinees were selected from 3 different skill levels of inspectors: one from the most experienced group who boasted the highest defect-detecting skills (a.k.a. the advanced level inspector), one from the semi-experienced group (a.k.a. the intermediary level inspector), one untrained inspector (a.k.a. the novice inspector).

In the experiment, examinees were asked to conduct inspections for 5 trays wearing the Eye Cameras. Operators' eye movement was tracked during the tests. The movement was then analyzed for defect-free trays and trays with defective parts.

The trajectories shown in Figure 3 show the tracked eye movements from each of the 3 examinees. They also show the fixation points during the inspections of the defect-free trays. For the advanced level inspector, fixation points moved at an almost certain angle, as is shown in the trajectory. Studying the field of view monitor, the inspector did not move the head, and instead, handled the tray skillfully so as to minimize the inspection scanpaths. Movement speed was almost constant and not as variable as the others. The inspector's eyes moved at a certain, constant speed.

On the other hand, the intermediary level and novice inspectors moved their eyes at winding angles, and checked the inspection items, as shown in their inspection scanpaths. Looking at the trajectory from the novice inspector in particular, the inspection scanpaths were not followed and adjusted according to the focal point distance, which resulted in overlooking certain defects, and thus this inspector's visual inspection method was not consistent, as can be analyzed from the results. This seems to be because the inspector was poorly-trained. The field of view monitor showed that this

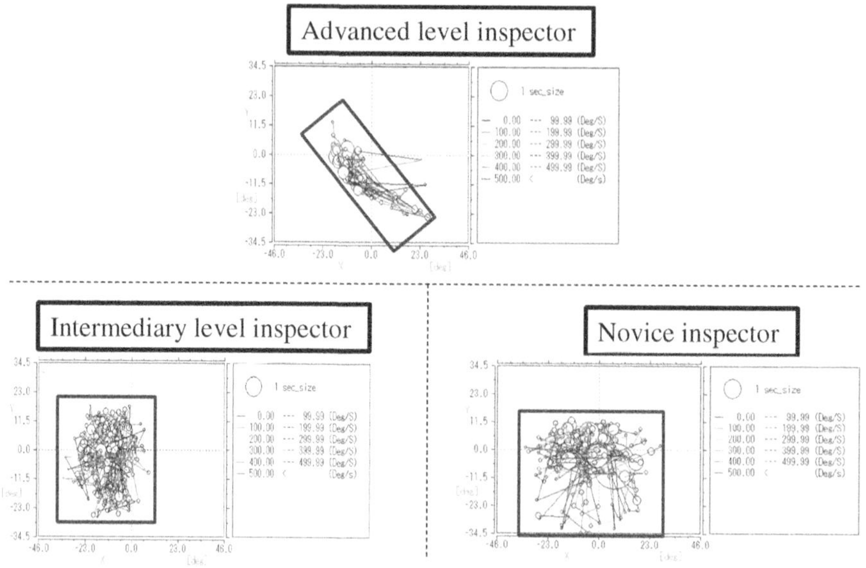

Fig. 3. The tracked eye movements from each of the examinees

inspector did not move the head or the trays and only moved the eyes, so the eyes' moving range was rather wide in this case.

Studying the visual inspection method of the advanced level inspector, it became clear that the ocular movement or the eye movement followed the targets with saccades. It was also found that the targets that the inspector's eyes followed were not single inspection items but rather several inspection items located at the lighted portion of the tray, using the peripheral visual field. In other words, the advanced level inspector used peripheral view to detect anything improper with the parts, and with saccadic eye movements, identified defects with the parts. Furthermore, when the defects were identified, the inspector's fixation points were not fixed on the defective positions themselves, although the defects were within inspector's peripheral visual field.

As for the intermediary and novice inspectors, their eyes moved to scan for detects on the targets upon inspection. When they moved from one target to another, one saccade was not enough for processing the necessary visual information to detect defects, thus their eye movement involved a series of quick saccades. However, as the inspectors continued inspection, their eye movement involved less and less saccades because of eyestrain, and their movement speed became unstable. Eyes that are fixed for a shorter period for each target resulted in overlooking defects or erroneous inspection.

5 Determining Inspection Items, Order and Implementation

Analyzing the results of the experiment above, proper visual inspection order should consider the elements for successful visual inspections identified in the experiment.

Upon determining inspection order, inspection angles at which inspection is to be conducted most efficiently for each inspection item must be clarified. At the same time, the number of handling steps must be considered to create efficient inspection flow. To consider these, visual inspection items and defective items to be checked during an inspection must be listed, and then the appropriate inspection order can determined.

First, visual inspection items and defective items to be dealt in each process of the inspection must be clarified. Possible defective items to be detected for the flash lamps were listed studying the product's specifications, interviewing inspectors, and considering properties of materials used for the each inspection position of the part. The items were cross-checked with the items to be covered in the specification, and an exhaustive list which includes all the possible defective items was then formed. As for the materials of the parts, plastic and other 3 types of metal (copper, aluminum, and lead) were used. Possible defects were considered for each type of material, such as defects uniquely observed for reused parts, deterioration-related defects such as rust or stains, and deformation defects such as dents or flexions.

From the created exhaustive list which covers the defective items, visual inspection items and defective items to be inspected by automatic testers were eliminated, and the existing visual inspection and defective items were streamlined. Automatic testers mainly deal with deformation defects, although visual inspection items undetectable

Table 2. Finalized inspection and defective items

Item	Defect	Part picture
Synchro armature	Stain Change of color Rust 砂付着 Flaw Lack of shape Substance	
Armature(+)	Stain Change of color Rust 砂付着 Flaw Lack of shape Substance	
SM armature	Stain Change of color Rust Flaw Lack of shape Substance	
Board	Stain Change of color Rust Flaw Lack of shape Substance	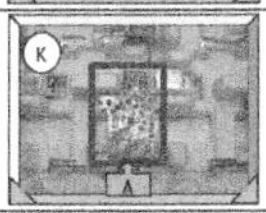
Protector / Reflector	Stain Change of color Flaw Lack of shape Substance Erased logo	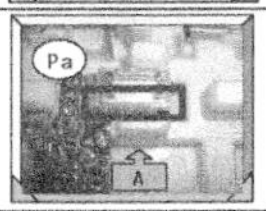
Protector / Reflector	Stain Change of color Flaw Lack of shape Substance Erased logo	
Synchro armature	Stain Change of color Rust Flaw Lack of shape Substance	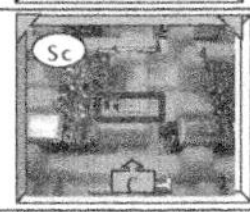

by a camera can only be inspected by the human eye. Such special cases were also taken into consideration and after reviewing the list, a new list covering visual inspection and defective items to be inspected was created.

Next, inspection order for visual inspection items to be covered for each scan is to be determined. There are 7 angles to be scanned in an inspection, and by conducting inspection from the same scan angle successively, the number of handling steps can be reduced. Possible inspection orders were thus considered and then listed in a matrix. Upon creating the matrix, 3 types of constraints were imposed. (1) Inspection items which can be inspected from the same angles should be covered successively. The angle at which the tray is originally placed is the first inspection angle. (2) Protectors and reflectors require 2 scans for their inspection items; inspection items serve as switching points of inspection angles. (3) Important inspection items that are most likely to have defects must be checked first. Importance was determined based on the previous data, calculated from the number of detected defective parts in all of the parts put into the visual inspection processes. Using the matrix, inspection order was finalized, and the finalized inspection order is shown on Table 2.

The inspection order and the visual inspection method discussed in the previous chapter were adopted at the targeted inspection process. Prior to the implementation of the new inspection method and order, all the inspectors were given a training session using a new manual which explains the new inspection items. Inspections were then conducted following the new inspection order. Now, we are examining the undetected defect probabilities in the visual inspection processes.

6 Conclusions

This study developed an effective visual inspection method for products using reused parts at Recycling Plant N, for assuring product quality. First, a survey was conducted to investigate the current situation at the subject plant. After the survey, problems were identified, and in order to solve the problems, another set of surveys was conducted for the 13 inspectors (12 experienced inspectors with a high defect-detecting rate and one novice inspector). The survey included a video analysis, a questionnaire survey analysis, and an analysis of eye movements using an eye-tracker, and after the analyses, the differences were specified. By considering the inspection method and inspection orders used by the experienced inspectors, appropriate inspection items, inspection order, and inspection method could be devised. The devised inspection method was applied for the target processes.

The authors would like to examine the result of the devised inspection method and further develop effective training methods for activating the peripheral visual field as experienced inspectors do upon inspection, as an issue in the future.

References

1. Akutsu, M.: Ergonomic Design Approach for Improving Industrial Visual Inspection - Study on Structure and Optimum Feeding Rate in Visual Sorting Inspection. The Japanese Journal of Ergonomics 31, 426–427 (1995)

2. Chan, A.H.S., Courtney, A.J.: Effects of object size and inter-object spacing on peripheral object detection. International Journal of Industrial Ergonomics 25(4), 359–366 (2000)
3. Kubota, S., Ohkura, M.: IIDA Hiroyasu: Eye Movement in Visual Inspection. The Journal of science of labor 57(12), 623–631 (1981)
4. Moeck, M.: Lighting design based on luminance contrast. International Journal of Lighting Research and Technology 32(2), 55–63 (2000)
5. Sasaki, A.: A method of Visual Inspection with Peripheral Vision [2]. IE Review 46(5), 61–68 (2005)
6. Sasaki, A.: A method of Visual Inspection with Peripheral Vision [1] – Introduction of Outline. IE Review 46(4), 65–70 (2005)
7. Takeoka, K., Tagawa, S., Akutsu, M.: Study on Optimum Speed of Work in Visual Inspection Conducted by Tact Inspection System -Study on Structure and Optimum Speed of Work (Part III). Journal of Japan Industrial Management Association 42, 1–7 (1991)

Practical Use of Task Models for Building and Modeling Operations Chart in the Industrial Production

Tomasz Mistrzyk[1] and Alexander Redenius[2]

[1] OFFIS Institute for Information Technology, Oldenburg, Germany
[2] Salzgitter Mannesmann Forschung GmbH, Salzgitter, Germany
tomasz.mistrzyk@offis.de, a.red@gmx.de

Abstract. In this paper, we will show how hierarchical task modeling can give beneficial support for the specification and validation of a production process in its early development stages. This will be done with respect to the growing challenges coming from the market environment. In particular, it is a substantial benefit to use an adequate modeling environment to be able to take specific customer requirements process during the production planning into account. For our methodology, we use AMBOSS, a task modeling environment designed for safety-critical socio-technical systems.

Keywords: Hierarchical Task Models, Expert System, Process Modeling, Production process.

1 Introduction

The production of standard mechanical products is no longer cost-effectively possible in Western Europe with the present market, due to strong competition from Asia and Eastern Europe. The trend therefore goes at industrial enterprises in the area of a customer's individual production needs. In the area of the steel industry, using pipe manufacturers as an example, standard products are still an important part of the production process, although development is moving towards customer solution products.

At these enterprises the biggest business volume can still be found at these enterprises in the area of standardized products. Hardwired work plan configuration systems are therefore frequently used for the design of operation charts in the production process. Using a work schedule configuration system the product is specified exactly on the basis of defined features. With the help of a firm set of rules, a qualified operation chart can be prepared automatically. This way is problematic for the production of customer individual products, whose market share will increase in the future as expected. These products cannot be described in the same way with the same features since the specification of the features or the combinations of the different features cannot be represented in the hardwired configuration system. The individual customer products are also manufactured on the same aggregates as the standard products. However additional parameters are necessary for the application of the aggregates.

Semiautomatic interventions or manual inputs are therefore necessary for the preparation of the operation chart, and it is often not possible to use the previous

M.J. Smith and G. Salvendy (Eds.): Human Interface, Part I, HCII 2009, LNCS 5617, pp. 140–148, 2009.

process knowledge in hardwired systems. This procedure is necessary even when only a small part of the process chain is different from a standard product, because it is not possible to configure one parameter in a way which was not considered in advance.

Also, to be able to use the collected process knowledge for the production of customer individual products, a concept is introduced which documents the process knowledge from a manufacturing plant on the basis of task models. Therefore, a part of the process chain of the production is represented with the help of hierarchical task models.

2 Approach Modeling Environment

Hierarchical task models represent a well establish approach for showing dependencies between tasks in a socio-technical system. The hierarchical representation of a task model is obtained through the decomposition of the higher level tasks into lower level tasks with respect to the temporal relationship between the investigated subtasks. The result is a tree-like structure with temporal dependencies between the tasks. The primary goal is the modeling of tasks, analyzing them with visualization, and a simulator to understand what an actor does or should do to accomplish his target. Since the task trees are easy to understand and build, such a concept proves itself as a powerful method for communication with domain experts, especially in the early phases of system development.

Specifying a task model for a production process documents the task order and the rationale behind the planning, along with structuring the tasks which have to be performed. These cognitions are useful for analyzing an existing socio-technical system, or to start the design of a new system which is not yet existing, which could be a production process. A task model of a particular production process can be helpful in detecting potential problems created by, for example, inadequate task order, disproportionate distribution of workloads between actors, or lack of time in critical phases of task execution.

Task models are already widely successful when used for supporting the constructive design of interactive systems. Their use as part of the product development process, however, is not yet widespread between practitioners.

There are various approaches that aim to specify tasks [1], [2], [3], [4]. They differ in aspects, such as the type of formalism they use, the type of knowledge they capture, and how they support the design and development of interactive systems.

In this paper, we consider task models that have been represented using the AMBOSS notation. AMBOSS [5] is a free modeling environment following the traditional approach of hierarchical task structure with temporary relations between the subtasks.

AMBOSS was developed at the University of Paderborn which can be freely downloaded from the homepage [5].

The software provides the typical tree-based editing functions, such as editing a child node or a sibling node; apart from that, however, AMBOSS allows direct editing and manipulation of nodes and connections for simple structural manipulation tasks. AMBOSS allows describing tasks at different abstraction levels in a hierarchical manner, represented graphically in a tree-like format. Figure 1 shows the notation used.

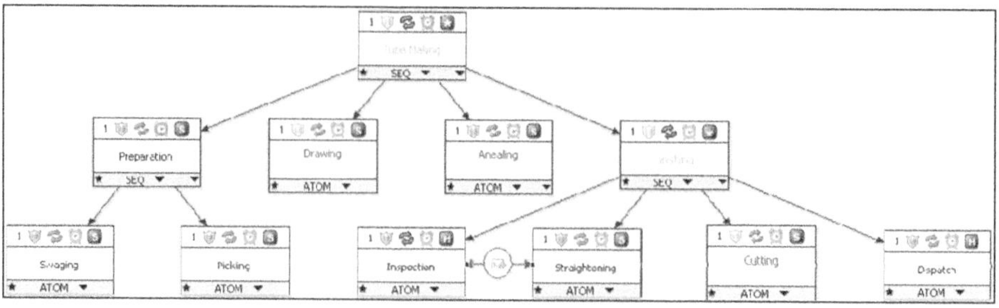

Fig. 1. Example for notation of AMBOSS models

It provides a set of temporal relations between the tasks such as:

- **sequential:** The subtasks are performed in a fixed sequence,
- **serial:** the subtasks are executed in an unsystematic sequence,
- **parallel:** in this relation the subtasks can start and end at random relation to each other,
- **simultaneous: t**he subtasks start in an arbitrary sequence with the constraints that there must be a moment when all tasks are running simultaneously before any task can end,
- **alternative:** just one randomly selected subtask can be executed

There are almost the same temporal relations that can be found in TOMBOLA [4] or in CTTE [3]. A task node without any subtasks is automatically noted as an **atomic** task.

The modeling environment has additional distinct views of a task model, which can be used for inspecting particular attributes of the tasks. For example, if an analyst likes to observe what kind of objects are manipulated in a system by a particular task, he can switch to the object view, take a look over the model, and analyze the dependencies between tasks and objects. It is also possible to review what kind of object is associated to a particular task, what kind of access rights (read or write) the task has and in which room are the objects are located.

AMBOSS makes it possible to handle topological aspects and relationships that are important in the analysis of a production process. Frequently, there are situations where tasks are performed in a certain position of the work space. To capture dependencies linked to the topology of a system, AMBOSS is able to build a relation between a room and a task which is performed in that room. An expert can specify the relationship either from the point of view of a task or a room. Additionally, every object of the task model can be assigned to rooms in a similar way as the tasks.

Analogous to other environments [3], [2] the concept of roles is implemented in AMBOSS. Basically there are three types of actors used: human, system, and abstract. The user can specify additional actor subtypes, (such as engineer or manager), and instances of actors, (such as Ms. Lee, the manager of marketing), as refinements of the human and system roles. If an actor is linked to a task, his role can be also specified as being the person who is also responsible for a particular task. This concept helps to organize areas of responsibility for each actor involved in a task model.

The main purpose during the development of AMBOSS is to provide a hierarchical task modeling environment that provides support for developing and analyzing task models in safety critical domains. For modeling tasks in such an environment, the model needs to be enhanced with more adequate parameters. This benefit can be directly used also in other domains as this paper shows. For the safety purposes, AMBOSS contains the concept of barriers. This concept allows specifying parameters which helps to protect human life and/or computer infrastructure. In a production process, barriers could be used to protect workers from hot temperatures. Barriers are a special kind of object which can be activated or deactivated by tasks. This can be directly "observed" during the simulation which is a part of the environment.

One of the challenges during the development of AMBOSS was to involve communication into a task model. The modeling environment allows precise description of communication with parameters describing the physical conditions, using options with respect to the medium of communication, form of message as well as type of transfer. Hence, the user is able to describe a communication flow while taking into account how the actors communicate with each other, who is communicating with whom, and what kind of role a piece of particular information plays regarding to a specific task.

By using such parameters like objects, barriers, risk factors, roles, timing, and communication, it is possible to describe a task model in more detail, and to have a good impression of the tasks by utilizing different views of the same model.

Similar to other modeling approaches, [3] AMBOSS is able to simulate a task model. The simulator shows to the user exactly what happens in a task environment at a particular moment. A finished task model can be simulated by taking into account the task hierarchy, temporal relations providing the task execution order, and communication flow showing messages along with their parameters. Additionally, during the simulation the user is able to observe the activation and deactivation of barriers.

After the preparation of the task model, the already named parameters of the single task can be interactively proved on reciprocal interdependency with the help of a simulation environment. The interdependency of single parameters can be observed during the simulation. If discrepancies are detected, they can be immediately corrected. In that way, we can validate our model or its scenario, and consult other experts. Therefore, it is also a suitable medium of communication for the cooperating members of the production process.

3 Methods and Techniques Applied

As already described in the previous chapter, the visualization of processes presents a suitable method for the modeling in a cross-domain environment. This approach becomes more and more important in the context of the industrial production in Europe, because the rigid production planning systems are no longer adequate.

In the past, tubes were ordered and produced in a standardized form. The customer could only order a set number of different material grades and standardized dimensions.

These predefined configurations of the tubes were hard-wired in the implementation of an order entry system in form of keys. The keys can be directly used for the order entry and for the production planning system. It was a difficult process to represent; the process knowledge from the precision tube production process into an extensive set of rules. This set of rules is needed to enable the automated planning of the manufacturing process, and to assure that nobody could order an impossible tube configuration. Deviations of the defined specifications were not accessible.

Because of the enforced competition situation in Europe, established tube manufacturers have to offer the customer individual products. Standard products can be produced more economically outside Europe. So it is essential for European tube manufacturers to offer their customer an added value to get the additional charge. This can be achieved by the production of exotic products with special material grades or special treatments.

The planning of the production process for these products is problematic due to the fact that the production routes cannot be encoded in the planning system automatically for these special features. A manual encoding of these special tubes is very effort intensive and fault-prone.

A detailed description of this type of tube with a complete specification is often created in close interaction with the development department. The manufacturing process has also been developed, because the properties of a new steel grade are not always known in detail. It is an iterative process, and the manufacturing process is designed in close interaction between the logistic and development department [6].

It was noticed in analysis of different development processes that it is possible to divide a complete development process into a restricted number of process blocks [7]. It is possible to use these process blocks to combine them into a new development process for a system, which was not developed before [8]. It is also possible to use this knowledge for the planning of manufacturing processes.

The process of the precision tube manufacturing can be divided into four main parts, as you can see in figure 2.

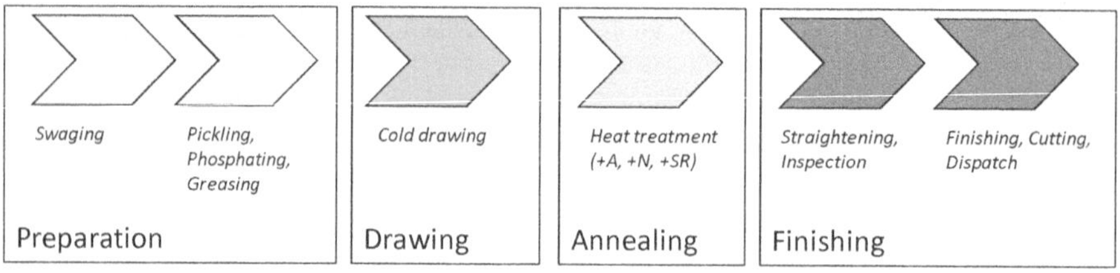

Fig. 2. Schema of precision tube making process

The following parts are:

1. **Preparation:** the seamless tube (Mannesmann process) has to be prepared for the following production steps. The tubes will be cleaned, pickled, phosphated, greased, and one end of the tube will be formed into a metal tang. This metal tang is used for the following drawing process.
2. **Drawing:** The tube will be drawn cold through a die. There can be an inner tool inside of the tube to plane the inner surface. It can be necessary to repeat this process several time, depends on the target dimension of the precision tube.

3. **Annealing:** The drawing causes a strain hardening effect into the tube. The strength of the tube gets higher and it is sometimes necessary to anneal the tube to fits the properties to the planed values.
4. **Finishing:** It is always the last step of the production process. The tube can be straightened, checked and prepared for the shipping.

For every part only finite configurations are possible. Every precision tube manufacturing process contains several of these parts (patterns), but stringently not every part has to be involved in such a process. Some tubes, for example, do not need the annealing part, while others need the combination of the preparation, drawing and annealing parts several times. For every of these four parts exist a limited number of possible realizations.

In figure 3 the different parts of the manufacturing process are colored in four different colors.

pr	ps 1	ps 2	ps 3	ps 4	ps 5	ps 6	ps 7	ps 8
50	AWS 100	Pickle 1	50 to Drawing 1	50 to Drawing 1	furnace A	flow line I	test line A	
51	AWS 100	Pickle 1	50 to Drawing 1	furnace A	flow line II	test line A	leveling line	
52	AWS 100	Pickle 1	50 to Drawing 1	flow line I	test line A	leveling line		
53	AWS 40.1	Heater 1	Pickle 1	50 to Drawing 2	furnace A	flow line II	test line B	leveling line
54	AWS 40.1	Heater 1	Pickle 1	50 to Drawing 1	furnace A	flow line III	test line B	leveling line
55	AWS 100	Pickle 2	50 to Drawing 1	flow line IV	test line A	leveling line		
56	AWS 100	Pickle 2	50 to Drawing 1	furnace A	flow line II	test line A	leveling line	
57	AWS 40.1	Heater 2	Pickle 1	50 to Drawing 2	50 to Drawing 2	flow line I	test line B	
58	AWS 40.1	Heater 2	Pickle 1	50 to Drawing 1	furnace B	flow line I	test line B	leveling line
59	AWS 100	Pickle 2	50 to Drawing 2	furnace B	flow line III	test line B	leveling line	
60	AWS 100	Pickle 2	50 to Drawing 2	50 to Drawing 2	flow line IV	test line B		
61	AWS 100	Pickle 2	50 to Drawing 1	furnace A	flow line II	test line A		

Fig. 3. List of different process routes for precision tubes

On the left side is a list with different process routes (pr) in green. Every column stands for one production step (ps). To the right to every process route is the sequence of the needed process blocks, and it is divided in to several production steps (columns). The yellow blocks, for example are preparation steps, the blue ones are drawing steps, the light blue blocks are annealing steps and the purple blocks present finishing blocks.

As you can see in the figure 3, some combination of process steps occurred in different process routes. For example the combination of "AWS 100" and "Pickle 1" can be found in process route 51 and 52. There are several examples in the whole list for all production steps. In figure 1 is just a small section of the whole list. There are a limited number of different process blocks for the four production steps.

It is possible to combine a new process for a special tube with these process blocks. The knowledge of a completed process can be reused. The different combinations of these process blocks make it possible to build up almost every special tube manufacturing process. Beside the description, a number of boundary conditions are stored for each of these process blocks. This special information is required, for example, for the processing or for temporal information.

In addition, for the definition of the individual aggregates of the process chain the dependence between the aggregates is also defined. For instance, there are special conditions for the sequence of the working steps. The use of some blocks is only possible in a special sequence. A preparation step has to be executed before a drawing process, to form the metal tang, and to soap and pickle the tube for the following drawing process. It is also often necessary to anneal the tube after a drawing process with a high cross-section diminution, because the stiffness of the tube got to high.

The planning of a manufacturing process can be supported by these process blocks with additional information about their usage and their boundary conditions. The graphical representation of the blocks supports the close interaction between engineer and logistic planer and helps to model the needed process in an iterative method under the usage of the knowledge from finished processes. The logistic planner has the possibility to combine the individual process blocks of the process chain in a dialog with the product engineer and they are supported by a "light" system of rules. This system of rules suggests a possible configuration on the basis of the input and output, or shows possible compatibilities. So it is just possible to insert aggregates in a certain order. For example is the pretreatment in front of the drawing process.

This procedure is described in figure 4. On the left side there are three libraries with process blocks for the preparation, the drawing and the annealing. In every library, are a number of different process blocks with a description and with information about their boundary condition. The process planner can use the process blocks out of the library and combine the manufacturing processes with the help of a system of rules, which evaluate consider the background information for every process block.

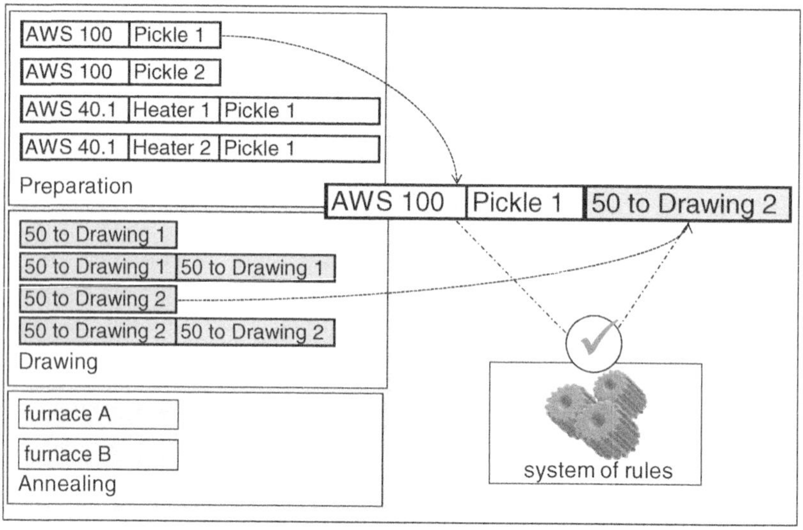

Fig. 4. Combination of different process steps supports by system of rules

In addition it is possible to reuse the knowledge of finished manufacturing processes. The finished process models are labeled with their fundamental boundary conditions. So it is in principle possible to access the process knowledge from a completed precision tube production process.

4 Evaluation

Hierarchical structure of task models together with the temporary relations between the task nodes makes it possible for the flexible handling of a production process. Such structures are difficult to represent in a table. The individual operations can be grouped to bigger sections and combined with each other.

We tested our approach in a case study. The work plans were modeled and stored in an adapted version of the modeling environment. Now, the created work plans can be reused in other case studies. AMBOSS allows use of textual descriptions for each node. In this way, the collected individual experiences of employees can be saved. Thus, the internal knowledge is not lost, and future operations can quickly be created.

The work schedules were already modeled and stored with an adapted version of AMBOSS in a case study. These work schedules can be used consistently. It is possible to store the collected experiences from the employees in the description field of the individual process block. So the internal knowledge is not lost and future work processes can be created faster. The process blocks are stored in a library and can be easily reused. The process blocks in the library, along with possible process routes, were generated from the production planning system. For every process block in the list, a representation of it was created in Amboss. The engineer and process planer can use it to model the required process. The graphical interface helps both sides in this iterative process.

The library is divided into the four parts of the manufacturing process (preparation, drawing, annealing, finishing), as described above. For every part, there exists a limited number of process blocks with possible realizations of the manufacturing part. It is possible to combine these process blocks to a complete process. The process planer and product engineer are supported during this process by a system of rules, which are based on the divided boundary condition and other information.

The provided background information for the process blocks in AMBOSS and the information about the already modeled process blocks assure that only feasible sequences can be arranged. With this background information, it is also possible for the system to suggest possible process blocks for the next steps. It is also possible for special requirements to be searched for in the process library with a full text search tool.

After modeling the process model in AMBOSS, it is possible to start an export to the production planning system. The interface is based on the XML-notation. But at the moment it is only possible to use this export in one direction. Changes have to be modeled in the graphical process model, and in the next step they have to transfer again into the production planning system.

5 Conclusions

The task modeling environment was used in our approach to specify a production process in its early phases. AMBOSS offers elements like objects, barriers, topology, actors, and the communication flow of the system, which can be used to build up a process model. Additionally, AMBOSS provides a concept of views which help to visualize product relevant factors as well as parameters for production facilities or resources on different abstract levels in the enhanced task models.

With AMBOSS the user get a flexible modeling environment which allows editing as well as the direct manipulation of the task structure in an easy and intuitive way. The production process can be modeled, analyzed, and simulated using different views or by taking into account different production parameters.

Since the software has been developed with reference to the analysis and modeling of complex socio-technical systems, the developers paid special attention to the creation of big models which can be easily visualized and used by engineers. Within the product development phase the task modeling is an interesting approach which provides effective support especially for the design of operation charts for customized products.

For this reason, it is a suitable medium to consult between experts of the production process and the product development. By using our approach, the experts are able to reuse process knowledge for the planning of customer individual products. The knowledge needed to develop customized products can be saved into elements of task models and be reused if needed in a more flexible form than classical PPS systems can offer.

References

1. Baron, M., Lucquiaud, V., Autard, D., Scapin, D.L.: K-MAD: un environment pour le noyau du modèle de description de l'activé. In: Proceedings of the 18th French-speaking conference on Human-Computer Interaction, Montreal, Canada (2006)
2. Biere, M., Bomsdorf, B., Szwillus, G.: Spezification and simulation of task models with VTMB. In: CHI 1999 Extended Abstract on Human Factors in Computing Systems, Pittsburg, Pennsylvania. ACM Press, New York (1999)
3. Mori, G., Paternò, F., Santoro, C.S.L.: CTTE: support for developing and analyzing task models for interactivesystem design. IEEE Transactions on Software Engineering 28(8), 797–813 (2002)
4. Uhr, H.C.: TOMBOLA: Simulation and USer-Specific Presentation of Executable Task Models. In: HCI International (2003)
5. Giese, M., Mistrzyk, T., Pfau, A., Szwillus, G., von Detten, M.: 7th Int. Workshop on Task Models and Diagrams TAMODIA 2008. Springer, Berlin (2008), `http://mci.cs.uni-paderborn.de/pg/amboss/`
6. Pahl, G., Beitz, W.: Engineering Design: A Systematic Approach. Springer, Berlin (1995)
7. Dörner, D.: Errors in planning and decision making and the nature of human information processing. Universität Bamberg, Lehrstuhl Psychologie II (1994)
8. Redenius, A.: Verfahren zur Planung von Entwicklungsprozessen für fortgeschrittene mechatronische Systeme. HNI-Verlagsschriftenreihe, Paderborn (2006)

A Model for Measurement and Analysis of the Workflow Processes

Pavel Ocenasek and Miroslav Sveda

Brno University of Technology, FIT, Bozetechova 2, 612 66 Brno, Czech Republic
{ocenaspa,sveda}@fit.vutbr.cz

Abstract. This paper presents an approach based on 3-layered model that may be useful for analysis and measurement of workflow processes. The workflow system can evaluate a state of the products or services in each point of processing and is divided into three layers (sources, workflow management system and workflow processes), whereas each layer covers the processing of its specific tasks.

Keywords: workflow, process, metrics, layer, model, analysis.

1 Introduction

There exist two ways of how to manage a the processes in collaborative workflow systems. The first of them is a function management method where main idea of this method is a dividing work among particular function units. These function units are generated on their functions and knowledge. These units perform only their part of the task and they don't take the total result.

The second way of management is a Workflow management. The main differences between both methods can be found in different views to company. The workflow management is focused on result of work produced by employed instrustrial embedded systems. Whole Workflow system is managed by customer requirements, generally by interactions and interfaces. The Workflow system can evaluate a state of the products or services in each point of processing. If the products or services show some abnormal values during the processing, the process method can be adjusted according to the measured values, even the processing can be halted [5].

2 Layered Model

Workflow system appears to be a complex problem for beginner. Therefore we can use the idea from computer network and that is a layered model like a TCP/IP or ISO/OSI. We can divide the complex Workflow system into particular logical layers, whose content is such difficult. The objects from the upper layer can use objects or services from the lower layer.

In our layered model we provide the use of three layered model. The lowest layer is called "*Resources*". This layer includes hardware, operation system and his services, users, data and applications. Second layer is called „*Management system*",

M.J. Smith and G. Salvendy (Eds.): Human Interface, Part I, HCII 2009, LNCS 5617, pp. 149–153, 2009.

which take care of synchronization among particular workflow processes and their initiation in proper sequence. In the last layer, there are particular workflow processes. Therefore, this layer is called "*Processes*".

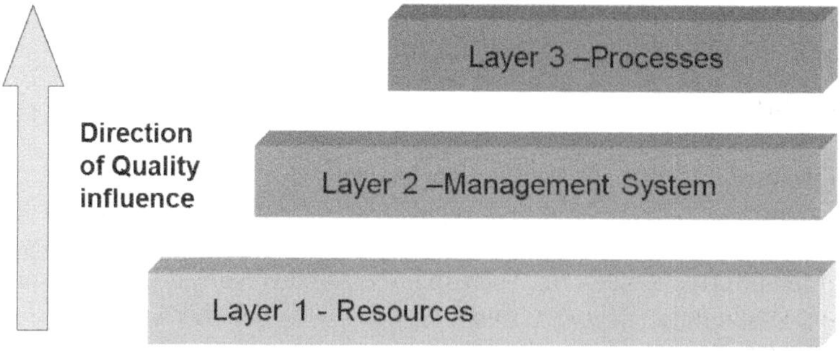

Fig. 1. Layered model

For implementing a high-quality Workflow system, it is necessary to determine chance to monitor running processes. Quantified processes make possible to consider different possibilities and make perfect processes. Quantified processes provide possibilities to compare processes among rival products.

In this case performance metrics for measurements of workflow processes are time, cost and quality.

3 Performance Metrics

3.1 Time

Time is a common and universal measure of performance. For workflow systems, it can be defined as the total time needed by an instance to transform a set of inputs into outputs. The first measure of time is task response time (*T*). Task response time corresponds to the time, which the instance takes to be processed by a task. The task response time can be divided into two major components: delay time and process time. Delay time (*DT*) refers to the non-value-added time needed in order for an instance to be processed by a task:

$$T(t) = DT(t) + PT(t) \tag{1}$$

The delay time can be further divided into the queuing delay and setup delay. Queuing delay is the time, which instances spend with waiting in the task list before the instance is chosen for processing. Setup delay is the time the instance spends with waiting for the task to be set up. Setup activities may correspond to the warming process carried out by a machine before executing any operation, or the execution of self-checking procedures [6].

$$DT(t) = QD(t) + STD(t) \tag{2}$$

3.2 Cost

Task cost represents the cost associated with the execution of workflow tasks. Cost is an important factor, since organizations need to operate according to their financial plan. Task cost (*C*) is the cost incurred when a task t is executed; it can be divided into two major components: fixed cost and realization cost.

$$C(t) = EC(t) + RC(t) \tag{3}$$

The Cost of task contains a fixed cost (*EC*) and realization cost (*RC*). The realization cost includes direct labor cost, machine cost, direct material cost and setup cost.

3.3 Quality

It is possible to interpret the Quality of the product like a set of abnormalities at predefined parameters of the product or service. The Quality consists of two basis items: reliability and fidelity. Task Reliability (*R*) corresponds to the likelihood that the components will perform for its users on demand; it is a function of the failure rate. The failure rate is given as the ratio of successful executions/scheduled executions.

$$R(t) = 1 - \left(\frac{SuccessfulExecutions}{ScheduledExecutions} \right) \tag{4}$$

We see the fidelity as a function of effective design; it refers to intrinsic properties or characteristics of a good produced or service rendered. Fidelity reflects how well a product is being produced and how well a service is being rendered. Fidelity is often difficult to define and measure because it is subject to judgments and perceptions.

$$\begin{aligned} F(t) = {} & \left| f_1(F(t).a_i) \right| \cdot wi_1 + \ldots + \\ & + \left| f_n(F(t).a_n) \right| \cdot wi_n \end{aligned} \tag{5}$$

Workflow tasks have a fidelity (*F*) vector dimension composed of a set of fidelity attributes ($F(t).a_i$), that reflect and quantify task operations. Each fidelity attribute refers to a property or characteristic of the product being created, transformed, or analyzed. Fidelity attributes are used by the workflow system to compute how well workflows, instances, and tasks are meeting user specifications.

4 Performance of the System

The particular layers of Workflow system are ordered so that lower layers provide elementary services to upper layers. The qualities of services from lower layer are transferred to upper layer. Therefore the lower quality of resources takes effect into quality of workflow process. In ideal workflow system, the quality on workflow process layer is equal to 1. When we use for example hardware of poor quality, than quality of workflow layer goes down. For determination of quality of workflow system we use a function *f* with *n* variables $q_1, q_2, \ldots q_n$, where *n* is a dimension of quality. Vector of total quality layer $Q = [q_1, q_2, \ldots q_n]$ of this dimension is a vector of quality of whole workflow system. Variables of $q_1, q_2, \ldots, q_n$ take the value from 0 to 1. The function of total performance workflow system can be defined as:

$$f(Q) = \frac{\sum_{i=1}^{n} a_i q_i}{n} \tag{6}$$

In our case of three layered workflow model, the function is defined as:

$$f(Q) = \frac{a_1 q_1 + a_2 q_2 + a_3 q_3}{3} \tag{7}$$

Relevance of participant layer is impressed with a weigh coefficients a_1, a_2, ... a_n. These coefficients participate in determination of importance each layer. For the coefficients, the following figure should be valid:

$$\sum_{i=1}^{n} a_i = 1 \tag{8}$$

4.1 Total Performance of the Task

The total performance of task (process, activity or service) is a result of factors like time, cost and quality. Though, to calculate the total efficiency of the task, we need a mechanism, which serves for calculating both time and costs. Possible way of this solution is an implementation of conversation coefficient (*CC*) for each task. This conversation coefficient determines the organization or company and it is a rate between costs and spent time during a processing task. Conversation Coefficient *CC(t)* determines costs per unit of time for processing task t. The total cost (*TC*), which is expended to process task t by an organization, can be defined like:

$$TC(t) = CC(t) \cdot ElapsedTime(t) - C(t) \tag{9}$$

where *C(t)* is the cost of task *t*. This cost contains fixed cost (*EC*) and realization cost (*RC*).

Total performance of task *PT* is computed in general as the ratio of quality of process per expended costs. The quality of task *t* can be divided into fidelity *F(t)* and reliability *R(t)*. The total performance of task *PT* is defined like:

$$V(t) = \frac{R(t).\ F(t)}{TC(p)} \tag{10}$$

$$V(t) = \frac{R(t).\ F(t)}{CC(t).\ ElapsedTime(t) + C(t)} \tag{11}$$

With increasing failure rate of task *t*, the performance of the task t goes down even if the quality, namely fidelity has a high level.

4.1 Total Performance of the Layer

The quality of each layer is determined from weight average of total performance of tasks, which are provided by particular layer. The total performance of layer is defined as follows:

$$p_i = \frac{a_1 . V(t_1) + \ldots + a_n . V(t_n)}{n} . (1 - z) \quad (12)$$

For each tasks of $t_1, \ldots, t_n$ there is assigned a coefficient $a_1, \ldots, a_n$ which expresses a weight of services in a layer. The sum of these coefficients is equal to 1. Coefficient z determines a dependence measured layer with lower layer. Vector quality of all layers can be defined in normalized form as:

$$Q = \left(\frac{p_1}{s}; \frac{p_2}{s}; \frac{p_3}{s}; \right) \quad (13)$$

where p_i is total performance of layer and s is defined as:

$$s = \sum_{i=1}^{3} p_i \quad (14)$$

5 Conclusions

This abstract briefly presents a workflow layered model from the view of quality of processes and services. Moreover, the final version of the paper also includes the case study in the area of human processes and the results of the application of the proposed model in the real collaborative environment.

Acknowledgement. The research has been supported by the Czech Ministry of Education in frame of the Research Intention MSM 0021630528: Security-Oriented Research in Information Technology, MSM 0021630503 MIKROSYN: New Trends in Microelectronic Systems and Nanotechnologies, and by the Grant Agency of the Czech Republic through the grant GACR 102/08/1429: Safety and security of networked embedded system applications.

References

1. Carda, A., Kunstova, R.: WORKFLOW: Nástroj manažera pro řízení podnikových procesů. Grada Publishing, a.s. (2003) ISBN: 80-247-0666-0
2. Cardoso, J., Miller, J., Sheth, A., Arnold, J.: Modeling quality of service for Workflows and Web Service Processes (2002)
3. Fiala, J., Ministr, J.: Průvodce analýzou a modelováním procesů, VŠB – Technická univerzita Ostrava, OFTIS-Ostrava (2003) ISBN: 20-248-0500-6
4. Muehlen, M., Rosemann, M.: Workflow-based Process Monitoring and Controlling - Technical and Organizational Issues, International Conference on System Sciences, Hawaii (2000)
5. Racek, J.: Workflow spravnich procesu v oblasti zivotniho prostredi. FI MU Brno, disertacni prace (2002)
6. Trchalik, R.: Metrics in the Informatics and their using in Workflow Systems. Diploma thesis, Masaryk University, Faculty of Informatics Brno (2005)
7. Ucen, P., et al.: Metriky v informatice. Grada Publishing, Praha (2001)
8. Fisher, L. (ed.): Workflow Management Coalition. Workflow Handbook (2004) ISBN: 0-9703509-6-1
9. Zhao, L.J., Stohr, A.E.: Temporal Workflow in a claim handling system (1999)

Tactical Access to Complex Technology through Interactive Communication (TACTIC)

Alexei Samoylov, Christopher Franklin, Susan Harkness Regli, Patrice Tremoulet, Kathleen Stibler, and Peter Gerken

Advanced Technology Laboratories, Lockheed Martin, 3 Executive Campus, Suite 600
08002 Cherry Hill, United States
{asamoylo,cfrankli,sregli,ptremoul,kstibler,
pgerken}@atl.lmco.com

Abstract. Tactical Access to Complex Technology through Interactive Communication (TACTIC) is a user interface prototype developed under a technology effort designed to provide personnel deployed in the field with access to resources that typically require intercession by stationary experts with extensive training and experience. We present rationale behind a computer system that judiciously and beneficially exposes expert knowledge to novices, outline the domains in which such a system is applicable and describe a prototype interface implementation in the domain of military expeditionary forces.

Keywords: expert knowledge, interface, workflow, expeditionary forces.

1 Introduction

Technology that enables remote, mobile personnel to tap into intellectual, computing, and other informational resources from the field could improve performance in domains ranging from installing and maintaining communications (telephony, cable, networking, satellite) technologies, deploying and troubleshooting crucial infrastructure services (electricity, plumbing, HVAC systems), providing up-to-date intelligence to military expeditionary forces on patrol, to first responders in disaster situations.

Tactical Access to Complex Technology through Interactive Communication (TACTIC) technology is designed to provide a bridge from human to information in any situation for which the following conditions are true:

1. A complex information environment (including enterprise services, databases, manuals, policies, etc.) exists in which information requests are typically composed and executed by an expert who is extremely familiar with the data sources, required values, parameters, and expected results that relate to an information query.
2. Relative novices must be able to use this complex information environment to acquire operationally relevant information in near real-time with minimal assistance.

In such situations, mobile personnel are limited in the scope of their situational awareness, while the completion of their task depends on knowledge of conditions

M.J. Smith and G. Salvendy (Eds.): Human Interface, Part I, HCII 2009, LNCS 5617, pp. 154–162, 2009.

beyond their current environment. Therefore, timely and correct assistance from a remote resource is typically required to understand the bigger picture. For example, an ambulance might take a different route if an operator can warn them ahead of time that the current road is blocked by a traffic jam several miles ahead. Expert operators will even go beyond simple reactionary measures and proactively monitor for problems. Operators will also be the most likely to suggest best solutions due to the greater access to data about current conditions (in the case of the ambulance, these would include weather conditions, geography, blocked roads, etc.). Unfortunately, operators can become overloaded with incoming data as well as with the number of tasks and therefore, can devote only limited time to each request [9].

A novice operator may have access to the same information, but because of their lack of experience in the domain, they do not know what information is required for a useful query. In the example of the ambulance, a novice operator may not know to inform the driver about detours on their route because it never occurred to them to check for that information. In the case of a novice driver, they may want to know what is happening several blocks ahead, but this question may be completely useless to the expert operator without a specification of what type of information the driver wants to know (traffic conditions, one way streets, construction, etc.).

2 Related Works

Research has explored how to determine what an expert knows [12], as well as how to create external representations of that information or knowledge [1, 3, 4, 11]. Although difficult to define, an expert in a domain is generally someone with substantial training and/or practical experience as well as someone who performs better than average in their domain and has developed excellent skills in that domain [12]. One way to represent an expert's knowledge is via Cognitive Structure Analysis (CSA). CSA states that what an expert knows can best be represented by one of five knowledge structures: memory organization packets, production rules, semantic networks, object frames and mental models [11]. Some knowledge can only be accurately represented by one of these structures while other knowledge could be represented by all of them.

One of the main reasons for representing the knowledge of an expert is to enable non-experts (either human or machine) to accomplish tasks currently performed by experts. Efforts to create external representations have been used to inform the development of tutoring systems [5, 6, 10]. In these cases, the goal is to have the non-expert understand or embody enough of the expert knowledge to enable them to use and manipulate it themselves.

In contrast to the systems above, search mechanisms attempt to enable someone who generally knows something about the domain in which he or she is searching to find the desired information through the use of key phrases and reference information [2, 7]. The "searcher" presumably knows enough about the domain of interest to choose viable search terms in that domain.

In the work reported here, we assume that a representation of expert knowledge exists (we have other ongoing research efforts that address how to obtain expert knowledge) and concentrate on the situations in which a novice in a domain needs to ask a question of the representation *without understanding or fully being able to visualize* the domain space (usually because of a lack of knowledge, but occasionally because of time pressure).

As outlined above, experts use their deep understanding of a domain to create a map of meaningful relationships between data. One could say experts are looking at a domain and associated data through a set of filters that allow them to efficiently extract information necessary to fulfill queries. A novice who is unfamiliar with the domain can benefit from a modified version of the expert's view that exposes a simpler and more limited set of data and relationships that is easy to grasp yet yields correct results.

In essence, while an expert needs a complicated set of filters to examine data in a given domain, a novice posing a query only needs to know what characteristics of that data are important to capture in a query so that an expert can provide meaningful results. Sample characteristics could be types of data to return (e.g., images, text, audio, etc.), time window in which results are meaningful (e.g., two hours ago until now), objects of interest and any useful attributes (e.g. red cars, friendly civilians, parked busses, etc.) and areas of interest (e.g., one block south and two blocks west of current location) among others. Our effort focused on finding effective translations from the expert's view to a novice's view such that the most salient characteristics of the data and domain are preserved, but the rest of the details are abstracted away (i.e., novice does not need to know that cars can also be labeled as vehicles in a database).

3 User-Centered Design

The TACTIC interface assists users in formulating requests for information by producing a set of templates that deployed personnel can easily configure based on their current information needs. TACTIC template configurations provide a generic description of the information need. Prior to being capable of performing an action on behalf of the user, the configuration must first be transformed into an executable agent. The transformation process is a systematic mapping between end-user workflows[1] and the service interaction workflows[2] produced as part of domain analysis required to build a TACTIC prototype for a particular application.

We applied the Interaction Design and Engineering for Advanced Systems (IDEAS) process [8] to extract end user and service interaction workflows as part of analyzing a domain of interest for TACTIC. IDEAS is a proven, iterative methodology that places the human operator at the center of systems engineering activities. This innovative interdisciplinary approach combines standard best practices from

[1] **End user workflows** describe the process that a user follows to complete a task (e.g. a traveler specifying to either a travel agent, or to a travel application's interface, information about desired destination(s), dates, etc). These workflows detail the individual steps performed by a non-expert user to enable an expert (human or machine) to assist the user in achieving a desired outcome.

[2] **Service interaction workflows** are schematic representations of sequences of operations performed by systems and expert users. These workflows specify what is necessary, from both systems and expert users, to accomplish a specific goal. Service interaction workflows are used in conjunction with corresponding user workflows to translate an existing process into a series of tasks that can be completed by a user aided by an advanced system.

User-Centered Design (UCD) theory with a cutting-edge systems engineering approach to account for human cognitive and task processes throughout the design of the system architecture and human-computer interface. While UCD practices have traditionally focused on methods to understand user needs, IDEAS ensures that user functional and cognitive requirements underlie design and development throughout the entire systems engineering process.

IDEAS goes a step farther than traditional UCD by directly facilitating discussion and translation between engineering subject matter experts and domain subject matter experts. The methodology brings engineering SMEs, operational SMEs, and stakeholders together through documentation, conversation, and hands-on brainstorming sessions to facilitate the exchange of information critical to designing technology solutions. This facilitation enables engineers to communicate what their technology is capable of doing while operators communicate what is worth doing with technology; the result is that both types of experts to think in terms that are both visionary and achievable.

To support our user interface prototype development effort, we conducted three interviews with a subject matter expert (SME) familiar with Army and Marine intelligence activities and tactical operations to develop the information structure. These interviews were geared toward understanding the needs of an operator in the field as well as the structure of requests that can be answered intelligently by an analyst. The goal of the interviews was to understand how a non-intelligence specialist could compose actionable intelligence requests.

4 User Interface Prototype

Although our prototype development effort focused on the domain of military intelligence, we started by creating a generic framework that may be used to develop user interfaces for any of several complex information domains where remote personnel may need to obtain data 'in the field'. The interactions associated with our framework are based on the concept of applying relations to entities. Entities are the nouns of the domain—people, places, and things. All entities have characteristics. A person, for example, has the characteristics of name, age, gender, and nationality. Relations are the functions describing how entities are to be compared. Relations are n-ary in form. For example, "simple search" is a unary relation that will find entities of a particular type (e.g. people) that have specific characteristics (age that is greater than 30). Another example is "proximity," a binary relation that compares the closeness of two entities.

Once this framework had been established, we leveraged it to create a user interface prototype called TACTIC, which is based on the information we gathered during SME interviews on military intelligence. The prototype is designed to be a tool used by personnel while on patrol to ask questions about new threats that arise in a changing operational situation.

To prototype interface interactions we created a domain involving military patrol operations based on SME input. A patrol in an urban environment might frequently request information such as locations of possible or previous improvised explosive devices (IEDs) within a zone or in a radius around a certain position or whether vehicles of a particular type are present in an area to which the patrol does not currently have a visual. To match the domain terminology the abstract concepts of entities and relations have been renamed to objects and conditions. However, the style of interaction is the same: mobile personnel can combine one or several objects with conditions to compose an intelligent actionable query. As the query is composed, the system accesses background information available about the entities and relations in the domain and facilitates requests for more detailed information if necessary (e.g., a vehicle might have a make, model, and color associated with it).

In our prototype, the main working area of the screen is comprised of three side-by-side panes that contain the graphical trees depicting objects (Fig. 1, area A) and conditions (Fig. 1, area C) as well as the "Request Specification Area" (Fig. 1, area B).

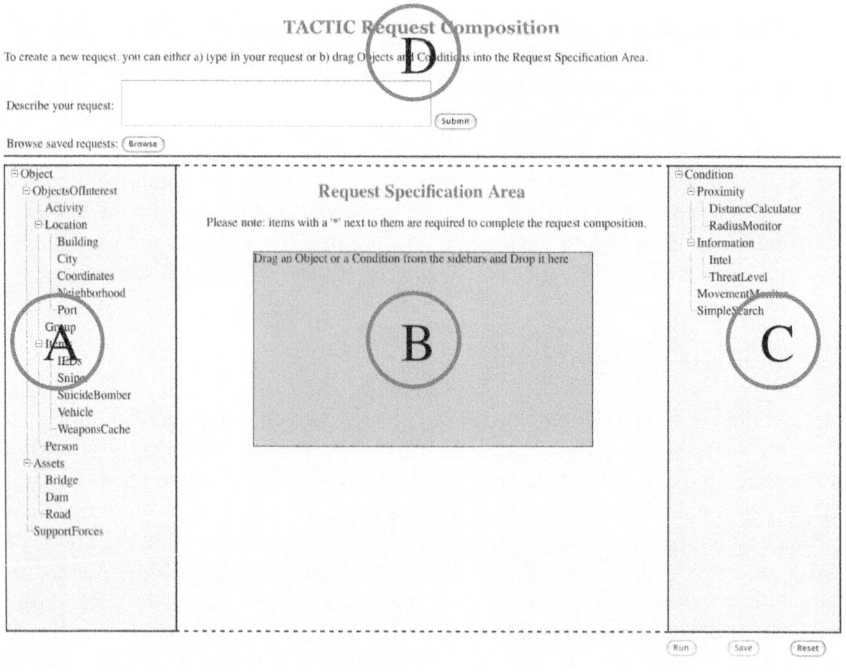

Fig. 1. Complete view of the TACTIC interface. User graphically composes a query by dragging and dropping objects and conditions into the Request Specification Area.

The Request Specification Area is the primary workspace of the user. Through click-and-drag interactions the user can express his desire to know about a particular type of vehicle in a graphical way. If he wants to know about vehicles around a given building, he might start with a "Building" object. Once dragged to the middle of the screen, the object is placed into an accordion-style GUI element that exposes a number of optional and required (outlined in red) characteristics that prompt the user to provide more details. As the information is entered, the user is aided in a number of ways depending on the richness of the ontology. Most fields will automatically validate entries to avoid erroneous input. Additionally, when appropriate, the interface will provide auto-completion functionality to aid with difficult-to-remember terms (Fig. 2). At the same time, the relation tree is pruned to prevent selection of incompatible items (e.g., buildings do not move so Movement Monitor condition is omitted, Fig. 3).

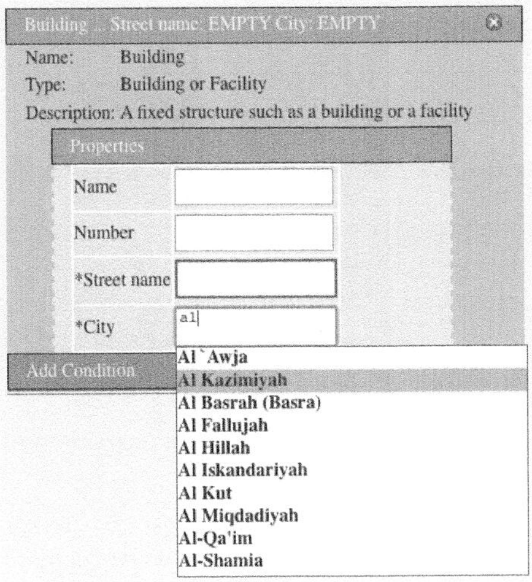

Fig. 2. Building object with exposed characteristics. City field provides lookup functionality to aid the user in recalling city names.

Once the user has specified Building characteristics, he will select a condition. In this example we use the Radius Monitor (recall that this is a binary relation) to demonstrate the interface's capability of coupling entities. After the Radius Monitor is dragged and dropped into the Request Specification Area, the view is updated with an accordion menu containing the Radius condition, the Building object, and a placeholder for another object that still needs to be coupled to make this query complete.

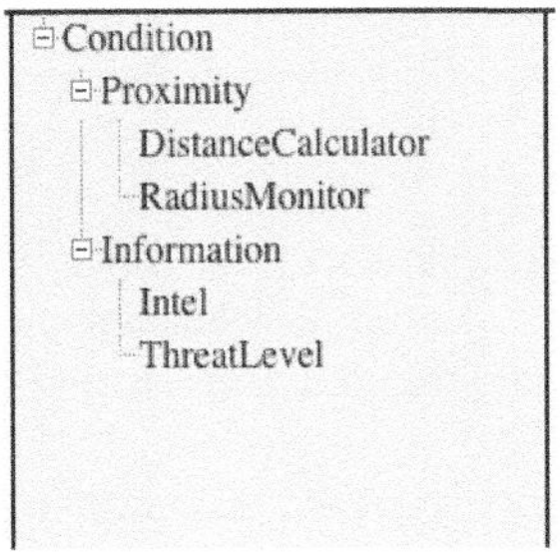

Fig. 3. Relations list is filtered based on the choice of entity to which they relate

An important feature of the interface is the visual feedback of the progress of the query. Even when portions of the accordion are collapsed (i.e., characteristics of items are hidden) the interface provides an at-a-glance status of that entity. The accordion header is filled with a partial summary of the object's most salient characteristics while a green checkmark or a red "X" denotes the validation status (Fig 4).

Finally, we have also provided a tool for advanced users to rapidly create queries. The users who are familiar with the objects and relations that comprise a given domain can use the natural language parsing interface (Fig. 1, area D) to generate partial or full queries.

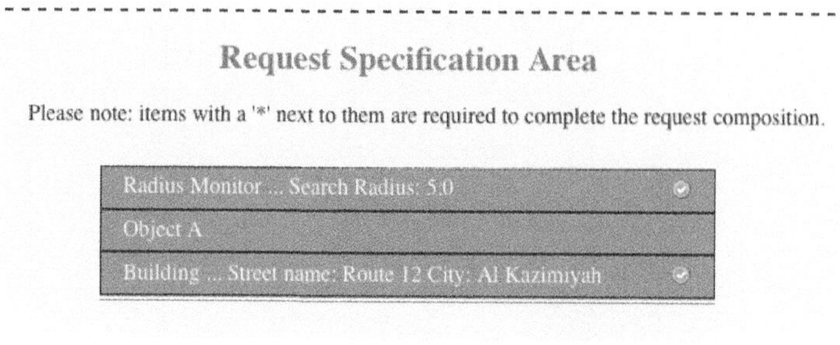

Fig. 4. At-a-glance summary of query progress

In summary, through the use of dropdown boxes, text fields, and supporting validation ontology, TACTIC's interface enables a powerful agent search and monitoring tool that is usable by novices.

5 Context-Based Workflows

One of the areas of research associated with our effort is applying context-based workflows in an agent environment. Agent-based software helps to break apart various reusable tasks and can make it easier to troubleshoot when there are problems. Agents usually have a predefined workflow that contains a list of tasks that execute in a predefined way. Each task takes a certain amount of time to execute and is generally written to execute quickly in its own thread and not block other work from being completed. Agents can also be complex with many paths to different tasks within the workflow. Tracking the planned flow through the agent's tasks to what actually happens when the agent is executed can also be a challenge and usually is done by watching log messages on the console.

One new approach leverages the workflow concept from agent-based software and adds layers of built in context awareness when the agent is executed. The first layer includes behind the scenes software that collects and calculates performance metrics at the task level. As each task is executed, the time it takes to complete the task is measured as well as how many times the task was used and a running tally of minimum, maximum and average times for the task. All of the metric data collected can help a developer understand how the agent is performing during its execution. If a task was expected to run fast, but actually runs slow, the metric data would be able to pinpoint which task is taking to long to run.

The second layer makes use of the metric data that is being collected and calculated to influence the flow of the agent's workflow. There may be more then one way to perform a task, but each method may have a set of pros and cons. (i.e., pros: task runs fast, task returns "better" data; cons: task runs slow, task may not finish) The first pass through the workflow might use historical metric data on how well a task has performed in the past. Each time the task is executed (i.e., retrieve data from database A in 1 second), the historical data is updated and if a certain negative threshold is found (retrieve data in 60 seconds or no response from database A), then the workflow might change to use a similar task (retrieve data in 10 seconds from database B) with "better" metrics.

The third layer makes use of external contexts within the domain of the project. These may include finding new data sources when a new database appears online. When the context-based workflow is made aware, it may automatically decide to use the new database. Or if an agent-based query is created while the user is physically moving (i.e., convoy or patrol), the results coming back may be filtered based on the user's current location.

Determining which metrics are useful to collect and how to apply those metrics to a dynamically changing workflow is a topic for future research.

6 Future Research

Many operational domains present complex information environments in which queries are composed and executed by experts familiar with data sources, required values, parameters and expected results. The primary long-term goal of our research effort is to enable novices to ask for and receive information from these complex information environments without the assistance of experts. The work presented in this paper focuses on the "ask for" side of the equation: exposing an expert's understanding of a query space in a way that a novice would be able to create a valid query in a complex domain. Future research will need to take a critical step toward the "receive information" challenge by addressing what is necessary for those queries to be executed.

To do this, we will need to understand what an expert does to execute a query. Taking a concrete example, TACTIC addresses how a query for intelligence information can be expressed in operational language such as "Has there been any hostile activity within a five mile radius of my location?" and then translated into a potentially executable query. To answer this query, an expert intelligence analyst would create sub-queries that address context (where is the soldier, is he moving), details (what constitutes hostile activity) and available information sources (where is information about IEDs, how should information from various sources be summarized). Some of these sub-queries have been exposed in the expert query expression (e.g., the query creation interface exposes some of the details about what could be considered hostile activity), but there is a complex expert workflow that executes between expression of the query and the availability of worthwhile results.

Our TACTIC prototype serves as a proof-of-concept for a new genre of technology: systems that provide remote, mobile, relative novices access to complex information environments, enabling them to acquire situationally relevant information in near real-time with minimal assistance. Future research will need to further our understanding of how to find out and automate what an expert does to execute a complex query, addressing advanced workflow framework technologies and expert knowledge elicitation techniques.

Acknowledgments. We would like to acknowledge the research efforts of Dr. Lisa Anthony and Dr. Susan Lawrence for the work they have done along with our team in the area of expert knowledge elicitation.

References

1. Aguirre, J.-L., Barron, O., Brena, R., Garcia, M.-N.: Using a High Level Knowledge Representation for Expert System Knowledge Acquisition and Prototyping. In: Proceedings of Symposium of Applied Computing (1993)
2. Bates, M.J.: The Design of Browsing and Berrypicking Techniques for the Online Search Interface (1989), `http://www.gseis.ucla.edu/faculty/bates/berrypicking.html`
3. Davis, R., Shrobe, H., Szolovits, P.: What is a Knowledge Representation? AI Magazine 14(1), 17–33 (1993)
4. Grigorova, D., Nikolov, N.: Knowledge Representation in Systems with Natural Language Interface. In: Proceedings of International Conference on Computer Systems and Technologies (2007)
5. Hakulinen, J., Turunen, M., Salonen, E.-P., Raiha, K.-J.: Tutor Design for Speech-Based Interfaces. In: Proceedings of Designing Interactive Systems (2004)
6. Najjar, L.J., Thompson, C., Ockerman, J.J.: Using a wearable computer for continuous learning and support. Mobile Networks and Applications 4, 69–74 (1999)
7. Raab, M., Johnson, J.G.: Extertise-Based Differences in Search and Option-Generation Strategies. Journal of Experimental Psychology: Applied (2007)
8. Regli, S.H., Tremoulet, P.D.: The IDEAS Process: Interaction Design and Engineering for Advanced Systems. In: APA Division 21, Division 19, & HFES Potomac Chapter Annual Symposium on Applied Research. George Mason University, Fairfax (2007)
9. Sutcliffe Kathleen, M., Weick Karl, E.: Information Overload Revisited. In: Handbook of Organizational Decision Making. Oxford University Press, Oxford (2006)
10. Syn, S.Y., Spring, M.B.: Can a system make novice users experts? Important factors for automatic metadata generation systems. In: Proceedings of International Conference on Dublin Core and Metadata Applications (2007)
11. Van Geenen, E.W., Witteman, C.L.: How experts reason: the acquisition of experts' knowledge structures. The Knowledge Engineering Review 21(4), 335–344 (2006)
12. Venable, J.R.: Teaching Novice Conceptual Data Modellers to Become Experts. In: Proceedings of Software Engineering: Education and Practice (1996)
13. Vidal, J.C., Lama, M., Bugarin, A.: A Framework for Unifying Problem-Solving Knowledge and Workflow Modeling. Technical Report SS-08-01. AAAI Press, Menlo Park (2008)

A Method for Evaluating the Relationship among Four Perspectives of the Balanced Scorecard

Yumiko Taguchi[1], Takumi Kaneko[2], and Tsutomu Tabe[2]

[1] Department of Business Administration and Communication, Shohoku College, 428 Nurumizu, Atsugi, Kanagawa 243-8501 Japan
[2] Department of Industrial and Systems Engineering, College of Science and Engineering, Aoyama Gakuin University, 5-10-1 Fuchinobe, Sagamihara, Kanagawa 229-8558 Japan
taguchi@shohoku.ac.jp, c5605046@cc.aoyama.ac.jp, tabe@ise.aoyama.ac.jp

Abstract. The purpose of this study is to clarify a method for quantitatively evaluating the relationship among the four perspectives of the BSC. In order to establish the purpose of this study, we (1) clarify the methodology to quantitatively understand the vision and relationship among the four perspectives, (2) analyze the KPI data of a corporation model which is similar to practical business with the evaluation method shown in this study and (3) verify the practical effectiveness of the method based on the obtained results.

Keywords: Balanced scorecard, four perspectives, balance, high-order factor analysis, covariance structure analysis.

1 Introduction

Changes in various perspectives of business management are strongly requested due to recent economic globalization and increased competition. The balanced Scorecard (BSC) is gathering attentions as one of these management systems [1]. In the BSC, specific targets (strategic objectives) required for executing strategies (company strategy, business strategy, etc) to accomplish the vision of a company constructed from the status of the future industry and development of the company are categorized into four perspectives: "Financial," "Customer," "Internal Business Processes" and "Learning and Growth." The strategic objectives categorized into these four perspectives further evolve into each Critical Success Factor (CSF), which accomplishes the four perspectives respectively, and the Key Performance Indicator (KPI) to quantitatively evaluate CSF.

By the way, David P. Norton, who is one of the developers of the BSC uses the term "Unbalanced Scorecard", meaning "bad operation status of the BSC." It is a state of the BSC in which the balance between the financial perspective and the other three perspectives is completely lost. There are many cases in which the financial perspective is significantly stronger than the other perspectives and benefits of the BSC, and thus "balance" is not received at all. Therefore, an organization should be evaluated and

M.J. Smith and G. Salvendy (Eds.): Human Interface, Part I, HCII 2009, LNCS 5617, pp. 163–170, 2009.

improved by keeping balance among the four perspectives while focusing on vision and strategy when executing the BSC.

Previous studies reported methods for quantitatively understanding the causal relationship among strategic objectives and the pervasion level of strategy [2][3]. However, no study has reported a method for evaluating and improving the balance among the four perspectives.

Hence, the purpose of this study is to show a method for evaluating the relationship among the four perspectives of the BSC. In order to establish the purpose of this study, we (1) present the methodology to quantitatively understand the vision and relationship among the four perspectives, (2) analyze the KPI data of a corporation model which is similar to practical business with the evaluation method proposed in this study and (3) verify the practical effectiveness of the method based on the obtained results.

2 A Method for Quantitatively Understanding the Relationship among the Four Perspectives

At first, a model expressing the status of the balance is constructed. Then, an analysis method applied to the model is discussed. Finally, we propose a method for evaluating the balance through the analysis method.

2.1 A Model Expressing the Status of the Balance of the BSC

Balances in the BSC can be categorized into the balance between short and long-range objectives, the balance among the past, present and future, the balance between financial and non-financial perspectives, and the balance between external and internal perspectives.

The balance has no fixed criterion (i.e., distribution by percentage). It is important to build a relationship as expected and predicted by a business manger or the person who prepared the BSC.

Therefore, we evaluate the relationship among the four perspectives in this study. These four perspectives were created from vision and strategy. It is difficult to make an evaluation based only their relationship. Thus, we "evaluate the relationship among the four perspectives" by "quantitatively indicating the relationship among the four perspectives."

We construct a model for the relationship between the vision, strategy, four perspectives and strategic objective, and the KPI. This model is called an evolutionary model of the BSC.

We should note that the KPI (key performance indicator) is quantitative data, while the strategic objective, four perspectives and vision are qualitative data (i.e., concepts) in this study. The method used to analyze this model will be specified while paying attention to this respect.

2.2 Discussion of the Analytical Method

We should consider the following four issues concerning the method for analyzing the evolutionary model of the BSC: We can (1) freely construct the model to be analyzed.

(2) Analyze the model including concepts that cannot be directly measured. (3) Distinguish the relationship among variables. (4) Analyze the relationship among qualitative data.

We use high-order factor analysis in order to consider these issues in this study. High-order factor analysis is "a covariance structure analysis and a model assuming higher concepts exist behind the factors supposed to be at the back of quantitative data." We can construct a model specific to the data according to the data format by the covariance structure analysis. A variable that can be measured is called an "observed variable." A factor that cannot be directly observed is called a "constructive concept." We can analyze the causal relationship by including constructive concepts using this method.

In other words, by using high-order factor analysis, we can quantitatively evaluate the relationship among factors that cannot be directly explained by observed variables and analyze models by utilizing the method to evolve the BSC without modification. More specifically, when the evolutionary model of the BSC is applied to high-order factor analysis, as shown in Fig. 1, we can quantitatively evaluate the relationship between the vision and the four perspectives from the path coefficient between the second and third-order factors. Because the model is applied by assuming the KPI data to be observed variables, the strategic objective to be the first-order factor, the four perspectives to be the second-order factors, and the vision to be the third-order factor.

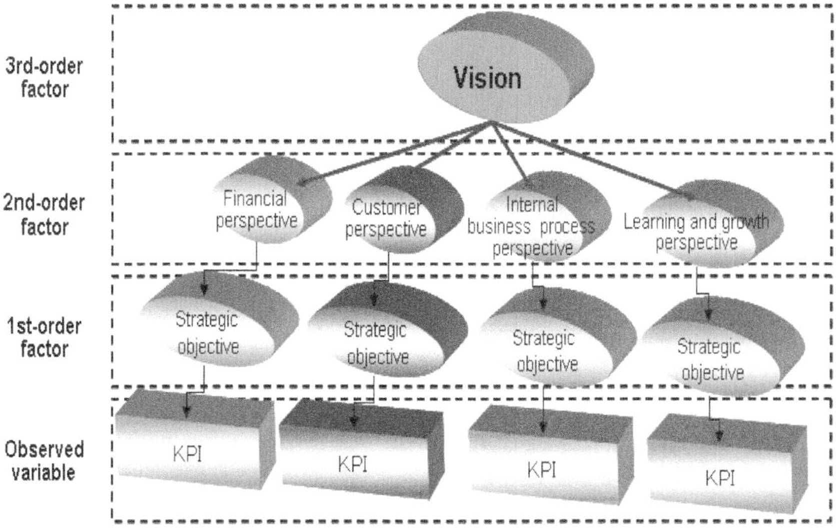

Fig. 1. Setting evolutionary model and high-order factors

2.3 A Method for Evaluating Balance

In this study, balance of each perspective is a ratio obtained from the following equation (1) based on the path coefficient of the vision and four perspectives.

$$X_i = \frac{x_i}{x_1 + x_2 + x_3 + x_4} \times 100 \quad (1)$$

where

X_i :balance of perspective i (%)

x_i :Path coefficient of vision and perspective i

i=1: Financial perspective

i=2: Customer perspective

i=3: Internal business process perspective

i=4: Learning and growth perspective

We used high-order factor analysis in the evolutionary model of the BSC. As a result, we could not always construct the relationship between vision and the four perspectives as expected.

If no relationship was constructed as expected, approaches after that were very important. As shown in Fig. 2, we prepare a chart to interpret the analysis results and trace the relationship between the four perspectives and strategic objectives, and the strategic objectives and KPIs in the chart to identify causes of the imbalance.

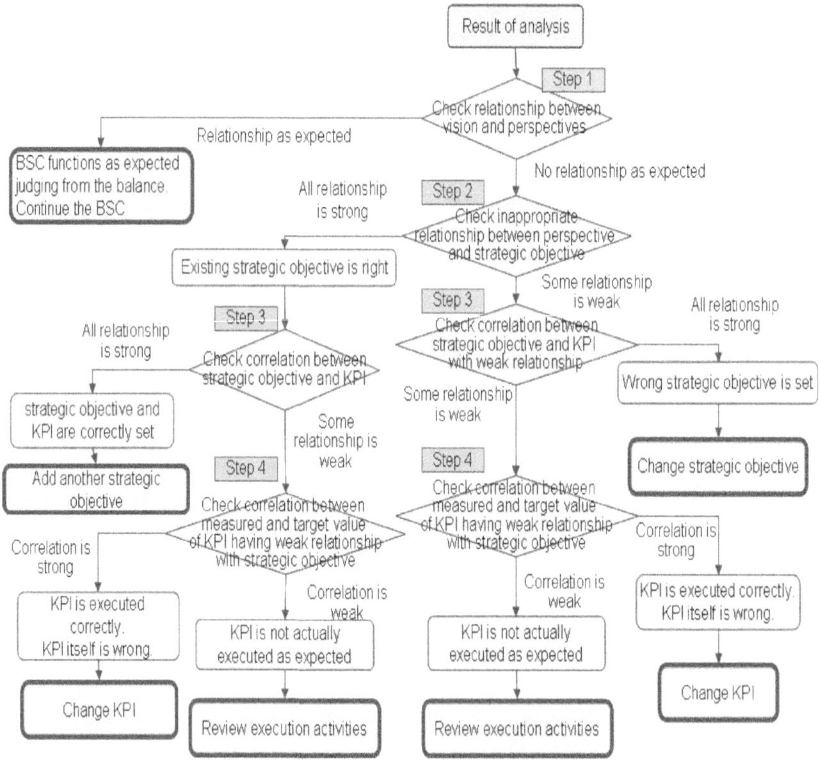

Fig. 2. Chart to interpret results of the analysis

The interpretation chart consists of four steps: (1) checking the relationship between vision and perspectives. (2) checking the inappropriate relationship between perspectives and strategic objectives. (3) checking the relationship between strategic objectives and KPIs. (4) checking the correlation between the actual value and target value of KPIs which have a weak relationship with strategic objectives. The strength of the relationship of (1) through (3) shown above is determined by the path coefficient obtained from the high-order factor analysis.

3 Discussion of the Effectiveness

We prepare the BSC of a corporation model simulating actual business conditions in order to discuss the effectiveness of the evaluation method shown above. We verify the following two issues based on the KPI data of the BSC. First, we construct a "balanced model" and "unbalanced model" and perform a high-order factor analysis. We determine whether there is an expected relationship between the vision and the four perspectives.

Second, we construct "a model in which inappropriate KPIs are set" by design. We verify whether the result affects the relationship between the vision and perspectives. We further determine whether we can extract "inappropriate KPI" which are the cause.

3.1 Corporate Model

We use data consisting of actual data provided by a corporation, as well as virtual data. We asked a businessperson to check the virtual data prepared in this study. The businessperson considered the virtual data to be useful.

The corporate model X is a manufacturer of resin. They have a vision "to provide good products and solutions aiming at business development with customers." The prepared BSC consists of a company-wide BSC, a BSC of the production department, and a BSC of the sales department.

3.2 Verification 1

As shown in Fig.3, we applied high-order analysis to the evolutionary model of the company-wide BSC for the "balanced model." Meanwhile, we applied high-order factor analysis to the evolutionary model of the BSC of the production department for the "imbalanced model."

It is estimated that the four perspectives are well-balanced in the evolutionary model of the company-wide BSC. On the contrary, it is also estimated that the "internal business process perspective" and "learning and growth perspective" account for a larger ratio than the "financial" and "customer" in the evolutionary model of the BSC of the production department. We applied high-order factor analysis on these two models by using AMOS, which is covariance structure analysis software. We obtained the following results:

In the evolutionary model of the company-wide BSC, financial, customer, internal business process, learning and growth had an effect of 25%, 25%, 26% and 24%, respectively, showing an almost equal ratio. Meanwhile, financial, customer, internal business process, learning and growth accounted for 19%, 24%, 28%, 29%, respectively, in the evolutionary model of the BSC in the production department. The "internal business process" and "learning and growth" perspectives had larger ratio in their effect. As shown above, the high-order factor analysis can show the relationship among the four perspectives in a quantitative manner.

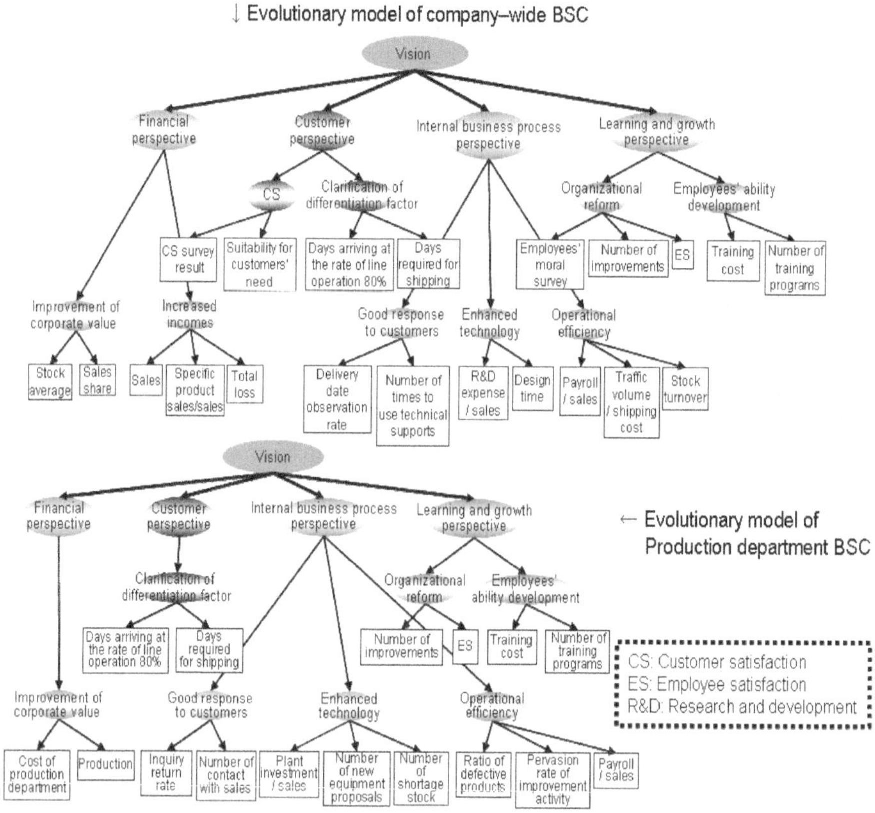

Fig. 3. Evolutionary BSC models company-wide and of the production department

3.3 Verification 2

As shown in Fig. 4, we set the "training cost", supposed to be a KPI of learning and growth perspective, in the KPI for the strategic objective of customer perspective in the evolutionary model of the BSC of the sales department. We confirm whether the setting is reflected in the results or not. Normally, it is estimated that the "financial" and "customer" perspectives account for a larger ratio than the "internal business process" and "learning and growth" perspectives in the evolutionary model of the BSC of the sales department.

We performed a high-order factor analysis as in verification 1 shown above. The following results were obtained: The four perspectives account for 50%, 8%, 36% and 6% respectively. The four perspectives were imbalanced. We tried to identify the cause using the "Chart to interpret analysis results." We obtained the results shown in Fig. 5.

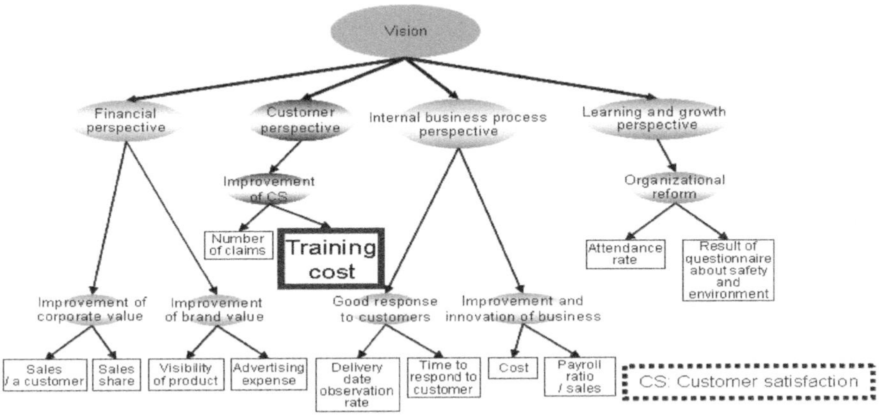

Fig. 4. BSC evolutionary model of the sales department

In step 2 of the interpretation chart, we check the relationship between inappropriate customer perspective and the improvement of CS, which is a strategic objective. The path coefficient between the two factors is -0.99, showing a strong relationship. Therefore, the improvement of CS is felt to be appropriately set for customer perspective.

Then, in step 3, we check the relationship between improvement of CS, and the number of complaints and training cost which are KPIs of improvement of CS. The path coefficient between the improvement of CS and the number of complaints is -0.99 (strong relationship). Meanwhile, the path coefficient between improvement of CS and training cost is 0.34 (weak relationship). Consequently, in step 4, we check whether or

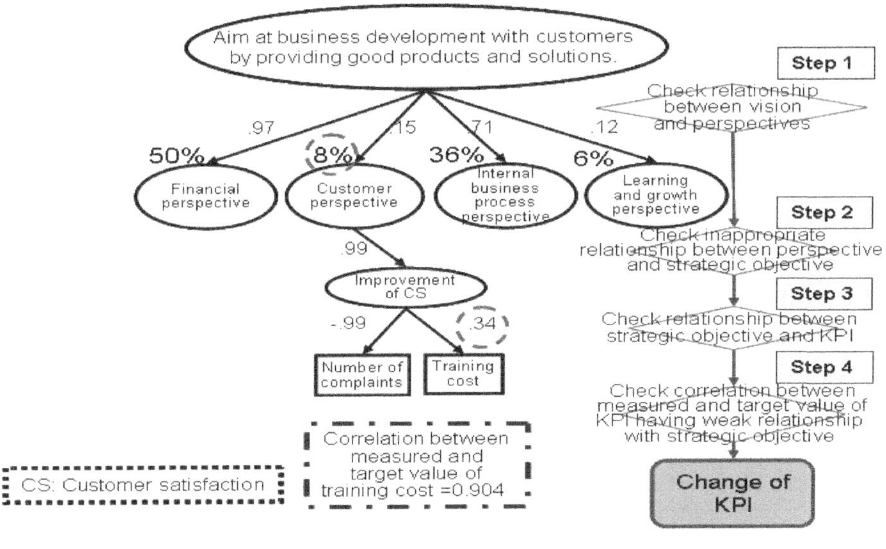

Fig. 5. Analysis results of BSC in the sales department

not the setting of the training cost itself was appropriate by using the correlation between the measured value and the target value of the training cost. As a result, the value is 0.9, showing a strong correlation. Therefore, we understand that the imbalanced customer perspective in the BSC of the sales department was caused by an inappropriate setting of KPI (training cost).

4 Conclusion

We demonstrated the usability of an evaluation method using high-order factor analysis as a methodology to quantitatively understand and evaluate whether or not a balanced BSC is constructed in the relationship of the four perspectives in this study. In the future, we should further investigate how to apply data including the relationship between the target value and measured value as the handling of observed variable data, perform analyses and extend our knowledge.

References

1. Kaplan, R.S., Norton, D.P.: The balanced scorecard—measures that drive performance. Harvard Business Review 70(1), 71–79 (1992)
2. Kaplan, R.S., Norton, D.P.: Using the Balanced Scorecard as a Strategic Management System. Harvard Business Review 85(7-8), 150–161 (2007)
3. Taguchi, Y., Ota, M., Tabe, T.: A method for quantitatively appraising strategy penetration in BSC. In: 2nd International Conference on Applied Human Factors and Ergonomics, CD-ROM (2008)

Exploring Employee Perspectives on Information Privacy and Security in the Mobile Environment

Peter Tarasewich, Benjamin Ngugi, and Mansi Sanghavi

ISOM Department, Sawyer Business School, Suffolk University
8 Ashburton Place, Boston, MA 02108 USA
tarase@suffolk.edu, bngugi@suffolk.edu,
mansi.s.sanghavi@gmail.com

Abstract. Maintaining information privacy and security in the mobile environment, an issue having personal and organizational implications, remains a challenge because the context of a mobile device can change rapidly. In response to this, the authors have been investigating methods of maintaining the privacy of sensitive information accessed in mobile environments (i.e. potentially in public places). This particular study, still in progress, will collect and analyze data on employee information privacy and security practices and perceptions. A complicating factor is that the classification of information as "private" may not be absolute. Privacy levels of organizational information will be dictated by company policy, while privacy concerns with personal information are based on the employee's own comfort level. Hence potential conflict arises if an individual uses a single mobile device for both personal and organizational information access. A summary and status of design activities and the planned implementation of the study is provided.

Keywords: privacy, security, information, mobile, context, organization.

1 Introduction

Maintaining information privacy and security in the mobile environment remains difficult because the context of a mobile device or application can change rapidly and without notice. This is in sharp contrast to a fixed environment, where people can control the way that information is handled to minimize the chance of divulging sensitive information. In an office, computer screens can be pointed towards users such that they cannot easily be read by others [8]. People are not intentionally careless when it comes to protecting information in public places, but normal human behavior can lead to unsafe conditions [4]. For example, laptop computers are often used whenever and wherever needed or desired (e.g., on airplanes). In this situation, the user can become more focused on the task at hand rather than the fact that information might be overseen or recorded by someone close by. While current technology makes it easy to access information anywhere and anytime, protection of that information is not always adequate. If this does not improve, users must accept tradeoffs between the pervasive availability of information and a potential loss of privacy and security [8].

M.J. Smith and G. Salvendy (Eds.): Human Interface, Part I, HCII 2009, LNCS 5617, pp. 171–180, 2009.

While privacy and security is often maintained through methods (e.g., encryption) that keep data from being read by unauthorized parties, this research looks at the relatively unexplored problem of maintaining the privacy of displayed information by employees in the mobile environment. The overall goal of this work is the creation of technically sound but practical methods of maintaining privacy of sensitive information that is accessed and displayed in potentially public spaces. Furthermore, the classification of information as "private" may not be absolute. Privacy levels of organizational information will be dictated by company policy, while the privacy levels of personal information are customized to a user's own comfort level and requirements. For example, an individual may not be concerned with displaying names and addresses of friends, but company policy may restrict disclosure of customer information to unauthorized parties. This can potentially lead to conflicts when devices are used for both personal and business use [21].

This research-in-progress continues to build upon the authors' previous work in the area of information privacy and security, and seeks to extend it to address additional issues related to information use by employees and organizations while mobile. Specifically, the ultimate goals of this research are to:

- determine what types of information employees and organizations consider private;
- establish the relative importance to each party of keeping that information private and secure;
- determine whether employee security and privacy concerns are related to information sensitivity and the context of the information; and
- identify potential conflicts between personal and organizational privacy concerns.

These goals will be studied under assumptions that information can be accessed at any time with different devices in different contexts (i.e. users are mobile).

An organization's privacy concerns potentially impact all of its stakeholders - including employees, customers, and suppliers. Information privacy needs might be documented through an organization's official privacy policy [13] and/or may be dictated by its executives. While understanding organizational privacy concerns is an overall goal, this research starts by studying employees' perspectives of information privacy to gain insight into 1) the policies of the organization, and 2) any conflicts that exist between personal and organization privacy concerns. This will be accomplished by gathering data that measures the importance and understanding of information privacy to employees, and documents how employees operate under existing organizational privacy policies and security expectations in the mobile environment.

2 Background

Since technology is no longer restricted to the home or office, users can now access information in virtually any sort of public setting, including city streets, trains, restaurants, theaters, shops, hotel lobbies, and airports. This creates situations where sensitive information, organizational or personal in nature, can be seen and captured by people and technology in the immediate vicinity. Information such as product designs, customer data, sales transactions, and personnel files can easily be used for harmful purposes in the wrong hands. People may just unintentionally see another person's sensitive information on a PDA or laptop computer, but there is little to stop people from purposely trying to collect information in public places for criminal gain. Technology has made the gathering of visual information easier through devices such as digital cameras, now common in mobile phones and often used for public surveillance. Consequences of seeing or stealing another's private information can include misinformation, corporate espionage, monetary loss, and identity theft [1].

2.1 Context

In mobile computing, context is often equated simply with location but is actually more complex. Mobile application use can vary continuously because of changing circumstances and differing user needs. A context model can be defined using three broad categories of context - environment, participants, and activities [18]. The "environment" category is concerned with the properties of objects in the physical environment. "Participants" includes the status of the user(s) and other participants in the environment. "Activities" covers user, participant, and environmental activities. The model also includes interactions or relationships that exist between categories, and context history is tracked over time. This model can be narrowed down and simplified to several aspects that are most important to an employee's (as well as any individual's) use of mobile information systems. One is the location of the employee when the system is being used. For example, is the employee in his/her office, at a client's business, or on a train? Another is co-location, or whom the employee is with. In a conference room, an employee can be alone, with colleagues, or with clients. Related to the co-location of people is the co-location (or availability) of resources, such as communications networks (e.g. cell or Wi-Fi), output devices (e.g. printers), and even power to keep devices running for longer periods of time.

2.2 Information Sensitivity

Information sensitivity is a measure of the effects that lost or misused information can have on an individual or organization. It also assesses how unauthorized access to or modification of information might adversely affect these same parties. Organizations will often create categories of information sensitivity, with specific rules and procedures for each one. For example, some US government security organizations classify information into the categories unclassified, restricted, confidential, secret, and top secret [5]. As another example, medical organizations, using guidelines based on the Health Insurance Portability and Accountability Act (HIPPA), can classify information as public, office, confidential, and restricted confidential [10].

2.3 Information Privacy and Security Concerns of the Employee

Employees potentially work with two types of information in the mobile environment personal information and organizational information. On a personal level, the comfort associated with using or viewing information in public will vary by person and with context, although some categories of information most people seem to want to always keep private. A recent study by the authors examined individuals' concerns when accessing personal information in public settings [13]. Results of a questionnaire showed that subjects felt most strongly about protecting the privacy of medical and financial information.

As an employee, an individual may also have to deal with the privacy and security of information of the organization. its workers, and clients. While losing or divulging personal information about oneself can be damaging, doing so on an organizational level can potentially harm thousands of people at one time. One difficulty in this regard is that the sensitivity of information may be viewed differently by employee and organization. For example, an employee might freely browse names, addresses, and phone numbers of friends without worrying about that information being intercepted and misused. But that same behavior with an organization's client list could prove detrimental. Such situations can be compounded if employees use mobile devices for both business and personal work. Even with rules in place, switching between personal and organizational tasks may not always signal that the level of security has changed as well.

2.4 Protecting Information in Public Spaces

In general, there are many factors that need to be taken into account when using mobile devices to send and receive data in the mobile environment [e.g. 23]. Any device should be secured against someone accessing its data without permission. The risk of losing the entire device should be assessed. When connecting to a communications network, the security of the network and the location from which the device is used may need to be evaluated and verified. Can someone intercept the data, or take data from the user interface (e.g. screen) itself?

The authors have been looking specifically at the problem of protecting the privacy and security of displayed information. The authors initially investigations the design and use of pixel-based displays, which convey information using one or more individual lights [19]. One of the most significant benefits of these displays is that information can be personalized such that only the user knows what it means, even if shown in public. For example, three blue lights on a person's ring, even when noticed by other people nearby, could convey a message only understood by the wearer. This led to the development of privacy blinders [20], which mimic the use of yellow sticky-notes to cover parts of a larger document so that they are not viewable by others. Blinders can be used to provide a mixed display in which sensitive information is hidden (covered) but other information is displayed normally. A user can view sensitive information by temporarily removing the blinder, perhaps by a stylus tap.

Furthermore, privacy blinders can automatically respond to a predefined organizational and/or personal "privacy policy," which specifies what types of information are covered under different circumstances. An organizational policy might be dictated by the company a person works for, while a personal policy is customized to a user's own privacy requirements. This method can also account for user context changes; if a person moves to a less public space, they might turn off the blinder feature. Context might also be taken into account automatically by the system [18]. For example, moving from a private office to a conference room might modify privacy settings by design - adapting to the changing environment of the mobile device user.

Other steps can be taken to protect sensitive data stored on mobile devices. It may be in the best interest of the user (and of anyone else whose data is stored on a device) to destroy stored data or permanently disable a device if an invalid login is attempted. These types of security measures create additional overhead and potential operating problems for the user, so a risk analysis may be appropriate to determine the level of security desired or needed. For example, it may not make sense to encrypt data on a device that is solely being used to play games or music. But if the game is one that is under development, or the music consists of new songs by an artist being considered for distribution by a record label, then security should be a top concern.

2.5 Organizational Mobile Device Policies

Given that mobile devices are more vulnerable to loss or theft than their desktop counterparts, mobile devices and applications need to squarely focus on securing data that is stored on the device, and transmitted from or received to the device [2, 12]. This may involve data encryption, validation of the user's identity through passwords or biometric techniques, and validation of the sender or receiver's identity (in the case of transmitted data) through electronic signatures. What becomes more critical to mobile (versus desktop) application design is the implementation of features that make it difficult or impossible to access data on (or with) the device should it fall into the hands of an unauthorized user. Some organizations have instituted policies such as requiring all data to be stored on servers with no data residing on the mobile device.

Another design concern is the security of communications pathways and security at a systems level. Implementing proper levels of security for data as it is transmitted over wireless pathways is critical. Wireless access points need to be secured. Both ends of the communication pathway need to prevent and detect viruses and spyware.

Security and privacy policies for mobile device use and information access must be set and enforced (e.g., 2, 12, 14]. Security policies of organizations often have to take many stakeholder requirements into account, including customers, governments, and regulatory bodies. An example can be found in the brokerage industry, which requires mobile devices and applications to handle sophisticated tasks without jeopardizing security and regulation compliance [3].

2.6 Designing Mobile Systems for Both Organizational and Employee Use

Mobile devices used to access the information resources and systems of an organization may be used for both personal and business purposes, and may also be used for multiple application types. Levels of required security and the classification of information as "private" may not be absolute. Privacy and security levels of organizational information are based on company policy, and they usually differ from privacy and security levels of personal information. Hence, conflicts can arise.

Contradictory forces between personal and business needs of users may influence patterns of user behavior in the mobile environment [6]. Hence, organizations need to design applications with both individuals and the corporation in mind as traditional boundaries between work and non-work become blurred with increased worker mobility. To maximize effective use of mobile applications, and to minimize possible dilemmas faced by employees, it may become necessary to resolve any conflicting concerns of the individual and the organization. This can involve design issues such as boundary management settings between work and personal activities, and managing organizations' priorities relating to empowerment and control.

3 Methodology and Research Activities

This research is based on the premise that privacy is valued and expected by most people to varying degrees. Usually an employee expects reasonable access to personal information while restricting strangers' access to this same information. Privacy requirements will also vary based on the type of information, the information's context, and the preferences of the information's owner (e.g., employee or organization) [e.g., 11]. Along with privacy, security is also valued to protect information from being misused.

The information privacy perspectives of employees will be studied in this research project. While this will not provide a complete picture an organization's privacy concerns, an employee will be influenced by both individual (personal) and organizational privacy issues when dealing with information in the mobile environment. Two hypotheses will be tested as part of this project:

H1: For employees, the amount of privacy and security concerns in the mobile environment will depend on the sensitivity of the information.

H2: For employees, the amount of privacy and security concerns in the mobile environment will depend on the usage context

Hypothesis 1 investigates the variation in employee concern over privacy and security with changing information types. Information is categorized by some level of sensitivity, such as secret, confidential, or unclassified. Hypothesis 2 investigates the variation in individual concern with changing context. Context is categorized by the conditions under which the information is being accessed, such as in a public setting, a private setting, or a setting where the conditions may be mixed (or unknown).

Initial exploratory studies are underway to provide additional background, support, and direction for this work. The status of these activities is summarized below, along with the results of an exploratory study involving students taking an upper-level class at the authors' institution. Section 3.2 describes a survey to gather data on employee information privacy and security practices and perceptions, which will be used to formally test hypotheses H1 and H2.

3.1 Initial Research Activities

The authors are currently conducting a set of exploratory studies and other research activities as the ground work for a formal study of employee information privacy and security practices and perceptions. Studies have already been conducted on the importance of data privacy to individuals [e.g. 9, 20], although further analysis is underway. Information on organizational privacy policies and security procedures is being collected from academic studies [e.g. 17], corporate case studies, and existing survey data [e.g. 22]. In preparation for the formal testing that is central to this study, a preliminary list of questions was created to address 1) employee use of information in the mobile environment, 2) employee perceptions of privacy and security while accessing different types of information in different contexts, 3) employee privacy concerns, and 4) employers' privacy and security policies. These questions were used as the basis of a class activity to further refine the questions and begin designing the survey instrument for formal testing. A summary of that activity follows.

Exploratory Discussion of Mobile Information Security and Privacy Concerns. The goals of this class activity were to 1) discover types of potentially sensitive information kept by employees on different mobile devices and 2) explore privacy and security concerns associated with use of that information. The class consisted of sixteen evening graduate and undergraduate students, most of who were working full-time. Students were first asked what types of mobile devices they carried. Next students were asked about data storage on those devices, security measures taken, use of the devices for work-related activities, and any privacy concerns while using the devices. Data was gathered by counting a show of hands, or by taking notes of students' comments during the discussion. Of these students, all of them carried mobile communication devices (including cell phones and Blackberry-type devices) with varying information storage capabilities; all carried USB memory sticks; fourteen had laptops, four carried PDA devices; and eight had iPods. When queried about security measures taken with these devices, only seven students were found to be using some type of access control method, such as a login and password, to prevent unauthorized access to the device and its information.

Seven of the students carried or could access potentially sensitive information related to their employment on their communication devices, USB memory sticks, or laptops. Such information included customer, employer, or employee contact lists, text messages, emails, schedules, and appointment calendars. These students reported an occasional need to exchange potentially sensitive information with members of their organization while mobile, via documents, spreadsheets, email, and text messages.

Nine students acknowledged having privacy and security concerns when carrying and accessing information on their mobile devices. One issue raised was the fact that Blackberries and similar devices lack the security features inherent in desktop devices. For example, three students raised the issue of the Blackberry's capability to download and store emails, documents, and spreadsheets from organizational servers. The net result is that sensitive information can be stored in a multitude of disparate places. This mobile data is usually not encrypted, and there are no audit trails to help identify if an intrusion has occurred. These same students were also concerned about the security and privacy of sensitive information when accessing it in public places. Two students stated that they would not view sensitive financial (e.g., banking) or organizational information in public places such as Wi-Fi hot spots.

Eight students worked for employers that have specifically enacted policies and guidelines on the use of mobile devices to carry or access company information. One student described a strictly-enforced policy which prohibits the transmission of any potentially sensitive information by email or even storing it in employee computers (mobile or otherwise). On the other hand, four students mentioned that their organizations had privacy and security policies related to mobile device use, but that these policies did not seem to be enforced. One of these students stated that even company management seemed to pay little attention to stated mobile device privacy and security policies and procedures. The other students were not certain if their organization had formal policies.

3.2 Survey of Employee Information Privacy and Security Practices

The objective of the survey will be to measure employee concerns with the privacy and security of different types of information being stored and accessed under different contexts. A questionnaire will be developed in iterative phases as recommended by [15]. The development phases will be (1) conceptual specification and definition of constructs, (2) construction of items, (3) data collection, and (4) measurement purification. Phases 1 and 2 are currently being completed, and plans have been made for phase 3.

In terms of phase 1, the dependent variable to be measured will be the "amount of privacy and security concern" for a given type of information under a given context. This is measured in terms of the level of perceived risk in accessing, using, storing, or transmitting the information. This could range from no risk at all (i.e. no concern over the data being compromised) to potentially catastrophic risk (i.e. information is critical to the well-being of the organization and should not be accessed under the given circumstances). The first independent variable, type of information, will use three military-like classifications of secret, confidential, or unclassified, but may still potentially be modified to fit more generic organizational information classifications. The second independent variable, context, will consist of three categories stating the conditions under which the information is being accessed - in a public setting, a private setting, or a setting where the conditions may be mixed (or unknown). At least four questionnaire items will be defined for each of the variables.

The "security and privacy concern" constructs has many facets that will need to be captured in constructing the questionnaire items. The first decision that an employee needs to make is whether to use a mobile device for a work related task. This may depend on the organizational policies, on the sensitivity of information etc. If the employee decides to use a mobile device, then the next decision is which type of mobile device to use. Such a decision may be based on the kind of tasks that the employee expects to execute while on the go, and may also depend on the reputation of the mobile device and the manufacturer in terms of security and intrusion protection. Another concern would be whether to store information in the device, since the device protection mechanisms may not be as good as those of the organizational servers. Equally important is whether to transmit sensitive information while on the go. Wireless security is still a big challenge and data in transmission is at its most vulnerable state since it is outside the control zone of the employee's organization. The employee might be equally worried of people tracking their position. All these are aspects that need to be addressed when constructing the questionnaire items. The considerations will be tempered by the sensitivity of the information and the context of usage.

For phase 3, the surveys will be administered initially to Executive MBA students, and then to other organizational employees. Surveys will be followed up with interviews to obtain more specific information about the respondents and their activities, and inquiries will be made to the organizations themselves about any formal policies in place concerning the use of mobile information systems. In addition, the authors will seek out other local firms willing to participate in the study and discuss their needs for privacy and security in the context of mobile information applications.

Phase 4, measurement purification, will involve refining the measurement items set after the survey is administered. The measurement items will be tested for; (1) Content validity, (2) Individual items reliability, (3) Construct reliability, and (4) Discriminant validity. Content validity requires that questionnaire items represent the materials (or content areas) that they are supposed to represent [16]. Individual items reliability tests the convergence of each manifest question item on the latent construct that they are measuring. Construct validity refers to the degree to which the questionnaire scales measure the psychological characteristics of interest [7]. Discriminant validity checks whether a construct differs from other constructs. Measures that are supposed to be related should correlate well together and not correlate to dissimilar measures. Items not meeting expected criteria will be eliminated from the measurement.

4 Expected Findings and Contributions

This research will add to the existing knowledge on maintaining privacy and security of information used in public spaces. Specifically, this study is expected to provide:

1. Further insight into what types of information people (specifically employees) and organizations consider private, and the relative importance of keeping that information private and secure.
2. Greater insight into the conflicts that may arise between personal and organizational privacy concerns.

This study will also document current practices of securing information use in public places by organizations and their workers, and try to see if such actions follow expected norms. In addition, it is expected that the findings of this study will guide the development of potential solutions to address the access of both personal and organizational information through a single mobile device. It is expected that the test instrument for the study will soon be finished and pilot tested, and data will be collected during the spring 2009 semester. Results of this study will be integrated into the paper's presentation at HCII 2009.

References

1. Berghel, H.: Identity Theft, Social Security Numbers, and the Web. Comm. of the ACM 43(2), 17–21 (2000)
2. Britt, P.: Mobile Security for Those On-the-Go. Information Today 23(9), 32 (2006)
3. Bienfang, M., Egan, B.: Handheld Device Trends In the US Retail Brokerage Advisory Industry. TowerGroup Research (2006)
4. Cooper, G.: The Mutable Mobile: Social Theory in the Wireless World. In: Wireless World: Social and Interactional Aspects of the Mobile Age, pp. 19–31. Springer, Heidelberg (2002)
5. Committee on National Security Systems, National Information Assurance (IA) Glossary, National Security Agency (2006), `http://www.cnss.gov/Assets/pdf/cnssi_4009.pdf`
6. Cousins, K.C., Robey, D.: Human Agency in Wireless World: Patterns of Technology Use in Nomadic Computing Environments. Info. and Org. 15, 151–180 (2005)
7. Cronbach, L.J., Meehl, P.E.: Construct Validity in Psychological Tests. Psychological Bulletin (52), 281–302 (1955)
8. Dourish, P., Grinter, R.E., Delgato de la Flor, J., Joseph, M.: Security in the Wild: User Strategies for Managing Security as an Everyday, Practical Problem. Personal and Ubiquitous Computing 8, 391–401 (2004)

9. Grimes, A., Tarasewich, P.: Investigating Privacy-Augmented Displays for Mobile Devices. In: Proceedings of HCI International 2005 (2005)
10. Guidelines for Classifying Data, CAL HIPPA (2008), `http://www.calhipaa.com/main/classification_sample1.htm`
11. Hawkey, K., Inkpen, K.M.: Keeping Up Appearances: Understanding the Dimensions of Incidental Information Privacy. In: Proceedings of CHI 2006, pp. 821–830 (2006)
12. Herold, R.: Top 10 Mobile Device Privacy Policies, InformationWeek (September 1, 2005)
13. Karat, J., Karat, C.-M., Brodie, C., Feng, J.: Privacy in Information Technology: Designing to Enable Privacy Policy Management in Organizations. International Journal of Human Computer Studies, 153–174 (2005)
14. Kavanagh, J.: Plans Into Action. Computer Weekly, pp. 52–54 (September 13, 2005)
15. Malhotra, M., Groover, V.: An Assessment of Survey Research in POM: From Constructs to Theory. J. of Ops. Mgmt. 16(4), 403–423 (1998)
16. Rosenthal, R., Rosnow, R.: Essential of Behavioral Research:Methods and Data Analysis. McGraw Hill, New York (1991)
17. Smith, H.J., Milberg, S.J., Burke, S.J.: Information Privacy: Measuring Individuals' Concerns about Organizational Practices. MIS Quarterly 20(2), 167–196 (1996)
18. Tarasewich, P.: Towards a Comprehensive Model of Context for Mobile and Wireless Computing. In: Proceedings of AMCIS 2003, pp. 114–124 (2003)
19. Tarasewich, P., Campbell, C., Xia, T., Dideles, M.: Evaluation of Visual Notification Cues for Ubiquitous Computing. In: Dey, A.K., Schmidt, A., McCarthy, J.F. (eds.) UbiComp 2003. LNCS, vol. 2864, pp. 349–366. Springer, Heidelberg (2003)
20. Tarasewich, P., Gong, J., Conlan, R.: Protecting Private Data in Public. In: Adjunct Proceedings of CHI 2006, pp. 1409–1414 (2006)
21. Tarasewich, P., Gong, J., Nah, F., DeWester, D.: Mobile Interaction Design: Integrating Individual and Organizational Perspectives. Information Knowledge Systems Management 7, 121–144 (2008)
22. The State of Information Security 2007, a Worldwide Study by CIO, CSO and PricewaterhouseCoopers. CIO Magazine (September 15, 2007)
23. The Year of Working Dangerously: The Privacy Implications of Wireless in the State Workplace–Part I, NASCIO (2005), `http://www.nascio.org/publications/documents/NASCIO-WirelessPartI.pdf`

Organization Diagnosis Tools Based on Social Network Analysis

Takanori Ugai and Kouji Aoyama

Fujitsu Laboratories Limited
4-1-1 Kamikodanaka Nakaharaku Kawasaki Kanagawa 211-8588, Japan
ugai@jp.fujitsu.com, aoyama.kouji@jp.fujitsu.com

Abstract. Many organizations have challenges such as inter-organizational barriers and motivation of employees. However, these kinds of problems are not easy to visualize, and it is even more difficult to derive, implement and assess appropriate measures to deal with them. We developed a tool to visualize the dynamic structure of cooperative relationships between employees in organizations based on questionnaires given to employees of those organizations. This tool is used for visualizing barriers between teams and the effects of measures. In this paper we explain some features of this tool and verify its capabilities and effectiveness with a case study. The case study is some field research based on interviews that we conducted in which we applied measures to improve the employees' communication. We collected a set of data about relationships in an organization with questionnaires before and after implementing the measures. And we compared the observed result produced by the visualization tool with the result from the field research.

1 Introduction

It is very important to have a company that is growing and whose entire workforce is activated. To have such a company, one of most important disciplines is to collect a lot of useful information and effectively use the knowledge of individuals within that organization [1]. We have proposed a mathematical model for transferring knowledge with the purpose of making a KM mechanism or system take root in an organization and obtaining guidelines to make that system functional [2]. We have applied this model to some cases and produced guidelines for appropriate systems to manage knowledge [3].

We have found that many organizations have communication problems such as inter-organizational barriers. Mistakes occurred because of a lack of communication. The efficiency of work does not improve because information and knowledge do not spread in the organization owing to this lack of communication.

However, this kind of problem is not easy to visualize, and it is even more difficult to derive, implement and assess appropriate measures for it.

In many cases, people in the organization realize there is a lack of communication after an accident has occurred, but it is often too late to take measures. Generally, measures to eliminate a lack of communication cannot be expected to produce short-term

M.J. Smith and G. Salvendy (Eds.): Human Interface, Part I, HCII 2009, LNCS 5617, pp. 181–189, 2009.

results. Because organizations change and improve gradually, using indicators such as the decreased number of mistakes or ROI is inappropriate.

We developed a tool to visualize the dynamic structure of cooperative relationships between employees of an organization based on questionnaires given to employees. This tool visualizes not only the organization's static structure but also its dynamic structure. It can clearly visualize barriers in an organization and the effect of measures which the organization has taken.

In this paper, we propose tools which show a layout of employees as nodes and the cooperative relationships between them as edges and a layout of employees' position among organizational vision. Vision gap visualizing tools show relationships between organization's vision and personals'. Relationship tool visualizes the static structure of cooperative relationships and the dynamic structure of relationships as a sequence of static structures and shows any difference between those static structures. We also describe how these tools work with a case study. The case study is based on our field research on an organization that develops software. We compare the field research with findings from the tools' visualization based on a questionnaire given to the employees.

There are such visualization and analysis methods and tools in existence today [4,5,6]. However, they are general analysis tools, and it is difficult to understand problems which happen in the field from the diagrams which those tools draw. The users have to analyze and interpret the meaning of the diagrams and numbers by making investigations in the field. Our tools target the managers of an organization. It should show what kinds of communication problems their organization has in an intuitive way.

In chapter 2, we describe the functions of the visualization tool of employees' relationship. In chapter 3, we describe how the gap-visualizing-tool layouts employees and their personal vision among organizational vision. Chapter 5 describes a case study to show the tool works in practical manner.

2 Visualizing the Cooperative Relationship

This tool graphically represents answers to a questionnaire given by employees of an organization regarding their cooperative relationships. We developed the questionnaire by combining an Interpersonal Solidarity Scale [7] and Bales' Interaction Process Analysis [8], and simplified it as much as possible. The questionnaire asks the employees to classify the other employees into any of the following five categories.

1. I don't know him or her.
2. I know his or her by sight.
3. I know his or her work and responsibilities.
4. I have talked about business with him or her.
5. I have worked with him or her.

The tool produces a matrix between employees which expresses their relationships.

Static Structure: The tool renders employee's names as nodes and shows the cooperative relationships between employees as edges using a spring layout algorithm [9]. The spring strength is proportional to the value of the cooperative relationship. The edge is Omni directional and the strength of an edge between the nodes is the average value of the relationships. A diagram shows that the cooperative employees are collocated and unknown employees are located far from each other. This tool can decorate nodes and edges in the following ways.

- The thickness of the edges can be proportional to the length. Users can easily see which relationships are strong in the diagram.
- Details of members of the group including their names and affiliation to the group are shown by color in each node.
- The nodes can be cauterized based on edge betweenness[10] and the nodes in a cluster are painted with a color. Users can identify isolated members.

Dynamic Structure: The tool computes differences within the organization and draws a comparison using two matrixes.

- It is possible to place two diagrams of cooperative relationships side by side, or switch over two drawings with a click.
- For each diagram, branches that show differences between cooperative relationships can be shown in red if their value is larger than the other branches or in blue

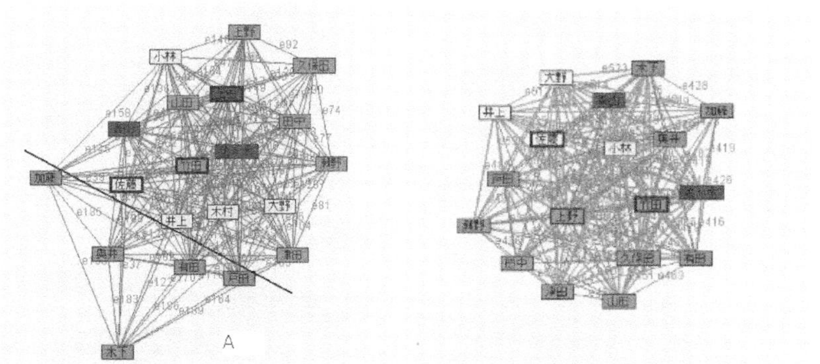

Fig. 1. Pre-poll cooperative relation and Post poll cooperative relation

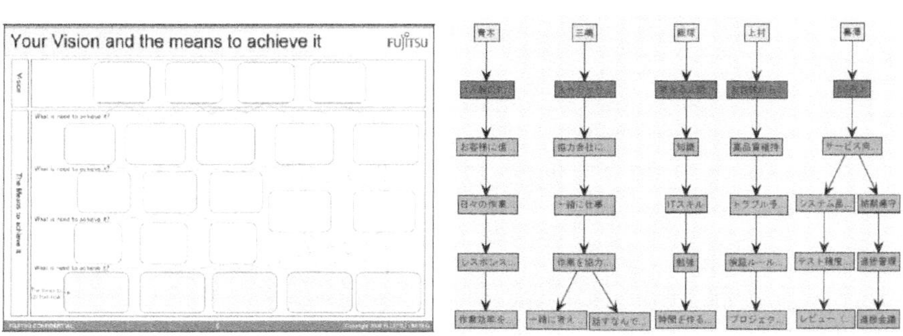

Fig. 2. Vision-Mean Analysis Worksheet and Examples

- if their value is smaller than the other branches.
- A diagram showing only the difference between cooperative relationships can be displayed.

If nodes are divided between each group as shown in figure 1, it is expected that the distance between groups is large and there are no cooperative relationships between those groups. On the other hand, if employees of each group are uniformly spread as shown in figure 2, it is expected that there are cooperative relationships between the groups.

3 Visualizing the Gap between Organizational Policy and Personal Vision

We use a work-sheet form, as shown in Fig. 3, to have the members fill in their personal visions and the means to achieve them. The work-sheet is derived from a motivation theory described in the section 4.3.1.

In the right-hand figure in Fig.4 is made by unifying the same or similar description as a single box.

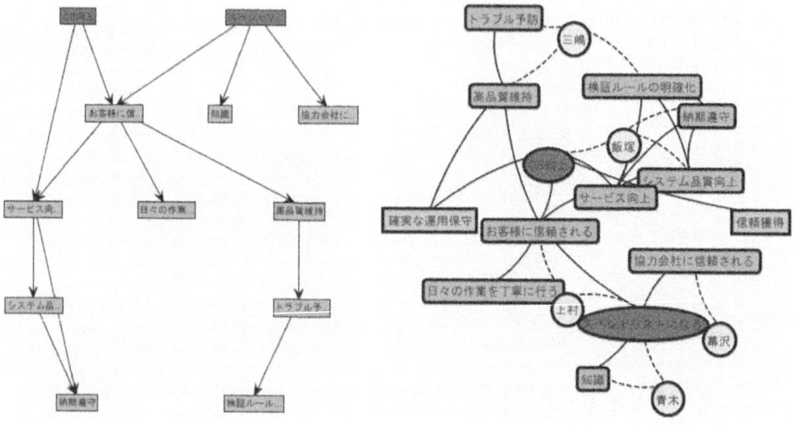

Fig. 3. Eliminated Tree Graph and Anchored Map of Vision-Mean Analysis

The left-hand figure in Fig.3 is drawn by Anchor Map [13]. Anchor Map has two kinds of node: "anchors" and "free nodes." The anchors are arranged on the circumference at equal intervals and in the figure two squire boxes are anchors. The free nodes are arranged at suitable positions in relation to the adjacent anchor nodes using a spring layout algorithm [9]. The edge is directional and the default length of an edge is the half length between two boxes. In the left-hand figure of Fig.3 all node but the two organizational policies which are squire boxes are free nodes. The five small circles stand persons and the locations stand for the position among the organizational vision. The figure says that all five members are located almost middle of two visions but none of them are connected to the visions.

4 A Case Study by Field Research and Questionnaires

4.1 Research Field and Research Methods

The organization we surveyed implements software development projects based on various pieces of package software. It has about 200 employees in 7 groups, and we studied 4 of those groups. Of the employees, 23 people from 4 groups were interested in our research and we decided to do this research with them. The four groups deal with different work packages for their customers. Two or three people from one group worked together on a system development project, though usually they don't work together on projects and they don't talk or chat with employees of a different group in general.

We conducted two-hour interviews with 5 people from the 23. And we analyzed them, derived measures to improve their communication, and helped them to implement those measures. We have continued to observe them. Independently from the interviews and observations, we gave questionnaires to all 23 employees and got answers from 21.We decided to use those 21 employees and their answers for our analysis and to make diagrams.

4.2 Cooperative Relations

Cooperative Relations Based on Prior Surveys. Figure 1 shows result of the questionnaire was taken in prior surveys. Three teams B, C and D have a relatively uniform spread, but team-A is located too far from the other three teams. We found a barrier between the team-A and the others. The average value was 2.7, and the average value with different team employees was 2.3. This means they know other team employees by sight but don't know what other team employees are doing in their work.

We found that the organization doesn't place much value on knowledge exchange and employees don't get much profit from providing knowledge. A lot of background knowledge is required to share knowledge of their work. And they don't know each other by sight. So we can say the barrier between employees is very large. We can suppose that there is insufficient communication for knowledge transfer and knowledge sharing.

Casual Information Exchange Meeting. From the aforementioned situations, we suggested that the employees should have meetings for casual information exchange to develop relationships between employees in which they can help each other. In order to reduce barriers we tried the following variety of strategies to implement.

- **System:** Following the instructions of managers, employees participate as part of the work. By positioning this work as work done during work hours, it becomes authorized work and the employees are not working as volunteers, and we thought this would reduce barriers in the system.
- **Trust:** the purpose of this meeting is anxiety awareness where employees create relationships in which they can consult with each other, and we combined this meeting with self-introductions and icebreakers to build a relationship of trust.
- **Sense of camaraderie:** We selected topics for the meeting very carefully, such as facilitation and reflection. These topics related to all employees because the topics

are not related to their work and are what all employees require in general. All employees can empathize with each other and feel a sense of camaraderie. We took what was bothering them as the theme, and had them share their problems and also share a sense of camaraderie.

The information exchange meeting was held once a month for seven months. Figure 2 was made using the results of the questionnaire which we collected after this period. And also we got the following feedback from the employees.

- There were relationships in which people can consult with each other.
- It was good to know that employees in other groups had the same problems a slight angle and to the right of you.
- We want to continue with implementation but we cannot do that yet by ourselves.

Validation of Findings from the Visualized Cooperative Relationships. In the interviews, some employees said they know other employees only by sight. In figure 1, there is a gap between the upper right group and lower left group. This gap is backed by results from the interview. Also, the edges in Fig. 4 have values of more than 3.0. The gaps between the groups are shown in the diagram. The value of 3.0 means that they know what they are doing as their work. The diagram shows that they don't know about the other groups' work. The findings from this diagram are consistent with the fact that the employees said they required lots of background information to understand each other.

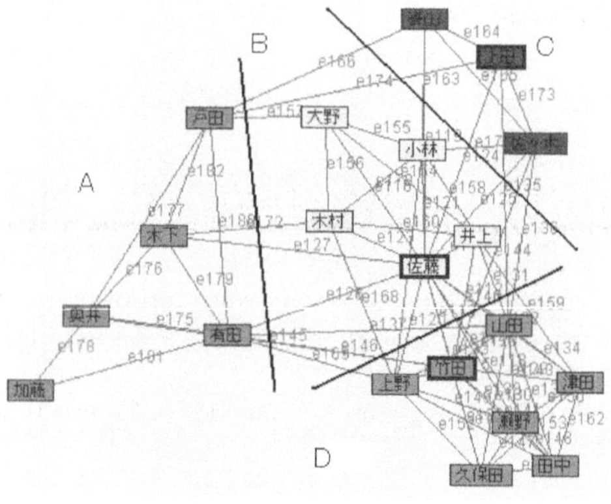

Fig. 4. The edges have values of more then 3.0

The number of samples was 21 and the average points rose from 2.7 (they knew each other by sight) to 3.1 (they know what each other are doing in terms of work). In the t-test, the value of p was $0.000012 < 0.0001 (0.01\%)$, which means the rise of the average value was statistically significant.

Each employee increased the value of his or her relationship with 6.8 employees on average. This means each employee built a deeper relationship with about 7 employees compared with before. The size of the circumcircle of employees in figure 2 is 17% smaller than the size of that in figure 1. This also backed the feedback from employees in section 4.4.

The purpose of the meeting was to develop relationships in which employees could consult with each other and we believe that purpose was achieved. This was also backed by the feedback from employees mentioned.

4.3 Gap between Organizational Policy and Personal vision

Theory of Motivation. In this section, we describe the method to increase members' intrinsic motivation. We discovered a theory of motivating members for knowledge sharing and built it up as a method.

In the interviews, we asked members about their own visions and the means to achieve them. Then, we discovered the means to achieve their own visions were in close agreement with the knowledge sharing of organizational policies. The purposes of knowledge sharing were improvement of profit and loss or improvement of efficiency. The members recognized these purposes as the organization's issues. On the other hand, the personal visions of the members were concerned with working more easily or developing a relationship built on trust with customers. When they analyzed their own visions, the means to achieve their visions were reusing someone else's documents or standardizing development processes. These means were organizational policies themselves. We discovered that when members analyzed the means to achieve their personal visions, they discover that means are connected to their organizational policies. We considered this theory was useful to promote the members' motivation for knowledge sharing. And we built it up as a method to increase intrinsic motivation. We called it Vision-Mean Analysis.

Experiment. We did an experiment to verify the Vision-Mean Analysis and whether it motivates the members' for knowledge sharing. We applied the method to fifteen system engineers of two different organizations and showed them the diagrams like Fig. 4.

As a result, it was affirmed by all fifteen members that the means to achieve their own visions were connected to the organizational policies. For example, one member held up increasing customer satisfaction as his own vision. And that member thought that providing a high-quality system was a means to achieving his vision. Furthermore, the member thought that knowledge sharing and explicating his own tacit knowledge was a means to achieving his vision. So, these means were the organizational policies in themselves. And we interviewed the members of one organization after the experiment, and got the following positive comments.

- "I couldn't usually understand the intention of the organizational policy; however I can understand the relationship between my own work and the policy."
- "I usually make some documents for customers by myself. I will try to re-use other people's knowledge from now on."
- "I think it is important to motivate the members for knowledge sharing. So we should have an opportunity to consider knowledge sharing like this."
- "I found the means to achieve my own goal were connected to the means to achieving the organization's policy. So I can recognize that the organization's policy is my own problem."
- This result shows that the Vision-Mean Analysis and the diagrams could motivate the members for knowledge sharing.

5 Summary

In this paper, we described a tool which gives a layout of employees as nodes and shows the cooperative relationships between them as edges. It visualizes the static structure of cooperative relationships and the dynamic structure of relationships as a sequence of static structures and the difference of static structures. We also described how the tool works in a case study. The case study was based on our field research on an organization that develops software. We compared the field research results with findings from the tool's visualization based on a questionnaire given to the employees. The time needed to answer the questionnaire was between three and five minutes on average and this is much shorter than the time taken for observations and interviews. But the many findings from the diagram drawn up using the results of the questionnaire back the findings of the survey results based on interviews.

As shown in section 4.4, in an analysis of the field research, we thought it was a big issue that employees knew each other only by sight. But such employees were in the minority and they did know each other, but they didn't know each other's work well. That is an example of how field research based on interviews can be misleading. The figure showed the barriers between the groups more notably. Therefore, managers in the field can use this tool and get suggestions on how to improve communication in their organization. Looking at changes in diagrams is easier to understand than listening to opinions in the field through interviews, such as "I can now consult other team employees".

We will continue to investigate this organization to help the measures and guidelines take root. In addition we would like to apply this tool to other organizations to enhance the features of this tool and improve its accuracy.

References

1. Senge, P.M.: The Fifth Discipline: The Art & Practice of the Learning Organization. Currency (2006)
2. Ugai, T., Aoyama, K., Arima, J.: A mathematical model of knowledge transfer and case studies. The International Journal of Knowledge, Culture & Change Management 7, 9–17 (2007)
3. Aoyama, K., Ugai, T., Arima, J.: Design and evaluation a knowledge management system by using mathematical model of knowledge transfer. In: Apolloni, B., Howlett, R.J., Jain, L. (eds.) KES 2007, Part II. LNCS (LNAI), vol. 4693, pp. 1253–1260. Springer, Heidelberg (2007)
4. Wasserman, S., Stanley, K.: Social Networks Analysis: Methods and Applications. Cambridge University Press, Cambridge (1994)
5. Borgatti, S., Everett, M., Feeman, L.: UCINET IV Version 1.0 User's Guide. Analytic Technologies, Columbia, SC (1992)
6. Carley, K.M., Pattison, P., Breiger, R.L.: Dynamic Social Network Modeling and Analysis: Workshop Summary and Papers. Natl. Academy Press (2004)

7. Rubin, R.B., Palmgreen, P., Sympher, H.E. (eds.): Communication Research Measures. Larence Erlbaum Associates, Mahwah (2004)
8. McGrath, J.E.: GROUPS: Interaction And Performance. Prentice-Hall, Englewood Cliffs (1984)
9. Eades, P.: A heuristic for graph drawing. Congressus Numerantium 42, 149160 (1984)
10. Girvan, M., Newman, M.E.J.: Community structure in social and biological networks. Proc. Natl. Acad. Sci. 99, 8271–8276 (2002)
11. McClelland, D.C.: Human Motivation. Cambridge University Press, Cambridge (1988)
12. Misue, K.: Anchored Maps: Visualization Techniques for Drawing Bipartite Graphs. In: Jacko, J.A. (ed.) HCI 2007. LNCS, vol. 4551, pp. 106–114. Springer, Heidelberg (2007)

Designing for the Next Generation: Generation-Y Expectations

Patañjali Venkatacharya, Sean Rice, and Lulit Bezuayehu

Oracle USA Inc., 200 Oracle Parkway
Redwood Shores, CA USA 94065
{Patanjali.Venkatacharya,Sean.J.Rice,
Lulit.Bezuayehu} @oracle.com

Abstract. In designing applications for the next-generation workforce, we considered the requirements of future workforce members (Generation Y). A study was conducted using a wants and needs analysis to analyze a consumer trend toward the use of social networking tools. A significant proportion of future workforce members expected to see consumer social-networking style features in their enterprise applications. The data highlighted specific requirements that provided some further insight into the expectations of future enterprise application users.

Keywords: Design, User Experience, Generation-Y, Enterprise Applications.

1 Introduction

McKinsey and Forrester, two noteworthy high technology industry analysts, conducted studies of workplaces worldwide in 2005, 2007, and 2008, and found that the very nature of work has changed dramatically, as technology has become a central point in people's lives, both at work and at home [1, 2, 3, 4]. The increasing use of personal computers (PCs) and mobile devices outside the workplace, namely at home, has changed the expectations of all end users in recent years. Attempting to accurately classify today's users vs. future users is likely to be controversial – we may never arrive at an answer that is satisfactory to everyone [5, 6]. But what is clear from studies by McKinsey and Forrester is that today's users, however they may be classified and to whatever generation they may belong, collectively expect much more from their enterprise applications than they did just 15 years ago [1, 7, 4].

This commonality among user expectations across multiple generations has motivated us to seek answers to questions around the future direction of enterprise applications.

2 The Changing Face of Work

Just twelve years ago, less than 25% of adults in the United States used the internet at home, work, or school [8]. Today, the number of "wired adults"[1] in the United States

[1] Pew Internet & American Life Project refers to these "wired adults" as those making use of one or more new communications technologies, including internet, email, and mobile phone. A full 73% utilize all three basic tools regularly.

M.J. Smith and G. Salvendy (Eds.): Human Interface, Part I, HCII 2009, LNCS 5617, pp. 190–196, 2009.

alone is a staggering 96% of all working adults (156 million), of which more than 55% have broadband access in the home [9].

When enterprise applications emerged in 1995, most end users were exposed to computers only in the workplace. And these early enterprise applications had one important and common goal: to provide more control for management over things such as employee benefits, cost of operations, and overall corporate expenses – what McKinsey calls "transactional work" [7]. Today, almost half (45%) of those employed work at least sometime from their home [9]. And, unlike the early enterprise applications, most popular consumer applications today turn "command and control" on its edge. They instead focus on empowering end users to do more, take greater ownership of their methods as well as their results, and do things such as "share" and "relate" with one another [10, 11]. McKinsey refers to this kind of work as "tacit work" [7]. In this study McKinsey determined that: "*... the way companies deploy technology to improve the performance of the tacit workforce is very different from the way they have used it to streamline transactions or improve manufacturing.*" [7]

Will the increased use of consumer internet applications in the home, and the reality that companies are deploying different tacit technologies at work, have a direct impact on the enterprise applications expectations of these users? If this is true, in just a few short years, these gaps may grow even greater, as even more students from colleges and universities across the world enter the workforce for the first time – expecting their enterprise applications to look, feel, and behave more like their everyday consumer applications. It seems appropriate for us to understand the true expectations of these users and determine what impact (if any) their expectations might have for future considerations in enterprise application design.

2.2 Generation Y

In considering these shifts in the workplace [2, 11], we asked these questions: How do we design for the next generation? Have users' core requirements and expectations changed so much that we need to start looking closely at the consumer space for indications of what to do in the enterprise?

In order to address these questions, we needed to first take a closer look at this future workforce and understand what their expectations might be. If we can understand what this future workforce will want and need in their enterprise applications, we will be able to design for the *majority* of the next generation of enterprise users[2]. As a result, we conducted a wants and needs analysis of Generation-Y users – those born between 1978 and 1986 [12] – to better understand what their true expectations are in terms of enterprise applications.

The way that Generation Y views and uses technology is different from the way earlier generations have used technology. Generation Y is an example of a primarily tacit workforce [7]. Characteristics of this cohort include a sense of entitlement, a desire for customization, a team orientation, and a high value on peer opinion. Many in this generation are natural multitaskers, capable of efficiently using multiple technologies at the same time [13]. They are constantly surrounded by technology and

---

[2] This in turn assumes we will see similar trends in common expectations of both future users and today's users, as McKinsey outlines in its changing workplace studies 2005-2008.

desire immediacy. They also want and expect access to tools that will fill those needs: cell phones, Facebook, and instant messaging [3]. Generation Y expects to communicate within the workplace across a flattened hierarchy, in the same way they communicate outside the workplace. They want to not just inform, but to increase engagement. This generation expects not just directions, but dialog [14].

Workplaces are generationally diverse, and different generations have different values and needs in the workplace [12]. The sheer size of Generation Y, with 70 million members (compared to 76 million for baby boomers), will certainly have a strong influence on the future workplace [15].

3 Generation-Y Social Networking Case Study

One technology trend, social networking, will have a tremendous impact on the workplace as Generation-Y workers enter the workforce. According to the Wall Street Journal, "social networking is just one of many consumer technologies, including blogs, wikis, and virtual worlds, to cross over into the corporate world" [16]. The entry of Generation Y into the workforce has pressured employers to enable communication tools such as online social networks in the workplace. In response to this trend, enterprise software companies have introduced social networking products such as KickApps, Passenger, and hi5 [17]. The market intelligence firm IDC predicts that the market for enterprise social networking applications will grow to $428.3 million in 2009 [18].

To understand the requirements of the Generation-Y worker population, particularly with respect to social networking, the Oracle Application User Experience team conducted a study targeting this group. A wants and needs study was conducted in December 2007 and involved Generation-Y participants from three geographic locations. Wants and needs analysis is a structured brainstorming method used to elicit user requirements for a product or service [19]. Our findings revealed that there is indeed an impact of social networking technology in the workplace. Generation-Y workers want and need some of the same social networking features that they are accustomed to using outside of work. They often expect these same features and functions to be available in their enterprise applications.

3.1 Participants

Three wants and needs sessions were conducted (Denver, Colorado; Redwood City, California; and Reading, UK). Four to eight people participated in each session, and a total of nineteen people participated in the study overall. To qualify for the study, participants had to:

- Be adults under 30 years of age (members of Generation Y) who use consumer social networking sites or applications.
- Use a self-service human resources (HR) application.

Each session lasted two hours, and participants were compensated for their time with a $100 value gift card.

3.2 Procedures

Participants were asked to sign informed consent and nondisclosure forms, then the scope of the study was explained: a brainstorming session regarding social networking in a work context. Social networking was defined as the "creation of an online community where a person can gather information, build contacts, and interact with others." After social networking was defined, participants were asked what consumer social networking Web sites they currently use and how they use these products.

Participants performed a brief 5-minute practice brainstorm, regarding the characteristics of their ideal toaster. This practice session accustomed participants to the flow of brainstorming.

Each brainstorming question concerned a different aspect of social networking: tasks, features, information, and system characteristics (see Table 1). Participants were given the first brainstorming question. After the participants had run out of ideas for the first brainstorming question, they were given the second brainstorming question. This process was repeated for the remaining questions. The ideas generated were recorded in a text document, which was projected onto the screen in the front of the session room for all to see.

Table 1. Brainstorming Questions

Number	Question
1	What kinds of social networking / collaboration **tasks** would you like to perform in a work context?
2	What social networking **features and information** would you want?
3	What are the **characteristics** of your ideal system?

After participants completed the brainstorming for all questions, they were asked to individually pick their top five items from the pool of brainstormed ideas and identify the social networking characteristics, features, and tasks they personally would like to have in a work context. After all of the participants had identified the five items that they wanted most in a work context, the process was repeated to identify the top five social networking characteristics, features, or tasks that participants would like to have in their HR applications.

3.3 Analysis

Data from the three sessions were pooled, and each of the items in the top five worksheets for the work context exercise was given a participant identifier and sorted into groups based on verbatim content. If two or more of these groups were similar or used different wording for the same idea, they were combined into one group. Duplicate items were removed to avoid counting multiple votes from a participant. The percentage of participants wanting or needing each item was calculated.

This process was repeated for the top five worksheets from the HR prioritization exercise. Items were ranked to produce two prioritized lists, one for social networking items desired in a work context and another for social networking items desired in HR applications.

3.4 Results

Top Wants and Needs for Social Networking Features in the Workplace. The following items were identified as desirable features to have in a work context by greater than 35% of all participants:

- The ability to see skills and qualifications of colleagues
- The ability to view and schedule meetings
- Mini-feeds (RSS feeds[3])—information of interest to the employee in his or her role, that is, "what's new" information, training information, and project updates

Top Wants and Needs for Social Networking Features in HR Applications. The following items were identified as desirable features to have in HR applications by greater than 35% of all participants:

- A personal profile (including data such as contact information, skill set, role in the company, and personal interests or hobbies)
- Security—the ability to secure information from unauthorized individuals or parties outside the company
- Training recommendations—the ability to view training that others have taken
- Reminders and updates

Based on the study and the findings, Oracle Application User Experience confirmed that a crossover of personal technology use to the workplace will indeed occur. We determined that Generation-Y workers expect and want to have the same features and functionalities in enterprise applications that they are accustomed to using for their social interactions and connections outside of work.

4 Implications for Enterprise Applications

The next-generation workforce is accustomed to always being connected to peers and expects to have the same technology available in the workplace that they have outside of work. In order to enable knowledge workers to work smarter, they need to have access to the right information to complete their tasks and need to be able to identify the right people to collaborate with to meet their objectives. The findings of the wants and needs study indicated that participants wanted to have access to the information and people they needed to complete work both in the workplace and outside work.

[3] RSS (Really-Simple Syndication) is a family of Web feed formats used to publish frequently updated works—such as blog entries, news headlines, audio, and video—in a standardized format. (From: http://en.wikipedia.org/wiki/RSS_feeds)

Specifically, participants wanted their tools to not just provide more information, but to provide the right information based on context. They wanted to see who was best qualified and available to help them complete a task, and wanted tools that enabled them to easily collaborate with others. Applications that provide easy access to these capabilities could benefit not just members of Generation Y, but also baby boomers and other demographic groups.

4.1 Next Steps

One area for future research is identifying enterprise social networking features that users would want to have available from a mobile device. Some participants in the wants and needs study indicated that a key characteristic of their ideal enterprise social networking application would be mobile capability. Oracle plans to investigate potential mobile applications user requirements as part of a set of ongoing studies on mobile device use in the workplace.

Another question that could be addressed with future research is whether the results of the Generation-Y wants and needs study, which was conducted in the United States and United Kingdom, apply across cultures. A recent global study predicts that within just a few years China will surpass the United States as the "most wired" country on the planet [20]. Perhaps, after all, we may see similar patterns of interaction across much of the world.

5 Conclusions

End-user expectations of enterprise applications have dramatically changed in the last two decades. The changing work landscape throughout the world, brought on primarily by the increasing use of PCs, the internet, and mobile devices at home as well as at work, has created a major shift towards more consumer-driven experiences. The main questions here have been: Who is this shift really affecting the most? Is this shift impacting the requirements for designing next-generation enterprise application user experiences? Our own studies of Generation-Y users utilizing social networking capabilities indicate that the landscape for all users has indeed changed the design requirements for enterprise applications. The enterprise is now playing catch-up to the consumer space, as end users of all kinds are expecting similar experiences in their everyday enterprise applications.

References

1. Beard, D., Schwieger, D., Surendran, K.: Preparing the millennial generation for the work place: how can academia help? In: Proceedings of the 2008 ACM SIGMIS CPR Conference on Computer Personnel Doctoral Consortium and Research, pp. 102–105. ACM, New York (2008)
2. Business-iTnews Australia, Social networking not just a Gen Y 'thing' (2008), http://www.itnews.com.au/news/78059,social-networking-not-just-a-gen-y-%E2%80%98thing%E2%80%99.aspx

3. Courage, C., Baxter, K.: Understanding Your Users. A Practical Guide to User Requirements Methods, Tools and Techniques. Morgan Kaufmann, San Francisco (2005)
4. Forrester Research.: Designing a Social Computing Strategy to Attract Gen-Y Users, Forrester Research Report (2008)
5. Hicks, J.: Looking for the magic formula: evaluating our IT communication mix. In: Proceedings of the 35th Annual ACM SIGUCCS Conference on User, SIGUCCS 2007, pp. 169–171. ACM, New York (2007)
6. Howard, C.: Tips for Social Computing in the Enterprise. The Burton Group, CIO.com. Midvale, Utah (2008)
7. HR Magazine, After X Comes Y - echo boom generation enters workforce (2008), http://findarticles.com/p/articles/mi_m3495/is_4_46/ai_73848314/print?tag=artBody;col1
8. IDC (2008), http://www.idc.com/getdoc.jsp?containerId=prUS20852407
9. Johnson, B., Manyika, J., Yee, L.: The Next Revolution in Interactions. The McKinsey Quarterly 4, 20–33 (2005)
10. Lowell, B., Claudia, J.: The 21st-century organization. The McKinsey Quarterly 3, 24–33 (2005)
11. Lowell, B., Matson, E., Weiss, L.: Harnessing the power of informal employee networks. McKinsey Quarterly (2007)
12. Mac an Airchinnigh, M.: Classifying the user. SIGCHI Bulletin 14(2), 3–8 (1982)
13. McIntosh-Elkins, J., McRitchie, K., Scoones, M.: From the silent generation to generation x, y and z: strategies for managing the generation mix. In: Proceedings of the 35th Annual ACM SIGUCCS Conference on User Services, SIGUCCS 2007, pp. 240–246. ACM, New York (2007)
14. McKinsey & Co. Building the Web 2.0 Enterprise Global Survey (2008), http://www.mckinseyquarterly.com/Information_Technology/Management/Building_the_Web_20_Enterprise_McKinsey_Global_Survey_2174_abstract
15. Pew Internet & American Life Project Surveys. Percentage of U.S. Adults Online. Pew Research Center for People & the Press Reports, Washington, DC (2000—2006)
16. Pew Internet & American Life Project. China's Online Population Explosion: What It May Mean for the Internet Globally... and for U.S. Users. Washington, DC (2007)
17. Pew Internet & American Life Project. Home Broadband Adoption 2008. Washington, DC (2008)
18. Pew Internet & American Life Project. Networked Workers. Washington, DC (2008)
19. Potosnak, K., Hayes, P., Rosson, M., Schneider, M., Whiteside, J.: Classifying users: a hard look at some controversial issues. In: Proceedings of the SIGCHI Conference on Human Factors in Computing Systems, pp. 84–88. ACM, New York (1986)
20. Wall Street Journal dot Com. Social Networking Goes Professional (2008), http://online.wsj.com/article/SB118825239984310205.html

A Study on a Method of Support for Improving the Motivation of Employees

Daisuke Yatsuzuka[1], Yumiko Taguchi[2], and Tsutomu Tabe[1]

[1] Department of Industrial and Systems Engineering, College of Science and Engineering, Aoyama Gakuin University, 5-10-1 Fuchinobe, Sagamihara, Kanagawa 229-8558 Japan
[2] Department of Business Administration and Communication, Shohoku College, 428 Nurumizu, Atsugi, Kanagawa 243-8501 Japan
c5607155@cc.aoyama.ac.jp, taguchi@shohoku.ac.jp, tabe@ise.aoyama.ac.jp

Abstract. The purpose of this study is to clarify a method for generating a "priority order for improvement" and "specific ideas for improvement" to support the improvement of the motivation of employees, and to verify the effectiveness of the method by a questionnaire given to employees of an actual company.

Keywords: Employee, improvement of motivation, graphical modeling, text mining.

1 Introduction

Recently, more and more Japanese enterprises are expanding throughout the world, while at the same time overseas enterprises are expanding and increasingly coming over to Japan due to the globalization of business administration. Therefore, it is anticipated that competition between Japanese and overseas enterprises will further increase. Under these circumstances, the tendency of younger employees to change occupations, the loss of the aging baby boomers who are retiring, and a lower awareness and energy of employees have become subjects of discussion. In this social context, enterprises should improve performance with a limited labor force. Therefore, management should form and maintain higher motivation in order to increase the ability and work efficiency of existing employees. Currently, many studies on the improvement of motivation are being published. They are confined to an understanding of the mechanism and structure of motivation. As a result, with only these studies, the priority order for improving motivation within a company and specific contents of these improvements are hardly understood. Therefore, it is very difficult to improve motivation with these studies. Additionally, the establishment of a method to support the improvement of motivation in order to constantly correspond to requests from employees is required.

Consequently, this study clarifies a method for generating the "priority order for improvement" and "specific ideas for improvement" in order to support the improvement of the motivation of employees. In addition to that, effectiveness of the method is discussed by using a questionnaire given to employees of an actual company. This

M.J. Smith and G. Salvendy (Eds.): Human Interface, Part I, HCII 2009, LNCS 5617, pp. 197–204, 2009.

study proceeds through the steps shown below in order to establish this purpose. 1) Discussion of motivation and analysis methods through a bibliographic survey, 2) Clarification of methods for generating the priority order for improving motivation and specific ideas for improvement (hereinafter referred to as "motivation improvement support information"), 3) Analysis of a questionnaire given to employees of a company by a method proposed in this study, and 4) Discussion of the effectiveness of the method based on the results of the analysis.

2 Factors Forming Motivation and Its Analytical Method

At first, factors forming motivation in this study are discussed. Then, based on this discussion, an analytical method to generate motivation improvement support information is discussed.

2.1 Factors Forming Motivation

Herzberg proposed a two factor theory that the factors by which people are made satisfied (motivator factors) are different from the factors by which people are made dissatisfied (hygiene factors). However, motivator factors sometimes cause dissatisfactions and hygiene factors sometimes cause satisfactions. Therefore, this theory is not always applicable to the present time [1].

In this study, both motivator factors and hygiene factors are assumed to be factors which improve motivation. Both of them are considered to be motivation formation factors in this study.

Motivation formation factors are extracted from motivator factors and hygiene factors consisting of eight factors which Herzberg proposed as being closely–linked to motivation and interviews which Herzberg performed to investigate motivations.

Fig. 1 shows the contents of an interview about "recognition", which is one of the motivator factors and hygiene factors, and formation factors used in this study. The same process is repeated for seven other factors used to obtain formation factors. As a result, factors shown in the cause-effect diagram (Fig. 2) are used in this study. Major categories (ellipses) are called characteristic factors. Minor categories (rectangles) are called component factors. Characteristic factors consist of eight factors set up based on the motivator factors and hygiene factors (i.e., pay, growth, achievement, recognition, work, shop, working conditions, and personal relationship). Component factors are called detailed formation factors defining the characteristic factors. Component factors consist of 27 factors.

2.2 Analytical Method for Motivation Improvement Support Information

The analytical method for generating motivation improvement support information will now be explained.

Analytical method for calculating the priority order for improvement. The following three points should be considered for deriving the priority order for improvement. 1) Factors forming motivation include factors which may be difficult to

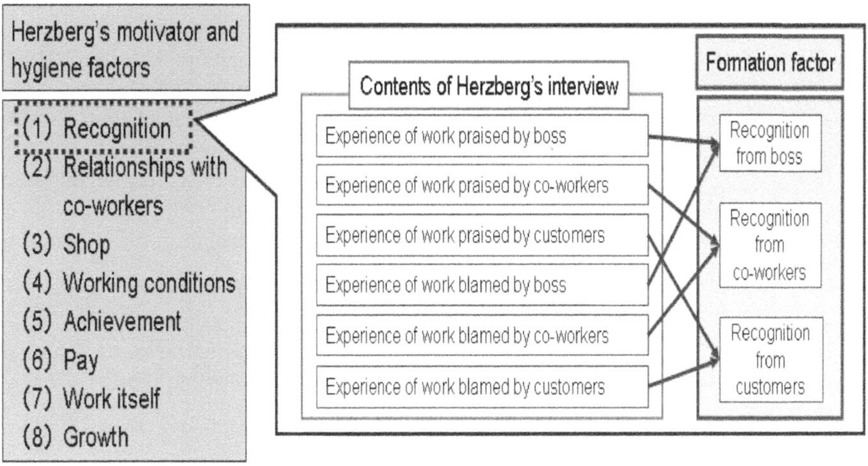

Fig. 1. Example of extraction of motivation formation factor

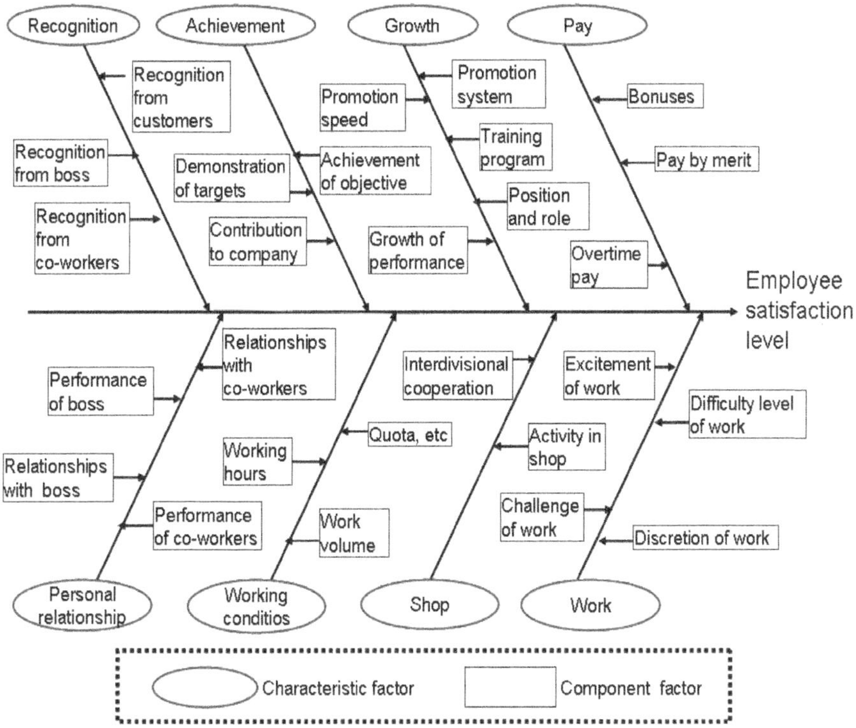

Fig. 2. Motivation formation factors

be observed directly as quantitative data. 2) The model should be set up freely because each company has different factors that form motivation. Therefore, the relationship among factors should be quantitatively analyzed considering these issues. 3) The relationship among factors should be considered carefully because factors forming motivation are intricately interrelated.

Covariance structure analysis allows 1) and 2) shown above [2]. Graphical modeling is a method which allows 3) shown above. Graphical modeling can organize the relationship among factors, as shown in Fig. 3, by partial correlation coefficients in an exploratory manner [3].

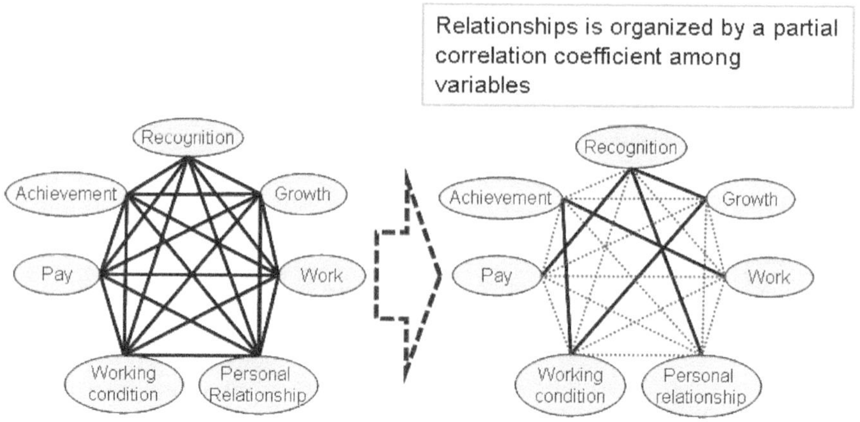

Fig. 3. Image of graphical modeling

Consequently, organized motivation formation factors are extracted by using graphical modeling. Based on the results, the relationship among motivation formation factors is quantified by covariance structure analysis. Eight characteristic factors that are hardly measured by covariance structure analysis are used as latent variables. Component factors are used as observed variables.

Analytical method for deriving ideas for improvements. The priority order for motivation formation factors shows factors to be improved. Opinions of employees should be considered to devise ideas for improvements concerning these factors. If the ideas for improvements reflect the opinions of employees in the field, the ideas will be received smoothly by those employees in the field and motivation will be improved.

Therefore, text mining that can quantitatively catch opinions from open questions is used in this study to consider local input. In this study, the questionnaire, in an open question format (Fig. 4), is performed on employees concerning factors with higher priority for improvement. Possible ideas for improvements are derived based on the results. The questionnaire consists of a definition of requests (specific images) and a description of the reasons to extract broader opinions.

What do you think of requests regarding "work volume"? Give concrete examples.	
Specific requests	Reason
( )	( )
( )	( )
( )	( )

Fig. 4. Example of the questionnaire regarding requests of employees

3 Method for Generating Motivation Improvement Support Information

In this study, the method for generating motivation improvement support information is divided into two phases, which are the priority order for improvement of motivation factors and ideas for improvement.

3.1 Priority Order for Improvement

The priority order for improvement is calculated by the following three rough steps. 1) Questionnaire regarding satisfaction level towards work. 2) Quantification of each factor affecting motivation improvement by using covariance structure analysis and graphical modeling. 3) Derivation of motivation improvement priority order.

Questionnaire regarding satisfaction level towards work. The level of satisfaction towards work is investigated by a six-level response for 27 component factors, as shown in Fig. 2.

Level of the effect on motivation improvement. At first, a confirmatory factor analytic procedure is performed based on the questionnaire data obtained shown above in order to understand the relationship among latent variables. Then, graphical modeling is performed by using a simple correlation coefficient obtained by the confirmatory factor analysis in order to organize the factors.

Calculation of the priority order for improvement. At first, a model is constructed based on the relationship organized shown above and covariance structure analysis is performed. Then, the total effect towards motivation formation factors is calculated based on the obtained path coefficient and the priority order for improvement is calculated based on the results.

3.2 Ideas for Improvements

Ideas for improvements are derived by three steps. Firstly, the questionnaire (Fig. 4) is performed concerning requests by employees for three factors which rank higher in the priority order for improvement shown above. Then, text mining is performed on the results of the questionnaire to execute a frequency analysis on certain words in the requests from the employees. Finally, specific ideas for improvements are derived from the extracted requests.

4 Discussion of the Effectiveness

The effectiveness of the methodology proposed in this study is verified in the following four steps:

- Step 1: The questionnaire is given to employees of an actual company. In this study, the questionnaire was conducted at a restaurant business. Business hours are from 17:00 to 24:00, seven days a week. The subjects included 31 employees working at the restaurant. The investigation was performed from November 21 to December 5, 2008.
- Step 2: The methodology proposed in this study is applied to the data obtained from the questionnaire to generate a "priority order for improvement" and "specific ideas for improvements" as motivation improvement support information.
- Step 3: From the viewpoint of goodness of model fit (GFI and AGFI), the generated priority order for improvement and specific ideas for improvements are then compared to a model which does not consider the relationship among the factors.
- Step 4: Motivation improvement support information is then proposed to the manager of the company to which the questionnaire is performed. The manager is asked to judge the effectiveness of the ideas for improvement.

4.1 Motivation Improvement Support Information Based on Questionnaire Data

Factors making up the top three priorities were "work volume," "achievement," and "labor hours", from highest to lowest.

The frequency of "reason" data in the questionnaire regarding employee requests was analyzed concerning "work volume", which ranked first. Fig. 5 shows the results. The frequency analysis shows that the labor amount is different depending on the section. Therefore, ideas for improvements include that busy sections should be given some help to eliminate this variation in labor amount by section.

4.2 Comparison to a Model Which Does Not Consider the Relationship among the Factors

Our model was compared to a model which did not consider the relationship among the factors. Fig. 6 shows the results. The model used in this study shows a rather higher GFI and AGFI, and has more validity as compared with models which do not consider the relationship among the factors.

Priority order for improvement was compared between our model and a model which does not consider the relationship among the factors. As a result, the factors which ranked first were different between the two models. We asked employees about "challenge", which rank first in the model which does not consider the relationship among the factors. They said, "We think the labor amount is a problem. However, there is no large problem concerning the challenge. A priority order which considers the relationship among the factors mirrors reality better."

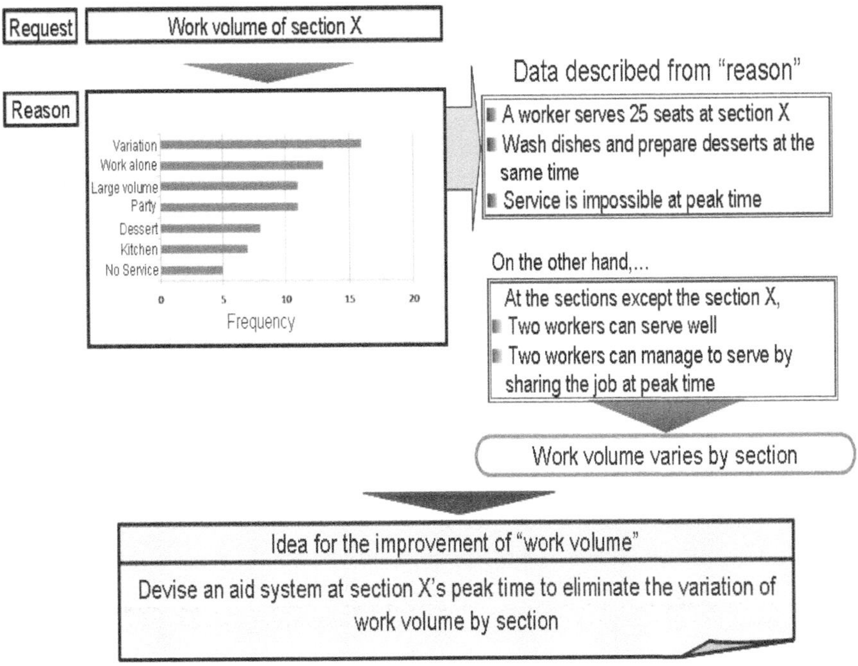

Fig. 5. One idea for the improvement of work volume

Therefore, the priority order for improvement considering the relationship among the factors proposed in this study is "a priority order for improvement with higher reliability."

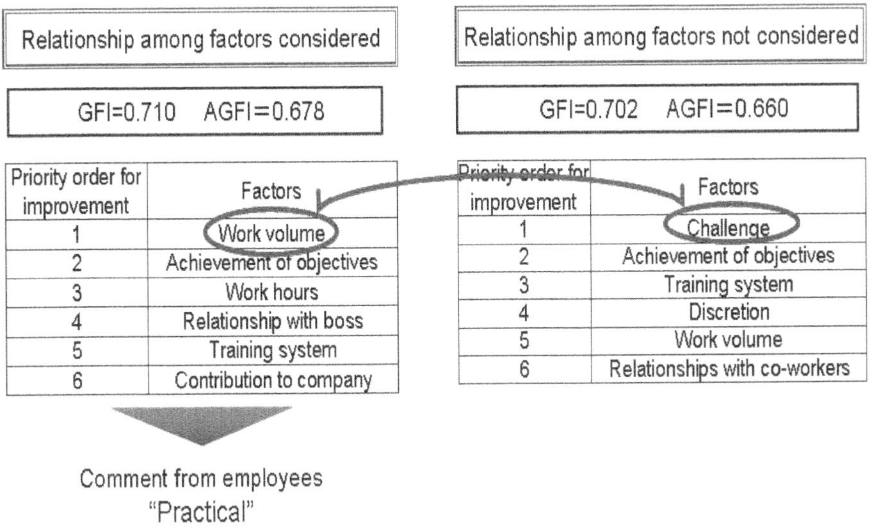

Fig. 6. Comparison to a model not considering the relationship among the factors

4.3 Motivation Improvement Support Information Evaluated by the Manager

The motivation improvement support information generated by the method proposed in this study was presented to the manager. The manager commented as follows:

- That is an idea of improvement derived from the questionnaire given to employees. Therefore, it is an idea of improvement considering the characteristics of the restaurant. It is expected that the idea would improve the actual situation.
- It is an interesting idea for improvement because we can understand relationships between co-workers which are not conceivable.
- Some ideas for the improvements are impossible in the current situation.

As shown above, there are some problems in that feasibility of the ideas for improvement is not considered. However, derivation of the ideas for improvements may support improvements that enhance motivation.

5 Conclusion

A questionnaire was given to employees of a collaborating company by using the methodology proposed in this study based on the discussion of effectiveness shown above. As a result, priority order for improvement and ideas for improvements are expected to have an effect.

As mentioned above, the method to conduct a "presentation of priority for improvement" and "proposal of specific ideas for improvement" to improve motivation has been clarified.

References

1. Herzberg, F.: The motivation to work. Wiley, Chichester (1959)
2. Joreskog, K.G.: Structural analysis of covariance and correlation matrices. Psychometrika 43, 443–477 (1978)
3. Dempster, A.P.: Covariance selection. Biometrics, 157–175 (1972)

Research on the Supplier Promise Management Based on the Lean Six Sigma

Wang Ying-chun[1,2], Wang Shi-jie[2], Liu Yong-xian[2], and Zhang-Xinmin[2]

[1] Sch. of Mechanical Engineering & Automation, Northeastern University, China
[2] Mechanical Engineering College, Shenyang University of Technology, China

Abstract. In the supplier management, the supplier promise management is one important area. The scope of the supplier promise is very widespread, such as the promise of HSE (health environmental safety), the promise of quality, the promise of price, and the promise of delivery performance etc. As one character of one enterprise the promise has become the most significant evaluation factor which was used to select the supplier for the upstream firm. So during the modern production and management, how to help suppliers to improve their performance to promise has become one key problem. Evidently the nature of this problem is usually complex and unstructured. The paper tried to utilize Lean Six Sigma to solve the problem of the supplier promise management. Such creates a new point of view to study and control the supplier promise management. Lean Six Sigma is an new advanced and high efficient method of management which combined the virtues of Lean production and Six Sigma Management. It focus on the subtle production and the strategic management simultaneously. The three key elements which were attach importance of the modern companies are velocity, cost and benefits, as well as the modern supplier promise management. To use the thinking of the lean Six Sigma in the supplier promise management will bring greater economical benefits than before. In this paper one example which is about the supplier performance to promise date is discussed. The results show that to apply the Lean Six Sigma into the supplier promise management is more practical and effective than the other managing ways.

Keywords: Lean Six Sigma, Supply Chain, Supplier Promise Management, Procession.

1 Introduction

In the modern production and management of the supply chain, the supplier promise management is an area of tremendous importance. The material and equipment supplied from the suppliers play an important role in the effective management of a company. Many issues in the supply chain are influenced by the supplier promise management. In the logistics decisions of a firm, the supplier promise management has a great influence on the supply chain design in terms of transportation and distribution planning and production planning. It is important to manage the supplier promise so that different objectives of the supply chain are achieved. Similarly, because the reliable supplier promise management can minimize the risk associated with the

M.J. Smith and G. Salvendy (Eds.): Human Interface, Part I, HCII 2009, LNCS 5617, pp. 205–213, 2009.

purchase, the associated costs increase with this approach, can minimize the waste associated with the production, the associated costs increase with this approach, can minimize the waste associated with the production, the associated benefits increase with this process, to make a study of the supplier promise management is very necessarily. Hence, the supplier promise management is to identify better performing suppliers in a supply chain. The supplier promise management has been considered as a complex problem in the literature due to the several following reasons:

1. Individual suppliers may have different performance characteristics for different criteria [1];
2. Suppliers may impose constraints on the supplying process so as to meet their own minimum order quantities or maximum order quantities that may be based on their production capacity;
3. There may be time constraints on the delivery of items. Within these time constraints, some criteria for the supplying the items may become important, while other criteria may not be the dominant ones [2];
4. Constraints related to suppliers' quota allocation, number of suppliers to employ, minimum and maximum order quantities, use of minority suppliers etc [3].

Supplier promise management is complex in nature and invites strategic decision of long-term implications.

2 Lean Six Sigma

Face on the commercial competition, the enterprise must focus on the velocity, cost and benefit simultaneously, as well as the supplier promise management. In the supplier promise management, the firm should attach importance to the key elements which were mentioned upper. Lean Six Sigma just is such a management method which takes account of these sides. Now more and more enterprises adopted Lean Six Sigma to manage their supply chain, include the production, inventory, logistics and suppliers. Lean Six Sigma is the most advanced managing method in the managing field. It combined the Lean Production and Six Sigma management into a new useful managing system. Originally, some large companies utilized the Six Sigma to manage their producing operation and acquired brilliant achievements. For instance, Motorola Company, GE, etc. These companies have founded their own enterprise kingdom with powerfully competitive strength. With the development of the production and management, firms found that Lean Production is more convenient than Six Sigma on some subtle or delicate manufacture procession. Gradually Lean Production and Six Sigma were combined into one management way. GE even had created its new Lean Six Sigma firm culture. The "promise" becomes the symbol of this great company. In the early years, the firm uses the Lean Production to carry the producing process out and to solve the problems emerged during the manufacture. The Six Sigma was continually used to manage the upper or overall management or some important projects managing. The Lean Production makes up some missed issues of the Six Sigma, for example, partial production and so on small and subtle problems.

1. **Six Sigma Management.** Six Sigma management was defined firstly by American Motorola Company. It is a set of proxess improvement method and system which

developed basing on many advanced theories and modes. It core of Six Sigma Management is DMAIC, that means: Define, Measure, Aanlyze, Improve, Control. Firms often use it to solve large and overall problems.

2. **The Lean production.** Lean Production was defined firstly by Japanese Toyota Company. It's core thinking is to eliminate any kinds of waste include each step of manufacture and management. Firms usually use the Lean Production to control the producing process and to solve the manufacturing problems. It's "JIT producing', "Zero stocks" made the production more effectively and more economically than before.
3. **Lean Six Sigma.** Lean Six Sigma is developed based on the Lean Production and Six Sigma Management. In the practice of the Lean Production and Six Sigma Management, the company found their respective superiority, then integrated them into one method to manage the enterprise. Its essence is to eliminate any waste and to diminish defects [5]. This integrated way has more strong power to operate the production and management, furthermore it can creat more astomishing benefits.

Many large companies adopted the Lean Six Sigma Management into their companies' circulation. The Lean Production and Six Sigma Management combine their own virtues, then develop stage by stage till become an advanced popular managing method.

The "DMAIC" means: define, measure, analogy, improve, control. As usual, the company uses the Lean production into the producing process to solve manufacturing problems, and used Six Sigma management into the upper or overall management or some important projects managing. The core of the Lean Six Sigma Management is the "JIT producing", "zero stocks", "3.4 defects percent million" [6].

In the supplier promise management, the enterprise uses the theory of Six Sigma Management to manage the suppliers' promise of the quantity, cost, lead time and transportation and so on. When the high degree of Lean Six Sigma Management is involved, due to its perfections and advancement, the supplier promise management based on Lean Six Sigma Management becomes one of the most effective management [4].

3 One Example of the Supplier Promise Management

This example is about the supplier promise dates and the data are collected form a firm. The supplier promise dates are very important to a firm, because as usual one company will make a serial decisions and production or process steps according to the supplier promise dates. If the promise dates are very accurate, the company will save huge stocks and enormous wealth and a large number of production-time even to considerable quantities people. So, to the company how to manage and improve the supplier promise dates accuracy is the most important matter [7].

3.1 Project Summary –Define the Project

1. **Problem statement:** According to the data from the company, the 7- day forecast accuracy of the supplier promise dates is 45%. This accuracy number is too low to enable the managers of the company to be satisfied. Accordingly this kind of production and management can't bring great beneficial result for the firm. To

improve the supplier promise management what steps will the firm take? Such transforming works which will be adopted are as follow:

2. **Project goal:** The goal of this project is to improve and control the accuracy of the supplier promise dates to 100%.
3. **Business case:** Inaccurate supplier promise dates affect internal planning and delivery commitments to the end customers of the company. These will decrease the benefits and destroy the relationship between the company and the customers.
4. **Project scope:** Includes the supply department, the quality department of the studied firm
   Includes PO's forcasted to be received during the current week
   Excludes POs created in the last 2 working days
5. **CTQ of the project:**
   7- day forecast accuracy
6. **Additional requirements to close this project:**
   Supplier Utilization of e-PO >= 95%
   At least data of last 4 consecutive weeks

3.2 Measure the Project

1. Baseline Data taken for performance to promise in the firm:
   DPMO = 545,455; Z-Bench = 1.39; Yield = 45% On Time to Promise

With the Fig.1, we can know the DPMO, Z-bench and the general situation of the company. The quantity information about the company is very value than any other things.

STARTING DPMO

Line	D	U	OP	TOP	DPU	DPO	DPMO	Shift	ZB
1	6	1	11	11	6.000	0.545	545.455	1.500	1.39

Fig. 1. The starting DPMO of a firm

2. The firm had 11 POs promised for that week
   In the week which was took to be analyzed, the firm had 11 promised POs.
3. Delivered 5 of those 11 promised

3.3 Analyze the Project

1. Initial Process to Update/Verify Promise Dates of the firm

From Fig. 2, we can find that the mortal wound of the initial process is: steps not followed, ePO not utilized properly (training issue). The process is too simple and many factors which will affect the firm aren't disired conscientiously. If we exploit the Lean Six Sigma Management into the process, we will design new production process and develop new management way. These will bring brand-new appearance and vigor to the company or enterprise.

2. The firm Causes Analysis

The fishbone chart is the popular tool to display and found the causes to the production and management of the firm. It can tell the causes into different varieties [8].

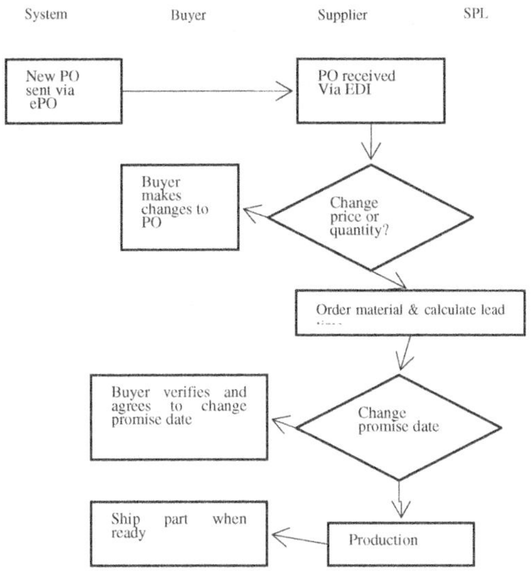

Fig. 2. The initial process not use Lean Six Sigma

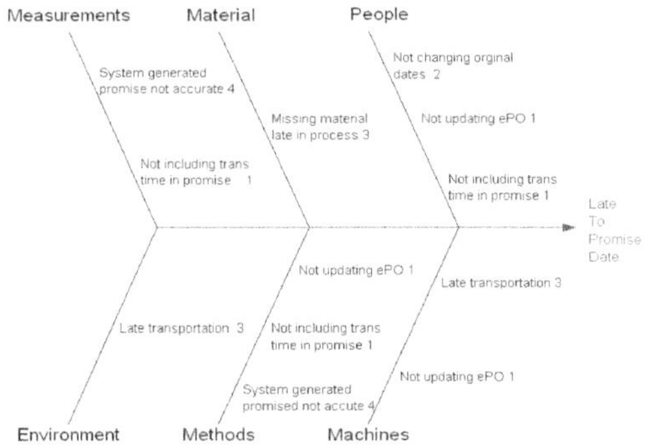

Fig. 3. The fishbone chart of "analyze the root the cause"

Table 1. The affection of the causes

	High Impact	Low Impact
Easy Implement	1	2
Hard Implement	3	4

By the upper fishbone chart of Fig. 3, we can get the causes that make the firm face the low forecast date. So we can focus on every cause to take the useful steps to improve the bad production and management. Table 1 proved the each cause made different influence.

3.4 Improve the Project

The firm's Improvement Plan:

1. Get the firm engaged and using ePO daily
2. ePO trained
3. Follow Fig. 4 for changing Promise Dates early-on in the timeline
4. Include 7 calendar days for delivery in original and updated Promise Dates

After the upper improvement steps, we amend the working flow process. Just reference the Fig. 4. The improved process looks more comprehensive and more specific. Evidently it's more effectively than the upper one (Please refer to Fig. 2).

How do we utilize the new process (Fig. 4) correctly? We can use the process map (Fig. 5). It is just used to present what is the process map for correct utilization of ePO. In the process map, we can find that the change is made at seventh step. Between fourth step to fifth step, do not propose change because the opportunity is not mature. The improvement phase is progressed step by step.

If promise dates are updated in this way, they will be accepted immediately in this management system. If promise dates are updated in any other way, they will not be accepted until the buyer acknowledges, and then the firm will get another amended PO from the buyer that the firm will have to acknowledge again. The firm will have to do double work!

By using the up process flow diagram (Fig. 4) and the process map (Fig. 5), we find: ePO Utilization started at 53% in Aug-02, >95% since Trajning Sep-02. The role is very quickly and obviously.

Table 2. The result of utilizing ePO

Supplier	2-Aug	2-Sep	2-Oct	2-Nov	2-Dec	2-Jan
The firm	53%	94%	100%	98%	100%	100%

3.5 Control the Project

In this phase the firm select and adopt such ways:

1. the final data:
   DPMO:0; Z-Bench:6.0; Yield:100%
2. the controlling way:

Table 3. The report of promise dates

12/16-12/22				12/23-12/29			
-1WF	Ship	Yield	Open	-1WF	Ship	Yield	Open
2	2	100%	0	7	7	100%	0
12/30-1/5				1/6-1/12			
-1WF	Ship	Yield	Open	-1WF	Ship	Yield	Open
6	6	100%	0	13	13	100%	0
1/13-1/19				1/20-1/26			
-1WF	Ship	Yield	Open	-1WF	Ship	Yield	Open
6	6	100%	0	7	7	100%	0

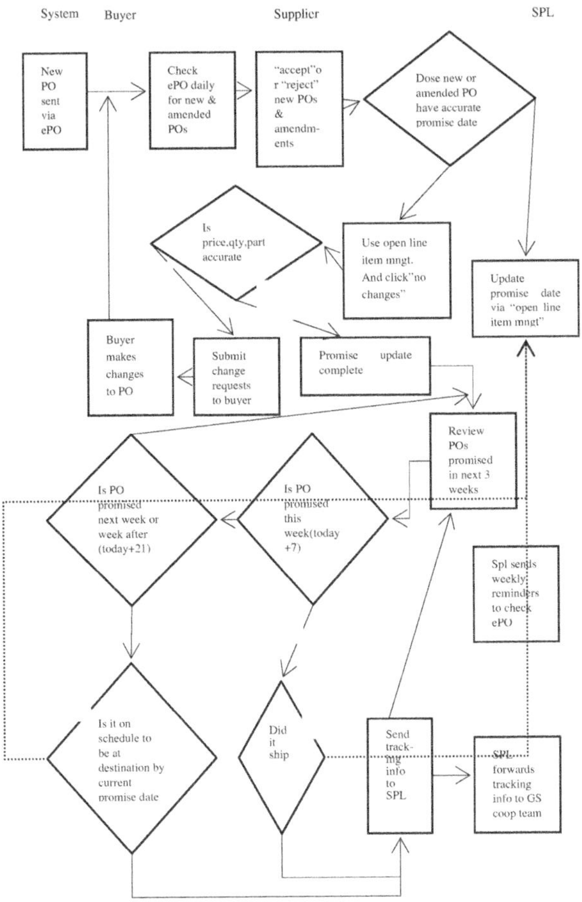

Fig. 4. Weekly/Bi-Weekly Flow after using Lean Six Sigma

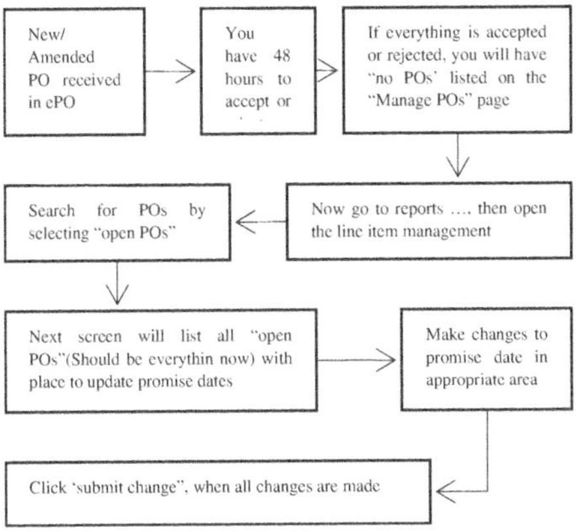

Fig. 5. The process map for the correct utilization of ePO

3. The performance after using the new management method

Table 4. The report of product performance

Characteristic	Defs	Units	Opps	TotOpps	DPU
1	6	11	1	11	0.545
2	0	34	1	34	0.000
Total	6			45	
Characteristic	DPO	PPM	Zshift	Zbench	
1	0.545455	545455	1.500	1.386	
2	0.000000	0	1.500	6.000	
Total	0.133333	133333	1.500	2.611	

4. The verification about the statistics

Table 5. The Statistical Verification

Unira	Opps	Defects	Good	Bad	Profess
11	1	6	5	6	Before
34	1	0	34	0	After

5. The control plan
   a. Daily e-PO usage
   b. Promise Date update according to improvement plan and process map
   c. Short-term: SPL will continue to send weekly PO reports and verify delivery to Promise Date for 30 days out;
   d. Long-term:
      GEPS Sourcing (Buyer/SPL) monitor e-PO usage
      Metric will continue to be monitored on sample basis
      Process will NOT be in Control if:
         >2 continuous weeks supplier is NOT at 100% VTP Yield;
         >2 weeks in a 5 week term supplier is NOT at 100% VTP Yield.

4 Conclusion

The supplier management is the key problem of supply chain management, so companies couldn't ignore the problem of the supplier promise management. To apply the Lean Six Sigma into the supplier promise management will develop a new method and way to manage the supplier promise. Such will bring the enterprise more and more benefits or economical results. The case that the paper quoted and discussed just proved the same points: to adopt Lean Six Sigma into the supplier promise management is very effectively and practicefully. The cole thinking of Lean Six Sigma: saving and making the defects minimum was used into the supply chain will bring large changes and enormous benefits. This kind of study and demonstration is very significant and useful.

Acknowledgements

The paper is completed after long time hard study and work. The authors gratefully acknowledge the financial support of the Provincal Science Council, Liaoning Province, China, under Project number 2006BAForA19.

References

1. Noman, E., Hotinton, S.H.: Customer centered Six Sigma, pp. 3–17. China Machine Press (2003)
2. Yao-qiu, W., xian-liang, S.: Supply chain management, pp. 92–102. China Machine Press (2006)
3. Wei, C.C., Chien, C.F., Wang, M.J.J.: An AHP-based approach to ERP System. Int. J. Prod. Econ. 96, 45–68 (2005)
4. Chopra, S., Meindl, P.: Supply chain management: Strategy, Planning, and operation, pp. 67–89. Prentice-Hall, Upper Saddle River (2001)
5. Erol, I., Ferrel, W.G.: A methodology for selection problem with multiple, confliction objectives and both qualitative and quantitative criteria. Int. J. perod. Econ. 86, 187–199 (2003)
6. Hui, L.: Complementary Lean production snd Six Sigma management. The Way on the Spot and Practixe 5, 94–97 (2004)
7. Lin, J.T., Chen, T.L., Tsai, T., Lai, T.J., Huang, T.C.: A SCOR-based methodology for analyzing and designing supply chain. Int. J. Electron Bus Manage 3, 4–12 (2005)
8. Naumann, E., Hoisington, S.H.: Customer Centered Six Sigma: Linking Customers, Process Improvement, and Financial Results, vol. 2, pp. 134–141. China Machine Press, Beijing (2004)

Part II

Product Design and Development

Supporting Scenario-Based Product Design and Its Adapters: An Informal Framework for Scenario Creation and Use

Irene Anggreeni and Mascha van der Voort

University of Twente, Laboratory of Design, Production & Management,
P.O. Box 217, 7500 AE Enschede, The Netherlands
{i.anggreeni,m.c.vandervoort}@utwente.nl

Abstract. This paper proposes a support tool for designers who have realized the potential benefits of using a scenario-based approach, yet need a more concrete guidance for its implementation. As an informal framework for scenario creation and use, the tool supports three areas: (1) documenting the large amount of design information, (2) making sense of and summarizing the information by creating scenarios, and (3) sustaining an integrated use of the scenarios throughout the design process. By making explicit and supporting the process of using scenarios, we expect that designers can better structure their design information, communicate their ideas and confirm their rationales. Our ongoing work includes the verification of the tool concept and the development of a testable prototype.

Keywords: scenario based design, product design, scenario based product design, scenario generation, design tool.

1 Introduction

Designing consumer products nowadays is becoming more complex. In the effort to accommodate diverse users and their use situations, designers have to take into account a product's dynamic use situations [1] to ensure a good usability in all possible settings, for varied user characteristics, etc. To cope with different design aspects, design teams often have to integrate different disciplines and backgrounds, which in turn poses a challenge in communication. Stories have been an inseparable part of communication as it is natural for us to communicate by telling them. Scenarios in this research context refer to a more structured, exploited version of stories. Whether expressed or not, scenarios are inherent in design processes and play a special role as a communication tool. Their use has been made more explicit with the advance of scenario-based design (SBD) within software development [2-5] as well as within product design domain [6-8]. Nevertheless, SBD as a methodology remains largely unguided and is perceived as an eclectic combination of existing design techniques and appropriate scenario uses. Consequently, industrial adapters of SBD are often discouraged because they have to speculate their own eclectic approach in applying

M.J. Smith and G. Salvendy (Eds.): Human Interface, Part I, HCII 2009, LNCS 5617, pp. 217–226, 2009.

scenarios. A concrete and practical guidance in carrying out scenario creation and use will support them to conduct SBD effectively and efficiently.

There exist formal frameworks that integrate scenarios in requirements engineering (e.g. [9-11]) as well as tools that explicitly address the generation and management of scenarios (e.g. [12-15]). Nevertheless, these frameworks and tools tend to focus on a specific activity or phase within the development process, and not the complete design project. To actively support SBD adapters, there needs to be a framework that sustains an all-round scenario-based approach throughout the design process, with the in-between results being optimally used. This framework should also fit in with the storm of creativities and dynamics of product designers. Such support is currently not available to the industry. We propose an informal framework that guides sustainable information documentation and scenario generation. This paper will explain how a scenario generation support tool could help an easier transition to a scenario-based design practice.

2 Approach: Adaptation of SBD Theories into Design Practice

Existing scenario-based design frameworks are rather theoretical and not proven in practice. Dealing with different types of design projects, practitioners nowadays have to take an eclectic approach to apply scenarios where they see fit. As the first step in our research, we have studied literature to get acquainted with the developing state of scenario-based design. An overview of scenario uses in the design domain has been formulated into a scenario classification [16]. This classification has helped us to pinpoint the often-not-explicit process of using scenarios in our target group users. By making the knowledge transparent to both researchers and designers (our target users), we were able to elicit their underlying reasons and concerns in using scenarios in real projects. The designers' involvement has given insights into their practice-grounded requirements that enable the extraction of criteria for the support tool.

Our proposed tool targets product designers in small- to medium-sized design companies. Interaction among designers in this target group is direct and transparent, which makes it easier for us to observe and subsequently support the information exchange. Contacts with designers from different backgrounds (roles, experiences, project types) have been established. From a series of workshops and questionnaires involving our contacts, we found the following challenges in design practices: (1) documenting the vast amount of relevant design information, (2) organizing the information into manageable and meaningful pieces and (3) making sure important and relevant information is traceable and useful to support design decisions. Creating scenarios is a partial answer to these challenges. However, due to a lack of guidance the created scenarios are often scattered and their use not fully integrated in the process. Furthermore, designers also wish for a more efficient and stimulating way of documenting information and composing scenarios.

We therefore focus on the backbone activities of SBD, i.e. the documentation of design information as scenario building blocks and the creation of scenarios to keep this information alive. Designers wish 'guidance' instead of 'prescribed steps' in conducting a scenario-based approach. The flexibility of our proposed framework will give room for a living approach which is strongly connected to each design phase.

3 Functionality

A structure for documenting design information is an essential part of the tool. It could then guide designers to create meaningful scenarios that give a better picture of the various use situations. Furthermore, dealing with scenarios and requirements, the design team and stakeholders should be able to get an overview as well as detailed views of information during decision making.

To introduce the actors involved in our support tool, we summarize their roles and interests in a design project. We assume that these typical roles in a design project are generic enough and can be recognized in different project settings.

- *Project manager* (PM) is the one who manages the course of a design project. During the initial stage of the project, PM is responsible for defining an agreement with the client and possibly other stakeholders on subjects such as project time-frame and budget. An important part of this agreement is the design approach which the design team will use to accomplish the assignment. As it directly relates to cost, the client could wish to remove parts of the proposed approach to cut expenses.
- *Client* is the one giving an assignment to the design company. The client has the biggest interest in the design project in terms of business profit. During the initial stage of the project, the client defines a clear orientation and project scope with the design team. He also has rich information, for instance about the target users/buyers, a rough product idea and relevant technologies, potential sources of information (e.g. people, documents, standards or regulations) and competitor products of similar line. Therefore, even in the early stage predefined requirements are often already constructed. This initial information however could have been assumed by the client and therefore needs to be verified through further research.
- *End-users* are the persons or professionals who are going to use the product. They have the domain skill and knowledge that is resourceful towards the design. The design team needs to obtain this tacit information to be able to design a useful product. A variety of participatory design techniques/methods can be used to involve these people in design activities.
- *Designers* are members of a design team and could be of diverse functions such as engineers, software developers, usability specialists, business analysts, ergonomic experts, etc. They conduct design activities to fulfill the assignment as agreed by the PM and client. Even with the design approach already defined, designers are more stimulated when they have freedom to employ their multitude of skills, creativity and perspectives in executing design techniques. Therefore despite their different backgrounds and functions, we refer to these creative problem-solvers in a design project as simply *designers*.

The set of scenarios below explains briefly the context for which the tool offers support. The functions described in this paper are select ones that highlight the scenario identification and generation, and therefore do not show the complete functionality. In each section, a set of illustrations and intended use scenarios are presented to portray the interaction with the tool. Figure 1 shows the first step using the tool, in which its components are introduced by means of an example dealing with the design of a bike accessory.

A new project Urban Mobility kicks off
The project 'Urban Mobility' has reached its final agreement with client Speeda Inc. The design team, lead by Bob, is quickly preparing a set of action plans to tackle the initial phase. For instance, Marie is visiting an exhibition of bicycle latest gadgets/technology to survey the market. John, another designer in the team, is recruiting current users of bicycle accessories for carrying luggage, to get insights on their way of using existing products.
During the focus group session...
John invites two users to a focus group session. The session takes place in an informal setting at the users' place. The discussion deals with the users' daily activities, situations and events which occur around these activities and the users' wishes and needs for improvement.
At the exhibition...
Marie collects brochures, takes pictures and notes comments about the latest bicycle gadgets and technologies that may be relevant with the project. She also learns about the benefits and limitations of the latest products from talking with the people at the exhibition.

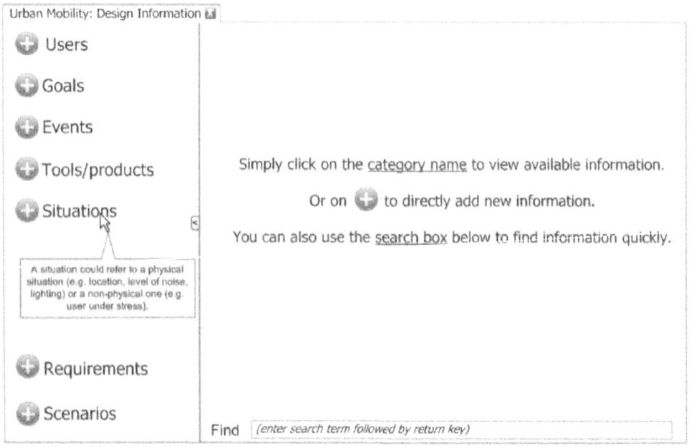

John tries out the tool for the 1st time
John sees neat categories to group information on the left. He familiarizes himself with each one of them. When he hovers the mouse pointer on a category name, a brief explanation shows up. Now John understands what kind of information should be documented under each category.

Fig. 1. The welcome screen provides simple categories that guide the design team to document relevant information

3.1 Design Information Documentation: Identifying Scenario Elements

During the initial phase of a design process, understanding the global view of information and making sure the validity of all information up to now are essential. A combination of common design methods/techniques such as ethnography, observation, focus group, interview, participatory design can be used to capture all elements of product use, e.g. users, their goals, current products they use, their tasks/activities and events that might happen during product use.

With the various sources of information, it can be overwhelming for design team members to grasp all relevant aspects of the use situations. The diverse media (e.g. video/audio recording, photos, brochures, mockup models) in combination with vast amount of information could result in a lack of organization, which makes information less traceable. Designers wish for a more flexible and stimulating way to sort and record information pieces while still keeping them coherent.

The tool provides a template to guide designers to classify their information in categories inspired by scenario elements. The existence of a template will inform design team members of the priority information. Fig. 2-4 show how the designers register information of different categories. By organizing the information early on, the tool could sustain designers' motivation and enthusiasm while documenting.

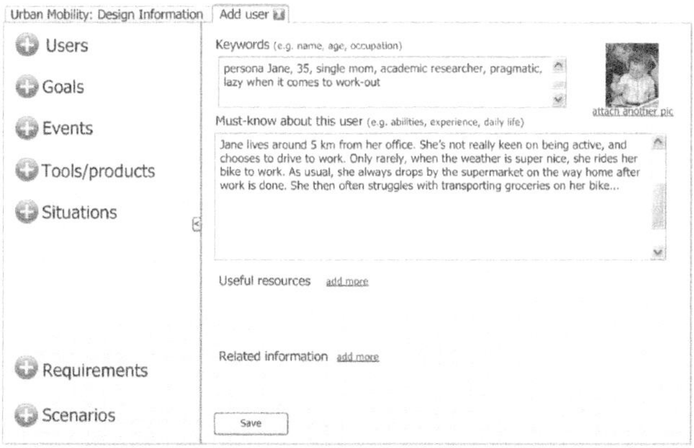

John creates a user profile using the scenario support tool
John summarizes what he has learned from the focus group session. Both participating users are quite similar in the way they are using bike pannier. Therefore John creates a fictive user Jane that combines their characteristics.

Fig. 2. Adding a user element without much prescribed steps; the designer is free to include whatever information he or she feels necessary

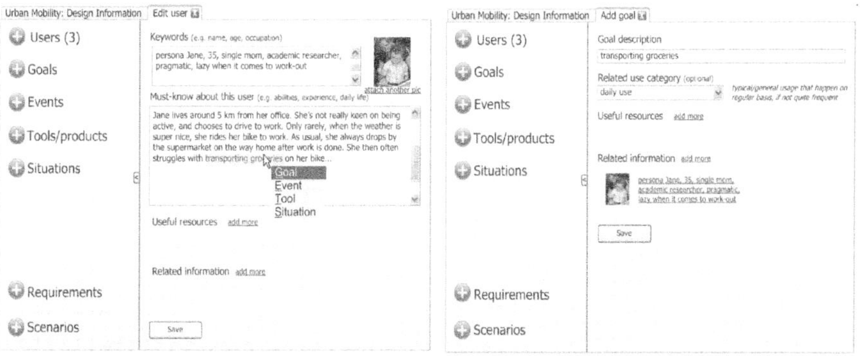

John identifies and extracts other scenario elements from a user profile
John revisits the user profile Jane and breaks it down further by scanning for other useful information. For instance, he notices a goal 'transporting groceries' is implicit in the user description. He marks the piece of text and by right-clicking, could register it as an explicit goal. The new goal keeps a reference to the user profile Jane as its source of origin.

Fig. 3. New scenario elements can be denoted from existing ones; their relations are explicitly maintained for better traceability

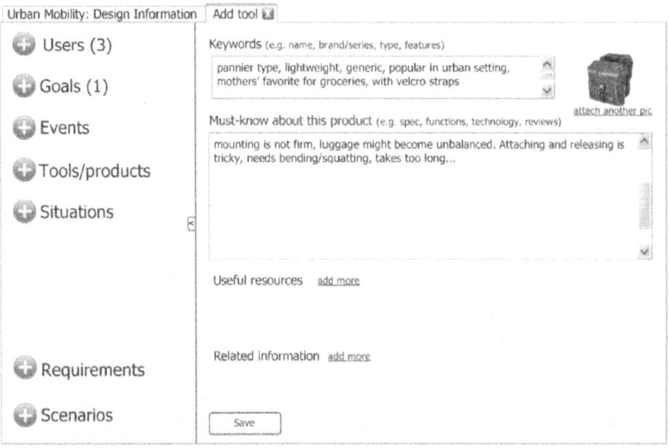

Marie adds product profiles and completes an existing product profile

Marie finished scanning the brochures from the fair and uploading pictures to the company's network drive. She adds new information in the 'Tools' category along with a summary of her notes. Based on her observation at the fair, she composes reviews on the pannier product.

Fig. 4. An existing product can be added as a tool scenario element; the designer can also freely add a product concept in progress to make mention of its characteristics

3.2 Design Information Processing: Scenario Building and Requirements Extraction

A design team is responsible to the client and the users. For this reason any design approach needs to be reliable, which means that it has to take into account all possible uses as well as hazardous/critical situations. The designers are required to understand the full extent of design information. To better structure the information, formal analysis frameworks are often used. However, the use of formal frameworks could also be inefficient because they tend to be rigid and impose high mental load to the designers.

A design team needs somehow a mechanism to keep everyone on the same level of knowledge, no matter what his or her function is. Every team member needs to know the big picture and the scope of the design assignment. In dealing with the large amount of information, a concrete guidance in making sense of the information is necessary. As a potential solution, different types of information could be combined into situations that give a reasonable coverage of the product use. Designers could then think of these vivid situations as a frame of reference. Nevertheless, they may not know where to start, which combinations to prioritize, whether they have covered all necessary ones, also how to systematically document these combinations of information for easy recall later on.

Creation of scenarios narrows down the information into its essence considerably. Scenarios are a coherent assimilation of information in a form that is easily understood. The tool helps to identify and record combinations of information that make the basis of scenarios (Fig. 5 illustrates this). We also propose to make explicit the relations between the different types of information, as can be seen in Fig. 6 and 7. This includes the relatedness between the elemental information and scenarios, among scenarios, and between scenarios and requirements. During information use later on,

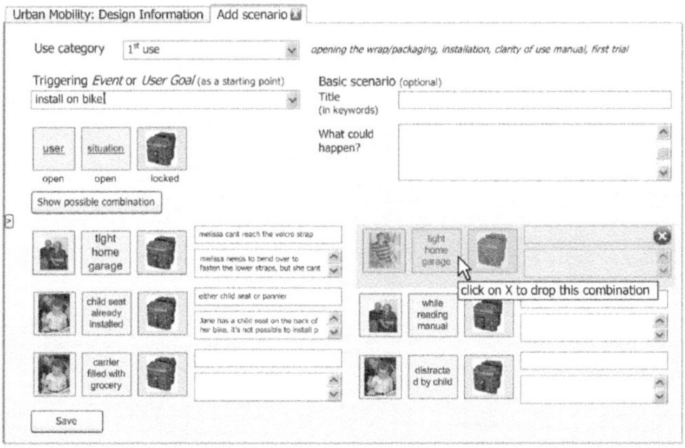

John generates scenarios

John starts identifying scenarios that contain important issues. He chooses a common user goal 'installation of product'. Then he pulls out other scenario elements: his main interest being the pannier product type. As John is not sure what other elements would make interesting scenarios, he asks the tool to show him possible combinations of existing information. A few of the combinations trigger John to scribble stories about what could happen.

Fig. 5. Combination of various elements could inspire different scenarios, for which the designer could register short narratives

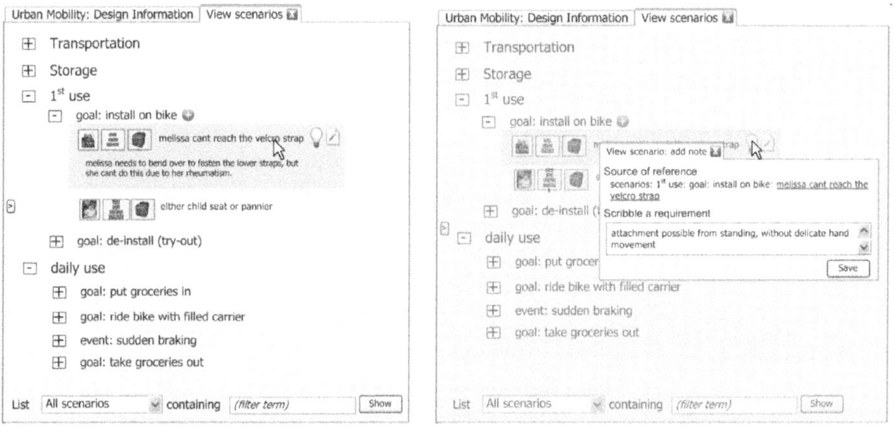

Marie scribbles requirements inspired by the scenarios

Marie scans through the scenarios related to 'installation of the product'. A scenario provokes her mind and she wants to immediately write down her thought. She clicks on the title to highlight the scenario. A set of icons appear next to the scenario title: the light bulb for note-taking and the pen & paper for editing.

The tool keeps a record on which information has inspired this note, so that Marie can easily find it again for her own reference. After writing down her note, Marie continues scanning the scenarios for inspiration. She knows she can always come back to her scribbled notes later.

Fig. 6. When a scenario inspires the designer, the tool accommodates him or her to readily make notes for tentative requirements

we expect that designers can effortlessly find out relevant information and rationales for scenarios and requirements that have been created.

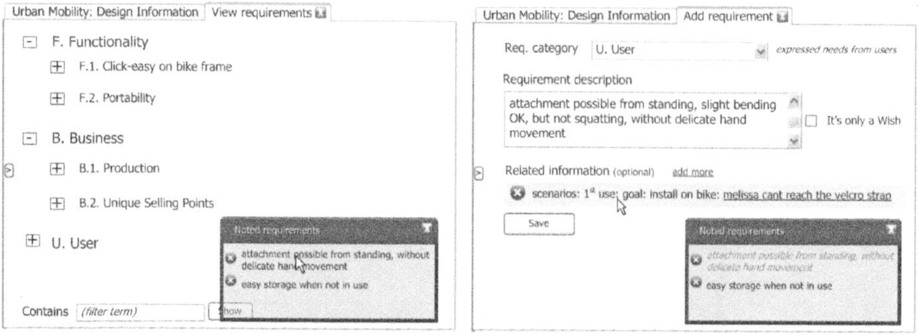

Bob registers a scribbled note as a formal requirement

Bob opens the requirements overview to check out how the team is doing with the analysis. He sees that the 'noted requirements' box now contains new tentative requirements. He clicks one of them about 'attachment (should be) possible from standing' to edit its details. He modifies the description and assigns a proper category for the new requirement, and then saves it. Bob notices that the scenario about Melissa that has inspired this requirement is saved as related info.

Fig. 7. The registration of requirements could take place whenever it is convenient for the designers, therefore not disrupting the creative process by forcing formal tasks in the middle of it

3.3 Design Information Use Guidance: Integrating Scenarios in the Process

Scenarios and requirements together are useful as a valid evaluation and reflection tool. While design requirements are often expressed in high-level manner, scenarios identify concrete use situations with explicit elements that advise a proper validation setup. With these potential uses of scenarios and requirements, there needs to be an effortless mechanism to skim, find and relate them to one another.

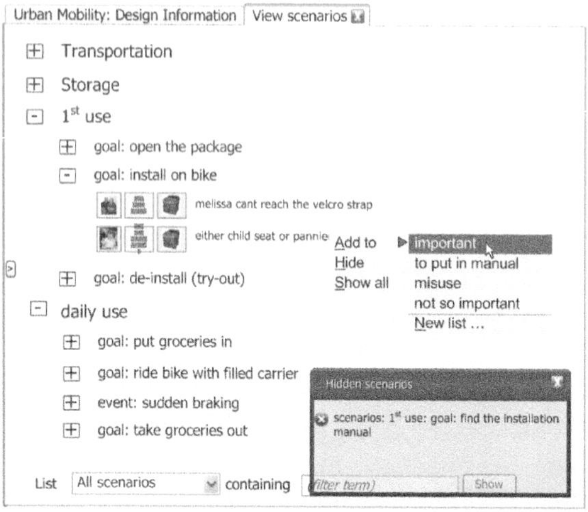

Marie wants to group scenarios on her own categories

Marie studies the scenarios overview. Scenarios about 'finding the installation manual' are not interesting for her; she already hid this scenario group. A small window keeps track of these hidden scenarios.

Marie adds a scenario to the 'important' list; the design team will have to pay special attention to it. Anyone in the design team can easily see the custom lists by choosing from the dropdown in the left-bottom corner.

Fig. 8. Filters and custom sets of scenarios help designers, individually as well as a team, to select and prioritize scenarios according to their main tasks or interests

Since scenarios are never meant to be the end result of a design process, they need to maintain a connection with other design artifacts. These artifacts could be for instance user requirements, engineering requirements, design concepts, ergonomic or safety standards. Scenarios address mainly the aspects of product use. To relate them with the rest, the design team needs to see the big picture of all the information it has, while at the same time also be able to quickly find a group of related information to examine more closely. This has not been easy in practice, and information sharing among designers and between designers and stakeholders often happens on the spot.

Our tool aims to present an overview of scenarios that is easily grasped by designers. There is a possibility to mark scenarios with specific keywords and priority ranks, to create semi-customized search results for individual design team members (see Figure 8). There should also be different perspectives to view the information as there are multiple functions working together in a design team, as well as varied interests and purposes. The hierarchical structure makes it possible to display compact yet resourceful information.

4 Discussion and Future Work

This paper has described our proposal for a tool that serves as an informal framework in a scenario-based product design practice. To conclude our proposal, we reflect upon the suitability of our proposed functionality to solve real design challenges. The first challenge concerns *the large amount of design information*. The tool puts forward the intention of building and using scenarios throughout the design process, and accommodates this by providing a structure. Designers, as the users, will be guided to single out information that is important as scenario building blocks. Secondly, with all important information neatly documented, the actual benefit lies in the designers being more guided in *making sense of the vast amount of information*. Designers need to understand a decent multitude of product use situations, without neglecting the critical or extreme situations. By explicitly prompting the designers the combinations of information pieces, the tool inspires them to identify concrete use situations that may lead to crucial design issues. Thirdly, scenarios are not the final results of a design process, rather a helping medium that sums up the designers' knowledge of the design domain. Scenarios therefore cannot be independent from the rest of design artifacts. *Integration of scenario uses within the design process* however is still largely experimental. Our tool aims to help designers to effortlessly find related information (including scenarios and other information) and to apply filters when needed. With information readily available and well organized, it is easier to employ scenarios and other information for specific purposes, e.g. reflection, validation and decision making. An additional benefit is that designers could use the processed information to generate formal documents e.g. deliveries to third parties.

Our ultimate goal is reached when designers use the proposed tool within their practice because it helps them structure their design knowledge, develop and confirm their rationales, and communicate their ideas and concerns. In achieving this goal, we make sure a close interaction with our stakeholders (i.e. designers and design companies) during the development and later during the evaluation of the tool.

References

1. van der Bijl-Brouwer, M., van der Voort, M.C.: Designing for Dynamic Usability: Development of a Design Method that Supports Designing Products for Dynamic Use Situations. Design Principles & Practices: An International Journal 2(1), 149–158 (2008)
2. Carroll, J.M.: Making Use: Scenario-Based Design of Human-Computer Interactions. MIT Press, London (2000)
3. Rosson, M.B., Carroll, J.M.: Usability Engineering: Scenario-based Development of Human-Computer Interaction. Morgan-Kaufman, San Francisco (2002)
4. Alexander, I.F., Maiden, N. (eds.): Scenarios, Stories, Use Cases: Through the Systems Development Life-Cycle. John Wiley and Sons, New York (2004)
5. Carroll, J.M. (ed.): Scenario-based Design: Envisioning Work and Technology in System Development. John Wiley and Sons, New York (1995)
6. Moggridge, B.: Design by story-telling. Applied Ergonomics 24(1), 15–18 (1993)
7. Suri, J.F., Marsh, M.: Scenario building as an ergonomics method in consumer product design. Applied Ergonomics 31(2), 151–157 (2000)
8. Tideman, M., van der Voort, M.C., van Houten, F.J.A.M.: A Scenario Based Method for Product Design. In: 15th International CIRP Design Seminar, Shanghai, China, pp. 127–133 (2005)
9. Cockburn, A.: Using Goal-Based Use Cases. Journal of Object-Oriented Programming 10(7), 56–62 (1997)
10. Sutcliffe, A.G., Maiden, N.A.M., Minocha, S., Manuel, D.: Supporting scenario-based requirements engineering. IEEE Transactions on Software Engineering, Special Issue on Scenario Management 24(12), 1072–1088 (1998)
11. Penna, G.D., Intrigila, B., Laurenzi, A.R., Orefice, S.: An XML environment for scenario based requirements engineering. Journal of Systems and Software 79(3), 379–403 (2006)
12. Rolland, C., Souveyet, C., Achour, C.B.: Guiding goal modelling using scenarios. IEEE Transactions on Software Engineering, Special Issue on Scenario Management 24(12), 1055–1071 (1998)
13. Maiden, N., Minocha, S., Manning, K., Ryan, M.: CREWS-SAVRE: Systematic Scenario Generation and Use. In: 3rd International Conference on Requirements Engineering, Colorado Springs, USA, pp. 148–155 (1998)
14. Shin, J.E., Sutcliffe, A.G., Gregoriades, A.: Scenario advisor tool for requirements engineering. Requirements Engineering 10(2), 132–145 (2005)
15. Lim, Y., Sato, K.: Describing Multiple Aspects of Use Situation: Applications of Design Information Framework to Scenario Development. Design Studies 27(1), 57–76 (2006)
16. Anggreeni, I., van der Voort, M.C.: Tracing the Scenarios in Scenario-Based Product Design: A Study to Understand Scenario Generation. Design Principles and Practices: An International Journal 2(4), 123–136 (2008)

How to Learn from Intelligent Products; The Structuring of Incoherent Field Feedback Data in Two Case Studies

Renate de Bruin, Yuan Lu, and Aarnout Brombacher

Eindhoven University of Technology, Department of Industrial Design, Business Process Design Group, Den Dolech 2, 5600 MB Eindhoven, The Netherlands
{R.d.Bruin,Y.Lu,A.C.Brombacher}@tue.nl

Abstract. A growing number of products - particularly highly innovative and intelligent products - are being returned by customers, while analysis shows that many of these products are in fact functioning according to their technical specifications. Product developers are recognizing the need for information that gives more detail about the reason for these product returns, in order to find the root cause of the problems. Traditionally a lot of information from the field is stored in departments like sales and service (helpdesks and repair centers). Combining these data sources with new field feedback data sources, like customer experiences on the internet and product log files, could provide product developers with more in-depth and accurate information about the actual product performance and the context of use. Case studies were done at two different industries: a company from consumer electronic industry and a company from professional medical system industry. The Maturity Index on Reliability (MIR) method, a method to assess capability for businesses to respond to product reliability related issues, was combined with Contextual Design, a method for user context mapping and requirement definition, and explored as a means to analyze the structure and capability of the current field feedback process in supporting the development of complex and innovative products .

Keywords: Field Feedback Data, Product Failure, Root Cause Analysis, Business Information Flow, MIR Method, Contextual Design.

1 Introduction

A vast and growing number of products are being returned by customers, while analysis shows that many of these products fully function according to their technical specifications [1, 2]. There are several trends in product development and user emancipation that may contribute here [1, 3, 4], the fact is that a lot has changed since the beginning of the digital era. First of all the products themselves have changed tremendously. What first were simple, mono-functional products have become complex, multi-functional and adaptive products (take for example the remote-less television of the 70s and compare this with the totally adaptive digital television of today). Secondly the product development organizations have changed from having simple, monolithic business chains to complex, multi party chains as they have been driven by logistic efficiency [5].The consequences of these changes have been that;

M.J. Smith and G. Salvendy (Eds.): Human Interface, Part I, HCII 2009, LNCS 5617, pp. 227–232, 2009.

- users experience a lot of (new) problems with complex products, not only the typical hardware malfunctions, but in addition also because of software faults, bad usability, or incorrect expectations about functionality [1, 2]
- companies do no longer receive feedback about their product from first-hand, but only through a sequence of service delivering companies, and sometimes not even at all [5].

So while users are experiencing more difficulties than ever before, the platform to handle these complaints and to learn from them has been getting more diffuse and more difficult to manage. Reconnecting the multi party chains in business organizations again through merging of their separate field feedback data sources is thought to offer a potential solution [6]. In addition to the traditional field feedback sources (functionality tests, sales orders, helpdesk loggings and repair job sheets) new sources of information (internet websites, web-communities and product data loggings [4, 7]) can add rich information about actual product use and the context of use. Merging incoherent field feedback data into prioritized design information is the objective of the Data fusion project [6]. This project is financed by the Dutch ministry of Economic Affairs and carried out by a consortium of academic and industrial partners. All are committed to form an improved field feedback information process to generate prioritized consumer complaints information to support product improvement in future products (Data fusion Information System).

In order to elicit requirements for this Data fusion Information System the following research questions need to be addressed first;

- How is the current field feedback information process structured?
- How does the current field feedback information support the different stake holders in developing complex and innovative products?

This paper will be addressing these questions. The obtained insight will support the (re)design of the field feedback process. The paper is organized as follows; in section 2 the chosen research methodology is discussed, followed by a discussion of the research results in section 3 and conclusion in section 4.

2 Research Methodology

Considering the nature of the research questions the choice for an explorative case study as research methodology was considered appropriate [8].

2.1 Case Study Selection

Two different product developing industries were chosen for this case study research: Company A, a manufacturing company from consumer electronics industry, and Company B, a manufacturing company from professional medical systems industry. Although their product scales, markets and end-users largely differ, they both focus on the development of complex, high quality and innovative products. In these case studies the focal point has been the field feedback information process from customer service and trade departments to the strategy, design and development departments.

2.2 Data Collection, Documentation and Analysis

In the Data fusion Information System the multiple field feedback sources will be merged and information that in a smart way has been compressed and focused will be fed back to the different stakeholders in the product development process.

In the current business practice often improvements can be made regarding the flow of information coming from the field to the stake holding business departments [9]. In one way the Data fusion Information System will be the facilitator of such improved information flow and therefore will affect the design of business processes directly. From a requirement definition point of view the Data fusion Information System therefore defines the information process on 'macro' level. At the same time the Data fusion Information System aims to provide information to support different stakeholders in developing complex and innovative products. How to define the required information from the stakeholder point of view touches just exactly the information processes on 'micro' level. It makes sense that different data collection methods are applicable when studying these nearly opposite subjects. A combination of the two would suit this project however.

The Maturity Index on Reliability is a method that has proved to be very useful for analyzing information flows in -macro scale- business processes. The maturity index on reliability is a method to assess, in a relatively short period of time, business processes with respect to their effectiveness to control product quality and reliability [9, 10, 11]. By interviewing people operating both on management and execution level, the MIR method aims to generate a valid and representative overview of the different activities of stakeholders and the interactions between them [9]. Eventually major bottlenecks and opportunities for improvement can be identified [10].

Contextual Design is a popular user-centered research and design method focusing on the -micro scale- information systems design [13]. It is based on ethnographic methods for studying product users and the context of product use and its principles are deducted from Activity Theory [12]. By conducting interviews and participant observation studies in the field, diagrammatic work models showing each an aspect of the workflow, in order to create a common understanding of the context of use and to build the system requirements from a human factors point of view [14].

The MIR method and the Contextual Design method both can be used in the Data fusion project, but neither of them suits the research objectives and project's resources perfectly. The Contextual Design method is a very time-consuming method, but does generate detailed results and give a multi-dimensional view on the context of use. With the MIR method a quick assessment can be made with regards to the information flows in a business process. The result however is showing this one dimension, disregarding the influence of different stakeholders in the processes. In this research a well-balanced combination of the two methods (see table 1) is used. In this combined method the basic approach is that of the MIR method, though extended with field observations with representatives at different stakeholders, for example, working with different software applications, in the company. These applications -seen from an activity theory point of view - are mediators of human thought and behavior [13] and form practical illustrations of how different stakeholders can influence the systems they work with and systems can influence the way different stakeholders work.

Table 1. Comparison of MIR, Contextual Design and combined method

	MIR	Contextual Design	Combined
Discipline / objective	Business process design	Information system design	Business process and information system design
Data gathering tools	Interviews, Document analysis	Field interviews, Field observation, Application analysis	Interviews, Document analysis, Field interviews and observation Application analysis
Interviewed / observed	Department representatives of management and workforce	Representatives of the typical system user roles, covering the diversity	Department representatives of management and workforce, as well as representative users of mainly used SW applications.
Synthesis	Consolidation with management of client company	Consolidation with system users at site and team members afterwards	Consolidation with system users at site, and afterwards cons. with interviewed and management
Visualization tools	Activity diagram showing main business information flows	Work models showing the multi dimensions of workflow	Activity diagrams and work models showing the multi dimensions of information flow

3 Results

In 49 1-hr-meetings 43 different persons have been interviewed (Company A: 13 meetings, 14 different persons; Company B: 36 meetings, 29 different persons). The software applications that support the main work have been discussed in these interviews, though only in 2 instances (Company A) and 3 instances (Company B) more extensive field observations have been made. During the interviews and field observations notes were taken, which were later consolidated in a high level information flowchart (MIR diagram), that was extended with an overview showing the different stakeholders involved in every phase, their objectives in work and the (software) tools they use.

So far this has lead to the following cross-case findings on the 'macro' level of information flow:

- A logistics oriented service process leads to an ineffective feedback process from (external) service and trade organization to the product development process in supporting new product development or product improvement.
- Stake holding departments are focusing on their own departmental needs with regards to the collection and processing of field feedback data. Hence departments are not always fully aware of the needs of other departments with regards to the required quality of field feedback data.

On the 'micro' level of information flow a few preliminary findings are:

- The logistics oriented characteristics of software applications at the service organization and trade organization have an influence on the quality of field feedback, especially in the data gathering stage.
- The obstructions in the information flow caused by software applications are originating often from historical grounds (once a good solution for mechanical products, but not adapted to the present 'digital' situation, for example, there is insufficient IRIS code to describe software faults), misusage (applications that are used but not designed for the specific purpose e.g. customer service-applications that are also used for problem trace activities) or an imperfect fit (applications that are developed for a larger public, not entirely adapted to the specific needs of the company or department).

Due to the small amount of field observations the work models of contextual design were not made, though the observations on 'micro' level that did take place offered already very valuable insights and showed a negative effect on the quality of information flow on 'macro' level. In short, in both case studies it can not be concluded that the current field feedback process already supports the different stakeholders in the product development process to its full potential.

More field observations are necessary to gain insight on the 'micro' level, which can directly contribute to the redesign of the field feedback process. These case studies therefore will continue and a next round of interviews has been planned in the course of 2009.

4 Conclusion

In this paper, field feedback information processes in two different manufacturing companies are analyzed on their structure and capability in supporting the different stakeholders in development complex and innovative products by using a combined approach based MIR method and Contextual Design. The preliminary results show that it is possible to get a quick overview of the main information flows within a company on macro level, and to address not only the business process aspects in these, but also the 'micro level' role and influence of the different stakeholders on the information flow. The analysis of the results is still ongoing and more field data will be collected. Further results will be subject of future publication.

References

1. Den Ouden, E.: Development of a Design Analysis Model for Consumer Complaints, Revealing a New Class of Quality Failures. Eindhoven University of Technology, Eindhoven (2006)
2. Overton, D.: Investigating the mobile No Fault Found Phenomenon. Media Bulletin - White paper. WDSGlobal (2006)

3. Brombacher, A.C., de Graef, M.R.: Anticiperen op trends. In: de Graef, M.R. (ed.) Betrouwbaarheid van technische systemen, anticiperen op trends. Stichting Toekomstbeeld der Techniek, vol. STT64, pp. 392–417. Den Haag (2001) (in Dutch)
4. Pitt, L.F., Berthon, P.R., Watson, R.T., Zinkhan, G.M.: The Internet and the birth of real consumer power. Business Horizons 45(4), 7–14 (2002)
5. Visser, I.M.: Analyzing user perceived failure severity in consumer electronics products. Incorporating the user perspective into the development process. Eindhoven University of Technology, Eindhoven (2008)
6. Lu, Y.: Merging of incoherent field feedback data into prioritized design information. A project proposal for the IOP Integrated Product Creation and Realization. Eindhoven University of Technology, Eindhoven (2007)
7. Füller, J., Bartl, M., Ernst, H., Mühlbacher, H.: Community Based Innovation: A Method to Utilize the Innovative Potential of Online Communities. In: Proc. Proceedings of the Proceedings of the 37th Annual Hawaii International Conference on System Sciences (HICSS 2004), vol. 7, p. 10. IEEE Computer Society, Washington (2004)
8. Yin, R.K.: Case Study Research. Design and Methods. Applied social research method series, vol. 5. Sage Publications, California (2002)
9. Sander, P.C., Brombacher, A.C.: Analysis of quality information flows in the product creation process of high-volume consumer products. International Journal of Production Economics 67(1), 37–52 (2000)
10. Sander, P.C., Brombacher, A.C.: MIR: The use of reliability information flows as a maturity index for quality management. Quality and Reliability Engineering International 15(6), 439–447 (1999)
11. Brombacher, A.C.: Maturity index on reliability: covering non-technical aspects of IEC61508 reliability certification. Reliability Engineering & System Safety 66(2), 109–120 (1999)
12. Beyer, H.R., Holtzblatt, K.: Contextual design: defining customer-centered systems. Morgan Kaufmann Publishers Inc., San Francisco (1998)
13. Nardi, B.A.: Activity Theory and Human-Computer Interaction. In: Nardi, B.A. (ed.) Context and Consciousness: Activity Theory and Human-computer Interaction, pp. 7–16. MIT Press, Cambridge (1996)
14. Notess, M.: Using Contextual Design for digital library field studies. In: Proc. ACM/IEEE-CS Joint Conference on Digital Libraries workshop. ACM/IEEE-CS Joint Conference on Digital Libraries workshop Studying digital library users in the wild: theories, methods, and analytical approaches, Denver, CO, June 10-11 (2005)
15. Holtzblatt, K.: Contextual Design: Experience in Real Life. In: Oberquelle, H., Krause, R.O.J. (eds.) Proc. Mensch und Computer, vol. 1, pp. 19–22. B.G. Teubner, Stuttgart (2001)

Successful User Experience in an Agile Enterprise Environment

Melissa Federoff and Catherine Courage

Salesforce.com, The Landmark at One Market St., San Francisco, CA 94105
ccourage@salesforce.com, mfederoff@salesforce.com

Abstract. Salesforce.com is a leader in the enterprise Customer Relationship Management (CRM) marketplace. In 2006, salesforce.com's Research and Development (R&D) organization transitioned over 30 product teams from a waterfall development process to an agile one. The R&D department is responsible for producing all products offered to salesforce.com customers. After the transition, it was clear that User Experience (UX) team members were dissatisfied. When asked 6 months into the rollout if agile was making their work life better, only 24% agreed. This paper discusses how the team and management responded to this data and as a result reached an 85% satisfaction rate a year later.

Keywords: Agile, waterfall, enterprise, user experience, design, research.

1 Transition from Waterfall to Agile Development Process

R&D teams at salesforce.com are functionally organized into program management, user experience, product management, development, quality engineering, and documentation. Prior to agile, these teams leveraged a waterfall development process (Figure 1).

Program management oversaw projects and coordinated feature delivery across the various functions. Product management created business requirement documents that specified what was to be built. User experience produced and evaluated feature prototypes. Development wrote technical specifications and coded based on prototypes. The quality assurance team tested and verified the feature functionality. The documentation team documented the functionality. The system test team tested the product at scale. These functions were performed in a serial fashion.

As the company grew and the salesforce.com application gained complexity, development became increasingly unable to accurately estimate time and scope for new features using the waterfall method. Product teams suffered from feature creep, redesign work, extended development times and compressed testing schedules during development cycles. In October 2006, it had been almost a year since salesforce.com's last major release - a release that had been rescheduled five times.

In an effort to increase the number and accuracy of releases, the R&D organization decided to move all 30 plus development teams from a waterfall development process to an agile one. Agile is a philosophy toward software development that was

M.J. Smith and G. Salvendy (Eds.): Human Interface, Part I, HCII 2009, LNCS 5617, pp. 233–242, 2009.

established in an Agile Manifesto [2]. There are many tenets to this approach; its core lies in methods that are adaptive and people-oriented. One common agile methodology is 'Scrum' which is the process salesforce.com decided to espouse during the company's self termed "big-bang rollout" in 2006.

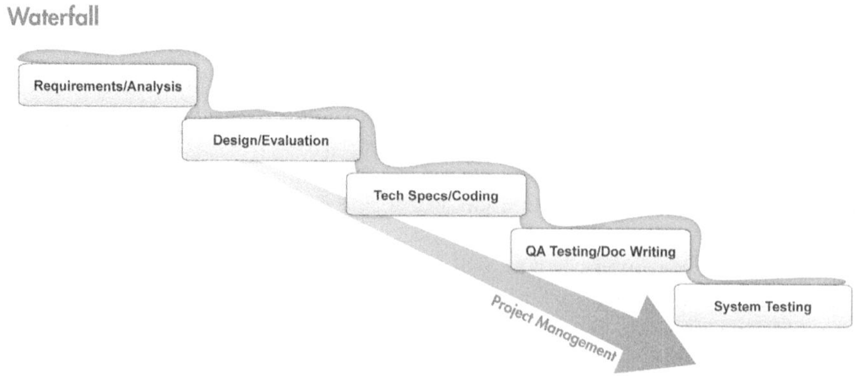

Fig. 1. A visualization of the Waterfall process used at salesforce.com

With Scrum, projects progress via a set number of time-boxed iterations called sprints (typically 2 - 4 weeks in duration). The goal at the end of each sprint is to have fully functioning code that has been tested and that could be released (Figure 2).

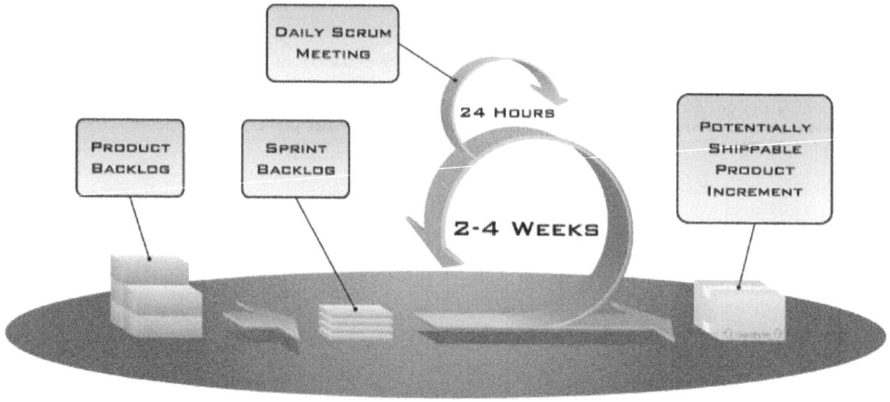

Fig. 2. A visualization of the Scrum process, taken from Mountain Goat Software (2005)

The procedure of the Scrum method at a high-level is as follows [3]: A Product Backlog is created by a product owner (a product manager typically fills this role). This backlog is a list of all desired changes to the product for the release.

At the start of each sprint a planning meeting is held with all members of the Scrum team (product management, user experience, documentation, quality assurance, development, project management). During this planning meeting the product owner prioritizes the product backlog and the Scrum team selects a chunk of the backlog that it can complete during the coming sprint.

Items selected for the sprint are then moved from the Product Backlog to the Sprint Backlog. Because the time-frame is short this allows the team to effectively estimate and commit to work that it can reasonably accomplish. During the sprint, all items in the sprint backlog need to reach completion: at salesforce.com this means they will be designed, validated with users, coded, tested and documented.

During the sprint the Scrum team conducts a brief daily meeting called the Daily Scrum, which helps the team stay on track and creates visibility into any barriers to work completion. Program Managers take the roll of Scrum master which entails facilitating all meeting and planning sessions and ensuring that any roadblocks are removed.

This process repeats for a pre-defined number of sprints, and then work is released. Salesforce.com completes three sprints and then releases new features to customers quarterly.

1.1 Impact of the Transition to Agile on the R&D Organization

In the transition to agile, some successes were seen immediately. The most significant change for salesforce.com was that the R&D organization met the February 2007 release date. At the one year anniversary of moving to agile, the rollout team calculated overall improvements. Changes included a 61% improvement in mean time to release for major releases, a 94% increase in feature requests delivered in major releases, and a 38% increase in feature requests delivered per developer.

1.2 Initial Impact of the Transition to Agile on the User Experience Team

While moving to agile was clearly a success for the R&D organization at salesforce.com, the initial transition period was far from smooth for the User Experience team. In 2006, the vast majority of literature on the topic of agile did not include guidance on the inclusion of User-Centered Design (UCD) processes [4]. In agile literature, 'design' typically referred only to coding or system design [1]. The User Experience team was left with many unanswered questions; some included:

- How can we properly identify the target users and ensure that their needs are met within agile's just-in-time process?
- How can we accomplish holistic design when agile teams are planning and building features in a piecemeal fashion?
- How do UX professionals succeed across multiple teams when agile espouses one team per person?

The unhappiness of the UX team was illustrated clearly in the results of a survey sent out to the R&D organization about 6 months after the rollout. Eighty percent of total respondents felt that agile was making their Scrum team more effective. However, when looking at the data by functional area, only 30% of User Experience team members agreed with this statement. When asked about several other impacts of agile, the same trend was seen (Figure 3).

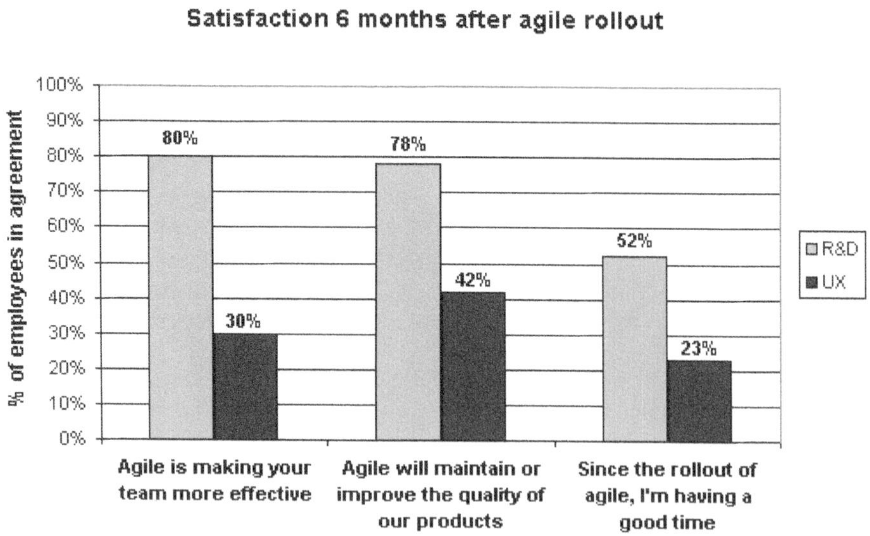

Fig. 3. UX satisfaction with agile as compared to all of the R&D Department

Some of the reasons cited by User Experience for their dissatisfaction with agile included:

- Assigned to too many teams
- Spending too much time in meetings
- Not enough time to complete work
- Lack of focus on the big picture

User Experience team members supported four product teams on average, while developers and other disciplines only supported one, as dictated by the Scrum process. For team members that are only assigned to one Scrum team, attending planning meetings, daily Scrums and retrospectives is a reasonable time investment. For those on multiple teams, attending all of those meetings meant having little time to do anything else, such as research or design work.

Tighter timelines were also mentioned. In Scrum, teams select what they will work on at the beginning of a sprint and then build those items to completion over the course of that sprint; in the case of salesforce.com that is a one month period of time, start to finish. These aggressive timeframes can cause UCD to be compromised, since investigating and iterating designs often takes longer than weeks, particularly for complex enterprise applications. In this initial phase, the User Experience team struggled to find ways to achieve the success that they had reached using waterfall. The R&D division made significant progress using agile, though, so the UX team had no choice but to adapt and evolve.

2 Strategies for Success

The road to agile success within an enterprise space was not an easy one for the User Experience team at salesforce.com. The key factors that drove the success included:

- A New Resource Plan
- Design Transformation
- Getting Usability RITE

2.1 A New Resource Plan

Management made the decision to reduce the number of Scrum teams per User Experience team member from four or more to a maximum of two. The process of determining which teams would receive assigned resources involved the UX managers meeting with the VPs of the different product areas to determine the priorities of the new features. UX management took these priorities and assessed them against the complexity of the features (e.g., would the feature require a brand new user interface). Assignments were made accordingly. The Scrum teams that did not have an assigned UX resource became responsible for working as a team to create the user interface as best they could.

This solution to de-support approximately 35% of teams was a situation that no one was satisfied with. Certainly, the UX team members were pleased with their new, realistic workload, but the decision to strand some Scrum teams was not accepted. Ease of use is a core value at salesforce.com. Placing the burden on the Scrum teams put that value at risk.

In the spirit of agile, UX team members brainstormed an alternate solution, and the concept of Office Hours (OH) emerged [5]. The UX team members determined that by giving two hours of their time per week, they would be able to assist Scrum teams that did not have assigned UX resources without spreading themselves too thin.

One of the goals was to make OH easy for everyone involved. In the spirit of agile, OH slots are self-service. Scrum teams are not scheduled by UX team members, but instead, Scrum teams are encouraged to make use of the time and schedule it when most appropriate for them.

Scrum teams are asked to bring user stories, design objectives, and a design artifact to OH to maximize the effectiveness of the session. A typical agenda looks like this:

- Project review – < 15 min – Cover the user stories and design objectives
- Artifact presentation – 10 min – Show the UX team member(s) the proposed design artifact
- Discussion – 35 min – Converse about the artifacts and objectives, and assess the design

The OH program has been successful on several fronts: job satisfaction is higher among UX team members; Scrum teams with minimal user experience needs are supported very well with OH; teams that might never have received any support are receiving some (e.g., departments outside of R&D such as Marketing). However, the program is a stop-gap and complex features do suffer if not given full design support. We have noticed a slight decrease in the overall consistency of deliverables for complex features, and have found OH slots to lack adequate time to address Scrum teams' needs fully.

Overall, at salesforce.com, the need for the Office Hours program is decreasing over time. When management supported the cut of Scrum teams supported by each UX member, they also supported headcount for rapid growth of the UX team. The UX team in two years has gone from less than ten, to more than 30 full-time staff. This larger team can cover more product areas, so fewer Scrum teams are left without an assigned design resource. The Scrum teams that are not given a resource now are typically the ones that have simplier features and therefore are served well through the OH program.

2.2 Design Transformation

Fitting the complex design of enterprise features into the agile process proved to be a major challenge. Developers were frustrated that they did not have UI designs that they could start to code in the first sprint and designers felt the one month sprint model did not leave them with enough time to create and evaluate their design. Also, in waterfall, the designer could focus on the design in its entirety. In agile, the UX team was forced to take the design and break it into pieces that could be tackled by development in one month segments. This can be difficult for complex enterprise features such as analytics or workflow which are often impossible to design and evaluate as component parts.

Some of the approaches that have improved the process include:

- Moving to parallel development and design [7], [8]
- Working a release ahead
- Utilizing interactive prototypes for usability testing and communication of designs to developers
- Design Studios

Parallel Development and Design. Parallel development and design (Figure 3) is beneficial in the enterprise applications space because features are typically too big to be designed, usability tested, iterated, validated, built, translated and documented in one sprint. Sometimes single sprint design and validation is actually possible to achieve, when the feature being built is an enhancement of something that already exists. However for new features or for features that require new interface components to be designed, sometimes it is not possible to accomplish a finished design in that time period. Parallel tracks between design and development can help solve this problem (see Figure 4).

At salesforce.com, when a new feature or interface component needs to be designed, the UX team member will often choose to work in parallel with development. The UX professional will create and verify designs during the first or second sprint, while developers are working on back-end features that require little to no user interface. Designs are then handed off to development to be completed in the final sprint(s) of that release.

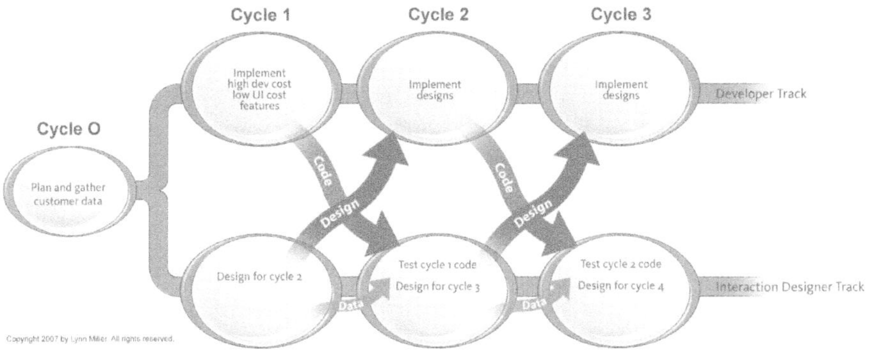

Fig. 4. Lynn Miller's depiction of parallel tracks at Alias [7]

Designing a Release Ahead. For very large scale projects that do not have a precedent in the product or for large features where the design cannot be chunked into sprints, the UX team will work on designing and iterating the prototypes an entire release ahead, while the development team is working on minor feature enhancements or features that do not require front-end design.

Interactive Prototypes. The prototypes that are created by UX become the means for expressing designs to the development teams. They essentially replace the need for written specifications. This method is not only more efficient, it is often more effective at conveying detailed interactions that are hard to describe in words and sometimes left open to interpretation.

Design Studio. We have recently had success with the Design Studio approach [9]. This process gets the entire Scrum team together for a single-day participatory design session. During this design session, all attending members are able to critique concepts and contribute to design that will be embarked on during the course of the sprint. Some key advantages are:

- Rapid exploration of design alternatives and the creation of an initial design
- The entire Scrum team, rather than the UI designer alone, owns the design and, as a result, feels very invested in the design
- Provides an opportunity for User-Centered Design education
- Provides the Scrum team an alternate way to view the functionality proposed. The resulting design from the exercise can help the team take a more realistic and informed look at whether the functionality they have proposed is feasible for the release.

2.3 Getting Usability RITE

Once the UX team has designs ready for feedback, they are swiftly taken into the lab for testing with users who represent those who will actually use that product. The UX team conducts an extraordinary amount of formative testing due to the nature of our business. Since salesforce.com is an enterprise company with millions of existing customers, we have to be extremely diligent about making changes to existing paradigms, to ensure that these changes are worth the learning curve to the user. Also, due to salesforce.com's software-as-a-service model, all of our customers are on the same version of our product at all times. We do not have the luxury of A/B style testing (testing where multiple versions are released at the same time to see which is has the best impact). As a result, it is most effective for us to do most of our testing in a formative way, prior to release. Luckily, since we have this existing customer base, we also have a huge pool of users to pull from easily when we need to conduct testing quickly.

Since moving to agile, the UX team at salesforce.com has only on rare occasion conducted a standard usability test. For formative testing, Rapid Iterative Test and Evaluation (RITE) is almost always the method utilized to evaluate designs. RITE-style testing is based on the principle of iterating the design as you test each participant. This is in contrast to a traditional usability test, where changes are only made after a full set of participants has evaluated the design [6]. In RITE, the designer is empowered to change the interface if: (1) the participant is believed to be a representative user for that feature; (2) the problem is believed to be understood and; (3) a solution is proposed. RITE testing acknowledges that initial designs will be flawed, but that a successful design will be achieved through user input and iteration.

RITE and agile development go hand in hand. RITE is fast, iterative, inclusive of the team, collaborative, and in the end produces a prototype that acts as a proven design specification for development. No time or resources are wasted; every hour is utilized; every problem is acted upon. At salesforce.com, the UX team often makes changes after seeing one participant have a problem. Sometimes those changes are even made during the session.

In the field of usability, many researchers have a difficult time getting product teams to want to attend usability tests. Typically, by the third or fourth user, everyone, including the researcher, knows what most of the problems are, so there really is little value in continuing to attend. In RITE, the lab setting changes from a stagnant place where problems are repeatedly rediscovered, to a dynamic environment where problems are solved in real time. The lab becomes an active design space.

During UX training at salesforce.com, certain mindsets are taught to enable designers and researchers to engage in RITE successfully. Designers are encouraged to embrace the following principles:

- Collaboration: Use the lab to discuss ideas
- Focus: Don't do anything other than watch and iterate
- Flexibility: Be willing to try new ideas at a moment's notice
- No ego, no blame: Find out what's wrong, don't validate that you're right

Likewise, researchers are taught to support RITE by valuing the following:

- Collaboration: Keep attendees involved and inspire brainstorming
- Flexibility: Be willing to change your protocol to support design iteration
- Report Immediately: Do continual lightweight reporting that keeps the team iterating

Using these principles and combining that with the evolution of our prototyping tool set over time, the UX team has found a way to be able to iterate designs as fast as agile requires.

3 Success!

With the advent of these changes, UX team members have reported greater job satisfaction. The most recent agile survey showed a significant improvement in how much UX team members are enjoying their jobs (Table 1).

Table 1. A comparison of job satisfaction survey responses from UX team members in March 2008 and March 2007

Survey question and responses: Since the rollout of agile, how much fun are you having?	**% that agreed in March 2008**	**% that agreed in March 2007**
The best time	24	0
A good time	62	23
Not much fun	14	30
A terrible time	0	46

The recent survey also revealed improvements in UX team members' attitudes toward agile (Table 2).

Table 2. Comparison of survey responses from UX members in March 2008 and 2007

Survey question	**% that agreed in March 08**	**% that agreed in March 07**
Is agile making your team more effective?	73	30
Agile will maintain or improve the quality of our products.	86	46

The job satisfaction of UX members reflects the team's belief that UCD is being achieved again at salesforce.com. All customer-facing features undergo usability testing and design iteration until all serious usability issues are resolved, just as they did within waterfall. This transition from standard development to agile wasn't easy for UX, but with adaptation, the team has experienced great success.

References

1. Beck, K.: eXtreme Programming Explained: embrace change. Addison-Wesley, Reading (2000)
2. Beck, K., et al.: Manifesto for Agile Software Development (2001), `http://agilemanifesto.org`
3. Cohn, M.: Agile Estimating and Planning. Prentice Hall PTR, Englewood Cliffs (2005)

4. Fowler, M.: Put Your Process on a Diet, Software Development, v.8, 12. Expanded version (December 2000),
`http://martinfowler.com/articles/newMethodology.html`
5. Leszek, A., Courage, C.: The Doctor is "In" – Using the Office Hours Concept to Make Limited Resources Most Effective. In: Proceedings of the Agile 2008 conference 2008 (2008)
6. Medlock, M.C., Wixon, D., Terrano, M., Romero, R., Fulton, B.: Using the RITE Method to improve products: a definition and a case study. Usability Professionals Association, Orlando, FL (July 2002), `http://www.microsoft.com/downloads/details.aspx?FamilyID=3b882eb1-5f06-41d9-baba-d39ad13bc3ff&displaylang=en`
7. Miller, L.: Case Study of Customer Input For a Successful Product. In: Proceedings of the Agile 2005 conference (2005),
`http://agileproductdesign.com/useful_papers/miller_customer_input_in_agile_projects.pdf`
8. Sy, D.: Adapting Usability Investigations for Agile User-Centered Design. Journal of Usability Studies 2(3), 112–132 (2007),\
`http://www.upassoc.org/upa_publications/jus/2007may/agile-ucd.html`
9. Ungar, J., White, J.: Agile user centered design: enter the design studio - a case study. In: CHI 2008 Extended Abstracts on Human Factors in Computing Systems, CHI 2008, Florence, Italy, April 5-10, pp. 2167–2178. ACM, New York (2008),
`http://doi.acm.org/10.1145/1358628.1358650`

From Research to Product: Integrating Treemaps into Enterprise Software

Joseph H. Goldberg, Jonathan I. Helfman, and John Beresniewicz

Oracle USA, 500 Oracle Parkway, MS 2op2, Redwood Shores, CA USA
{Joe.Goldberg,Jon.Helfman,John.Beresniewicz}@oracle.com

Abstract. The difficult journey of introducing a new treemap visualization into enterprise software products is described, highlighting interactions among a research team, product teams, and management. Successful product integration ultimately required multiple iterations of prototyping, review, and redirection, as products were cancelled and modified. Several lessons learned are provided, including the need to build flexible and generic prototypes, cultivate champions, and be tenacious.

1 Introduction

Introducing new user interface (UI) ideas into products at large enterprise software companies can be challenging. Development teams, product teams, executives, and others often view innovation as risky and therefore sometimes resist proposed changes to successful, money-making products. This resistance usually stems from a lack of understanding about the trade-offs between the risks of innovation and the benefits of increased customer satisfaction with associated sales revenues and profits. However, communication between research and product teams may be poor, and corporate methods or processes for introducing innovation may be undefined.

This paper highlights the journey of a treemap visualization concept into enterprise software products at Oracle. The focus is on the interplay between a corporate UI research group and product teams. Design decisions were made at each step, and several rounds of prototype development were required. Lessons learned from this experience are provided to facilitate streamlining the adoption of innovative visualizations and other technologies.

1.1 Treemap

A treemap is a graph that shows hierarchical datasets as nested rectangles, with areas of rectangles conveying a quantitative (or numerical) dimension. The color of the rectangles may also represent an additional data dimension [2]. Treemaps gained significant popularity following Smart Money's introduction of the Map of the Market (Fig. 1, http://www.smartmoney.com/map-of-the-market/), which enables a user to simultaneously monitor hundreds of stock prices [3].

Oracle has a UI research group that investigates new concepts and interaction techniques. Several Oracle product teams expressed interest in treemaps following an

M.J. Smith and G. Salvendy (Eds.): Human Interface, Part I, HCII 2009, LNCS 5617, pp. 243–252, 2009.

early-2004 UI research group proposal. The research group began collecting treemap requirements for enterprise users and obtained demo treemaps from commercial vendors, to determine which commercial version best satisfied these requirements. In early 2005, the research group conducted a usability evaluation in which 10 network administrators used treemaps and hierarchical tables to monitor transactional data. The administrators completed their tasks more slowly when using a larger dataset than when using a smaller dataset, and this difference was twice as great when the administrators used tables than when they used treemaps. The network administrators also missed more information when it was presented in tables rather than in treemaps. Subjective impressions strongly favored treemaps over tables for completing network administration tasks [1].

Fig. 1. Smart Money Map of the Market, with area indicating relative market capitalization and color indicating stock price movement for the current day. A two-level hierarchy is shown, with individual stocks grouped by sector. (Used with permission. To license Tree Map software from SmartMoney, email licensinginfo@Smartmoney.com.)

1.2 Interactivity and Enterprise Technology Decisions

User interface innovation, such as a new treemap, carries many potential design tradeoffs in areas such as interaction capabilities and choice of technology platforms. Product requirements and results from user performance testing should ultimately determine these design decisions. Treemap technology decisions include issues such as thick versus thin client, and client- versus server-based data aggregation and rendering. In addition, enterprise software must support large and complex queries on both relational and OLAP databases, providing challenges for applications with user interfaces that include interactive data visualizations.

User experience can differ markedly among different treemap designs with different levels of interactivity. At one extreme, a treemap may be a static image, with or without (1) tooltip feedback, (2) user-defined configuration of data to color and area, and (3) clickable master/detail drilling. Static treemaps are intended for more casual or occasional users, who primarily consume information. At the other extreme, a treemap may be a dynamic view with or without (1) in-place drilling, (2) hierarchical, area and/or color filtering, and (3) automated updating. Dynamic treemaps are intended for more advanced enterprise users such as analysts or administrators.

A visualization such as an enterprise treemap must be both vertically and horizontally scalable. Vertical scalability refers to the ability to investigate deep data hierarchies that may pivot about multiple dimensions. Horizontal scalability refers to the ability to see and compare many objects (e.g., databases) within a single view. Treemaps can be used to navigate, filter, and show deep, multidimensional data. Their space-filling property enables a user to simultaneously view, prioritize, and interact with hundreds or even thousands of objects at typical screen sizes.

Enterprise applications are developed using a combination of standard components and custom code. A standard visualization component is always a compromise that attempts to implement a majority of the highest-priority requirements from the product teams that will use it. Ultimately, the design of an enterprise visualization component such as a treemap must address significant design trade-offs that are specific to applications. Dynamic updates and interactive features may be sacrificed in support of faster performance. The need for real-time filtering may supplant certain performance requirements in some applications. Two versions of a component may be required in some cases: one that implements basic functionality very efficiently and another that implements sophisticated features but requires more memory. These decisions can also be complicated by difficulties in obtaining useful and accurate information from multiple product teams about how they would prioritize requirements for a visualization component with which they may have little experience.

The treemap discussed here is not yet a standard component in any of Oracle's development toolkits. The entry path into product was through custom-coded components. While product integration is easier using standard components, product teams may be willing to include nonstandard components because their end users can benefit from customized functionality.

2 Oracle Business Intelligence: First Treemap Attempt

In early 2005, the Oracle team that developed a product to create ad hoc queries and business intelligence reports was interested in providing treemaps as data visualizations to their business analyst end users. The team wanted the treemap in their next release, which was to include both a thick-client product for creating and editing reports, Workbook Builder, and a thin-client product for reading reports, Report Center. Intended users of the thick-client product had different interaction requirements and were more technically oriented than intended users of the thin-client product. Workbook Builder users were expected to choose and configure graphs for reports, while Report Center users were expected to mostly read reports that had been prepared for them. Report Center requirements included minimal interactive capabilities, while Workbook Builder had extensive and dynamic interaction, including in-place drilling and interactive filtering.

Differences in data storage between the thick- and thin-client treemaps dictate important architectural differences in treemap layout and rendering. As an example, layout and cell color assignments should logically run on the server because client-side data storage is discouraged in thin-client treemaps. From a usability perspective, this can cause a frustrating and slow visual refresh for interactive tasks such as filtering data.

The UI research group worked with both product teams to design a flexible treemap component with Java classes to support a rich API that could be used to load hierarchical data into treemaps, compute layouts, and associate treemap cells with colors. The component was designed to render to Java (Workbook Builder) or HTML/JavaScript (Report Center). The product teams and their management approved the design.

The research group subsequently helped to integrate the interactive Java applet treemap and the HTML/JavaScript treemap into Workbook Builder (Fig. 2) and Report Center, respectively. The Java applet implemented business intelligence requirements to control the visibility, coloring, and text labeling of the treemap cells.

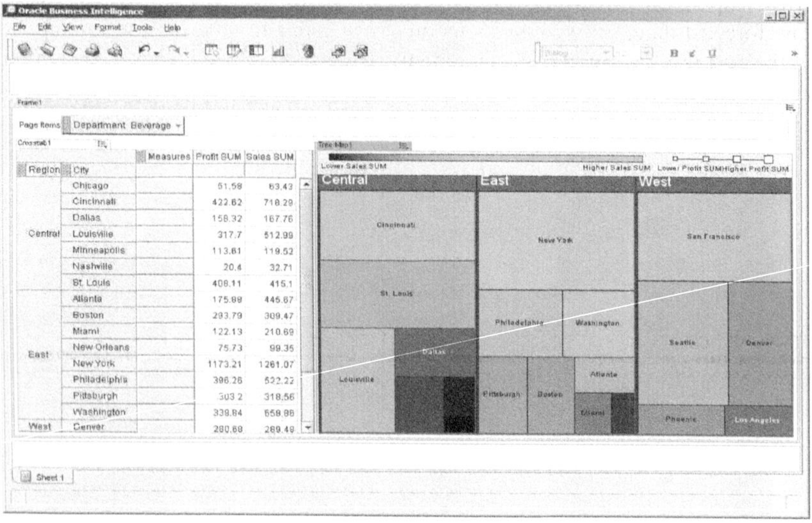

Fig. 2. Workbook Builder screen, merging a flexible hierarchical table with a treemap separated by a vertical splitter bar. Selecting a cell on either component highlights the same data on the other component. Dragging and dropping the table's row and column headers pivots and reaggregates the data, causing the treemap to recalculate its layout and re-render.

It had deep interaction, enabling end users to select dimensions, filter dimensions based upon color and area, customize graphical notifications, and trim the depth of the visible hierarchy. The Report Center treemap cells were implemented as HTML elements with cell borders, background colors, and textual content specified by using cascading stylesheets. JavaScript code was developed to provide a tooltip and to handle selection events on cells.

In late 2005, Oracle purchased Siebel Systems, which had a competing ad hoc reporting product, and both Workbook Builder and Report Center were cancelled. In response, the UI research group worked with both product teams to amend and reprioritize the treemap requirements. In the first half of 2006, the research group developed a JavaScript treemap that used animation while zooming and helped define dialogs for configuring treemaps. The product teams were then reorganized and development priorities changed. The thin-client code was to be integrated into a Siebel product, but later fell out of scope.

Although the treemap was not successfully integrated into a product at this point, the extensively featured thick-client prototype completed by the UI research group provided a flexible testbed for other teams exploring the use of treemaps in their products. One of these teams was Oracle Enterprise Manager.

3 Enterprise Manager: Second Attempt

Oracle's Enterprise Manager (EM) is a software tool through which IT administrators can monitor and manage the availability and performance of IT infrastructure, applications, and databases. EM is a complex, three-tier Java application comprising thousands of pages and hundreds of use cases.

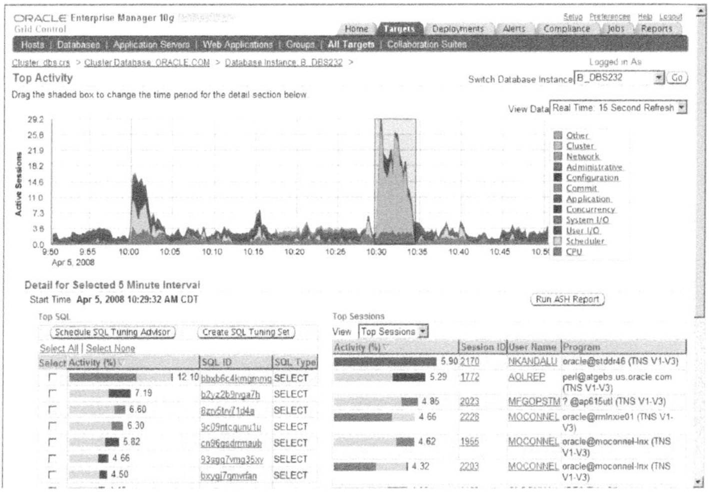

Fig. 3. Enterprise Manager 10g Database Top Activity page. The top chart graphs DB Time over time as a set of stacked line charts by wait class, with preassigned colors for each class. The bars below show the highest accumulators of DB Time by other dimensions within the user-draggable highlighted area in the upper chart.

The end users are highly technical and frequently select and filter data dimensions as they diagnose system issues. Release 10g of the Oracle database introduced the concept and instrumentation for measuring Database Time (DB Time) as the primary internal measure of database activity and performance. Stripchart visualizations of DB Time were included in the user interface to support database tuning and performance diagnosis, where an end user identifies and reduces the largest sources of DB

Time accumulation. In these screens, larger amounts of DB Time display with a larger visual footprint, attracting user attention for subsequent drill-down analysis. This usage metaphor ("click on the big stuff") is easily explained to and adopted by users. Two important visualization properties of DB Time are its inherent multidimensionality (some 30 dimensions of potential interest are instrumented and captured) and scalability within dimension (dimensions could potentially include thousands of values).

The EM Top Activity page (Fig. 3) has been both commercially successful and popular with users. The page facilitates interactive diagnosis, enabling the user to see unusual accumulations of DB Time then isolate specific time periods for details. The details show ranked contributors to DB Time during the selected time period as color-coded stacked bars for correlation with the main chart. User-selectable dimensions and navigation drill-down to dimension-specific tuning tools are also available.

3.1 Prototyping DB Time Treemaps

EM product architects started exploring treemaps for visualizing DB Time after reading the 2004 UI Research group report. The Top Activity page displays two independent dimensions as lists, but requires scrolling to see all elements. The ability of a treemap to integrate different dimensions and quickly identify large DB Time accumulations would combine these lists into a single nonscrolling view. The research group's Java treemap provided a testbed to conduct visualization experiments of production DB Time data (Fig. 4).

The Java applet enabled users to explore different dimensional hierarchies, coloring schemes, and simulated real-time updates. Several diagnostically useful treemaps

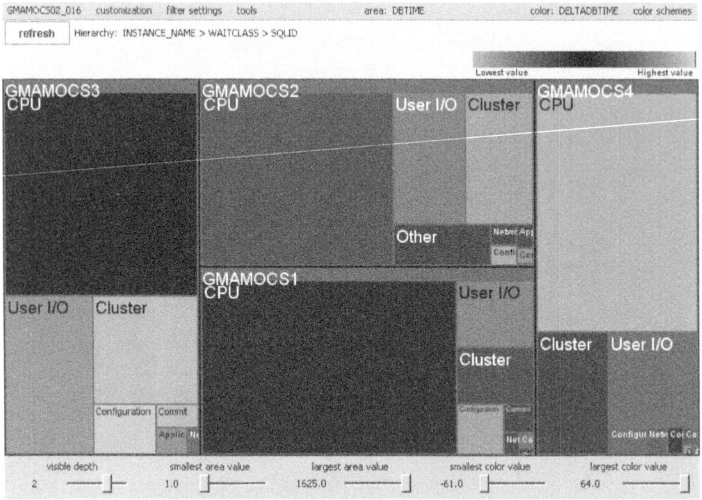

Fig. 4. Java treemap experiment showing breakdown of DB Time spent in an Oracle database using the INSTANCE_NAME>WAITCLASS>SQLID hierarchy. The visible depth level is set to show the instance name and wait class but not SQLID. Sliders enable filtering by low/high color values, large/small areas, and visible depth level.

using DB Time for cell size were generated and a presentation of results was circulated, resulting in significant interest.

Extending from this initial enthusiasm, a prototype treemap viewer was developed in early 2007. Preconfigured sets of DB Time dimensions were loaded into the treemap using predefined queries in order to demonstrate its usefulness for performance analysis on production databases. After starting the prototype, further work would require formal project staffing and resources. Despite substantial interest in the treemap, the existing EM Top Activity page was considered more than adequate for users. In addition, the availability of resources to integrate and maintain the treemap was unclear. The potential benefits of a new treemap view didn't appear to outweigh its potential costs. At this time, treemaps were considered an interesting research area, but not cost-justified for production development within EM.

3.2 DBA Console: A New Opportunity

A decision was made in early 2008 to have administrator-specific login pages for the next version of EM, including a homepage console for database administrators (DBAs) that would summarize database performance across the enterprise. A proposal was developed that showed each database as a treemap cell, sized by DB Time accumulation. This treemap ("Enterprise Loadmap") would be the centerpiece of the DBA console page, which would also show a list of databases that are "down" (not bearing load and thus not appearing in the treemap) as well as a list of recent database incidents. The treemap would be implemented as an Adobe Flex component to ease the integration process with the existing product, which already included other Flex components.

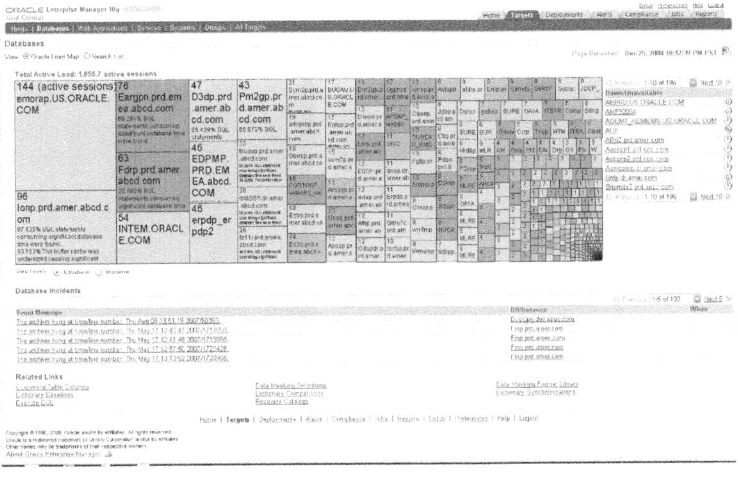

Fig. 5. Enterprise Database Loadmap, a treemap showing DB Time accumulation across an enterprise. Cell size is proportional to database load, and color indicates dominant time component: CPU, I/O, or wait. Cell text shows recent diagnostic findings, and cell click-through navigates to the Top Activity page (Fig. 3) for that cell.

Whereas the earlier treemap DB Time experiments focused on analyzing how time accumulates within a single database, the console Enterprise DB Loadmap (Fig. 5) shows where DB Time is accumulating across an entire enterprise of databases. Databases accumulating more DB Time have visually larger cells and this real estate is used to display database-specific diagnostic information. Busier databases are usually more important to watch, so the visualization favors this natural prioritization. Cell color indicates the dominant component of DB Time (CPU, I/O, or wait). The treemap's efficient use of space scales well to the 10 to 500 (or more) databases characteristic of larger data centers.

The treemap solved the problem of how to show many database targets in a single view, organized so the most important are most accessible and information-rich. Decision makers clearly understood these advantages over list-and-search oriented solutions, and approved the design for product inclusion. The DB Time treemap was once again scheduled to be included in a product.

A final challenge for the treemap surfaced in mid-2008 when the scheduled EM release (including the DBA Console) was suspended in favor of developing the next incremental version of the prior release, which did not include the console. The team quickly developed a proposal for including the treemap into this incremental release as a user-selectable alternative to an existing screen listing all databases in the enterprise. The proposal for inclusion was accepted, receiving very positive reception from an audience of EM executives. The version of EM including the Enterprise Loadmap is currently scheduled for release. Other EM groups have also expressed a desire to explore similar visualizations over other datasets.

4 Discussion

The preceding examples used a variety of approaches for showcasing treemaps in enterprise products. Use cases include monitoring databases, responding to slow database instances, querying sales data, and reporting sales data. Technologies include Java, HTML/JavaScript, and Flex. Each of the treemaps had tooltip feedback and selection capability to expose details about a cell. Otherwise, interaction levels differed in the various prototypes. The business intelligence thick-client treemap supported interactive filtering, in-place drilling, and coordinated highlighting with a hierarchical table. The Enterprise Loadmap enabled users to control depth level, but didn't support filtering to choose a subgroup of the results.

While the level of interaction needs to be matched to the expected user tasks, this matching becomes quite a challenge for a general component that is to be used across multiple applications with multiple user roles and different tasks. We found that having at least one prototype that implements a very large and representative set of the requirements is invaluable for allowing product teams to experiment with the features. Product teams need to be able to configure the prototype with representative customer datasets to see how the features apply to actual customer tasks. The more product teams that prioritize the requirements, the more likely the design of a reusable component will meet the interaction requirements for most of the products.

Why did the Enterprise Loadmap get into product, but not the earlier DB Time treemap prototype? The difference was that the Enterprise Loadmap addressed an unresolved issue: it provides the DBA with a comprehensive overview of the enterprise, with easy access to details or drill-down into specific databases. It is intuitive and easily readable by anyone familiar with the problem domain because databases are what matters most to a DBA, and each database can be seen as a treemap cell, the most important databases having the largest cells.

Product innovation is essential for companies to compete and grow; yet the path for incorporating new ideas into products is not always clear or efficient. The present treemap case studies illustrate how an established concept can be included in different enterprise products, while adapting and morphing to suit individual product and user requirements. Successfully growing organizations should strive toward eliminating the technical and social barriers to include new ideas.

The present exercise showed that both great ideas and attention to interpersonal relationships are needed to integrate a new, innovative concept into an enterprise product. While great ideas are timeless, significant effort must also be spent convincing decision makers that there is an added value from the concept. Though several ups and downs were encountered in the present case, a product team eventually adopted the ideas.

Lessons learned from this project can be divided into those related to developing realistic concepts that provide solutions to real problems, and those related to gaining uptake from product teams.

Concept and Prototype Development

- Ensure that the new concept provides a solution to a well-articulated problem.
- Remain flexible on technology details, as the version that makes it into product will likely require a different implementation.
- Separate ideas from underlying technology. Changing technology may make previously impractical ideas feasible.
- Create a prototype implementing as many requirements as possible so different teams can use it to understand which capabilities they find most useful.

Gaining Product Uptake

- Large companies may move slowly when implementing new solutions, due in part to the complexity of integrating and communicating across many diverse teams.
- Even though potential benefits of new ideas can outweigh their cost of development, constant juggling of project leadership and release schedule priorities may prematurely terminate an innovative project.
- If one implementation avenue fails, try another. Once one team uptakes an idea, other teams may follow.
- Integration of great ideas into existing products requires great timing, in addition to good usability and marketability. This requires management of relationships, not just management of ideas.
- Find and cultivate champions that can shepherd the new concept through the ups and downs of product adoption.

- Product and development teams are most likely to understand a complex visualization if it shows their own data, with demonstrations that solve their problems. Many need to be able to see meaning in their own data to recognize the value of an unfamiliar visualization technique.

References

1. Goldberg, J., Helfman, J.I.: Enterprise Network Monitoring Using Treemaps. In: Proc. 49th Ann. Hum. Factors & Ergo. Soc., pp. 671–675. Santa Monica, HFES (2005)
2. Shneiderman, B.: Tree Visualization with Tree-Maps: 2-d Space-Filling Approach. ACM Trans. on Graphics 11(1), 92–99 (1992)
3. Wattenberg, M.: Visualizing the Stock Market. In: Proc. CHI 1999, ACM/CHI, pp. 188–189 (1999)

How Product Differentiation Affects Online Shopping

A. Walkyria Goode

ESPAE Graduate School of Management at ESPOL,
Campus Las Peñas Malecón No 100, Guayaquil – Ecuador
wgoode@espol.edu.ec

Abstract. Online product comparison sites aid consumers in making purchase decisions by providing information regarding product attributes, reviews and retailers. Despite the enormous research on information processing and consumer behavior, little has been directed at investigating how information display impacts online purchase choices. Cognitive theory is used to evaluate whether the order in which consumers receive information influence their decision making. An experiment with 60 participants was performed. Results indicate that no order effects mediated choice among alternatives with equal utility. However, preference was observed for a particular type of product. While all the products had the same overall rating, this product had both a larger number of high and low ratings. It is suggested that this was a sufficient condition to differentiate the product listed in this position from the other products.

Keywords: consumer choice, product differentiation, order bias.

1 Introduction

Online product comparison sites aid consumers in making informed purchase decisions. Online shopping provides the capability for consumers to obtain more information about both price and non-price attributes [1]. Information regarding product attributes, reviews and retailers is presented for assessment. Despite the enormous research on information processing and consumer behavior, little has been directed at investigating how information display impacts online purchase choices. This research will help stipulate factors that influence online comparisons and shopping decisions.

One of the strengths of the Internet is its wide availability of products. It is also one of its challenges. The Pew report [2] states that 19% of online shoppers feel overwhelmed by the amount of choices offered. Choices are made from a set of very similar alternatives. Presentation formats that provide shopping assistance can provide services that maximize the advantages of the Internet. A relevant issue to consider is whether choice is influenced by the order in which information is displayed. Research that predates the Internet indicates that the order in which individuals receive their information affects their decision making and judgment [3]. Better understanding of how this order affects purchase decisions is important for both consumers and merchants. Consumers ought to know that the layout of the information they are researching may unconsciously affect their final decision. Merchants could take advantage of these possible effects and display information that can help maximize their sales. This

M.J. Smith and G. Salvendy (Eds.): Human Interface, Part I, HCII 2009, LNCS 5617, pp. 253–262, 2009.

study examined the effect of information display in situations that compared products with the same global utility.

1.1 Consumer Decision Making

Consumer preference and choice result from evaluating different products against a set of criteria. The subjective value for each option is derived by integrating its attributes' weights and values. Comparison between dissimilar options is possible since the subjective value of each alternative is represented in a same interval scale unit [4], although the weights given to each attribute vary for each decision maker. The values of all the criteria are processed for each option and a preference structure is built.

A systematic strategy used to evaluate alternatives is the Multi-Attribute Utility Theory (MAUT) [5]. MAUT is an additive linear strategy. First, the importance of each attribute is assigned a weight (w_j). Afterwards, each alternative is assessed by examining the value (v_j) each attribute has; these values are weighted by their importance. As per Equation 1, global utility (U_i) is calculated for each alternative by adding the weighted values. The alternative with the highest global utility is chosen.

$$U_i = \sum_{j=1}^{n} w_j v_j \quad (1)$$

1.2 Comparison Matrix

Consumers, particularly online, are demanding higher levels of product information before making purchasing decisions. Users need to be able to scan, evaluate, and select items easily and effectively. The choice and structure of the information presented has a major impact on the purchasing decision [6]. The usual method of presentation for comparison shopping sites is the comparison matrix, which is based on a table format. It organizes attribute information about multiple products in an alternatives-by-attributes (rows-by-columns) matrix [7]. This format maximizes the comparison of attributes across products.

1.3 Order Effects

Research has shown that different properties of an information display influence decisions [8]. The order in which items or groups of items are displayed often determines the order in which information is read and processed. If the decision maker decides to reduce cognitive effort and select the first acceptable alternative, the options listed first in the sequence have a higher probability of being selected. Primacy is defined as a bias toward selecting the first object considered in a set [9]. Confirmation bias [10] is one possible explanation for primacy. Confirmation bias is the tendency to search for information that validates one's decision rather than information that invalidates it. Satisficing theory [11] is another possible explanation for primacy effects as decision makers select the first acceptable solution they encounter. If they follow the sequence presented, they would be biased towards the first items in the list. Recency is defined as a bias toward selecting the last object considered in a set [12]. This bias can be explained when the selection process is the opposite of confirmation bias. Due to a lack of positive information, negative information drives choice. Decision makers survey a list of options by looking for reasons not to select an alternative, creating stronger negative attitudes towards the first alternatives in the sequence and therefore selecting the latter alternatives. Research has found numerous examples of order effects in judgment and decision tasks. Following some examples found in the literature.

Examples of Order Bias in Judgment and Choice. Research in judgment of experienced personnel observed both primacy and recency effects [13]. Patriot air defense operators were to decide whether an approaching aircraft was friendly or hostile. Two different information sequences were used: cues started confirming and then disconfirming the initial cue and vice versa. Primacy effects were found if the context could rationalize the last cues presented, recency effects were found if it did not.

Analysis of the 1998 Democratic primary in New York City showed order effects in 71 of the 79 precincts favoring the first position listed in the ballot. In 17 precincts, the lead the first position received was higher than the winner's margin of victory [14]. The Democratic primary rotated the candidates' names within ballots, so primacy did not affect the outcome. The authors caution that rotation is performed in only 14 states in the U.S. statewide and two additional states do so in some jurisdictions.

A study in healthcare judgments found that patients considered proposed low- and medium-risk treatments less favorable when the list of risks was presented last. This recency effect was not observed when benefits were presented last or with the high-risk treatment [15].

These last three examples involve both judgment and choice and result in the presence of primacy, recency or both. If serial position influences these types of tasks, it is foreseeable that it would affect tasks that involve judging and choosing products. The following are a list of studies that investigate the presence of order effects in consumer tasks.

An experiment – involving a consumer scenario with two products – was performed in a physical store with a salesperson presenting product information to actual customers [16]. Results indicated a primacy bias. The author argued that if a product had a small advantage, it would not trump order bias. A study of online grocery choices also found an effect of primacy [17]. Participants preferred products presented in the first screen especially when the product assortment was difficult to evaluate. The experimental method was not described in detail.

A study that compared vendors in a comparison shopping site [18] resulted in the presence of a primacy bias. Although, it is questionable whether comparison shopping took place in the study. As cited in the paper: "Some participants carried out product comparisons, but product comparison was ignored" (p. 480). In addition, the study did not control for list length or price. Authors propose more strictly controlled replications. Their study is a foray into investigating the impact of serial position on online comparison shopping.

These last three examples involve both consumer judgment and choice and result in the presence of a primacy bias only. However, this does not necessarily imply that recency will not be observed in the present study. Note that both [16] and [17] are different from online comparison shopping where consumers are simultaneously presented with several products. [18] did not control for list length. The present research offers a systematic empirical investigation of order effects in online product comparisons.

1.4 Predictions

Research for judgment and decision tasks has found both order biases. It is predicted that order biases will be observed for product choices performed in an online comparison site. It is not suggested that order biases will always appear during online product comparison and shopping. Descriptive information allows the decision maker to differentiate between choices. However, there are occasions when individuals need to select an option from a set of very similar alternatives, which is quite common in online purchases. Order biases are predicted to occur in decision tasks where information does not differentiate products within the alternative set.

This study will use a consumer decision task. The information layout will be organized by means of a comparison matrix, which allows for a multi-product by multi-attribute representation. If there is no product differentiation, a primacy bias is predicted to occur.

2 Experiment

This experiment required participants to select a restaurant (product) based on restaurant ratings. These ratings were in the form of 1-5 star rating scale, provided by fictitious food critics. This information was organized as a 5x5 comparison matrix that presented 5 restaurants and 5 ratings per restaurant. Products and attributes (i.e. restaurants and food critics) were listed in the rows or columns of the matrix respectively.

Hastie and Dawes [5] state that the most thorough and systematic strategy to evaluate alternatives is to use Multi-Attribute Utility Theory (MAUT). Note that if all the attributes are considered equally important, the weight factor is no longer required in Equation 1. Thus, the value for each alternative is the sum of its attribute values as per Equation 2. In this study, it is assumed that there was no preferential bias for a particular critic since they were fictitious. Decision makers would consider all attributes equally important.

$$U_i = \sum_{j=1}^{n} v_j \quad (2)$$

It should be noted that the MAUT strategy is one among several strategies that a decision maker can use. In fact, this strategy is considered to be one that requires high mental effort. Research has shown that people select choice strategies by means of an effort-accuracy tradeoff [19]; this implies that MAUT would be used only when high accuracy is needed. However, the elimination of the weight factor for the decision task reduces the amount of cognitive effort usually associated with MAUT. The global utility for each alternative is the sum of its ratings. This mathematical operation is quite simple, since it involves single digits (ratings: 1 - 5). It is assumed that participants would use an MAUT strategy; it associated with high accuracy and the experimental conditions entail nominal effort.

Although each restaurant had distinct ratings, the distribution of ratings was manipulated so all the products had the same sum of rating. As per Equation 2, all alternatives have the same global utility (U_i=18). Thus, there is no differentiating information that could elicit preference for a particular restaurant. If order effects do not influence choice, all alternatives should have equal probability for being chosen.

2.1 Method

Participants perform a consumer decision task after observing each stimulus. A test of memory recognition follows each decision. This experiment is part of a larger research program aimed at investigating the influence of information display on decision making.

Participants. Sixty undergraduates from the University of Maryland participated in the study as a partial fulfillment of a course requirement. All subjects had normal or corrected-to-normal vision. Participants consisted of various demographic groups representative of a typical urban university. Participants were run individually in single sessions lasting approximately 35 minutes.

Materials. A 5x5 (restaurants x critics) comparison matrix was created (See Figure 1). It was necessary to create a matrix were all the labels elicited the same awareness. If the matrix compared market products and price was among the attributes compared, the "price label" could elicit attention due to its connotation and not due to its position on the matrix. Thus restaurants were chosen as the compared product and food critic ratings were selected as the compared attributes. It should be noted that all restaurants and critics were fictitious. To avoid preference for a certain type of cuisine, the restaurants compared within each trial were always of the same type (all Italian, all Mexican, etc.).

	food.com	restaurants.com	foodexpert.com	dining.com	chef.com
The Alps	4	5	4	2	3
Little Switzerland	3	2	5	4	4
Bern	2	3	4	5	4
Geneva	5	4	2	2	5
Matterhorn	4	4	3	5	2

Fig. 1. Comparison Matrix

Restaurants and food critics appeared in different order for every comparison. The ratings were presented in 5-point scale, similar to the common 5-star rating system. The rating scale was explained to the participants before the experiment began. Each restaurant had the same utility in order to control for preference and the ratings for each critic always summed up to the same value so that none would be regarded as either a harsh or lenient critic.

Procedure. The participant was greeted and the general instructions for the test were provided. A pre-test questionnaire and consent form were then administrated. The pre-test questionnaire gathered demographic data that showed non-significant results. The task consisted of selecting a restaurant based on the food critics' recommendations. Participants could look at the comparison matrix as long as they wanted. In order to observe the ratings, the participant had to click each particular row. After pressing a continue button, participants entered the preferred restaurant's name and select from a list the critic that influenced their decision. A memory test followed, in which participants where asked to recall the ratings for all products. Participants were informed that a memory test would take place before starting the experiment. Participants were not allowed to go back and see their previous entries or the previous comparison matrices. This task was repeated for four trials. It was the participants' understanding that the first trial was a practice session to get familiar with the experiment and no data would be collected during practice.

2.2 Results

Statistical analysis showed that the counterbalancing conditions did not have significant effects on the results.

Preference. A meta-Chi-Square[1] was calculated, it showed significant preference for Rows ($\chi^2(8)=17.26$, $p<.05$) but not for Columns ($\chi^2(8) = 5.66$, $p>.05$) (See Figure 2). Results indicate that no order effects were present.

Results were analyzed in order to evaluate whether the stimuli influenced preference. One assumption of the experimental design was that the information describing the products would not differentiate them. If no product differentiation was present, all products should have equal probability of being selected. This was not the case as Row 4 was preferred over the other rows. In fact, participants significantly chose the fourth row during both trials ($\chi^2(16)=27.04$, $p<.05$). Even though all restaurants had the same average rating, it was noted that restaurants presented in Row 4 had a higher distribution of 5-star ratings.

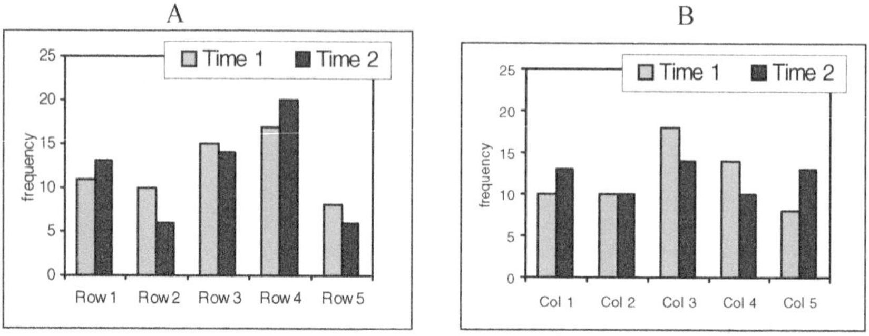

Fig. 2. Preference– (A) By Row. (B) By Column

Recall. Recall was scored as the ratio of correctly recalled cells over the total number of cells per row or column; i.e. if the participant recalled 4 cells in Row 1, the recall score for Row 1 would be 0.80.

[1] Both Chi-Square distributions and degrees of freedom possess additive properties. The meta-Chi-Square was calculated by summing –for both trials– χ^2 variables and their corresponding degrees of freedom.

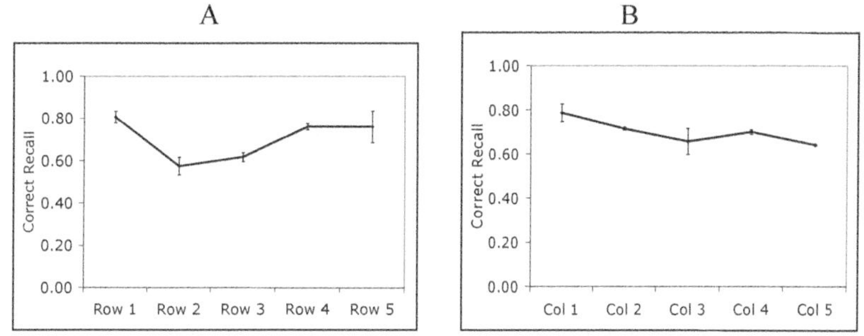

Fig. 3. Recall – (A) By Row. (B) By Column.

A repeated measures ANOVA was conducted, there was a main effect of position for recall in Rows ($F(4,236)=11.31$, $p<.001$) and Columns ($F(4,232)=3.04$, $p<.05$). A Bonferroni correction for multiple comparisons was used, the first row was recalled significantly more ($p<.001$) when compared to the other rows – with the exception of the last row – and the first column was recalled more ($p<.05$) when compared to the other columns (Figure 3). A primacy effect for both rows and columns was found. There were no interactions between rows or columns ($F(16,960)=1.10$, $p>.05$).

Probability of Recall given Preference. A new variable was coded to examine the probability of recall given preference (i.e. If Row 2 was preferred, what percentage is recalled for Row 2?). The probability of recall given preference was significantly higher than probability of recall given non-preference for both rows and columns. See Table 1 and Figure 4. There was a significant positive correlation ($r_b=.11$) between recall and preference ($t(60)=2.11$, $p<.05$).

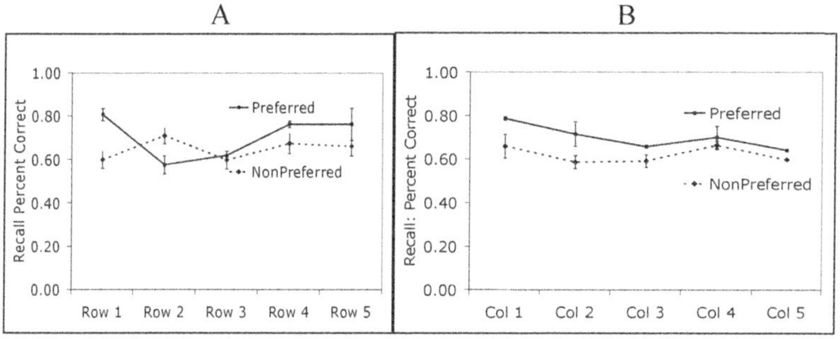

Fig. 4. Recall for Preferred vs. Non-Preferred Items– (A) By Row. (B) By Column.

Table 1. p(recall I preference) vs. p(recall I non-preference)

	p(recall I preference)		**p(recall I non-preference)**		**t(60)**
	M	**SD**	**M**	**SD**	
Rows	0.69	0.23	0.64	0.22	1.78 *
Columns	0.71	2.29	0.63	0.21	2.70 **

* p<.05
** p<.01

3 Discussion

No order effects – primacy or recency – mediated choice. The study predicted the presence of order effects in online consumer choice under a condition with no product differentiation. Stimuli were designed so that product attribute values would not allow for differentiation between the alternatives. In addition, all restaurants had the same global utility as per MAUT. An alternative's value is calculated by summing the ratings assigned to it. If all alternatives have an equal sum, they all should have the same subjective value and there should be no preferences based on distribution of ratings. However, a particular distribution of ratings seemed to favor one specific product. Preference was observed for the product listed in Row 4. All rows had the same average rating; however, the fourth row had the higher combination of 5-star ratings. It appears that consumers seek differences that help them justify their choice. Conditions that presented alternatives with equal global utility and characteristic information showed a preference for such singular products. Participants perceive this specific distribution of ratings as the differentiating information that guides choice.

Porter [20] describes differentiation as developing a unique position on an attribute that is "widely valued by buyers" (p. 14). This is not the definition used in this study. Differentiation is defined as the strategy that helps separate a product from a set of similar alternatives. In the present scenario, the successfully differentiated product was the one with more 5-stars albeit more 2-stars.

Research has shown that choice among options is context dependent and is conditional on how the choice set is represented [21]. Context dependence implies that an option's value is not solely determined by the attributes it possesses; but also by the attributes of the other available options. Row 4 represented restaurants with combinations of the rating composition 2-2-4-5-5. The other products in the comparison matrix had combinations of the rating composition 2-3-4-4-5. It appears that a confirmation bias influenced choice. Those with a larger number of high ratings were preferred despite the experimental design requiring those products to also include a larger number of low ratings. Participants discounted the low ratings in favor of the high ratings.

A primacy bias was observed in recall. Participants recalled the first position significantly more than the other positions. Research in memory has persistently found a relationship between the position in which an item is studied and the accuracy of recall for that item. If recall is immediate, both primacy and recency effects are observed. If there is a delay between study and recall, the recency effect is reduced [22]. The present study was not a learning study, although participants were aware that a memory test would be part of the procedure. The observation phase of this study could represent a learning phase of a memory study and the decision phase would be the equivalent of the delay before testing.

Participants were able to recall correctly most of the items they preferred. They were not selecting options randomly but rather paying attention to the decision task. Recall was higher for preferred items when compared to non-preferred items.

The results of the current research carry interesting theoretical and practical implications. It appears that participants were not following MAUT but some satisficing algorithm. In fact, it is hypothesized that the satisficing strategy is based on confirmation bias. Participants preferred products with the highest ratings even though these same products included the lowest ratings. The results of this study suggest that a successful differentiation strategy for online commerce is to focus on attaining great ratings on a set of attributes rather than obtaining good average ratings in all attributes. Further evidence for this premise is found in [23], a regression analysis of the craft beer industry found a positive significant correlation between the standard deviation of ratings and sales growth. The authors hypothesize that having good average reviews is less important than occupying and extreme position. A consequence of occupying such position is an increase in rating variance since extremes could lead to some unfavorable reviews.

A possible criticism to this study would entail that actual consumer behavior does not incorporate comparisons among products with equal weighted attributes. The empirical design determined that all attributes should be weighted equal. Most purchases usually involve attributes with unequal weights. However, there are exceptions; for example, a consumer not knowledgeable with the product. If there are no *a priori* product requirements and the alternatives are similar in price, all attributes would have equal weights.

To further strengthen the findings from this experiment, two additional studies can be undertaken. First, the absence of order effects in online consumer choices should be evaluated. The hypothesis stated that both primacy and recency biases appear if there is no differentiating information between alternatives. It is necessary to create a comparison matrix with similar products (same utility) where the ratings would not be able to be considered as differentiating information. Second, the effect of the differentiation caused by rating distributions needs to be examined. A study could be developed that specifically manipulates different rating distributions and analyzes how they affect choice.

References

1. Degeratu, A.M., Rangaswamy, A., Wu, J.: Consumer behavior in online and traditional supermarkets: The effects of brand name, price, and other search attributes. International Journal of Research in Marketing 17, 55–78 (2000)
2. Pew Internet & American Life Project: Online Shopping, `http://www.pewinternet.org/ppf/r/237/report_display.asp`
3. Hogarth, R.M., Einhorn, H.J.: Order effects in belief updating: The belief-adjustment model. Cognitive Psychology 24, 1–55 (1992)
4. Norman, K.L.: A solution for weights and scale values in functional measurement. Psychological Review 83, 80–84 (1976)
5. Hastie, R., Dawes, R.M.: Rational choice in an uncertain world: The psychology of judgment and decision making. Sage Publications, Thousand Oaks (2001)
6. Pereira, R.E.: Factors influencing consumer perceptions of Web-based decision support systems. Logistics Information Management 12, 157–181 (1999)
7. Häubl, G., Trifts, V.: Consumer decision making in online shopping environments: The effects of interactive decision aids. Marketing Science 1, 4–21 (2000)
8. Kleinmutz, D.N., Schkade, D.A.: Information displays and decision processes. Psychological Science 4, 221–227 (1993)
9. Krosnick, J.: Response strategies for coping with the cognitive demands of attitude measures in surveys. Applied Cognitive Psychology 5, 213–237 (1991)

10. Koriat, A., Lichtenstein, S., Fischhoff, B.: Reasons for confidence. Journal of Experimental Psychology: Human Learning and Memory 6, 107–118 (1990)
11. Simon, H.A.: A behavioral model of rational choice. Quarterly Journal of Economics 69, 99–118 (1955)
12. Miller, J., Krosnick, J.: The impact of candidate name order on election outcomes. Public Opinion Quarterly 62, 291–330 (1998)
13. Adelman, L., Bresnick, T.A., Christian, M., Gualtieri, J., Minions, D.: Demonstrating the effect of context on order effect for an army air defense task using the Patriot simulator. Journal of Behavioral Decision Making 10, 327–342 (1997)
14. Koppel, J.G.S., Steen, J.A.: The effects of ballot position on election outcomes. The Journal of Politics 66, 267–281 (2004)
15. Bergus, G.R., Levin, I.P., Einstein, A.S.: Presenting risks and benefits to patients, the effect of information order on decision making. Journal of Internal Medicine 17, 612–617 (2002)
16. Scarpi, D.: Effects of presentation order on product evaluation: An empirical analysis. International Review of Retail, Distribution and Consumer Research 14(3), 309–329 (2004)
17. Breugelmans, E., Campos, K., Gijsbrechts, E.: Shelf sequence and proximity effects on online grocery choices. Marketing Letters 18, 117–133 (2007)
18. Xu, Y., Hee-Woong, K.: Order effect and vendor inspection in online comparison shopping. Journal of Retailing 84(4), 477–486 (2008)
19. Bettman, J.R., Johnson, E.J., Payne, J.W.: A componential analysis of cognitive effort in choice. Organizational Behavior Human Decision Processes 45, 111–139 (1990)
20. Porter, M.E.: Competitive Advantage. The Free Press, New York (1985)
21. Bettman, J.R., Johnson, E.J., Luce, M.F., Payne, J.W.: Correlation, conflict and choice. Journal of Experimental Psychology: Learning, Memory, And Cognition 19, 931–951 (1993)
22. Postman, L., Phillips, L.W.: Short-term temporal changes in free recall. Quarterly Journal of Experimental Psychology 17, 132–138 (1965)
23. Clemons, E.K., Gao, G.G., Hitt, L.M.: When online reviews meet hyperdifferentiation: A study of the craft beer industry. Journal of Management Information Systems 23(2), 149–171 (2006)

Design and Evaluation of the Customized Product Color Combination Interface Based on Scenario Experience

Ying-Jye Lee[1], Cheih-Ying Chen[2], and Fong-Gong Wu[3]

[1] Department of Cultural Business Development, National Kaohsiung University of Applied Sciences, 415 Chien Kung Road, Kaohsiung 807, Taiwan, R.O.C.
[2] Department of Multimedia Design, Fortune Institute of Technology, No.1-10, Nwongchang Rd., Daliao Township, Kaohsiung County 831, Taiwan, R.O.C.
[3] Department of Industrial Design, National Cheng Kung University, 1, Ta-Hsueh Road, Tainan, Taiwan 70101, R.O.C.
yjlee@cc.kuas.edu.tw, cychen@center.fotech.edu.tw, fonggong@mail.ncku.edu.tw

Abstract. The customized product color interface based on scenario experience is defined as the experienced marketing model in this study. There are 48 color combinations of the spatial image resulted from four scenario styles and 12 popular sofa colors. The image compositing technique is adopted to appear the 48 color combinations of the spatial image on computer screen. This study compares the difference between the experienced marketing model and traditional marketing model by using the evaluation items of Personal Involvement Inventory. Results show that eight evaluation items including interesting, exciting, means a lot to me, appealing, fascinating, valuable, involving, and needed for the experienced marketing model are significant better than the traditional marketing model. Besides, two evaluation items including important and relevant doesn't appear significance between the two models. Therefore, the entrepreneur who wants to display the color primarily commodity should design the customized color combination interface with scenario experience for consumers to take opportunity to find the appropriate products to meet with consumers' needs, so as to shorten communication time between entrepreneurs and consumers.

Keywords: Customized product; Color combination; image compositing technique; Personal Involvement Inventory.

1 Introduction

In new age, the new consumption attaches great importance to consumer experience. Consumers gradually take to expend in the process that brings the unforgettable experience; and the traditional marketing have some changes. Nowadays, consumers have already regarded product function, benefit and quality as essentiality, and they need the product stimulating their perceptions and touching their thoughts. Schmitt [1] announced that the experience economy time has already approached. The experience economy for the enterprise indicates the service as the core and the product as the material to devise the sense experience, change consumer behavior, and create the feeling for the consumer which is worth recollecting [2].

M.J. Smith and G. Salvendy (Eds.): Human Interface, Part I, HCII 2009, LNCS 5617, pp. 263–270, 2009.

Customization is one kind of excellent experience economy tools. Consumers can customize their own products by means of customization modules. While a product is customized, consumer goes through the active experiences at the same time. Regarding customized products, it is more easy produce to vary product's color than its form for meeting the need and preference of individual consumer. Different color combination create vary image and provide different needs for consumers [3]. Lin et al. [4] addressed the color as a key attribute in object recognition because object colors correlate strongly with object identify. Lai et al. [5] demonstrated that the product color has a greater affect on product image than product form. Wu et al. [6] noted different color products can achieve different visual effects and create more pleasing and stylish product image.

Consumers can interact with the designer or the entrepreneur by customized way. Besides, providing the diverse color component to involve the consumer in the design process, continue product life cycle, reduce product development cost and gain a higher profit [7]. Therefore, different color combinations by applying image compositing technique to the customized design interface; it will give the consumers different color fascination and design experience. However, the ordinary process of product color selection always consumed most of time in customizing and communication. The end product may fall short of consumers' expectations, since only sample pictures of the catalogue were shown to the consumers during the selection process, the finished effect of a completed product could not be effectively pictured. This vast discrepancy in the end product and consumers' expectations causes as much frustration to the consumers as well as to the manufacturers. The kind of selecting process is called the traditional model in this study.

The main objective of this study is to design the customized product design interface and to compare the difference between the experienced marketing model and traditional model. The study plans to design the customized product color interface based on scenario display via the marketing approach of experience economy. The study takes the leather sofa as an example. The sofa belongs to the large-scale commodity, and all different colors sofas can't be displayed completely in the shop. Moreover, the each different color sofa matches the different scenario to build the different spatial image for meeting consumer's need. Considering above descriptions, a customized sofa color combination interface with scenario experience is designed in this study.

2 Image Compositing Technique

The image compositing technique can be used in a wide variety of applications such as virtual reality which requires the scene to be displayed from different viewpoints and stereo matching [8]. Virtual environments may also be produced through the image compositing process. Definitions and descriptions in regards to virtual environments can be found in the research conducted by Bayliss et al. [9]. A good discussion on virtual reality has been presented by Machover and Tice [10] and Ellis [11]. Jayaram et al. [12] and Connacher et al. [13] also developed a virtual assembly design environment through the concept of virtual environment. From the above, it can clearly be seen that the concept of virtual environment and image compositing has already been widely adopted in many fields. This study adopts the image compositing technique to assist in the color combination of the sofa by letting consumers preview a virtual image of the completed work on a computer monitor. Specially, descriptions and procedures with regard to image compositing technique can be found in the research conducted by Wu et al. [14] [15].

3 Methods

3.1 Scenario Style and Interface Arrangement

In order to achieve the objective of this study, the scenario style of the study refer to the sofa style image constructed by factor analysis [16], so as to design four different scenario styles including grand, leisure, taste and vogue styles (Fig. 1) In addition, there are 12 popular sofa colors including red, orange, light yellow, yellow, light blue, blue, green, purple, black, white, brown and dark brown (see Table 1) which are suggested by five interior designers from leather samples provided by sofa industry professionals [16], and thus there are 48 (4*12) color combinations of the spatial image. This study applies image compositing technique to design the customized color combination interface with scenario experience to appear 48 color combinations of the spatial image on computer screen. For example, if the scenario style is vogue and the sofa color is orange, then the customized color combination interface with scenario experience of the spatial image appears as Fig. 2.

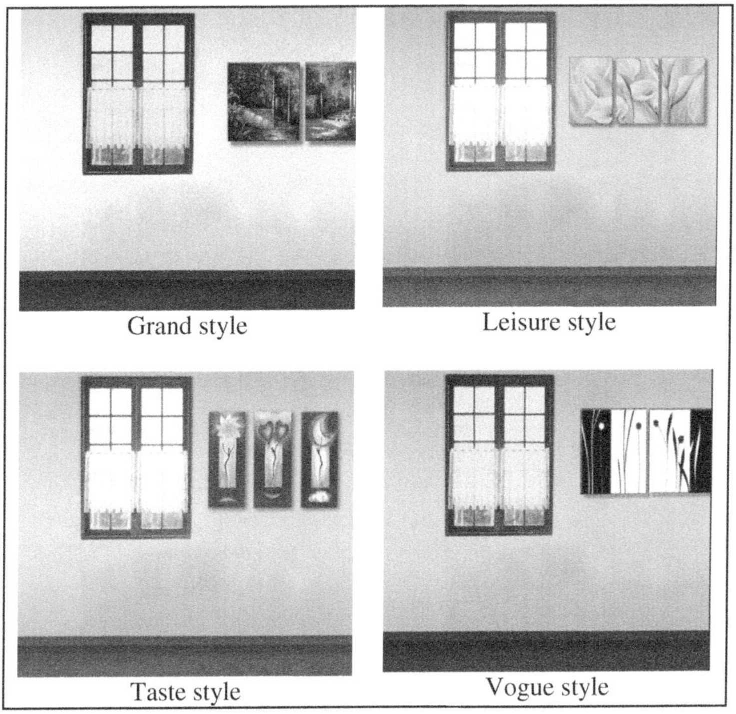

Fig. 1. The four different scenario styles in this study

Table 1. The 12 popular sofa colors

Color name	Color Chip	Lab Color space
White		L93, a1, b2
Red		L45, a46, b18
Orange		L69, a35, b54
Light Yellow		L90, a-1, b41
Yellow		L86, a7, b75
Green		L37, a-13, b2
Light Blue		L78, a-17, b0
Blue		L38, a2, b-47
Purple		L31, a31, b-36
Light Brown		L72, a5, b29
Dark Brown		L36, a7, b9
Black		L40, a0, b3

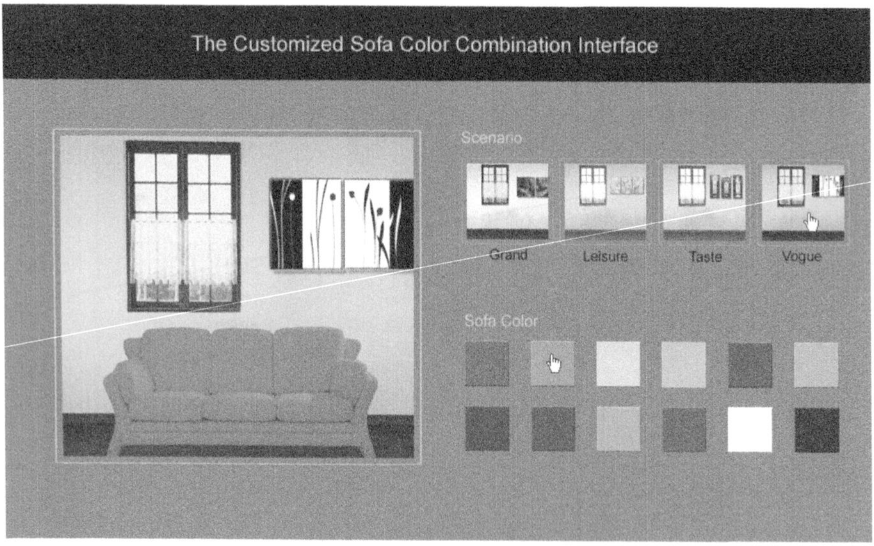

Fig. 2. The customized color combination interface with scenario experience

3.2 Participants

To attain effective results, this study invites 30 consumers who visit or purchase some house fitting from the furniture shop to participate in interface evaluation, 16 males and 14 females, with ages ranging from 26 to 48 (mean=35.7, SD=5.2), and all possessed a normal or after correction, eyesight of 1.0., and requests the participants to operate the customized color combination interface of the scenario experience.

3.3 Apparatus/Stimuli

A multimedia computer was used here to assist in the process of interface evaluation. Additionally, a high-resolution monitor (19") with 1440 (horizontal) × 900 (vertical) pixels resolution and 60HZ refresh rate was used to display the experiment stimuli. Stimuli used in the evaluation process were processed composite images of the customized color combination interface, each with a dimension of 1000 (horizontal) ×600 (vertical) pixels, and subtended a visual angle of 47.6°×29.6° from a viewing distance of 40cm (Fig. 3).

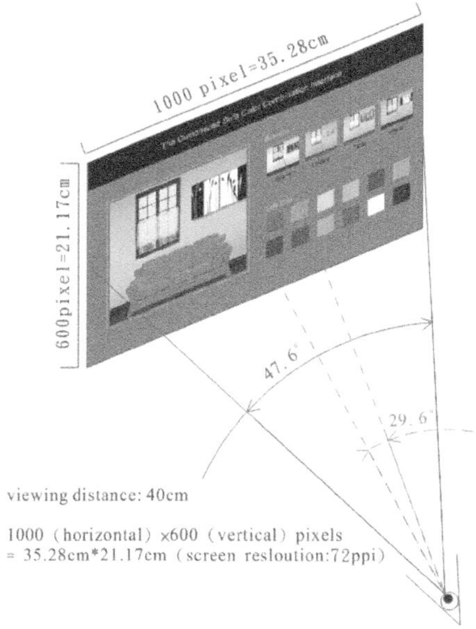

Fig. 3. The stimuli subtending a visual angle of 47.6°×29.6° from a viewing distance of 40cm

3.4 Procedure

Participants were seated in front of a monitor during the evaluation process. Keyboards and chairs were adjusted individually for participants to achieve a viewing distance of 40 cm from the screen to ensure a comfortable experimental environment. Before commencement of the interface evaluation, participants were introduced to the procedures. All the evaluating time regarding experienced marketing model for a participant is about 10 minutes. Afterwards, the study asks the participant's comments regarding the experienced marketing model (i.e. the customized color combination interface in the study) and traditional marketing model to fill in Personal Involvement Inventory [17]. Furthermore, participants were asked to make ticks on a 100mm measuring scale in accordance to their evaluation of the 10 evaluation items of the PII by means of the measuring scale. Take the item "Interesting/boring" as an example (Fig. 4); the extreme left of the measuring scale corresponds to "exceedingly boring", and the extreme right "exceedingly interesting". After evaluation, the ticks on the degree of realistic effect marked on the measuring scale were further quantified into 0 and 10 values (0 representing the left side of the scale and 10 the right side). At the ending of the experiment, data collected on the reaction of participants was further statistically analyzed.

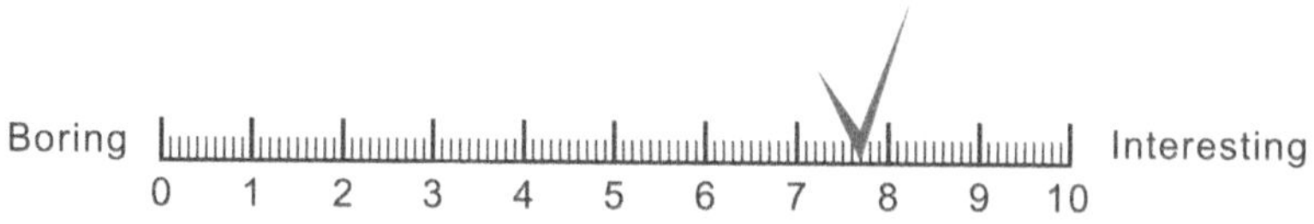

Fig. 4. The measuring scale for evaluating experienced marketing model and traditional marketing model

4 Results and Discussion

This study compares ten evaluation items in Personal Involvement Inventory for the experienced marketing model with traditional marketing model by T-test. Results show that the eight evaluation items including interesting, exciting, means a lot to me, appealing, fascinating, valuable, involving, and needed for the experienced marketing model are statistically significant better than the traditional marketing model (Table 2).

Table 2. T-test for comparing the experienced marketing model with traditional marketing model using the 10 evaluation items of Personal Involvement Inventory

Evaluation item	Mean of Evaluation values		*P* value
	Experienced model	Traditional model	
Important/unimportant	6.72	6.99	0.209943
Interesting/boring	8.06	5.55	0.000000**
Relevant/irrelevant	6.14	5.70	0.050951
Exciting/unexciting	7.90	6.00	0.000000**
means a lot to me/means nothing	6.40	5.37	0.000498**
appealing/unappealing	8.02	5.70	0.000000**
Fascinating/mundane	7.90	5.13	0.000000**
valuable/worthless	7.81	5.85	0.000000**
involving/uninvolving	7.47	5.42	0.000000**
needed/not needed	7.56	6.62	0.000158**

*: P <0.05; **: P <0.01

Besides, the rest evaluation items including important and relevant doesn't appear significance between the two models. According to the analysis results, consumers have high favor on the experienced marketing model (i.e. the customized color combination interface with scenario experience). Therefore, the entrepreneur who wants to display the color primarily commodity should design the customized color combination interface with scenario experience for consumers to take opportunity to find the appropriate products to meet with consumers' needs, so as to shorten communication time between entrepreneurs and consumers.

5 Conclusion

The customized product color interface based on scenario experience is designed as the experienced marketing model. After evaluation, the experienced marketing model is statistically significant better than the traditional marketing model. Furthermore, the image compositing technique gives consumers the advantage of viewing the customized product color interface on a computer screen, thus enabling them to effectively select suitable spatial image that best suit their needs. The experienced marketing model based on the image compositing technique would improve on the efficiency in product color combination selection; minimize communication time between entrepreneur and consumer. The proposed experienced marketing model in this study may also be extended to similar researches in other furnishing related industries.

Acknowledgments

This research was partially supported by the National Science Council in Taiwan through Grant NSC 96-2411-H-151 -001 and NSC 97-2410-H-151-010.

References

1. Schmitt, B.H.: Experiential Marketing: How to Get Customers to Sense, Feel, Think, Act. Relate Free Press (1999)
2. Pine II, B., Joseph, Gilmore, J.H.: The Experience Economy: Work Is Theatre & Every Business a Stage. Harvard Business School Press (1999)
3. Chronicle Books Inc., The color book (1997)
4. Lin, Z., Wang, J., Ma, K.K.: Using eigencolor normalization for illumination-invariant color object recognition. Pattern Recognition 35(11), 2629–2642 (2002)
5. Lai, H.H., Lin, Y.C., Yeh, C.H., Wei, C.H.: User-oriented design for the optimal combination on product design. International Journal of Production Economics 100(2), 253–267 (2006)
6. Wu, F.G., Chen, C.H., Lee, Y.J.: Optimum product color combination using the FAHP approach combined with the image compositing technique. In: HCI International 2005, Las Vegas, USA (2005)
7. Chen, C.Y., Lee, Y.J., Wu, F.G., Su, C.F.: Screen Layout on Color Search Task for Customized Product Color Combination Selection. In: Jacko, J.A. (ed.) HCI 2007. LNCS, vol. 4551, pp. 32–40. Springer, Heidelberg (2007)

8. Zhang, Z.Y.: A robust technique for matching two uncalibrated images through the recovery of the unknown epipolar geometry. Artificial Intell. J. 78, 87–119 (1995)
9. Bayliss, G.M., Bowyer, A., Taylor, R.I., Willis, P.J.: Virtual manufacturing. In: Proceedings of CSG 1994: Set-Theoretic Solid Modeling Techniques and Applications, pp. 353–365 (1994)
10. Machover, C., Tice, S.E.: Virtual reality. IEEE Computer Graphics and Applications 14(1) (1994)
11. Ellis, S.R.: What are Virtual Environments? IEEE Computer Graphics and Applications 14(1) (1994)
12. Jayaram, S., Connacher, H.I., Lyons, K.W.: Virtual assembly using virtual reality techniques. Computer-Aided Design 29, 575–584 (1997)
13. Connacher, H.I., Jayaram, S., Lyons, K.: Virtual assembly design environment. In: Proceedings of 1995 Computers in Engineering Conference, Boston, MA (1995)
14. Wu, F.G., Lee, Y.J., Chen, C.H.: Evaluation of the Realistic Effect of Image Compositing to Assist in Curtain Selection. International Journal of Industrial Ergonomics 32(1), 1–12 (2003)
15. Wu, F.G., Lee, Y.J., Lin, M.C.: Using the fuzzy analytic hierarchy process on optimum spatial allocation. International Journal of Industrial Ergonomics 33(6), 553–569 (2004)
16. Ma, M.Y., Chen, C.Y., Wu, F.G.: A design decision-making support model for customized product color combination. Computers in Industry 58(6), 504–518 (2007)
17. Zaichkowsky, J.K.: Measuring the Involvement Constructure. Journal of Consumer Research 12(3), 341–352 (1985)

Distribution of Human-Machine Interfaces in System-of-Systems Engineering

Sandro Leuchter and Dirk Mühlenberg

Fraunhofer Institute for Information and Data Processing (IITB)
Fraunhoferstr. 1, 76131 Karlsruhe, Germany
{sandro.leuchter,dirk.muehlenberg}@iitb.fraunhofer.de

Abstract. System-of-systems integration requires sharing of data, algorithms, user authorization/authentication, and user interfaces between independent systems. While SOA promises to solve the first issues the latter is still open. Within an experimental prototype for a distributed information system we have tested different methods to share not only the algorithmics and data of services but also their user interface. The experimental prototype consists of nodes providing services within process portals and nodes realizing services with software agents. Some of the services were extended with WSRP (web service remote portlet) to provide their own user interface components that can be transmitted between separated containers and application servers. Interoperability tests were conducted on JBoss and BEA Portal Workshop. Open questions remain on how the layout of one component should influence the internal layout of other GUI-components displayed concurrently. Former work on user interface management systems could improve todays tools in that respect.

Keywords: HMI, portlet, wsrp, web clipping, interoperability.

1 Preface and Problem Statement

Generally integration of software systems means to use functions or data from one system in the other. Enterprise application integration (EAI) deals with frameworks and middleware to integrate business data sources of different software systems. Another important field of data integration is how to deal with different user login identities and access rights. The problem is how to identify users from one system for the other (authentication) and how to map their access rights (authorization). Solutions have to address not only the technical side but also organizational issues. The third important field of system integration is how to distribute functionality between integrated systems. To achieve this functions have to be identified that are to be shared between systems. The system that provides such a function has to be extended with a new interface that can be accessed by function consumers over a network. The current technological approach is using a communication middleware like an enterprise service bus with different adaptors as interfaces to the services providers (e.g. via web services). A further area in integration is discovery of such services.

M.J. Smith and G. Salvendy (Eds.): Human Interface, Part I, HCII 2009, LNCS 5617, pp. 271–278, 2009.

It remains open how to interactively use services by consumers: Normally a service provides a specified functionality that is accessed via a certain interface over which parameters are given to the underlying process. The nature and semantics of the parameters depend on the service. But how to specify the parameters interactively remains to the consumer that has to build input fields in it own user interface. From an engineering point of view it is not useful to share functionality but not user interface elements. Thus there are some different methods emerging how to share user interface elements besides business data, user data, and functionality. In the following section web service remote portlets, web clipping and smart clients are presented.

2 Methods for Sharing Human-Machine-Interfaces

2.1 Technological Approach

In a context of a network of SOA nodes (like in a System of systems ad-hoc approach) services may be exchanged on a simple syntactic contract base i.e. by exchanging W3C web service signatures via WSDL (web service description language) files for integration in a business process coded in BPEL4WS (business process execution language for web services). This approach lacks any (semantic) information on how to use the selected Web service, the responsibility for the correct usage of the service falls to the consumer of the remote service. If the service is simply an endpoint for a deployed BPEL process, the advertisement via WSDL is enough for a correct operation. But in an end user scenario for a real business process in a big enterprise portal, the user needs the semantics of the service usage to avoid erroneous or useless operation. The requirement for usage semantics covers the range of provision of simple constraints on input parameters like valid numerical ranges over the more elaborated networked dependencies between parameters - in case of input space of the service is not normalized – to the necessity of user guidance through a wizard – in case of input parameter sets, which need a deep understanding of the sequence of parameter input and the appropriate expert knowledge for parameterising the underlying process like an automatic target recognition.

The lack of usage semantics may be easily remedied by not simply supplying only input / output data, but annotating the method's signature by usage handles in the form of declarative or real user graphical elements. This approach of UI (user interface) surfacing not only improves end user operation, but also avoids error prone code duplication of UI element generation in each portal, and makes the integration problem of portal fragments not a programming but a real management task performed by the administrator of the consuming portal. An additional essential advantage of this responsibility shift is, that the administrator is the one role which manages user authorization and authentication information, thus the management of the portal fragments goes hand in hand with SSO and access right configurations and management. Three roles are identified in the context of surfacing:

- the producer owns the service of interest and hosts it via a network-enabled protocol
- the consumer accesses the web service and provides the UI to its registered client

- the client accesses the consumer, which is a proxy to the UI of the service on the producer, displays the graphical elements and relays user interaction to the proxy.

The concept of UI surfacing is technically realized in different approaches, from which the most common are presented here in the following three sub-chapters.

2.2 WSRP

WSRP is the acronym for Web services for remote portlets, which summarizes the technology blend realizing this approach. In fact, reviewing the conceptual needs for UI surfacing, this solution implements the idea of consolidation of service data and the required GUI elements in a smart and direct fashion. The main idea is, to bring existing industry standards together to realize a new standard which fits seamlessly into existing portal and SOA technologies. In 2003, the OASIS [1] organization published their version 1.0 of the WSRP standard, which is adopted by all relevant portal vendors. One of the crucial factors was, that the main technical contributors (participants like Microsoft, IBM, ORACLE, ...) tested their implementations from the beginning against each other, to assure the WSRP interoperability between portals based on different portal vendors (see Interoperability SC on OASIS website). Two technology base concepts, namely portlets and common W3C web service stack were brought together to constitute a new web service interface which published not only the data, but the complete graphical user interface in the context of portal fragments (named parts or portlets). WSRP service descriptions are published by standard WSRP files, but instead of coding the functional signature of the underlying process, the WSRP WSDL contains a series of well-defined technical service endpoints which realizes the WSRP framework. From these services two a mandatory:

- Self-description: this web-service allows the consumer to reflect the producer's capabilities and the portlets hosted on producer site inclusive their meta data,
- Access to HTML markup: this service allows the consumer to access the markup of the portlet running on a selected producer.

There are two optional service ports which expand the functionality of the consumed portlet:

- Registration: instantiates a binding between producer and consumer for accounting purposes or auditing, this binding allows the consumer to parameterise attributes of a portlet,
- Portlet management: this service gives access to the life cycle management of a portlet and some persistent state saving.

Each vendor has the possibility to publish and implement additional services, which are only significant between portals of this vendor.

The portlet specifications (JSR-168, 286) are understandable as an extension of the Java servlet specification (JSR-154), which in fact realizes rectangular non-overlapping areas in a standard web page. They are visualized as discrete windows of independent mini applications, this technique is preliminary intended for visualizing a bulk of diverse data in a compact and dense manner (like graphical charts or tables for a stock exchange page). A portal page based on portlets aggregates the data by a compact view through many small windows. Nevertheless, portlets may exchange data or events by vendor-specific extensions, if this is necessary. Portlets may be implemented in any language, which is supported and understandable by the portlet container and the underlying web (or application) server, the UI elements may be coded in simple HTML up to elaborated usage of JSF (java server faces) or dynamic features like AJAX (asynchronous Javascript and XML).

With portlets as the main structuring building blocks of a portal page, one is able to aggregate an application by simply arranging the desired mini applications on the page and configuring the attributes of these portlets. This task may be performed by an administrator for main portal pages and their core applications, but also by registered users on their own pages (so called dashboards or community pages with accompanying user spaces). This is a dynamic approach, each addition, removing, rearranging and configuring of such pages is done during normal server uptime. To achieve such a dynamics, it is vital to provide a framework by the portal software, which allows these different modifications during runtime. These frameworks are vendor-specific with proprietary APIs, which are generally hidden by comfortable tools in the management workspace of the administrator or the community tools for the dashboard handling by normal portal community members.

2.3 Web Clipping

The web clipping concept stems from the need to display filtered Web content on small mobile devices like PDA (first implemented in Palm OS 3.5) to avoid overloading these crippled devices by the rich internet content designed mainly for the more powerful desktop computers. The main idea is to filter heavy static data from the overburdened internet pages (images, banner, videos, big audio streams), then fit these information blocks to the capabilities of the device, to cache the adjusted data only once for a page and to update only dynamic data during online time. The major difference to programmable web filter is, that the filter intelligence is generally hosted on a proxy server or at least a separate process, with which the client (some browser) interacts over an eventually proprietary (i.e. in the case of the Palm) protocol. This technique is adopted to some portal implementations and stand-alone frameworks (like Kapow RoboSuite Web Integration Platform [2]), thus enabling standard desktop web application to clip from existing remote portal pages. The main motivation in the context of desktop applications is the same as in the case of WSRP, to easily aggregate own portal pages by leaning some clipped content (and thus the underlying applications) from remote portals (known under the keyword "enterprise mashup via presentation level integration").

Summarizing the existing implementations in the enterprise context, one finds three solutions to integrate web clipping in existing portal software:

- through specialized (web clip enabled) portlets (like portlet bridge [3])
- through browser extensions (like google notebook [4])
- as a remote service (like openkapow with a web clip robot [5])

A similar technique to web clipping is web scraping or harvesting (through web crawler) with the main motivation to focus, filter or sample information into a new often more condensed form.

One noteworthy aspect is the notice of legal issues which results in access or copy restrictions due to (often only printed) copyrights of the source pages.

2.4 Smart Clients

The third variant of sharing human-machine interfaces (HMI) presented here is a technique, which is not restricted to Web content. The metaphor of smart clients results from a very new technology blend of the frameworks Spring (J2EE abstraction [6]) and OSGI (open services gateway initiative [7]). These technologies brought together allow for a completely new client-server concept named as smart or rich client in contrast to thin and fat client known from the web respective swing fraction. The main technical advantage stems from the symmetrical conceptualization of using the same interfaces and data structures on client and server side; this symmetry allows a more dynamic assignment of responsibilities of the functional logic building blocks between server and client along the actual needs or constraints (like network availability) without changing one source code line. One possible realisation of this approach is available in the context of the Eclipse Server-Side framework (Rich Server Platform – RSP [8], see Fig. 1). Each client type mentioned above has its own pros and contras, but the smart client resembles the "copy & run" paradigm plus the rich responsiveness and comfort of a fat client as its best. One highlight is a smart client may run from a simple memory stick or any other removable media, allowing a sales representative to carry its own secure and actual version of his catalogue browser on a DVD ready for use on the client's site.

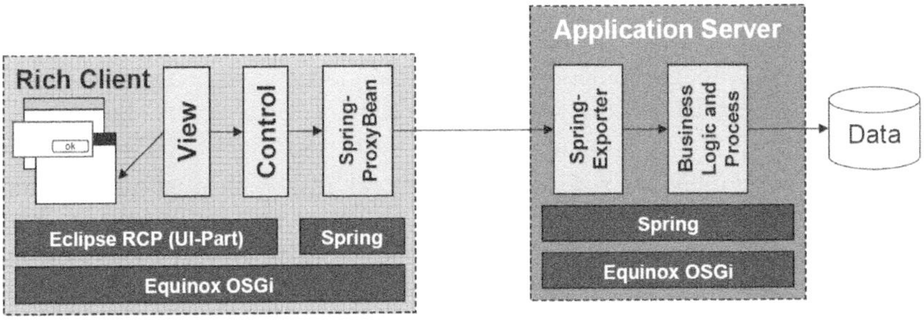

Fig. 1. Possible implementation of RSP [8]

The main disadvantage of this new technology is its unavailability in commercial or open source portal implementations, so actual implementations of a smart client concept have to coexist aside existing enterprise portals on their own platforms using server-side eclipse or the software from compeople [9].

3 Interoperability Tests

3.1 Experimental Setup and Procedure

For the experimental prototype we chose the WSRP framework due to the excellent integration and support into the portal software existing from earlier phases of the project. The OASIS Interoperability sub committee (SC) is the primary source for interoperability questions referring to the portal implementations of the participants in this SC. In the experimental setup regarding distributed HMI two nodes built on different portal software were tested against each other. The used implementations were the BEA Weblogic Portal 9.2 and the JBoss Portal 2.6.

On the BEA side, there is a desktop with one book containing different pages. The desktop and its content is built solely with the existing management tools of BEA portal and workshop. BEA portal allows for the installation of a community site with accompanying tools, so registered users in this group may manage their own pages. In addition to the vendor provided tools, the prototype contains a special portlet, which allows dynamic adding and removing of portlets on consumer site. This portlet allows the existing prototype portal to aquire new portlets hosted by remote portal nodes.

The setup included a login portlet, the whole desktop is SSO enabled via SAML (Security Assertion Markup Language). The application portlet that was used in the tests is a content based image retrieval (CBIR) application running as a software agent on an agent platform (JADE). The portlet has access to the user's filesystem (user space), which may reside on producer, consumer or client site. The access to the client's data is done via up- or download from respective to the client's computer, the representation of the user space in a portal is implemented by an enterprise bean, which is a façade to a virtual filesystem residing on the bean's host node. The user space allows the delivery of data from one step in a workflow to the next one.

The application portlet is implemented as a pageflow portlet, which realizes a strict separation between control (Java) and view (JSP) in a web adopted model-view-controller manner. As a prototype for stress testing the WSRP concept and the different implementations, this portlet is overloaded with many web artifacts, which are candidates for producing problems on the remote consumer site. Some of these features are Javascript, URLs referencing local resources on a portal, different AJAX implementations, a tree representation of the user space and a HTTP file upload, which forces form submission more demanding. The portlet was tested on its producer site in all aspects regarding the functional behaviour and the portlet specific features like personalisation and management to assure correct functionality. The tests on the different consumer sites exposed many obstacles in programming real interoperable portlets.

On the JBoss site, we used the standard unmodified JBoss portal downloaded from jboss.org. This portal comes with a ready to use portal and community site, the only management tasks are to add users and to enable SSO via SAML. Portlets from a remote producer site are easily managed and arranged by the tools of the admin portal or the community customization tools.

4 Results

4.1 JBoss

The CBIR portlet can be displayed in the JBoss portal. The actual site displays the dashboard of user "admin". The portlet has a different visual style due to other style sheets delivered by this portlet container. Beside these visual differences the handling of the portlet content and the portlet controls is totally alike. One aspect which can be a show stopper is the fact that the registration of a consumer with a producer on the

BEA side has to provide a special attribute which is BEA specific. The consumer has to know it per se, or to avoid erroneous registration cycles, one can enforce a so called out-band registration on producer side with a protocol which handles in its first step an agreement about the next registration step. This agreement is achieved in any communication context, which may be a phone call or a simple letter.

4.2 BEA Portal Workshop

The first tests with the application and management portlet were done on two BEA Weblogic portal implementations to discover implementation issues regarding the WSRP concept in a homogeneous environment related to vendor specific incompatibilities.

5 Discussion

Open questions remain on how the layout of one component should influence the internal layout of other GUI-components displayed concurrently. A working mechanism for notification of a layout rearrangement of another portlet to accommodate the own data layout is the event messaging for portlets. The WSRP V1.0 specification makes no proposition, how to implement such a event mechanism for portlets, so this feature is vendor-specific, a standardisation is scheduled for the WSRP 2.0 paper, which is not yet released, not even as a draft. The work on standardisation of inter-portlet communication is done by the WSRP Cross portlet coordination SC. A generalisation of the event paradigm between arbitrary components of one portlet with one of another portlet is not available, the set of event types to react on is restricted to portlet mode or window state changes, the actions then raised are restricted to mode or state changes, page activation or a generic user event action. Events may be accompanied by a payload, which is simply user defined data. Thus the influence of a state change of an arbitrary element in one portlet is not easily communicable to another portlet, not with the supplied event framework. The proprietary implementations of events permit the usage of such a feature in the WSRP context.

One of the next steps in the project workflow is the inclusion of the Web clipping technique. This approach allows a reuse of nearly every part of an existing web page in a remote portal, not only the reuse or dissemination of portlets. First of all we have to test the possible integration types, portlets as a vehicle for web clipping seem to have much restrictions regarding the clipping functionality.

One major question in using distributed HMI components or even a simple web service is where to find the appropriate service for the user's problem. The retrieval of a service matching the user's requirements is the precondition for the overall system acceptance by the user. To achieve a successful matching, the system must provide a registry for services (directory service, yellow pages) and a convenient information model for the service and its capabilities. The information model is the publish-find-bind abstract model of WSRP. It states in its own data structure named "businessEntity" how to publish portlets and producers as own services in the registry. In W3C web service context a UDDI (Universal Description, Discovery and Integration) is responsible for managing the registrations and responses to search requests in the context of the installed service model (the so called tModel). Version 1.1 of WSRP includes the concept of publishing the WSRP WSDL files to UDDI to manage the advertised WSRP service descriptions in a network-enabled repository. The information model of producers and portlets in the service model is very sparse and restricted to direct properties of the modelled components. Meta data is only accessible indirect via the service description web service. There is a potential need to enhance the model with semantic annotations to achieve an appropriate level for quality of service analogous to normal W3C web services lacking any semantic information model.

6 Conclusion

In this paper we have presented a study on distributed human-machine interfaces. We used a portlet approach to integrate not only functionality but also parts of graphical user interfaces over a SOA. Interoperability tests suggest that this standard is a promising way to integrate interactive software systems over a SOA. Depending on the nature of the applications to be integrated web clipping is also a relevant standard.

References

1. Internet portal of the Organization for the Advancement of Structured Information Standards (OASIS), http://www.oasis-open.org
2. Internet page Kapow RoboSuite Web Integration Platform, http://www.kapowtech.com
3. Internet page Portlet Bridge, http://www.protletbridge.org
4. Internet page Google Notebook, http://www.google.com/notebook
5. Internet page Openkapow, http://www.openkapow.com
6. Internet page Spring Framework, http://springframework.org
7. Internet page of OSGi Alliance, http://www.osgi.org
8. Internet page Rich Server Platform, http://www.infonoia.com/en/rich_server_platform.jsp
9. Internet page Compeople, http://www.compeople.de

Determination of Inspection Threshold Using Perceptive Sensitivities of Experienced Panel

Masao Nakagawa[1], Hidetoshi Nakayasu[2], and Tetsuya Miyoshi[3]

[1] Shiga University, Faculty of Economics, Center for Information Processing, 1-1-1 Bamba, Hikone, Shiga 522-8522, Japan

[2] Konan University, Faculty of Intelligence and Informatics, 8-9-1 Okamoto, Higashinada, Kobe 658-8501, Japan

[3] Toyohashi Sozo University, Faculty of Business and Informatics, 20-1 Matushita, Ushikawacho, Toyohashi-shi, Aichi 440-8511, Japan

mnaka@biwako.shiga-u.ac.jp, nakayasu@konan-u.ac.jp, miyoshi@sozo.ac.jp

Abstract. A determination of the threshold of inspection has been proposed based on the sensitivities by perception of panel. As the simulation in the proposed methods based on theory of signal detection are applied to determine the threshold level of experienced panels. It was found from the results that the value of index that evaluated the degree of experiensed skill enabled one to determine the threshold level of inspection.

Keywords: Theory of signal detection, Experienced panel, Stimulus threshold, Psychometric function.

1 Introduction

In the practical situation of quality control on visual inspection, a panel must decide the judgment of the quality by using one's sensitivities for products whether the quality of object is good or not. On the other hand, it is required in the field of automated visual inspection system with knowledge management to measure the characteristics of sensitivities for products and to evaluate the detective probability when the experienced skills were used for the judgment of inspection. In this case, it is an important viewpoint to obtain the skill of the professional sense.

The psychometric function method has been researched by the authors [3], though the classical methods of constant stimuli, such as limit and adjustment methods are applied until now [1]. A sorting performance by human vision is very flexible; therefore such a soft information processing has been regarded unfit on the computer which is good at digital information processing. The outline of the objectives in this study is drawn in Fig. 1. The approaches for knowledge management were illustrated in Fig. 1. The former management is two kind of stimulus as a mixture of signal and noise. On the other hand, the latter is concerned with the two kind of human response by knowledge and skill of experienced panel. However, it is hardly so far that defect inspection process using two kind of stimulus. To obtain the psychometric function

M.J. Smith and G. Salvendy (Eds.): Human Interface, Part I, HCII 2009, LNCS 5617, pp. 279–286, 2009.

for detecting the deference on the modalities of graphics, a prototype research was developed based on the constant stimulation method, has been proposed by the authors [2, 3, 4].

The aim of this work is to discuss the potentiality of identifying human perception and skill of experienced panel, it is a new approach. In this paper, the simulation in the proposed methods based on theory of signal detection are applied to determine the level of threshold whether the quality of industrial object is good or not by perception performance of experienced panel. The computer simulation was performed to determine the thresolds for inspection.

2 Measurement of the Degree of Skills

2.1 Strength of Stimulus and the Human Response

The psychophysical method advanced by authors enables one to measure the perceptual skill related to the defect inspection. It seems to be possible to add the knowledge and the experience by requesting information obtained from the visual performance of the experienced panel. It is assumed in this paper that the difference of the figure attribute is detected in the defect detection work as shown in Fig. 1. As for this method, the discrimination capacity of man who inspects it by the paired comparison is based on the method of constant stimuli. Therefore, it is assumed in this paper that the human response is identified by the mathematical function of the strength of stimulus.

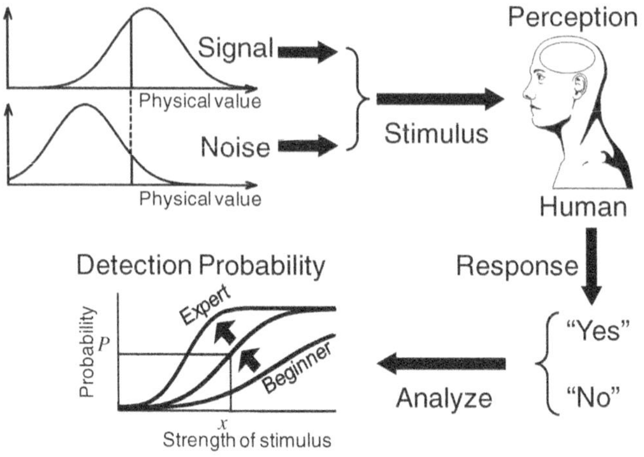

Fig. 1. Outline of concept

2.2 Theory of Signal Detection

As mentioned in 2.1, human response is identified by the mathematical function of the strength of stimulus. From this point of view, the model of human response is formulated by the theory of signal detection. In this formulation, there are two kinds of response and stimulus. The former consists of "yes" and "no", the latter consists of "signal" and "noise" as shown in Table 1. Even if the detection subject is a machine or human, the signal is detected from among the noise that always exists in the background. The level of the noise in the background is assumed to be random, it is generated from the outside or inside of the detection subject.

Table 1. Response of human by stimulus

		Response of Human	
		Yes (*SN*)	No (*N*)
Stimulus	Signal + Noise (*sn*)	A: hit $P(SN \mid sn)$ **OK**	C: miss $P(N \mid sn)$ **NG**
	Noise (*n*)	B: false-alarm $P(SN \mid n)$ **NG**	D:correct-rejection $P(N \mid n)$ **OK**

Table 1 shows the combination of the responses such as yes and no to the stimulus such as signal and the noise. The response to the stimulation of the SN distribution is said, "hit" for "yes" and "miss" for "no". When each probability is $P(SN \mid sn)$ and $P(N \mid sn)$ the relationship is as follow:

$$P(SN \mid sn) + P(N \mid sn) = 1 \tag{1}$$

If SN shows "signal-plus-noise", N shows "noise", the response to the stimulation N (noise) is called "false-alarm" when the response is "yes" and "correct-rejection" when the response is "no". If each probability is $P(SN \mid n)$ and $P(N \mid n)$, the relationship is

$$P(SN \mid n) + P(N \mid n) = 1 \tag{2}$$

Therefore, it is enough only by the calculation of $P(SN \mid sn)$ of "hit" and $P(SN \mid n)$ of "false-alarm".

The participant first receives stimulation, and the judgment from perception. They should judge whether stimulation is a signal or a noise in each trial. However, the judgment is difficult. One of the reasons is a random change of the noise that it strengthens and weakens. This situation can be shown in the Fig. 2 as two probability distributions that show a random change of the noise (N) and the signal-plus-noise (SN) distributions. As for the perception, the SN distribution is always larger than N distribution, because the signal is added to the noise. When the strength of the signal weakens, the difference of the mean value of two distributions becomes small. If SN distribution corresponds to N distribution then the judgment is very difficult. A relative distance of the mean value of the SN distribution and N distribution becomes the index of the sensitivity of the detection that is called d'. Let us designate by $\overline{x}_{sn}$, $\overline{x}_n$ the mean of the SN and N distributions. The distance d' is designated by the next equation.

$$d' = \overline{x}_{sn} - \overline{x}_n \tag{3}$$

When d' is small, the detection by human becomes difficult.

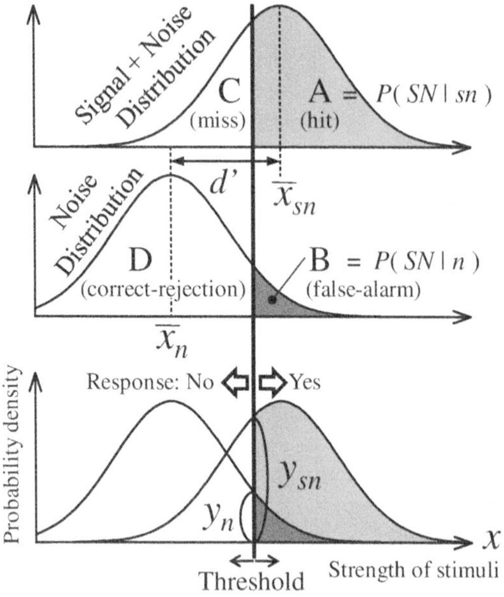

Fig. 2. The noise (N) and the signal-plus- noise (SN) distributions

The participant perceives when stimulation is strong or weaker than the standard values by method of constant stimuli, and responds with "Yes" or "No". They should respond yes-no for the stimulation that mixed the signal and the noise was received at random. In the physical value x of stimulation, the probability density of N distribution is defined $y_n(x)$, the probability density of SN distribution is defined $y_{sn}(x)$. The ratio $l(x)$ is shown by the next equation.

$$l(x) = \frac{y_{sn}(x)}{y_n(x)} \tag{4}$$

This $l(x)$ is called a likelihood ratio. In the case where $l(x)$ is greater than 1.0, the probability of responding for the SN distribution is larger than that of N [1].

2.3 Measurement and Analysis

It is known that in the event of the natural world, there is something that takes the distribution of the amount according to normal distribution because of central limit theorem. If one assumed that N distribution and SN distribution are among according to normal distribution, these probability density function are given by eq. (5).

$$\begin{aligned} y_{sn}(x) &= \frac{1}{\sqrt{2\pi}\sigma_{sn}} \exp\left[-\frac{1}{2}\left(\frac{x-\mu_{sn}}{\sigma_{sn}}\right)^2\right], \\ y_n(x) &= \frac{1}{\sqrt{2\pi}} \exp\left[-\frac{1}{2}x^2\right] \end{aligned} \tag{5}$$

The likelihood ratio is the following.

$$l(x) = \frac{1}{\sigma_{sn}} \exp\left[-\frac{1}{2}\left(\frac{x-\mu_{sn}}{\sigma_{sn}}\right)^2 + \frac{1}{2}x^2\right] \tag{6}$$

It becomes the following when assuming that the variance of N and SN distribution is equal.

$$l(x) = \exp\left[x\mu_{sn} - \frac{1}{2}\mu_{sn}^2\right] \tag{7}$$

This is monotonically increasing function. Therefore the $l(x)$ is in proportion to strength of stimulus by physical value x.

Applying the probability function to the above equations, it is found the following results:

1) It is possible to evaluate the relationship between the signal and noise.
2) d' and $l(x)$ is able to calculated.
3) The validation methodology is given which of a sensitivity or a threshold influences.

3 Practical Application

3.1 Design of Simulation for Sensory Tests

The decision method of the inspection threshold is examined by applying the theory of signal detection about the relation between the judgment of the inspection panel and the received stimuli by signal of physical value. It assumes that it is stronger than the threshold, when the human is received the stimulation of SN and N distribution, and judges the defect.

In the case where the inspection panel judges the defect, there is probability $P(SN \mid sn)$ that correctly judges in the SN distribution, and is probability $P(SN \mid n)$ that makes a mistake in the N distribution. When you apply the signal detection theory to the design of the inspection standard value, it is necessary to enlarge $P(SN \mid sn)$ and to reduce $P(SN \mid n)$. It is suggested in the paper to calculate the best inspection threshold from the relation of this trade-off. Fig. 3 is the flowchart of the inspection threshold decision.

3.2 Results and Discussions

It is shown in Fig. 4 that the relation between the likelihood ratio by eq. (6) and the strength of stimulus x is dependent on the ratio of $\overline{x}_n/\overline{x}_{sn}$. It is seen from the Fig. 4 the tendency of graph shows monotonically increasing function. However, it is also found that the likelihood ratio does not increases when the ratio of $\overline{x}_n/\overline{x}_{sn}$ becomes to be close to 1.0.

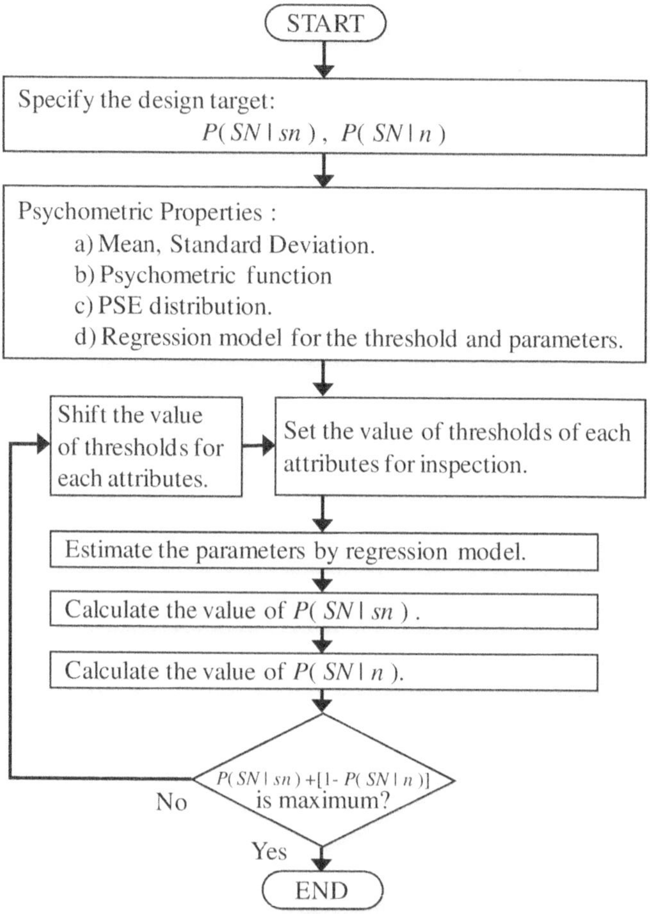

Fig. 3. Flowchart to determine the threshold in inspection

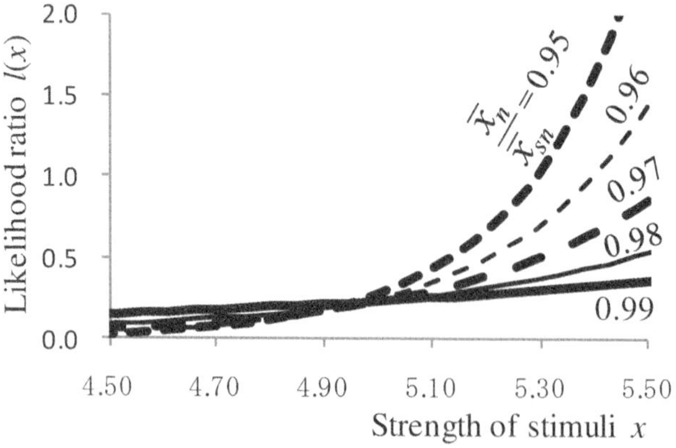

Fig. 4. Likelihood ratio in each $\frac{\overline{x}_n}{\overline{x}_{sn}}$

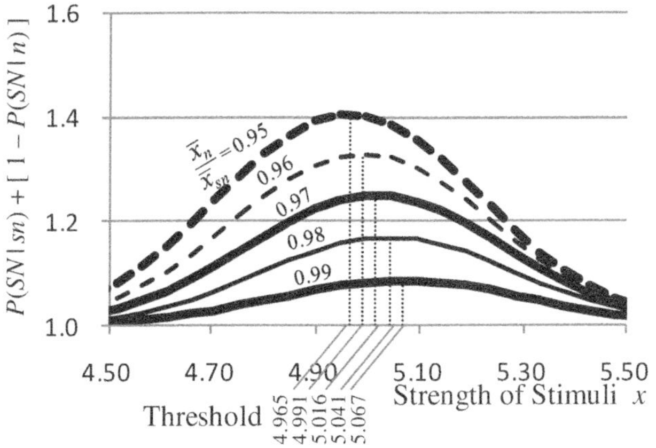

Fig. 5. Decided Threshold by $P(SN \mid sn) + [1 - P(SN \mid n)]$ in each $\frac{\overline{x}_n}{\overline{x}_{sn}}$

Fig. 5 shows the summaries the algorithm of updating the new values of threshold level for visual inspection by iterating processes. These steps consists of

1) Experiment of paired comparisons
2) Analysis of psychometric function
3) Set the values of thresholds for each attributes and estimate the parameters of psychometric functions.
4) Estimate the parameter of psychometric functions for each attribute by regression model
5) Calculate $P(SN \mid sn)$ and $P(SN \mid n)$
6) If the value of $P(SN \mid sn) + [1 - P(SN \mid n)]$ is not maximum, shift the value of threshold and go to step 3). Otherwise, the process is end.

Fig. 5 shows that the threshold level is dependent on the ratio of $\overline{x}_n/\overline{x}_{sn}$ that is strongly correlated to the value of d'.

Two main features in this study are as follows. The first is the likelihood ratio $l(x)$ means the judgment index of the human performance in visual perception, and the second is the distance d' between $\overline{x}_{sn}$ and $\overline{x}_n$ means the sensitivity of the human performance in visual perception,. This discussion will focus on the measurement for the skill level of experienced panel apply $l(x)$ and d'.

The theory of signal detection can be applied to various kinds of fields. Because the sensitivity of the detection and the response to stimulation are individually appreciable, it was impossible in traditional psychophysics. It should be noted that the method of determining the level of threshold proposed in this paper is very efficient and improves the accuracy of the measurement of perceptive sensitivities of experienced panel for visual inspection.

4 Conclusion

The following conclusions were derived from the results and discussions.

1) The judgment index clarified showing by the likelihood ratio.
2) It is found that d' means the sensitivity of the human performance in visual perception.

In the future it will be necessary to develop the experimental process for measure the likelihood $l(x)$ and sensitivities d' experienced panel in visual inspection.

Acknowledgments. This research was supported in part of IICT, Konan University by MEXT ORC JAPAN (2004-2008)

References

1. Ooyama, T., Imai, S., Wake, T. (eds.): Shinpen Kankaku Chikaku Handbook, Seishinshobo, pp. 19–41 (1994) (in Japanese)
2. Hayashi, H., Nakagawa, M., Nakayasu, H.: Psychometric evaluation with correlation for multiple perceptions. In: Proceedings of the First World Congress of the International Federation for Systems Research, CD-ROM, pp.S3-5-2. JAIST Press (2005)
3. Nakagawa, M., Nakayasu, H.: Evaluation of detection probability for flaw by multiple modalities. Journal of the Society of Materials Science 54(1), 2–7 (2005) (in Japanese)
4. Hayashi, H., Nakagawa, M., Nakayasu, H.: An evaluation method of human property using psychometric function in sensory test. Journal of the Chinese Institute of Industrial Engineers 24(4), 259–267 (2007)

Reversing the Simon Effect with Prior Practice of Noncorresponding Location Words

Andrea Rottermann and Kim-Phuong L. Vu

California State University Long Beach, Dept of Psychology
1250 North Bellflower Blvd, Long Beach, California
{arotterm,kvu8}@csulb.edu

Abstract. A benefit for spatial correspondence, called the Simon effect, is typically obtained in choice-reaction time tasks when the stimulus location is irrelevant to the task. Reversal of the Simon effect to favor noncorresponding stimulus-response locations has been obtained for physical-location stimuli after minimal practice (84 trials) with an incompatible spatial mapping. After practice with location-word stimuli, the Simon effect for physical locations is not reduced. The present study evaluated whether practice with "incompatibly" mapped location words can reverse the Simon effect when the practice session emphasizes color-to-response mappings rather than spatial mappings. Two conditions were tested in which the proportion of noncorresponding to corresponding trials was manipulated in the practice session. A full reversal was evident when all trials in the practice session were noncorresponding. Implications for interface design are discussed.

Keywords: Simon effect, stimulus–response compatibility, practice-transfer paradigm.

1 Introduction

One goal of designers is to make products easy to use. For example, when using a stove to cook food, users expect to be able to turn on a specific burner without having to receive any instructions. This expectation is satisfied with well-designed stoves that set the burners and control knobs in layouts that spatially correspond. That is, users can expect a top left burner to be turned on with the knob located in the top left position, top-right burner with the knob in the top-right position, and so on. In addition, to the stimulus-response compatibility effects brought about by mapping of display and control elements, population stereotypes regarding operation of individual controls could also be included in the design to make the product more usable. With the stove example, a right/clockwise rotation of the control knob is expected to turn the burner on and a left/counterclockwise rotation of the knob control is expected to turn the burner off. Both natural response tendencies brought about by stimulus-response compatibility and population stereotypes come from our experience interacting with display and controls in our daily lives.

M.J. Smith and G. Salvendy (Eds.): Human Interface, Part I, HCII 2009, LNCS 5617, pp. 287–295, 2009.

1.1 SRC and Simon Tasks

Stimulus-response compatibility (SRC) refers to better performance for compatible than incompatible mappings of stimulus locations to response locations [1]. In a two-choice task where the stimulus could occur in a left or right location and the response is a left or right keypress, the compatible mapping of left stimulus to left response and right stimulus to right response yields better performance than the incompatible mapping of left stimulus to right response and right stimulus to left response. Spatial compatibility effects not only apply to physical locations, but they are also obtained with conceptually similar stimuli and responses. For example, compatibility effects occur when the meaning of the stimulus (e.g., the word "left") is compatible with the meaning of the response (e.g., turn left). In general, compatibility effects occur when there is similarity, or dimensional overlap, between the stimulus and response set [2]. Thus, compatibility effects are obtained with arrow stimuli that point in left-right directions and "left"-"right" location-word stimuli. Similarly, compatibility effects occur when the responses are verbal (e.g., saying the words "left"/"right") as well as when they are manual (left-right keypresses; joystick movements; steering wheel rotations) [1].

Compatibility effects also occur when stimulus location is nominally irrelevant to the task. When responses are to be based on a non-spatial feature, such as stimulus color or shape rather than its location, performance is better when the stimulus and response locations correspond compared to when they do not. For example, if a participant is instructed to press a left button whenever a circular stimulus appears in green and a right button whenever the stimulus appears in red, then responses will be faster if the green stimulus appears on the left and the red stimulus on the right than vice versa. This spatial correspondence effect is called the Simon effect after its discoverer, J. R. Simon [3]. As with the SRC effects, the Simon effect also occurs for physical-location, arrow-direction, and location-word stimuli. The size of the Simon effect varies as a function of the stimulus and response modalities. For example, the Simon effect is usually twice as larger for auditory than visual stimuli [1].

Kornblum et al. [2] developed a dimensional overlap model to explain compatibility effects. For stimulus and response sets that have a dimensional overlap (perceptual, conceptual, or structural similarity), response selection occurs via two routes: direct and indirect. The direct route is based on automatic response tendencies. Kornblum et al. define automatic, "as the process that leads to the activation of a congruent response" (p. 262). The level of activation varies with the amount of dimensional overlap between the stimulus and response sets, such that the more overlap, the greater the benefit for congruent responses and the greater the cost for incongruent responses. Degree of overlap can explain why compatibility effects are larger with specific combinations of stimulus and response modes. Before response selection is complete, information goes through a verification stage to determine whether the automatically activated response is the correct response to be selected based on the task instructions. If the automatically activated response is correct, then the response is executed. If not, then the automatically activated response is inhibited, and the correct response is retrieved and executed. There is a delay for inhibiting the automatically activated response along with retrieving the correct response. According to the dimensional overlap model [2], compatible stimulus-response mappings yield shorter reaction time than incompatible mappings because the compatible mapping benefits from automatic activation via the direct route and more efficient response translation via the indirect route. For incompatible spatial mappings, there is no benefit in response selection via the direct route, but response selection via the indirect route is more efficient if a response-selection rule can be applied (e.g., respond at the mirror-opposite location) than if the individual stimulus-response pairings are arbitrary. Although both routes contribute to the effects of SRC proper, only the direct route contributes to the Simon effect.

1.2 Practice and Transfer

S-R compatibility and Simon effects have been of basic and applied interest because they have been shown to be robust, persisting even after extensive practice. Studies that have had participants complete between 900-3,000 trials over multiple blocks, days, or weeks have shown that participants become faster and more accurate all mappings, but the benefit for compatible mapping over the incompatible one is not eliminated [4, 5]. Dutta and Proctor [6] had participants complete 300 trials a day for 8 days (total of 2,400 practice trials) with a compatible or incompatible spatial mapping for two-choice and four-choice reaction tasks. For all mappings and tasks, participants showed a practice effect in which RT decreased by roughly 30 ms by the 4th day. For the two-choice reaction tasks, performance leveled off at that point, but for the four-choice task, there was an additional benefit of practice: RT continued to decrease by another 11 ms by the 8th day. The additional benefit of practice for the four-choice task is usually attributed to the additional number of response choices [2]. However, the benefit for the compatible mapping is not eliminated. Similar practice effects have been found with the Simon task as well [3]. Thus, studies of practice with SRC proper and the Simon effect show that activation of the spatially corresponding response cannot be overridden easily.

However, studies that have employed a practice-transfer paradigm to examine the influence of prior spatial mappings on the Simon effect show that compatibility effects are more malleable than that suggested by the aforementioned practice studies [4, 5, 7]. In these studies, participants practiced with as few as 72 trials with a spatially incompatible mapping of stimulus locations to responses and then were subsequently transferred to a Simon task where responding was to be based on stimulus color or shape. After practice with the incompatible mapping, the inherent advantage for corresponding stimulus-response relations giving rise to the Simon effect was eliminated or even reversed. The elimination of the Simon effect supports the notion that natural response tendencies can be overridden and that the underlying mechanisms of the transfer effect are powerful enough to prevail over the long-term associations. Moreover, the transfer effect has been shown to persist over a 1-week delay [7].

In the initial studies, the stimuli varied in left-right physical locations and responses were made with left-right keypresses in both the practice and transfer sessions. Comparing the similarities and differences between stimulus modes within a practice-transfer paradigm could provide insight on the nature of the stimulus-response associations that are being acquired and transferred to the Simon task. Proctor et al. [8] examined all possible combinations of practice and transfer for physical location, arrow-direction, and location words stimuli mapped to keypress responses.

In the practice session, participants performed with an incompatible mapping of locations, arrows, or words. Within each of these conditions, different groups of participants were transferred to a Simon task in which they responded to the color of the location, arrow, or word stimuli. With as little as 84 practice trials with an incompatible spatial mapping, the Simon effect was eliminated for location and arrow stimuli, regardless of whether the same stimulus mode was use in the practice and transfer sessions or not. The complete transfer of associations between locations and arrows suggests that these stimulus modes share visual-spatial codes. With practice of location word stimuli, there was no transfer to physical location stimuli after 84 practice trials, suggesting that verbal-spatial associations are distinct from visual-spatial ones.

One reason for the separation of verbal-spatial and visual-spatial codes in this task is that there is a salient distinction between how the physical locations and location words are displayed in the practice and transfer task, as illustrated in Fig. 1. Physical locations stimuli are presented in left and right locations and overlaps with the physical locations of the response keys. Location words are centrally presented and the meaning of the words overlaps with the left-right physical location of the response keys. This difference in the presentation of the stimuli may lead participants to treat the location word practice task as separate from the physical location Simon transfer task. Vu [9] showed that with only 84 practice trials, the associations for horizontal and vertical locations are kept distinct, but after extended practice, general response-selection procedures are learned and transferred across dimensions to the Simon task. Proctor et al. [8] showed that even after extended practice with incompatibly mapped location words, though, there was no transfer to the Simon task for physical locations. One question that arises then is, what do participants learn in the practice session? Is it the spatial associations of left stimulus-right response and right stimulus-left response, or is it more specific associations of left circle-right key and right circle-left key / the word "left"-right key and the word "right"-left key. Current evidence suggests the latter. However, if there is a way to make the practice and transfer task more similar, then the spatial associations may transfer between location words and physical locations. In the present study, the practice task using location word stimuli was made to be more similar to the transfer task using physical location stimuli by manipulating the task instructions for the practice session.

Instructions on the practice task can either explicitly describe an incompatible spatial mapping, or they can refer to a dimension such as color with the incompatible spatial mapping implemented implicitly. In previous studies, participants were explicitly instructed to respond to the stimulus location with an incompatible spatial mapping in the practice session. In the present study, the participants were given instructions to respond based on the color of the stimulus in both the practice and transfer sessions. Although both tasks are nominally Simon tasks, the proportion of noncorresponding-to-corresponding stimuli in the practice session was 1.0 to 0 in Condition 1, making it equivalent to an incompatible spatial compatibility task. Condition 2 use a standard Simon task (proportion of noncorresponding-to-corresponding trials is .5 to .5) in both the practice and transfer sessions to serve as a control group.

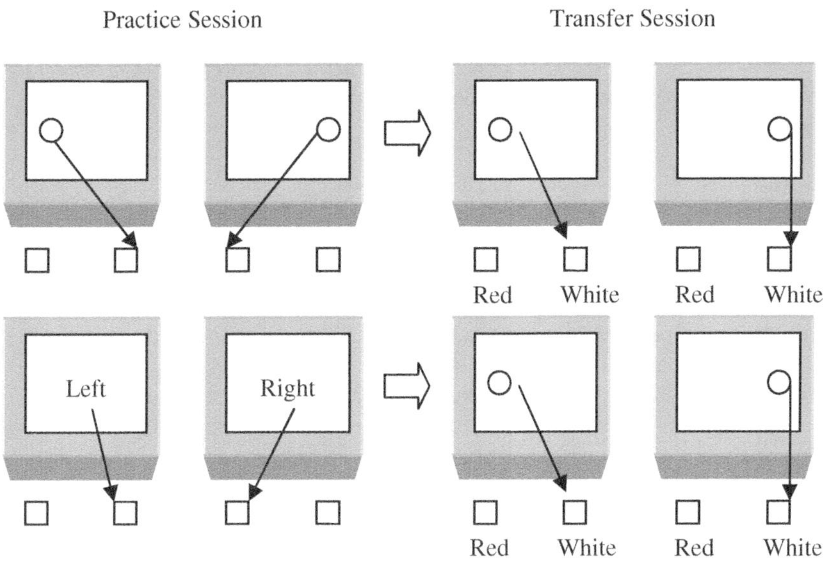

Fig. 1. Illustration of the practice and transfer task when physical locations are used for both the practice and transfer sessions (top) and when location word stimuli are used for the practice session and physical location stimuli for the transfer session (bottom). For the transfer session, only one color-to-response mapping is illustrated.

1.3 Present Study

This study was conducted to evaluate whether practice with non-corresponding mappings of location-words can reverse the Simon effect with as few as 84 practice trials when the practice session emphasizes color-to-response mappings rather than spatial mappings. If the Simon effect is reduced in the transfer session then this would indicate that specific, color-response associations transfer between stimulus modes, but the spatial relations do not. Two conditions were employed: condition 1 consisted of 0% to 100% corresponding to noncorresponding trials, and condition 2 50% to 50% corresponding to noncorresponding trials (which is the composition of a standard Simon task). Condition 1 is equivalent to a spatial incompatible mapping, where the stimulus and the response can be summarized by a rule to respond opposite to the spatial locations. At the end of the session, participants were asked if they were aware of the any patterns or differences between the two sessions to determine whether learning of spatial associations were implicit or explicit [10].

2 Method

2.1 Participants

A total of 80 students from California State University Long Beach participated for experimental credits towards an Introductory Psychology course requirement. There were 14 males and 66 females, age ranging between 18 to 39 years (M = 19.35, SD = 3.02). All subjects reported having normal or corrected-to-normal vision. Forty participants were assigned to one of the two practice conditions that varied in the proportion of noncorresponding-to-corresponding trials.

2.2 Design

A practice-transfer paradigm was used in which subjects practiced with a location-word Simon task in which the percent of corresponding to noncorresponding responses was 0%-100% (condition 1) or 50%-50% (condition 2/control) and then subsequently transferred to a standard Simon task using physical location stimuli. In the transfer Simon task, the percent of noncorresponding-to-corresponding trials was the standard 50%-50%. Although Simon stimuli were used in practice sessions, the fact that all trials were noncorresponding in condition 1 makes it equivalent to a spatially incompatible mapping if responses were to be based on stimulus location and not color.

To analyze the effects of the practice session on the Simon effect, the study employed a 2 (Condition: 1 and 2) x 2 (Correspondence: noncorresponding and corresponding) mixed design. Condition was the between-subjects factor. The dependent measures were reaction time and percent error.

2.3 Apparatus and Stimuli

Micro Experimental Laboratory (MEL v 2.01) was used to program all components of the experiment, including stimulus presentation, timing of events, recording the responses, and presenting the instructions. The program was run on a personal computer with a 14" VGA color monitor. The participants were tested individually in a dimly lit room directly in front of the monitor at a viewing distance of 60 cm. Participants responded with the number pad of the keyboard that was aligned to the center of the monitor. Responses were made on the computer keyboard's number pad by pressing the "4" and "6" keys with the index finger of each hand. Stimuli in the practice sessions for both conditions were the words "left" and "right" presented in lowercase letters at the center of the screen (approximate size of 12 mm x 5mm and 15 mm x 5 mm with visual angles of 1.56° x 0.52° and 1.24° x 0.52°, respectively) occurring in either red or green (MEL color codes 4 and 2, respectively). In the transfer sessions, filled red and green circles of 5 mm (visual angle 1.43°) diameter were presented to the right or left location approximately 3 inches from the center of the screen.

2.4 Procedure

For the practice session, participants were instructed to respond to the color of the location word stimuli, while ignoring the meaning of the word. Half of the participants were told to press the "4" key when the location word was presented in the color "red" and the "6" key when the location word was presented in the color "green". For the other half of the participants, the color-to-response assignment was reversed. Participants were instructed to respond as quickly and accurately as possible. The practice sessions included 72 trials plus 12 warm-up trials. Every trial started with a fixation point that remained on screen for 1,000 ms, and then the target stimulus was presented for 1,500 ms or until the participant responded. If the allotted time lapsed without a response or if an incorrect response was made, a 400-Hz error tone was presented for 500 ms, followed by a blank inter-trial interval of 1,000 ms.

Upon completing the practice session, the participant notified the experimenter, who then set up the program for the transfer session, allowing the participant to take a break. The participant then completed the transfer session in which they were instructed to respond to the color of the circles while ignoring the spatial location. The transfer sessions included 144 trials and 12 warm-up trials. The color-to-response assignment, trial timing, and all other constraints remained the same as in the practice session. At the end of the experiment, all participants filled out a quick survey that collected demographic information along with a question that asked if they noticed a pattern and if so, to identify the pattern. The pattern recognition data are presented in the Discussion section.

3 Results

Reaction time (RT) was measured as the time between stimulus onset and the depression of a response key. For the RT analysis, only correct responses were used, and percent error (PE) was analyzed separately. Trials in which RT was less than 200 ms or greater than 2,000 ms were excluded as outliers (< 1% of all trials); the first 12 trials were also excluded and considered warm-up trials.

3.1 Practice Sessions

A one-way analysis of variance (ANOVA) was run comparing overall RT and PE as a function of condition. The effect of practice condition was not significant for either RT or PE, $Fs < 1.0$ (see Table 1 for means). This finding indicates that participants were performing at similar levels prior to the transfer session. An additional analysis was performed to determine the Simon effect for location words in condition 2. A repeated measures ANOVA was conducted on mean RT and PE for corresponding and noncorresponding trials in condition 2. The effect of correspondence was significant for RT, $F(1, 39) = 5.32$, $p < .03$, with the Simon effect being 17 ms. The effect of correspondence for PE was not significant, $F(1, 39) = 1.68$, $p > .20$.

Table 1. Means for both practice sessions

Condition (C-NC trials)	Mean	
	RT	PE
1. 0-100%	532	2.70
2. 50-50%	548	3.09

3.2 Transfer Sessions/Simon Effect

A 2 (Correspondence: corresponding or noncorresponding) x 2 (Practice Condition: 1: 0-100% or 2: 50-50%) mixed ANOVA was performed on mean RT and PE. The main effect of correspondence was marginally significant for RT $F(1, 78) = 3.37$, $ps = .07$, but not PE $F < 1.0$ (see Table 2 for means). The main effect of practice condition was not significant for either measure, $Fs < 1.0$. The interaction between correspondence and practice condition; however, was significant for both measures, $Fs(1, 78) = 47.00$ and 8.42, $ps = .00$ and .005. Follow-up analyses were performed to determine the effects of practice condition on the correspondence effect. The Simon effect reversed in condition 1, where the corresponding to noncorresponding trials was 0-100%, for both RT and PE, $Fs(1, 39) = 32.78$ and 4.31, $ps = .00$ and .04, respectively. For condition 2, in which the corresponding to noncorresponding trials was 50-50%, the Simon effect was significant for RT but not significant for PE, $Fs(1, 39) = 10.98$ and 1.05, $ps = .002$ and .31, respectively. Simon effect sizes can be seen in Table 2.

Because condition 2 was intended to serve as a control condition, an ANOVA was performed to determine if the Simon effect varied as a function of practice and transfer session. There was no significant difference in Simon effects across the two sessions in condition 2 for RT or PE, $Fs < 1.0$.

Table 2. Means and Simon Effect for both transfer sessions

Condition (C-NC trials)		Mean		Simon Effect	
Correspondence		RT	PE	RT	PE
1. 0-100%	C	553	3.31		
	NC	524	2.06	-28**	-1.2%
2. 50-50%	C	532	1.84		
	NC	548	2.77	+16**	+1.0%

Note: C = corresponding, NC = non-corresponding
$* p < .05$, $** p < .002$

4 Discussion

When all the trials were non-corresponding in the practice block (condition 1), the Simon effect was significantly reversed in the transfer block. Thus, prior practice with noncorresponding mappings of location words to keypresses can influence the Simon effect for physical location stimuli in a subsequent session. The Simon effect was present in condition 2, which was the control condition. This implies that practice with the color-response associations alone does not eliminate the Simon effect.

To determine whether the impact of the spatial correspondence relation was implicit or explicit, participants were asked if they had noticed any patterns during the experiment and, if so, to specify the pattern. This question was not asked until after the conclusion of the experimental sessions when participants filled out the demographic form, so there was no possibility that the question could have primed the participants to look for patterns during the experimental sessions. For condition 1, the majority of participants, 77.5%, responded with a correct pattern recognition response, indicating that they had indeed noticed the complete noncorresponding relation (Yes = 31/40). Only 22.5% of the participants responded that they either had not noticed a pattern or proceeded to elaborate on an erroneous pattern (No = 9/40). Because the majority of participants in condition 1 did in fact notice the pattern that stimulus assignment color was always presented on the side opposite to the response, we cannot attribute the transfer effect to learning implicitly [9]. To determine whether the impact of the transfer effect was affected by the explicit recognition of the spatial incompatible relation, reaction times were compared for those who noticed the pattern and those who did not notice a pattern. No significant differences were found between the two groups. This finding requires further investigation to evaluate the differences that are encountered when the participant is explicitly informed of the pattern, compared to when they notice it themselves.

In daily life we switch between verbal and spatial tasks such as when we follow directions in a navigation task. For example, when driving, we may be looking at our Global Positioning System (GPS) to determine where to go. We might see that there is a left turn coming up and so we automatically switch on our left blinker and start moving to the left lane preparing for the turn. GPS directions provided while driving could be commands presented verbally, visually, or both. Currently, the GPS is designed to give directions that are compatible with the response (e.g., turn left or turn right), but more advance systems can give information about locations to avoid (e.g., traffic jams). In this case, a spatially incompatible response is made to the location to be avoided. This study showed that people can be influenced by a spatially incompatible mapping that is no longer relevant to the current task. Thus, designers should take care in making sure that the impact of spatial incompatibility of subsequent tasks is minimized.

References

1. Proctor, R.W., Vu, K.-P.L.: Stimulus-Response Compatibility Principles: Data, Theory, and Application. Boca Raton, CRC Press, Boca Raton (2006)
2. Kornblum, S., Hasbroucq, T., Osman, A.: Dimensional Overlap: Cognitive Basis for Stimulus-Response Compatibility-a Model and Taxonomy 97(2), 253–270 (1990)
3. Simon, J.R.: The Effects of an Irrelevant Directional cue on Human Information Processing. In: Proctor, R.W., Reeve, T.G. (eds.) Stimulus-Response Compatibility: An Integrated Perspective. North-Holland, Amsterdam (1990)
4. Proctor, R.W., Lu, C.-H.: Processing Irrelevant Location Information: Practice and Transfer Effects in Choice Reaction Time Tasks. Mem. & Cog. 27(1), 63–77 (1999)
5. Vu, K.-P.L., Rabas, A., Roberson, R.: The Effects of Practice and Speed Stress with Different Stimulus-Response Mappings. In: Smith, J., Salvendy, G. (eds.) Human Interface, Part I, HCII 2009. LNCS, vol. 5617. Springer, Heidelberg (2009)
6. Dutta, A., Proctor, R.W.: Persistence of Stimulus-Response Compatibility Effects with Extended Practice. J. of Exp. Psy.: Learning, Memory, & Cognition 18(4), 801–809 (1992)
7. Tagliabue, M., Zorzi, M., Umiltà, C., Bassignani, F.: The role of LTM links and STM links in the Simon effect. J. of Exp. Psy.: Human Perception and Performance 26, 648–670 (2000)
8. Proctor, R.W., Yamaguchi, M., Zhang, Y., Vu, K.-P.L.: Modulation of Spatial Correspondence Effects by Learned Incompatible Stimulus-Response Association: Transfer Within and Between Different Visual Stimulus Modes. J. of Exp. Psy.: Learning, Memory, & Cognition (in press)
9. Vu, K.-P.L.: Influences on the Simon effect of prior practice with spatially incompatible mappings: Transfer within and between horizontal and vertical dimensions. Mem. & Cog. 35, 1463–1471 (2007)
10. Destrebecqz, A., Cleeremans, A.: Can Sequence Learning be Implicit? New Evidence with the Process Dissociation Procedure. Psychonomic Bulletin & Review 8, 343–350 (2001)

Object-Oriented Interactive Processes in Decentralized Production Systems

Thomas Schlegel

Institute for Visualization and Interactive Systems, Universität Stuttgart, Universitätsstr. 38, 70569 Stuttgart, Germany
Thomas.Schlegel@vis.uni-stuttgart.de

Abstract. Supervision and controlled execution of processes and parameters are core requirements in Digital Production. But even in rather homogeneous systems, deviation from defined workflows and even process change occur daily. To be able to react flexibly to changing market conditions and to respond to increasing individualization of products and their adjacent processes, even more and ad-hoc variants have to be created and supported by production planning and control systems. We solve this problem by introducing a semantic process model, which can be executed and extended using a decentralized, message-based system. In situations where unforeseen deviations and needs for interactive intervention occur or ad-hoc input from users is needed, a generic, model-based user interface is used to create correct interactions in such a context. Our contribution makes processes much more flexible and opens up a new way of creating correct interactions in unforeseen process contexts.

Keywords: Interactive Processes, Process Models, Meta-Models, Production Systems, Semantic Models, Process Execution, Workflow, User Interface, User Interface Generation, Decentralized Systems, Process Instances.

1 Introduction

Complexity and flexibility are the prevalent driving forces and challenges of European Production Systems. Today, a production system and even a factory represent concepts that reach far beyond what is generally defined as an enterprise. Virtual factory networks and reconfigurable Production Systems demand flexibility and interactions at all levels – from a single machine to complex supply webs.

These huge Production Systems and their processes quickly become too complex and dynamic for the use of a fixed, standard user interface. Thus, many adaptations and process changes would lead to high costs for system development making user interfaces either simple or inadequate for the growing number of specific tasks to be accomplished on an interactive system that runs parts of the processes.

In this paper, we therefore present an approach for modeling and executing processes that are:

- interactive (provide interactive steps, gather missing data interactively from users by deriving correct dialogs from the type of input needed),

M.J. Smith and G. Salvendy (Eds.): Human Interface, Part I, HCII 2009, LNCS 5617, pp. 296–305, 2009.

- object-oriented (instantiated for execution, variant-enabled and using a classification and inheritance structure) and
- executed without central control in a decentralized peer/agent-based environment, i.e. process instances traveling from peer to peer.

Digital Production Systems and even organizations as a whole become more and more dynamic. Virtual enterprises and dynamic supply webs in today's turbulent economy are not restricted to former and formal organizational borders anymore. This requires to integrate data and processes crossing systems and company borders. Even within factories, different companies work together. This makes it necessary to support clear separation and profiles of stakeholder organizations as well as individual stakeholders by intelligent models and decentralized execution of processes.

The main goal is therefore to integrate static structural models and dynamic flow models into a joint model for object-oriented processes in production, which allows users to interact with processes executed in a complex, dynamic and decentralized environment. (see fig. 1)

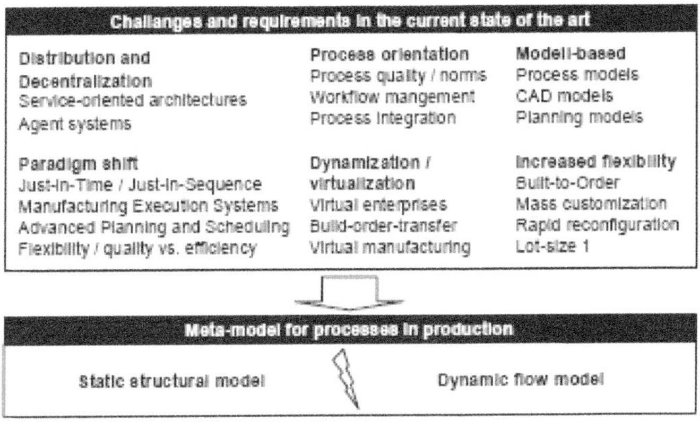

Fig. 1. Overview of state of the art, challenges and approach of decentralized processes in Digital Production

2 Decentralized Process Model and Execution

2.1 The Smart Connected Control Platform SCCP

The research project INT-MANUS has developed a communication layer to integrate various machines into a peer-to-peer based decentralized network infrastructure. [2]

Processes are executed within the Smart Connected Control Platform (SCCP) using a semantic model for definition and interpretation of organizational structures and processes. A concept for message interpretation and routing as well as decentralized process execution and the network join mechanism of peers into a decentralized system using a SuperNode concept is detailed in [1].

This decentralized platform is message-based to ensure communication even when direct connections are (often temporarily) not possible. Messages can be delayed (queued) and are able to carry a specific payload with them. In the case of SCCP, this contains the model in the form of a model fragment describing the whole contents of the message.

As the underlying communication paradigm and technology has already been detailed in [2], we concentrate in this publication on how the process model can be partitioned and executed on decentralized control system structures.

2.2 A Distributed Object-Oriented Process Model

Actions in a production system – even in rapid reconfiguration scenarios – have to be based on specified processes executed to perform in a reliable, traceable and repeatable way. This is only possible using a process execution system based on a process model that describes sequences of tasks in the form of relations between actions and interactions of processes.

While creative processes, for example, often do not follow a predefined process model, production processes and other highly defined and structured processes have to be executed strictly as they are described to ensure correct and repeatable results. This is often referred to as "achieving product quality through process quality". Creating the same product each time with different processes would directly have an impact on control of the whole system and the quality of its products.

The implication is that processes have to be predefined and repeatable to ensure quality and control, while nowadays processes also need to be flexible to answer needs of reconfiguration and customization.

In research carried out in INT-MANUS project and more detailed in the frame of a dissertation on runtime-modeling of object-oriented interactive processes in production [3], a concept and realization for consistent use of object-oriented principles in process modeling and interactive execution (OMICRON) have been designed.

OMICRON is a conceptual framework that enables decentralized modeling and execution of object-oriented processes. It has been implemented prototypically in the frame of INT-MANUS as a JAVA-based prototype system that has proven to be able to control production workflows and NC machines decentrally at FIDIA in Turin based on OMICRON model and workflow instances.

2.3 Meta-Model-Based Runtime Adaptation

When reconfiguration or unforeseen change happens during runtime in the workflow system or even during the execution of a process instance belonging to the type of process affected by the change, these changes have to be reflected in the process definition in order to maintain control over the process execution and prevent a deviation of the model from reality. This is also necessary to be able to provide the same workflow and context information related to a variant process V also to process instances of V created in the future.

[5] and others have described the adaptations of workflow definitions after deviations in single process instances have occurred. To correct the process in order to comply with the instance's actual state, the process definition is changed. While this allows for flexible execution of process instances even if deviations occur, the process definition becomes non-mandatory, if deviations from the predefined processes

become possible. Also, older instances cannot be executed and interpreted anymore, because the old process scheme has been adapted.

For this reason, we use inheritance as a powerful mechanism to incorporate changes in a process model without touching the basic process definition. By inheritance, it is possible to create variants in the form of child process models of an interactive standard process. These child processes inherit from their basic process to which additional process steps and coherent interactions can be added.

Each child inherits all properties (that is attributes, types and signatures) from the basic process. This means that an unchanged child process will be executed exactly the same way as the basic process, it is inheriting from, showing the same interactive behavior.

Once changes are made to the child process, they overwrite the basic process in only the specific part changed – leaving all other parts unchanged and still executable.

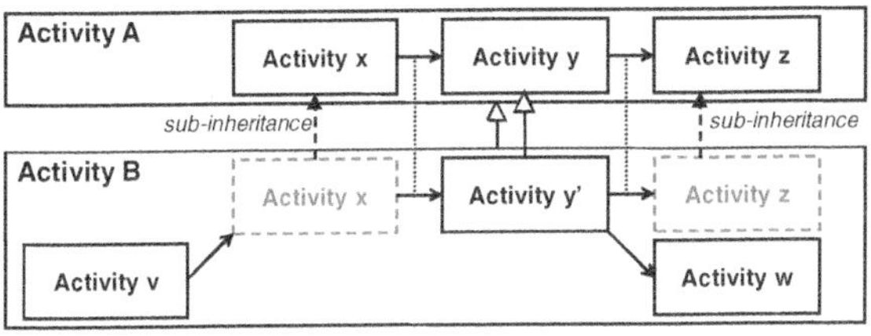

Fig. 2. Creation of a variant B of activity A by inheritance and extension

In figure 2, Activity B is created as a child specialization of Activity A, because additional steps have become necessary to produce a specific variant B. This consists of an additional initial step v (e.g. order special material, fill in a specific quality form) and a parallel final activity w (e.g. perform additional quality measures, add a special sticker). Additionally the activity y is being modified to become y'.

While these changes are made to B, no changes are applied to A. Because of the inheritance concept, B inherits all unchanged activities and fields from A – in this example the activities x and z. While they would be copied in usual workflow systems, in OMICRON they are inherited as "sub-inheritance". The sub-inheritance is applied to all elements recursively that form a part of a component from which a variant or specialization is being created.

The newly created process B is a subtype and specialization of process A. This means, in the sense of object-orientation, that everywhere a process instance of process type A is requested, also an instance of process type B can be used.

This is possible because B delivers at least all information that is delivered by A, performs at least all the steps (or specializations) of A and also consumes all input.

While correct output and steps are already covered by the inheritance concepts, for input parameters, specific measures have to be taken. The reason is that specialization of parameters often involves a reduction of scope of atomic types and the addition of new elements to complex types.

For this reason, specialized input parameter slots of B would require more than is delivered by parameters of type A or reject values, which would be possible with A. For this reason, specialized input signatures of processes can only be allowed, if they are filled interactively, leading to a need for dynamic request for values at runtime. When additional steps are added, which require or allow for input by the user, interactions can be derived from the type of information or commands needed.

Input, process steps and output of a complex component are characterized by the input, steps and output of their sub-components. These interface components are also specified in an object-oriented manner, with the implication of being fully classified in the semantic model created by the component structure and the relations between the components. This allows for directly recognizing required input for a process.

2.4 Distribution of the Object-Oriented Process Model

To achieve an execution of such flexible interactive processes even in decentralized environments, processes can be instantiated at runtime in OMICRON. Instantiation is possible for each process described in the OMICRON common model.

This means that every process instance in a whole OMICRON system is an object-oriented instance of its process model. It carries with it its informational context such as type of the process (and therefore of the instance, too), types of the sub-components, required input, created output and relations between this and other instances, e.g. triggers that initialized another process instance.

Sharing the process model information across the whole distributed OMICRON system, enables all stake-holders to interpret process instances received from other stakeholders in the production system.

If the next process step cannot be accomplished on the current peer, the process instance is wrapped into a semantic message as model fragment on the instance layer. Required model fragments from the process model layer (class layer) can be injected into the message directly to enable execution on the remote peer, or the remote peer requests them to be sent in case the model information is not present on it.

The object-oriented OMICRON model allows for a distributed extension of it by all authorized stakeholders. This way, variants, as described above, can be created on a single machine using the powerful semantic model infrastructure already present in the system. With the instance mechanism, it is assured that an instance of such a variant can be interpreted and executed everywhere in the system.

3 Interactive Processes

The demand for greater flexibility as well as for adaptability of control and monitoring systems on runtime leads to situation, where user interfaces for interaction with processes change rapidly. This implies the necessity of user interface adaptation on runtime to comply with changes in the processes executed and their context.

The scientific community in the field of HCI has experienced strong interest in automated, model-based and generative system in the 1990s [9]. Model-based User Interface generators started to create a semi-automated or even automated bridge from analysis and design models as used in Software Engineering and Task Modeling to executable software or User Interface (UI) models interpreted by User Interface Management Systems (UIMS).

While the Software Engineering community still discusses and uses generator concepts for example in the frame of Model-Driven Architecture (MDA) and Domain Specific Languages (DSL), manufacturing industry has only very rarely adopted automated and generative models from the field of HCI.

When it comes to the type of distributed processes described above, the frequency and impact of changes is too high to create fixed user interfaces for process control on design time of the system. Talking about rapid reconfiguration of processes with a high frequency, for example in Mass Customization and beyond, it is even impossible to rewrite or manually produce new software versions each time changes in the processes occur.

This issue leads directly to the use of models that can be adapted to changes and also allow for consistent adaptations of the user interfaces that connect people and processes.

The semantic process model described above, lays the foundation for providing users with interactive integration into a complex process landscape of model-based processes interfering with each other. As it is impossible to use pre-programmed, fixed user interfaces, interactions have to be created according to the actual process model, using generic user interface concepts triggered by the semantic model.

3.1 Explicit Interactions

In classical User Interface Generation interactions are often generated using descriptive models. [9] provides an overview on concepts and approaches and introduces a flow-based model that is able to describe not only static data structures and input components but also interactive workflows that can be transformed into an interactive software.

Explicit interactions in the semantic process model are as well described explicitly and classified to allow an interactive workflow engine to generate interactions from the descriptive model. This interaction concept is important, where specific interactions have to be ensured in an executed process, for example introducing the four-eyes-principle, show error messages in an exceptional workflow, or entering parameters not available from earlier process steps or model information.

3.2 Implicit Interactions

In a complex and flexible production environment, semantic processes are often dynamically connected depending on context and situation. For example a special order can trigger a configuration process that is normally started from an administrative process, which provides all data needed. When started from the special order process, some parameters may be missing or unknown because of the different context of the process execution. It is impossible to avoid or foresee such situations in dynamic, model-based and decentrally controlled environments.

When a process needs further information, it is possible to derive interactions implicitly from the semantic model information available with this process. As all information processed in a workflow is classified in the model, this model information can be used to generate correct interactions to fill the vacant parameter slots interactively. This enables the process to be executed with a complete parameter set in such a situation.

We use this concept and model information to develop a visual and interactive user interface for process control that uses explicit and implicit interactions for generating the needed interactive elements.

4 Visual and Interactive Process Control

In workflow modeling, for example Business Process Modeling, often models are used as pure graphical representation without the ability to execute and partially automate them. When decentralized process execution is described, the focus often lies on the pure automation without human involvement – with some exceptions like WS Human Task and BPEL4people. This leads to a gap between models of processes, execution of processes and interactions with these processes.

Complex networks like decentralized and multi-stakeholder production systems, Virtual Enterprises or whole supply webs cannot be run automatically nor be managed completely manual due to their complexity and dependencies. This kind of complex system needs to provide optimized interaction with the processes executed, including a context-based, automated execution of partial processes that are triggerd manually or by other processes in the decentralized control structure.

To achieve this, visual interaction with and feedback from the processes executed is needed, which enables the human users to understand and control them visually using a dashboard-like interface. While visual control of agent-based supply webs [4], production systems [7] and management systems [8] has already been proposed, real-time interaction with model-based processes dynamically executed in a decentralized environment goes a step further: it requires generic software that interprets complex semantic model infrastructures to derive the necessary interactions from them. This implies that the user interface has to be created dynamically on runtime, using concepts from model-based user interface generation [9]. While such models and systems full specifications of the intended user interface are provided, interactions for an OMICRON environment dynamically change with the processes, specific processes instances and their context. This makes runtime generation from a partial semantic model in a decentralized agent environment a complex problem, which occurs in all dynamic decentralized environments, for example user interfaces for service-oriented architectures (SOA) [6].

To allow for user interface generation from the OMICRON environment, three model layers are defined and explained in the following: the User Information Model layer, the User Interface Element Model layer and the User Interface or Dashboard Model layer.

The *User Information Model (UIM)* connects the local User Interface to the OMICRON model and processes. It defines the type of information visualized or requested, its structure and conditions. If for example a "latest delivery date" is to be selected, the UIM defines that it consists of a day, a month and a year, as well as an optional time in the form of hour and minute. A rule will ensure that the date entered is after the current date.

Once a process instance is executed and requires a parameter which is not supplied to it by the caller, the OMICRON runtime system has to create an interaction to request the data from the user, e.g. the supervisor. If the interactive request is already foreseen in the process, the task will be similar but may yield already more information to the user interface system. If the contracts applicable to an order for example have not been given as parameters to the process, they have to be selected interactively. The UIM then defines type of selection to be used, e.g. "m out of n", and how the options are generated.

For the selection of the correct interaction element to be used, the *User Interface Element Model (UIEM)* is required. It defines interaction elements and their semantic input and output characteristics as well as their (classified) capabilities to be used. In the example of selecting the applicable contracts, in the form of an open m out of n selection, a multi-select list box could be chosen from the UIEM using the specification from UIM. This matching of specifications from UIM to elements of UIEM is done by the local user interface system. Also interconnectivity of the user interface elements on the complete dialog or dashboard interface is defined by the UIEM. This is needed to form semantic groups and relate the interaction elements to each other.

The integration and setup of a complete user interface dashboard is accomplish by the interpretation of the *User Interface or Dashboard Model (DM)*, which defines the setup of a dashboard using elements (E) derived from the UIEM. Each element E from UIEM is referenced in the DM and made available on the user interface. This way, the DM connects the user interface to information channels from OMICRON system via the UIM and assembles the elements into a dashboard profile that matches the requirements of the actual user. User profiles for different user roles provide specific user interfaces for different tasks associated with the user, e.g. distinguishing between supervisors, managers and workers. Figure 3 shows the process of generating a new element for interaction on the user interface.

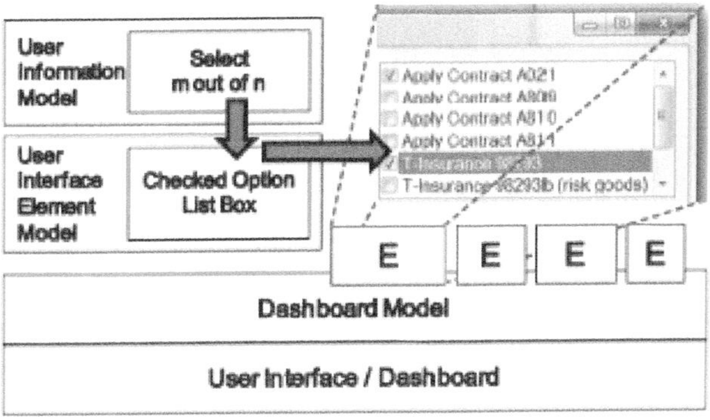

Fig. 3. Population of the Dashboard with User Interface Elements (E)

5 Summary, Conclusion and Outlook

With the OMICRON framework, we have provided a model, modeling concepts and runtime system able to execute variant processes type-safely and decentrally using object-oriented concepts. It allows for connecting machines, systems and even companies with fully automatable, model-based processes. When user input is foreseen in the process, parameters are missing of unforeseen process contexts occur, with the Interactive Process concept described in this publication we additionally provide an approach on generating interactions from the OMICRON model. Currently we are creating an improved implementation for the OMICRON model, runtime and instance management as well as an implementation for the Interactive Process concept, which builds on the model and concepts of OMICRON to derive interactions from models explained.

The decentralized, flexible processes described show their strength in production environments that are determined by a high amount of variants and a high degree of process automation (see fig. 4):

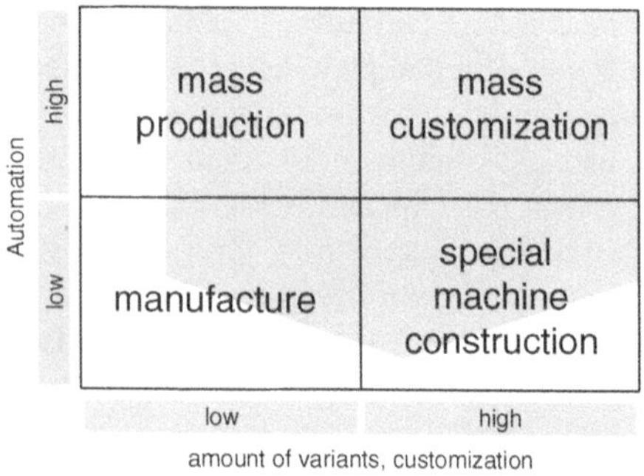

Fig. 4. Automation and flexibility: applicability of object-oriented process models (grey)

For environments where every product is different, the application of a model-based process framework only makes sense if two factors are prevalent: high automation requires a clear model-based and executable definition of processes, but many variant processes are also required for the different products.

In fields with no process variants and a very limited number of product types, cost of such a complex system will often be higher than the benefit of it. In all other fields, especially where mass customization shows a high degree of automation, but also requires flexibility for customization, object-oriented processes can be of great benefit for flexibility and dynamic changes as well as well-defined execution and operation.

The concept described, supports the model-based provision of interactivity, as a key factor for the integration of humans into a complex system landscape of globalized production systems and especially their distributed processes. Also, the object-oriented distributed model provides model-based runtime system integration for the information layer as well as the interaction layer. Taking this work as a basis, further research will be necessary to achieve fully dynamic interaction with decentralized processes in the field of production and beyond.

Acknowledgement. Most of this work has been done in the frame of my work at Fraunhofer IAO. INT-MANUS has been funded by the European Commission.

References

1. Schlegel, T.: Interactive Processes and Semantic Models in Production Systems. In: Proceedings of the First International Conference on Business Sustainability, Ofir, Portugal (2008)
2. Schlegel, T., Thiel, S.: Semantic Models and Processes for Information Gathering and Interactive Workflow Execution in Decentralized Production Environments. In: Intelligent Production Machines and Systems – 3rd I*PROMS Virtual International Conference (2007)
3. Schlegel, T.: Laufzeit-Modellierung objektorientierter interaktiver Prozesse in der Produktion. Jost-Jetter, Dissertation, Heidenheim (2008)
4. Rahman, M.A., Sadik, S., Ali, A., Ahmad, H.F., Suguri, H.: Formal model of agent-based supply chain framework in a virtual enterprise for reliability. In: Proc. IEEE INDIN 2008, pp. 1407–1412 (2008)
5. Rinderle, S.: Schema Evolution in Process Management Systems. Dissertation, Universität Ulm, Fakultät für Informatik (2004)
6. Beinhauer, W., Schlegel, T.: User Interface for Service Oriented Architectures. Intelligent Production Machines and Systems, 129–134 (2005)
7. Li, X.J., Schlegel, T., Rotard, M., Ertl, T.: A model-based graphical user-interface for process control systems in manufacturing. Intelligent Production Machines and Systems, 89–94 (2006)
8. Spath, D., Nostdal, R., Göhring, M.: Visual Management - Concept for the support of strategy, planning and control in real-time. In: Kuhlin, B. (ed.) The practical real-time enterprise: Facts and perspectives, pp. 321–332. Springer, Berlin (2005)
9. Schlegel, T., Burst, A., Ertl, T.: A flow centric interaction model for requirements specification and user interface generation. In: Proceedings of the 7th international conference on WWCS, bridging diversity at work. Damai Sciences, Kuala Lumpur (2004)

A Retrospective and Prospective View of Information Technology Professionals' Use of Tools: Maturing the User Experience

Candace Soderston

Microsoft Corporation, One Microsoft Way, Redmond, WA 98052
csoders@microsoft.com

Abstract. The Information Technology (IT) professional's user experience has evolved both in terms of software tool attributes and the process workflow within which the IT tasks exist. Based on user research with IT professionals over the past two decades, including ethnographic observational field studies, focus group discussions, and hands-on user tests with analyses of errors and perceived complexities, this paper presents a set of hypotheses about user interface design aspects that across the industry most impede the workflow of typical IT professionals. It discusses how this relates to typical information workers' behavior, as well as what steps we as tools and systems designers can take to better support the IT professionals' user interface needs.

Keywords: IT professional, enterprise software, user interface, user experience, UI design, command line interface, CLI, graphical user interface, GUI.

1 Introduction

This paper started with an intent to re-examine and report on the state of the art in understanding and designing software to enable the most productive and satisfying workflow and habits of IT professional (hereafter referred to as "IT pro") audiences. Specifically, my objective was to challenge our typical answers to these 2 questions:

1. How far have we come in the last decades in extending IT management capabilities beyond the realm of the dedicated scientists who were deeply rooted in creating the technology to meet their own needs, to a more generalist though highly skilled IT pro end user audience?
2. Are IT pro audiences fundamentally different from the larger set of end users of information productivity software and, if yes, in what ways do these differences require fundamentally different UI designs, principles, and paradigms?

Although I started with the intent to answer these questions, my review did not end with this. In this paper I posit an answer to the first question, while decomposing the 2nd question into parts. Evidence leads me to claim that IT pro audiences are NOT *fundamentally* different from other end user sets, but rather, that they sit in different places on a continuum of user experience, and they carry their learning forward from their information worker role to their IT pro role. This understanding could help us

M.J. Smith and G. Salvendy (Eds.): Human Interface, Part I, HCII 2009, LNCS 5617, pp. 306–315, 2009.

change UI design approaches from the continuing typical "either/or" propositions (e.g., whether to optimize for novice or for expert) to a more synergistic approach that enables both, in concert. Training wheels don't have to be removed if they're not in the way, aren't heavy and are needed from time to time.

Admitting this common "humanity", as well as identifying those aspects that evolve with further learning, puts us in good position to more insightfully merge the well-proven benefits of the common graphical user interface (GUI) -- where users can explore and recognize landmarks – with the also well-proven benefit that expert fast paths provide, such as those offered by command lines and scripts.

1.1 How Far Have We Come in Extending IT Management Capabilities beyond the Realm of the Dedicated Scientists Who Created the Technology?

It is a commonplace understanding today that the computer was initially a tool for select scientists, to run their own analyses, to support their scientific research and lines of inquiry through sharing the time/cycles offered by a mainframe computer. At this stage in history, the people creating the technology and the interfaces to it, and the people using them, were essentially the same individuals.

As computing capabilities grew beyond the "glass house" (those working inside the raised-floor, water-cooled IT center), to support other audiences (e.g., word processing for professional authors, financial accounting for professional accountants, searchable databases for librarians, and all these for everyone), so did the number of IT producers grow and diverge from the targeted end users. In the United States alone, even during today's economic slow-down, there are over 5000 postings for "Information Technology management" jobs in online databases[1]). These skills are in high demand, even though they represent a large cost category for enterprises in managing their businesses (along with capital and energy).

Up through the 1970s, computer science and software engineering degree programs didn't typically exist as major areas of study, but were an adjunct to the mainline scientific degrees that used them (math, physics, chemistry, biology, and all the "engineerings": e.g., electrical, civil, mechanical.). Early generalists, hired to support the scientists, were trained on-the-job to write in binary, "machine language"

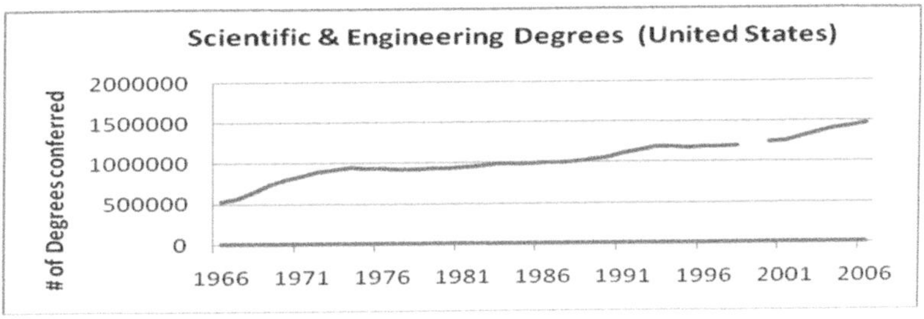

Fig. 1. National Science Foundation Statistics see http://www.nsf.gov/statistics/nsf08321/content.cfm?pub_id=3785&id=2

[1] For example, www.monster.com (on 2/9/2009)

or other low-level code to automate some of the more common aspects of programming, and came from other main-line disciplines, most notably from both music and math majors. Today, degrees in software engineering and computer science are offered by most universities and colleges, and the numbers are large (see Fig.1).

Graduates coming out of the computer science and software engineering programs today and working in the IT management jobs differ in many keys ways from their earlier counterparts. These graduates are highly technical, but they are also not dedicated to one particular scientific field, research focus, program, task, or language. They are much more horizontally distributed in focus than were yesterday's technocrats and they are not typically representative of their own end users.

1.2 In What Ways Do IT Professionals Require Fundamentally Different UI Designs, Principles, and Paradigms from Other End Users?

Today's IT pros' lives are filled with multi-tasking, with learning and using multiple programs, with diagnosing and fixing problems that occur at multiple layers in the stack. The wider heterogeneity in the end users being supported, the greater span in the applications being supported, and the more rapid rate of progress in technology development all contribute to the challenge an IT pro has in juggling knowledge … what gets left in short-term memory, constantly being wiped out and refilled, and what gets stored in long-term memory for continued, repetitive access.

1.3 The UI from "Yesterday" – Blank Screens, Programmed Function Keys, Keyboard Templates, "Run Books"

Command line sessions, blank screens. The command line, blank screen user interface was the only user interface in the early days of computing, enabling scientists to run their few dedicated programs and get their resulting data in "real-time" or to "batch" the programs up for running later during the slowest times, typically overnight. Command line access depends on ability to recall and produce the appropriate syntax.

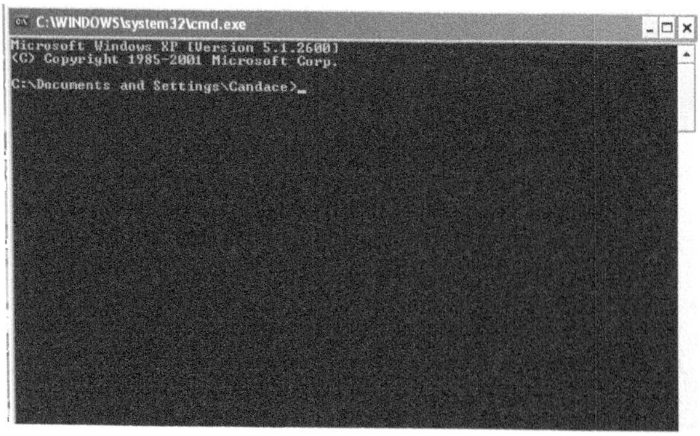

Fig. 2. Command Prompt

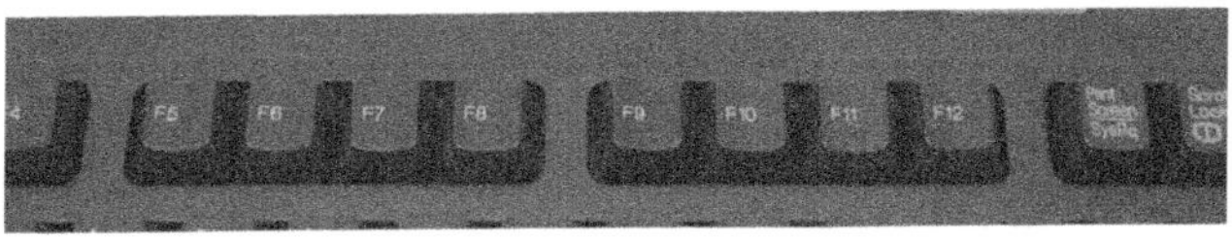

Fig. 3. Function keys

Programmed function keys. PF keys, shortened to F keys in most keyboards today (Fig 3), were created to enable IT staff and other users to tie their most frequently used sequences of commands (scripts) to particular function keys, so that a single key press could do what formerly required more typing. Remembering a single key rather than a string with exact sequence/syntax was much easier and cut down on errors and time.

Keyboard template overlays and paper "run books". Both of these user interface aspects -- dedicating a key to a frequent-use command, macro/script, or application, and making this visible via a keyboard template overlay – were created to help with the recall problem that people face when they are not dedicated to and frequently using the functions. Today many applications and most users don't rely on these.

Through cognitive science we know that recognition is more robust than recall, and this applies to all humans, including IT pros and other end users. The UI paradigm shifted over the last decades from blank screen command prompt UIs to graphical UIs providing explorable landscapes, aimed to enable users to find features when they are multi-tasking rather than performing a single task repetitively.

1.4 Today – GUIs, Multi-tasking, Multiple Programs, Scripts, Commands

In the late 1980's - early 1990s, I led a small team at IBM in benchmarking early GUI shells that were developing, before there was a shake-out of the alternatives into a set of common principles and standards. Users were crying for consistency and predictability in common mechanics across applications and environments, so they could transfer their learning from task to task. (As a backdrop, the average number of software packages they reported using was 3; the packages listed most often included Lotus 1-2-3, WordPerfect, MacWrite, and DOS.)

In this work, we tested 95 information workers who used productivity software at home and work, in different GUI environments, and logged their behavior:

- 34 people did tasks across a full day, half of the people in Apple's MAC System 6 and half in Microsoft's Windows 3.0 environment.
- 31 people did the same tasks as above and other similar tasks across 2 full days, about half of the people in DOS and half in IBM's OS/2 environment.
- 30 people did the same tasks as above and other similar tasks across 3 full days, one-third of the people in Apple's MAC System 6, one-third in Tandy's DeskMate (a DOS shell), and one-third in IBM's PS/1 (another DOS shell).

We codified the errors people made within these environments and studied this across time to understand the learning curve characteristics: which errors persisted over time in a longitudinal study and which others dropped off as people became more experienced and repeated tasks. Out of this lengthy and rigorous benchmarking effort (see Fig.4), we codified the top 10 user error types that occurred across ALL of the GUI environments and that persisted across days. (See Table 1.) We developed a faster heuristic evaluation method based on these aspects and used this to compare HP'sNewWave 3.0, Sun's OpenWindows 2.0 OPENLOOK, NeXT's NeXTStep 2.1, Metaphor, Microsoft's Windows 3.0, and OS/2. [1]

Fig. 4. Benchmark test representation

Table 1. Top 10 User Errors Associated with Graphical User Interfaces

1. Ambiguous menus and icons – Users were unable to tell from cryptic word labels what categories of tasks they represented. Object/action and noun/verb ambiguities contributed. For example, a user might be looking for a way to view a file, and would look under "view", but had to instead look under "file".
2. Single direction language – People chose actions without first selecting objects. This resulted in "no-operation": no action and the user was given no feedback. In communication between human beings, people can talk in either active or passive voice ("the ball was hit by the boy" = "the boy hit the ball").
3. Complex linkage between and within applications – Extraneous data conversions and mode switches were required to complete tasks.
4. Unclear step sequences -- People chose wrong actions, omitted required actions, and performed unnecessary actions – all contributing to losing their way in the sequence of steps to complete their overall task goal.
5. More steps to manage interface than to do common tasks (in users' model) -- For example, accomplishing "simple" cut-copy-paste goals required 12-15 steps in the GUI platforms evaluated, whereas participants tried all kinds of shortcuts that didn't work. This is highly related to the previous category but this is a subset that relates only to the common GUI tasks, not to the end-to-end step sequencing to achieve the users' overall task goal.
6. Inadequate feedback and confirmation – People were not confident that their requests were acted upon by the system so, for example, sent notes redundantly and saved data, then exited and re-entered files in order to make sure the changes had been made. This type of user operation resulted in no feedback from the system.

Table 1. (*Continued*)

7. Lack of system anticipation and intelligence – Users were forced out of their task by lack of anticipation on the part of the software. E.g., when trying to save a file to an unformatted diskette, most systems simply issued a dead-end error message. On the other hand, some systems (the MAC, notably) issued an error *dialog* listing the alternatives that could be behind the error condition (could be unformatted, formatted for another system, or damaged) and asked the user if he/she wanted to try formatting the diskette. If user chose 'yes', the system formatted the diskette and saved the user's file.
8. Input and direct manipulation limits – People had difficulty with operations related to positioning the mouse-cursor correctly: to pick up small targets, such as window borders; to place an insert cursor within text, etc., and related to timing: double-click speed. User statements indicated frustration with these miscellaneous recoverable errors and requested support for alternatives to direct manipulation.
9. Highlighting/selecting limitations – People tried selecting multiple discontiguous text strings and spreadsheet cells to operate on and were frustrated when their 2nd selection caused the 1st to be de-selected. E.g., this happened when trying to make bold all proper nouns in a document and when trying to add a specific value into several spreadsheet cells. People also tried to refine selected areas, for example, to stretch one of the end-points out to cover more text or to shrink it to cover less.
10. Inadequate error messages, help, tutorials, documentation – Users referred to these support items and were unable to locate helpful information. Problems lay in retrieval aids and in information content that was descriptive rather than procedural.

Yes, these problems were identified by studying typical information workers rather than IT pros, and the GUI paradigm has improved since the time these data were collected. However, from observing many IT pros over the past decade doing their tasks in the field and in usability tests in the laboratory, there is reason to believe that these same UI design aspects remain top contributors to user errors – all users.

2 Do IT Professionals Require Fundamentally Different UI Designs, Principles, and Paradigms Than Information Workers?

Over the past 4 years I and other user experience professionals in the teams I've worked with here at Microsoft collaborated in conducting ethnographic field research at customer sites[2]. This research incorporated typical "follow me to work / day in the life of" observational study procedures, where we spent a minimum of a day and an average of 2 days at each site. User experience researchers "shadowed" selected participants over several hours, intruding as little as possible so as to observe the typical work day/work flow, and concluded with a structured interview at the end.

Included in these studies were IT pros and particular types of information workers (data analyzed separately), including Exchange messaging administrators, SQL

[2] Lin, A., Huang, D., Engelbeck, G., Mueller, L., Stempski, M., Gunderson, N., Mings, S., Soderston, C., (2006/2007)

Server database administrators, database developers, PBX / voice messaging administrators, Compliance / security administrators, business analysts, financial analysts, administrative assistants, and other highly mobile information workers. Following are three of the hypotheses developed from these studies that have already been incorporated into Microsoft solutions for these audiences in the field.

2.1 Hypothesis 1: IT Pros Spend Most Time in Information Worker Tools

These IT pros spend the lion's share of the time that they are at their computer in their email program, their internet browser, and their spreadsheet program, e.g., Outlook, Internet Explorer, and Excel. They multi-task, they have many programs open at the same time and they shift their attention back and forth across windows as needed to support the tasks that pre-empt, interrupt and integrate with their on-going work flow. They like having multi-windowing capability, they like being able to do another task while one is running in the background. They collaborate, they delegate, and there is workflow that they support outside of their typical IT tools.

This mirrors what we observe in the typical office information workers' behavior. The IT pro end users are very familiar with and are spending the lion's share of their computer time in the same UI constructs as the information workers. This supports the proposition that UIs don't have to be *fundamentally* different for IT pros from what they are for information workers. IT pros will be using both "classes" of tools side-by-side and transfer learning at the UI "primitives" level from one to another if they are consistent at this level. For a "caveat" on this, see the next items.

2.2 Hypothesis 2: IT Pros Expect Full GUI & Scripting Support for All Operations

Most messaging and database administrators do not write scripts in their daily jobs today. They either write scripts infrequently or use existing scripts that were written by others for infrequent tasks (re-installing), and for frequent bulk, repetitive operations. Like information workers, the IT pros expect to be able to do all tasks through a GUI. Unlike information workers, however, they also expect all tasks to be "scriptable".

This finding was addressed in a novel way in a "version 1" sense in Microsoft's Exchange Server 2007, in the design of the Exchange Management Console, and in Windows Powershell (an extensible command line shell and scripting capability – see http://en.wikipedia.org/wiki/Windows_PowerShell)[3]. The management console in Exchange Server 2007 was the first Microsoft administrator console written entirely in Powershell. All GUI operations done by the user are executed by the GUI "under the covers" in Powershell.

Early usability testing showed that Exchange administrators would have a learning hurdle to overcome in scripting in Powershell, though they loved the value proposition. To help in the initial learning curve as well as in continuing memory refresh, the team designed a way for the GUI to show what Powershell scripts were being run to do the tasks (see Fig.5). People can ignore this or attend to it in order to "learn by doing". In addition, they can copy/paste to create and modify their own scripts.

[3] Lin, A., Langowski, A., Clark, B., Frijlink, N., Mings, S., Sharma, V., Soderston, C. (2006).

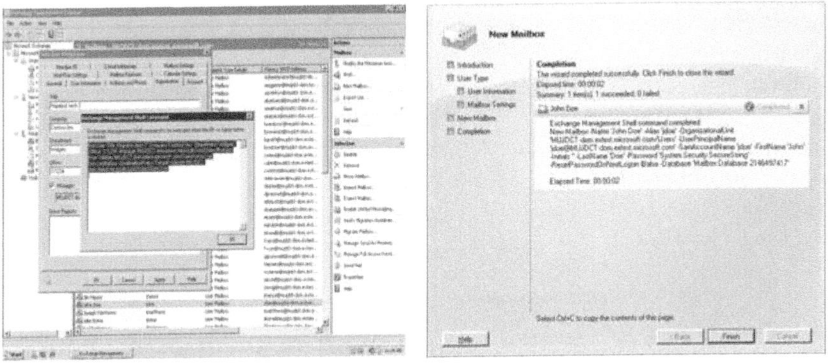

Fig. 5. Script display in dialog confirmation and in wizard

Information workers don't typically encapsulate their GUI tasks into macros, but IT pros appreciate the capability to do so.

2.3 Hypothesis 3: IT Pros Need Task Orientation in UIs

The last hypothesis on the commonality between IT pros and average information workers has to do with transfer of learning and navigation challenges. IT pro roles today are separated across individuals, but their roles are constantly evolving. The way responsibilities are allocated across IT pro team members shifts continually, the task responsibilities transfer across individuals, and the trend is that the pace of change in this regard is accelerating.

Being able to transfer learning from one tool to another and to find function (if authorized) from a common launchpoint, is highly valued by both IT pros and their companies and is still a significant challenge today. They are stressed to find function as quickly as possible, and many times this is in front of their customer/end user.

IT pros, like other users, think in task terms, in verb/action language, and have similar challenges fitting this aspect of their interaction approach into the object-oriented, single direction language environment. They're challenged to find where the first step for tasks lies in deeply nested object hierarchies, and what order to perform steps when their task crosses objects and hierarchies without any sequencing built into the software. In a "version 1 and 2" sense, this proposition was addressed in the UI design of Microsoft's BizTalk Server 2006 and Exchange Server 2007.

Microsoft administrator consoles, like others, are object-oriented, laid out with a navigable tree structure in a left pane. IT pros navigate to the objects of their choice and then perform their chosen tasks applied to those objects. In BizTalk Server 2006, the team applied a task overlay to the console, enabling the IT pros to locate their task start points through navigating a task hierarchy – *in addition to locating through the object tree* (see Fig.6). This console "root hub" of the tree hierarchy provides a display in the right pane of the major task categories, with procedural information on how to complete them. This was a user experience-led initiative, which quickly garnered great support and collaboration across the technical teams and user education staff [4].

In Exchange Server 2007, the team took this concept even further, to support learning the OO structure "by doing". In Exchange Server 2007, the tree structure was significantly simplified, an actions pane was added on the right that is context-sensitive and "bubbles up" the tasks that can be applied to the selected objects, and a console root hub page similar to BizTalk Server's was added to integrate all configuration tasks (see Fig. 7).

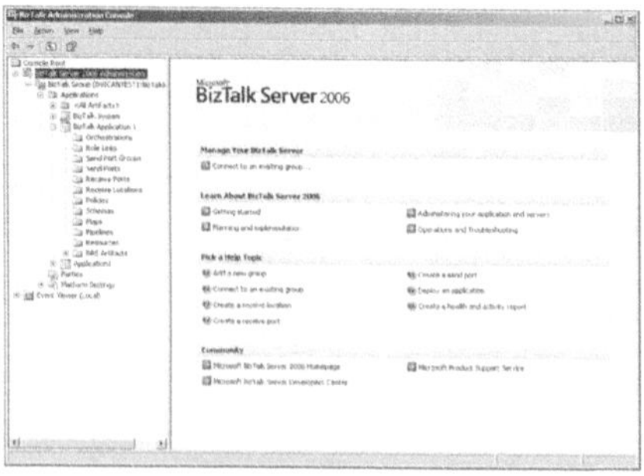

Fig. 6. BizTalk Server 2006

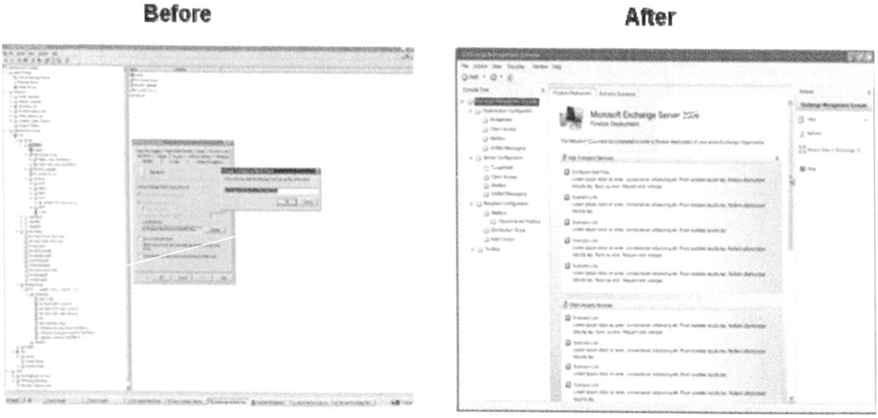

Fig. 7. Exchange Server 2003 vs 2007

This "start page" took the design to the next level, creating a tab structure to make it extensible to other task categories, and adding what we light-heartedly code-named a "post-its" construct to provide procedural task information and links to the appropriate places in the object hierarchy. These persist and float on top, even when the task flow replaces the middle pane start page information with the particular object's contents. These start page structures and the floating task instructions/links enable IT

[4] Vican, D., Scott, P., Christman, C., Hanswadkar, K., Perry, N., McElroy, P., Casey, T., Soderston, C., (2005).

pros to learn the object hierarchy while they're doing tasks. This was a relatively lightweight way to superimpose task structure on top of an existing object hierarchy, without having to re-architect the object hierarchy in a more substantial way.

3 Conclusion

Observational studies of IT pros and information workers in their environment show the many ways their behavior is similar [See also 2,3,4]. IT pros typically spend the lion's share of their working day in email, browser, and spreadsheet programs (e.g., Outlook, IE, Excel). Being able to transfer learning from one tool to another and to find function (if authorized) from a common launchpoint, is highly valued by both IT pros and information workers. Both IT pros and information workers think in task terms, in verb/action language. They have similar challenges fitting this aspect of their interaction approach into a single-direction UI, one in which objects must always first be chosen. Ideally they want to be able to traverse through software in both object/action and action/object sequences, depending on which better fits the immediate goal.

Like information workers, the IT pros we studied report that they expect to be able to do all tasks through a GUI. Unlike information workers, they also expect all tasks to be "scriptable". When we moved the lion's share of user interface interactions from blank screen command line interfaces, first to page/panel driven text UIs and then to full GUIs, we didn't consider how to build in command line and query capability seamlessly in the GUI paradigm. Instead, these are in separate windows and work doesn't typically flow seamlessly between the two. This lost opportunity is recoverable: it is time to more insightfully craft the best conventions for converging command line access into the standard GUI paradigm and for enabling task navigation/structure in object-oriented frameworks.

References

1. Soderston, C., Lanzetta, T., Scanlon, J.: Problem-Guided Heuristic User Interface Evaluation Method: Identifying Nuggets of Opportunity & Fool's Gold in User Interface Design. IBM Technical Report 36.0017 (1992)[5]
2. Intel Ease of Use/PC Quality Roundtable white paper: Improving PC Ease-of-Use for IT Professionals (2001), `http://www.eouroundtable.com/files/ITFiles/ITWhitePaper.pdf`
3. Wichansky, A.: Customers Reveal Top Enterprise Software Usability Issues to Oracle (2008), `http://usableapps.oracle.com/design/pages/OUAB.html`
4. Haber, E.M., Bailey, J.: Design Guidelines for System Administration Tools Developed through Ethnographic Field Studies. Paper presented at the 1st symposium on Computer Human Interaction for Management of Information Technology (CHIMIT) Cambridge, MA (2007), `http://chimit.cs.tufts.edu/ap-papers.html`

[5] This work was first cited externally by Theo Mandel in his 1995 book entitled "The GUI-OOUI War: Windows vs. Os/2: The Designer's Guide to Human-Computer Interfaces". The full report is not available externally; is now dated, not sensitive. The results of this work helped to inform what became IBM's Common User Access architecture, which was licensed in early versions of Microsoft Windows, and other environments. See http://en.wikipedia.org/wiki/Common_User_Access

Customer Boards as Vehicles of Change in Enterprise Software User Experience

Anna M. Wichansky

Oracle USA, 500 Oracle Parkway, MS2op2, Redwood Shores, California 94065 USA
Anna.Wichansky@oracle.com

Abstract. Traditional user-centered design processes do not leverage long-term customer-vendor relationships as a means of improving product usability. While designing a next-generation applications software suite, Oracle reached out to its most-involved customers for creative solutions to common user-experience issues. The mission of the Oracle Usability Advisory Board was to take enterprise software to a whole new level in usability. The board consisted of executives and senior managers primarily in information technology positions in different types of organizations. The board identified three major areas where it wanted to improve usability: consistency and design, integration and performance, and Web 2.0. Through various working groups, the board has developed tools for obtaining customer feedback on product usability, online seminars on technical topics, and outreach mechanisms to other customers. The board has effectively become Oracle's partner in ensuring product understanding and use, thus setting the stage for improved usability in the next-generation product.

Keywords: customer boards, enterprise software, user experience.

1 Introduction

To improve product usability, major enterprise software vendors have interacted with customers to obtain information on customer work processes as input to task flow modeling, and to get user feedback on prototypes of products under development. This type of user-centered design process generally involves end users in the target product market as participants in vendor ethnographic research, focus groups, surveys, and usability testing studies.

The methods in the user-centered design cycle are fairly labor intensive. A relatively small number of end users are often observed one at a time, various parameters of productivity and acceptance are measured in usability lab studies, and generalizations are made about how the users think and act in the course of work. Agile methods have speeded up this process, largely by using a minimalist approach [1].

In reality, how enterprise users work and what software they use is heavily influenced, if not largely determined, by purchase decision makers in their information technology (IT) departments. IT departments provide base image software installed on worker computers. This software is heavily customized and configured for the jobs that users are doing. Document and Web templates exist and are used throughout the organization. These templates dictate the content and appearance of deliverables, promote corporate consistency, and improve department productivity.

M.J. Smith and G. Salvendy (Eds.): Human Interface, Part I, HCII 2009, LNCS 5617, pp. 316–322, 2009.

So why aren't usability experts talking to IT departments and other corporate decision makers about usability, productivity, and process roadblocks in modern enterprise software? In most enterprises, these entities have real control over end user work processes and software.

Previous attempts to reach out to customers on behalf of usability efforts have been reported. Some good examples were the User Partnership Program at Siebel Systems and the Global Design Partners at Oracle [2]. These programs provided vendors with access to customers for three basic purposes: test recruiting, customer site visits (ethnographic style research), and verbal feedback about existing products. The goal of these programs was primarily to access end users (rather than decision makers). The author concludes that positive relationships are important for repeat sales; however, he does not detail exactly how this happens.

At the end of 2007, the Oracle Applications User Experience (UX) department was completing the design phase of an integrated suite of next-generation enterprise applications software. UX had flourished and grown at Oracle since 1994, when the first department and labs were founded. Numerous customers were tested—all of whom were in the participant database and some of whom had prior contacts in the department. It was time to access these long-term relationships with our most involved customers in a systematic way, not only to promote our new product, but also to educate customers about included Web-based technologies. Oracle wanted to see if customers expected new usability issues to crop up with the next-generation products, and if so, we wanted to address these issues early in development. These types of dialogues were necessary not only with end users but also with decision makers.

The Applications UX department sought to establish a new way of tapping into customer experience to drive usability a quantum level higher in future generations of software. The department encouraged open communication of experiences with its customers as a means of accessing customer strengths and problem-solving abilities at an executive level. The goal: maximize the usability of enterprise software in an atmosphere of mutual trust. The department also considered enlisting consultants, usability experts, and industry partners to facilitate, educate, and otherwise contribute as needed to reach solutions.

2 Founding of the Oracle Usability Advisory Board

Late in 2007, the Oracle Usability Advisory Board was formed to identify and address enterprise software usability issues in current and future products. This approach was intended to deal with usability on a macro level to determine and influence industry trends. The board reviewed and provided feedback on future technologies, shared best practices, and developed use cases, industry guidelines, standards, and requirements. The ultimate goal was to bring enterprise software usability to a whole new level through industry, government, and university collaboration.

The board was recruited from Oracle's most involved customers—those with a history of support for user-centered design activities and with a willingness to work on joint projects with Oracle UX groups. The goal in founding the organization was to identify 10-15 active customers that were enthusiastic about the prospect of a joint venture to improve enterprise software usability.

The composition of the board is shown in Table 1. Participating on the board were representatives from government, university, and financial, pharmaceutical, and manufacturing industries using Oracle, PeopleSoft, Siebel, and JD Edwards enterprise applications. The distribution of members represented small, medium, and large customer entities across various organization types and user groups. Members were typically chief executives or senior managers with direct accountability within their organizations for user effectiveness, efficiency, and satisfaction with enterprise software. Select independent UX consultants under nondisclosure agreements with Oracle advised and educated the board on an as-needed basis.

Table 1. Oracle Usability Advisory Board Demographics

	Executive	VP/Dir/ Chief IT/	VP/Dir Apps	Architect/ Analyst	Other
Manufacturing	1	3		1	1
Government		4	1	1	
Education	1		1	1	
Pharmaceutical/Medical Devices	2	1	1	2	
Financial/Insurance/ Real Estate		3	1	1	
Consulting	3	1			

Board member organizations were required to:

- sign a Customer Participation Confidentiality Agreement (CPCA),
- attend three working meetings per year,
- perform necessary pre- and post-meeting assignments (typically collecting data within their organizations),
- contribute actively during working meetings and online conferences,
- communicate with and support Oracle usability executives and senior staff driving action items, and
- facilitate Oracle access to users in their organizations for studies that would advance the goals of the group.

Later requirements were to use the Oracle Councils secure Web site for communication and social networking and to attend Web and telephonic conference calls to advance the progress of working groups and to organize activities between meetings.

It was emphasized to new board members that this was a *working* board. Board members were expected to stay engaged during and between meetings, prepare for meetings, present and advocate for positions representing their organizations, and actively collect and contribute usability information relevant to the board.

As of this writing, the board included 30 representatives from 25 customer organizations, an Oracle chair and cochair, and three Oracle working group facilitators. Recent additions to the board were value-added resellers and database consulting services.

Two UX consulting firms familiar with the Oracle user-centered design process also sat on the board. A steering committee was formed in 2009 to help lead the board.

3 How the Board Operated

Meetings with customers before we had a board typically consisted of a review of customer UX problems with their applications systems. It was often hard to understand and even harder to fix such problems. Enterprise software tends to be highly customized, so we had trouble even recognizing the user interfaces in some cases. Also, customers tended to relegate almost all problems to the category of usability, including integration, installation, performance, and documentation.

The paradigm shift that we were looking for in establishing the board was to engender open sharing, communication, and problem solving at an executive level around topics in enterprise usability that were pertinent to all customers. Organized meetings therefore became a series of dialogues between Oracle and customers and among customers themselves. Presentation of specific screen shots and use cases was encouraged.

Meetings occurred three times in the first year. Meeting themes were determined based on what was of interest to customers, based on their usability issues. Topics for the first year were collaboration tools, business intelligence, and mobile computing applications. Subsequent agendas have been determined in board planning meetings and through online board surveys.

Before meetings were held, board members completed assignments relevant to the meeting that involved reading articles, watching videos, preparing slides, or collecting data within their organizations.

Meetings were rotated geographically to accommodate as many board members as possible. Meeting sites for the first year included the Oracle Conference Center in Redwood Shores, California; the Oracle Usability Labs in Burlington, Massachusetts; and the Oracle Usability Labs in Denver, Colorado. Future meetings may be held at customer sites and in Europe.

Meetings began with an informal dinner hosted by Oracle the evening before a full day agenda. All speakers, facilitators, customers, and lab staff were invited. This helped customers get to know staff throughout Oracle who could serve as future resources. It also allowed catching up between customers, introduction of new board members in an informal setting, and setting the stage for discussion of important topics that would be featured the next day.

Day-long meetings typically consisted of presentations by new members on their top-three usability issues, reports on interim activities and statuses of working group projects, a keynote address on the meeting's major theme by a major Oracle executive, and an afternoon spent in the working groups. These events were followed by new business and a general summary of what happened in the meetings.

The working groups were the heart and soul of the customer board. Groups were originally organized on three key topics, which were based on customer representations of their worst usability problems. The three areas were: user interface consistency and design, integration and performance, and Web 2.0. Customers initially joined a group in which they were most interested, but they could switch groups. New board members sampled the discussions of multiple groups. Oracle senior management facilitated the groups. Oracle scribes, often designers and usability engineers, took notes during working group sessions and prepared minutes and action items to be addressed between meetings. Outcomes of these groups were report formats for usability issues, Web conference training on technical topics, and sharing of opinions about the importance of Web 2.0 features.

The issues selected by the working groups were both tactical and strategic. The consistency and design effort was largely based on the desire to provide feedback on current software, pointing out places where a feature was called by two different names or a feature operated differently between application platforms. The Web 2.0 and integration and performance groups were more strategic, generally addressing future users, enterprise software architectures, and functionality.

UX consultants contributed to the working groups in various ways. They generally wore the hats of usability experts but not Oracle experts. Therefore, they were credible to customers for their UX knowledge but not aligned with Oracle in terms of product agenda. They could provide examples from various vendors to illustrate their points and findings from published research on other software platforms.

Between meetings, additional work was scheduled and performed. The primary means of communication was the Oracle Councils secure Web site. This Web site was used by more than 60 Oracle Councils and Customer Advisory Boards to archive and share documents, provide distribution list service, and host polls and surveys. The Web site was not available to the general public, so online communications could be carried out while maintaining confidentiality. The board started out using e-mail, which was too cumbersome and not secure enough for the work that we were trying to accomplish. The polling feature made it easy to take a quick head count on small logistical questions such as potential attendance at a user group meeting or interest in a panel or presentation. Secure online survey tools were also used with the board to collect data.

Working groups conducted conference calls and Web seminars between meetings. Working groups also met in online conferences without Oracle UX facilitation to discuss work between meetings.

More in-depth visits between board organizations and Oracle UX teams were also scheduled between meetings. These meetings typically involved Oracle headquarters lab tours and discussions with customer staff (other than the board members) and Oracle staff selected to represent topics of interest to the customer. These visits took place at the customer sites as well.

4 What the Board Has Accomplished

In the first year, there have been three meetings; three Web seminars on secure enterprise search, applications integration, and Web application development tools; a panel at Oracle OpenWorld 2008 (Oracle's annual technical conference and tradeshow); organization of three working groups; and planning of the agenda for the second year of operation.

One working group developed a spreadsheet tool to collect information on usability issues in customer implementations of Oracle software. This spreadsheet was designed to give sufficient detail for Oracle developers to address the issues and the working group facilitators to track the issues.

Subtle changes in our relationships with some of our most involved customers resulted. Due to the board meeting exposure, we had greater familiarity with these customers, and we felt more comfortable with each and every encounter. We were able to immediately get down to business and accomplish more work in fewer meetings on customer issues.

As we met these partners at conferences and user group meetings, we continued our agendas and developed new ideas for projects that we could accomplish together. The Oracle UX group met between board meetings with UX groups of board customers to share best practices and concerns.

We have achieved the start of a paradigm shift in the way usability is addressed between customers and the Applications UX department. In general, customers are taking more initiative than they were previously. Some examples:

- Customers have gained a greater understanding of the potential use of the next-generation product. This understanding is a result of communication of features and functions by UX facilitators, Oracle developers, and product managers. We have moved away from selling these customers on new features and functions to customers seeking information.
- Instead of customers providing anecdotal descriptions of their usability issues, there is now a clear process and form to enable Oracle to actually address the problems affecting multiple customers.
- Customers can now participate meaningfully in discussions ranging from UX enhancements in point releases of existing software, to features and functions that need to be developed in future software.
- Presentations of Web 2.0 technology and discussion with customers to overcome unfamiliarity may translate to earlier and greater acceptance of the next release.
- In addition to Oracle UX speakers at industry conferences, board members are now talking to other customers about usability.
- Usability has been promoted from a one-sided vendor initiative to a joint project in which customers can participate.

5 Future Goals

The board is now working on a set of customer-driven goals for 2009. These goals include evaluating the efficacy of the inputs that it has made so far on Oracle products. Meetings are planned in conjunction with user groups and hosted by customers. An industry magazine is researching a story about UX with testimonials from board members. Board organizations participate increasingly in customer feedback sessions on pre-released products at industry conferences.

Beyond 2009, the board may sponsor projects such as participating in beta-style product trials at customer sites, inviting vendors to participate in themed meetings, and hosting a usability conference.

6 Conclusions

The success of the Oracle Usability Advisory Board resulted from having faith in customers as equal partners in the effort to improve enterprise software usability. The customers have come through with virtually everything that we expected of them and more. These were our most involved customers, who became even more interested and active as they spent time and effort with us on joint projects.

Some lessons learned:

- Customers appreciate rotating meetings to different geographical sites because it minimizes travel expenses. Rotation also enables a variety of local labs and technical talent to host and participate in meetings with lower cost and effort.
- The board needs a quorum of active customers in good standing for its opinions to convince Oracle development. UX needs to be able to "name names" when providing feedback to development and marketing. Chief executive opinions are particularly valuable.
- Customers are willing to pay their own expenses and spend time and effort to help vendors achieve the next level of usability. They are also willing to read articles and take online training to help themselves understand the latest trends in technology and usability.
- To conduct successful working groups, there must be a core set of customers who are willing to work on a project until it is done.
- It is helpful to have customers on the board who like to present to others. "Customers talking to customers" was a particularly compelling model in convincing audiences of the importance of usability.

References

1. Detweiler, M.: Managing UCD with Agile Projects. interactions 14(3), 40–42 (2007)
2. Lee, W.O., Solorzano, S., Harrison, C.: Creating and Managing User Partnership Programs. In: Proc.CHI 2005, p. 2123. ACM Press, New York (2005)

Fashion Support from Clothes with Characteristics

Yuri Yonezawa and Yoshio Nakatani

Ritsumeikan University
1-1, Noji-Higashi 1, Kusatsu, Shiga 525-0058, Japan
cc017043@is.ritsumei.ac.jp
nakatani@is.ritsumei.ac.jp

Abstract. Fashion can be a source of daily enjoyment, as well as source of anxiety, in our lives. In this paper, we propose a new system for supporting a person's fashion coordination. This system produces pleasant coordination between a person and his or her clothes when the person stands in front of the closet each morning and picks out the day's attire. Each item of clothing has its corresponding software agent with a specific characteristic decided by color, based on psychological findings. The scenario is as follows: First, the user enters the total fashion image for the desired fashion profile for the day, such as "today's image is red." When the user goes to the closet, clothing that is red or verging on red start to appeal. The user hears messages such as "Choose me because you had a good day last time you chose me," and "No, choose me because it is hot today." An Integrated Circuit (IC) tag is put on each hanger, and the chosen article of clothing is identified by the IC tag reader affixed to user's finger. The episodic database and the database of basic fashion knowledge are connected to each as a knowledge source for conversation.

Keywords: fashion, coordinate, software agent, IC tag.

1 Introduction

This proposes a support system aimed at easily coordinating a large number of articles of clothing. As compared with the past, we own many more articles of clothing. However, we cannot wear all these clothes. There are too many items of clothing, and sometimes we worry about coordinating all the items. If there is no time to think about our choices, we will often wear the same clothes over and over. Some clothes are almost never worn or taken out of the closet.

In response to this problem, many systems for coordinating our wardrobes have been developed. We introduce two examples. First are the systems that recommend combining clothes by color, balance, and other details. These systems make recommendations based on fashion rules. Second are the systems that display an image of the person wearing the clothes. These systems project the user's image, using virtual reality technology to portray the user wearing the articles of clothing, seen as if reflected in a mirror. Almost all the systems display a result in front of a terminal. These are not providing decision-making support while the user thinks about coordination.

M.J. Smith and G. Salvendy (Eds.): Human Interface, Part I, HCII 2009, LNCS 5617, pp. 323–330, 2009.

We thought at first that fashion coordination is what is performed in front of the closet. The most important part of coordination is not what is "correct" but what is fun. What you would like to wear changes with your feelings at that time. In this paper, we aim at coordinating with the user's decision-making through "dialog" with the clothes while the user stands in front of the closet. We propose the structure for the system needed for realizing this goal.

For the realization of such a system, we believe that two elements are required. One is an intuitive interface and the other is the structure for having fun. We propose using an IC tag as an interface. We propose the view of sense engineering as structure for having fun.

Section 3 explains these ideas in detail. We devised the new structure of the system using these ideas, and we named the structure the "talkative closet." We describe related research below and we explain the outline of the system and, finally discuss the future.

2 Related Research

2.1 Virtual Stylist Project

Kansai University and Kansai University Graduate School researched the coordination of a clothing and accessory support system [1]. The user can obtain data about the clothing through a network from the clothes data server. The search terminal searches the clothing that the user likes. If there is a picture of an article of clothing on hand, they can automatically coordinate other clothing that go well with it. This coordination is performed as follows: First, the system coordinates a harmonious color scheme and a color scheme image judging is performed. Second, this is suited to the user's image and coordinates with the color scheme that was able to take color harmony. The coordination system can perform an imaginary fitting, as if the user were actually wearing the clothing. We base these steps on the data that were extracted from the image of the user and the image data of the clothes.

This system works with color judging and virtual reality (VR) fitting. Even if a user does not have knowledge of color, he or she can get the coordination of the harmonizing color scheme. If the user extracts liking, s/he can also prevent the system from acquiring inaccurate information.

In this system, the user can perform a virtual fitting using VR. The user can do a more fascinating coordination in liking or color adjustment. Thus, VR has the ability to study future clothing purchase. However, we would like to explore the problem of how a user chooses clothes through intuition based on the clothes that s/he has.

2.2 Virtual Mirror

Figure 1 shows a virtual fitting of shoes in the showroom of the Avenue des Champs Elysées, which Adidas opened in Paris. Fraunhofer Telecommunications Institute and Heinrich-Hertz-Institute in Berlin jointly developed this virtual mirror. Unlike a conventional mirror, the virtual mirror does not project a reflection. A system processes the image of the leg photographed with the camera in a 3-D image and projects the image wearing shoes. Since the processing speed is quick, in order that it may unite

Fig. 1. Virtual fitting of the shoes in Avenue des Champs Elysees

with a visitor moving and an image may also move, the feeling is as if you were looking at an ordinary mirror.

The virtual mirror has also projected the photograph and a short movie of shoes or clothes in addition to the image of the individual wearing the shoes. If the individual points at the items, s/he can purchase them. This technology differs from a touch screen. Even if the individual does not actually touch the screen, the system recognizes the direction where the individual pointed out the goods, and the individual can make an article selection [2].

Unlike the system in Section 2.1, the user's physique and size are compensated for by image processing. You think that it is actually you who is wearing the clothes. However, a special mirror, the camera used for image processing, are required. Considering the cost and space required, the difficulty at this time is having a virtual mirror at home.

2.3 My Stylist

My Stylist® is game software used by the handheld game machine PSP (PlayStation Portable). It is a tool that supports synthetic coordination for women's fashions [3]. My Stylist® can take a photo of clothing with an attached camera and then it can manage the photo. It can simulate coordination and can recommend coordination based on information about a schedule and the weather. My Stylist® has a fashion dictionary and supports enjoyment of fashion synthetically and it performs personal color diagnosis. Figure 2 shows My Stylist®.

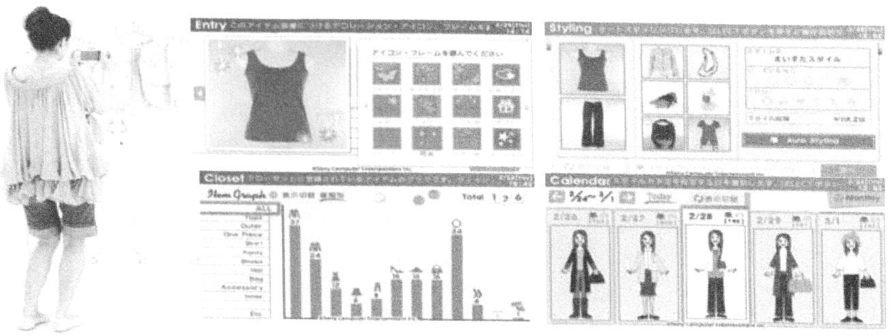

Fig. 2. sample display of My Stylist [3]

2.4 Example of an IC Tag

Hankyu Department© Stores and Dai Nippon Printing© jointly developed an over-the-counter-sales support system using an IC tag. Image information, such as a function of the goods that attached the tag is inputted beforehand. If it detects that goods have been removed from the showcase, an image will flow automatically from the small screen attached to the shelf. In retail sales, an IC tag has come to be used for sales promotion as well, although an IC tag had been utilized for inventory control until now.

The system consists of personal computers that input merchandise information, the touch-sensitive screen, the tag-reading antenna that is included in a showcase, and the card IC tag attached to goods of the card size. The input of information is patterning arrangement of a picture or a letter, and the salesperson can easily do it using a computer mouse. A sound effect and animation are also possible.

3 The System Concept

We wear various kinds of clothing. When we select our clothing, we present ourselves. We choose our clothes in accordance with various environmental changes, for example, in response to our culture, the seasons, a particular situation, and our community. We also choose our clothes in accordance with how we feel. We are concerned about own presentation of self in connection with clothes. In this section, we propose a way of shifting these feelings of concern into feelings of enjoyment.

3.1 Touch and Enjoy

Our system makes the pleasure that conversation is born by touching clothes. We believe that we do not feel concerned about coordination if coordination is enjoyable. However, support of the coordination that we touch and enjoy is difficult on a screen. We propose an interface that uses an IC tag. With an interface, we can access a system by touching our clothes.

3.2 People Talk with an Object

We communicate with our clothes. We project a sense of character onto our clothing. We have associations with colors. For example, the feeling of excitement is red. Although such ideas differ from country to country, for our paper, we refer to a color study of Japan. We determine the character of our clothing with color. We determine the character of the items of clothing from the association we have with their color.

We build the following structure in a system. This system gets a color from a picture and changes the color into a form from which it is easy to extract an image. The system then decides a conversation tendency in the extracted image. When we talk about clothes, the system decides the conversational content.

3.3 The New Recognition of Value

The value of clothing decreases day by day. We have many clothes, and when we get bored with them, we buy new clothes. Even if a style is no longer in fashion, we do not throw our favorite clothes away. The most typical examples are clothes that we wear on a memorable day, such as a wedding dress. However, the clothes of a memorable day are unique examples.

We explain why we do not throw away my clothes. These clothes are a trigger for recollecting special memories. We may choose our clothes in accordance with our recollections. For example, on a examination day, we wear auspicious clothes that will bring luck for an examination. However, too many records are annoying for us. Then we will use a part of record.

4 The System Method

In Section 3, we explained the concept of the system. In this section, we discuss a system scenario, realizing the system, and the effect expected.

4.1 A System Scenario

For our scenario, we use choosing clothes in the morning. The user explains her/his image of the day's fashion to the system. The user goes to the closet. Then, if clothes find an image that matches the user's explanation, they claim user to choose themselves. The user hears the opinion of the clothes and makes a decision based on the system's recommendations.

The ideal process is described below. In this case, the information for the user's clothing is a premise previously registered into the system. The picture of the clothes, a color, an image word, and other details are registered. We presuppose that the recollections about clothes are also separately registered through the recollection arrangement system under development.

1. If a user stands in front of a mirror and opens the closet, the clothes inside enter an opinion in the log. For example, "Recently, you have not worn me; wear me today."

2. The user accepts these opinions or states his/her own opinion. For example, "Today I have an orange feeling."
3. The system receives the user's reaction and the clothes analyze their information and express their revised opinion. For example "I am orange" or "When your feeling is orange, I believe I am suitable for you because ‘xx.’"
4. The user chooses clothes from those opinions, or the user chooses another set of clothing. Other clothes insist that they unite themselves. The clothes that assert themselves at this time refer to a past record with the decided clothes. For example, "Had you set me at the time of an orange feeling?" or "Although the clothes are not related to orange, I am orange. I suit you."
5. Repeating the above processes several times, the user makes decisions about what to wear.
6. The user registers into a recollections arrangement system the comments of the clothes worn on that day. The user inputs recollections at a terminal, such as a cell phone.
7. This system accumulates registration information as a log and performs registration and updates as new rules.

In this research, in order to communicate on a screen as a prototype system, our plan is as shown in Figure 3.

Fig. 3. Prototype system

4.2 System Features

This system does not support the knowledge of a fashion. Even if you input a target, this system does not display the result. In this system, we support the fun of coordination. For example, we support the surprise that clothes assert or we support the fun in which a choice spreads. Our primary goal is not good coordination. Our goal is that you become pleased and that you will become fond of love your clothes.

4.3 The Step of a System

1. Use IC tag
   The first step in this system is environmental construction for using IC tag between the closet and a terminal [4]. An IC tag is one of the wireless-communications standards RFID. We use the combination of the IC tag and the reader writer (henceforth R/W) that reads and writes information on IC tag by non-contact. Various forms in an IC tag and R/W. Figure 4 shows an example.

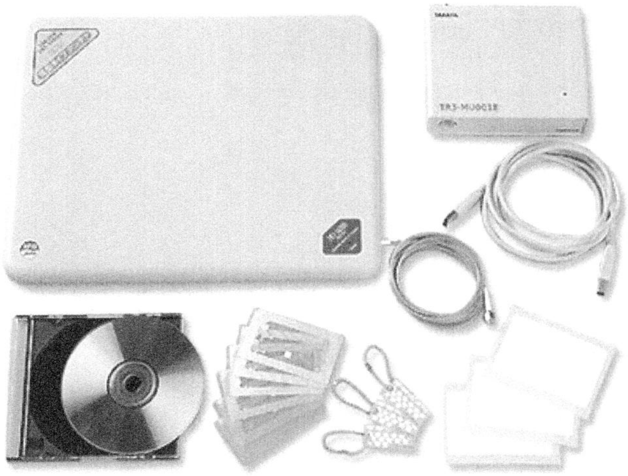

Fig. 4. ICtag and R/W

An IC tag is weak in water. Therefore, we cannot tag the clothes themselves. From such a problem, we gave up attaching an IC tag to clothes. At the present stage, we tagged the hanger and confirmed that we could connect an IC tag with a reader.

2. Color study use
   Color study use is classification using the color information. We get hue saturation value (HSV) from a picture. HSV is the evaluation of color decided by hue, chroma saturation, and brightness. We decide character using hue and tone that mixes S and V [5].

5 Future Work

For future work, our aim is to develop an application based on the structure described in Section 4. Then, we plan to conduct an actual experiment and examine the usefulness of a system.

References

1. Tokumaru, M., Muranaka, N., Imanishi, S.: Virtual Stylist Project. In: 2002 IEEE World Congress on Computational Intelligence, CD-ROM (2002)
2. Article of Adidas opens world's most advanced showroom in Paris, http://www.gizmowatch.com/entry/adidas-opens-worlds-most-advanced-showroom-in-paris/ (access date 2009.2) (in Japanese)
3. PlayStation (Japan): My Stylist, http://www.jp.playstation.com/scej/title/mystylist/ (access date 2009.2) (in Japanese)
4. Yanagida, T.: RFID technology, All of the wireless IC tags: Nikkei BP, 2004.4 (in Japanese)
5. Nationwide dress federation text revised edition edit committee: Color authorization, AFT project Ltd. 2005.3 (in Japanese)

Part III

Interacting with Information, Documents and Knowledge

Exploring History with Narrative Timelines

Robert B. Allen and Sumanth Nalluru

College of Information Science and Technology, Drexel University,
Philadelphia PA, 19104, USA
{rba,sumanth}@drexel.edu

Abstract. We develop novel timeline interfaces which separate the events in timelines into threads and then allow users to select among them. This interface is illustrated with five threads describing the causes of the American Civil War. In addition to selecting each of the threads, the sequence of events it describes can be played. That is, the user can step through the sequence of events and get a description of each event in the context of its thread. In addition, many of the events have links to more focused timelines and to external resources such as Wikipedia.

Keywords: Causation, Events, History, Interactivity, Narrative, Timelines, Visualization.

1 Introduction

Timelines are a common graphical technique which present ordered sets of events. But, the events are often jumbled with several different threads interlaced. For instance, a timeline of the events leading to the American Civil War might include constitutional debates, Abolitionist protests, and changing agricultural patterns. While simple timelines help people to understand the temporal relationships among events, we believe that disentangling the threads among those events would make the entire sets of events more comprehensible. Indeed, visualization could enhance texts and be a sort of cognitive organizer.

This paper extends a series of interface techniques for interaction with timelines. Allen [1] described several ways timelines can be interactive such as by filtering the displayed items by attributes. Allen [2] introduced focus-context timelines which have two segments. One of these presented focus events and the other displays context events. That interface provided menus to the user for controlling specifically which sets of events are presented. The Lifelines project [3] developed a hierarchical timeline viewer for presenting significant life events such as the duration of medical conditions and treatments in patients.

Other interfaces have described techniques for presenting narrative and causal threads. For instance, Allen and Acheson (e.g., [4]) describe narrative lines which are sets of connected events that form the plot of stories. The connections are the causal links which give coherence to the plot. In narrative theory, this interface could be said to explore the story's fabula. Allen et al. [5] developed an interface for presenting causal explanations of scientific models. Specifically, that work described a geological

M.J. Smith and G. Salvendy (Eds.): Human Interface, Part I, HCII 2009, LNCS 5617, pp. 333–338, 2009.

hypothesis as a set of state changes. Events are central for most of these interfaces and events have been of broad general interest (e.g., [6, 7, 8]).

In a different tradition, it is also worth noting hypertext guided tours (e.g., [9, 10]) which provide a linear framework for presentation of texts. These do not claim any particular connection among the pages, but it seems likely that the connections would often be used to connect a causal or narrative thread.

The current project explores the combination of narrative and causal links with timelines. Specifically, we present causal threads which connect coherent sets of events. Such interfaces should be helpful for supporting navigation through explanations of historic events as well as comprehension of and memory for the relationships of those events. Thus, these interfaces could improve history education.

2 Implementation

2.1 Content

As sample content, we used an analysis of the causes of the Civil War.[1] This analysis proposed five threads which lead to Secession of the South from the Union. We slightly refocused some of the threads to emphasize narrative relationships among the events. Specifically, we used: Cultural/Conflict, States Rights, Abolitionists, Slavery in the New Territories, Abolitionists., and Lincoln as the five Causes of the Civil War.

In some cases, the events which composed each of these threads were simple causal chains. For instance, the development of the Cotton Gin directly changed agriculture in a way that increased the number of slaves. However, the events in some of the other threads suggested more of a trend than a simple causal chain. Thus, while we can say that Abolitionism led to the book *Uncle Tom's Cabin*, the effects of that book, although powerful, were too diffuse to easily represent with the current model.

2.2 Interface

We have developed a prototype interface for presenting causal links and threads as a stand-alone Java application. The main screen of the interface is a focus-context timeline framework related to [5]. However, the focus view is subdivided into separate threads. Initially, the threads are in a compressed or closed format (Fig. 1). Each of them can be opened by clicking on its label. Fig 2, shows the full expansion of all five threads.

As noted above, the threads are like narratives or guided tours. They generally provide a coherent presentation through the sequence of events. In our interface, each thread may be activated with the "Play thread" button which is below the thread label. When that button is selected, the "Play thread" button is replaced with "forward" and "back" arrows and the thread title is highlighted. The events in the active thread are highlighted one-by-one and a text box about it is presented.

---

[1] http://americanhistory.about.com/od/civilwarmenu/a/cause_civil_war.htm

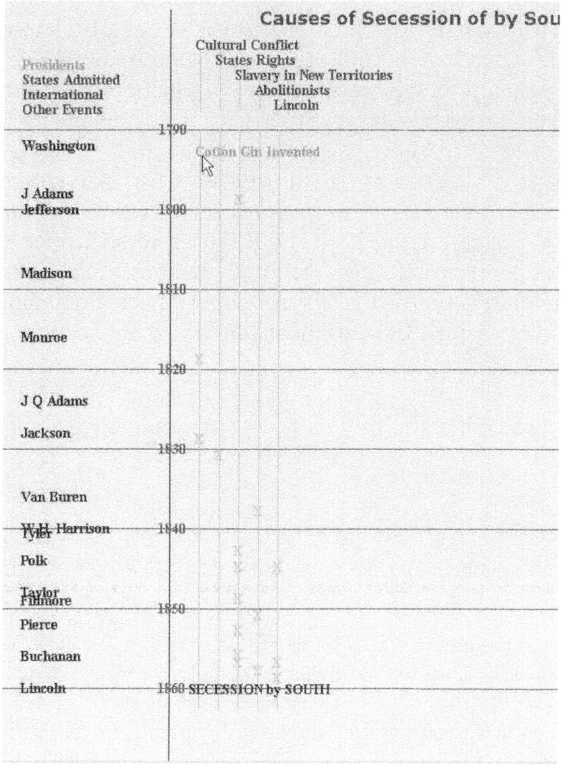

Fig. 1. A portion of the narrative timeline page I shown. On the left, a context timeline shows presidents. Other options are also able to be selected. The historical threads are shown on the right in the closed configuration. The events for each thread are indicated with an "X". Hovering over the "X" pops up a tooltip-like label for the event. Thus, the user can quickly scan a set of events. In addition, each of the historical threads can be opened by clicking on the label associated with it.

Clicking on the event label when a thread is being played produces a short menu which lists other resources associated with that event. In the current implementation, there are up to three resources: links to a sub-thread, links to Wikipedia, and the audio description. The sub-threads are focused timelines which present finer-grained subevents. For instance, the "Amistad" sub-thread elaborates the mutiny, trial, and convictions. The links to Wikipedia provide an even richer resource for background such as biographies of people and documentation.

We have focused thus far on the thread interface. However, we envision a broader system composed of many such thread interfaces. As a step in that direction, we developed a home page (Fig. 3) which displayed links to various sets of narrative timelines. (In the current implementation all but the "Causes of the Civil War" were stubs). Originally, we had developed a text-only outline for the home page, but many of the threads did not fit the hierarchical outline structure. The graphical approach in the figure allows multiple overlapping threads.

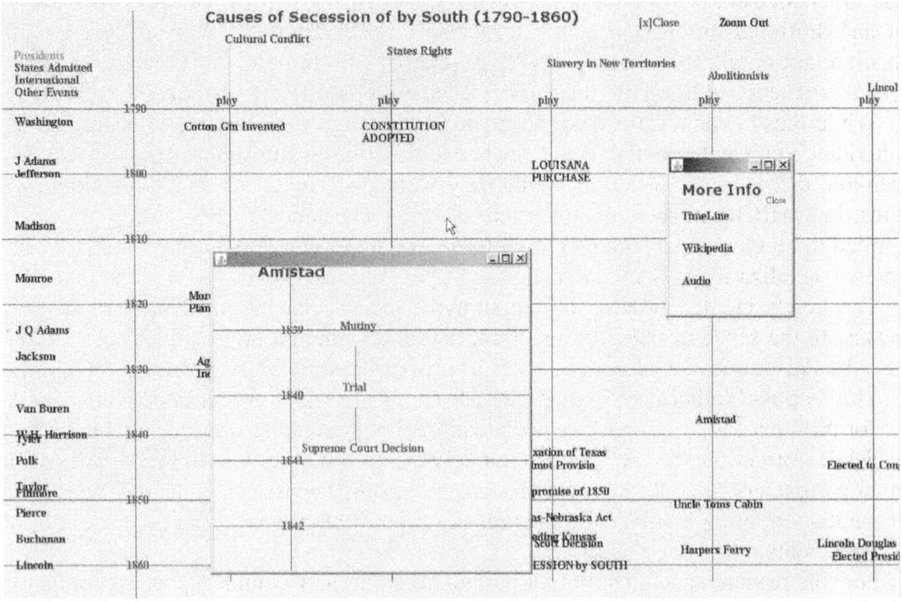

Fig. 2. In the open configuration, each of the threads has been expanded with full labels presented. A menu indicating links for additional information for the "Amistad" has been opened. Furthermore, a timeline with the detailed timeline has been opened.

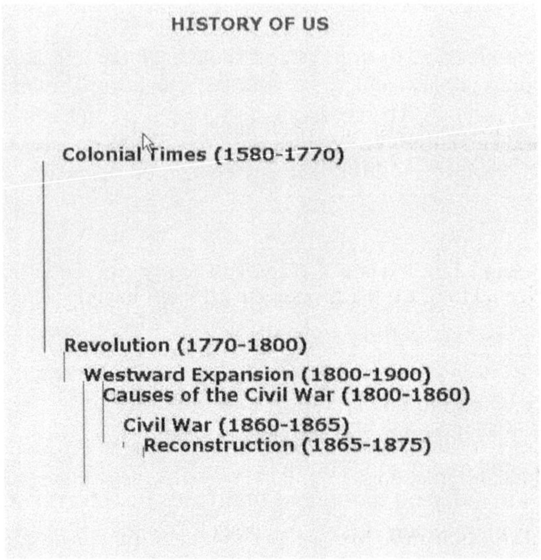

Fig. 3. This approach can be applied to major episodes of US history. The history browser home page is shown. At present only the "Causes of the Civil War" link is functional, as described above.

3 Conclusion

We have developed a prototype interface for incorporating narrative and causal threads in an interactive focus-context, interactive timeline environment. Moreover, each of the threads can be "played". That is, the user can step through the events in the thread.

4 Extensions

Several useful features could be added to this system. For instance, we could incorporate aspects of adaptive hypertexts (e.g., [11]). For instance, when playing multiple threads, the system could adapt the node descriptions based on what other nodes had previously been presented to that same user. We might also allow individuals to add commentary and even arrange their own versions of the causal associations of links.

However, there are also several unresolved issue with the design which may affect its generality. First, some historical reasoning is based on trends and these are not easy to represent as events. For instance, the impact of the Abolitionist movement would have been based more on public acceptance of its arguments than in the direct effect of the events associated with it. A second limitation is that many event threads are not easily linearized. Rather, they are probably best represented as tangled webs,

Moreover, some of the events get deeper and could be expanded into their own full thread. For instance, the "Cotton Gin" node could be expanded into an elaborate story of how the cotton gin was developed. Similarly, "Secession of the South" could be followed by the whole story of the attack of Fort Sumter and a large network of links dealing with the Civil War. This is amplified by the inclusion of external resources. Though we could also invert the viewpoint and consider the multi-timeline system we have developed to be a useful adjunct for providing structure to those external resources. Alternatively, these threads could be used as a way to create guided tours through Wikipedia.

Presumably, many of these issues can be mitigated. We are currently developing formalisms for describing events and the relationship among them which extends the model in [5]. Such a framework should facilitate developing larger and richer applications.

4.1 Applications

Large scale applications should help history students understand both temporal and causal relationships in history. Such a system could also support access to the millions of pages of historical newspapers which have are currently being digitized [12,13,14]. However, to apply these interfaces to such a large number of newspapers would entail the automatic extraction categorization of events from the text and such techniques still need to be developed.

Acknowledgements. This work was funded by an NEH Digital Humanities Start-up Grant. We thank Abhijeet Ganachari for assistance.

References

1. Allen, R.B.: Timelines as Information System Interfaces. In: Proceedings International Symposium on Digital Libraries, pp. 175–180 (1995)
2. Allen, R.B.: A Focus-Context Timeline Interface for Browsing Historical Newspapers. In: Proceedings Joint ACM-IEEE Digital Libraries Conference, pp. 260–261 (2005)
3. Plaisant, C., Milash, B., Rose, A., Widoff, S., Shneiderman, B.: Life Lines: Visualizing Personal Histories. In: ACM SIGCHI, pp. 221–227 (1996)
4. Allen, R.B., Acheson, J.A.: Browsing Structured Multimedia Stories. In: Proceedings ACM Digital Libraries Conference, pp. 11–18 (2000)
5. Allen, R.B., Wu, Y.J., Luo, J.: Interactive Causal Schematics for Scientific Explanations. In: International Conference on Asian Digital Libraries, pp. 411–415 (2005)
6. Allen, R.B.: Developing a Query Interface for an Event Gazetteer. In: Proceedings Joint Digital Libraries Conference, pp. 72–73 (2004)
7. Shaw, R., Larson, R.R.: Event Representation in Temporal and Geographic Context. In: European Conference on Digital Libraries, pp. 415–418 (2008)
8. Pustejovsky, J., Castaño, J., Ingria, R., Saurí, R., Gaizauskas, R., Setzer, A., Katz, G.: TimeML: Robust Specification of Event and Temporal Expressions in Text. IWCS-5. In: Fifth International Workshop on Computational Semantics (2003)
9. Marshall, C.: Guided Tours and On-line Presentations: How Authors Make Existing Hypertext Intelligible for Readers. In: ACM Hypertext (1989)
10. Zellweger, P.T.: Scripted Documents: A Hypermedia Path Mechanism. In: ACM Hypertext (1989)
11. Hitzeman, J., Mellish, C., Oberlander, J.: Dynamic Generation of Museum Web Pages: The Intelligent Labelling Explorer. Archives and Museum Informatics 11, 107–115 (1997)
12. Allen, R.B., Japzon, A., Achananuparp, P., Lee, K.-J.: A Framework for Text Processing and Supporting Access to Collections of Digitized Historical Newspapers. In: HCI International Conference (2007)
13. Allen, R.B., Waldstein, I., Zhu, W.: Automated Processing of Digitized Historical Newspapers: Identification of Segments and Genres. In: International Conference on Asian Digital Libraries, pp. 380–387 (2008)
14. Allen, R.B., Hall, C., Zhu, W.: Automated Processing of Digitized Historical Newspapers beyond the Article Level: Finding Regular Features, Sections, and Genres of Images (submitted)

Facilitating Search through Visualized Results

Takahisa Ando, Satoko Shiga, Tomoya Iwakura, and Seishi Okamoto

Fujitsu Laboratories Ltd.
1-1, Kamikodanaka 4-chome, Nakahara-ku, Kawasaki 211-8588, Japan
{andou.takahisa,shiga.satoko,iwakura.tomoya,
seishi}@jp.fujitsu.com

Abstract. This paper presents a search engine interface enriching search results with thumbnails of documents such as reports, presentation materials, spreadsheets, and so on. Usual search results consist of text information, such as the title and its snippet of each document. Our interface adds the thumbnails of documents listed in search results for assisting the search processes of users. We evaluate the interface on a web site search engine. The experimental results show that our interface enables users to find their target documents more efficiently than using an interface without thumbnails.

1 Introduction

Conventional search results basically consist of text information, such as the title and snippet of each document. For example, major web search engines, such as Yahoo! 1 and Google 2 return text-based search results. We can find target information by using text-based search results. However, text information is not always enough for users to find documents.

For example, when we find documents in business, we have to deal with many types of documents such as presentation materials, spreadsheets, and so on. If we use a conventional search engine returning text-based search results, we have to perform several steps to judge whether documents are relevant for our purpose. These steps include downloading each document, opening document with its corresponding application, turning pages on the application and so on.

We present a search engine interface that generates search results with visual information to save these steps. Our search results include the thumbnails of documents in addition to text-based information used in conventional search results. Our interface also provides users with four types of browsing method to facilitate understanding of the documents. Users can check the contents of documents without downloading.

2 System Description

We describe our system as follows. Let *Q* be a search query given by a user. Our interface posts the query *Q* to a search engine. The interface receives the search-result *R(Q)* obtained with *Q* from the search engine. *R(Q)* consists of *N* results. We assume that each result consists of the title, URL, and its snippet. We denote *Ri* is the i-th result in *R(Q)* $(1 \le i \le N)$. *Ti*, *Ui*, and *Si* are the title, URL and snippet of *Ri*.

M.J. Smith and G. Salvendy (Eds.): Human Interface, Part I, HCII 2009, LNCS 5617, pp. 339–344, 2009.

Next, our interface generates the thumbnail of each result in *R(Q)*. The thumbnail of each document consists of the image of each page in the document. If a document *Di* located in *Ui* consists of |*Di*| pages, our interface generates the thumbnail *THi* for *Di* by generating the image of each page in Di. *Hi* consists of |*Di*| images.

Finally, our interface generates a search result *RT(Q)* including the text-based information of documents and their thumbnails. Our interface includes the thumbnail of the first page of each document in *RT(Q)*. Fig.1 shows a snapshot of a search result generated by our interface.

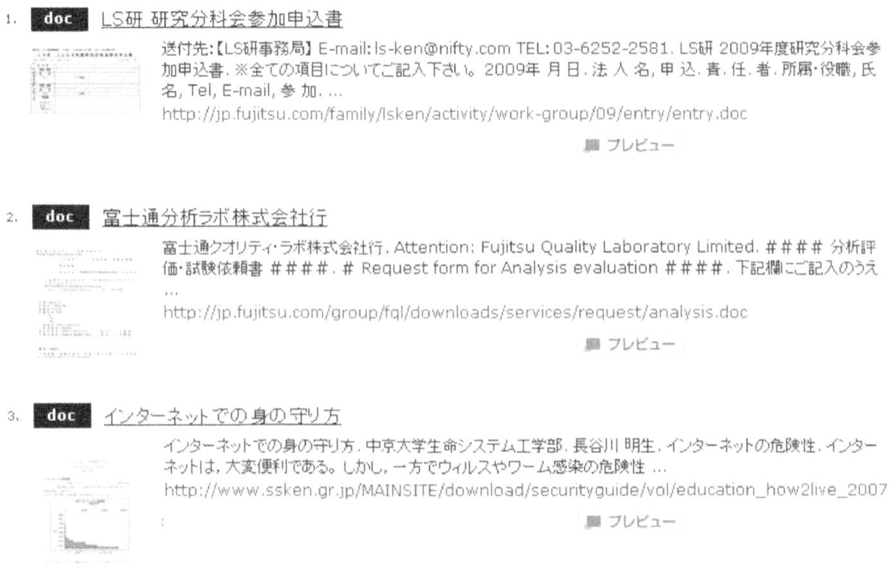

Fig. 1. A snapshot of our interface

Our interface has the following four types of preview functions for thumbnails.

- **NORMAL:** This preview function enables us to check the pages in the thumbnail of each document page-by-page. If we click one of the arrows for moving previous or next page, users can see previous or next preview of the current page. If we click the button for automatic page turner, we can see all the images for the pages without clicking the arrows. The figure in the left side of Fig.2 shows an example of NORMAL browsing.
- **BOOK:** This preview function enables us to check the images of pages in each document like reading by two-page spread. Consecutive two pages are displayed in this view. We can see a next or previous two pages by clicking the edge of each view. We can check each thumbnail like flipping a page. The figure in the right side of Fig.2 shows an example of BOOK browsing.
- **MEADERING:** This preview function enables us to check several images in the thumbnail of each document simultaneously. A set of images for a document are displayed from top to bottom curving like snake movement. We can grasp of the overview of each document by MEADERING. The figure in left side of Fig.3 shows an example of MEADERING browsing.

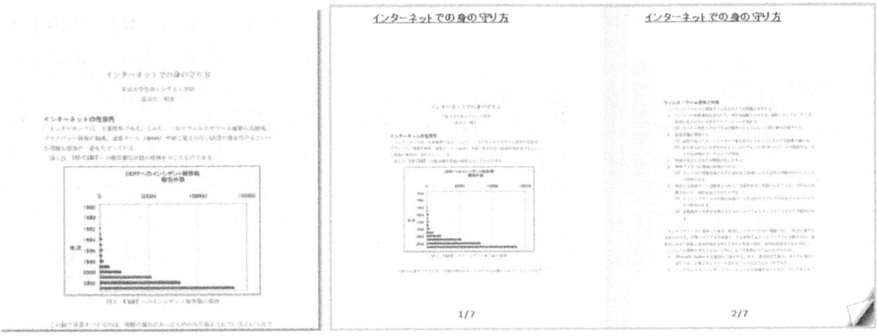

Fig. 2. Snapshots for NORMAL and BOOK: The left figure is a snapshot for NORMAL. The right figure is a snapshot for BOOK.

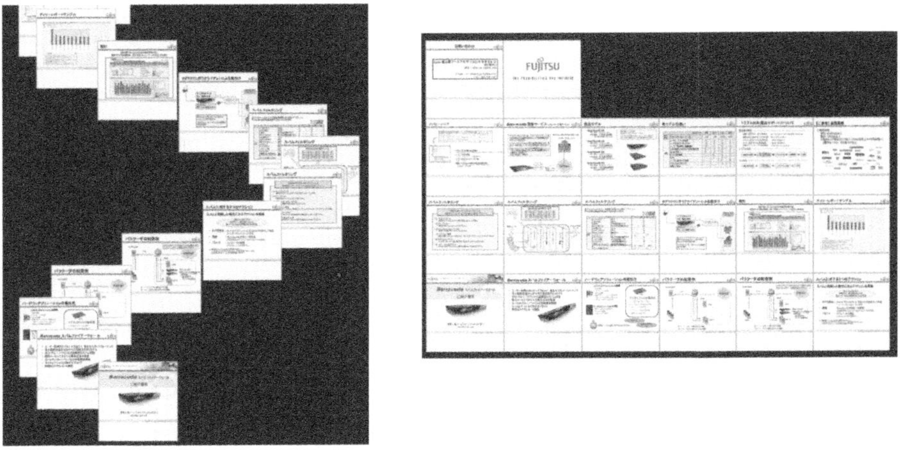

Fig. 3. Snapshots for MEADERING and TILE: The left figure is a snapshot for MEADERING. The right figure is a snapshot for TILE.

- **TILE:** This preview function also enables us to check several images in the thumbnail of each document simultaneously. A set of images for a document are displayed like tiles. We can also grasp of the content of each document as well as MEADERING browsing. The figure in right side of Fig.3 shows an example of TILE browsing.

We can start the NORMAL preview of each document by clicking the thumbnail listed in *RT(Q)* or the button for starting the preview of the document. Then we can choose MEADERING, TITLE or BOOK.

3 Experimental Settings

We evaluate our interface with the following questions.

- Q1: How many inquiries by emails did CERT have in 2000?
- Q2: What is the function of Barracuda for spasm other than segregation and block?

– Q3: Which company is in charge of receiving orders for machine glazed paper?

We prepare the search-results including a document that have the answer of each question.(We use the site search for Fujitsu 3). To evaluate the effect of thumbnails on each file types, we control the file types of search-results for each question.The file type of Q1 is documents such as reports and prepares created with a word processor application. The file type of Q2 only consists of presentation materials, and the search-result for Q3 only consists of spreadsheets.

Users firstly try a sample question for studying the usage of our interface. Then the users start to find the answer of each question from the search-results. We prepare two types of search-results for each question that are with thumbnails and without thumbnails. We randomly reordered results in each search-result to alleviate the effect of the order decided by a search engine ranking schema.[1]

Six users evaluate our interface in this experiment. To evaluate our interface, we divide users into two groups. Each group consists of 3 people. The users in the first group (Group1) start the first evaluation with the search-results including thumbnails. Then the Group1 start the second evaluation with the search-results including only text-based information. The users of second group(Group2) evaluate our interface in reverse order, firstly using search-results only including text-based information, and secondly using search-results including thumbnails. So the users solve the questions two times. We compare time to find the answer of each question in this evaluation.

4 Experimental Results

Table 1 shows the experimental results for evaluating our interface. The top table in Table 1 lists the experimental results. In first evaluation, we see that users of Group1 found the answer of each question faster than the users of Group2. For example, the average time for Q3 expended by the users of Group1 show about 2.7 times shorter than that of the Group2. The average times for Q1 and Q2 expended by the users of Group1 also show shorter than those of the users of Group2. These results show that our interface contributes to improved time any file types.

In the first evaluation, the users of Group1 were using preview functions (mainly normal preview function) to select the document. Finally they downloaded the document to check the detail. All of the users of group2 couldn't get enough information of document from title or snippet. Thus, they downloaded each document from high-order to low-order and checked it. The answer document was ranked 1st or so, user could find the answer quickly, but if it is ranked low-order, it takes much time to find the answer.

The bottom table in Table 1 lists the experimental results in second evaluation. All users had already known the answers of all the questions, because they had already answered them in the first evaluation. Despite all users knew the answers, the users of Group1 showed about 2.7 times faster speed for finding the answer for Q3 than

[1] Previous researches 45 have shown users are more likely to click on the first displayed result regardless of its relevance. If the result including the answer of a question is displayed on the top of the search-result, we guess users are likely to check the result and find answer without checking the other results. To evaluate the effectiveness of thumbnails, we reordered the results in each search-result randomly to alleviate the affection of ranked order.

Table 1. Time (second) for finding the answer for each question. The top table shows the results for first evaluaton. The bottom is the results for second evaluation.

First evaluation

Query/ User	With thumbnails (first group)				Without thumbnails (second group)			
	A	B	C	AV.	C	D	E	Av.
Q1	147	284	28	153	162	180	249	197
Q2	43	66	53	54	104	45	41	63.3
Q3	26	26	61	37.66	240	15	50	101.66

Second evaluation

Query/ User	Without thumbnails (first group)				With thumbnails (second group)			
	A	B	C	AV.	C	D	E	Av.
Q1	60	23	14	32.33	26	90	50	55.33
Q2	25	30	9	21.33	18	80	20	39.33
Q3	40	10	51	33.66	14	10	13	12.33

* Av. (Average time of each question).

Group2. Average times for Q1 and Q2 expended by the Group1 are little bit slower than those expended by users of Group2. However, the users of Group1 in the second evaluation show much faster speed than speed in their first evaluation.

5 Related Works

There exist services that provide rich information other than text-based information. Searchme 6 and ManagedQ 7 provide users with search results of web pages as a set of thumbnails of pages. MARSFLAG returns search results including thumbnails in addition to their text-based results like our interface 8.

Compared with these services using thumbnails, our interface gives users four types of browsing functions. Users of our interface can choose browsing methods. Our browsing methods aim at improving procedures of browsing of business documents such as presentation materials and spreadsheets.

6 Conclusion and Future Work

This paper presents a search engine interface enriched with thumbnails of data such as documents, presentation materials and spreadsheets, and so on. Our interface adds the thumbnails of documents included in search results for assisting the search processes of users. We evaluate our interface on a web site search engine. The experimental results show that our interface enables users to find their targets information more efficiently than an interface without thumbnails.

We have evaluated our interface with small number of users in this experiment. We should evaluate our interface with more users in future work. We should also evaluate our interface with different aspects such as effect of number of results, mixed file types, and so on.

Acknowledgements. The authors thank the users help us to evaluate our interface.

References

1. Yahoo Top Page, `http://www.yahooo.com/`
2. Google Top Page, `http://www.google.com/`
3. Fujitsu Japan Top Page, `http://jp.fujitsu.com/`
4. Guan, Z., Cutrell, E.: An eye tracking study of the effects of target rank on Web search. In: Proceedings of CHI 2007, pp. 417–420 (2007)
5. Joachims, T., Granka, L., Pan, B., Hembrooke, H., Gay, G.: Accurately interpreting click-through data as implicit feedback. In: Proceedings of SIGIR 2005, pp. 154–161 (2005)
6. Serachme Top Page, `http://www.searchme.com/`
7. ManagedQ Top Page, `http://beta2.managedq.com/`
8. MARSFLAG Top Page, `http://www.mitsubishielectric.co.jp/`

The 7 Basic Functions of a Digital Library - Analysis of Focus Groups about the Usefulness of a Thematic Digital Library on the History of European Integration

Eric Brangier, Jérôme Dinet, and Laurent Eilrich

Université Paul Verlaine – Metz. 2LP, Laboratoire Lorrain de Psychologie. Faculté des Sciences Humaines et Arts. BP 30309. Île du Saulcy – 57006 Metz, France
Centre Virtuel de la Connaissance sur l'Europe,
L-4992 Sanem – Grand Duchy of Luxembourg
{brangier,dinet}@univ-metz.fr, laurent.eilrich@cvce.lu

Abstract. This paper presents the main results of a study involving an original user-centred design approach to modify and improve a specific digital library (DL) related to the history of European integration. The ultimate goal of the project is twofold: (1) to find ways to improve end-users' use of this thematic DL; (2) to develop an original method to measure real users' needs and mental representation. This user-centred approach is based on focus groups, this technique being a powerful means to evaluate services or test new ideas. In our study, more precisely, focus groups were set up by conducting interviews with 58 users (researchers, professionals in documentation, journalists, historians, teachers, students, ergonomists, politicians, computer engineers, etc.), but including four to five people at the same time in the same group. Each of these 14 focus groups was videotaped for a 3-hour period and all verbalisations (needs, viewpoints, etc.) were transcribed. Our analysis showed that the users' needs can be expressed through seven basic functions which match the users' expectations. This paper brings together the process by which these functions were identified and also discusses the interests, limitations and possible generalisations of these functions in the context of all digital libraries.

Keywords: digital library, functions, function analysis, user-centred design, user requirements, user needs, ENA, European integration.

1 Introduction

Although there has been considerable investment by organisations, academics and researchers targeting online resources [11], little attention has been paid to the effective use and usability of these digital systems [17]. Moreover, in updating and introducing a new version of digital library (DL) services, end-users should be comfortable with the new version and should not struggle to complete their tasks and to be able to "find what they need, when they need it and in a form they want it" [2]. Even if some specific problems have been identified in prior empirical studies, e.g. the quality of

M.J. Smith and G. Salvendy (Eds.): Human Interface, Part I, HCII 2009, LNCS 5617, pp. 345–354, 2009.

sources [3], diversity of the users' knowledge [13], users' perceptions [12] and usability [17], work remains to be done by designers [9]. In this paper, we intend to analyse the real and effective needs of users of a specific DL [16] and to deal with the problem of what is a "good" DL [10]. In the next section, we will introduce the social construction of needs in the field of digital libraries. In section 3, we will describe the problem and related method to solve the key issues. Section 4 will discuss the results.

2 Social Construction of Needs of Digital Libraries

From a general end-user point of view, technological environments are the product of their activities. Nevertheless, if we tend to perceive these environments as an "external" world of obvious things, this can be an error of assessment. Technology does not exist by itself; technology is primarily a social construct, even if many designers reject and/or deny this social constructivism.

2.1 Limits of Prior Studies

Design approaches are often deterministic. These approaches support the idea that an artefact is designed in an autonomous or independent way. In this case, technological progress is seen as imperative, determining many dimensions of the user's needs and a large number of the social dimensions involved in the use of the technologies. In other words, this progress is perceived as the result of scientific discovery applied to technology and social change.

However, sciences do not generate technological innovation; in our view, specific groups of individuals (i.e. end-users) build a social representation of what they desire to use. These individuals have complex and interdependent relationships, and take part in innovation by building interactions between society and technology.

2.2 Principles for an Innovative Approach Based on the Social Construction of Users' Needs

Our approach is both based on the theory of the social construction of technology (SCOT [2]) and on methods to produce the social construction of the user's needs [6].

According to the model elaborated by Pinch & Bijker, the social construction of technology is a theory within the field of sociology of technology and society. The SCOT theory argues that technology does not determine human action but rather that human action shapes technology. The way to understand and design technology is first to refer to the social context of the technological use. This approach highlights the importance of three social constructivist concepts:

- Interpretative flexibility: each technological artefact has different meanings and interpretations for various groups. Every social group has different priorities and regards the artefact's features in different ways. These distinct interpretations create different problems to be solved. For example, what should be prioritised: performance, aesthetics, convenience, usability or robustness?
- Relevant social groups: end-users and producers of the technological artefact are the most suitable groups to address the issue of understanding its features. Many

subgroups can be often delineated. But sometimes there are relevant groups who are neither users nor producers of the technology, such as journalists, students, multicultural specialists, politicians and teachers. Distinctions may be made between the various groups based on their shared or diverging use and assessment of the technology in question.
- Design flexibility: there are many ways of designing technologies. A design is the result of a decision process, this process trying to construct technical possibilities to reflect the different points of views of relevant social groups.

So a new technology cannot begin to be designed without understanding how that technology is embedded in its social situation. Comprehension of the social context depends on methods used to establish the user's needs and to analyse them.

From a methodological point of view, previous studies [5, 6, 8, 9] insisted on the fact that need cannot exist as such outside humankind, outside history and outside the society that generates it. The user's need is a social construction. Need does not build itself in an isolated manner. Need is the result of complex transactions between a user, a designer and an environment where imitation, learning, co-construction of knowledge and sharing of representations play an essential role; it involves reciprocal process validations. Needs emerge in and through social interactions and through the mediation of language. When designing an interface, if the user and the designer are not able to solve their own problems, they will have more chance of achieving their objectives by means of cooperation and social interactions. Therefore, needs emerge from collaborative efforts where users and designers mutually enrich their knowledge by being confronted with the knowledge of others. This knowledge, which ultimately shapes the representation of needs, can be obtained by using participatory and creative methodologies. The target of these methods is to explore the intellectual creative works generated by relevant social groups. Within the scope of participatory methods, verbalisations are produced concerning new forms of ideations which might be useful for users. In short, the benefits anticipated by implementing participatory and creative methods are the development of technologies which are useful, usable, acceptable and adaptable for communities of users.

2.3 Implications for the Design of a Digital Library

The approach centred on the social construction of users' needs involves taking different dimensions into account in order to understand users' requirements. This approach draws on the following ideas, which it seeks to develop.

We consider that a DL is a social reality before becoming a technological reality. This reality is not a fixed construct. It is built on the basis of social interactions and sharing (or not) inside social groups. This reality is essentially built from conversations (face-to-face or technologically mediated) between individuals who co-produce representations about the use of a DL. They share and disseminate these representations. Each social group will develop a flexible interpretation of the DL, its functionality, its aesthetics and its overall usage. We must therefore look at the most relevant communities of users to promote flexible interpretations and understand the degree of familiarity of each group with the future of the DL. The design of an artefact, such as a DL, must be viewed from the "perspective" of each relevant group, even if, and because, these perspectives are different. But the views of the social groups are ultimately processed in the same way by the designer.

The design has to take into account variation in the flexibility of interpretation, which converges into a common representation of the DL. When there is convergence of views, this leads to a socially constructed innovation. Thus, the contribution of a group to a specific technology (and vice versa) is strongly related to the degree of familiarity of its members and their shared knowledge.

Because language is the primary way to construct, share and disseminate representations about the future DL's functions and attributes, analysis of verbalisations produced by relevant social groups must be central.

3 Problem and Method

The challenge of the ENA (European NAvigator, www.ena.lu) project is to preserve digital content on the history of European integration [1, 7]. For this project, our main problems and issues were the following: How can we generate a situation that provides conditions favourable to the social construction of users' needs with regard to the DL? How can it be monitored? Does this situation provide a relevant expression of users' requirements? Does this situation accelerate the ideation of usable features for the DL? As we have already pointed out [6], users' needs are social constructions: the aim is not exactly to analyse these needs, but to create a social situation in which they may become apparent. To do this, we establish a method of user comprehension based on the principles of (1) social construction of users' needs and (2) participatory and creative design.

Because experimental studies, laboratory testing or job analysis are not relevant for obtaining real needs, producing in-depth exchanges between end-users or obtaining qualitative information such as emotions [8], it was decided to use focus groups for our study. These focus groups were used in order to extract qualitative information from the end-users taking part and to investigate whether participants felt that they shared similar problems and experiences. It was felt that this technique would be an effective way to probe the problems they were facing, in particular identifying barriers to their use of the DL, their perceived benefits and their needs. Table 1 presents the steps related to our focus groups.

Table 1. General organisation of the process for collecting users' needs with a socio-constructivist approach.

• **Definition of communities of practice:** What communities are involved in the project? The proposed solution is to map communities in order to identify those directly or indirectly involved in the use of the DL. – Principles: The purpose of defining communities of practice is to choose relevant social groups to express needs regarding the evolution of the DL. The aim is to gather as much information as possible on people who may be affected by the project and to determine a profile of users of the DL. – Results: 14 communities of practice involved in the use of a digital library on European integration were defined and approved by the head of the DL (researchers/historians, lawyers, professionals in documentation, journalists, teachers, cross-cultural teachers, students, PhD students, software ergonomists, politicians, computer engineers, experts in intercultural studies, experts in new digital leisure, members of historical associations).

Table 1. (*Continued*)

• **Identification of communities' experts:** Who are the experts in this community? Who can speak for this community? Who are the legitimate representatives for this community? The proposed solution is to contact institutions to produce a panel of experts. – Principles: A laboratory engineer contacts a number of individuals to find community experts and uses social networks to identify the experts' communities. They are invited to participate in a filmed focus group for 3.5 hours. An expert of a community of practice is an individual who can speak for his/her community. He/she is a legitimate representative for this community. – Results: 58 experts (from France, Germany, Luxembourg and Belgium) agreed to participate. The list of experts was validated by the initiator of the study.
• **Organisation of 14 video-recorded focus groups:** What do they think? Who are they? What do they do? How do they see the future? The proposed solution is to organise a group working session using expert "focus groups". The working session will be video-recorded and conducted with the aid of various tools. The leader of the focus group will target expertise and the community. – Principles: 14 focus groups were organised in three phases: (1) participants were asked to speak freely about subjects relating to European integration; (2) participants were asked to discuss some sections of the existing DL; (3) participants were asked to organise knowledge about Europe by carrying out a short card exercise. Each session should last 3 to 3.5 hours. Because focus groups are basically multiple interviews, many of the guidelines for conducting focus groups are similar to those for conducting interviews. Focus groups were conducted with four to five communities of experts who were given equal status in the programme. – Results: 50 hours of video records; representation of ideas of each relevant social group; ideas for improving the functionality and usability of the DL; better representations of users (direct, indirect, primary, secondary, etc.); considerations on the future of the DL (content, status, organisation of information).
• **Results analysis:** What knowledge is useful for designing the DL? The proposed solution is to summarise the results of each focus group, to define all ideations discussed in each social group, and briefly to explain the main function of the DL. – Principles: A content analysis was carried out on the 14 focus groups. The verbalisations of each focus group were summarised and a list of major improvements was drawn up. – Results: 53 new ideations were listed and the descriptions of the 7 basic functions of the DL were described.
• **Consensus workshop:** How should the knowledge produced by the different groups be put to use? How should the different results be prioritised? The proposed solution is to hold discussions with the initiator of the study and try to reach a consensus. – Principles: A "consensus workshop" is a standard method to conduct a process of collective reflection to discuss controversial issues and reach agreement on joint recommendations. The main objective of consensus methods is to develop recommendations modelling the opinion of experts to strengthen the objectivity of the requirements produced. It is particularly relevant in cases where the subject matter is controversial. – Results: share and validate the seven basic functions.

4 Results and Discussions: "The Seven 'A' Functions"

The focus groups highlight expectations, needs, gaps and desires. Corpus analyses stress different types of features that people find useful as a whole. We will show that a DL is not only a system archiving relevant information, but that users want to enjoy a total of seven useful functions, as we explain below:

1. **Function: To Archive resources:** "to give efficient access to relevant data".
   a. **Definition:** To sort rational, reliable and organised resources and to make them easily accessible and usable for users by specifying their usage rights.
   b. **One example of archive needs constructed by focus groups**

User expressions	"We need disasters, errors, things that have gone wrong." "What didn't work is also interesting for the historian." "We should not be limited to the official version." "There should be horizontal links for more in-depth information."
Ideas & solutions	Archive documents as widely as possible, including those which may be controversial. Include links to "less official" websites and comment on the links. Explain to the user what might be found on other sites.
Focus group	Researchers, historians, multicultural experts, teachers

2. **Function: To Accredit the information:** "to improve the credibility of the DL"
   a. **Definition:** To officially recognise the DL as a credible institution with credible expertise. The DL must be an authoritative source of knowledge.
   b. **One example of accreditation needs constructed by focus groups**

User expressions	"Have comments made by credible historians." "To be accompanied by criticisms on other aspects." "We cannot be limited to facts."
Ideas & solutions	Involve facts and documents with contextualised explanations written by European historians. Comments by expert historians are important. The authors of academic papers must be cited.
Focus group	Teachers

3. **Function: To Actualise knowledge**: "to update the knowledge"
   a. **Definition:** To update the information and provide up-to-date knowledge is an ongoing user need.
   b. **One example of actualisation needs constructed by focus groups**

User expressions	"Have home pages linked to the news." "Europe is not over but under construction." "We need information on Europe relating to the news." "There should be a link with the news of Europe." "Reflect the fact that the story does not end every day."
Ideas & solutions	Making homepage more attractive. Zoom on a point of relevance of a European country. Give news on the various European countries.
Focus group	All.

4. **Function: To Analyse the data**: "to help the user to interpret the archives"
   a. **Definition:** To help the user to analyse data. Users express the need to have usable systems to analyse the archives. The DL should promote

understanding of facts, comparing the resources, giving its cultural referents or contextualising ideas. The DL has to offer evidence of analysis to identify the constituent historical, geographical, cultural, artistic, social, psychological and political facts and archived events.

b. **One example of analysis needs constructed by focus groups**

User expressions	"Have modes of entry other than chronology, institutions and data formats or search engine information." "The history is seen as too monolithic and official."
Ideas & solutions	Give other modes of entry (geography, countries, capitals, maps, articles of law, treaties, values, identity, etc.). Link the archives to software data analysis.
Focus group	All groups.

5. **Function: To Affirm an identity**: "to express a good, positive and relevant image"

a. **Definition:** To assert, point out or affirm the corporate identity behind the DL; this is an important background for interpreting the archives based on the nature of the DL. The objective is to affirm the DL's identity and therefore to seek to differentiate it and put forward a specific identity.

b. **One example of identity affirmation needs constructed by focus groups**

User expressions	"A digital library on the history of Europe should be a place of cultural democracy." "Must express European values."
Ideas & solutions	User interface and interaction design in line with European values.
Focus group	Experts in new cultural practices, multicultural experts.

6. **Function: To Associate**: "to help users to connect with specialised social networks"

a. **Definition:** Involving various forums (individual or collective, private or public) in developing common knowledge.

b. **One example of association needs constructed by focus groups**

User expressions	"Have a discussion forum under the supervision of a moderator." "Create a network of exchanges between citizens or between businesses on specific European topics."
Ideas & solutions	Have an access code and identifier to monitor users' credibility. Build European forums. Boost friendship groups.
Focus group	Multicultural experts, politicians, ergonomists, legal experts, experts in new cultural practices,

7. **Function: To Animate:** "to increase user interest by developing digital events"

a. **Definition:** To stimulate the users of the DL by encouraging them to produce and exchange knowledge.

b. **One example of animation needs constructed by focus groups**

User expressions	"Relationships between researchers are important but not sufficient." "It is necessary to facilitate the relationships between research scientists to solicit papers, publications, memoranda, etc." "Much work is performed in history research centres but this is not properly valued."
Ideas & solutions	Developing the relationships between research centres in European history. DLs might help identify historical research, i.e. lists of theses written on the history of Europe, dissertations, etc.
Focus group	Experts in new cultural practices, historians.

Technology has often overshadowed social practice in DL design. DLs are complex, heterogeneous social entities that are difficult to understand without considering their social implications. Our rule of the seven basic functions underlines the fact that users would like to have large patterns of activity which are almost socially oriented.

Generally, a DL offers only the first three functions: (1) Archiving knowledge in different forms and formats; (2) Ensuring the credibility of its information; (3) Actualising knowledge by keeping it up to date. However, DLs have not yet focused their efforts on new features that facilitate analysis of records by increased visualisation, intelligent sorting or statistics analysis. DLs don't make much effort to assert the consistent identity of their contents and thus to enhance their external image. DLs generally do not look to engage users and develop social networks of users. Finally, DLs do not desire to stimulate users willing to contribute to their development by promoting the exchange of information, or by directing users to participate in one-off virtual events. There really are new needs to explore!

5 Conclusion and Prospects

The design of a digital library requires the identification of the needs of users who interact with this system. Need is often the starting point for technology projects. Understanding it is complex and delicate, particularly when this determines the success or failure of the digital library. Far from a linear design point of view where the need is seen as a finished object, this paper shows that users' needs are social constructions which can be extracted from relevant conversations with users. As a result, the understanding, development and formalisation of users' needs involves establishing a user-centred design process to produce new knowledge built in conjunction with experts from relevant user communities. Placed in a focus group, communities of experts have specific needs, requirements, expectations and desires – all statements that we classified into seven categories, "the seven A functions" (to **Archive,** to **Accredit,** to **Actualise,** to **Analyse,** to **Affirm,** to **Associate, and** to **Animate**).

Even if this research highlights new functions for digital libraries, as explained by Wilson [17], the discussion on the enormous problems of defining "information need" and how "information" can satisfy the end-user is still open [8]. These seven rules have to be operational and efficient in order to turn them into features for designing human-computer interactions. But that is the subject of another publication!

Acknowledgments. The authors would like to thank the CVCE (Centre Virtuel de la Connaissance de l'Europe – Chateau de Sanem – Sanem – Grand Duchy of Luxembourg). The achievements of this project are greatly indebted to Marianne Backes, Director of the CVCE, and Ghislain Sillaume, Project Manager, and we want to thank them for their involvement.

References

1. Backes, M., LeBrun, M., Sillaume, G.: A multilingual, multisource and multilingual digital library, safeguarding the European heritage. BRICKS: The European solution for Digital Libraries services, Lisbon (2006), http://www.brickscommunity.org/BRICKS-Lisbon
2. Bijker, W.E., Hughes, T.P., Pinch, T. (eds.): The Social Construction of Technological Systems: New Directions in the Sociology and History of Technology. MIT Press, Cambridge (1999)
3. Bishop, A.P., Van House, N.A., Buttenfield, B.P. (eds.): Digital Library Use: Social Practice in Design and Evaluation. MIT Press, Cambridge (2003)
4. Borgman, C.L.: The Invisible Library: Paradox of the Global Information Infrastructure. Library Trends 51(4) (2003)
5. Brangier, E., Bastien, J.-M.-C.: L'analyse de l'activité est elle suffisante et/ou pertinente pour innover dans le domaine des nouvelles technologies? In: Vallery, G., Amalberti, R. (eds.) L'analyse du travail en perspectives: influences et évolutions. Toulouse: Octarès, Collection Entreprise, Travail, Emploi, pp. 143–156 (2006)
6. Brangier, E.: Besoin et interface. In: Akoka, J., Comyn-Wattiau, I. (eds.) Encyclopédie de l'informatique et des systèmes d'information, pp. 1070–1084. Vuibert, Paris (2007)
7. Consalvi, M., Eilrich, L., Sillaume, G.: Rôles et perspectives des bibliothèques numériques dans l'économie de la connaissance: une étude de cas. In: Colloque En route vers Lisbonne, Luxembourg (2008),http://www.tudor.lu/cms/lisbonne2008/publishing.nsf/0/FDECF548D12BC30BC12575140048AB73/$file/15h30_CONSALVI_EILRICH_SILL.pdf
8. Dinet, J., Vivian, R.: Comportements et stratégies des utilisateurs dans les bibliothèques numériques. In: Papy, F. (ed.) Usages et pratiques dans les bibliothèques numériques, pp. 241–268. Hermès Lavoisier, Paris (2007)
9. Dinet, J. (ed.): Usages, usagers et compétences informationnelles au 21e siècle. Hermès Science Lavoisier, Paris (2008)
10. Goncalves, M.A., Moreira, B.L., Fox, E.A., Watson, L.T.: What is a good digital library? - A quality model for digital libraries. Information Processing & Management 43(5), 1416–1437 (2007)
11. Jones, M.L.W., Gay, G.K., Rieger, R.H.: Comparing Evaluations of Digital Collection Efforts. DLib Magazine 5(11) (1999), http://www.dlib.org/dlib/november99/11jones.html
12. Kani-Zabihi, E., Ghinea, G., Chen, S.Y.: User perceptions of online public library catalogues. International Journal of Information Management (2008) (in press)

13. Komlodi, A., Hou, W., Preece, J., Druin, A., Golub, E., Alburo, J., Liao, S., Elkiss, A., Resnik, P.: Evaluating a cross-cultural children's online book community: Lessons learned for sociability, usability, and cultural exchange. Interacting with Computers 19(4), 494–511 (2007)
14. Pinch, T.J., Bijker, W.E.: The Social Construction of Facts and Artefacts: or How the Sociology of Science and the Sociology of Technology Might Benefit Each Other. Social Studies of Science 14, 399–441 (1984)
15. Silva, L.V., Goncalves, M.A., Laender, A.H.F.: Evaluating a digital library self-archiving service: The BDBComp user case study. Information Processing & Management 43(4), 1103–1120 (2007)
16. Tsakonas, G., Papatheodoroua, C.: Critical Constructs of Digital Library Interaction (2007), http://eprints.rclis.org/archive/00011654/01/PCI2007_TsakonasPapatheodorou_eprints_.pdf
17. Wilson, L.A.: If we build it, will they come? Library users in a digital world. In: Lee, S.H. (ed.) Improved access to information: portals, content selection, and digital information, pp. 19–28. Haworth Information Press, Binghamton (2003)
18. Xie, H.I.: Users' evaluation of digital libraries (DLs): Their uses, their criteria, and their assessment. Information Processing & Management 44(3), 1346–1373 (2008)

Exploiting Browsing History for Exploratory Search

Wei-Lin Chen and Wei-Guang Teng

Department of Engineering Science,
National Cheng Kung University,
Tainan, Taiwan
realseven@gmail.com, wgteng@mail.ncku.edu.tw

Abstract. With the advance of information technologies, web search has become a necessary activity for most Internet users. Although current search engines are powerful enough to respond required search results in seconds, information seekers may still feel arduous to digest them when conducting the exploratory search. In other words, when users are unfamiliar with the domain of their goals or unsure about the ways to achieve their goals, they may need to read numerous pages before fully understand what they are searching for. In this work, we propose to fully exploit the browsing history to help users clarify their thoughts and discover new insights during the process of exploratory search. Specifically, interactive user interfaces of two different modes, i.e., the timeline mode and the relevance mode, are devised to provide users a vivid impression of their browsing history. Hopefully, the feasibility and usability of web search for users can be further improved.

Keywords: Exploratory search, information seeking, interactive interface, timeline.

1 Introduction

As the data amount gathered on the Internet increases at a very rapid speed, web search is becoming one of the most frequent tasks performed by the users. With proper query terms, one may reach required information or perceive some knowledge by exploring web pages in the search results. Moreover, as is investigated in previous studies, user goals in web search can be classified in to three types, i.e., navigational, transactional and informational search [6,14,21]. For users conducting either a navigational or a transactional search, there are always specific targets, e.g., a website to be visited or an item to be purchased, in their mind. On the other hand, for users conducting an informational search, they may experience a learning process to discover desired information from a number of web pages from corresponding search results.

It is thus observed that user activities in web search vary with their different goals [7]. With current search engines, a simple search activity can usually be completed in seconds. Nevertheless, when conducting the search in an unfamiliar domain, a user tends to browse a few search results and then recursively issue the next query for obtaining some further information. Such a process of *exploratory search* is a time-consuming and tentative task for most of the users since they may need to read numerous pages before fully understand what they are searching for. In other words, if a

M.J. Smith and G. Salvendy (Eds.): Human Interface, Part I, HCII 2009, LNCS 5617, pp. 355–364, 2009.

user does not know which keywords to use or looks for several answers, he usually needs to spend much time on querying and browsing activities to accomplish an exploratory search.

When users try to investigate, evaluate, compare and synthesize information pieces grabbed during their exploratory search processes, they may encounter the data overloading problem due to the limited short-term memory of human beings. In view of this, an interesting concept which is inspired from the Pensieve in the famous series of novels "Harry Potter" is noted in this work. Specifically, the Pensieve is a fictional item used to store memories and to provide a near-omniscient perspective of the events to be relived later in the novels. To realize such a concept in practical applications, we propose to develop a proper approach for users to retain and to organize necessary information.

Furthermore, we note that search sessions of a user are not totally irrelevant, especially when the goal of web search is for obtaining more and more knowledge, or more specifically, *learning* something new. Consequently, the browsing history of users can help them on not only reviewing previous search activities but also clarifying next search targets. Nevertheless, to the best of our knowledge, the web browsing history is simply recorded in sequence and is usually loosely organized in previous works. For supporting the user needs of retrospecting their previous activities in exploratory search, we propose in this work to present and to organize browsing history of users in more feasible forms. Specifically, our goals are twofold. First, we expect to provide users a novel interface to support their historical retrospect. Moreover, relationship among search sessions should be identified to help users clarifying their thoughts and discover new insights.

The rest of this paper is organized as follows. Relevant works on exploratory search and usage of browsing logs are generally reviewed in Section 2. The concept and advantages of our approach are explored in Section3. In Section 4, implementation details of our approach are presented. This paper concludes with Section 5.

2 Preliminaries

The process of exploratory search and relevant information theories are generally discussed in Section 2.1. Moreover, in Section 2.2, we show how browsing logs can help to improve the user interactivity in web search.

2.1 Exploratory Search

Search can be considered as a specific type of information seeking behavior. Generally speaking, there are different stages for this goal-oriented and problem-solving process, i.e., problem recognition, problem definition, problem resolution, and solution statement [27]. With the advance of information technologies in recent years, web search has become an increasingly important part of most computer users. There are several previous works on improving search results or achieving personalized search. Typical techniques include reranking search results based on the personal profile [2,22,24], comparing search results of different search engines [18], and offering faceted search interfaces [12]. In addition, there are more and more works focusing on the visualization of search results [7,8].

A search engine can handle numerous user requests at a same time and returns more than a few matching results to the users, respectively. On the other hand, for users who are unfamiliar with the domain of their goals or unsure about the ways to achieve their goals, they may need to spend much more efforts to explore and to digest these documents on the web. An illustrative example is that if a user does not know much about classical music, how should he even begin to find a piece that he might like. This reveals that typical keyword search scenarios are insufficient. Specifically, relevant studies on exploratory search emphasize the circumstances that users start with vague information needs [15,17,25].

When conducting an exploratory search task, a user acquires not only specific documents but also the knowledge discovered during the interaction process with a search engine since the user may keep revising his query terms and thus the search engine returns some new results. When facing a large amount of retrieved data, users need to spent time and effort filtering useful information and organizing their own thoughts. Thus, a good tool supporting in exploratory search should help users to discovering new association and kinds of knowledge, resolving complex information problems, or developing an understanding of the terminology and the information space structures [25].

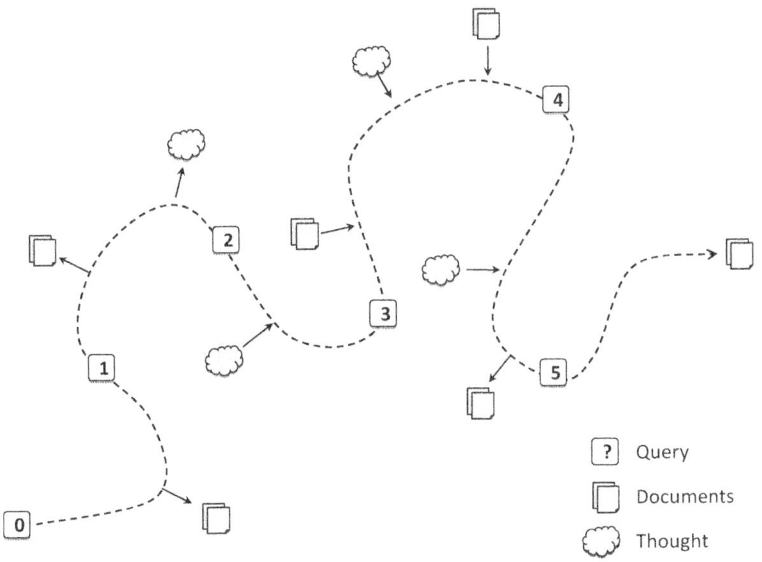

Fig. 1. An evolving search path as described in the berrypicking model (adapted from [5])

An informational model of *berrypicking* [5] best describes the actual behavior of information searchers, especially for the process of exploratory search. Unlike the comments made in conventional models, the search behavior is not merely a single movement for users to reach the best retrieved result, but a berrypicking or an evolving pattern. As shown in Fig. 1, the path of a searcher is usually a curve instead of a straight line. Also, searchers may iteratively change their direction of the query after gaining new knowledge from retrieved documents and clarifying their own thoughts.

In other words, the process of exploratory search usually consists of several sessions of keyword search and is significantly influenced by the current state of the searcher.

2.2 Usage of Browsing Logs

Logs of user activities can be feasible in discovering usage patterns and other applications. For seeking legal information, user interface tools based on search histories are developed [13] and the corresponding results show that the log information is useful for many information-seeking tasks. On the other hand, although current web browsers are usually capable to offer users simple logs of their browsing history, there are only a few basic functions provided for users to access their logs. Usually the most recently-viewed websites are listed so that a user can easily revisit a site even if he forgets the corresponding URL. Also, a user can search the log to find some sites that he may have forgotten he visited.

Note that time is the explicit information recorded with all entries in the browsing log. Thus, the timeline is a common approach to present time sequences [16,6] so that corresponding events can be organized in order. To provide a better visualization form, the utilization of timeline is adopted in many relevant works so that users can understand the relationship among information segments in a much easier way [1,3,6]. For example, news photos on Flickr can also be visualized as a time flow to demonstrate the evolving history [9]. Research papers can also be presented as a timeline with the tool SIMILE [23]. Generally speaking, how to use the temporal information effectively is already a crucial issue in the field of information retrieval [4,20]. For our purpose of exploratory search, the use of time for clustering and browsing online documents may provide information seekers a much distinct way than simply returning some documents in response to a query.

To the best of our knowledge, the records of web browsing history are usually loosely organized in current web browsers. Thus, it is difficult to recognize dissimilarities among browsing records, not to mention to discover the valuable relevance among them. By analyzing the browsing history, browsing patterns of users can be used to recommend websites which users might be interested in [10]. Moreover, for users to get more personalized search results or to follow interesting trends in their web activities, Google also provides similar services for users to review their search history online [11]. In this work, we thus propose to exploit two significant types of information embedded within the browsing history, i.e., record time and document relevance, to provide users better interactivities in conducting the exploratory search.

3 Exploiting Explicit and Implicit Information in Browsing Logs

As users may feel lost in the large quantity of gathered data in exploratory search, we propose to organize the search activities based on both the explicit and the implicit information in browsing logs. Our approach of utilizing the explicit record time is illustrated in Section 3.1. Moreover, the discovery of implicit relevance among search sessions and the corresponding usage are presented in Section 3.2.

3.1 Usage of Timeline and Temporal Granularities

As mentioned in previous sections, logs offered by current web browsers are usually simple and are not well-structured. The arrangement of records is simply ordered by browsing time in conventional approaches. Moreover, the basic unit of the browsing log is either a website or a web page. Nevertheless, for the purpose of exploratory search, we notice that a search session should be a more appropriate unit for tracking the browsing history. Also, we believe that more efforts can be elaborated on utilizing the corresponding temporal information.

Once a user's browsing history is tracked, search sessions along with some timestamps can be defined. Specifically, each keyword search and the corresponding browsing behavior of the result pages are regarded as a *session* in this work. Also, each search result is considered as an *item*. Also, the start timestamp and the duration of each session are crucial in our approach. Instead of presenting a long list of previously visited websites, we choose to organize previous search sessions on a timeline with variable temporal *granularities*.

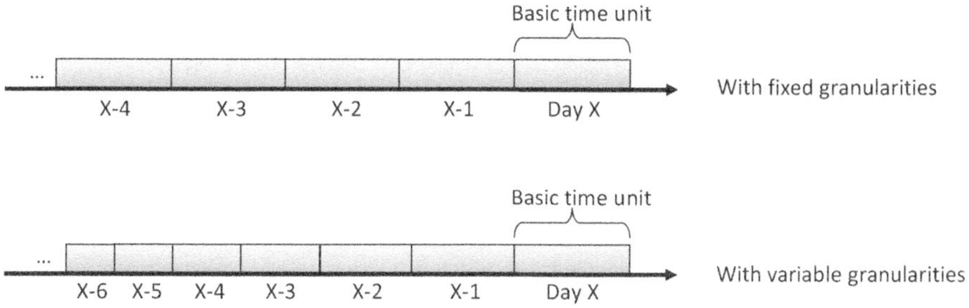

Fig. 2. Timeline with either fixed granularities or variable granularities

Granularities of a timeline can be adjusted according to the temporal distance between a previous search session and the present time. It is observed that a more recent event usually results in a higher impact to people. In other words, the search session issued yesterday is considered more impressive than the one issued a month ago for the same user. The difference of using fixed granularities and variable granularities on a timeline is depicted in Fig. 2. Note that although every time unit stands for a day in Fig. 2, the corresponding width is not fixed in the timeline with variable granularities.

We further incorporate the concept of *tilted-time window* to present the whole timeline in different granularities. In general, recent search sessions are presented in a finer granularity while older ones are presented in a coarser granularity. In addition to such a recent-biased arrangement, search sessions of more interestingness to the user can also be presented in a finer granularity. Such an extension of using variable granularities can be even more illustrative for users. This is similar to the memory of human beings since some significant events can be remembered for a quite long time.

3.2 Usage of Relevance among Search Sessions

Browsing logs not only reveal the behavior of users, but also help to establish the fundamental knowledge of users. Through reviewing the histories, users can reformulate their thoughts to confirm their next directions in searching. In other words, relationship among previous sessions may help to facilitate the exploratory search since it can be a long-term learning process. This process is just like playing jigsaw puzzles, a user may gradually obtain some clues to place more pieces properly or suddenly encounter a bottleneck preventing him from moving forward. Consequently, search sessions of a user may not occur in series during a complete process of exploratory search. For example, many users search for the business affairs in office hours and search for their casual hobbits after work. Thus, two successive search sessions can be quite irrelevant.

Obviously, if the implicit relationship among search sessions can be identified from browsing logs, the search history can be further organized. Note that since each search session contains the corresponding query terms and a number of items, i.e., web pages in the search result, two relevant sessions may contain duplicate items. In other words, the number of duplicate items between two different sessions is used to measure the strength of their relevance. Moreover, browsed items and non-browsed ones should be of different interestingness to the user. Specifically, if there exists a duplicate item among relevance sessions and is not browsed by the user, then it is regarded as an interesting item.

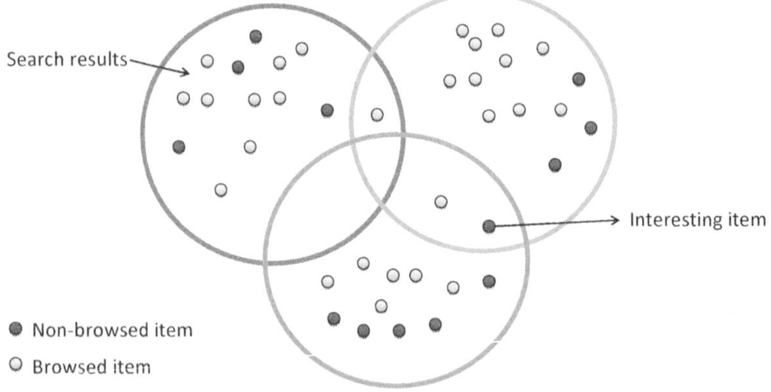

Fig. 3. Duplicate items is used for measuring the relevance among search sessions

As shown in Fig. 3, there are three different search sessions represented as hollow circles in the example browsing history of a user. Grey and red dots represent the browsed items and non-browsed items, respectively. Dots in the overlap region stand for duplicate items. Therefore, the relevance between the yellow and the green sessions is higher than that between the yellow and the blue ones. Also, an interesting item of this example is indicated in Fig. 3.

4 Implementation of Proposed Interfaces

To make previous concepts into practice, we propose new user interfaces to present the organized browsing history. Specifically, users can review their previous search activities in either the timeline mode or the relevance mode. Details of these two modes are illustrated in Section 4.1 and Section 4.2, respectively.

4.1 Timeline Mode

The interface of the timeline mode is as shown in Fig. 4. Search sessions are specified with different colors. On the left is the overall browsing history presented on a timeline with variable granularities. After the user specifies a desirable period to review his or her browsing history, details of all corresponding search sessions are presented on the right.

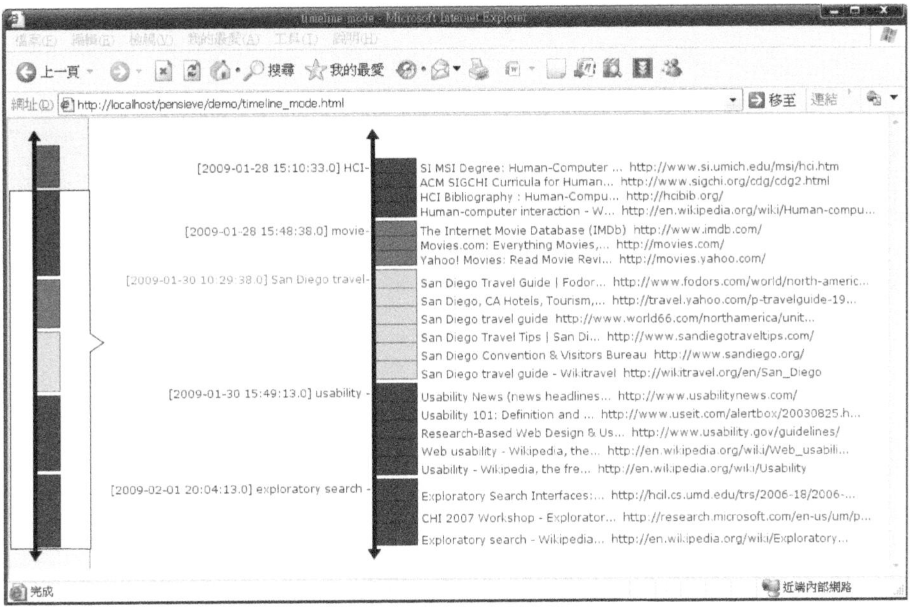

Fig. 4. Visualization of the browsing history in the timeline mode

In the timeline mode, all search sessions are arranged according to their timestamps. The granularities of the timeline vary with the recency and the interestingness of search sessions. Besides, it is noted that relevant sessions are in similar colors so that the user can distinguish them easily. As compared to the usage of a typical timeline of fixed granularities, our approach in timeline mode offers users a vivid impression of their browsing history.

4.2 Relevance Mode

The interface of the relevance mode is as shown in Fig. 5. Our implementation is built on the basis of an interactive visualization toolkit, i.e., prefuse [19]. Search sessions are represented as circles marked with the corresponding query terms. Also note that relevant sessions are in similar colors as the way in the timeline mode.

In the relevance mode, the relationship among search sessions can be easily identified since relevant sessions are connected. Note that user behavior can affect the strength of a relationship since not all search results are browsed by the user.

Generally speaking, our approach in the relevance mode breaks through the hedge of the timeline. Hence, users may obtain more insights about their own browsing history. For the purpose of exploratory search, in addition to what users already know, they can extend the current knowledge base by creating more search sessions of relevant query terms.

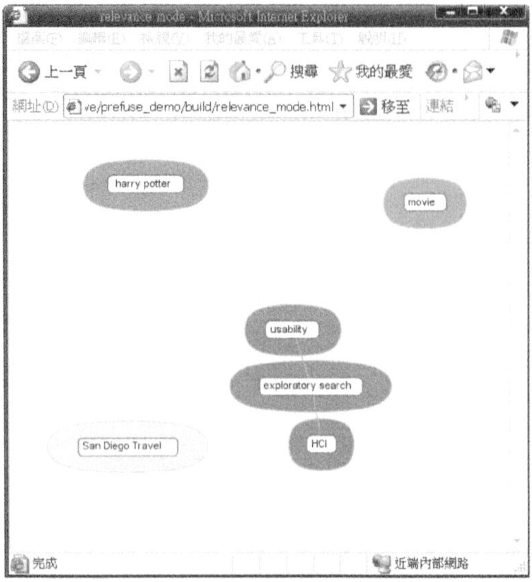

Fig. 5. Visualization of the browsing history in the relevance mode

5 Conclusions

As the tasks and goals involved with exploratory search are probably undefined and unpredictable, it is difficult and subjective for users to obtain insights during the search process. Much effort has been spent on studies in relevant fields, e.g., information seeking and human-computer interaction. In this work, we have proposed to fully exploit the browsing history to help users clarify their thoughts and discover new insights during the process of exploratory search. With one of the motivations being to support users when keyword search is not enough, we have focused on developing novel user interfaces and interaction models that support the user in different ways. Hopefully, the feasibility and usability of web search for users can be further improved as both the explicit and the implicit information of browsing history are properly utilized.

Acknowledgments. The work was supported in part by the National Science Council of Taiwan, R.O.C., under Contracts NSC95-2221-E-006-319-MY3.

References

1. Aigner, W., Bertone, A., Miksch, S., Tominski, C., Schumann, H.: Toward a Conceptual Framework for Visual Analytics of Time and Time-oriented Data. In: Proceedings of the 2007 Winter Simulation Conference, December 2007, pp. 721–729 (2007)
2. Agichtein, E., Brill, E., Dumais, S.: Improving Web Search Ranking by Incorporating User Behavior Information. In: Proceedings of the 29th Annual International ACM SIGIR Conference on Research and Development in Information Retrieval, August 2006, pp. 19–26 (2006)
3. Alonso, O., Baeza-Yates, R., Gertz, M.: Exploratory Search Using Timelines. In: Proceeding of the 25th Annual SIGCHI Conference on Human Factors in Computing Systems, April 2007, pp. 23–26 (2007)
4. Alonso, O., Gertz, M., Baeza-Yates, R.: On the Value of Temporal Information in Information Retrieval. ACM SIGIR Forum 41(2), 35–41 (2007)
5. Bates, M.J.: The Design of Browsing and Berrypicking Techniques for the Online Search Interface. Online Information Review 13(5), 407–424 (1989)
6. Broder, A.: A Taxonomy of Web Search. ACM SIGIR Forum 36(2), 3–10 (2002)
7. Chen, T.-W., Chang, P.-L., Teng, W.-G.: Supporting Informational Web Search with Interactive Explorations. In: Proceedings of the 4th International Conference on Signal Image Technology and Internet Based Systems, November 2008, pp. 153–160 (2008)
8. Dörk, M., Carpendale, S., Collins, C., Williamson, C.: VisGets: Coordinated Visualizations for Web-based Information Exploration and Discovery. IEEE Transactions on Visualization and Computer Graphics 14(6), 1205–1212 (2008)
9. Dubinko, M., Kumar, R., Magnani, J., Novak, J., Raghavan, P., Tomkins, A.: Visualizing Tags over Time. In: Proceedings of the 15th International Conference on World Wide Web, May 2006, pp. 193–202 (2006)
10. Fu, X.-B., Sudzik, J., Hammond, K.J.: Mining Navigation History for Recommendation. In: Proceedings of the 5th International Conference on Intelligent User Interfaces, January 2000, pp. 106–112 (2000)
11. Google Web History, `http://www.google.com/searchhistory/`
12. Hearst, M.: Design Recommendations for Hierarchical Faceted Search Interfaces. In: Proceedings of the ACM SIGIR Workshop on Faceted Search, August 2006, pp. 26–30 (2006)
13. Komlodi, A., Soergel, D., Marchionini, G.: Search Histories for User Support in User Interfaces. Journal of the American Society for Information Science and Technology 57(6), 803–807 (2006)
14. Lee, U., Liu, Z., Cho, J.: Automatic Identification of User Goals in Web Search. In: Proceedings of the 14th International Conference on World Wide Web, May 2005, pp. 391–400 (2005)
15. Marchionini, G.: Exploratory Search: from Finding to Understanding. Communications of the ACM 49(4), 41–46 (2006)
16. Müller, W., Schumann, H.: Visualization Method for Time-dependent Data - an Overview. In: Proceedings of the 2003 Winter Simulation Conference, December 2003, pp. 737–745 (2003)
17. Park, H., Myaeng, S.H., Jang, G., Choi, J., Jo, S., Roh, H.: Interactive Information Seeking Interface for Exploratory Search. In: Proceedings of the 10th International Conference on Enterprise Information System, June 2008, pp. 276–285 (2008)

18. Petratos, P.: Informing through User-centered Exploratory Search and Human-computer Interaction Strategies. Proceedings of the Informing Science and Information Technology 5, 705–727 (2008)
19. Prefuse - Interactive Information Visualization Toolkit, `http://prefuse.org/`
20. Ringel, M., Cutrell, E., Dumais, S.T., Horvitz, E.: Milestones in Time: the Value of Landmarks in Retrieving Information from Personal Stores. In: Proceedings of the 9th IFIP TC13 International Conference on Human-Computer Interaction, September 2003, pp. 184–191 (2003)
21. Rose, D.E., Levinson, D.: Understanding user goals in web search. In: Proceedings of the 13th International Conference on World Wide Web, May 2004, pp. 13–19 (2004)
22. Shen, X., Zhai, C.: Exploiting Query History for Document Ranking in Interactive Information Retrieval. In: Proceedings of the 26th Annual International ACM SIGIR Conference on Research and Development in Information Retrieval, July 2003, pp. 377–378 (2003)
23. SIMILE Timeline Toolkit, `http://simile.mit.edu/timeline/`
24. Speretta, M., Gauch, S.: Personalized Search Based on User Search Histories. In: Proceedings of the 2005 IEEE/WIC/ACM International Conference on Web Intelligence, September 2005, pp. 622–628 (2005)
25. White, R.W., Kules, B., Drucker, S.M., Schraefel, M.C.: Supporting Exploratory Search. Communications of the ACM 49(4), 36–39 (2006)
26. Wild, P.J., Johnson, P., Roast, C., Czerwinski, M.: The Temporal Aspects of Work for HCI. In: Proceeding of the 22nd Annual SIGCHI Conference on Human Factors in Computing Systems, April 2004, pp. 1735–1736 (2004)
27. Wilson, T.: Human Information Behavior. Informing Science the International Journal of an Emerging Transdiscipline 3(2), 49–56 (2000)

A Proposal of Awareness Services for the Construction of Quality Community Knowledge Supported by the Knowledge Management System KnowCat

Ruth Cobos[1], Ivan Dario Claros Gómez[2], and Jaime Moreno-Llorena[1]

[1] Computer Science Department, Universidad Autónoma de Madrid, 28049 Madrid, Spain
{Ruth.Cobos.Jaime.Moreno}@uam.es
[2] IDIS Research Group, Universidad del Cauca, FIET, Sector Tulcán, Popayán, Colombia
iclaros@unicauca.edu.co

Abstract. It is well known that members of work groups need awareness about one another, about shared elements, and about the group process. We present a generic framework of awareness services for groupware systems. The proposed approach has been tested through the implementation of a prototype composed by several awareness services for the Knowledge Management system called KnowCat. KnowCat is a groupware that supports the collaborative construction of quality community knowledge. We carried out a research study with a group of students enrolled in a graduate course at Universidad Autónoma de Madrid (Spain) to test our approach. The study outcomes corroborated that providing to the KnowCat users the most useful information about how the classmates have interacted with the system through the awareness services, can help them to know how they should solve their own work in the system and to be aware that they are working in a collaborative way.

Keywords: Awareness, CSCW, Groupware, Knowledge Management, Learning communities, Social Presence.

1 Introduction

Nowadays, knowledge in contemporary organizations is getting more important and relevant due to the new paradigms associated with the Information Society and the New Economy based on knowledge. For this reason, companies have made extensive attempts to manage this new value with the aim of attending to the needs of an increasingly demanding market. From the organizations' viewpoint, knowledge can be defined as information that has value to it [14], in other words, ones that allows to generate action associated with meeting the demands of the market, and support new opportunities through the exploitation of the core competencies of the organization [13]. However, it is important to emphasize that the organizations by themselves can't create knowledge, however, the people and even more, the results of their action, who set the new insights and experiences with which will form the Organizational Memory [5].

M.J. Smith and G. Salvendy (Eds.): Human Interface, Part I, HCII 2009, LNCS 5617, pp. 365–374, 2009.

Related to this introduction, the generation of effective organizational knowledge lies in the support for the potential sources of knowledge: individuals, groups, teams, projects, areas, departments, among others. Therefore, the organization should support the creativity of each source through the development of means and measures allowing the interaction between individuals and thus providing an appropriate environment for the generation of new knowledge or ownership of the existing one. Where the media are channels of communication that allow the efficient and effective exchange of information and actions, all policies and protocols that frame strategies to deal with a problematic situation, share a creative idea, registering an experience, etc.

The knowledge generated should be managed to support the development, acquisition and application of concepts and experiences that the organization needs to address its dynamics. And here is where the Information and Communications Technology (ICT) can help making some of the conditions associated to the implementation of the Knowledge Management, supporting the generation of competitive advantages based on the processes of innovation and exploit the capabilities of each participants of the organization.

However, the implementation of a system for Knowledge Management by itself does not guarantee that knowledge flows efficiently within the social networks of the organization. It is required then to establish an environment conducive to learning and dissemination of this knowledge, which suggests the adoption a technological environment based on the principles and concepts of engineering collaboration, which allows the efficient and effective exchange of information to and from each individual.

It is known that members of work groups need awareness about one another, about shared elements, and about the group process [9]. From this perspective, the media services platform for awareness in support of collaboration, seeks to reduce the effort needed to achieve a natural and efficient communication during the execution of a process and allows the creation of an environment of fluid communication, which clearly favours the establishment of an environment conducive to learning and creative chaos.

Since 1998, at Universidad Autónoma de Madrid (UAM, Spain) we have used with several student communities a fully consolidated and thoroughly tested and validated Knowledge Management system called KnowCat (acronym for "Knowledge Catalyser") [1][3], in order to create among students educational material of high quality. During these last ten years, these students communities have generated collaboratively several knowledge areas about “Operating Systems”, “Computer Systems”, “Artificial Intelligence”, etc. [2][4][6][7][8].

The main aim in this paper is to show how KnowCat, which supports the collaborative construction of quality community knowledge, can be improved through the provision of some group awareness to its users. More in detail, we are interested in providing to the system users the more useful information about how the classmates have interacted with the system in order to help to the others useful information about how they should solve their work in the system.

In the next section, a generic proposal of Awareness Services for Groupware is presented. In Section 3, the KnowCat system is presented. In Section 4, the details of the proposed approach in presented, which has been implemented as a prototype of awareness services for KnowCat. In Section 5 we present the experimentation and results obtained with the proposed prototype with a student community at UAM. And we conclude this paper with a Conclusion and Future Work Section.

2 Proposal of Awareness Services for Groupware

We can find different proposals of awareness approaches for groupware systems [9] [10] [11]. However, most current groupware systems are not well prepared to handle heterogeneity perspectives and information overload at the moment of providing awareness information to their users.

In order to afford a general solution to this limitation a generic framework of awareness services for groupware systems is proposed in this paper. Due to the heterogeneous and distributed nature of the information in groupware systems, the framework architecture is composed by three interconnected elements: the data model, the agent model and the manager model.

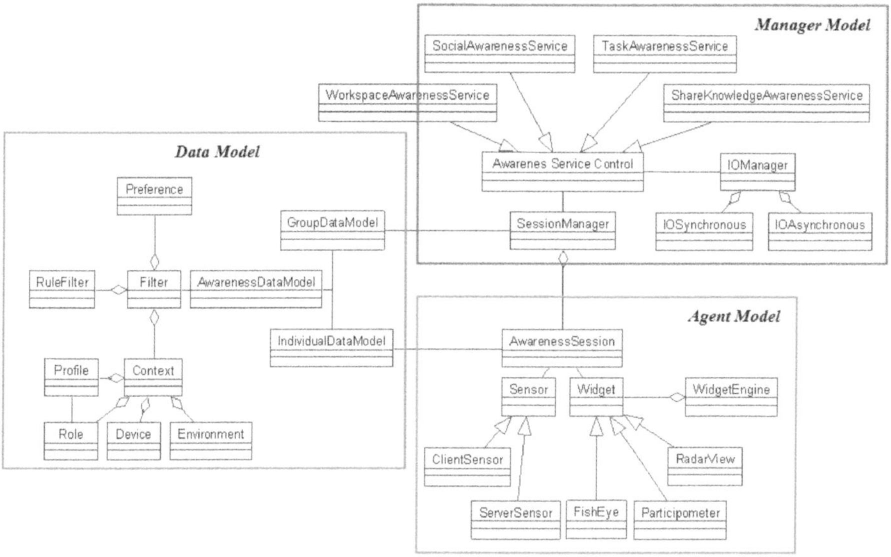

Fig. 1. General Architecture of the Awareness Services for Groupware

Data model. It organizes the data in the following layers: i) individual data model: user's interaction, such as, mouse movements, clicks, scrolling, etc., ii) awareness data model: previous information is transformed into domain objects of the application taking into account the information about the type of awareness that the system supports (i.e context awareness, group awareness, task awareness, change awareness, etc. [9][15]); and finally iii) group data model: a global model of users' collaboration in the system.

Agent model. It provides to the users adapted awareness services. They are managed as widgets [12]. All widgets are composed by three layers; the following two ones are common in all widgets. The first one is called "sensor" and its mission is to collect all the interactions of the users with the system interface. Immediately after, the agent model transfers this information to the individual data model layer. The second layer of the widgets is called "contextualization", its mission is to provide the information adapted to the user characteristics and in the user context. Furthermore, each widget has its own specification in a third layer. These specifications are related with its purpose and with the representation of its provided information (graphical or textual).

Manager model. It controls the provided awareness services, e.g., it decides what kind of information has to be provided to the agents taking into account group awareness, task awareness, context awareness, etc. It manages the communication channels among the models.

3 KnowCat: Knowledge Catalyser

The KnowCat system, an acronym for "Knowledge Catalyser", was developed and in active use since 1998 at the Universidad Autónoma de Madrid (UAM). KnowCat is a distributed non-supervised system for structuring knowledge whose purpose is to enable the crystallization of collective knowledge as the result of users' interactions without an editor managing the task. The main aim of this system is to generate quality educational materials as the automatic result of the students' interactions with the materials, by catalysing the crystallization of knowledge [1][3].

KnowCat enables us to build up knowledge sites, known as "KnowCat sites" or KnowCat nodes. These knowledge sites can be accessed through a specific URL using a Web browser. These knowledge sites are organised in these knowledge elements: a) a knowledge tree: a hierarchical structure of topics which facilitates the organisation of the community knowledge; b) a set of documents contained in each topic which provides alternative descriptions of the topic and c) a set of annotations contained in each document which express explanations, comments and opinions about the content document.

At any given time, all documents contained in the same topic compete with each other to be considered as the "best" description of the topic. This competitive environment is achieved by the Knowledge Crystallisation mechanism of KnowCat, which is supported by virtual communities of users.

The Knowledge Crystallization mechanism takes into account the user's opinions about the documents and the evolution of the opinions received to determine what documents are socially acceptable, in which case they remain in the knowledge site, and which of those are found unsatisfactory, in which case they are removed from the knowledge site. Whether or not a document is socially acceptable is determined by its "degree of acceptance" as calculated by the Knowledge Crystallization mechanism. More specifically, the degree of acceptance of a document is formulated using the explicitly received opinions concerning the document: the received votes, how these votes were received, the received annotations and their respective types (see below the descriptions of votes, annotations and annotation types); and the implicitly received opinions regarding access to the document.

Moreover, we have taken into account in this mechanism the "quality" of the users. In other words, we prefer to give more credibility to opinions from experts than those from occasional users. KnowCat establishes categories of users through the same means as the scientific community establishes its member's credibility, that is, by taking into account past contributions. Therefore, this system deals with "virtual communities of experts".

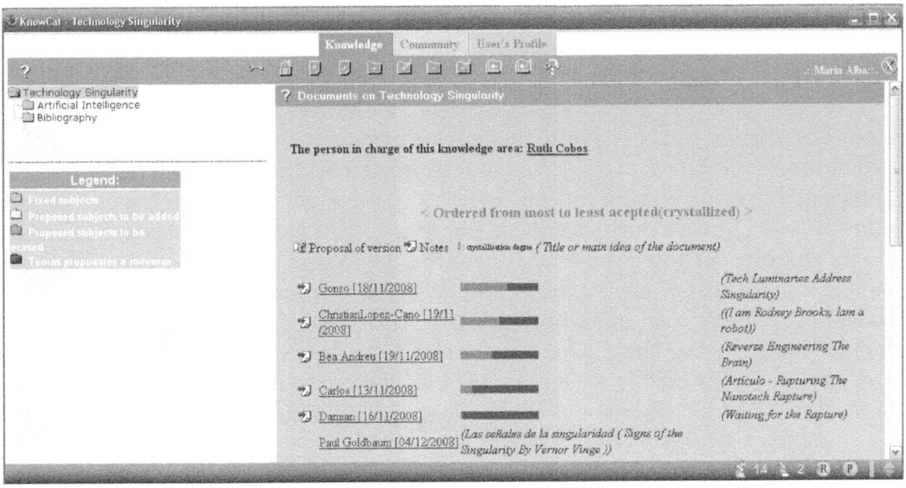

Fig. 2. Example screenshot of the "Technology Singularity" KnowCat node

An example screenshot of a KnowCat node is shown in Figure 2. This system screenshot shows the ''refinement'' of a selected topic and its contents. The left side of the screen shows the knowledge tree on the knowledge area "Technology Singularity". The right side shows the documents added to this topic. Documents are identified by the author's name, arrival date and title. They are ordered by their degree of acceptance, which is shown to the right of the identification heading of each document (with the green-red bar). On the left side of the identification heading of each document are the icons indicating whether a document has received annotations and whether a new version of the document is available. For example, the document identified by "Gonso [18/11/2008] (Tech Luminaries Address Singularity)" shows the highest degree of acceptance in the topic, and this document has received annotations which are shown with the corresponding icon.

4 Experimental Case: Implementation Proposal in the Context of the KnowCat

The proposed architecture for the Awareness Services for Groupware has been tested through the implementation of a prototype composed with several awareness services for the groupware system called KnowCat. The proposed prototype deals with the following KnowCat elements: i) space data, such as topics, documents, etc., ii) user interaction data, taking into account what the last interactions are and what have the users done and iii) task data, such as, when has a user added a document to a topic, and what annotation has added a user to a document, etc.

The proposed prototype of awareness services appears in a console in the bottom part of KnowCat. The Awareness console is composed by the following five awareness widgets. Due to the widgets' nature, the users' interactions with these widgets can interact with the KnowCat system at the same time.

Registered Users. It provides brief information about the registered users in KnowCat. This widget displays contact data (e-mail, occupation, etc.) of a selected user and a list of the last executed tasks in the system by this user.

On-line Users. It provides the following information of the on-line users in KnowCat: contact data (similar to the provided one in the previous awareness service) and the current location of a selected user. A user can be located visiting a topic of the knowledge tree, or visiting a document or visiting a note. This information about a user location allows a direct access in KnowCat. In Figure 3, the contact data and the current location of the on-line user identified as "Maria Alba" is displayed. At this moment, the actual KnowCat user, who is seeing this mentioned information in the awareness console, can click in the Maria Alba's location (Inicio -> Technology Singularity ...) in order to access in KnowCat to the note identified as "Guiomar-2008-11-21 [7]", which is located on the document named as "ChristianLopez-Cano [19/11/2008]", which is contained in the "Technology Singularity" topic.

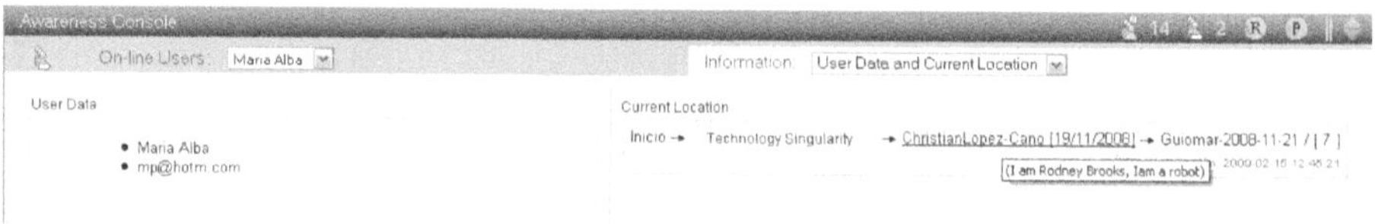

Fig. 3. Example screenshot of the On-line User awareness widget

Radar View. It provides the locations of the on-line users through a replicate knowledge tree of the KnowCat site. Moreover, in each topic of the replicated knowledge tree the number of the on-line users, who are interacting in each topic, is shown. In Figure 4, we can see that the user "Maria Alba" is located in the root topic of the knowledge tree, where she is visiting the document identified as "Carlos [13/11/2008]".

Fig. 4. Example screenshot of the Radar View awareness widget

History View. It provides chronically and graphically information about the realised tasks of a selected user. In this view, see Figure 5, there is a time line vertically; in the left-hand of this time line the deadlines for the different programmed tasks by the group instructor are shown. For example, the instructor programmed that the students had to submit their documents to the KnowCat site before 2008-11-13. On the right-hand of the time line it is possible to see the realised tasks, presented by icons, that a previous user has done. Each icon corresponds with a task type. For example, this user collaborated with 3 notes and visited 4 documents at 2008-11-17.

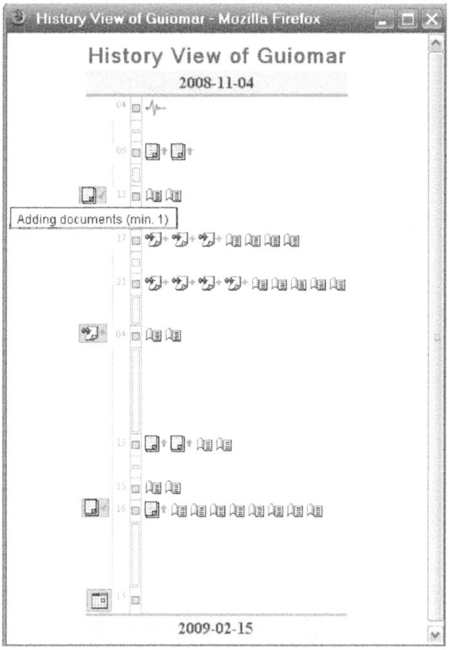

Fig. 5. Example screenshot of the History View awareness widget

Notes View. It provides graphically information about interaction among users in the annotating task, i.e., with this widget we can know who has annotated the other's document. This widget provides this information in these two ways: i) a summary of the notes view: it displays how many notes have added each user to documents of other users; and ii) detailed notes view: it displays the identification of the users, of the annotated documets (which is identified by its author's name and arrival date) and the realised annotation (which are the arcs of the graph and show the note type, moreover we can see more information about the annotation when we click it).

5 Experimentation and Results

In order to test the proposed approach, we carried out the following research study with a group of students enrolled in a graduate course about "Artificial Intelligence" of the Computer Science Engineering studies at Universidad Autónoma de Madrid. This course took place during the first semester of the academic year 2008/2009, i.e., the actual academic year.

Every year, the enrolled students in this course have to elaborate individually or in groups own documentation about an actual subject related to the contents of the course. This information is useful to the students in order to facilitate them their participation in a debate that takes place at the end of the course.

In the academic year 2008/2009 the debate subject was "Technology Singularity". The students were in groups of 10-12. Each group has 2 persons with the coordinator role, who had to coordinate the collaborative work of creating a good common corpus of knowledge about the debate subject. Two of these groups used the KnowCat system with the Awareness console. These students knew KnowCat, because they used it in other courses, such as "Operating Systems" and "Information Systems" in previous academic years, but in these previous years they used KnowCat without the Awareness console, due to this console was used at first time in the actual academic year.

The students' tasks during the mentioned period were the following: i) in the first month each student had to read documentation (papers, books, etc.) about the debate subject, ii) the group coordinators assigned to each member of their group an specific topic of the subject, iii) individually each student had to elaborate a document about the assigned topic and add it into the KnowCat site, iv) during the second month they had to annotate their classmates documents and v) finally, in the third month, they had a face-to-face meeting in order to organise the documents and annotations for their utilization in the debate.

As we mentioned, the students knew the use of KnowCat without the Awareness console, and they did similar collaborative activities (generation of community knowledge) in previous academic years. Therefore, the main difference between their work with KnowCat from previous years and the actual year was the use of the awareness services presented in this paper.

At the end of the semester the students were asked to do a questionnaire in relation to their work with the system. The main aim of this activity was to obtain information about how the collaborative work was for the students taking into account that they could use the awareness widgets of the Awareness console.

All students recognised that thanks to the Awareness console they had all the time the perception that they were working with other classmates collaboratively. Moreover, they were aware at any time of their contributions and of other classmates' contributions in the construction of the community knowledge. Therefore, they recognised that in general the awareness services are useful in order to inform them about when and who has contributed with what and where.

The awareness widget catalogued by the students as the most useful was the "History View". They argued that this widget displays complete and organise information about the contributions of each person in a convenient way. However, the "Notes View" was the least used awareness widget and it was the least useful ~~in~~ according to the students' opinion.

The group coordinators emphasized that the Awareness console helped them in the execution of their role in the group. They used mostly the History View in order to check if the group rhythm was good or not. And they used the contact data displayed by the Awareness console in order to communicate with the classmates who were later in the tasks.

To sum up, the students have declared that the information about how the classmates have interacted with the system that are provided by the mentioned awareness widgets is really useful information about how they should solve their work in the system. Furthermore, these students' opinions about the awareness widgets were positive and emphasized that their work with KnowCat with the Awareness console was more successfully that with only KnowCat (without the console) in previous academic years.

6 Conclusions and Future Work

We have presented a generic framework of awareness services for groupware systems, which architecture is composed by three interconnected elements: the data model, the agent model and the manager model.

In order to test and validate the proposed framework we have implemented a prototype of our presented approach for the Knowledge Management system called KnowCat. KnowCat (acronym for "Knowledge Catalyser") is a fully consolidated and thoroughly tested and validated Knowledge Management system which has been developed and in active use at Universidad Autónoma de Madrid (Spain) since 1998. The main aim of KnowCat system is to support the crystallization of collective knowledge as the result of user interactions [1][3].

The proposed prototype of awareness services appears in a console in the bottom part of KnowCat. This prototype offers the following awareness widget services: brief information about registered users (what have these users done?), brief information about connected users, a radar view (where and what are the connected users doing?), participation-meter (How many times have the registered users done each task?), a fish eye view (when, where and what has each registered user done?) and a map of interaction among users in the annotating task (who has annotated the document of whom?). Due to the widgets' nature, the users' interactions with these widgets can interact with the KnowCat system at the same time.

We carried out a research study with a group of students enrolled in a graduate course about "Artificial Intelligence" of the Computer Science Engineering studies at Universidad Autónoma de Madrid. The students in groups of 10-12 had to elaborate proper documentation about an actual subject related with the contents of the course, i.e., they had to create a good common corpus of knowledge about the subject.

These students have declared that the information about how the classmates have interacted with the system that are provided by the mentioned awareness widgets is really useful information about how they should solve their work in the system. Furthermore, these students' opinions about the awareness widgets were positive.

Nowadays, we are starting new research studies in order to corroborate the previous mentioned outcomes with groups of students at Universidad Autónoma of Madrid, and at Universidad of Cauca (Colombia). Moreover, at Universidad of Cauca (Colombia) we have recently finished the following research study: a group of students used the traditional KnowCat system without the Awareness console, and a second group used it with the Awareness console, both groups work with the same aims and rules in this study. The initial obtained results corroborate hat the members of the group which used KnowCat with the Awareness consoles were more aware of their participation in a collaborative work than the members in the other group.

Acknowledgments. This research was partly funded by the Spanish National Plan of R+D, project number,TIN2008-02081/TIN; and by the AECI (Spanish Agency for the International Cooperation) project number A/017436/08.

References

1. Alamán, X., Cobos, R.: KnowCat: a Web Application for Knowledge Organization. In: Kouloumdjian, J., Roddick, J.F., Chen, P.P., Embley, D.W., Liddle, S.W. (eds.) ER Workshops 1999. LNCS, vol. 1727, pp. 348–359. Springer, Heidelberg (1999)
2. Cobos, R., Alamán, X.: Creating e-books in a distributed and collaborative way. Journal of Electronic Library on Electronic book for Education 20(4), 288–295 (2002)
3. Cobos, R.: Mechanisms for the Crystallisation of Knowledge, a proposal using a collaborative system. Doctoral dissertation. Universidad Autónoma de Madrid (2003)
4. Cobos, R., Pifarré, M.: Collaborative knowledge construction in the web supported by the KnowCat system. Computers & Education 50, 962–978 (2008)
5. Cross, R., Lloyd Baird, L.: Technology is not enough: Inproving perfomance by building organizational memory. Sloan Management Review, EEUU, 69–78 (Spring 2000)
6. Díez, F., Cobos, R.: A Case Study of a Cooperative Learning Experience in Artificial Intelligence. Journal Computer Applications in Engineering Education 15(4), 308–316 (2008)
7. Dumitrescu, D.A., Cobos, R., Moreno, J.: A Multi-Agent System for extracting and analysing users' interaction in a Collaborative Knowledge Management System. In: Cao, L., et al. (eds.) Data Mining and Multiagent Integration. Springer, Heidelberg (in press, 2009)
8. Gómez, M., Gutiérrez, A., Cobos, R., Alamán, X.: Collaborative Learning with computer-based support in the design of material for the development of abstract thought in Preschool Education. An Experiment in the teaching of mathematics. In: Proceedings of the International Symposium of Computers in Education, pp. 241–254 (2001)
9. Gross, T., Stary, C., Totter, A.: User-Centered Awareness in Computer-Supported Cooperative Work-Systems: Structured Embedding of Findings from Social Sciences. International Journal of Human-Computer Interaction 18(3), 323–360 (2005)
10. Gutwin, C., Greenberg, S.: Design for individuals, design for groups: Tradeoffs between power and workspace awareness. In: Proceedings of the ACM CSCW 1998, pp. 207–216 (1998)
11. Gutwin, C., Greenberg, S.: A descriptive framework of workspace awareness for real-time groupware. Computer Supported Cooperative Work (Kluwer Academic Press) (2002), `http://www.cpsc.ucalgary.ca/grouplab/papers/` (access: April 2002)
12. Kreijns, K., Kirschner, P.A., Jochems, W.: Supporting social interaction for group dynamics through social affordances in CSCL: Group awareness widgets. In: Paper presented at the 10th EARLI Conference, Italy (2003)
13. Prahalad, C., Hamel, G.: The Core Competence of the Corporation. Harvard Business Review, EEUU, 79–91 (1990)
14. Stewart, T.: Intellectual Capital: The new Wealth of Organizations, Doubleday, EEUU (1999)
15. Tam, J., Greenberg, S.: A framework for asynchronous change awareness in collaborative documents and workspaces. Int. J. Human-Computer Studies 64, 583–598 (2006)

Electronic Document Tracking System (EDTS): A Prototype

Rochin Demong, Lailatul Faizah Abu Hassan,
Tuan Badrol Hisham Tuan Besar, and Zulhaimi Zulkifli

Faculty of Office Management and Technology
Universiti Teknologi MARA
40450 Shah Alam, Selangor Darul Ehsan, Malaysia
laila@salam.uitm.edu.my, tuanbadrol@salam.uitm.edu.my,
zzulhaimi@yahoo.com

Abstract. With rapidly-changing technology and increasing emphasis in managing information properly, Electronic Document Tracking System (EDTS) has been developed as a solution system to improve the efficiency of retrieving the document online at anytime and tracking the movement of documents in and out of Faculty of Office Management and Technology (FOMT) in University Technology MARA, Malaysia. FOMT have tried to improve document handling and use through the application of information technology. EDTS was developed within windows environment using Open Source tools such as PHP and MySQL. The application relies on TCP/IP and internet to support server and client communications.

Keywords: Electronic Tracking System, Tracking System, Web Tracking System.

1 Introduction

This project is aimed to develop Electronic Document Tracking System through the web for Faculty of Office Management and Technology, UiTM. To do this, a number of forms available were employed to staff accessible locations throughout the Faculty of Office Management and Technology, UiTM. This project is becoming more crucial as the university steps up the utilization of paperless society to help staff especially delivering, maintaining and managing all documents efficiently towards excellence service quality.

The current condition for document management is managed manually by administrative staff of Faculty of Office Management and Technology, UiTM.

This project explores the scope and importance of EDTS in detail and illustrates how it expands our view of information management. It is designed to help structure the field by approaching it from three perspectives: technologies that are making EDTS possible, the application areas in which business value is being realized, and the roles and responsibilities of several personnel that will be involved in maintaining EDTS. The project suggests what administrators can do now to begin preparing for this major advancement in information management.

M.J. Smith and G. Salvendy (Eds.): Human Interface, Part I, HCII 2009, LNCS 5617, pp. 375–383, 2009.

The roles of the EDTS are: Each document can be logically registered, preserved, retrieved and renewed at high speed while tracing the required documents or files. Thus, efficiently performing an office work, reducing a volume of consumed paper and minimizing a work space in an office.

Improvement of functions and operations in the electronic document filing systems are been required with the spread of the system in many offices to assist the university community in the use of information technology.

1.1 Significance/Benefit of the Project

In general, the significant or benefit of this project is to manage the documents in Faculty of Office Management and Technology, UiTM through web-based application which enable the users ease in retrieving the documents.

Table 1. Substantial benefits the FOMT can expect to reap form the use of EDTS

Problem/Issue	EDTS Impact
Missing or lost files/documents.	Electronic files, if indexed and backed-up properly, will not get lost.
Take long time to retrieve required documents (lost documents).	The documents will be linked to cases as soon as they are scanned and indexed. Available to users immediately.
File available to only one user at a time.	Electronic files are available to multiple users at the same time.
Documents are copied to circulate.	The need for extra copies will be eliminated.

1.2 Object Modeling Technique (OMT) Methodology

EDTS was developed by using Object Modeling Technique (OMT).OMT is one of the most popular object-oriented development techniques developed by Rumbaugh et. al. According to Teo Xiu, X. [30], it is primarily used by system and software developers supporting full life-cycle development, targeting object-oriented implementations and has proven easy to understand, to draw and to use.

There are five main phases in OMT methodology namely analysis, system design, object design, implementation (programming/coding, installation) and testing as shown in Figure 1.

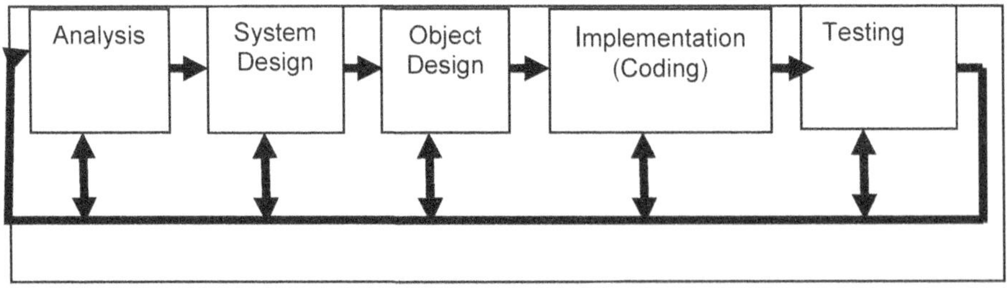

Fig. 1. OMT phases. (Source: [30]).

2 Analysis of Current System

This section will discuss about the data collection on current system of FOMT Electronic Document Tracking System which currently handled manually by the administration staffs. Issues regarding the current system were analyzed and explained with more detailed.

Category of Document

Type of document to or from the faculty includes memo, circulation, letter from outside organization or other department and others.

Current Document Management and Tracking System

The users of the documents consist of all staffs in the faculties and students. Two major transactions involved in the current system namely user record and manage the document received and user request for document. Each copy of the document needs to be kept in a file for future references.

Users Roles And Responsibilities

The roles and responsibilities of the related users that using this system are as follows:

Sender

The sender of the document can be categorized into three category namely staff (academic/administrative staffs), users from others organizations/departments and students.

Receiver

The receiver of the document can be categorized into two category namely staff (academic/administrative staffs), and students.

Current System Scope

FOMT is a faculty that has around 50 academic staff and 20 administrative staff who will be responsible for the document in and out to/from the faculty.

For this project, the OMT methodology with UML notation is used in order to analyze the system. The scope of the Electronic Document Tracking System in FOMT has been analyzed and is shown by using a use case diagram. Figure 2 shows a use case diagram of the current system of the Electronic Document Tracking System.

The description of the use case diagram is listed in the Table 2 below.

Table 2. Use case descriptions

Send document	The faculty will receive document from sender either from own staff or outside organization.
Record document	The staff (system) that responsible for the document will record the document and keep the document in a respective file.
Request for document	The receiver will request for the document and the staff responsible need to make sure that the documents are in a files for easy retrieval and can be access easily.

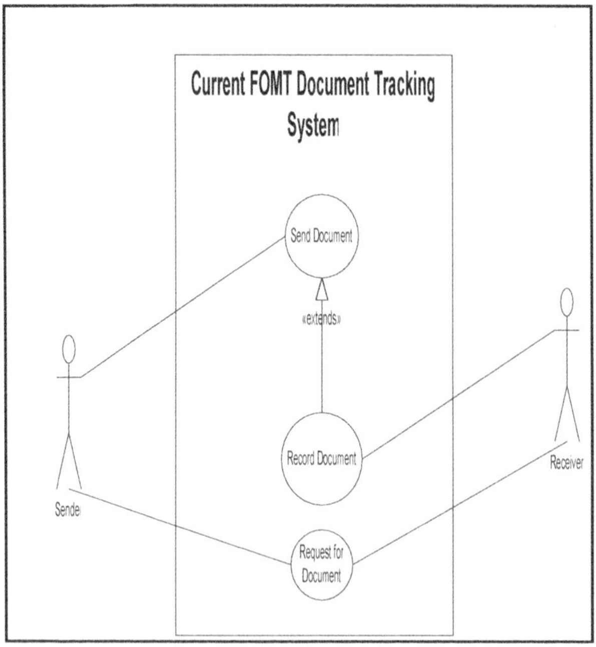

Fig. 2. Use case diagram showing the current document tracking and management system in FOMT

2.1 Issues and Challenges of Using the Current System

The current system used in order to record and tracking the document from the respective files faced many problems especially when the document needs to be retrieved in a short time.

Current system used generates some problems or issues that become as challenges for the management of the document. The issues or problems that related with the current system include:

More Space Needed To Kept A Copy Of The Document In A File, More Time Needed To Search And Retrieve Document Use A Lot Of Papers, Very High Cost For Papers, Ink To Make A Copy Of The Document, Missing Document Or Misplaced ,Duplication Of Document In Different File, No Sharing ,Separation And Isolation Of Documents,

2.2 Propose Electronic Document Tracking System through Web

All these issues and problems need to be solved to make sure every sources and information can be used effectively. Therefore, a systematically computerized or online document system need to be developed for benefits of all users namely the staffs of the faculty. The current practices of handling or recording the documents has to be automated with the availability of higher technology and higher specification software in the market.

3 Analysis of Document Tracking System

Analysis phase is an important phase in OMT methodology in developing a system. During analysis stage, information about a problem domain is captured. OMT allows a lot of flexibility as to what information to capture in the models and how to represent that information. The models are explained according to scope of the process of the system. The process of the current system is done manually and the issues occurred discussed in the previous chapter.

3.1 User Requirements

User requirements are the output from the process of collecting data and analyzing information about the part of the organization that is to be supported by the database application. The user requirements for this system are as follows:

User need to login to access into the system, user can retrieve the document through the system, user can change password of the system, user can change type of users by using the system. user can upload the document through the system. user can change the details of the documents for accuracy of document and indirectly easy for searching, user can delete document which are outdated to save space in the database, user can search for document according to different field and user can register to get their username and password.

Fig. 3. Use Case diagram of the system develop

3.2 Object Model

The object model is the most important model. It identifies the object classes in the system and their relationships, as well as their attributes and operations. It represents the static structure of the system. The object model is represented graphically by a class diagram. The object model describes the data structure that the dynamic and functional models operate on.

In addition, the operations in the object model correspond to events in the dynamic model and functions in the functional model. The OMT identifies the following for the system such as identifying objects and classes and prepare a data dictionary.

3.3 Identify Use Cases of the System

Figure 3 showing a use case diagram of the electronic document tracking system that consist of three actors namely super user, normal user and administrator.

4 Implementation

This section will discuss the design and implementation of the prototype system namely FOMT Electronic Document Tracking System. It will give an overview of object form related to the system, the architecture of the web application, and the prototype system architecture

In this phase, output of the design will be transferred in source code. Any programming language whether using object approach or not can be used during programming language process. For this FOMT Electronic Document Tracking System, PHP is used as a programming language with MySQL database.

4.1 FOMT Electronic Document Tracking System Programme

The programs of the system are as follows:

- Object Form for Recording/Uploading, Searching /Viewing, Updating And Deleting Document.
- Object Form for Registering New User, Searching/Viewing, Updating And Deleting User Details.

4.2 The Prototype System Architecture

The prototype model proposed in this project can be implemented on several other platforms. However for the purpose of this project, the prototype was developed using the relational database in the web environment.

This project chooses Apaches as the web server. Apache web server is one of the two-web servers that dominate the market. The other is Microsofts's IIS. Apache server is an open source, anyone with the skill can write code that extends the functionality of Apache [11].

For the middleware, this project chooses Hypertext Preprocessor (PHP). PHP will most often run as an Apache extension, known as the Apache module. PHP belongs to a class of languages known as middleware. This language work closely with the web server to interpret the requests made from the World Wide Web, process these requests, interacts with other programs on the server to fulfill the request, and then indicate to the web server exactly what to server to the client's browser. PHP is a cross-platform and it will run on Windows 2000/NT and UNIX and with both IIS and Apache. PHP also works on Netscape, Roxen and other wide variety of systems. For the database, MySQL is chosen because it is free. MySQL will be extremely fast for small-to-medium-sized databases.

There are three types of actors that will use the system. The actors are the users that consist of the administrator, super user and normal user. The system architecture contains two main subsystems and one relational database. The two main subsystems are the Document Tracking System (involve the process of Recording/Uploading new document, searching/viewing, updating document details and deleting document) and User Registration (involve the process of register new user, search, update and delete user) as discussed in chapter six. The relational database, which is used to store the information, is MySQL. The subsystems will interact with the relational database when storing or retrieving information. The actors will also interact with the systems and its architecture.

5 Conclusions and Recommendations

This section reviews the overall progress of this project. It includes the advantages, problems and limitations encountered during the development of the prototype of FOMT Electronic Document Tracking System. Recommendations for future enhancements will also be discussed.

5.1 Conclusions

As for the conclusion, this system is one of the major systems that the organization must have to handle the management and tracking of documents. It gives many benefits and helps the top level management to retrieve the document easily at anytime and anywhere with the support of the Internet connection.

One of the major roles played by administrative staff mainly the Personal Assistant is to provide and manage the documents in and out to/from the faculty. Therefore, it is very important to make sure the best way is applied in order to gain competitive advantage and effectiveness or efficiency in managing or handling the documents.

This prototype system could be useful because of the use of an open source package and it is also hoped that this project would enhance the organizations management information systems and the development of an effective FOMT Electronic Document Tracking System.

5.2 Recommendations

In order to overcome the limitations mentioned above, several suggestions are recommended.

The system can be extended to include other processes such as updating the document online where user can edit the document and save the updated copy in the system to enhance the EDTS. More reliable test should be created in order to validate the accuracy and correctness of the developed system model.

This project has achieved the target, which is to generate a web-based system model for managing the document in and out to/from the faculty or organizations. Its will become the major advantage of this system because the organization will work with documents everyday and all activities done need documentation. Several suggestions for the enhancement of this system have also been recommended for the future development of this project.

References

1. ARTiSAN Software Tools: Extending UML to design Object-Oriented Real-Time Systems, A WHITE PAPER form ARTiSAN Software Tools (1998)
2. Bennett, S., McRobb, S., Farmer, R.: Object Oriented Systems Analysis and Design using UML. McGraw-Hill, London (2000)
3. Butcher, D., Rowley, J.: The 7 Rs of information management. Managing Information 5(3), 34–36 (1998)
4. Daly, P.G.: Publishing Content from Your Document Management System. Internet Journal (2002), `http://www.intranetjournal.com/articles/200208/it_08_12_02a.html`
5. Dorey, J.: 10 steps to becoming nearly paperless. National Underwriter 45(107), 47–58 (2003)
6. Élise, M., Pierre, G., Jean-Pierre, P., Anne, D., Pierre, D.: A web information system for the management and the dissemination of Cultural Heritage data. Journal of Cultural Heritage (2007), `http://rss.sciencedirect.com/getMessage?registrationId=JCCDJDCDKICERCDFLCCLJEELJLDHMKKMMUELMLEFMG`
7. Eugene, M.: Integrating Business Process Models with UML System Models. A WHITE PAPER form Popkin Software (2001)
8. Gary, R., Diana, H., Thomsan, M., Mark, D.: A distributed Document Management System: TCS Programmers (2002), `http://ieeexplore.ieee.org/iel5/2218/26014/01161469.pdf`
9. Gragg, E.: Managing printing costs in the post-paperless office. Office Solutions 21(1), 28–31 (2004)
10. Greenemeier, L.: Fewer devices add up to savings. Information Week, No. 959, pp. 72–75 (2003)
11. Greenspan, J., Bulger, B.: MySQL/PHP database applications. M&T, USA (2000)
12. HCI Journal. The Ideal Document Management System (2001), `http://www.hci.com.au/hcisite2/journal/Ideal%20document%20management%20system.htm`
13. Hemphill, B.: Paperless or less paper. Journal of Financial Planning 14(7), 22–28 (2001)
14. Hind, Z., Farhi, M.: Document management systems from current capabilities towards intelligent information retrieval: an overview. International Journal of Information Management 19(6), 471–484 (1999)
15. Horton Jr., F.: Information resources management: an overview. Infotrends, 185–191 (1986)

16. Hrvoje, B., Edgar, P., Sven, L.C., Vedran, M.: An Electronic Journal Management System. In: Paper presented at 25th International Conference on Information Technology Interfaces (2003)
17. Jack, G., Kalid, K.: Effective Document Clustering for Large Heterogeneous Law Firm Collection (2005), http://delivery.acm.org/10.1145/1170000/1165513/p177-conrad.pdf?key1=1165513&key2=3599614021&coll=GUIDE&dl=ACM&CFID=56940527&CFTOKEN=48460680
18. Jesús, T.: Open source tools for content management. Hypertext.net 3 (2005)
19. Joseph, H., David, C., Binshan, L.: Intranet document management systems: Internet Research. Electronic Networking Applications and Policy 8(4), 338–346 (1998)
20. Lancaster, W.F., Warner, A.J.: Information retrieval today Information. Resources Press, Virginia (1993)
21. Magnus, B., Anssi, R.: Use of Document Management Systems - A Case Study of The Finnish Construction Industry (2003), http://www.itcon.org/data/works/att/2003_26.content.03446.pdf
22. Malhotra, Y.: Deciphering the knowledge management hype. The Journal for Quality and Participation, the Association for Quality and Participation (1998), http://www.brint.com/papers
23. Phelan, S.E.: A paperless success story. Journal of Accountancy 196(4), 59–68 (2003)
24. Raymond, K.W., Nicole, L.: Managing and Query Multi-Version XML Data with Update Logging (2003), http://delivery.acm.org/10.1145/590000/585073/p74-wong.pdf?key1=585073&key2=8699614021&coll=GUIDE&dl=ACM&CFID=56940527&CFTOKEN=48460680
25. Rumbaugh, J., et al.: Object-Oriented modeling and design. Prentice Hall, Englewood Cliffs (1991)
26. Shlaer, S.: A comparison of OOA and OMT. Project Tecnology Inc. (1992), http://www.projtech.com
27. Teo, T.S.H., Wong, P.K.: An empirical study of the performance impact of computerization in the retail industry. The International Journal of Management Science 26(5), 611–621 (1998)
28. Torkazadeh, G., Doll, W.J.: The development of a tool for measuring the perceived impact of information technology on work. The International Journal of Management Science 27, 327–339 (1999)
29. US Patent Issued.: System for managing resources (2007), http://www.patentstorm.us/patents/7188170-claims.html
30. Xiu, X.T.: Object-Oriented software development using object modeling technique (OMT) (1998), http://www.institut_fur_automatik/ET/omt.html

Adaptive Visual Clustering for Mixed-Initiative Information Structuring

Hakan Duman, Alex Healing, and Robert Ghanea-Hercock

British Telecommunications plc, Adastral Park, Martlesham Heath, Ipswich, IP5 3RE, United Kingdom
{hakan.duman,alex.healing,robert.ghanea-hercock}@bt.com

Abstract. Cyclone is a mixed-initiative and adaptive clustering and structure generation environment which is capable of learning categorization behavior through user interaction as well as conducting auto-categorization based on the extracted model. The strength of Cyclone resides in its integration of several visualization and interface techniques with data mining and AI learning processes. This paper presents the intuitive visual interface of Cyclone which empowers the user to explore, analyze, exploit and structure unstructured information from various sources generating a personalized taxonomy in real-time and on-the-fly.

Keywords: Information Visualization, Data Mining, Human-Computer Interaction, Machine Learning.

1 Introduction

The structuring of information is a fundamental step in order to deal with the sheer volume now common in the digital universe and increasing at an accelerating pace [1]. The World Wide Web has acted as a catalyst, making it easier for both information providers to publish and consumers to link themselves to a virtually infinite array of heterogeneous sources of information. With such a huge resource at the fingertips of practically anyone with an internet connection, never before has it been more relevant to investigate means to deal with the potential information deluge by employing intelligent approaches to automating the process. Data mining is capable of extracting patterns in large corpuses of data that would be unfathomable for a human to work with, yet automated processes alone are insufficient, as the way in which they work to retrieve "interesting" features for their human users must, by definition, be user-driven. Rather than for the prescription and processing stages to act distinctly, a mixed-initiative approach, integrating them simultaneously, has several advantages in communicating to the human the results of the automated process as well as allowing the user direct manipulation with both the data and the automated process itself [2].

The field of interactive data mining [3],[4],[5] focuses on interfaces which can facilitate this form of bidirectional communication between automated processes and human, primarily consisting of three key goals: (1) *data visualization* - exploiting humans' pattern-matching abilities aiding understanding and to discover "interesting"

M.J. Smith and G. Salvendy (Eds.): Human Interface, Part I, HCII 2009, LNCS 5617, pp. 384–393, 2009.

features; (2) *direct manipulation* - to allow the user to specify more easily what they seek in terms of the automated processes (e.g. selecting a subset of the dataset for the automated data mining processes to analyze at a more fine-grained level); (3) *visibility of both the process and result of the data mining to the user* - minimizing misinterpretation [3],[6]. We have identified a fourth important goal, closing the loop once more between human and machine: *offering a means for the user to feedback a performance measure of the data mining to the system* - thus tailoring the way that the automated processes operate and visualizing this *adaptation* in real-time. In addition to specifying which data for the automated mining to act upon, this fourth element influences *how* the automated processes take place.

There has been a significant amount of work on applying data mining techniques in the context of information structuring, however the focus of this research has often been on the performance of the data mining algorithms and less about the human-computer interface and user experience [7],[8],[9],[10]. Commercial off-the-shelf products such as Verity, Gammaware, Vivisimo and Inxight offer tools for automating taxonomy generation and maintenance, however, the majority of these products are relatively heavyweight in terms of their processing and the data mining and machine learning employed typically remains a "black box" to users. Although [11],[12] in particular provide learning the categorization behavior from the user, the way in which this is accomplished lacks the visibility and flexibility we argue is necessary to ensure users understand and trust the system during this process. The commercial approaches mentioned above all provide a visual interface to aid understanding of the relationships between information as well as allow the user to explore, analyze and exploit it. However, this visualization is limited to the visualization and manipulation of the resulting taxonomy and facilities to increase the users' awareness of the automated data mining and categorization processes are neglected.

We introduce an interactive data mining environment called *Cyclone* which has an emphasis on clustering and structuring the information inputted, generating a taxonomy. The primary method of visualization in Cyclone is similar to that of [3],[13],[14], consisting of a force-based graph of nodes representing information points; the forces and subsequent distances between the nodes reflecting their derived similarity. We evaluate to what extent this approach is an effective means to address the four elements identified above, conveying the result of machine-based data mining employed but also acting as a means to *adapt* the data mining process based on direct manipulation by humans. Using adaptive visual clustering, the interface presents to the user how the automated data mining is based on both unsupervised (statistical) measures of the data in combination with a history of user actions as learnt through their interaction. This is particularly relevant for the information structuring task we focus on, where the categorization of information might be highly user-specific, as well as the interpretation of the data dynamic as the task or information changes. By introducing a supervised learning component into our system, and this being trained by the direct manipulation of data by the user in an intuitive manner, our system is able to adapt its automated data mining processes so that they are increasingly accurate over time.

This paper is structured as follows. Section 2 describes the conceptual framework of Cyclone and presents the proposed visual clustering based on spring forces. In Section 3 we introduce the adaptive visual clustering paradigm, which is capable of reacting to variations in users' categorization behavior and adjusting the forces on the objects in real-time to reflect this. Section 4 presents initial results obtained from multi-user experiments. Finally, Section 5 concludes with a summary and future work.

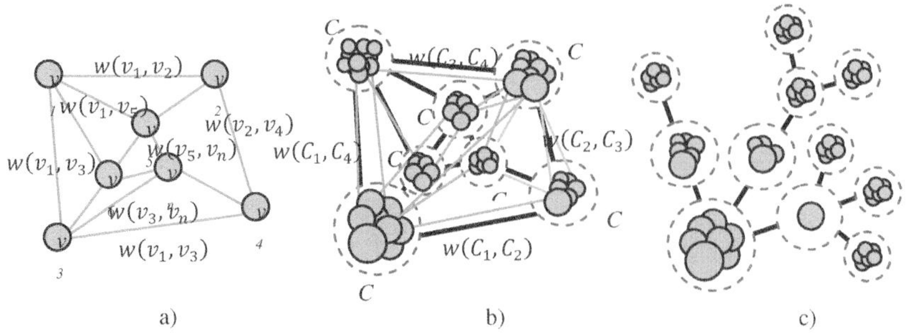

Fig. 1. The visual representations of a) an edge-weighted graph between visual objects, b) an edge-weighted graph between clusters and categories and c) a generated taxonomy T.

2 The Conceptual Framework

Cyclone is an intelligent agent-based visual framework offering a means for the user to exploit, analyze and categorize unstructured information from various sources into a more structured and manageable form [15]. The main strength of Cyclone is its capability to provide users with continuous visual feedback in real-time so as to increase their *understanding*, *confidence* and *trust* in the functional processes of the system, from clustering visualization to categorization.

The conceptual framework of Cyclone is based on the ideas provided by graph theory, which uses mathematical structures to model relations between a pair of visual objects, as well as visual data mining techniques, involving force-based real-time adaptive clustering and interface design. The adaptive learning system for clustering and automatic categorization is based on a single-layer feed-forward neural network utilizing a Hebbian rule style weight calculation.

The following subsections present and describe each of the aforementioned framework components.

2.1 Definitions

Definition 1. Let $G = (V, E)$ be a *graph of visual objects*, where V is the set of visual objects and E the set of edges, with $|V| = N$ and $|E| = M$. A visual object v_i has up to $N - 1$ edges (without self-loops) to visual objects of G. Each visual object v_i is equipped with a set of information, such as Name n, URL url, Description d as well as a set of metadata, i.e. tags t describing the content and context of the service or information it represents. An *edge-weighted graph* $G^* = (V, E, w)$ is a graph of objects, where V is the set of visual objects, E the set of edges and w the set of edge weights respectively. The edge weight is defined as $w: E \rightarrow \mathfrak{R}$, where $w \in [0,1]$ (see Fig. 1a).

Let $S \subseteq V$, the *subgraph* of G^* induced by S so that $G[S] = (S, E_s, w_S)$ where $E_s = \{e = (u, v) \in E \text{ such that } u \in S \text{ and } v \in E\}$. A subgraph S can also be regarded as a *cluster* or a *category* C_x.

Definition 2. Let $A = (V, w)$ be a *similarity matrix*, extracted from $G^* = (V, E, w)$, where V is the set of visual objects, w the set of weights of a pair of visual jects v_i, v_j, where if $i = j$ the $w(v_i, v_j) = 1$.

Likewise, let $A^* = (C, w)$ be a *cluster or category-based similarity matrix*, where C is the set of clusters/categories and w the set of weights of a pair of clusters/categories C_x, C_y where if $x = y$ the $w(C_x, C_y) = 1$ (see Fig. 1b).

Definition 3. A clustered graph (taxonomy) is an ordered quadruple $T = (V, C, A, A^*)$, where V defines the set of visual objects, C is the clusters set and A and A^* are the set of edges (and their corresponding weights) among the visual objects and clusters respectively (see Fig. 1c).

2.2 Visual Clustering

Cyclone's visual clustering component consists of the following steps [15]:

1. *Initializing the visual environment.*
   Cyclone loads the information and services from various given data sources and produces the initial display. The information is represented as a visual node and color-coded according to the data source it comes from. Cyclone offers a rich set of visualization and graph layout methods allowing the user to program and personalize their visual environment as seen fit.
2. *Calculating similarity among visual objects*
   Before Cyclone can start with the clustering process, it calculates the similarity among the visual objects based on the provided tag set. The cosine similarity function is used to obtain the degree of relevance between a pair of objects. The objects are considered identical if they both share the exact set of tags. The similarity and dissimilarity between objects in Cyclone is calculated using the following equation:

$$similarity\left(v_i, v_j\right) = \frac{\sum_{t_j \in T} v_i, v_j}{\sqrt{\sum_{t_j \in T} {v_i}^2}\sqrt{\sum_{t_j \in T} {v_j}^2}} \quad (1)$$

   Cyclone computes a similarity measurement for each pair of objects to obtain the similarity matrix A of the information space.
3. *Visual clustering based on Spring forces*
   Cyclone utilizes a force-directed layout to perform visual clustering where spring forces (simulated physical forces [16]) operate between the visual objects. Depending on the position and $similarity\left(v_i, v_j\right)$ of the visual objects, the nature of employed forces is determined, i.e. they may be either attractive or repulsive and to varying degrees. The strength of the forces for a pair of visual objects is derived from the similarity matrix A.

The following describes the forces applied to a pair of visual objects v_i and v_j (inter-cluster):

$$F_{v_{i,j}} = F^r_{v_{i,j}} + F^a_{v_{i,j}} \tag{2}$$

with

$$F^r_{v_{i,j}} = \sum_{j=1 \text{ and } j \neq i}^{N} (r - (similarity\,(v_i, v_j) * r) \tag{3}$$

and

$$F^a_{v_{i,j}} = \sum_{j=1 \text{ and } j \neq i}^{N} (similarity\,(v_i, v_j) * r) \tag{4}$$

where N is the total number of visual objects and r a predefined maximum radius of a pair of objects within G.

In addition to the inter-cluster forces, Cyclone defines the intra-cluster forces which act between groups of visual objects forming clusters, which is described as:

$$F^a_{C_{x,y}} = \sum_{x=1 \text{ and } y \neq x}^{N} similarity(\overline{C_x}, \overline{C_y}) * r \tag{5}$$

where N is the total number of clusters and $\overline{C_x}$ *and* $\overline{C_y}$ the centre of the clusters.

The equilibrium state of the forces is found when

$$\sum_{i=1}^{N} F_i = \sum_{i=1}^{N} (F_i^r + F_i^a) = 0 \tag{6}$$

Once the forces are calculated, Cyclone assigns them to each visual object and executes them in a physical model, allowing the interactions of the forces to determine the resultant positions of the objects. The objects interact and eventually settle into a low-energy steady state representing an emergent clustering.

4. *Manual or Auto-Categorization*

Cyclone's categorization process involves the specification of a hierarchical system of categories (see Fig. 1c) as well as placing information into the nodes of this hierarchy to ultimately generate a personalized taxonomy. Cyclone offers two alternative modes: (1) *Manual Categorization*, where the user explicitly selects and assigns a subset of visual objects into categories and (2) *Auto-Categorization*, where Cyclone's intelligent agent learns the categorization habits of the user and automatically performs the categorization of their behalf. For the latter, the agent employs a single layer feed-forward neural network utilizing a hebbian style weight calculation, which creates a feedback loop between the user and the system. The agent maps tags to categories by continuously updating the edge weights each time the user categorizes a set of objects into a category. The weight increases if the co-occurrence of tags within a category increases and conversely, the weights decrease over time if previous categorization patterns are not repeated or are frequently changed. The two modes of operation may be either used exclusively by users as they see fit or, more typically, in combination creating a mixed-initiative experience.

For a more detailed description on the intelligent automatic categorization learning process of Cyclone, please refer to [15].

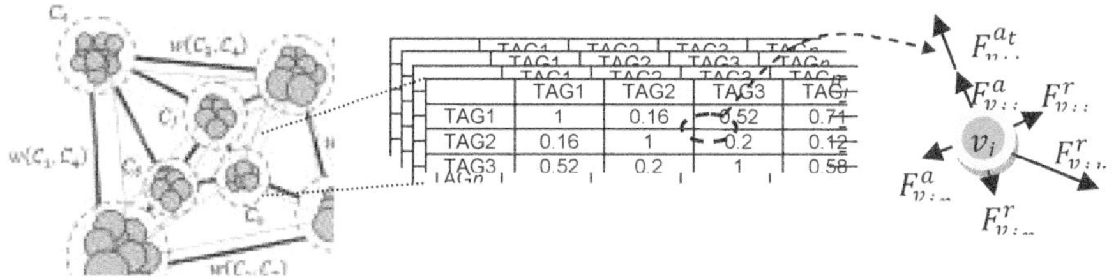

Fig. 2. The tag similarity calculation process for adaptive visual clustering

3 Adaptive Visual Clustering

A feedback loop between Cyclone and the user during the categorization process, taking a user's actions into account, updates the learnt model and visual clustering in an unobtrusive fashion. Adaptation of the model takes place each time the user assigns objects into categories and the visual clustering changes to reflect the categorization preferences of the user by updating and adjusting the forces on the visual objects.

The algorithm for the visual clustering adaptation is achieved by applying the following steps of *tag similarity calculation* (see Fig. 2) [15]:

- Step 1: Instantiate an edge-weighted graph for tags $G^* = (t, E, w)$ where t is the tags set, E the set of edges, and w the set of weights of the edges between a pair of tags. In addition, instantiate the similarity matrix $A^* = (t, w)$ based on G^*.
- Step 2: Set the edge weight $w_{i,j}$ for all possible pair of tags to zero, where $i \neq j$, and to 1, where $i = j$ (represents a self-loop which cannot be adjusted).
- Step 3: For categorization (manual or automatic), do the following:
  - o Extract from the selected objects of the *same category* C_x the tags set $t_{C_x} \subseteq t$. For every tag combination t_a, t_b ($a \neq b$) of t_{C_i}, use the following equation to calculate the w_{t_a, t_b}:

$$(w_{t_a t_b})^k = (w_{t_a t_b})^{k-1}(1 - \tau)\delta \tag{7}$$

    where $(w_{t_a t_b})^{k-1}$ is the degree of similarity (weight) before applying Eq. 7 and τ is the decay value which is set to 0.05. The decay τ is an empirical value, which prevents the weights to increase endlessly and δ the pre-defined (and empirically selected) learning rate (0.1).
- Step 4: Adjust the strength of the forces for every visual object F_i^a using Eq. 5:

$$F_{v_{i,j}}^{a^*} = \sum_{j=1 \ and \ j \neq i}^{N} similarity(v_i, v_j, w) * r$$

where $similarity\ (v_i, v_j, w)$ (8)

$$= \frac{\sum_{t_j \in T} v_i, v_j, w_{t_a, t_b}}{\sqrt{\sum_{t_j \in T} v_i w_{t_a, t_b}}^2 \sqrt{\sum_{t_j \in T} v_j w_{t_a, t_b}}^2}$$

The change in force strength causes the visual objects to rearrange and migrate to other positions so that the visual representations of the clusters adjust as a result.

4 Initial Experimental Results

We have conducted several multi-user experiments using real-data from different information sources and application context in order to validate the efficiency of Cyclone. Due to space constraints we focus on the qualitative aspects of the results in this paper.

The experiments were conducted on a group of researchers. Some of them were familiar with the task and others were non-experts and had no prior experience, however, neither group had used Cyclone before. Our hypothesis was that through using our proposed approach, in particular in its ability to visualize feedback in real-time, there wouldn't be a bias between the two groups in terms of their ability to familiarize themselves with the interface and use it efficiently to organize the data.

We chose a dataset consisting of 500 well-known songs (http://top500songs.blogspot.com) where each song was assigned 10 tags gleaned from the Audioscrobbler web service (http://www.audioscrobbler.net /data/webservices). A URL was also associated with each song such that the user could preview the song during the experiments.

The experiments were limited to 30 minutes, where the users were presented with the Cyclone interface (Fig. 3) showing a subset of the dataset read in and a visual clustering representation of it. The users were then asked to categorize the songs as they see fit through selecting and assigning of the visual objects.

An initial explanation stage was offered by the experimenter to ensure that the user was aware of the functional components of the interface. This stage, however, typically lasted only a couple of minutes as prescribed by how confident the users relayed they were with achieving their assigned task. It was observed that through successive categorizations by the users, the speed at which they categorized the objects increased. This effect is illustrated in Fig. 4 where the gradient of the curves gradually increase over time because of both an increased number of categorization actions, as well as a higher number of assignments per categorization.

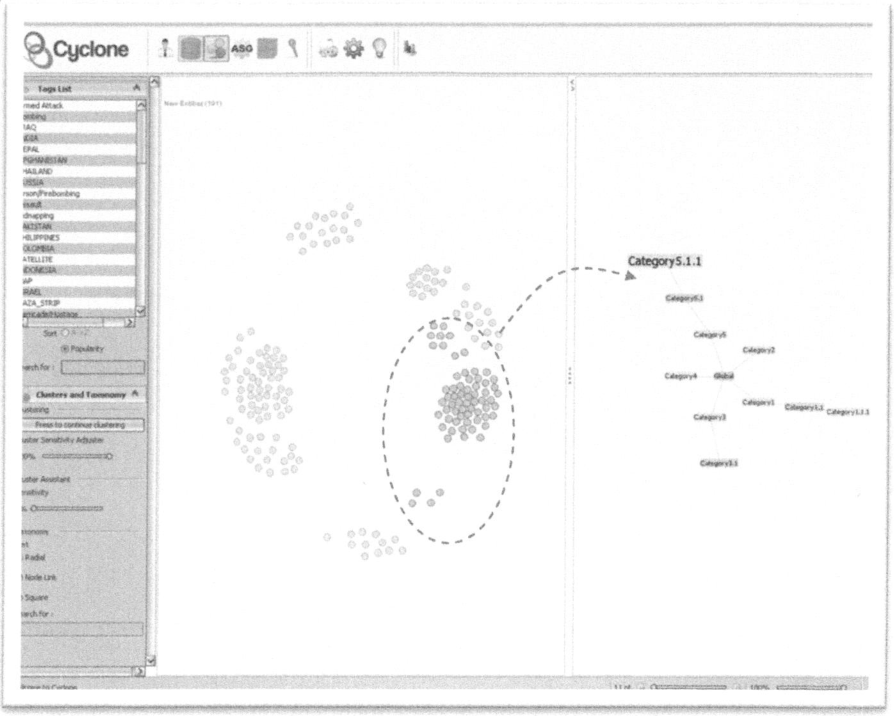

Fig. 3. The Cyclone Interface. It consists of three panels (incl. Tags Panel, Visual Clustering Panel and Taxonomy Panel) which the user can use to explore, analyze and exploit information. After the information are loaded (visual nodes), the user can initiate a categorization in a select-and-assign principle so that over time a personalized taxonomy is generated. Every time a categorization is performed and/or a change occurs, the visual clustering algorithm adjusts the force weights to adapt the visual clustering.

Clearly the effect above can be explained to an extent due to increased familiarization with the interface and dataset, but we argue that the speed at which this occurs is due in part by the real-time feedback offered by the interface. An important insight gained through observation and a post-experiment interviewing stage relates to the semantics of the categorization scheme chosen by users. A common attitude to dealing with categorizing the data was to, when encountering an object (song) which was deemed uninteresting, to assign it immediately to a category reflecting this. Unbeknown to the user at first, this behavior prompted the adaptive clustering to segregate those uninteresting objects from the interesting ones by visually migrating the objects on the screen. This allowed the users to concentrate more easily on the latter, creating more fine-grained categories for those objects which they were either more familiar with or valued more highly. This highlights the importance of the four primary goals for interactive data mining identified, but in particular the visualization of the system's interpretation of the data whilst incorporating users' preferences in real-time.

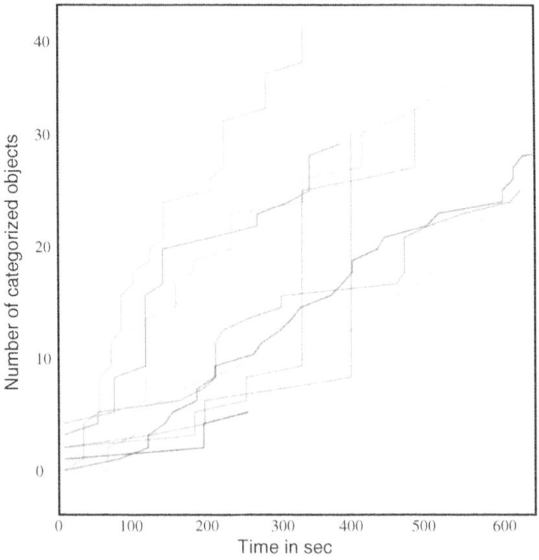

Fig. 4. User categorization process

5 Conclusions

The work presented discusses the case for introducing the provision to allow users to adapt the automated processes in an interactive data mining environment. In particular, we demonstrate the effectiveness of adaptive clustering to assist users and improve their ability to perform an information structuring task. Experimental investigation highlighted the importance of such real-time visual feedback to users and we believe that it is this visibility in particular which enables users of all skill levels to effectively tune the system to their benefit.

Current work involves applying the Cyclone system to various digital data sources such as Emails, web bookmarks, photo libraries and live video feeds to augment the information retrieval capabilities of users [17],[18].

Acknowledgments. The research described in this paper was conducted in the context of the Hyperion cluster project funded by the UK MoD Data and Information Fusion Defense Technology Centre (DIF DTC) research program and British Telecommunications plc (BT). We would also like to thank the members of the Centre for Information & Security Systems Research, in particular the Future Technologies Group at BT UK for their invaluable contributions and support, and also all the participants of the experiments presented in this paper for their involvement and constructive feedback.

References

1. Gants, J., Chute, C., Manfrediz, A., Minton, S., Reinsel, D., Schlichting, W., Toncheva, A.: The Diverse and Exploding Digital Universe – An Updated Forecase of Worldwide Information Growth Through 2011, IDC White Paper (2008)
2. Horvitz, E.: Principles of Mixed-Initiative User Interfaces. In: ACM conference on Human Factors in Computing Systems (CHI 1999), pp. 159–166 (1999)
3. Beale, R.: Supporting serendipity: Using ambient intelligence to augment user exploration for data mining and web browsing. J. of Human-Computer Studies 65, 421–433 (2007)
4. Ferreira de Oliveira, M.C., Levkowitz, H.: From Visual Data Exploration to Visual Data Mining: A Survey. IEEE Transactions on Visualization and Computer Graphics 9, 378–394 (2003)
5. Zhou, Y.: On Interactive Data Mining. In: Encyclopedia of Data Warehousing and Mining, vol. 2, pp. 1085–1090. Idea Group Inc. (2008)
6. Shneiderman, B.: Inventing Discovery Tools: Combining Information Visualization with Data Mining. Information Visualization 1, 5–12 (2002)
7. Xu, R., Wunsch II, D.: Survey of Clustering Algorithms. IEEE Transactions on Neural Networks 16, 645–678 (2005)
8. Gates, S.C., Teiken, W., Chen, K.-S.: Taxonomies by the numbers: building high-performance taxonomies. In: 14th ACM International Conference on Information and Knowledge Management, pp. 568–577 (2005)
9. Sanchez, D., Moreno, A.: Automatic Generation of Taxonomies from the WWW. In: Karagiannis, D., Reimer, U. (eds.) PAKM 2004. LNCS, vol. 3336, pp. 208–219. Springer, Heidelberg (2004)
10. Wetzker, R., Alpcan, T., Buckhage, C., Umbrath, W., Albayrak, S.: An unsupervised hierarchical approach to document categorization. In: 2007 IEEE/WIC/ACM International Conference on Web Intelligence, pp. 482–486 (2007)
11. Lan, S.C., Al-Hawamdeh, S.: Taxonomy-Building Tools: An Investigative Study. J. Informatoin & Knowledge Management. 2, 63–77 (2003)
12. Wolin. B.: Method for Automatic Categorization of Items. US Patent 1 160 52 (2001)
13. Rostoker, C.: Interactive Visualization of the Market Graph. Technical report, University of British Columbia (2005)
14. DesJardins, M., Ferraioli, J.: Interactive Visual Clustering. In: 12th International Conference on Intelligent User Interfaces (2007)
15. Duman, H., Healing, A., Ghanea-Hercock, R.: An Intelligent Agent Approach for Visual Information Structure Generation. In: 2009 IEEE Symposium on Intelligent Agents, Nashville, Tennessee (2009)
16. Eades, P., Huang, M.L.: Navigating Clustered Graphs using Force-Directed Methods. J. Graph Algorithms and Applications 4, 157–181 (2000)
17. Ghanea-Hercock, R., Gelenbe, E., Jennings, N.R., Smith, O., Allsopp, D.N., Healing, A., Duman, H., Sparks, S., Karunatillake, N.C., Vytelingum, P.: Hyperion - Next Generation Battlespace Information Services. Computer Journal 50, 632–645 (2007)
18. Healing, A., Ghanea-Hercock, R., Duman, H., Jacob, M.: Nexus: Self-organising Agent-based Peer-to-Peer Middleware for Battlespace Support. Defence Industry Applications of Autonomous Agents and Multi-agent Systems, pp. 1–13 (2008)

An Empirical Analysis of Personal Digital Document Structures

Sarah Henderson and Ananth Srinivasan

Department of Information Systems and Operations Management
University of Auckland
Auckland, New Zealand
s.henderson@auckland.ac.nz, a.srinivasan@auckland.ac.nz

Abstract. Hierarchies have long been used as useful structuring mechanisms for organizing and managing documents. This study looks at the problem of personal digital document management in the context of knowledge workers. We study and document strategies that users employ to manage the complexity imposed by the volume and variety of personal digital documents. Exploratory research was conducted, analyzing the file systems of 73 knowledge workers using Microsoft Windows in a university setting. The empirical results of this are presented, and compared to a previous study that examined the file systems of 11 users.

Keywords: Personal document management, personal information management, document organization, file system structure.

1 Introduction

Knowledge workers spend much of their time creating and using digital information. As well as being overloaded with a flood of information and data coming at them from all directions, they are also standing in a rising tide of information of their own making: the morass of reports, memos, articles, notes, presentations, graphics, contacts, web URLs, emails, tasks and appointments that they have been slowly but surely creating and accumulating on their computer. We refer to this collection as personal digital documents. While finding information in databases and on the web is becoming easier, finding information located on a local repository such as their own hard drive is becoming increasingly difficult as a consequence of user driven activity.

Many users will spend a great deal of time using software tools to locate, acquire, manage, communicate, process and otherwise interact with this growing plethora of digital information. Because these tasks occupy such a large amount of their time, it is important that these software tools are usable, that is, they are properly designed to effectively support information management activities. Given the ubiquitous nature of these activities, even small improvements in the usability of the tools could result in a large productivity gain for knowledge workers.

There are numerous digital information types that knowledge workers typically engage with: web pages, email, documents, images, sound, video, memos, contacts, appointments and tasks. Each of these different types of digital information has its

M.J. Smith and G. Salvendy (Eds.): Human Interface, Part I, HCII 2009, LNCS 5617, pp. 394–403, 2009.

own particular features and requirements. Due to the relative newness of web, email and multimedia technologies, management of these has been the focus of many research efforts. However, the older and more basic task of managing "ordinary" documents has gone relatively unstudied.

A hierarchical structure as a mechanism for storing and managing documents is a well entrenched systems paradigm. Most people store and manage documents through a user interface that exploits the structure of hierarchies [1]. These tools allow people to recursively create folders and place documents within folders by attaching meaning to the hierarchy thus created. Using this simple containment mechanism, people can build up a large hierarchical structure of folders. This basic paradigm has not changed in the decades since its introduction, although the user interface to it significantly improved with the widespread introduction of graphical user interfaces.

A basic principle of user interface design is that the design of a tool should be thoroughly grounded in an understanding of how the user works, what tasks they perform and how those tasks are carried out. However, with personal digital document management, very little research has been done with regard to investigating how people actually manage their documents and what the requirements are for document management tools. This study attempts to address this knowledge gap by empirically examining document structures that knowledge workers create for themselves.

2 Background

Previous research on personal document management (and personal information management in general) can be divided into two main approaches. The first strand of research examines how people manage various forms of personal information, and the second strand develops and tests new user interfaces and systems for the management of personal information.

In considering how people manage their documents, Lansdale [2] identified the trade-off that exists between the effort spent filing a document when it is first stored and the effort required to find it again later. Many subsequent studies of both email and paper filing systems have found there are two general filing strategies which people adopt in response to this trade-off: filing and non-filing [3-6]. A person adopting a filing strategy generally tries to create a folder structure, and makes an ongoing effort to try and file new information into this structure on a regular basis. They rely on the structure to help them locate documents again, typically using browsing (location-based search) in preference to using a search tool. People adopting a non-filing strategy tend not to maintain much of a formal organization structure. Instead they rely primarily on browsing fairly unstructured lists, or using search tools to locate information when needed.

These two strategies suggest two approaches to improving tool support for information management: improving the efficiency and effectiveness of using an organization structure or improving search tools.

To support the people inclined towards a filing strategy, another strand of research has worked to create new systems that are different from the currently predominant hierarchical containment approach. Most of these are based around a particular dimension of the information that is assumed to be primary. For instance, Lifestreams [7] is based on the premise that the most important dimension on which to organize things is time. TimeScape [8] also includes time as a primary dimension, but includes a spatial layout as well, while the Taskmaster system [9] is based on studies of email users that found that task or project is a common organizing principle. Along a similar vein, the Placeless Documents project [10] doesn't impose any type of structured organizing scheme at all, but allow the user to give attributes and attribute values to documents, which can then be used to search and group documents for viewing.

Future work on this strand of research could be helped by having more information about exactly how users currently structure their documents in the relatively unguided context of a hierarchical file system. Very little research has looked specifically at the document structures that people actually create to manage and organize their documents. Studies of paper filing systems in 1982 showed that people tend to create simple classification schemes, rarely more than 2 levels deep [3], however given the physical constraints of folders and filing cabinets, it is unlikely that a more layered system could be developed. The first study to examine computer file use and organization in 1995 [11] found that the study participants did not generally use directory structures at all, although some archived their files by placing them onto separate floppy disks. The only recent study to look specifically at file structures was conducted on 11 users of the Unix file system by Gonçalves and Jorge [12]. They looked at the total number of files and folders, the width and depth of the structure, as well as balance, and the distribution of file types.

To further this research, as part of a larger study into the personal digital document management practices of knowledge workers, we took snapshots of their file systems in order to analyse the document structures they created. The following section will describe how the study was conducted, followed by an analysis of the results. We then present a discussion of the relevance for the design of user interfaces for document management and give our conclusions and suggestions for further research on this subject.

3 Method

As part of a larger study into personal digital document management (including interviews and a survey), a snapshot of the file system of knowledge workers was taken (using custom-written software). The participants were all employees of a large university, drawn from all academic and supporting business units, and at all levels of the hierarchy. All were users of the Windows XP[1] operating system. We thought that the university setting was particularly helpful in understanding the dynamics of the problem given the proliferation in quantity and variety of digital documents that are typically found in such an environment.

The snapshot software instructed the participants to select all the locations where they store documents. The default locations were the My Documents and Desktop subdirectories, although participants could easily remove these and add other locations where they kept their document files. The snapshot was taken on their primary

[1] Microsoft Windows® is a registered trademark of Microsoft Corporation. Henceforth it will be referred to as Windows.

work computer, and could include network locations and flash memory devices, but not other desktop or home computers.

The information captured by the file system snapshot software includes the name, extension, date created, date last accessed and date last modified of every file and folder, as well as the structure of the folders and files. The data was checked and cleaned of any system folders or multi-user shared folders.

A total of 78 participants completed the file system snapshot. However, five of those only included the default locations of My Documents and Desktop, despite indicating in the other part of the study that their primary storage location was a network drive or removable drive. As a consequence, these participants only had a handful of files in the snapshot. These participant's snapshots were removed from this analysis, leaving a total of 73 snapshots for analysis. 34 participants were male and 39 were female. 48 had primarily academic responsibilities while 25 had general administrative responsibilities.

4 Results

4.1 Overall Size

The size of the document collection has an impact on the appropriate software support, since software to support the task of managing a few hundred files is going to be different from managing thousands or tens of thousands.

The mean number of files observed in the document folders was 5,850. However, the number of files ranged from a minimum of only 100 files, to a maximum of 33,902 (standard deviation of 7,605). As Fig. 1 shows, the distribution is significantly right skewed, with a median of 2,754 and a skewness statistic of 2.26.

The average number of folders was 628, with a standard deviation of 860. The distribution was also right-skewed, with the median number of folders being only 350. The smallest number of folders observed was 11, and the largest was 4,694. The participant with the highest number of folders was not the same person who has the highest number of files. As shown in Fig. 1 there are five participants with over 2,000 folders.

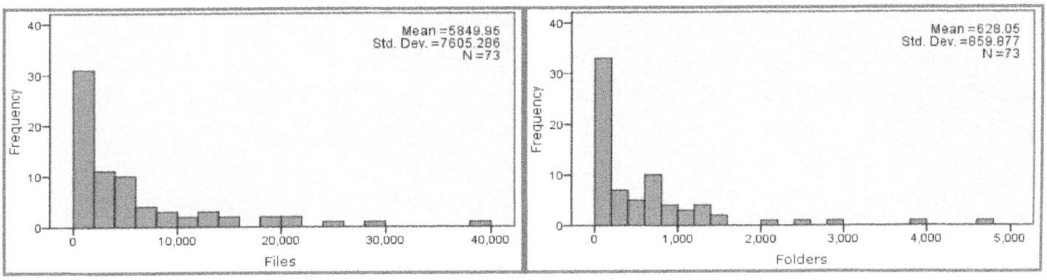

Fig. 1. Distribution of total number of files (left) and folders (right) in snapshot

As would be expected, there is a significant correlation between the number of files and the number of folders a person has in their file system (correlation coefficient

0.88) [2]. There is no correlation between the number of files or folders a person manages and any of the demographic data collected (age, gender, academic or general staff status, department, length of time they have been working in the same field, or duration of employment).

4.2 Tree Characteristics

Trees vary in several dimensions. Trees can be shallow or deep, broad or narrow, and can contain varying numbers of files in each folder. The maximum depth for each participant is the depth of the deepest folder in their document collection.

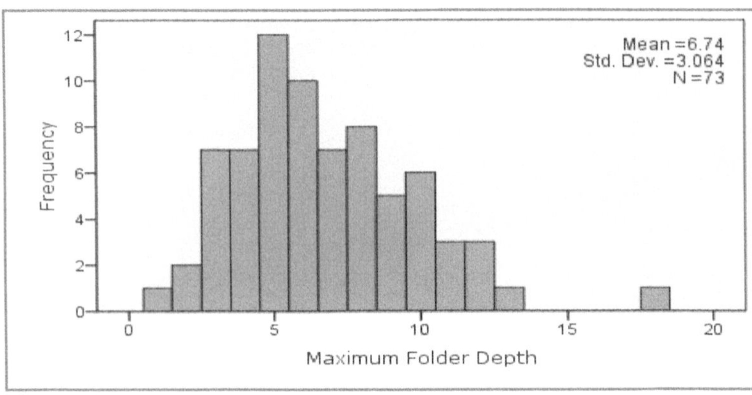

Fig. 2. Frequency distribution of the maximum depth of each document structure

There is significant variation in the (maximum) depths of folder structures. The shallowest structure was only 1 level deep, while the deepest was 18 levels deep. The average of the maximum depth across all participants was 6.8, with a standard deviation of 3.1. There is a significant correlation between the number of files a person has and the average depth of their file system ($r = 0.78$).

The width of a tree is determined by the average number of subfolders in each folder. On average 74% of folders did not contain any subfolders at all, only (possibly) files. These are considered leaf folders, and are not included in the average subfolders metric. The interior (non-leaf) folders by definition must contain at least one subfolder. The mean number of subfolders per folder was 4.1, with a standard deviation of 1.3. The highest average observed was 9.5, and the lowest was 1.8 subfolders per folder. The distribution of this metric is shown in Fig. 3.

There is no correlation between the average number of subfolders per folder and the total number of files and folders in the file system, so both small and large systems do not differ in their average breadth. There is also no significant correlation between the average depth of the tree and the average number of subfolders. Since depth does vary with the total size, this would imply that the bushiness of the tree varies independently of these factors.

[2] All correlations reported are statistically significant at the 0.01 level unless otherwise stated.

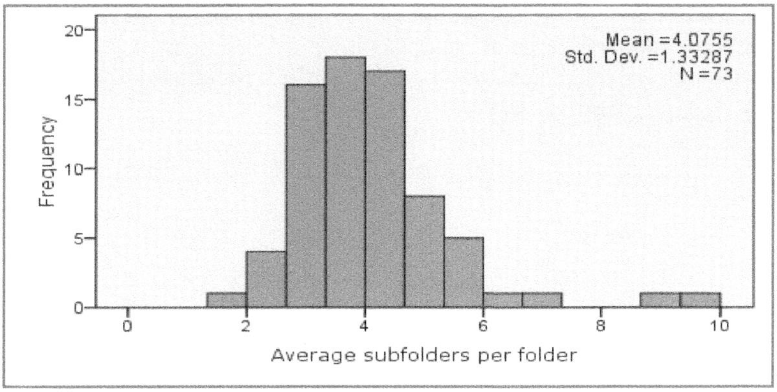

Fig. 3. Measure of bushiness - frequency distribution of number of subfolders per folder

Rather than comparing the average values for each user, we can also do the comparison at the individual folder level. There is a high average number of subfolders at the root of the tree (approximately 9), which sharply drops off (to less than 2) by two or three folders down.

In order to compare the result of this study directly to the study conducted by Gonçalves and Jorge [12], we also calculated branching factor as a metric of bushiness. The average branching factor in this study was 1.93. This ranged from 1.27 to 2.97 and had a standard deviation of 0.34.

As would be expected for two measures of bushiness, the branching factor and the average number of subfolders are correlated (r = 0.5). In common with the average subfolders metric of bushiness, there is no correlation between the branching factor and the total number of files and folders in the file system.

There is a significant negative correlation between the branching factor and the average depth of the tree (r = -0.36), indicating that wider trees tend to be shallower. There is also a positive correlation between the branching factor and the number of top level locations (r = 0.41). This is expected, since the locations essentially represent the top level of tree branching.

One of the key differences between the branching factor and the average subfolders is that branching factor assumes a perfectly even tree, whereas the average number of subfolders is affected by the tree's unevenness.

The more files people store in each folder, the more 'leafy' their folder tree becomes. Leafiness is the average number of files per folder. Higher leafiness indicates a denser tree. The average number of files per folder across all file systems was 11.1 (standard deviation 7.8). The highest number of files observed in a single folder was 1168.

The least leafy file system had an average of 4.5 files per folder, and the leafiest averaged 64.3 files per folder. However, this was a significant outlier, with the second leafiest file system averaging under 30 files per folder.

There is no significant correlation between the average number of files per folder and the overall number of files, nor with the average depth or bushiness of the document structure. However, similar to bushiness, there is a significant correlation between the *maximum* number of files per folder and total number of files (r = 0.56).

As shown in Fig. 4, the average number of files per folder is highest at the top levels of the tree, and then drops off sharply. It is fairly constant at levels 1 to 5 of the tree and then tapers off. Note that the average file system has a maximum depth of 6.8 levels.

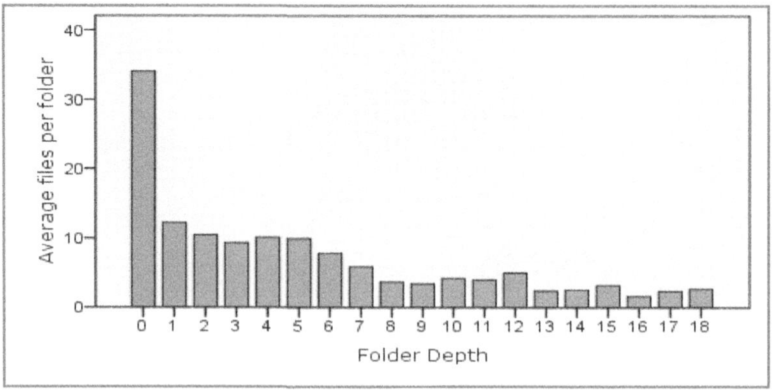

Fig. 4. Leafiness vs Depth - how the average number of files in a folder varies with the depth of the folder

To assess how even the distribution is across the tree, the standard deviation of the number of subfolders was used as a measure of balance. The lower the standard deviation, the more evenly balanced the tree. The average balance was 5.6, with a standard deviation of 3.5. While this indicates that most trees are fairly balanced, there is one significant outlier having a standard deviation of 27.9. This person has a relatively small file system of 628 folders in total. While most of their folders only have one or two subfolders, they also have folders that contain 105, 142 and 151 subfolders, giving them an extremely unbalanced tree.

There is no correlation between the balance of the tree and the overall size or the number of top level locations. Nor was there any relationship with the depth of the tree.

There is a statistically significant correlation between the balance of the tree and the bushiness, using both the average subfolder metric ($r = 0.73$) and the branching factor metric ($r = 0.42$). This would indicate that trees that are wider on average also tend to be less evenly balanced than narrower trees.

In addition to assessing how balanced the folder structure is, we can also examine how evenly distributed files are throughout the tree. The standard deviation of the number of files in a folder was used as a measure of balance. The average file balance was 23.4, with a standard deviation of 29.5. As with the folders, this was also considerably right skewed due to one outlier. This participant has 966 out of their total 1028 files in the My Documents folder itself. They have not created any folders to structure these documents, and do not appear to have made use of any of the system created folders to organise these documents.

There is no correlation between how evenly the files are distributed and the balance of the tree structure itself. There is also no correlation between the file spread and the overall size or depth of the document structure.

There is a statistically significant correlation between the file balance of the tree and the leafiness (r = 0.93). This would indicate that trees that are wider on average also tend to be less evenly balanced than narrower trees.

Empty folders are a potential sign of inefficiencies in the file structure, since the participant has expended effort created those folders but then has not made use of them. Only 3 out of the 73 file systems did not contain a single empty folder. The highest number of empty folders was 610. The mean number of empty folder was 37.8 (s.d 83.0). The distribution of the number of empty folder was extremely right-skewed (skewness 5.4), with a median number of empty folders being 13.

Since the number of empty folders can be expected to increase with the size of the file system, the proportion of empty folders is perhaps more important. The emptiest file system had 47.6% of the folders being empty.

Most file systems had only a small proportion of empty folders, with the mean proportion of empty folders being 7.9%.

4.3 Duplication

A very efficient system is likely to contain a low level of duplicated folder and file names. A measure of the proportion of duplication can be calculated from the number of non-unique files divided by the total number of files. Duplication can be calculated separately for files and folders. This measure of duplication will only reflect the fact that multiple files or folders are named identically. These files or folders may be exact copies of each other, or they may have entirely different content.

The mean level of file duplication was 21.8%. This means that on average, 21.8% of the documents in the file system have the same name as another file. The amount of file name duplication ranged from 0.4% to 60.4%. The level of folder name duplication was slightly higher, with a mean of 23.5%, and ranging from 0 to 73.4%.

There is a significant correlation between the level of duplication and the overall size of the file system. The correlation between total number of files and file duplication had r = 0.61, and the number of folders and folder duplication were correlated with r = 0.65.

There is also a significant correlation between the level of folder and file name duplication (r = 0.79). One explanation for this might be that entire folders and their contents are being duplicated together.

5 Discussion

It is surprising how similar the results of this study are to the study performed by Gonçalves and Jorge [12], despite the fact that many of their participants used different operating systems. The mean number of files that people are managing was slightly lower than their study, but we found similar levels of individual variation in file system size.

We also had fairly similar results in terms of the document structures. We found trees that were on average slightly bushier and leafier, a little deeper and a little less balanced. The larger the tree, the deeper and more unbalanced it tends to be. We found directory trees to be slightly deeper on average than Gonçalves and Jorge found - 9.65 compared to their 8.45. However, we also noted that people tend to average a lower depth value - only 3.4 folders deep on average. In fact, most people's maximum depth is about 3 subfolders deeper than their average.

Gonçalves and Jorge found an average number of top level folders per locus of 2.75. This study differed from theirs in only considering one locus (work computer), so therefore the number of locations in this study is comparable to the number of top level folders in a locus. Our value of 3.4 was probably inflated by the fact that the snapshot software automatically included the Desktop. Many users would probably not have added it themselves if it wasn't suggested to them, and thus the figure might be lower if the users had freedom to choose their top level folders themselves.

Our average branching factor was slightly higher than that found by Gonçalves and Jorge, although well within one standard deviation of their value. However, the branching factor metric eliminates all the variability in the tree and assumes the tree is completely uniform. The average number of subfolders in a folder is a better metric, since it eliminates the leaf (empty) folders, and better reflects the actual internal structure of the tree.

In terms of visualizing small sections of the tree, no extreme techniques are required, since the tree structures are not particularly bushy or leafy.

While there were some interesting correlations to emerge, what is perhaps more interesting is correlations that were not present that might have been expected. For instance, it might be expected that participants would have a tendency to create either wide tree or deep trees. Thus there would be an expected negative correlation between depth and either bushiness or leafiness. However, no such correlation was found.

6 Conclusion

The use of hierarchies as structuring mechanisms is an inherent part of how we approach document management. With digital documents, the volume and variety that knowledge workers typically encounter present a set of issues relating to how they ought to be managed effectively. We know that a significant number of computer users rely on a structuring strategy to enable effective retrieval. This study is an attempt to observe and document the strategies that individuals employ to assist them with this task. This study examines the behavior of knowledge workers who exploit a hierarchical structure to manage their documents. The results show that users vary considerably in terms of how the hierarchy is employed to manage the complexity of the problem. The results of such studies will help us improve tools that are integral to computer systems to help users be more productive. While this study is descriptive in that it examines various dimensions of usage, the real value of such work lies in our ability to construct predictive models of usage. Such models will form the foundation of improved usability of tools that support personal digital document management.

References

1. Faichney, J., Gonzalez, R.: Goldleaf Hierarchical Document Browser. Australian Computer Science Communications 23(5), 13–20 (2001)
2. Lansdale, M.: The psychology of personal information management. Applied Ergonomics 19(1), 55–66 (1988)
3. Cole, I.: Human aspects of office filing: Implications for the electronic office. In: Human Factors Society Annual Meeting, Seattle, Washington, USA (1982)
4. Ducheneaut, N., Bellotti, V.: E-mail as Habitat: An Exploration of Embedded Personal Information Management. Interactions 8(5), 30–38 (2001)
5. Malone, T.W.: How do people organize their desks? Implications for the design of office information systems. ACM Transactions on Office Information Systems 1(1), 99–112 (1983)
6. Whittaker, S., Sidner, C.: Email Overload: exploring personal information management of email. In: CHI 1996 Conference on Human Factors in Computing Systems, Vancouver, Canada (1996)
7. Freeman, E., Gelernter, D.: Lifestreams: A Storage Model for Personal Data. SIGMOD Bulletin 25(1), 80–86 (1996)
8. Rekimoto, J.: Time Machine Computing: A time-centric approach for the information environment. In: UIST 1999 Symposium on User Interface Software and Technology, Asheville, North Carolina, USA (1999)
9. Bellotti, V., et al.: What a To-Do: Studies of Task Management Towards the Design of a Personal Task List Manager. In: CHI 2004 Conference on Human Factors in Computing Systems, Vienna, Austria (2004)
10. Dourish, P., et al.: Presto: An Experimental Architecture for Fluid Interactive Document Spaces. ACM Transactions on Computer-Human Interaction 6(2), 133–161 (1999)
11. Barreau, D.K., Nardi, B.A.: Finding and Reminding: File Organization from the Desktop. SIGCHI Bulletin 27(3), 39–43 (1995)
12. Gonçalves, D., Jorge, J.A.: An Empirical Study of Personal Document Spaces. In: Jorge, J.A., Jardim Nunes, N., Falcão e Cunha, J. (eds.) DSV-IS 2003. LNCS, vol. 2844, pp. 46–60. Springer, Heidelberg (2003)

Search Mathematical Formulas by Mathematical Formulas

Yoshinori Hijikata, Hideki Hashimoto, and Shogo Nishida

Graduate School of Engineering Science, Osaka University
1-3 Machikaneyama, Toyonaka, Osaka 560-8531, japan
{hijikata,nishida}@sys.es.osaka-u.ac.jp

Abstract. Users cannot search information by mathematical formulas as queries in existing search engines. This is because mathematical formulas are not expressed as a sequence of characters. Some formulas are expressed in a complex structure like fractional numbers and index numbers. We present a search engine for MathML objects using the structure of mathematical formulas. The system makes the inverted indices by using the DOM structure of the MathML object. We proposed three types of index. One type is constructed from some paths of the DOM structure and expressed in XPath. The other type is constructed by encoding the nodes in the same level in DOM structure. The third type is a hybrid method of them. This paper describes the experiment conducted to study the effectiveness of those indices.

Keywords: MathML, search engine, mathematical formula, index, inverted file.

1 Introduction

The number of documents including mathematical formulas on the Web is increasing. However, people cannot search web documents by a mathematical formula as a search query in current search engines on the Web. The current search engines allow us to use a sequence of characters for a search query. Mathematical formulas are not necessarily expressed in a sequence of characters. They are sometimes expressed in a structure like fractional numbers and index numbers. For an example, xy and x^y have the same expressions if they are expressed in a sequence of characters. Both of them include x and y and they are arranged in the same order. However the meaning of xy and the meaning of x^y are totally different. Therefore people cannot search mathematical formulas by using a search engine which only uses the kind of characters and the order of them. If we want to search mathematical formulas, the system must consider their mathematical structures.

MathML (Mathematical Markup Language) [2] is being introduced worldwide as a standard to express mathematical formulas on the Web. People can express the structure of the mathematical formula by exploiting the hierarchical structure of XML. Nakanishi et al. have proposed a search method of mathematical formulas for MathML objects [3]. Its search type is the similarity search. It uses the vector space

M.J. Smith and G. Salvendy (Eds.): Human Interface, Part I, HCII 2009, LNCS 5617, pp. 404–411, 2009.

model, one of the popular search methods in the field of information retrieval. However the vector space model only has the information about the existence of characters and mathematical symbols. It does not express the structure of mathematical formulas. If two different formulas include similar characters and symbols, they are listed in the same search result. Furthermore, the search performance (speed) of the vector space model drastically becomes slow if the number of objects (documents) becomes large. The inverted file (or inverted index) is the better search method especially for commercial search engines. As adopted in major search engines like Google and Yahoo!, the inverted file provides a fast search performance.

In our research, we propose a search engine of mathematical formulas. We make an inverted index by using the DOM (one of the API to access the XML document [4]) structure of MathML of objects. We propose three types of indexing method of MathML objects. The first uses the vertical structure of the DOM structure. The second uses the horizontal structure of the DOM structure. The last uses the hybrid structure of the above. We conduct an experiment to evaluate the these indexing methods.

2 Overview of the Math Search Engine

Figure 1 shows the system structure of the proposed search engine of mathematical formulas. The system consists of two sub systems. One of them crawls the Web and collects web documents including MathML objects, and then it stores them with the search indices in an inverted file. The other system sends a query to the inverted file and outputs the search results when it receives the query from a user.

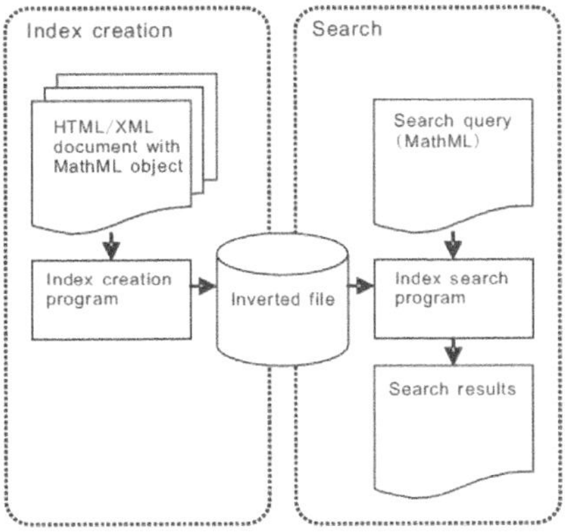

Fig. 1. Overview of our proposed system

Web documents which this search engine targets are HTML documents or XML documents which include mathematical formulas expressed in MathML like Figure 2-(a). <msqrt> represents the root, <msup> represents the square, <mi> represents the variable number, <mn> represents the numerical number, <mo> represents the mathematical operator.

When the system obtains a new web document, it retrieves each MathML object from the document. Then it acquires the DOM structure of the MathML object (see Figure 2-(b)). It generates indices by using a specific structure in the DOM structure. For giving an example of indices, we use a path in the DOM structure. XPath [5] are usually used for expressing a path in the DOM structure. Figure 2-(c) represents examples of XPath for all the paths of the MathML object in Figure 2-(a).

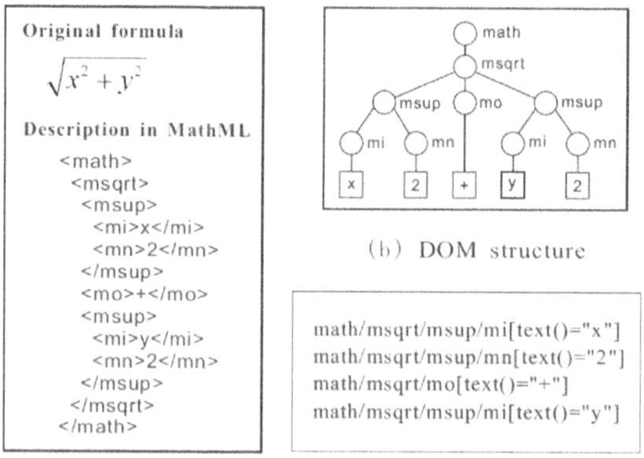

(a) MathML description (c) XPath expression

Fig. 2. An example of MathML description, dom and XPath expression

It stores the ID of the document with these indices as keys. In this step, the system does not include the following tags in the indices because they are not relating to the content of the formula: <mrow>, <mstyle>, <semantics>, <annotation>. Note that <mrow> represents the horizontal location of the formula at the time of rendering, <mstyle> represents the rendering attribute like color and font, <semantics> and <annotation> represents the additional semantic annotation to the formula. Table 1 shows an example of inverted indices. The first row shows that the document 1, 52, 70, 271, ... include the MathML object which has a fraction including the symbol " δ ."

Table 1. Examples of inverted index

XPath description	List of document ID
/math/mfrac/mo[text()="\delta"]	1,52,70,271,..
/math/msqrt/mfrac/mn[text()="1"]	2,16,55,102,..
/math/mfrac/mfrac/mi[text()="\pi"]	22,93,181,..

The system acquires the DOM structure of the given MathML object of the search query and acquires a key which is expressed as a specific path (Details are explained in the later section) in XPath. Then it acquires the document IDs whose indices match the key from the inverted file. For realizing a partial-match search, it just outputs these web document. For realizing an exact match, it checks all the paths (from the root node to the all leaf nodes) between the MathML object of the search query and the MathML objects in the acquired web document. It only outputs the web document in the search results which has a MathML object whose all paths match to all the paths of the MathML object of the search query.

3 Experiment

We think that using only one path for an index is not enough for narrowing down the search result. However, if we make an index from all the paths of the MathML object, the inverted file becomes large and many MathML objects are associated to one index. Therefore, we decided to select two paths from all the paths in the MathML object and generate an index by combining these paths. We built a search engine which introduces the above indexing method and conducted an experiment for the evaluation.

Table 2. Examples of created inverted index in the experiment

Description combining two XPath descriptions	List of document ID
/math/mi[text()="cot"],/math/mover/mi[text()="\infty"]	47206,47208, 47276,,,
/math/munderover/mo[text()="\Sigma"],/math/msup/mfrac/mn[text()="1"]	850,1211,1379,,,
/math/mo[text()="\int"],/math/mfrac/msqrt/msup/mi[text()="sin"]	18679,39824,39839,,,

Search query

$$\pi = 2\int_0^\infty \frac{\sin^2(t)}{t^2}dt$$

First path

/math/mi[text()="/pi"]

Deepest path

/math/mfrac/msup/mi[text()="sin"]

Search results

$$\pi = 2\int_0^\infty \frac{\sin^2(t)}{t^2}dt$$

$$\pi = \frac{8}{3}\int_0^\infty \frac{\sin^3(t)}{t^3}dt$$

$$\pi = 3\int_0^\infty \frac{\sin^4(t)}{t^4}dt$$

$$\pi = \frac{384}{115}\int_0^\infty \frac{\sin^5(t)}{t^5}dt$$

$$\pi = \left(2^n\int_0^\infty \frac{\sin^n(t)}{t^n}dt\right) \Bigg/ \left(n\sum_{k=0}^{\left\lfloor\frac{n-1}{2}\right\rfloor} \frac{(-1)^k (n-2k)^{n-1}}{k!(n-k)!}\right)$$

Fig. 3. Example (1) of the search result in the experiment

We used 87000 MathML objects provided by WOLFRAM RESEARCH [6]. We extracted all the paths in the DOM structure of each MathML object. We selected the first path and the deepest path and combined them as an index. By including the first path in the index, we can represent how the formula started (e.g. The formula expresses the definition or the equation). We think that the deepest path shows the most characteristic part in the formula. Table 2 shows examples of the created inverted file. The format is just a pair of the first path and the deepest path (expressed in XPath and separated by a commna). We selected one MathML object from the 87000 MathML objects and searched other MathML objects by our system. The search is conducted in the way of partial-match search. Figure 3 and 4 are examples.

In Figure 3, the XPath of the first path of the search query represents that the formula starts with π and includes the power of $\sin$. We can see that the search results include the formula with the square of $\sin$ and the cube of $\sin$. Also we can see that it includes the generalized formula that represents the n-power of $\sin$. The indexing methods using the first path and the deepest path can realize a similarity search by keeping the same structure with the search query. Figure 4 is an example with bad results. In Figure 4, the XPath of the first path of the search query represents that the formula starts with the square of cos. Actually, all the search results include the square (inverse) of cos. However, the whole mathematical structure of the third search result is different from the search query. The XPath of the deepest path of the search query represents that the formula includes the fraction with the square root of z. The third search result includes this structure, but there are two terms in the right-hand member. This is quite different from the search query. We notice that using the vertical structure in the DOM tree cannot solve this type of problem.

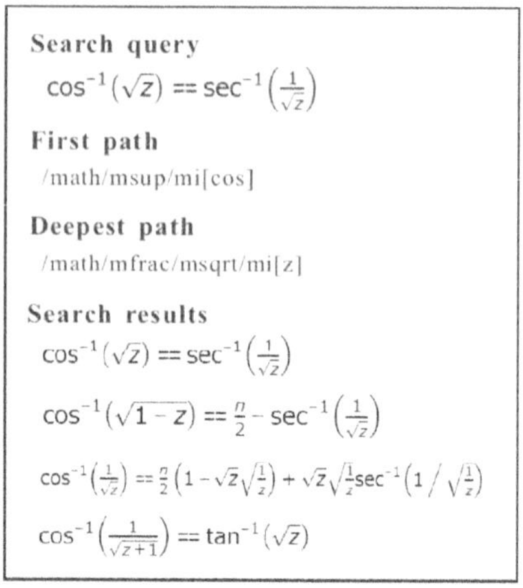

Fig. 4. Example (2) of the search result in the experiment

4 Incorporation of Breadth-First Search

We propose an indexing method of MathML object using the horizontal structure of the DOM tree. In detail, we use the tag names of the sibling nodes in one level of the

DOM tree (After here, "breadth-first search index (BFS index"). We compare the BFS index with the index which use the first path and the deepest path explained in the previous section (After here, "depth-first search index (DFS index)"). Ideally, BFS index represents the whole structure of the mathematical formula.

Figure 5 represents the sibling nodes in the same level in the DOM tree. In this experiment, we use the level which includes more than three nodes for the first time when checking the number of nodes in each level from the root level. In Figure 5, we use the sibling nodes for indexing this formula in the level surrounded by the red dashed line. Table 3 shows examples of BFS index.

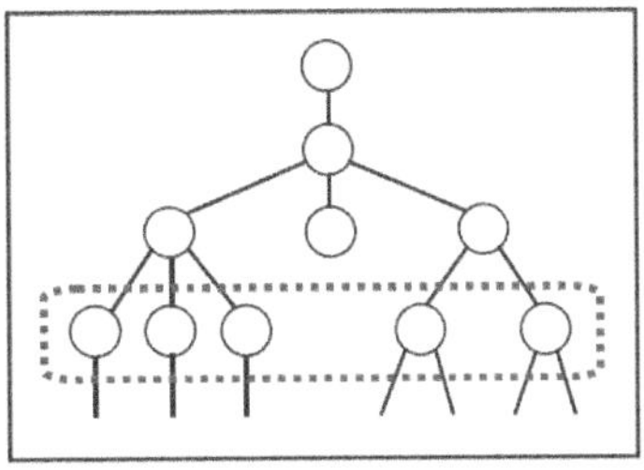

Fig. 5. Sibling nodes to be used for making a BFS index

Table 3. Examples of created BFS index

Description of sibling nodes	List of document ID
<msubsup><mo><mfrac><mfrac>	1,29,33,204,..
<mfrac><mo><mfrac><mo><mrow>	2,8,93,181,,,
<msup>Mmrow><mo><mrow>	3,64,226,,,

Search query

$\cos^{-1}(\sqrt{z}) == \sec^{-1}\left(\frac{1}{\sqrt{z}}\right)$

First path

/math/msup/mi[cos]

Deepest path

/math/mfrac/msqrt/mi[z]

Search results

$\cos^{-1}(\sqrt{1-z}) == \csc^{-1}\left(\frac{1}{\sqrt{z}}\right)$

$\cos^{-1}(\sqrt{z}) == \sec^{-1}\left(\frac{1}{\sqrt{z}}\right)$

$\cosh^{-1}(\sqrt{z+1}) == \mathrm{csch}^{-1}\left(\frac{1}{\sqrt{z}}\right)$

$\cosh^{-1}(\sqrt{z}) == \mathrm{sech}^{-1}\left(\frac{1}{\sqrt{z}}\right)$

Fig. 6. Example (3) of the search result in the experiment

Figure 6 shows the search results of BFS index whose search query is the same as Figure 4. We can see that formulas whose whole structure is different from the search query were deleted in the results. We think that combining BFS and DFS realizes the more efficient mathematical search.

5 Comparing Three Kinds of Indexing Methods

We compare the three kinds of indexing method: (i) DFS index, (ii) BFS index and (iii) a hybrid index of DFS and BFS. We use 550 mathematical documents which we collected from the high-school mathematical teaching material provided from Ikushinsya. For creating the correct dataset of the search result, we invite 6 users to give the correct search results to the query set (55 queries).

Table 4 shows the result. DFS method outputs many number of hits and achieves the high recall. Hybrid method narrows down the search result and achieves the high precision. When we see F-values, DFS achieves the best and hybrid method ahieves the worst. When we do not have a ranking algorithm, hybrid method is the best becaues the precision is the best. However, when we have a ranking algorithm, the results will be presented in the descending order of the ranking score. In that case, DFS is the best because the probability will be down that the users miss important mathematical documents.

Table 4. Results of comparing three kinds of indexing method

Indexing method	Number of hits	Number of matchces	Precision	Recall	F-value
DFS	61.5	2.18	0.10	0.58	0.34
BFS	59.7	2.9	0.13	0.46	0.30
Hybrid	17.0	1.2	0.16	0.30	0.23

6 Conclusion

In this research, we built a search engine for the MathML object by introducing the DOM structure and the inverted file. Firstly we investigated the paths in the DOM structures of MathML objects for generating an index. We found that combining the first path and the deepest path represents the characteristics of the formula well. We conducted an experiment to see what kind of results will be produced by using this index type. The results showed that in some cases it produces the search results which include only the mathematical formula with the same structure. However, it also showed that in some cases it produces the search results in which the whole structure is different from the search query. We also propse an indexing method which uses the horizontal structure in the DOM tree (sibling nodes in the same level in the DOM tree) and the hybrid method using vertical and horizontal structure. When we conducted an experiment, we found that the hybrid method achieves the best precision and the DFS achieves the best F-value. We will work on the ranking mechanism in the future.

References

1. Kopka, H., Daly, P.W.: A Guide to LaTeX, 3rd edn. Addison-Wesley, Harlow (1999)
2. W3C: M3C Math Home (1998), `http://www.w3.org/math/`
3. Nakanishi, T., et al.: An Implementation Method of Composite Association Retrieval System for Data of Mathematical Formulas. DBSJ Letters 4(1) (2005)
4. W3C: M3C DOM Home (1998), `http://www.w3.org/dom/`
5. W3C: M3C XPath Home (1999), `http://www.w3.org/TR/xpath`
6. WOLFRAM Research: MATHML CENTRAL (2007), `http://www.mathmlcentral.com/`
7. NICT: The EDR Electronic Dictionary (1999), `http://www2.nict.go.jp/r/r312/EDR/index.html`
8. Kohno, H., et al.: Search Engine and Agent. University of Tokyo Press (2002)
9. Kita, K., et al.: Information Retrieval Algorithm. Kyoritsu Publisher (2002)

Metadata-Based Reminder Classification in Reminiscence Engineering

Masashi Ishibashi and Yoshio Nakatani

Ritsumeikan University
1-1, Noji-Higashi 1, Kusatsu, Shiga 525-0058, Japan
cc001041@is.ritsumei.ac.jp, nakatani@is.ritsumei.ac.jp

Abstract. Our personal memory project studies a new framework which supports people construct the personal history on the computer. In this paper, we propose a framework of reminiscence engineering in order to classify reminders of personal memories. The reminders include photos, videos, gifts from significant people and so on. Today the reminders have been digitized, such as digital photos, and the amount of them keeps on increasing and the necessity of managing these digital reminders is also increasing. This system supports to manage reminiscence by classifying the digital reminders based on their metadata. The metadata in this research means the data about digital photos, such as the date and time at which it was taken, objects and the related memories. In order to classify these metadata, the system provides the vocabulary database that structures vocabulary necessary to classify the digital photos. We propose an effective framework which supports construct the vocabulary database through the entry of personal memory, as well as construct personal memory through the entry of new vocabulary.

Keywords: personal memory, classification, ontology, metadata.

1 Introduction

People meet various people and experience the various events every day. Memories of people and events are stored in each person and construct the personal memory ("Omoide" in Japanese). The personal memory is very important because it consists of the basis of the personality and people want to reconfirm their own personal history in this anxious society. Personal memories can be recalled by the related reminders, such as photos, videos and GPS data. Without these reminders, we are difficult to recall past memories. In recent years, some reminders are digitized and a vast volume of them are generated day by day. Personal computers, digital cameras, digital camcorders, camera-equipped cell phones, make it possible to easily record and preserve personal memories as digital data. The problem here is that people still waste time managing reminders. These digital data are not managed effectively and is not easy to reuse. Although they realize that management of the folders is more convenient as compared with before, they often feel it troublesome to arrange, edit, and continuously manage the digital data. In order to manage and reuse these digital reminders, the "Reminiscence Engineering" has been proposed in Japan [1].

M.J. Smith and G. Salvendy (Eds.): Human Interface, Part I, HCII 2009, LNCS 5617, pp. 412–418, 2009.

In this research, we propose a new framework to manage and reconstruct the memories on the computer by classifying so many digital reminders based on the vocabulary database about the metadata of the reminders. Our project includes a system which supports people construct the personal history on the computer. By integratedly using the vocabulary acquisition system and the personal history construction system, personal memories are expected to manage effectively.

2 Background

2.1 Importance of "Omoide Engineering"

When we recall memories, we usually count on the related reminders. Memories are in the memory of the individual person, and some of these memories are gradually weaken as time goes by and are difficult to recall. However by seeing reminders such as the photos, the home videos, and the diary, etc, the related memory is activated again and we are very easy to recall memories. In this sense, the reminders play an important role of recalling memories. Therefore, when people lose reminders in the disaster and the accident, they feel as if they have lost all memories and all of their pasts are denied. The person feels that he or she lost the goods, and lost memories though this was not lost of memories as the memory. More accurately, it is thought fear that the person might not be able to recall memories by having missed the chance. The reminders are the one attached importance to for the person because catching with "Reminders" = "Memories".

2.2 Related Research

In the field of memories engineering, the research how to manage goods of memories (digital photo) on the computer is advanced. Several studies have been proposed that "PPM" (Personal Portable Memory) has the function of an electronic mini album that can be easily carried[2], and "card type photo viewer" [3] . Moreover, there are proposed that input support to construct memories that exist on individual inside on computer and frame to arrange those memories is proposed [4]. Though various researches are performed from various aspects concerning memories, it has not come to still form a systematic field.

2.3 Questionnaire Survey

Three point questionnaire survey of memories (digital photography) was done for 20 students of Ritsumeikan University in January, 2008.

1. Do you have the digital photography?
2. How are you using the digital photography?
3. How have you treated the digital photography on computer?

Result of the questionnaire is as follows(figure.1).

In question 1, it was answered that 19 students(95%) had the digital photography, and nobody was using the film including the respondent of "No". In question 2 the use method is heard, it was answered that 10 students(53%) who answered with question 1 as "Yes" use both inspection with computer and the printout. It is thought

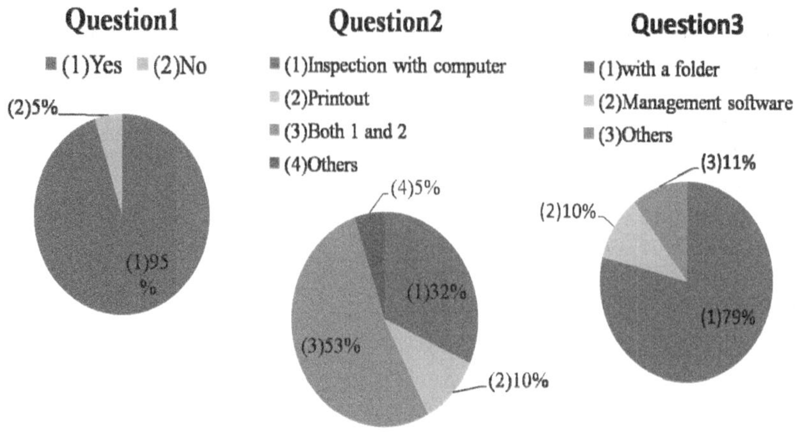

Fig. 1. Result of the Question

that a favorite part is printed out, because it costs this to print out all photographs too much. In question 3, it was answered that 14 students(74%) arranges the photography with a folder. Additionally, the answer such as use "Picasa" being offered from Google and management software attached to digital camera was obtained.

From the above-mentioned result, though the almost everyone uses the digital photography, it is a situation in which goods of memories are preserved as mere data, and it has been understood that the custom of devising it to manage the photography on computer doesn't take root. There are a custom of arranging an analog photography to the album. The reason for arranging analog photograph is that the possibility of getting scattered and lost at once and disappearing is large. However, the digitalized photograph is only divided into the folder on the computer. Because it has already been made to data, the person recognized like being managed on the computer, and the necessity for spending time any more is not felt strong. Moreover, the person takes a picture of a lot of number of sheets without the consideration of the development cost because the time of development is not necessary for the digital photography. Therefore, the number of sheets of the photograph increases more than time in an analog photograph, it seems that arranging the photograph became difficult, too. This cannot carry out the function of the photograph as the trigger that looks back on memories, and the digitalized advantage has not been made the best use of. Therefore it is thought that there is a necessity of the tool development that uses effective goods of memories.

3 Outline of system

3.1 Approach

For the arrangement of goods of memories, it is necessary to think about the arrangement of memories that coil round the goods at the same time.

This research pays attention to the meta data that accompanies memories. They are systematized as a peculiar vocabulary to the individual, and used to arrange goods of memories. The frame that supports the description and the arrangement of memories by using vocabulary is constructed.

So far, we have been researching the memories arrangement system that makes the community that has belong a classification frame[5]. The person belongs to the community(elementary school, district where it lives, friend group etc.)(figure.2). We construct the life history making system(Figure.3) in advance of the business service of the memories arrangement. This system displays the community on the axis of time in the Gantt chart style, arranges each memories that is related to the community. And, the effectiveness has been verified by the experiment. When you classify goods of new memories, the system proposed by this research mounts the following functions. When differing from the community that has been registered so far to which community information on the meta data on goods of memories has been registered, it is registered as a new vocabulary and it uses it to register goods of memories thereafter. At the same time, the system urges it on the user like registering as a new community for the memories input and the arrangement, and describing the content of memories.

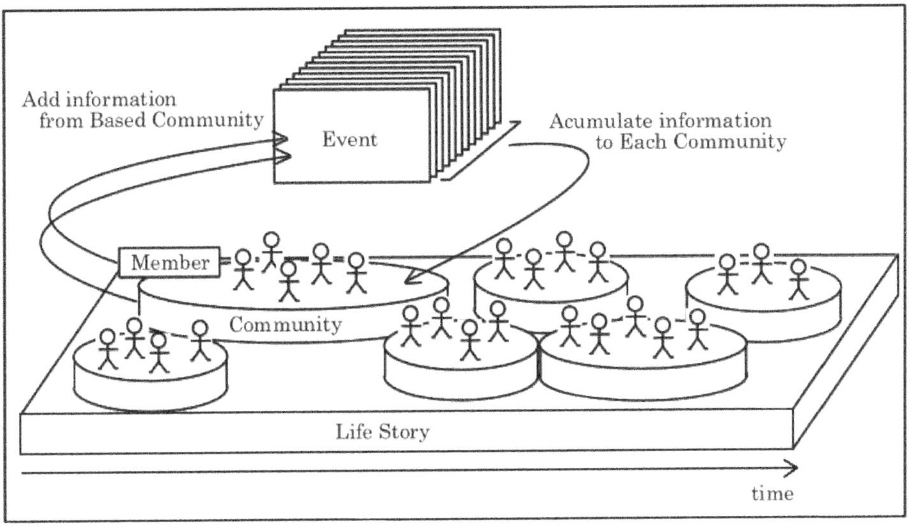

Fig. 2. Life History Based on Community

3.2 Flow of System

Figure 3 shows the flow of the system, from making individual vocabulary to displaying the community information.

The user inputs the basic data of community information.

1. The system arranges as an individual vocabulary, and accumulates in the data base.
2. The system outputs it in the Gantt chart style as community information.
3. Goods of memories are classified by using community information.

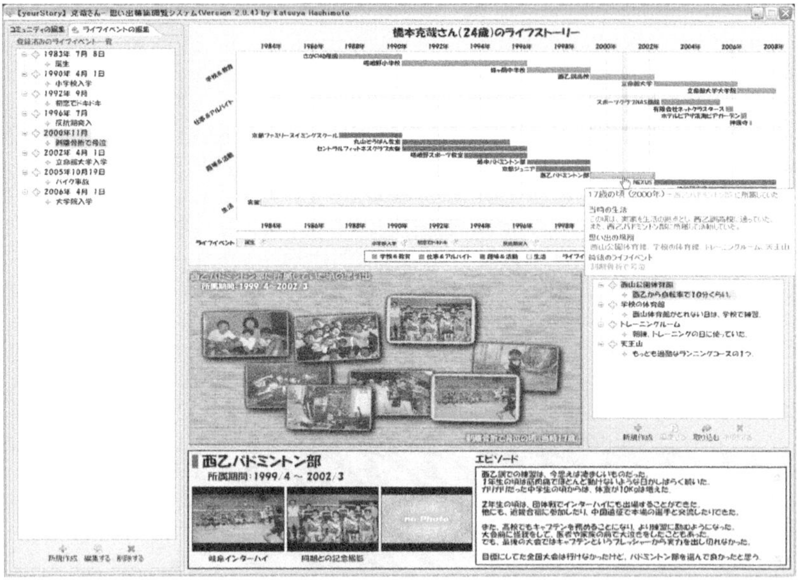

Fig. 3. A Screen of the Life History Making System

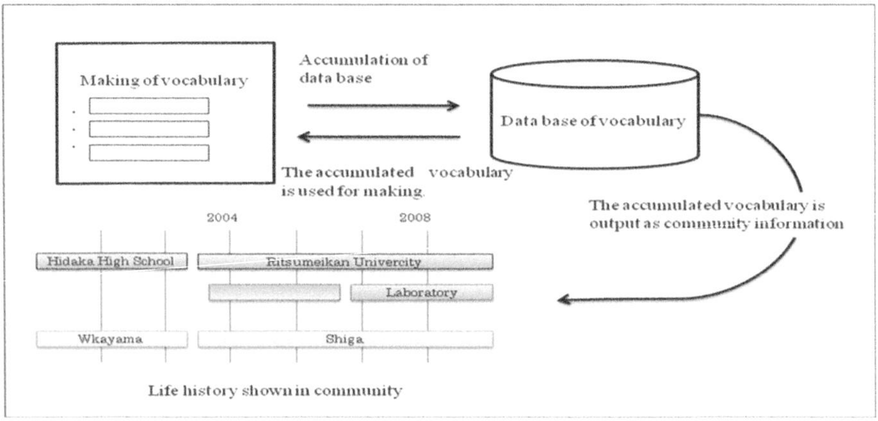

Fig. 4. Flow of system

3.3 Making of Community Information That Uses RDF

The system uses RDF(Resource Description Framework) / XML to systematize the meta data automatically as community information. Input information is output to RDF by using framework Jena of java. Community information that the individual belongs is shown by some layered structures. For example, we belong to Ritsumeikan University, and belong to the laboratory or the circle in Ritsumeikan University at the same time. Moreover, the content of (Table.1) is handled as a property element of community information.

Table 1. The Content

Name	*Example*
subject	Ritsumeikan Univercity
period	2004.04.01-2008.03.31
place	Shiga

The property element and the location of each meta data are provided for by the use of RDF/XML. The fragmentary meta data is automatically built in as community information on the life history. Moreover, a new vocabulary is located in the whole by using the vocabulary accumulated in the data base.

3.4 Goods Registration and Restructuring of Memories

Goods of memories are classified by using made community information. The candidate of the corresponding community is narrowed from the meta data concerning goods of memories, and goods of memories are registered. When corresponding community information is not registered, the meta data is newly input, and it registers as new community information. Moreover, individual memories are made visible by displaying registered community information in the Gantt chart style.

4 Experiments

After constructing the proposal system, we are scheduling the evaluation experiment by the testee. The evaluation item is the following three points.

1. Is the vocabulary accumulated whenever the system is used suitable assuming that individual memories are shown?
2. Is it suitable to arrange goods of memories by using community information as the frame that arranges memories?
3. Were goods of memories able to be arranged smoothly?
4. Can value be expected as goods of memories?

In (1) and (2), it is a question that asks whether it was the correct one as the frame that arranges goods of memories. In (3), it is a question that asks the usability of the system. In (4), it is a question that the memories package treated on the computer asks be worthy as memories.

5 Future Work

We proposed the above-mentioned system to goods of memories, to grope for the possibility of "Use" that invented "New value". We are developing systems now. We plan to execute the evaluation that by the testee in the future.

Moreover, it thinks about the following plans as an application of this system. This system makes a peculiar vocabulary to the individual that specialized in memories. We want to grope for standard memories vocabulary common to various people by using this data. We become easy to extract an experience and a community that is common or similar it, and want to construct the mechanism becoming communications support, by comparing differences of an individual vocabulary from this standard vocabulary.

References

1. Nojima, H., Harada, E.: <Inside in house> Cognitive science, Shin-yo-sha, pp. 269–288 (2004) (in Japanese)
2. Yamashita, K., Nojima, H.: Portable personal memory for communication. In: Human Interface Symposium 2002 (2002) (in Japanese)
3. Atsushi, T., Nojima, H.: Card type photo viewer for active of memories communications. In: 19th Japanese Cognitive Science Society, pp. 192–193 (2002) (in Japanese)
4. Nakatani, Y.: Support System for Reconstructing Fond Memories Lost in Disasters. In: Human Interface Symposium 2004, pp. 1–4 (2004) (in Japanese)
5. Hashimoto, K.: Evaluation of "yourStory" memories construction management system. In: 69th Information Processing Society of Japan, vol. 2, pp. 149–155 (2007) (in Japanese)

InfoScape: A Browser for User Behavior-Based Information Retrieval System

Masaaki Kawata[1] and Katsuhiko Ogawa[2]

[1] Graduate School of Media and Governance, Keio University
[2] Faculty of Environment Information Studies, Keio University
Endo 5322, Fujisawa-shi, Kanagawa, Japan
{kawatan,ogw}@sfc.keio.ac.jp

Abstract. We propose a system called InfoScape for exploring large histories of user behavior including location data (latitude and longitude). User behavior patterns are created from the histories of user behavior from data sources such as life logs. The browser of this system can find user behavior patterns by a query for a user behavior pattern. Currently, the histories of user behavior consist of 9,324 records. As a result, this study investigates the most effective conditions for the operation of this system working and demonstrates with factors are location and time for creating user behavior-based feature patterns.

Keywords: Context-aware computing, information retrievals, neural networks, ubiquitous computing.

1 Introduction

Recently, we obtained user histories of location data, including latitude and longitude, by using a mobile phone with a Global Positioning System (GPS) receiver. The amount of readily available user-generated content (UGC) with location information continues to grow. Life logs such as Twitter[1] for short messaging and Flickr [2] for photo sharing are examples of UGC sources. Although classic guide systems such as Active Badge [3], C-MAP[4], and Cyberguide [5] use location data, they do not accumulate histories of user behavior.

End-users benefit if they can explore content from other user behaviors. In this paper, we propose a browser called InfoScape, which we created to enable users to effectively explore user behavior histories. With the InfoScape browser, users can find similar user behavior patterns with a query for a user behavior pattern (Fig. 1).

The browser has access to 9,324 records of histories of user behavior, including location data. Currently, this research has been investigating the most effective conditions for operating this browser.

2 The InfoScape

In this section, we describe the system of InfoScape, which is composed of three building blocks (Fig. 1).

M.J. Smith and G. Salvendy (Eds.): Human Interface, Part I, HCII 2009, LNCS 5617, pp. 419–428, 2009.

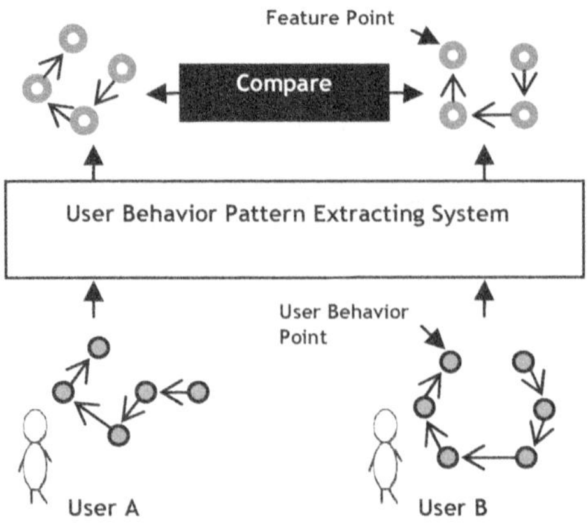

Fig. 1. Building blocks of InfoScape

The first block is the extraction of user behaviors. The extracting system creates user behavior patterns from the histories of user behavior. And, we called generated point is feature point from point of user behavior histories (Fig. 2).

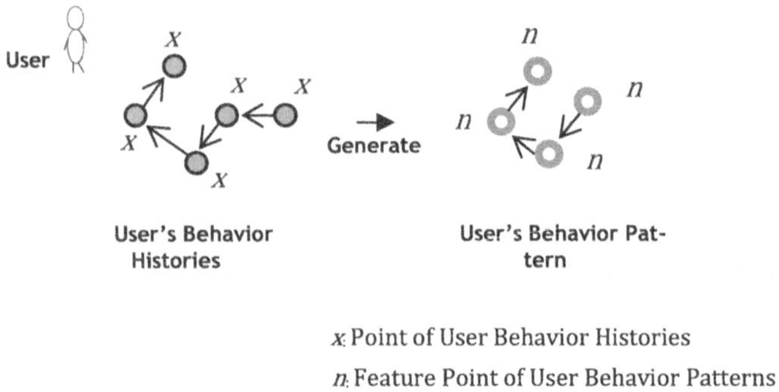

Fig. 2. Creating the User Behavior Patten from User Behavior Histories

These patterns are necessary because the number of user behavior items is different for each user or time span. The second block is the comparer. The comparer calculates the distance between items of each user. The third block is the user interface. This user interface provides visualization and selection of user behavior patterns.

2.1 The Algorithm for Creating a User Behavior Patten

User behavior patterns are created from the histories of user behavior through self-organizing maps (SOM) [6].

SOM's training rule is given by:

$$m_j(t+1) = m_j(t) + h_{ci}(\mathrm{t})\left[x(t) - m_j(t)\right] \tag{1}$$

h_{ci} is the neighborhood function is given by:

$$h_{ci} = \alpha(t) \cdot \exp\left(-\frac{\|r_c - r_i\|}{2\sigma^2(t)}\right) \tag{2}$$

First, we created input vectors from the histories of user behavior. An input vector is given by:

$$x_i = \begin{bmatrix} lat_i \\ long_i \\ t_i \\ g_i \\ a_i \end{bmatrix} \tag{3}$$

where lat and long are the latitude (4) and longitude (5), respectively, of the user, t is time (6), g is the user's gender (7), and a is user's age(8).

$$lat_i = \frac{(lattitude_i - \min lattitude)}{(\max lattitude \; - \min lattitude)} \tag{4}$$

$$long_i = \frac{(longitude_i - \min longitude)}{(\max longitude \; - \min longitude)} \tag{5}$$

$$t_i = \frac{(hours_i \cdot 60 + minutes_i)}{24 \cdot 60} \tag{6}$$

$$\begin{cases} 0, \text{if } gender \text{ is male} \\ \quad 1, \text{otherwise} \end{cases} \tag{7}$$

$$a_i = \frac{(age_i - \min age)}{(\max age \; - \min age)} \tag{8}$$

Then, using the input vectors, we obtain output vectors by training the SOM. The output vectors are the user behavior patterns.

2.2 The Compare Method for User Behavior Patterns

After creating output vectors, we compare the user behavior pattern M of a search query and a stored pattern N by the following equation:

$$dist = \sum_{\mathrm{i}=1}^{\mathrm{n}} \sqrt{\sum_{\mathrm{j}=1}^{\mathrm{m}} \left(M_{i_j} - N_{i_j}\right)^2} \tag{9}$$

2.3 The Select Method for User Behavior Patterns

The browser selects user behavior patterns by filtering the patterns by similarity. In addition, this method can manipulate the threshold of distance from far to near (or near to far) between the search query pattern and stored patterns (Fig. 3).

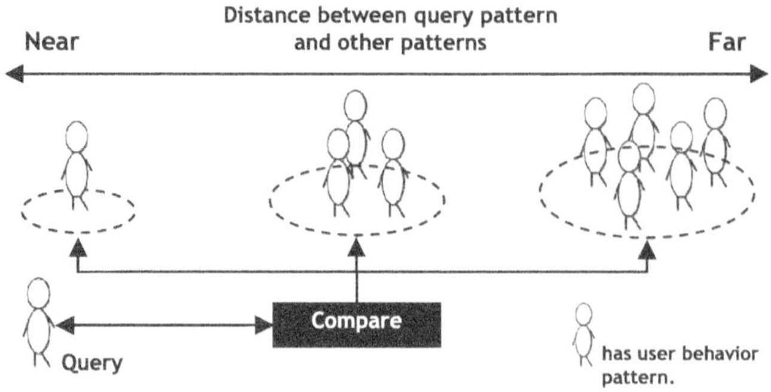

Fig. 3. Comparing user behavior patterns

3 Implementation

As stated above, we developed the InfoScape browser for exploring the histories of user behaviors linked to UGCs.

We developed an interactive InfoScape guide for a public space, such as a café, that consists of a graphical user interface (GUI) on a touch panel. This application provides exploring based on user behavior patterns (Fig .4).

Fig. 4. Browsing contents (UGC) on touch panel interface with user behavior-based information system

This system is built on MVC (Model-View Controller) architecture. InfoScape's GUI is View in MVC architecture. InfoScape's 'User Behavior Pattern Extracting and Comparing System (UBPECS)' is Controller and Model in MVC architecture (Fig. 5). Because, when creating the other application such as car navigation system, changing points is only View.

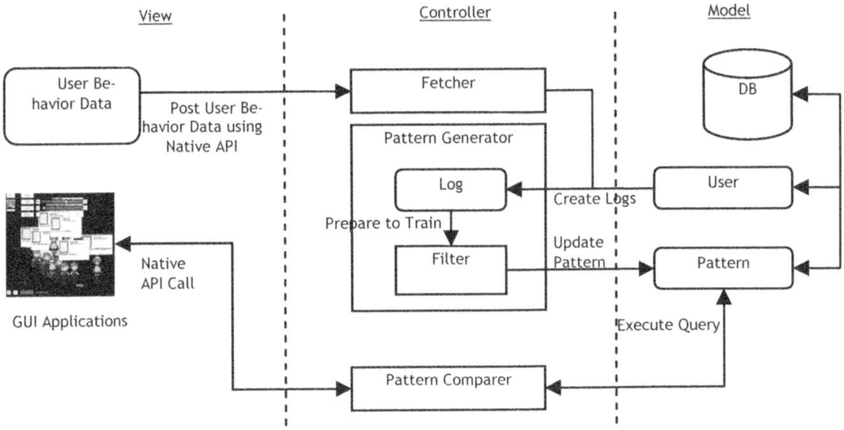

Fig. 5. System Overview

3.1 GUI

We now describe how InfoScape's GUI application works. (Fig. 6) First, an end-user posts a comment on short messaging service like Twitter, as an example, with location data. Second, the InfoScape's UBPECS generates the user behavior pattern from the histories of user behavior per day. Third, the end-user selects a user behavior pattern and sets a threshold on the slider control. (Fig. 7) Then, the application responds with similar comments. In addition, end-user can zoom in/out in this screen by using another slider control.

3.2 User Behavior Pattern Extracting and Comparing System (UBPECS)

Next, we describe how InfoScape's UBPECS works. First, when creating user behavior patterns from user behavior histories, send user behavior history to Fetcher. Second, Pattern Generator extracts a user behavior pattern from user behavior histories per user and a day. Finally, save user behavior patterns to DB.

When, an end-user search by using user behavior patterns, send a user behavior pattern to Pattern Comparer at first. Second, Pattern Comparer compares user behavior pattern of a search query and other user behavior patterns. Then, Pattern Comparer sorts by the distance between a pattern of search query and other patterns. Finally, Pattern Comparer responds user behavior patterns of search results to View in MVC architecture.

Therefore, InfoScape's UBPECS allows similar search by user using user behavior patterns.

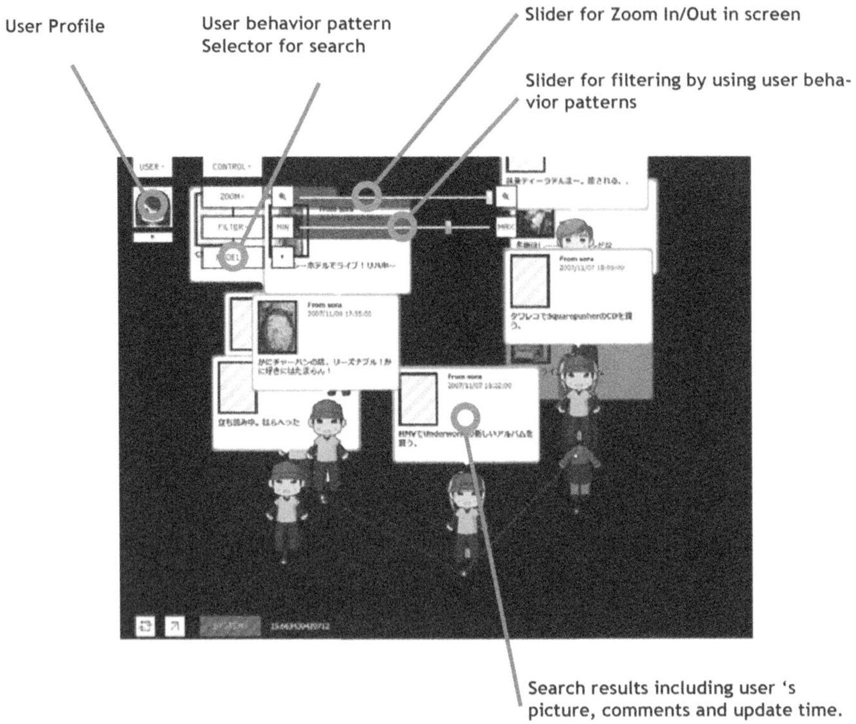

Fig. 5. Selecting contents on the user interface of InfoScape browser

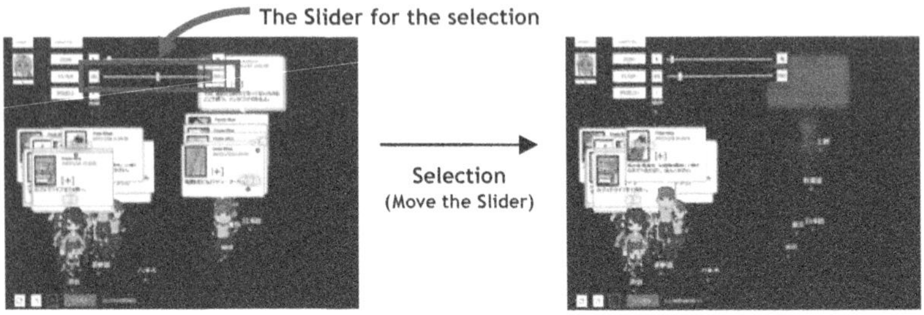

Fig. 6. Selecting contents on the user interface of InfoScape browser

4 Experimental Results

This section describes the experiments performed on InfoScape browser. First, we evaluate InfoScape browser's search accuracy by precision and recall. Second, we compare user status comments and search results of user behavior patterns linked user status comments for validation.

Further, we used histories of user behavior from 9,324 records gathered from 36 people for 13 days by Information Grand Voyage Project [7]. The overview of these user behavior histories is listed in Table 1.

Table 1. Overview of test users

Age	Male	Female	Total
18-29	6	7	13
30-44	8	6	14
45-60	4	5	9
Total	18	18	36

And, the number of user behavior data in 1 day is listed in Table 2.

Table 2. Overview of user behavior in 1day

MIN	AVG	MAX
2	10.7	34

Next, we looked at the most effective conditions of this system for creating user behavior-based feature patterns. After we found that the factors are location and time for linking patterns for finding user comments.

4.1 The Search Accuracy Using User Behavior Patterns

We evaluate the accuracy of searching by user behavior patterns. First, we created 445 user behavior patterns from user behavior histories for evaluation. Second, we chose a pattern in 445 user behavior patterns for a search query. Finally, we calculate small scale case and large sale case of precision and recall. The small scale case is small activities such as Shinjuku to Shibuya in Japan (Distance is less than 5 km). The large scale case is large activities such as Tokyo to Kanagawa in Japan (Distance is more than 50 km).

Further, when we calculate precision and recall, we defined results to averaged 5 frequently occurring search results.

We show the results from small scale case and large scale case in Fig. 6. These two scales are key features, which are frequently showed in the data of user histories. In the case where the first user behavior patterns are narrow, the high search precision is acquired according to the increasing of the feature points. The same tendency is provided, when user behavior patterns are broad.

This result predicts that there is a trend that the high search precision and recall can be acquired in the case where the user behavior patterns are narrow, because a low number of feature points are sufficient enough for the narrow range of the user behavior patterns.

As a result, over five feature points are required for acquiring of the high search precision and recall.

Additionally, since genders and ages are fixed, that data can not generate the difference in creating user behavior patterns by using SOM. Real-time varying values are suitable for SOM.

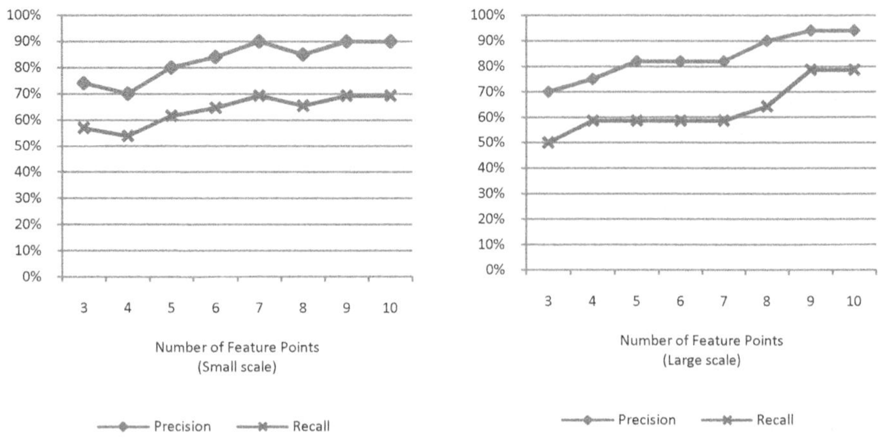

Fig. 7. Precision and Recall: User activities of small scale case (left) and large scale case (right)

4.2 Comparing User Comments and Search Result by User Behavior Patterns

Next, we show the results from comparing user comments and search result by user behavior patterns. In Table 3 and Table 4, we show status of user comments in the search query and results of small activities and large activities. The status of user comment is linked to user behavior pattern.

Table 3. Search query and results of small activities

Query	Result #1	Result #2	Result #3
Household chores/Child rearing/Care	Household chores/Child rearing/Care	Moving	Household chores/Child rearing/Care
Moving	Moving	Eating	Moving
Eating	Eating	Hobby/Entertainment	Eating
Hobby/Entertainment	Hobby/Entertainment	Breaking	Shopping
Waiting for people	Waiting for people		Hobby/Entertainment

The result, which the user behavior patterns are small activities, predicted that movements and eating are treated as a same behavior pattern in the query. This tendency is shown in the case where the user behavior patterns are large activities.

Additionally, the behavior of Hobby/Entertainment is predicted, because Shinjuku and Shibuya towns will be used for Hobby/Entertainment purposes.

Therefore, the location related to purposes of user, and user behavior histories frequently represented details of user's activities.

Next, which the user behavior patterns are large activities, predicted that movements and eating are treated as a same behavior pattern in the query. However, when a search query included working case is exists in undesirable search results are included Hobby/Entertainment status. This problem is necessary to solves by separating weekdays and holiday when create user behavior patterns.

Table 4. Search query and results of large activities

Query	Result #1	Result #2	Result #3
Working	Working	Eating	Eating
Moving	Household chores/Child rearing/Care	Moving	Moving
Eating	Moving	Free time	Hob-by/Entertainment
	Eating	Breaking	

5 Conclusions and Future Work

In this research, we created a flexible and personalized browser that is a retrieval system for user behavior patterns. We simultaneously developed a practical application of InfoScape's GUI and UBPECS. The slider control-based selection method retrievals enables important search results efficiently for the end-users, and UBPECS provides personalized search by using user behavior patterns. In addition, we looked at important factors, which are location and time, for the operation of this system.

The next challenging step is to evaluate this application by usability of the user interface. Also, we plan to build other user behavior-based information retrieval applications such as car navigation system.

Acknowledgements. This work was supported by Special Coordination Funds for Promoting Science and Technology of Ministry of Education, Culture, Sports, Science and Technology, Japan.

References

1. Twitter, `http://www.twitter.com/`
2. Flickr, `http://www.flickr.com/`
3. Want, R., Hopper, A., Smith, F.V., Gibbons, J.: The active badge location system. ACM Trans. Inf. Syst. 10(1), 91–102 (1992)

4. Sumi, Y., Asuyuki, E., Tani, T., Fels, S., Simont, N., Kobayashi, K., Mase, K.: Prototyping Context-aware Exhibition Guidance. Information Processing Society of Japan (IPSJ) 39(10), 2866–2878 (1998)
5. Abowd, G.D.: Cyberguide: A mobile context-aware tour guide. Wireless Network 3(5), 421–433 (1997)
6. Kohonen, T.: Self-Organizing Maps. Springer, Berlin (1995)
7. Information Grand Voyage Project, `http://www.igvpj.jp/index_en/`

The Roles of Profession and Gender in Some PIM Tasks

Guangfeng Song[1] and Chen Ling[2]

[1] Penn State Greate Allegheny, 4000 University Dr, McKeesport, PA 15132, USA
gus4@psu.edu
[2] University of Oklahoma, 202 West Boyd,Rm. 124, Norman, OK 73019, USA
chenling@ou.edu

Abstract. In this paper, the roles of selected individual difference factors were examined in some personal information management (PIM) tasks. A questionnaire was completed by 295 participants regarding their personal characteristics, retrieval of previously saved information, and selective use of multiple computers. Clustering of individual difference factors suggested further analysis of profession and gender as classifiers. Profession seemed to modulate the behaviors and attitudes of users in these PIM tasks. And certain gender difference in PIM tasks could be explained by different expectations or standards.

Keywords: Personal Information Management, Profession, Gender.

1 Introduction

Computers have become integral parts of modern life. They are playing increasing number of roles in people's lives. Computer users have also expanded from a few in scientific fields to almost everyone everywhere around the world. The phenomenal growth is accompanied by increasing diversities in computer usage and computer users. With recognition of the diversities, designers have strived to make computers easier to use for everyone. Usability is improved by reducing the knowledge or skills required to use computers. Universal usability is an effort to accommodate human limitations of all kinds so that everyone can use a well-designed computer system, despite difference in personal and social characteristics [1].

On the other side, the growing power and availability of personal computers resulted in a more and more complex computing environment for typical users. Multiple systems running on multiple computers may be readily available to the user. The user will inevitably have to choose from duplicate functionalities provided by these systems. Although this issue is a relatively new development in recent years when computers become increasingly affordable, it presents a growing challenge for users. We need to understand how users of different characteristics cope with ever more complex computing environment.

It is possible that individual difference of users can affect the ways they view and use computers, causing difference in attitude and activities. In particular, gender difference has long been a subject of studies [2,3]. The inquiry into individual difference continues to be an interesting topic due to the following competing effects. First, better usability tends to make individual difference less important. Second, today's

M.J. Smith and G. Salvendy (Eds.): Human Interface, Part I, HCII 2009, LNCS 5617, pp. 429–436, 2009.

users started using computers and received training or education much earlier than previous generations. It is possible the experience and education in computers reduce individual difference among them. Third, growing complexity of computing environment, along with growing list of activities performed with computers, are demanding more and more capabilities from users. Thus individual difference may be exposed and highlighted as a result of high demand.

The present study investigates several individual difference factors in some PIM tasks with multiple computers. This is unlike previous studies on individual difference that focused on statistics of computer usage and general attitude towards computers. By focusing on a common and popular set of tasks such as PIM tasks performed by almost every computer users, we intend to understand how users deal with the technological complexity of having multiple computers, and whether individual difference factors play a role in users' behaviors.

2 Individual Difference Factors

One of the most prominent individual difference factors is gender. Whitley concluded that gender difference exists in computer usage patterns but the effect sizes were small and probably of little practical significance [2]. In a more recent study of college students, Imhof could not detect gender gap in computer use, both in terms of time spent and activities in using computers [3]. However, Imhof did found male students continue to use computers more frequently than female, mostly for personal or non-study related activities. In other studies, females were found to email more than male but male searched the Internet more than females [4][5]. Males also utilized different type of sites compared to females [6]. Gender difference in computer experience was also reported in specific contexts such as video games[7] and online shopping [14]. Hence, the role of gender still deserves to be examined in specific context of computer usage despite hints from the literature that such role was diminishing.

Other important individual difference factors include age, ethnic background, education, and computer experience. We naturally expect some factors, such as computer experience, to affect attitudes and activities in using computers. Yet Garland and Noyes found that computer experience is a poor predictor of computer attitudes [8]. Inconsistent results regarding the roles of ethnic background were also reported. Digital divide among ethical groups has long been reported and studied [9,10]. On the other hand, some studies also found no effects of ethnic groups in attitudes and usage of computers [11,12].

Regarding the age factor, studies have compared the computer usage and attitude of older adults to young adults. Many studies were concerning senior adults of 55 years of older. Not surprising, senior adults had less access, less experience and skills in computers [13]. However, this finding might not apply to working professionals in their 30-50s. Instead of comparing the two extremes of young adults to senior adults, the present study compared college students around 20 to working professionals in PIM tasks.

3 Research Method

A survey of PIM was developed for the present study. The survey had 55 questions concerning the ownership and use of multiple computers, management and retrieval of information from the computers. These questions covered these issues regarding three types of information: textual files, bookmarks, and emails. We asked participants to rate the frequency of experiencing difficulty in retrieval of the three types of information. We also asked them to rate how satisfied they were with their management of the information. In addition, participants were asked to rate the importance of six possible factors in selecting a computer device to use, when they have more than one available computers. These factors are: overhead or how long it takes for the device to be ready; display; input or how comfortable is the input devices; software availability; purpose of usage; and length of usage. With these questions, we hoped to understand how computer users decide which computer to use.

The questionnaire also contained questions measuring several individual difference variables, including age, gender, ethnic classification, education levels, and profession. There were also questions regarding participants' computer ownership, experience, and frequency of usage.

Students and working professionals from two universities voluntarily completed the survey either online or by filling out the identical printout of the survey. A total of 296 completed surveys were collected. Participants indicated their profession as student, working professional, or others.

4 Individual Difference Factors

4.1 Clustering of Individual Difference Variables

We performed a two-step clustering to reveal the patterns of the following individual difference variables measured in the questionnaire: age, gender, ethnic background, education, and profession. The automatic cluster number selection method in SPSS was adopted. The result showed two clusters closely corresponding to the profession variable. 96% of working professionals belonging to cluster 1, and 98% of students belonging to cluster 2. Two participants indicated "others" as their profession and they belonged to cluster 1.

Table 1. Clusters found within individual difference variables

		Student		Working Professional		Others	
		Freq.	Percent	Freq.	Percent	Freq.	Percent
Cluster	1	4	1.6%	48	96.0%	2	100.0%
	2	239	98.4%	2	4.0%	0	.0%
	All	243	100.0%	50	100.0%	2	100.0%

Working professionals reported more available computers (p-value=0.006). This is because of greater number of laptops for working professionals than students. The number of available desktop computers did not differ significantly (p-value=0.677). Surprisingly, students reports significantly more mobile devices than working professionals. This is perhaps due to the ownership of many types of mobile devices popular among young college students, including mp3 player, cell phone, pda, game players etc.

We repeated the two-step clustering of these individual difference variables inside the student samples, which is much larger than the samples from working professionals. Six clusters were automatically selected by SPSS. No obvious patterns emerged from the result. However, the gender variable had an interesting distribution among the clusters as shown in Table 2. The distribution suggested that gender can be a meaningful classifier for our individual difference variables. Clusters corresponding mostly to female students (cluster 1-3) are characterized by less computer experience and less frequent use of computers.

Table 2. Clusters found within individual difference variables of student samples

		Female		Male	
		Freq.	Percent	Freq.	Percent
Title	1	58	55.8%	0	.0%
	2	18	17.3%	11	7.9%
	3	19	18.3%	15	10.8%
	4	9	8.7%	30	21.6%
	5	0	.0%	33	23.7%
	6	0	.0%	50	36.0%
	All	104	100.0%	139	100.0%

Hence, there appeared to be two classifiers, profession and gender, for the individual difference variables in our data. In addition, we did not find significant association between profession and gender as nominal variables (Chi-square test showed a p-value of 0.09). Our analysis is therefore focused on the difference in PIM variables across groups of participants identified by profession and gender.

4.2 Difference between Students and Working Professionals

Difficulty and Satisfaction in PIM. No significant difference was found between the satisfaction of working professional and students in management of files, bookmarks, and emails, although in the case of emails, the difference in satisfaction was approaching statistical significance (p-value=0.059).

Significant difference was found between working professionals and students in rated difficulty in finding files, bookmarks, and emails. Greater difficulty was reported in each category by working professionals.

Table 3. Difficulty and satisfaction in PIM, comparison by profession

	Students			Working Professionals			Difference		
	N	Mean	S.D.	N	Mean	S.D.	t	df	p
Difficulty in file[A]	243	2.34	1.324	50	3.28	1.294	-4.60	291	*<0.001*
Satisfaction in file[B]	243	4.82	1.281	50	4.54	1.541	1.22	64	0.229
Difficulty in bookmark[A]	243	1.73	1.233	50	2.12	1.438	-1.99	291	*0.048*
Satisfaction in bookmark[B]	243	4.65	1.547	50	4.62	1.652	0.14	291	0.888
Difficulty in email[A]	243	2.08	1.405	50	2.60	1.370	-2.40	291	*0.017*
Satisfaction in email[B]	243	4.98	1.372	50	4.56	1.593	1.89	291	0.059

A. Scale: 1: never, 2: every few months, 3: every month, 4:every week, 5:every few days, 6: every day
B. Scale: 1:very dissatisfied to 7: very satisfied

Selective Use of Computers. We measured 6 variables regarding the selective use of computers. We compared these variables between students and working professionals. Working professionals did not differ significantly in any computer-related variables (input, display, software, and overhead). However, working professional considered the other two variables (purpose and length of usage) less important than students. (p-values were 0.022 and 0.015, respectively).

4.3 The Effect of Other Individual Difference Factors

We examined other individual difference factors to identify possible factors responsible for the significant difference we found between students and working professionals in Section 4.2.

Ethnical Background. This is not affecting the significant difference between working professionals and students in selective use of computers. This is also not affecting the significant difference in reported difficulty between working professionals and students.

Age. When we limit the samples to older age ranges (18 students, 48 working professionals), the significant difference in reported difficulty between students and working professionals disappeared. However, the difference in means became greater and standard deviations of each group remained similar. Therefore, it is likely the difference in means were not significant because of loss of statistical power due to small samples. Similar situation was found for the two significant variables in selective use of computers.

Education. We tried to limit samples by requiring the education level to "Graduate school or higher" which resulted in 23 students and 42 working professionals. With the reduced samples, we found different results for variables in selective use of computers. "Overhead" and "input" differed significantly between professions while "purpose" and "length" ceased to be significant between professions. However, the difference in means for "purpose" and "length" remain similar to the larger sample, suggesting that the disappearance of significance was caused by lack of power. For the variables measuring difficulty in retrieval of files, bookmarks, and emails, similar difference in means were found in the small sample compared to the larger sample. However, only difficulty in retrieving files remained significant. The education factor appeared to play a big role in our data because it was associated with the age, computer experience, and nature of computer usage of our participants.

Number of Computers. We limited the samples to students only and performed one-way ANOVA with the factor being the number of computers. The number of computers was found to significantly affect the difficulty of retrieving emails, especially when comparing students with 3 or more computers to students with one or two computers. This result suggested that the difficulty in retrieving emails may be ultimately related to the number of computers people have to use, instead of as resulted from the profession.

Frequency of Usage. We limited the samples to students only and performed one-way ANOVA with the factor being the number of computers. The result was not significant for dependent variables difficulty in retrieving files, emails, and bookmarks. Frequency of use did affect the selection variable "display". Participants who use computers more frequently judged the variables to be more important in selecting a computer than those who use computer less frequently. No other selection variables were found to differ by frequency of computer use.

4.4 Difference between Male and Female Students

Our analysis in this section is limited to the samples from students only to eliminate the effect of profession. Male students have more desktops and overall counts of computers available to them than female students. On the other hand, the number of laptop did not differ significantly between the genders.

We measured the amount of computer usage by several questions: frequency of computer usage, number of files created, number of files received, and frequency of organizing files. A clustering of these variables produced two clusters corresponding perfectly to the two genders. This result suggested distinctive patterns of male versus female computer usage. A t-test of these variables revealed that significant difference existed between genders in frequency of usage, number of files created, and number of files received. Male students use computers more frequently than females. They also created and received more files than females.

A t-test by gender found significant difference in satisfaction regarding files and bookmarks management. Male students are more satisfied in their management of files and bookmarks than females. However, no significant difference was found between genders in difficulty of retrieving files, bookmarks, and emails.

Table 4. Difficulty and satisfaction in PIM, comparison by gender

	Female			Male			Difference		
	N	Mean	S.D.	N	Mean	S.D.	t	df	p
Difficulty in file[A]	104	2.26	1.344	139	2.40	1.311	-0.792	241	0.429
Satisfaction in file[B]	104	4.62	1.332	139	4.98	1.225	-2.202	241	*0.029*
Difficulty in bookmark[A]	104	1.83	1.340	139	1.65	1.147	1.078	241	0.282
Satisfaction in bookmark[B]	104	4.32	1.708	139	4.91	1.367	-2.986	241	*0.003*
Difficulty in email[A]	104	1.91	1.323	139	2.20	1.456	-1.586	241	0.114
Satisfaction in email[B]	104	5.01	1.451	139	4.95	1.315	0.336	241	0.737

A. Scale: 1: never, 2: every few months, 3: every month, 4:every week, 5:every few days, 6: every day
B. Scale: 1:very dissatisfied to 7: very satisfied

We also compared the importance of the 6 variables for selective use of computers. The data reported by male students did not differ significantly from female students. On the importance of the selecting computer with the purpose of usage, female students reported a mean of 4.81in a 1-to-7 scale. This was not significantly higher than the mean of 4.47 from male even though the p-value was only 0.075.

5 Discussion and Conclusion

We studied the roles of many individual difference variables in PIM. The effects of these variables were usually difficult to separate. We took a different approach in analyzing the data by clustering these variables. Profession and gender naturally emerged as candidate classifiers of individual difference. We tested whether behaviors and attitudes in some PIM tasks differed between the groups identified by the classifiers. For any significant difference between groups, we examined if any single variable could have explained the difference.

Our result suggested that people's professions made a difference in the frequency of difficulty they experienced in retrieving previously saved information. We also found that working professionals rated two usage factors (purpose and length) in selective use of computers more important than students. We could attribute the difference to having or not having a working environment that typically requires timely management of information and good organization skills. A working environment could make retrieving information a demanding task with time pressure, resulting in frequently experienced difficulty. It could also foster a purpose-driven behavior in computer usage. Students and working professionals in our study had different personal characteristics in variables such as age, education, and frequency of computer usage. Also it was possible for some of these variables to interact with profession, none of these variables alone could have explained the difference we found between students and working professionals. Therefore, the profession of people seemed to provide a powerful context that shaped the behaviors and attitudes of them in these PIM tasks.

The only difference we found between male and female students were their satisfaction of information management. On all three types of information, female students had lower satisfaction than male students. This was interesting because female students used computer less frequently and process less files than male students. We hypothesized that the low satisfaction of female students can be explained by high expectation or high standard of them to get organized in information management. The fact that female students did not differ significantly from male students in the frequency of difficulty in retrieving information provided support for this hypothesis. Further study on this issue is necessary to identify the specific difference.

References

1. Shneiderman, B.: Universal usability. Communications of the ACM 43(5), 84–91 (2000)
2. Whitley, B.E.: Gender differences in computer-related attitudes and behavior: A meta-analysis. Computers in Human Behavior 13(1), 1–22 (1997)
3. Imhof, M., Vollmeyer, R., Beierlein, C.: Computer use and the gender gap: The issue of access, use, motivation, and performance. Computers in Human Behavior 23(6), 2823–2837 (2007)
4. Durndell, A., Haag, Z.: Computer self efficacy, computer anxiety, attitudes towards the Internet and reported experience with the Internet, by gender, in an East European sample. Computers in Human Behavior 18(5), 521–535 (2002)
5. Jackson, L.A., Ervin, K.S., Gardner, P.D., Schmitt, N.: Gender and the Internet: Women Communicating and Men Searching. Sex Roles 44(5), 363–379 (2001)
6. Wasserman, I.M., Richmond-Abbott, M.: Gender and the Internet: Causes of Variation in Access, Level, and Scope of Use. Social Science Quarterly 86(1), 252–270 (2005)
7. Terlecki, M., Newcombe, N.: How Important Is the Digital Divide? The Relation of Computer and Videogame Usage to Gender Differences in Mental Rotation Ability. Sex Roles 53(5), 433–441 (2005)
8. Garland, K.J., Noyes, J.M.: Computer experience: a poor predictor of computer attitudes. Computers in Human Behavior 20(6), 823–840 (2004)
9. Hoffman, D.L., Novak, T.P.: Bridging the Racial Divide on the Internet. Science 280, 390–391 (1998)
10. Hacker, K.L., Mason, S.M.: Ethical gaps in studies of the digital divide. Ethics and Inf. Technol. 5(2), 99–115 (2003)
11. Campbell, N.J.: High School Students' Computer Attitudes and Attributions: Gender and Ethnic Group Differences. Journal of Adolescent Research 5(4), 485–499 (1990)
12. Tien, F.F., Fu, T.: The correlates of the digital divide and their impact on college student learning. Comput. Educ. 50(1), 421–436 (2008)
13. Cutler, S.J., Hendricks, J., Guyer, A.: Age Differences in Home Computer Availability and Use. J. Gerontol. B Psychol. Sci. Soc. Sci. 58(5), S271–S280 (2003)
14. Slyke, C.V., Comunale, C.L., Belanger, F.: Gender differences in perceptions of web-based shopping. Commun. ACM 45(8), 82–86 (2002)

Efficient Text Classification Using Best Feature Selection and Combination of Methods

M. Srinivas[1], K.P. Supreethi[2], E.V. Prasad[3], and S. Anitha Kumari[1]

[1] JNTUACE, Anantapur, India
[2] JNTUHCE, Hyderabad, India
[3] JNTUKCE, Kakinada, India
sreenu2521@gmail.com, supreethi.pujari@gmail.com,
drevprasad@yahoo.com, anitha.mahi7@gmail.com

Abstract. Lsquare and k-NN classifiers are two machine learning approaches for text classification. Rocchio is the classic method for text classification in information retrieval. Our approach is a supervised method, meaning that the list of categories should be defined and a set of training data should be provided for training the system. In this approach, documents are represented as vectors where each component is associated with a particular word. We propose voting method and OWA operator and Decision Template method for combining classifiers. In these we use an effective and efficient new method called variance-mean based feature filtering method of feature selection. Best feature selection method and combination of methods are used to do feature reduction in the representation phase of text classification is proposed. Using this efficient feature selection method and best classifier combination method we improve the text classification performance.

1 Introduction

Document retrieval, categorization, routing, and filtering systems are often based on text classification. A typical classification problem can be stated as follows: Given a set of labeled examples belonging to two or more categories (training data), classify a new test sample to a category with the highest similarity. Text classification has become a more and more important application of machine learning. Unlike the conventional machine learning domains, text classification has many special traits. Firstly, it is very often for a typical text classification application that there are hundreds of thousands features to be considered, while most of them are sparse in the document collections. Secondly, many features in the text classification tasks are redundant, which make classifiers prone to over fitting [12]. Text categorization is the problem of automatically assigning one or more predefined categories to free text documents. While more and more textual information is available online, effective retrieval is difficult without good indexing and summarization of document content. Document categorization is one solution to this problem.

M.J. Smith and G. Salvendy (Eds.): Human Interface, Part I, HCII 2009, LNCS 5617, pp. 437–446, 2009.

2 Previous Works

A number of classification methods have been discussed in the literature for document classification. These include, naïve Bayes classifier, decision trees [13], knearest neighbor classifier [3], linear discriminant analysis (LDA) [4], logistic regression [5] and neural networks [5], support vector machines [5], rule learning algorithms [6], relevance feedback [10], Lsquare, and neural networks [8]. Most of research in text categorization has been devoted to binary problems, where a document is classified as either relevant or not relevant with respect to predefined topic. In what follows we describe three algorithms for text categorization that have been proposed and evaluated in the past, and then describe our proposed algorithm, but first some general notation is given: Let d = $\{d_1, d_2 \ldots, d_M\}$ be the document vector to be classified and w_1, $w_2 \ldots, w_M$ are all possible words and let C = $\{c_1, c_2 \ldots, c_K\}$ be the possible topics. Further assume that we have a training set consisting of N document vectors $d_1, d_2 \ldots, d_N$ with true classes $y_1, y_2 \ldots, y_N$. N_j is then the number of training document for which the true class is c_j. In this paper, we use three different classification methods: Lsquare [20] and k-nearest neighbor classifier and Rocchio algorithm. Our approach is based on combining these methods by voting algorithms and OWA operator and Decision Template method. There is various feature selection methods are there such as Document Frequency (DF), Chi-square (CHI), Information Gain (IG), Mutual Information (MI), Term Strength (TS) GSS Coefficient, odds ratio. But in these we can use an effective and efficient new method called variance-mean based feature filtering method of feature selection to do feature reduction in the representation phase for text classification is proposed. This method is more efficient than remaining feature selection methods.

2.1 Feature Selection Variance-Mean Based Filtering Method of Feature Selection

While decreasing the dimension is the task of feature selection, how to select out the features that best characterize the text that belongs to a particular class is its main purpose. Consider the text class set $\{c_1, c_2, \ldots, c_n\}$ assuming each of the classes contains the same number of documents n in the training corpus1. Thus, we get the following matrix:

$$\begin{pmatrix} d_{11} & \cdots & d_{1n} \\ \cdots & \cdots & \cdots \\ d_{m1} & \cdots & d_{mn} \end{pmatrix} \tag{1}$$

Where d_{ij} is the jth document that belongs to the ith class. And for convenience of statement, let d_{ij} also stands for the weight $\text{weight}_w(d_{ij})$ of the candidate feature w (also called term) in that document denoted as in the training corpus. So, for every candidate feature in the original feature space, there is such a matrix. Each d_{ij} in the matrix (1) for a specified term w is computed by the formula defined in the weighting scheme, e.g.

$$d_{ij} = p_{ij}(w) \tag{2}$$

p_{ij} is the probability that term w will occur in document d_{ij}, which is approximated by the frequency of the occurrences of term w in document d_{ij}. In fact any (more complicated) weighting scheme can be employed here. And here we use the term frequency as the weight for the feature in a text's feature vector for further constructing our evaluation function just because of convenience. We then compute the mean and variance of the values for each row, producing the following vectors:

$$1)\ \begin{bmatrix} E_{r1} \\ \cdots \\ E_{rm} \end{bmatrix} \qquad 2)\ \begin{bmatrix} D_{r1} \\ \cdots \\ D_{rm} \end{bmatrix} \tag{3}$$

Where E_{r1} is the mean of the values in the ith row in the above matrix (1) and D_{r1} is the variance of those in the ith row. Thirdly, we compute the variance of the components' data in vector 1), denoted as D(E_{r1}), which shows the degree of dispersion among classes the term w can demonstrate and the mean of the components' data in vector 2), denoted as E(D_{r1}), which shows the average level of the degree of variability within every class the term w can show. The bigger the value of D (E_{r1}), the more distinguishable among classes using that term w is; the smaller the value of E (D_{r1}), the more cohesive within each single class averagely using that term w is. And the more distinguishable among classes and cohesive within each single class using that term w is, the more possible the term w should be remained. So D (E_{r1}) and E (D_{r1}) based criterion can be used to evaluate the importance of the candidate term w. Then the evaluation function could be:

$$1)\quad \beta * D(E_{r1})/E(D_{r1}) \tag{4}$$

Where β is the tuning parameter. If for term w, the F value is more than the threshold f, a.k.a. F>f, the term w is selected. And the threshold f is set according to the experiment. Or it could be

$$2)\quad E(D_{r1}) \tag{5}$$

The features whose E (D_{r1}) values are bigger than a threshold got by experiment are filtered. Or the filtering function

$$3)\quad D(E_{r1}) \tag{6}$$

can be applied, and the features satisfying D (E_{r1}) > d (where d is also a threshold) will be kept. The strategy of setting threshold on these evaluation functions' values to determine the size of the feature vector is called THR [2]. In fact another two strategies called PFC and MVS can also be adopted on these evaluation functions. PFC (Predefined Feature Count) selects the first several (the predefined feature count) features that have the largest or smallest evaluation function values. MVS (Mladenic' Vector Size) strategy proposed by Mladenic [2] decides the number of the features to be remained by defining its ratio to the total number of original feature items that appear in the training corpus. Any of the above feature evaluation functions and feature cutting strategies can be chosen to combine together to finish feature filtering in our method. And in our experiment, PFC and MVS strategies are used on feature evaluation function 3) in formula (6).

The computing complicity of the algorithm in accordance with our method is linear to the original feature space's size and can be largely cut down by using techniques in computing statistics. And even if it is time-consuming, compared with its effectiveness in better characterizing the text and its efficiency gained by decreasing the dimension of the vector, it is worthwhile especially when it is computed off-line and is once and for all. The classical methods of taking document frequency or document frequency incorporated measure as evaluation function to evaluate the importance of candidate features such as DF, CHI and IG, which are reported in [19] having better performance among all other methods. Overlook the fact that the properties of the candidate feature in different documents and different classes are different, because they only take whether the term w occurs in a text into consideration when computing the evaluation function, which is not enough to show that kind of differences of term w among different classes. For example, term t1 and term t2 may have the same document frequency, but they may appear different times in a single document. But the capability of a feature in characterizing a text as belonging to a class is closely related to its ability to express that kind of differences among different classes. Thus such classical methods reported having better performance in [19] seem to show no privilege. But if we take the variance of the average index values for each class as the evaluation function as our method does, the differences will be reflected, and the performance will be improved.

3 Proposed Algorithm

Many researchers have investigated the techniques of combining the predictions of multiple classifiers to produce a single classifier [1, 11]. By combining classifiers we are aiming at a more accurate classification decision at the expense of increased complexity. Voting algorithms and OWA operator and Decision Template are three classifier fusion methods that we use in this paper. In these paper we can proposed new an effective and efficient feature selection algorithm can be used. Using this feature selection with the efficient classifiers combination method (OWA) we can increase the classifiers performance. So our classification is more accurate and efficient

3.1 Classifiers Combining Methods

3.1.1 Voting Algorithms

Voting algorithms take the outputs of some classifiers as input and select a class which has been selected by most of the classifiers as output. We use three text classifiers, including Lsquare, Knearest neighbor and Rocchio. The output of these classifiers used as input for voting combiner.

Majority Voting

If two or three classifiers are agree on a class for a test document, the result of voting classifier is that class. But if each classifier has a different output, we select output of Lsquare classifier as output of voting classifier, because Lsquare has a better accuracy rather than the other classifiers. (voting).

Table 1. Classification rate for single classifiers based on test data

classifier	Accuracy
Rocchio	86%
K-NN	87.50%
Lsquare	89.60%

3.1.2 OWA Operators

Here we briefly review the class of aggregation operators called the OWA operators. An OWA operator defined on the unit interval I and having dimension N, is a mapping $F{:}I^n \rightarrow I$ such that

$$F(a_1,\ldots,a_n) = \sum_{j=1}^{n} w_j b_j \quad (14)$$

Where b_j is the jth largest of the a_j and w_j are a collection of weights such that $w_j \in [0,1]$ and

$$\sum_{j=1}^{n} w_j = 1$$

Note. If id(j) is the index of the jth largest of a_i then $a_{id(j)} = b_j$ and $F(a_1,\ldots,a_n) = \sum_{j=1}^{n} w_j a_{id(j)}$

Note: If W is an n vector whose jth components is w_j and B is an n vector whose jth components are b_j then $F(a_1,\ldots,a_n) = W^T B$. In this formulation W is referred to as the OWA weighting vector and B is called to ordered argument vector. The OWA operator is parameterized by the weighting vector W [2].

Minimum: $W = [0, 0,\ldots, 1]^T$

Maximum: $W = [1, 0,\ldots, 0]^T$

$$\text{Median: } W = \begin{cases} \left[\underbrace{0,\ldots,0}_{\frac{L-1}{2}},0,1,\underbrace{0,0,\ldots,0}_{\frac{L-1}{2}}\right]^T & ;L \text{ is odd} \\ \left[\underbrace{0,\ldots,0}_{\frac{L}{2}-1},.5,.5,\underbrace{0,\ldots,0}_{\frac{L}{2}-1}\right]^T & ;L \text{ is even} \end{cases}$$

$$\text{Average: } W = \left[\frac{1}{L},\frac{1}{L},\ldots,\frac{1}{L}\right]^T$$

$$\text{Competition jury: } W = \left[0,\frac{1}{L-2},\ldots,\frac{1}{L-2},0\right]^T$$

Table 2. Classification rate for OWA

Classifier	Accuracy
OWA1(W Maximum)	87.86%
OWA2(W Median)	89.57%

3.1.3 Decision Template

The idea of the decision templates (DT) combiner is to remember the most typical decision profile for each class w_j, called the decision template, DT_j, and then compare it with the current decision profile DP(x) using some similarity measure S. The closest match will label x [1].

$$DT_j = \frac{1}{N_j} \sum_{z_k \in w_j} DP(Z_k) \tag{15}$$

$$\mu_j(x) = S(DP(x), DT_j), \quad j = 1, \ldots, c. \tag{16}$$

Where S defined as below:

$$\mu_j(x) = 1 - \frac{1}{L \times C} \sum_{i=1}^{L} \sum_{k=1}^{c} \left[DT_j(i,k) - d_{i,k}(x) \right]^2 \tag{17}$$

4 Experimental Results

An experiment was performed to show the performance of fusion methods on real data. The dataset that we used consisted of Usenet articles collected from 20 different newsgroups (table 3). Over a period of time 100 articles were taken from each of the newsgroups, which make an overall number of 2000 documents in this collection. Each document exactly belongs to one newsgroup. The task is to learn which newsgroup an article was post to.

The documents in this dataset have the typical properties of Usenet articles. A random subset of 65% of the data considered in an experiment was used for training and 35% of the data considered for testing. The result has been shown in table 4.

Table 3. Usenet newsgroups used in newsgroup dataset

#	Newsgroup name	#	Newsgroup name
0	comp.design	10	eat.nonveg.fish
1	comp.sys.hcl.pc.hardware	11	sci.computers
2	comp.os.linux	12	sci.medical
3	comp.sys.del.hardware	13	sci.chemistry
4	comp.windows.x	14	sci.space
5	rec.cars	15	soc.religion.hindu
6	rec.sport.cricket	16	talk.politics.guns
7	rec.study.ms	17	cool.drinks.7up
8	rec.education.phd	18	Watch.movies.telugu
9	rec.movies	19	play.games.comp

Table 4. Classification rate

Classifier	Classification Rate
K-Nearest Neighbor	86.71%
Rocchio	87.57%
Lsquare	88.62%
Voting	88.90%
OWA1	88.99%
OWA2	89.59%
DT	88.75%

4.1 Best Feature Selection Result

The corpus used for training and testing is the "web text classification corpus" We divide the corpus into two non-intersected sets: a training set containing 10 categories with 100 texts in each and a test set containing the same 10 categories with another 100 texts in each also. Then we apply the open source lexical processing software made by ICT, (Institute of Computing Technology) to do the word segmentation task, which is part of the pre-processing. We then apply our method to do feature reduction. That is taking D (E_{r1}) as feature evaluation function and applying the MVS and PFC strategy to cut down the feature space's size. And, then we simply adopt the word frequency as the word's weight index in the vector. After vectors are got, the vectors corresponding to the texts in the training corpus are input into ten binary classifiers. Each binary classifier is trained for each of the 10 classes by using each class' texts at a time as positive examples with the rest of the data as negative examples. Each unknown vector from the test set is presented to all the 10 binary classifiers. The output of a binary classifier is positive when it decides that an unknown test vector belongs to the class it was trained for. Since there may be several simultaneous such claims or not a single such claim, a simple maximum selector is applied to the classifiers to make the final decision for an exclusive class [6]. Or multiple tags with the first largest selectors are assigned for the text in the multiple label condition. Every binary classifier is the Lsquare classifier. According to Thorsten Joachims, it is suitable to use Lsquare as classifier for text classification task. Because the theoretical analysis concludes that Lsquare acknowledges the particular properties of text: a) high dimensional feature space b) dense concept vector (most of the features are relevant) and c) sparse instance vectors. In all experiments we use a linear kernel and let C vary automatically accordingly. The performance evaluation data gained using our method in the experiment is depicted in the following diagrams (Table.5 and Table.6). The features are ordered by D(E_{r1}) the value decreasingly, using MVS and PFC feature cutting strategy by setting parameter RtoD=0.1,0.3,0.5,0.7,0.9,1; D=100,200,…,2800 correspondingly. "D" stands for the number of features we want to keep in the above-mentioned PFC strategy, a.k.a. the first several important features ordered by its D (E_{r1}) value decreasingly. "RtoD" means the ratio of the number of features to the total number of features originally extracted from the training corpus in the MVS strategy. Because every feature will be kept if the "RtoD" value in MVS is set to 1, this value of the parameter is set to get the performance result as using no filtering process for comparison. We also give the performance evaluation data gained by using the classical feature filtering methods introduced, the document frequency (DF) and χ^2statistic

(CHI) as a comparison, keeping other conditions being identical to those when using our method in the experiment.

Table 5. The performance evaluation data given by macro-precision M-P, macro-recall M-R and macro-f1 M-F1 using MVS strategy on filtering function $D(E_{r1})$ with the original dimension 28,260

RtoD	D	M-P	M-R	M-F1
0.1	2826	0.876	0.856	0.866
0.3	8478	0.912	0.898	0.905
0.5	14130	0.917	0.902	0.909
0.7	19782	0.938	0.9377	0.934
0.9	25434	0.938	0.9378	0.934
1	28260	0.67	0.67	7293

Table 6. The performance evaluation data given by macro-precision M-P, macro-recall M-R and macro-f1 M-F1 using the PFC strategy on filtering function $D(E_{r1})$

D	M-P	M-R	M-F1	D	M-P	M-R	M-F1
100	0.903	0.884	0.8934	1500	0.8498	0.83	0.8398
200	0.9126	0.894	0.9032	1600	0.8514	0.836	0.8436
300	0.916	0.9	0.9079	1700	0.8574	0.838	0.8476
400	0.9252	0.912	0.9018	1800	0.8617	0.842	0.8517
500	0.6434	0.668	0.6555	1900	0.8644	0.846	0.8551
600	0.6753	0.69	0.6826	2000	0.8654	0.846	0.8556
700	0.6624	0.681	0.6718	2100	0.8654	0.846	0.8556
800	0.7988	0.78	0.7893	2200	0.8679	0.848	0.8578
900	0.8089	0.788	0.7983	2300	0.8693	0.848	0.8585
1000	0.8138	0.796	0.8048	2400	0.8666	0.844	0.8551
1100	0.8274	0.808	0.8176	2500	0.8722	0.852	0.862
1200	0.8335	0.812	0.8226	2600	0.8722	0.852	0.862
1300	0.8421	0.822	0.8319	2700	0.8728	0.852	0.8623
1400	0.8405	0.824	0.8322	2800	0.8763	0.856	0.866

From the above diagrams, it can be seen that: 1) Feature reduction is really important because without it, the performance value is only 0.73 of macro-f1, much lower than the one gained with reduced feature space' size. 2) Our method shows a good property. The performance evaluation data are still high when the dimension is reduced to 100, as shown in Table 6. The performance still gradually goes up until the dimension reaches almost the original size, as shown in Table5. Although the performance gained when the dimension is 400 is not a real maximum, compared with the one gained at dimension 0.9*28260 shown in Table.5, Using feature vector of dimension 400 greatly decreases the computing time. With other conditions being identical, training a Lsquare using feature vectors of dimension 0.9*28260 costs about

25min while using dimension 400 only needs 10sec on Intel Celeron 2.4GHZ CPU with memory of 512M, and furthermore using vector of dimension 0.9*28260 to represent the incoming unknown free text for classification costs almost 120 times the time deciding the class of that text using vector of dimension 400. And the macro-f1 value gained at dimension 0.9*28260 is only slightly bigger. 3) Compared with the classical document frequency incorporated feature filtering methods such as DF, CHI used in our experiment, it shows in Figure.2, that our method can gain higher performance at a very low dimension, and quickly reach a peak, which means much less computing time and almost best performance than other methods.

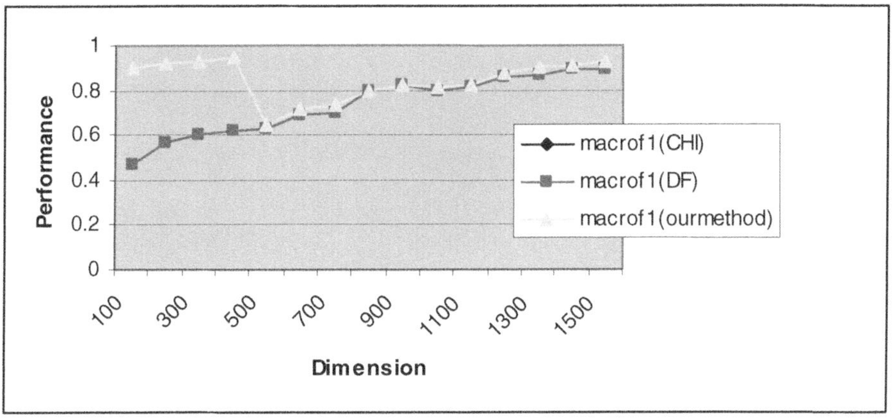

Fig. 2. Performance comparison among DF, CHI and ours. DF and CHI have almost the same performance curves.

5 Conclusion and Future Works

In this paper, we have proposed a novel approach for text classification. Our approach is based on combining classifiers. We combined Rocchio and k-Nearest Neighbour and Lsquare classifiers by voting algorithm and OWA operators and Decision Template method and achieve a better classification rate which experimental results show that the classification error decreased 15 percent. We use 2000 documents from 20 different newsgroups for testing our proposed methods. Good representation of a text is very important, of which the feature reduction has an obvious effect on the final performance of text classification. In these we can use efficient and effective feature selection method used. This method is called variance-mean based feature filtering method. Using this method we can also increasing the classifiers performance. We can also use an efficient combination method for combine the different classifiers. In these we can compare our method (OWA) with different methods. It will be give better result than compare to remaining methods. Using these two methods we can increase the classifiers performance more afflictive.

References

1. Kuncheva, L.I.: Combining Pattern Classifiers Methods and Algorithms. John Wiley, Chichester
2. Soucy, P., Mineau, G.W.: Feature selection strategies for text categorization. In: Xiang, Y., Chaib-draa, B. (eds.) Canadian AI 2003. LNCS (LNAI), vol. 2671, pp. 505–509. Springer, Heidelberg (2003)
3. Weiss, S., Kasif, S., Brill, E.: Text Classification in USENET Newsgroup: A Progress Report. In: AAAI Spring Symposium on Machine Learning in Information
4. Hull, D., Pedersen, J., Schutze, H.: Document Routing as Statistical Classification. In: AAAI Spring Symposium on Machine Learning in Information Access Technical Papers
5. Schutze, H., Hull, D., Pedersen, J.: A Comparison of Classifiers and Document Representations for the Routing Problem. In: SIGIR 1995, Washington, DC, pp. 229–237 (1995)
6. Salton, G., McGill, M.: Introduction to Modern Information Retrieval. McGraw-Hill, New York
7. Lewis, D.D.: Representation and Learning in Information Retrieval, University of Massachusetts Amherst, MA (1992)
8. Sahami, M., Dumais, S., Heckerman, D., Horvitz, E.: A bayesian approach to filtering junk e-mail, Learning for Text Categorization, pp. 55–62. AAAI Press, Menlo Park (1998)
9. McCallum, A.K., Nigam, K.: A comparison of event models for naïve Bayes text classification. In: Proc. of AAAI 1998 Workshop on Learning for Text Categorization (1998)
10. Dong, Y.-S., Han, K.-S.: A Comparison of Several Ensemble Methods for Text Categorization. In: Proc. of the 2004 IEEE International Conference on Services Computing (2004)
11. Kuncheva, L.I.: Switching between Selection and Fusion in Combining Classifiers: An Experiment. IEEE Transaction on Systems, Man, and Cybernetics - part B: Cybernetics
12. Douglas Baker, L., McCallum, A.: Distributional clustering of words for text classification. In: Proc. of the 21st Intl. ACM SIGIR Conf. on Research and Development in Information Retrieval, pp. 96–103. ACM Press, New York (1998)
13. Lewis, D., Ringutte, M.: A comparison of Two Learning Algorithm for Text Categorization. In: Third Annual Symposium on Document Analysis and Information Retrieval, Las Vegas, NV, pp. 81–93 (1994)
14. Bell, D.A., Guan, J.W., Bi, Y.: On Combining Classifier Mass Functions for Text Categorization. IEEE Transactions on Knowledge and Data Engineering
15. Dumais, S.T., Platt, J., Heckerman, D., Sahami, M.: Inductive Learning Algorithms and Representations for Text Categorization. In: Proc. Seventh Int'l Conf. Information and Knowledge Management (1998)
16. Felici, G., Sun, F., Truemper, K.: A Method for Controlling Errors in Two-Class Classification. In: Proc. 23rd Ann. Int'l Computer Software and Applications Conf.
17. Felici, G., Truemper, K.: A Minsat Approach for Learning in Logic Domains. Informs J. Computing 14(1) (Winter 2002)
18. Joachims, T.: Text Categorization with Support Vector Machines: Learning with Many Relevant Features. In: Nédellec, C., Rouveirol, C. (eds.) ECML 1998. LNCS, vol. 1398. Springer, Heidelberg (1998)
19. Yang, Y., Pedersen, J.O.: A Comparative Study on Feature Selection in Text Categorization. In: Proceedings of 14th International Conference on Machine Learning, San Francisco, pp. 412–420 (1997)
20. Al-Mubaid, H., Umair, S.A.: A New Text Categorization Technique Using Distributional Clustering and Learning Logic. Proc. of IEEE Transactions on Knowledge and Data Engineering 18(9) (September 2006)

Designing Sticky Knowledge-Network SNS for Japanese Science Teachers

Yoshihisa Wada[1], Masayuki Sakoda[1], Hiroshi Tsuji[1], Yuuki Aoki[2], and Kazuhisa Seta[2]

[1] Graduate school of Engineering, Osaka Prefecture University,
1-1, Gakuencho, Naka-ku, Sakai,
599-8531 Osaka, Japan
{Wada@mis.,Sakoda@mis.,Tsuji@}cs.osakafu-u.ac.jp
http://www.cs.osakafu-u.ac.jp/mis/

[2] Graduate school of Science, Osaka Prefecture University,
1-1, Gakuencho, Naka-ku, Sakai,
599-8531 Osaka, Japan
{aoki@kbs.cias.,seta@mi.s.}osakafu-u.ac.jp
http://www.mi.s.osakafu-u.ac.jp/e/

Abstract. To support knowledge sharing among science teachers on ICT (Information and Communication Technology), this paper proposes a knowledge-network SNS (Social Networking Service) based on stickiness, which have three factors to drive;(I) Relationship Capital, (II) Reputation and (III) Personalization. Building a prototype SNS based on these factors, this paper defines two kinds of quantification of stickiness which express how system is used and how user continue to use. Compared with the traditional knowledge-repository, the knowledge-network SNS transfers tacit knowledge learned in class from one to others. And more, the quantificated stickiness gives standards to strengthen teachers' knowledge-network; (1) picking up low-stickiness teachers for inviting into the network and (2) estimating stickiness of the system.

Keywords: ICT, Knowledge-Network, Stickiness, Social Networking Service.

1 Introduction

Many schools have introduced information and communication technology (ICT) as computer performance and the Internet have improved and become more affordable. In elementary or junior high schools, attention has been focused on applications of scientific digital content which include attractive subjects for experimentation or observation as shown in the Appendix Fig.5. Some teachers have *good practice* and others have valuable reviews; however, they do not always share their lessons and experiences of ICT classes. Therefore, an experience management system for sharing lessons and experience is necessary [1].

This research discusses a chain of knowledge based on personal networks. Previous research has focused on general knowledge spirals, a way to externalize tacit

M.J. Smith and G. Salvendy (Eds.): Human Interface, Part I, HCII 2009, LNCS 5617, pp. 447–456, 2009.

experience based on communities and is built on interpersonal interactions has been receiving much attention [2]. This is because epistemological learning, called social constructivism, is propounded. The purpose of this research is to build and discuss a prototype social networking service (SNS), based on a knowledge-network. This prototype deserves consideration because it differs from a repository system for digital content in which users query a database and obtain formatted answers. This prototype realizes to immediately share and trace knowledge learned in a classroom based on a chain of Japanese science teachers.

2 Knowledge-Network and Stickiness

Why human continue to query computer databases and how to promote human-computer interaction have been important questions. Suppose that a new knowledge-network is built. Why do human continue to interact with others via computers and how to promote such human-human interaction? This chapter describes the differences in knowledge-networks and knowledge-repositories, and the stickiness of these knowledge-networks.

2.1 Knowledge-Network vs. Knowledge-Repository

Tiwana claims that a knowledge-network is effective for lesson-learned-type knowledge management in the case of peer-to-peer relationships because it is similar to face-to-face communication in the real world [3]. A knowledge-repository type, which is a conventional document-type knowledge management system, differs from this in the way it obtains answers from users' queries as shown in Fig.1.

Repository types, which provide formatted answers from databases, require users to assign keywords for a query. Users in a knowledge-network, in contrast, directly obtain answers from other users in the same network.

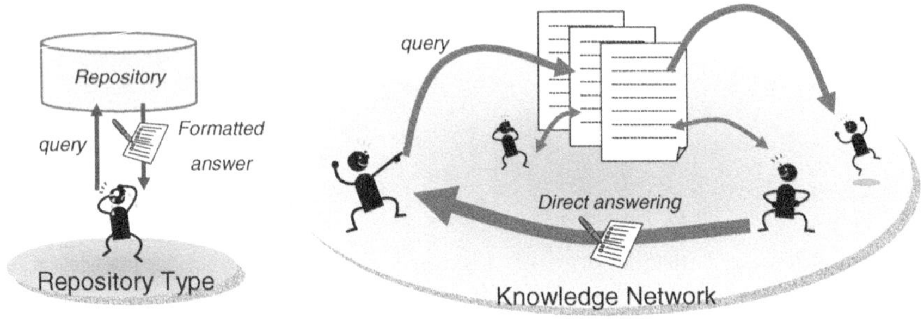

Fig. 1. Repository Type and Knowledge Network

2.3 Stickiness of Knowledge-Network

Because the value of knowledge-networks depends on how users continue to access it, Ashley et al. term the degree of this dependence as "stickiness" [4]. They also claim that there are three factors in making knowledge-network systems sticky; *Relationship Capital*, *Reputation* and *Personalization* as listed in Table 1.

Table 1. Three Factors to Drive Stickiness

	Driver	Reason and Design Features
(1)	Relationship Capital	Relationships provides privileged access Show network-relationships Develop relationship circles
(2)	Reputation	Reputation adds legitimacy and increases a user's influence Make reputation persistent Increase profiles' visibility Reduce transferability
(3)	Personalization	Increase context Reduce content/imitability of personalization

Relationship Capital focuses on the centrality of knowledge-networks for users. This factor forbids for other network users to access ideas, knowledge and opportunities, which raise reliability and affinity in the knowledge-network users. To develop this factor, it is important to show network-relationship and to develop relationship circles.

Reputation shows and quantifies users' contributions to a knowledge-network. If users continue to improve their reputation, a stickier knowledge-network is constructed. In addition, it enables users to share information with certainty because other users make decisions based on reputations. To implement this factor, it is necessary to make *Reputation* persistent, increase profile visibility and reduce transferability to other networks.

Personalization allows users to ignore irrelevant information and focus on their interests. Unlike repository types as mentioned in 2.1, users personalize their community by exchanging knowledge such as lesson learned and experience. This *Personalization* differs from personalization of content, which is recommended based on their records. The techniques for such personalization are discussed in other research [5] [6].

3 Proposal of Prototype and Quantified Stickiness

For Japanese science teachers, there has been a knowledge-repository called the "Rika Network", from which teachers download a large amount of digital content [7]. Although there is a variety of digital content on scientific education, it is a time-consuming task for teacher to use it in their classrooms because knowledge on how to use it has not been externalized, but remains as tacit knowledge. Our knowledge-network prototype, with a basic SNS, *Reputation*, *Personalization* and *pulling user in* is, on the other hand, interactive in real-time and provides two types of quantification for continually exchanging knowledge on digital content usage; quantification of the knowledge-network and that of the user.

3.1 Knowledge-Network SNS Based on Three Factors

For our research, the practice is rather indispensable than theory. The prototype is based on an open source SNS with the following functions for implementing and evaluating in a short period;

My Page: is a page for user who is invited by his colleague or system administrator. When he logs in this SNS, it shows his or her colleagues, his special interest group as a community and latest information.
My Friend: is a subset of SNS users. The user continues to build and show network relationships with acquaintances.
My Community: is a virtual space for commenting and browsing topics and writings with fellow science teachers. Once a user joins in a community of interest, he or she has a chance to exchange ideas and experience among community members.
Diary: is a kind of web log in which user select different privacy settings.
Others: including message exchanging such as e-mail, reviewing, and footprint, which shows users who has visited his or her *My Page* et al.

My friend enables the construction of a knowledge-network based on what Tiwana mentions as a peer-to-peer relationship, which is different from a knowledge-repository, for sharing information. User *X* might not be my friend but he or she is a friend of my friend (user *Y*). Such a chain-type relationship can be dynamically modified once user *X* becomes a friend of user *Y*. In other words, this function works as *Relationship Capital*, and users thereby acknowledge joining the knowledge-network.

My community is one way to realize social constructivism. Once a user joins a community, he or she can share information with other members of the community. Because some members might not be *My Friends* another type of knowledge-network is constructed. We believe that this function works as *Reputation* as Ashley mentions since an influential person who has a high reputation, frequently comments on topics.

Diary works as *Personalization* which is a factor driving stickiness. We believe that *Diary* and *Web Log* contribute to personalization because users bring casually information and then the effectiveness is claimed. According to our trial, it is confirmed that the basic SNS partially contributes to *Relationship Capital* but should be enhanced to be stickier as follows. We mention them later;

Reputation: improving reputation for users and content visualization.
Personalization: improving personalization of user to show user characteristics and advantages.
Pulling user in: developing a function to support non-sticky users.

3.2 Quantification of Stickiness

To express the degree on how users continue to use the knowledge-network, two approaches are considered when we quantify and build stickiness; "how stickiness is designed with the three factors; *Relationship Capital*, *Reputation* and *Personalization*" and "how sticky users are". We define stickiness of system $Sys(t)$ and user i $U_i(t)$ from the following equations;

$$Sys(t) = f(Rpt, Psnl, U_i(t))$$

$$U_i(t) = \int_{t-T}^{t} g(Frq, Utl, Act)dt + h(RC)$$

$f(x), g(y), h(z)$: arbitrary functions
Rpt: the net number of reputations which are visualized and quantified
$Psnl$: the net number of types for labeling
Frq: the visiting count in unit interval Δt
Utl: the viewing time in unit interval Δt
Act: the number of action in unit interval Δt
(Action indicates browsing and commenting articles, and messaging in this research)
RC: the total number of my friend and my community of user i

$Sys(t)$ declares variable *Reputation*, *Personalization* and $U_i(t)$. $U_i(t)$ is defined under the assumption that it varies over time. $U_i(t)$ also includes *Relationship Capital* which is one of the factors to driving stickiness since $U_i(t)$ is related to $Sys(t)$ and RC is dependent on the user.

There are two advantages in considering quantification. First, it enables evaluation of the degree of designing for *Relationship Capital*, *Reputation* and *Personalization* with an objective index. Second, it enables to determine non-sticky users, introducing the function to attract users as mentioned in 3.3.

3.3 Prototype SNS for Japanese Science Teachers

The Prototype SNS is designed for Japanese science teachers as shown in Fig.2. While the traditional Rika Network has been provided for science teachers to mainly search and stock digital contents prepared by experts, the SNS constructs a knowledge-network for sharing knowledge learned in classrooms based on the relationship between teachers. As such relationship makes a link one teacher to others, the links become a longer chain of teachers. Even though a colleague may not know the answer to a teacher's question, he or she has the opportunity to obtain the answer by following the chain. The SNS is developed with OpenPNE (http://esteem.center.osakafu-u.ac.jp/rikanet-sns/), which includes these four factors; basic SNS, *Reputation*, *Personalization* and *Pulling User in*.

(1) Basic SNS Since relationships between teachers and communities are shown on the *My Page* and lessons learned are available only for teachers who are in the networks, users thereby are conscious of the knowledge-network. Then it is thought that *My Friend* and *My Community* contribute to *Relationship Capital.*

Let us review the situation without an SNS on a knowledge network among teachers. Some teachers actively discuss their ICT class after that, the discussion still remains in the local area. The teachers in other areas cannot participate in the sharing of information. Even if guidelines on lessons learned are published after the discussion, a considerable time may have passed. For this situation, *Invitation*, a basic function for linking a user's colleague as his or her friend, is effective for discussing and giving feedback on a class.

(2) Reputation and Personalization *Reputation* and *Personalization* are available for inspection and contribution on *My Page* as shown in Fig.3. This is because

Reputation affects to *Personalization* and they are estimated by the user's actions. Recently, another study has proposed an interface for defining the maturity of science teachers and content [8].

Note that some externalized knowledge is available globally for all teachers and others may be available locally. These functions thereby enable other teachers to support knowledge internalization if content (teachers/digital contents/comments etc...) is characterized with labels and reputations. Since novice teachers (few experienced in ICT/normal classes) hope to learn from skilled teachers, they study contexts of these lessons learned. Uploaded content is judged for appropriate situations labeled and evaluated.

Reputation is expressed as stars, which visualizes and quantifies the degree of contribution to the knowledge-network, shows the effectiveness of externalized knowledge to *My Friend* and *My Community*. A high reputation is given if a teacher has many *My Friend*, *Diaries* and *comments* and if his or her externalized knowledge is referred to frequently. If teachers who have high reputations make an effort not to decrease their reputations and others make an effort to increase their reputations, they become stickier.

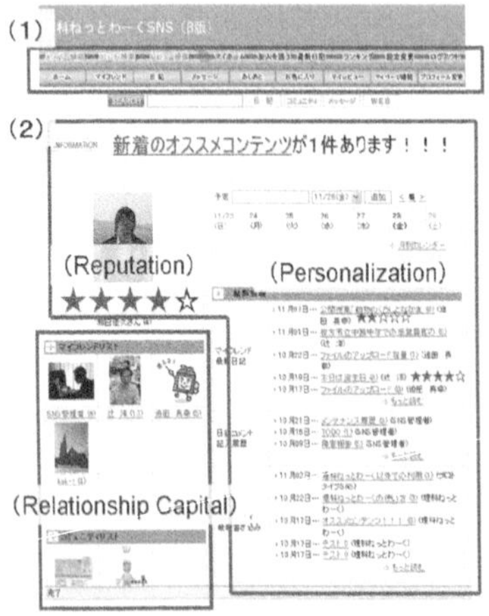

Fig. 2 Prototype SNS

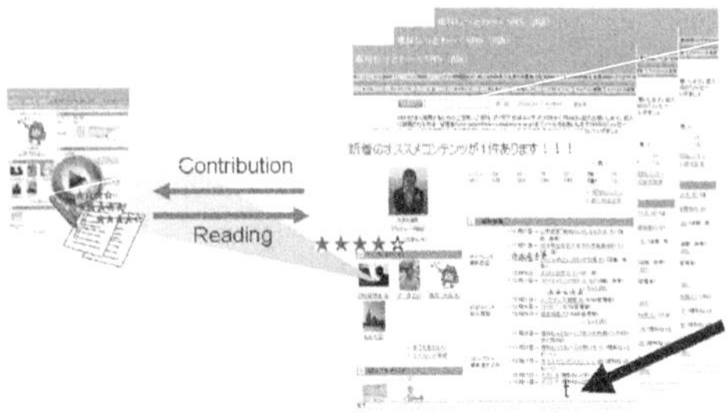

Fig. 3 Reputation and Personalization

Personalization, which gives label and function for individual users, is explained not only in terms of maturity as described previously, but also in accumulate ICT class research for providing an interface [9,10]. A chemistry teacher who has experiences in creating original digital content, for example, is labeled as a "professional developer of chemical content" and then other teachers become sticker by repeatedly referring to him.

(3) Pulling User in Quantification of stickiness enables a new function for attracting non-sticky users in as shown in Fig.4. This function identifies non-sticky users who are in the *My Friend* of an active user, and then sends an instant message to the latter user when the former acts. This is designed under the assumption that obtaining a

message from the same knowledge-network person is much more effective for non-sticky users compared to an automated system. Since the knowledge-network cannot function without sticky users, pulling users in as well as quantification is important.

In general, some non-sticky teachers do not repeatedly teach ICT classes once they fail nor they refer to as having good practices because of many troublesome tasks. Some sticky teachers, who contribute to the knowledge-network on the other hand, have incentives for non-sticky teachers. As a sticky teacher attracts non-sticky teachers, the SNS will continue to be effective.

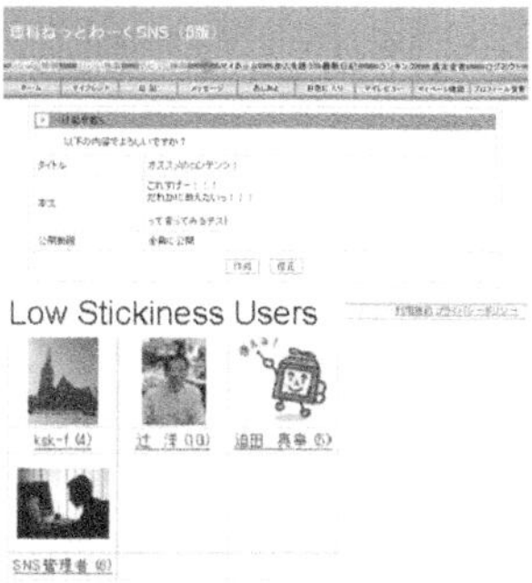

Fig. 4. Pulling non-Sticky User

4 Discussion

To confirm the effectiveness of the proposal, we discuss the infrastructure for knowledge sharing. It is mainly discussed that an SNS applied to a knowledge-network based on stickiness and quantified stickiness enables evaluation of the system and users.

Table 2. Comparison with Other Systems

System	Knowledge-Type Rep or Net)	Stickiness			Network Externality	Stock	Search	Main
		Relationship Capital	Reputation	Personalization				
Web Pages	Repository	bad	-	-	bad	-	bad	Download
Groupware	Network	-	bad	bad	-	good	-	Mail BBS Calender
SNS (Prototype)	Network	good	- (good)	bad (good)	good	-	-	Diar Comment Community Message
Wiki	Repository	bad	-	bad	bad	good	good	Edit

4.1 Comparison with Other Systems

Let us discuss four types of systems as listed in Table 2; standard web pages, groupware as a management information system, an SNS on which our system is based, and a Wiki, which can be accessed by anyone, such as Wikipedia.

For *Relationship Capital*, an SNS is superior to other systems because it is based on human relationships. Furthermore, *Reputation* and *Personalization* are not fully designed in any system. This is because designing approaches with *Reputation* and *Personalization* based on stickiness is not a concern for these systems. In addition, we found constructing an SNS for a knowledge-network is more beneficial because personalization often interrupts business affairs in groupware.

There is another concept, called *network externality*, for network services or products. The higher the number of user, the higher the value of the network services or products becomes [11]. For the prototype, the utility value is based on sharing information as frequently as possible in social networks [12]. From this viewpoint increasing the utility value, *Invitation* and *pulling users in* are effective.

4.2 Verification of Quantified Stickiness

To compare the value of stickiness, $Sys(t)$ and $U_i(t)$ are formulated for the prototype from the following equations;

$$Sys(prototype) = Rpt \times Psnl \times nU_{me}(t)$$

$$U_{me}(t) = (Frq,\ Utl \text{ and } Act \text{ in the last seven days}) + (RC \text{ in time } t)$$

n: the total number of user

The formulated quantification is strongly reflected by the number of users as listed in Table 3. Then $Sys(t)$ of the prototype is lower than that of Mixi with five million users, and that of Amazon with ten million reputation items. The $U_i(t)$ of the Mixi, on the other hand, is the highest in the author's case, because this function includes high RC and these variables are directly affected by the number of users.

It is considered that the "Pulling Users In function" increases $Sys(t)$ since this function is based on user stickier. In addition, precise quantification explains how these three factors change in our system change. This shall contribute to inspect not only knowledge-network but also related future systems.

Table 3. Estimated Stickiness

Example	$Sys(t)$	$U_i(t)$
mixi	7.8×10^7	155
Amazon	8.0×10^6	8
Prototype	900	9

Variables are shown in Apendix Table 4.

5 Conclusion

This paper has proposed a knowledge-network SNS for sharing knowledge and experience among Japanese science teachers. Stickiness can also be quantified when considering the three factors; *Relationship Capital*, *Reputation* and *Personalization*. The results showed Japanese science teacher hoped to use the proposed system to share their knowledge and experience concerning ICT classes. Therefore, introduction of this system into classrooms will be effective, though it needs more improvement and evaluation.

In the future, improvement and maintenance of the prototype SNS is recommended for disseminating knowledge and experience to many Japanese science teachers. A chain of knowledge and experience accordingly become longer as the number of user increase. Moreover, we have to precisely quantify the stickiness for designing systems and with sticky functions. Accordingly we can launch advanced engineering approaches to visualize and optimize. This knowledge-network SNS based on stickiness for sharing knowledge and experience will be available for other fields in the future.

Acknowledgement

This work is partially supported by Japan Science and Technology (JST). Some idea comes from discussion with Osaka Prefectural Education Center who gave opportunities to observe and review experimental classes in elementary or junior high school. The authors would like to be grateful to Mr. Kenji Kaibe and Mr. Hisanobu Tanabe for giving valuable chances and comments.

References

1. Ping, Z., Aikman, S.: Attitudes in ICT Acceptance and Use. In: Jacko, J.A. (ed.) HCI 2007. LNCS, vol. 4550, pp. 1021–1030. Springer, Heidelberg (2007)
2. Rushed, K., Malek, M.: Computing Social Networks for Information Sharing: A Case-Based Approach. In: Schuler, D. (ed.) HCII 2007 and OCSC 2007. LNCS, vol. 4564, pp. 86–95. Springer, Heidelberg (2007)
3. Tiwana, A.: Affinity to Infinity in Peer-to-Peer Knowledge Platforms. Communications of the ACM 46(5), 77–80 (2003)
4. Ashley, A.B., Tiwana, A.: Designing Sticky Knowledge Networks. Communications of the ACM 48(5), 67–71 (2005)
5. Resnick, P., Varian, H.R.: Recommender Systems. Communications of the ACM 40(3), 56–58 (1997)
6. Saga, R., Tsuji, H.: Sales Records Based Recommender System for TPO-Goods. The Transactions of the Institute of Electrical Engineers of Japan C 126(5), 661–666 (2006)
7. Japan Science and Technology Agency: Rikanetwork (2008), `http://www.rikanet.jst.go.jp/`
8. Sakoda, M., Wada, Y., Tsuji, H.: Maturity Model based Social Networking Service for Science Teachers. The Papers of Technical Meeting on Information Systems, IEE Japan, IS-08-46 (2008)

9. Suzuki, K.: Current Trends of Instructional Design Theories and Models for ICT Utilization. IPSJ Magazine 47(9), 993–1000 (2006)
10. Sato, S., Tanikawa, T., Hirose, M.: Proposal and Evaluation of an Interface to View SNS Data Accumulated through Practical Learning. Journal of Human Interface Society 10(4) (2008)
11. Katz, M.L., Shapiro, C.: Network externalities competition and comaptibility. American Economic Reviews 75(3), 424–440 (1995)
12. Mori, J., Matsuno, Y., Sugiyama, T., Ishizuka, M.: Real-world oriented Information Sharing System. In: The 19th Annual Conference of the Japanese Society for Artificial Intellingence, 1C2–03 (2005)

Appendix

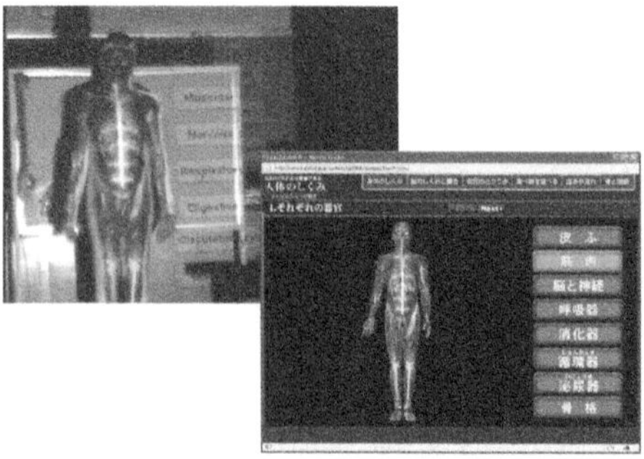

Fig. 5. An Example of Class with Digital Contents

Table 4. Variables to Estimate Stickiness

Example	*Rpt*	*Psnl*	*n*	**Frq*	**Utl*	**Act*	*RC*
mixi	5.0×10^5	1(text)		10	4	6	135
Amazon	1.0×10^6	2(text/star)	1	3	2	2	1
Prototype	100	**27			1		8

*Ratio when Prototype is assumed to be 1
**Year(Senior/-/Junior), PC(Good/-/Poor), DigitalContent(Good/-/Poor)

Part IV

Novel Devices, Interfaces and Interaction Environments

A Proposal of EMG-Based Training Support System for Basketball Dribbling

Seimei Abe, Takayuki Nozawa, and Toshiyuki Kondo

Tokyo University of Agriculture and Technology
2-24-16, Naka-cho, Koganei, Tokyo, Japan
seimei@livingsys.lab.tuat.ac.jp,
{tknozawa,t_kondo}@cc.tuat.ac.jp

Abstract. In the paper, we propose a novel HCI methodology for supporting human sports training. The proposed system utilizes electromyogram (EMG) signals as the metrics for evaluating the motor skill acquisition process, because it is well known that EMG signals measured from human experts are notably different from beginners in terms of timing and sharpness. According to this, we hypothesized that visualizing the difference of EMG signals between an expert and a learner, and providing the error information to the learner in real-time accelerates the learning process. The preliminary results show that the proposed method is effective especially in the early stage of training of beginners.

1 Introduction

In the field of sports science, body motions of human experts have been measured for motion analysis. In addition to motion capture systems, recently we can use various lightweight sensors (e.g. acceleration detector) by installing them in each part of body. The measured motion data are analyzed with computers, and it has been tried to clarify the experts' motor skill which is difficult for them to state [1, 2]. Moreover, in the area of motor learning or rehabilitation, these real-time measurements can be used as feedback (i.e. knowledge of results; KR) for acquiring motor skills [3]. Furthermore, nowadays these motion data are utilized for enhancing reality in entertainment applications [4].

When we begin to acquire some new body motion, it is generally accepted that we try to form an image of experts' motion in the brain as a model to imitate. The image we stated here corresponds to the position of body parts, the trajectories of joint angles, timing of motion, etc. It was reported that having the image to imitate accelerates motor learning, especially in the early stage of the learning [5]. However, visually observable information of experts' motor skills is restricted to kinematic one. Due to the lack of dynamic information (e.g. muscle tension), a learner needs to develop an internal model, i.e. a sensorimotor map between somatosensory stimulus and a motor command, via active trial-and-error.

On the other hand, electromyogram (EMG) is a biomedical signal which directly reflects motor commands from the brain to activate muscles. Therefore it has long been investigated in the sports science field. For example, Sakurai repeatedly

M.J. Smith and G. Salvendy (Eds.): Human Interface, Part I, HCII 2009, LNCS 5617, pp. 459–465, 2009.

measured the EMG data during smash motion of badminton, and compared the feature of beginners and experts' myoelectric signals [1]. According to the report, experts' myogenic potential immediately decreases after the impact, while the decrease cannot be observed in the beginners' EMG signal; however it gradually comes close to the experts' as the training goes on.

Based on the above observation, it is assumed that the difference of motor skill between beginners and experts can be quantitatively estimated from the discrepancy of the EMG signals. Thus we hypothesized that visualizing the difference of EMG signals between an expert and a learner, and providing the error information to the learner in real-time accelerates the learning process. Based on the hypothesis, in the paper, we propose a novel HCI methodology for supporting human sports training, in particular basketball dribbling.

2 System

The aim of the present study is to evaluate the effectiveness of the proposed training support system with regard to the basketball dribbling. First of all, we analyzed the EMG data measured from dribbling experts and beginners. In all experiments, we used the dry surface electrodes for EMG recoding (NM-512G, Nihon Kohden) and the multi-telemetry system (WEB-5000, Nihon Kohden). From the result of preliminary experiments, the positions for EMG recoding are determined at the *flexor carpi radialis* (forearm) and the *triceps brachii* (upper arm).

Figure 1 schematically shows a typical pattern of experts' EMG signals in one dribbling cycle (left: upper arm, right: forearm). The three peaks correspond to the following three tricks:

(1) in receiving the bounced ball, pulling up the forearm lightly and decelerate the ball,
(2) in pushing down the ball, pushing down the upper arm and stretching the elbow, and
(3) in sending down the ball, snapping the wrist.

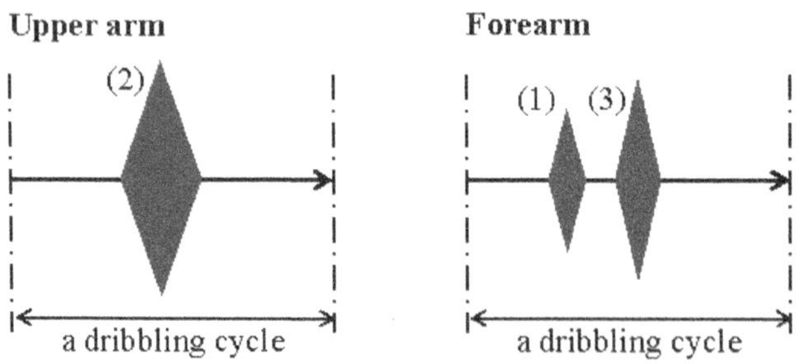

Fig. 1. Schematic shape of the EMG signals of dribbling experts

On the other hand, the EMG captured from beginners frequently lacked these features, especially the peak corresponding to (1). Thus we focused on the region that includes these three features.

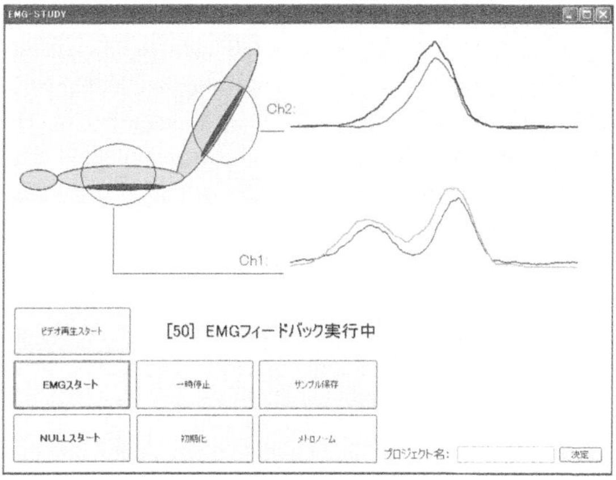

Fig. 2. Visual feedback of the upper arm and forearm EMG patterns. The patterns enable quick comparison of the learner's and the expert's muscle activity features during a cycle of dribbling.

As a measured EMG signal, i.e. raw data, has a lot of frequency components, the wave form itself is too complicated for a learner to understand quickly. To emphasize the three features previously described, we extracted schematic shape of the EMG signal in one dribbling cycle by using a first order differential filter, a full-wave rectification, and a smoothing by moving average with the time window 80 msec. The extracted EMG data is named *EMG patterns*.

In order to effectively provide the difference of expert and beginners' EMG signals during the training, we constructed the visual feedback system shown in Figure 2. As shown in the figure, on the expert's EMG patterns which are collected beforehand (blue and orange lines), the learner's EMG patterns (green and red lines), obtained via the above preprocessing, are superimposed and updated periodically.

During the training with the proposed system, a learner tries to tune his/her own EMG patterns to the expert's EMG patterns. Note that the learner's EMG pattern is updated every 5 sec.

3 Method

To evaluate the effect of the proposed EMG-based training support system, we executed a training experiment. Six healthy subjects (male, aged 21-24) participated in the experiment. None of them has any special training experience of the basketball dribbling task. They were randomly assigned to one of the two groups, i.e. the group of training with the proposed visual feedback system (hereafter, Feedback condition), or without any feedback information (Null condition). The difference of these training conditions was limited only to the presence of the visual feedback of the EMG patterns.

Before starting the training experiment, the experimenter who is a dribbling expert explained the three features observed in the expert dribbling to all subjects, and they were asked to keep these points in mind while the whole training session.

Additionally, the subjects in Feedback condition group were asked to tune their EMG patterns to the expert's ones as much as possible. Moreover, all subjects were made to hear an electronic metronome sound (1.5 Hz) to keep their dribbling period constant.

Each subject did the dribble training for approximately 30 minutes as a whole. It included first half training (15 min.), rest (5 min.), and second half training (15 min.). To evaluate the performance improvement through the training, three *performance tests* (30 sec.) were executed before and after the first half and after the second half, respectively.

4 Results

4.1 Qualitative Comparison of Raw EMG Data

Transition of the raw EMG signals, i.e. their changes from before to after the training, was compared between the two training conditions. In the case of upper arm, we could not confirm any difference between the conditions. Thus the results of forearm muscles are shown in the paper. Figure 3 represents transition of raw EMG signals measured from the forearm of a subject trained under Feedback condition, and Figure 4 shows the results of a subject trained under Null condition.

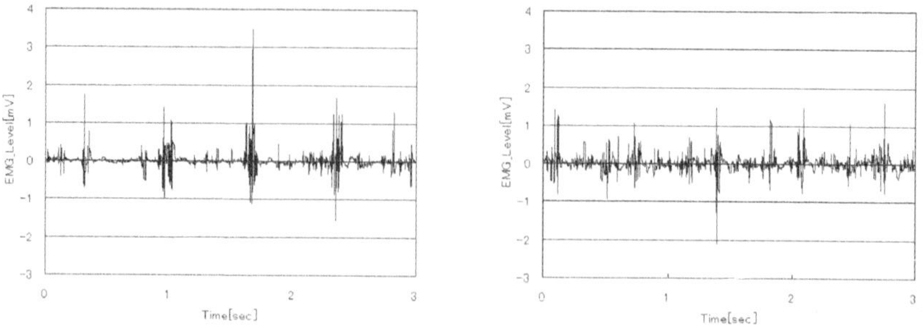

Fig. 3. Transition of raw EMG signals measured from a subject trained under EMG-Feedback condition (left: initial performance test, right: final performance test).

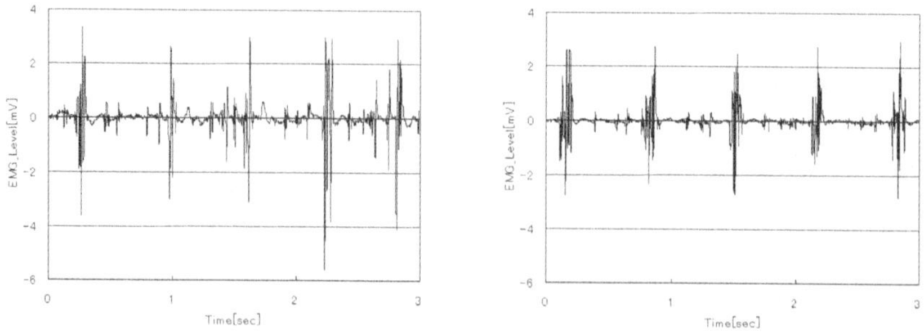

Fig. 4. Transition of raw EMG signals measured from a subject trained under Null condition (left: initial performance test, right: final performance test)

As shown in Figure 3 left and Figure 4 left, there was no distinct difference between the subjects of two conditions in the initial test. On the contrary, in the final test, one of the experts' EMG features—it is connected with the forearm pulling up motion when receiving the bounced ball—was observed in the subjects of Feedback condition (Figure 3 right). On the other hand, no acquisition of the experts' EMG features was observed in the subjects of Null condition (Figure 4 right).

4.2 Quantitative Evaluation of EMG Patterns

We conducted quantitative evaluation of the difference described above. First, standardizing the time interval of EMG patterns of one dribbling cycle as [0, 1], we identified that the muscle activity corresponding to the technique of pulling up forearm appeared in the region of [0.1, 0.45], which we named *technique region* (see Figure 5). Accordingly, we decided to use the amount of forearm muscle activity in this technique region as a measure of the forearm pulling up technique. The amount of muscle activity in a time interval was calculated by applying to EMG signals the following steps: (i) first order differentiation, (ii) full-wave rectification, (iii) standardization of a dribbling cycle into [0, 1], and (iv) integration over the given interval.

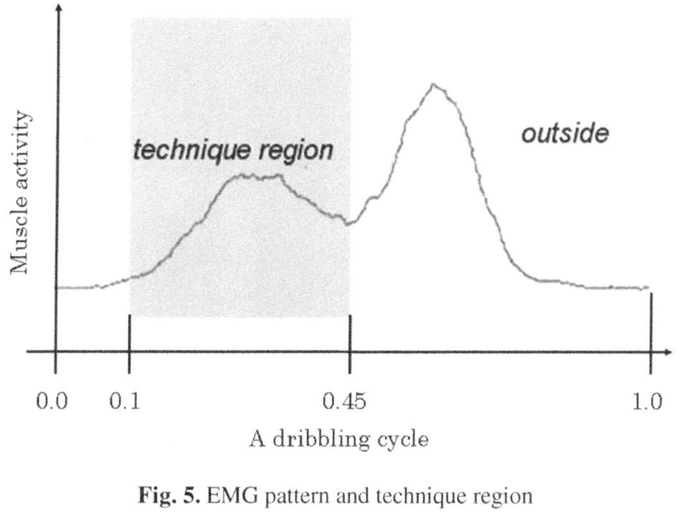

Fig. 5. EMG pattern and technique region

Figure 6 shows mean forearm muscle activity ratios of the technique region in the initial and final test, for each subject ("EMG_X" are the subjects under Feedback condition, and "NULL_X" are those under Null condition). To cancel the difference of muscle activity level between individuals, the amount of muscle activity in the technique region was normalized by the one over whole dribbling cycle. The dribbling cycles (number of samples) that contributed to the evaluation value were about 20 (for stability of evaluation, early parts of the tests were discarded).

T-test (P=0.05) of the mean value between initial and final test showed that for two of the three subjects who exercised under Feedback condition, the muscle activity of forearm at the timing of the technique region increased significantly after the training,

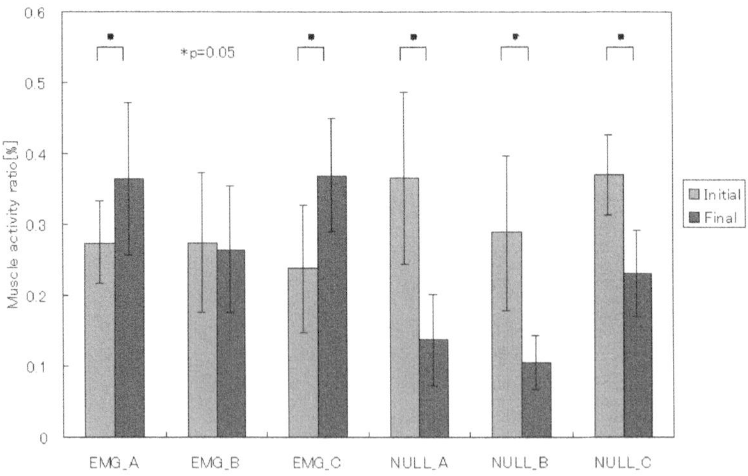

Fig. 6. Comparisons of muscle activities in the arm-pulling region (initial vs. final tests)

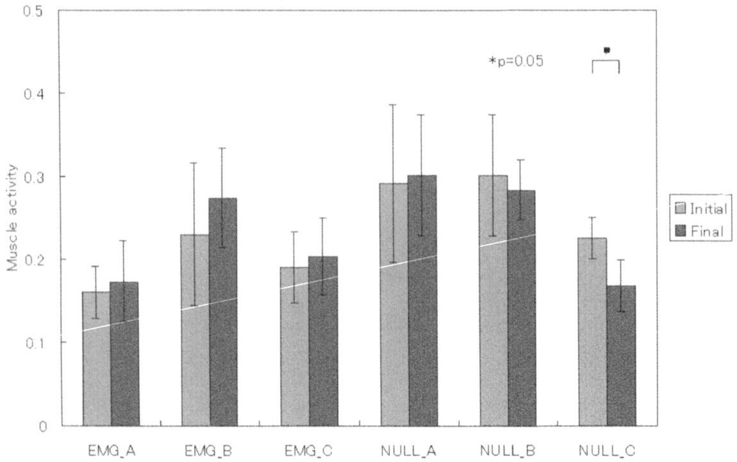

Fig. 7. Comparison of muscle activities outside the arm-pulling region (initial vs. final tests)

indicating the acquisition of the forearm pulling up technique. On the other hand, forearm muscle activity at the same timing significantly decreased for all the three subjects with Null condition, confirming their failure to acquire the technique. These results suggest that the proposed system is effective in supporting the acquisition of the forearm pulling up technique. A possible interpretation of these results is that the real-time feedback of EMG pattern made the learners conscious of their own muscle activity, and that promoted their learning process. In contrast, the learners under Null condition had no external information to judge whether they are achieving proper movement (desirable technique).

We also studied the change of muscle activity *outside* the technique region. As shown in Figure 7, for the subjects in both conditions, the amount of muscle activity outside the technique region indicated no significant change before and after training. This result confirms that Feedback condition worked to raise the muscle activity only at necessary point and timing, excluding the possibility that it just augmented general muscle stress and induced overstrain by making the learners too sensitive to their muscle activity.

In sum, our study supported the hypothesis that the EMG feedback in real time is effective in motor learning, especially with respect to timing and power (i.e. coordination).

5 Conclusions

In this study, we proposed a new motor skill acquisition support system. The proposed system utilizes electromyogram (EMG) signals as the metrics for evaluating the motor skill acquisition process, because EMG signals measured from human experts are notably different from beginners in terms of timing and sharpness. According to this, we hypothesized that visualizing the difference of EMG signals between an expert and a learner, and providing the error information to the learner in real-time accelerates the learning process. The preliminary results show that the proposed method is effective especially in the early stage training of beginners.

However, retainment of the motor skill acquired through the proposed system is not confirmed yet, because in our experiment, final performance test was executed immediately after the training. Furthermore, we should also confirm the causation that the similarity of EMG pattern results in improvement of the basketball dribbling skill of the subject.

Acknowledgments. This research was partially supported by the Ministry of Education, Culture, Sports, Science, and Technology, Grant-in-Aid for Scientific Research on Priority Areas (No.20033007), and "Symbiotic Information Technology Research Project" of Tokyo University of Agriculture and Technology.

References

1. Sakurai, S., Ohtsuki, T.: Muscle activity and accuracy of performance of the smash stroke in badminton with reference to skill and practice. Journal of Sports Sciences 18, 901–914 (2000)
2. Furukawa, K., Yoshinaga, S., Sawai, K., Shimizu, S., Kinjo, K.: On modeling bow arm movement in cello playing by whip motion. In: Proc. of 3rd European Medical and Biological Engineering Conference (2005)
3. Bilodeau, E.A., Bilodeau, I.M., Schumsky, D.A.: Some effects of introducing and withdrawing knowledge of results early and late in practice. Journal of Experimental Psychology 58, 142–144 (1959)
4. Nintendo Wii (2009), `http://www.nintendo.com/wii`
5. Mattar, A., Paul, G.: Motor Learning by Observing. Neuron 46, 153–160 (2005)

UbiSOA Dashboard: Integrating the Physical and Digital Domains through Mashups

Edgardo Avilés-López and J. Antonio García-Macías

Computer Science Department
CICESE Research Center
Km. 107 Carretera Tijuana-Ensenada
Ensenada, Baja California, México
{aviles1,jagm}@cicese.mx

Abstract. The current Web 2.0 stage of the Internet provided the basis for web-based communities and services aimed at collaboration and information sharing. Furthermore, Internet is now an application platform in which Web applications can be integrated to provide augmented services that could bring the basis for ubiquitous computing scenarios. Recently, the concept of mashups has been used to refer to applications built upon the integration and combination of public Web API's and data sources. Ubiquitous computing mashups go further by combining the functionality of both software and hardware components in an attempt to exploit computation and services provided by everyday objects. Typically, developing a mashup requires highly specialized knowledge in many topics (such as using different programming interfaces and languages). This problem is greatly magnified in developing mashups of both physical and digital services due to the various integration and communication issues. We exemplify these concepts through the use of UbiSOA Editor, a system that allows the creation of ubiquitous computing mashups through simple activities such as dragging and dropping graphical representations of the involved services in a desired scenario. Then we talk about the planning and execution of a sample scenario as a showcase of what can be easily accomplished.

1 Introduction

The vision of ubiquitous computing, as articulated more than 15 years ago in a seminal article by Mark Weiser stated that "the most profound technologies are those that disappear. They weave themselves into the fabric of everyday life until they are indistinguishable from it" [1]. Although this vision has not completely become a reality, we are getting closer. Nowadays we have GPS-equipped cars that will inform us about road conditions and propose driving routes, we can check our e-mail, make a phone call, listen to music and watch video clips with the same device (*e.g.*, Apple iPhone[1], which can be used for people-centric sensing applications [2]). All this is possible by micro-controllers embedded in the artifacts that we use everyday, some of which have truly embedded computers that we use without being aware of it. That is,

[1] Apple iPhone: http://www.apple.com/iphone/

M.J. Smith and G. Salvendy (Eds.): Human Interface, Part I, HCII 2009, LNCS 5617, pp. 466–474, 2009.

technology is weaving itself into the fabric of everyday life. Weiser also stated that "machines that fit the human environment instead of forcing humans to enter theirs will make using a computer as refreshing as taking a walk in the woods". Although using computers, embedded or not, is currently not as refreshing as a walk in the woods, there have been some advances in multimodal user interfaces promising better interactions that fit the human environment.

Some have suggested that integration is one of the main barriers to realizing the vision of ubiquitous computing [3]. Indeed, some of the appropriate infrastructural elements exist, and there have been significant advances in human-computer interfaces, but there are many ubiquitous computing scenarios still a matter of science fiction and not part of our daily lives [4]. We believe that an important factor is that, for the creation of the applications in these scenarios, we are accessing the infrastructural elements and integrating them at a very low-level. This forces developers to focus on unnecessary infrastructure-level details, instead of focusing on application-level issues.

In this article we start by discussing two of the key issues in the development of ubiquitous computing applications: access and integration (Section 2). We then present UbiSOA and discuss how it can help in accessing context providers (Section 3). We also discuss the integration issue and talk about the Web as a platform for ubiquitous computing by showing how mashups can be helpful for integrating different services and primary context sources. We exemplify those concepts through the development of a ubiquitous computing mashup application by using of UbiSOA Editor (Sections 4 and 5). Finally, we present some conclusions and future work (Section 6).

2 Access and Integration

Our current technological situation can be easily understood with an example: the development of a dashboard application where we can monitor what is going on at home, at the office, or at our grandparent's house. We would be interested to check out if some lights were left turned on at home, or maybe turn them on while we are traveling to deter burglars from breaking-in. Maybe we would like to keep an eye on our grandparents and monitor the air quality at their home to warn us of possible gas leaks, or want to have an overview of their motion patterns during the day. It would also be useful if this dashboard could integrate some services we already use, like traffic alerts to avoid the bottlenecks in the highway. All the elements to make this a reality exist: there are already in the market Zigbee-enabled[2] devices to control lights in a networked fashion; many networked sensors, including environmental monitors and motion detectors, are also in the market; traffic alerts services are also available from many providers. But there are two key problems that make creating this dashboard difficult: access and integration.

First, let us examine access. The intelligent network-controlled lights may use the Zigbee protocol stack. This implies having to learn some Zigbee concepts, including how to bind end-devices to controllers, how to create a network, etc. Coding the application may force the programmer to use a particular development environment

[2] Zigbee Alliance: http://www.zigbee.org/

(*e.g.*, Codewarrior[3] for HCS08 microcontrollers) and programming language (typically C). For the sensor network at our grandparent's home, there are already many wireless sensor network kits available. TinyOS [5] is a very popular platform in these kits, but it imposes programming in a particular programming language called nesC [6] which, not being a general purpose language will be unlikely already known by a developer. This language has a certain philosophy where everything is a component and in order to build an application we have to bind them using certain interfaces. Other embedded systems concepts, such as threads of execution and atomic sections may be in order. The part developers would be more familiar with is the traffic alerts parts, since companies like Google and others have public APIs[4] that can be used to invoke their services by means of familiar Web concepts. Now that we have access to the different platforms (the home automation network, the monitoring network, and the traffic alerts), there is yet another problem: the integration of them all. Since the dashboard must be accessed anytime, the home automation network and our grandparent's sensor network must be properly Internet-enabled. This means, writing some sort of Internet gateway for each of them, and since their code is written in different languages and they have different idiosyncrasies, reutilization is mostly not possible. A web page could be the integration element, since the gateways can be configured to output HTML and some JavaScript code would allow to access the services provided by Google Maps[5].

3 Better Access with UbiSOA

The concept of the Ubiquitous Web (UW) [7] has received a lot of attention lately, as it defines a pervasive web infrastructure where physical objects are integrated into the world of web information and services. Many features make the Web attractive as a platform for ubiquitous computing: it is device-independent and language-agnostic, it offers low barriers to entry, it has a content-centric model and it is, for all practical purposes, ubiquitous.

Regarding the access problem, providing better abstractions to developers has been a long-time motivation in software engineering. Therefore, we have witnessed advances in programming methodologies and paradigms ranging from sequential programming, modular programming, object oriented programming, component-based programming, and more recently service-oriented programming. This recent approach arises in response to modern needs and complexities such as distributed software, application integration, as well as heterogeneity in platforms, protocols and devices, including Internet integration. It features a very high abstraction level; any functionality is exposed as a service, so the user just asks for a given service without knowing or having to deal with any low-level issues required for the functionality to be completed. For example, if we have just installed a new wireless sensor node, and we want to know what the moisture reading is, we would have to go through the cumbersome tasks discussed earlier. In contrast, with the service-oriented model, we would

[3] Codewarrior Development Tools: http://www.freescale.com/codewarrior/
[4] Google APIs: http://code.google.com/apis/
[5] Google Maps: http://maps.google.com/

just ask what services the network provides and then we would invoke the service for reading the moisture level.

This is precisely the philosophy behind UbiSOA (which is a work in progress and has evolved from TinySOA [8]). As other authors [9], we think that the service-oriented and context-driven models are excellent means to implement successful ubiquitous applications. UbiSOA is a platform comprised of many different services, resources and tools to allow the creation of ubiquitous applications. Instances of such services are localization, environmental primary context providers (wireless sensor networks), identification context (RFID devices), and others. The access to those services is driven through standard and easy to use Web services (via the SOAP and REST protocol) and Web feeds (Atom and RSS feeds). An important advantage of Web services and feeds is that developers can use the same tools and languages they normally use for developing applications. This means, no hardware specifications needed, no need to learn or be tied to a particular language, and no wasting valuable time with these issues.

4 Better Integration with Mashups

In its origins, the Web provided a way to easily share information. Shared contents consisted of simple hyperlinked documents hosted on Web servers and were only maintained by their authors. One can refer to this stage as the Web 1.0, where the Internet's available content was on a "read-only" mode. As technologies evolved, server-side scripting languages (such as Perl, PHP, and others) gained attention. This made the Web servers to start changing dynamically the information they provided. This new "read-write" paradigm, gave the basis to web-based communities and services aimed at collaboration and information sharing. In this paradigm the user is no longer a consumer; he continuously contributes with new information (as seen on wiki systems). This still current stage is known as Web 2.0, where Web sites typically include technologies such as AJAX-enabled communications, syndication through RSS feeds, and a clear adoption of standards on presentation and content handling. Furthermore, we can see Internet as a platform in which many different Web applications are now providing public available APIs to manage and use the information they provide. This led to the introduction of mashups.

The mashup concept originated in music, where fragments of two or more songs are mixed and rearranged together creating new songs or compositions. A mashup in computer science refers to applications created by the integration and combination of services and content from different Web public APIs and sources. A very common example is a mashup that uses Google Maps to show the location of the most recent pictures of a Flickr[6] account. There are even many tools online that can help you on creating mashups: Microsoft Popfly[7], Yahoo! Pipes[8], Google Mashup Editor[9] and others. Those systems are ready to integrate information from some of the most used

[6] Flickr: http://www.flickr.com/
[7] Microsoft Popfly: http://www.popfly.ms/
[8] Yahoo! Pipes: http://pipes.yahoo.com/
[9] Google Mashup Editor: http://editor.googlemashups.com/

Web applications that already provide access to their functionality through Web services or other standardized means.

Recently, the term ubiquitous computing mashup is being used referring to the attempt to move computation off the desktop and integrate it with the artifacts of everyday life [10]. Ubiquitous computing application development can highly benefit from mashup technology. Common functionality can be isolated into single and independent components (service providers) that can be later combined to accomplish functionality needed for a specific scenario. Also, as the number of devices and other data sources grows, the need to filter the available information will be higher; mashups can then be used to delegate processing of the multiple sources (data or even functionality) generating a more manageable and scalable feed. Although, there is one problem, real-time needs could hardly be supported if they aren't carefully considered in the design and operation of each one of the service providers. Other aspects of Web 2.0 technology are being used to power ubiquitous scenarios, one of them is the model of sharing and annotating physical augmented devices in the same way wiki systems are [11].

Since UbiSOA's services allow us to retrieve primary context information from sensor networks and RFID devices (environmental and identification), we can go back to the monitoring dashboard we discussed earlier; we will see how UbiSOA can make building this type of applications a lot easier.

In order to showcase what can be easily accomplished, let us introduce the following scenario: let's suppose there are multiple sensor-networks deployed in some locations and that there are RFID readers in the door of multiple rooms on different locations that users use to report their arrival and departure. We need a Web application in which we can easily browse through all these locations (ideally using a map) and see the current status for each one, also, we need to be aware of many events including a user arriving/leaving a location. Clearly, developing this without using third-party services would be highly complicated, mainly because of dealing with the communication and data extraction of the many different involved devices, besides the problem of how to combine and remotely access all the data.

Second, we need to know what services we can use to accomplish the scenario and what we are going to integrate. In our case we will use Google Maps handled by an AJAX interface (for driving and selection of multiple locations), sensor networks (to fetch sensor data) and RFID services (to track people arriving/leaving locations and possibly other objects). This is going to be a static mashup; all the different providers will be manually selected to build the final application. Mashups can also be dynamic, allowing choosing on execution time which services could be used. This is highly useful, because for instance, we could use a backup service in case the main one is down or too busy to obtain timely responses.

Before discussing the third step which is the implementation of the scenario, we need to introduce UbiSOA Editor. In our infrastructure, a mashup application can be coded by means of a script that specifies the services involved, the connections between them, and the data flow though the configuration or assembled by using the UbiSOA Editor, a system to create mashups through simple activities such as dragging and dropping graphical representations of the involved services to accomplish the specification of configurations that support a desired scenario. In our scenario of use, a user must carry a device that provides him with personal services (such as localization, contacts, notification, and others), runs mashup applications, and allows

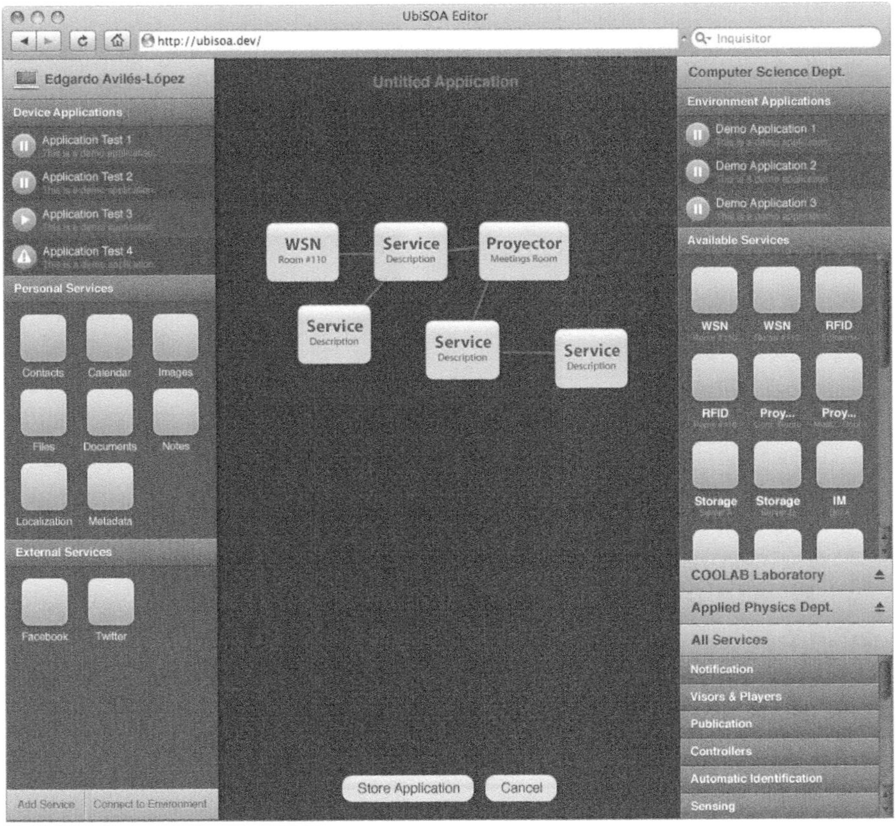

Fig. 1. Overview of the UbiSOA Editor

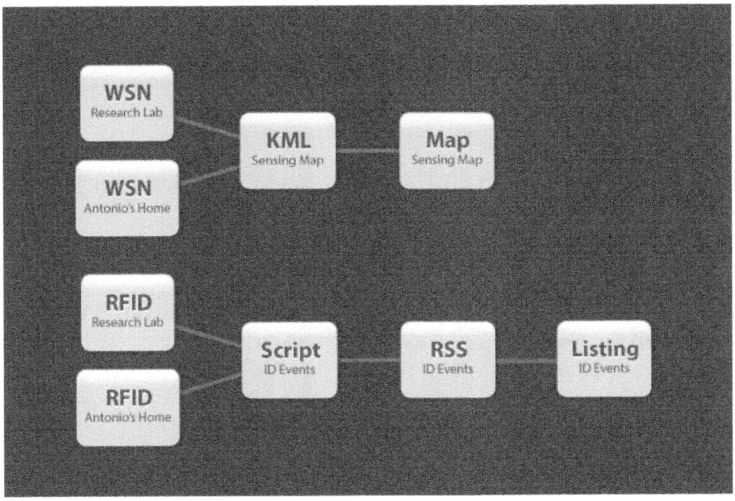

Fig. 2. Mashup configuration for the dashboard application

access to the infrastructure through the UbiSOA Editor. The personal applications and services are shown on the left sidebar of the editor (see Fig. 1). The UbiSOA infrastructure automatically discovers the services that are provided on the user environment and the applications that are currently running (those are shown on the right sidebar). Additionally to the basic personal services, the user can add external services such as his Facebook or Twitter account. The editor also allows interaction with remote environments. The area at the center of the editor is the instantiation area where the services are dropped, interconnected, and then stored into an environment or in the personal device as mashup applications.

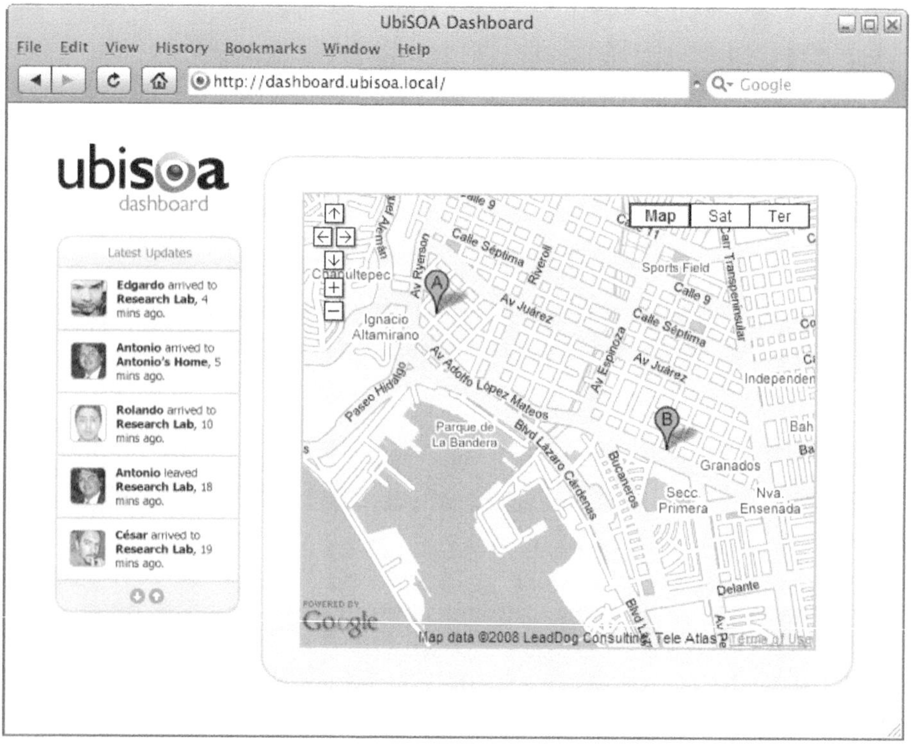

Fig. 3. The dashboard application build upon the embedded code from the mashup

Continuing with the dashboard implementation, we just have to drag, drop and interconnect the appropriate services for the scenario we desire. We need a sensor network service and a RFID reader service for each of the environments we want to monitor (we will first need to connect to them) and utility services that will provide our visualization needs. Is important to note that UbiSOA doesn't just introduce custom services, it also allows the integration with current Web resources and APIs such as Google's. In our sample dashboard we are going to support 2 locations, each with a WSN and a RFID service. As utility services, we need a KML maker service and a Script service that runs custom code for interpreting the RFID readings received from the readers, once interpreted we send the results to a RSS maker service. We could already use the generated KML and RSS files as data feeds for our dashboard application but UbiSOA also provides tools to embed special HTML+JavaScript code in our web applications that will help us on visualizing and updating those files when an infrastructure event is generated. Those tools are also services, in this case the Map and Listing services. The full configuration of this mashup is shown in Fig. 2. Once interconnected, we need to store this application in the environment or in our device to finally get the embedded codes of the sensing map and the identification listing. Figure 3 shows the finished dashboard application.

5 Towards the Ubiquitous Web

Almost by accident, or at least without being part of a grand plan, the Internet is playing a very important role in making the vision of ubiquitous computing a reality. What started as a medium for publishing and sharing information is now seen as a good platform for interconnecting all types of physical devices, integrating them into the information universe. The next stage in the evolution of the Web, already in progress and called by some Web 3.0, will be realized with semantic web developments. But beyond that, in what is called the Ubiquitous Web or the Internet of Things ([12, 13]), we need better means for accessing the services provided by physical objects and for integrating these services. Current Web 2.0 technologies, in the form of Web services for access and mashups for integration, provide a step ahead, although can only be seen as transitional technologies.

A decade ago, Weiser considered the Internet as a transitional stage between PC and UbiComp scenarios [14]. We can see proofs of this transition in current works such as the seamless integration of wireless sensor networks with IP networks [15]. More importantly, not only we have witnessed the needed changes on infrastructure and services, there are also changes on social aspects too; users are more prone to use a Web-based tool to do their tasks, for instance, increasingly more people are using webmail providers than institutionally hosted mail services, another notorious example is the adoption of instant-messaging systems.

6 Conclusions and Future Work

We believe that the benefits of mashup applications are important; they allow us to fulfill tasks and rapidly construct proof of concept applications which is highly valuable for ubiquitous computing research. Although there is a problem to overcome: developing a mashup application requires Web programming expertise due to the management of data representations, connectivity, and standards [16]. Higher-level abstraction tools that assist in constructing ubiquitous mashups are needed. These tools need to hide from the developer issues such as security, event management, real-time needs, communication, and others.

UbiSOA is still a work in progress; we are currently exploring communication and event notification models to allow a more transparent integration and cooperation between service providers. In this paper we have introduced the UbiSOA Editor which is an important part of the UbiSOA infrastructure. We plan to further improve the design by conducting a usefulness and adoption of technology experiment.

Acknowledgements. Financial support for this project was provided by the Mexican Council for Science and Technology (CONACyT).

References

1. Weiser, M.: The computer for the 21st century. In: SIGMOBILE Mob. Comput. Commun. Rev., vol. 3, pp. 3–11. ACM Press, New York (1999)
2. Miluzzo, E., Oakley, J.M.H., Lu, H., Lane, N.D., Peterson, R.A., Campbell, A.T.: Evaluating the iPhone as a mobile platform for people-centric sensing applications. In: Proc. of Intl Workshop on Urban, Community, and Social Applications of Networked Sensing Systems (UrbanSense 2008) (November 2008) (to appear)
3. Davies, N., Gellersen, H.: Beyond prototypes: Challenges in deploying ubiquitous systems. In: IEEE Pervasive Computing, Piscataway, USA, vol. 1, pp. 26–35. IEEE Educational Activities Department (2002)
4. Schmitz, M., Endres, C., Butz, A.: A survey of human-computer interaction design in science fiction movies. In: INTETAIN 2008: Proceedings of the 2nd international conference on INtelligent TEchnologies for interactive enterTAINment (January 2008)
5. Levis, P., Madden, S., Polastre, J., Szewczyk, R., Whitehouse, K., Woo, A., Gay, D., Hill, J., Welsh, M., Brewer, E., Culler, D.: TinyOS, an operating system for wireless sensor networks. Springer, New York (2005)
6. Gay, D., Levis, P., von Behren, R., Welsh, M., Brewer, E., Culler, D.: The nesC language: A holistic approach to networked embedded systems. In: PLDI 2003: Proceedings of the ACM SIGPLAN 2003 conference on Programming language design and implementation, pp. 1–11. ACM, New York (2003)
7. W3C: Workshop on the ubiquitous Web (2006)
8. Avilés-López, E., García-Macías, J.: Providing service-oriented abstractions for the wireless sensor grid. In: Cérin, C., Li, K.-C. (eds.) GPC 2007. LNCS, vol. 4459, pp. 710–715. Springer, Heidelberg (2007)
9. Yang, H., Jansen, E., Helal, S.: A comparison of two programming models for pervasive computing. In: SAINT-W 2006: Proceedings of the International Symposium on Applications on Internet Workshops, Washington, DC, USA, pp. 134–137. IEEE Computer Society, Los Alamitos (2006)
10. Hartmann, B., Doorley, S., Klemmer, S.R.: Hacking, mashing, gluing: Understanding opportunistic design. Pervasive Computing 7(3), 46–54 (2008)
11. de Ipiña, D.L., Vázquez, J.I., Abaitua, J.: A web 2.0 platform to enable context-aware mobile mash-ups. In: Schiele, B., Dey, A.K., Gellersen, H., de Ruyter, B., Tscheligi, M., Wichert, R., Aarts, E., Buchmann, A. (eds.) AmI 2007. LNCS, vol. 4794, pp. 266–286. Springer, Heidelberg (2007)
12. Broll, G., Siorpaes, S., Rukzio, E., Paolucci, M., Hamard, J., Wagner, M., Schmidt, A.: Supporting mobile service usage through physical mobile interaction. In: PERCOM 2007: Proceedings of the Fifth IEEE International Conference on Pervasive Computing and Communications, Washington, DC, USA, pp. 262–271. IEEE Computer Society, Los Alamitos (2007)
13. Michahelles, F., Thiesse, F., Schmidt, A., Williams, J.R.: Pervasive rfid and near field communication technology. IEEE Perv. Computing 6(3), 94–96 (2007)
14. Weiser, M., Brown, J.S.: The coming age of calm technology. Beyond Calculation: The Next Fifty Years of Computing 75(85) (1997)
15. Hui, J.W., Culler, D.E.: IP is dead, long live IP for wireless sensor networks. In: Proceedings of the 6th international Conference on Embedded Networked Sensor Systems (SenSys 2008) (November 2008)
16. Wong, J., Hong, J.: Marmite: end-user programming for the web. In: CHI 2006: CHI 2006 extended abstracts on Human factors in computing systems, pp. 1541–1546. ACM, New York (2006)

Internal Aspects of the Relationship between Pressing Force and Training Difficulty

Daisuke Doyo

Aoyama Gakuin University 5-10-1 Fuchinobe, Sagamihara-shi, Kanagawa, 229-8558, Japan
doyo@ise.aoyama.ac.jp

Abstract. In manual assembly work, parts are often assembled by applying force with a simple tool or by hand. A worker thus needs control the force he or she applies in working, as an appropriate level of force is requisite for minimizing work failures and improving efficiency. The object of this study is to clarify the relationship between the level of force (the "force") and level of training difficulty (the "training difficulty") in manual assembly work. Measurements of the training difficulty for 10 test subjects (persons being tested) at force gauges of 30, 40, 50, 60, 70, 80, 100, and 120N revealed a relationship between the force gauge internal to each test subject and the training difficulty.

1 Introduction

In the production of household electric appliances, office equipment, and information communication equipment, manufacturers assemble identical products in large-volume lots to satisfy the consumer demand for large quantities. Many companies adopt cell manufacturing systems, while others rely on older, more conventional forms of line production. Sometimes line production is the only choice available, as difficulties in hiring adequate numbers of multi-skilled workers rule out cell manufacturing. Product lifetimes, meanwhile, are being reduced as a consequence of diversifying consumer requirements and the frequent introduction of new products to the market. As a result, manufacturers are often forced to change their production lines. The obvious solution, automated production lines geared to accommodate frequent changes, are very costly and cannot be justified from an economical point of view. Thus, manual assembly work by workers remains the mainstream.

In a survey of the actual assembly work on a production line for office equipment, the assembly of parts using manual force, either by hand or with simple tools (e.g., tightening screws to attach sheet-metal, mounting E rings, or connecting connectors), accounted for 60% of the total work performed. Though some of these procedures required special skills with tools (e.g., caulking, soldering, or wrapping connections), most of them did not, and most people in the plant believed that the assembly work progressed smoothly as a whole. Yet by closely observing the actual work in this plant, frequent problems were identified. In some cases parts fell apart, for example, and assembly procedures were impossible to complete with a single maneuver. In other cases parts were assembled at an angle, or were dropped, or sprang out, or the

M.J. Smith and G. Salvendy (Eds.): Human Interface, Part I, HCII 2009, LNCS 5617, pp. 475–484, 2009.

assembly work wasn't properly finished due to a worker's failure to apply just the right force. Problems of this type not only wasted time and compromised quality, but also led to variations in the time required for work. The methods of work used, including the manual work by hand or with simple tools, were clearly flawed.

Two types of skill are considered necessary for accurate results in manual assembly work.

- A means to control the hand and arm in order to accurately position a part at the point where it is to be assembled, by controlling the direction, angle, and speed of the hand as the hand holds the part or holds a tool with the part attached.
- A means to control the pressing force required to assemble a part after the part has been positioned at the point it is to be assembled.

In an earlier study [1], the author developed a computerized training system which could convey the experience of "the ideal maneuver of a skilled worker" to a trainee. In testing with actual trainees, this system was found to be quite effective in improving control of the hand and arm motions, but not control of the pressing force applied through the hand motions. We therefore selected, as a target for the present study, a system which can train workers to keep the pressing force at necessary levels for the manual assembly work.

As suggested by Weber's law [2] or Stevens' power law [3], differences in the pressing force (hereinafter, the pressing force will be described as the "force") are expected to become more difficult to discriminate when the force gets larger, as the discrimination threshold becomes larger in parallel. Yet the experiments conducted to test these laws covered a relatively narrow range of stimulation. According to our survey of an assembly plant producing office equipment, the pressing force necessary for manual assembly was 30~120N, and the range of stimulation was much wider than that covered in experiments conducted on aesthesia. One study has even concluded that Weber's law and Stevens' power law are wholly inapplicable once the range of stimulation surpasses a certain point [4]. In view of the foregoing, we have an interest in clarifying the force required for manual assembly work and the level of training difficulty (the "training difficulty"). Such being the case, we have conducted this study to clarify the relationship between the force and training difficulty, and to obtain the knowledge necessary for training workers to control the pressing force in manual assembly work.

2 Work Maneuvers Studied and Evaluation Index

2.1 Work Maneuvers Studied

Manual assembly work involves various work maneuvers. With the tightening of sheet-metal screws and the mounting of E rings, for example, parts are assembled using simple tools such as motorized screwdrivers or E-ring holders. With the connecting of connectors, on the other hand, parts are assembled directly by hand. In this study we focused on processes reliant on manual force, a common requirement for many types of assembly work. To restrict the object of our study to maneuvers by hand, we eliminated any manual maneuver performed using simple tools. Thus, we selected a maneuver involving the simple application of force by the hand of the subject's primary arm (right arm for a right-hander, left arm for a left-hander), without the use of a tool (Figure 1).

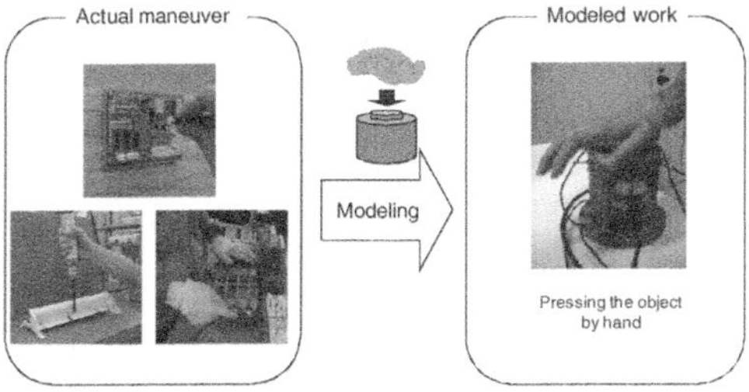

Fig. 1. Work Maneuver Studied

According to our survey of an assembly plant for office equipment, the minimum force required for any one maneuver was 20N, for tightening a sheet-metal screw, and the maximum force required was 120N, for mounting an E-ring. Thus, we set the range of force for this study as 30~120N.

In the training experiment described later, the subjects were trained to remember the force required for the maneuver studied (hereinafter called "target value"). Later, we tested whether or not the subjects could apply the target value repeatedly and accurately.

2.2 Experimental Setup and Layout

The experimental setup for this study is shown in Figure 2. The measuring module developed by Matsumoto and others was used to detect the force [5]. The detection module outputs the level of force as an electric signal via a load cell (Aiko Engineering: ultra miniaturized load cell for compression, CM-10K). The output from the load cell is amplified by an amplifier (Unipulse: Load cell converter, LC210), converted to a 12 bit digital signal by an analog/digital converter board (Contec: AD12-16S(98)H), and captured by a personal computer (IBM: ThinkCentre A51P, Pentium 3.6GHz). The voltages are continuously converted into force values, and the force values are automatically recorded every 0.015 seconds.

The minimum measurement accuracy is 0.0958N. Using the data captured by the personal computer, a bar which expands or contracts according to the force (the "feedback bar") is shown on a touch panel display (Iiyama: AX3844D) (Figure 2-5) to express the increase or decrease of force by the expansion or contraction of the feedback bar. In the training, the test subject learns the force of the target value by applying just enough force to the measuring device to keep the feedback bar as close as possible to the target value (indicated by a red line on the display). The software to display the feedback bar was developed using Visual Basic 6.0 from Microsoft.

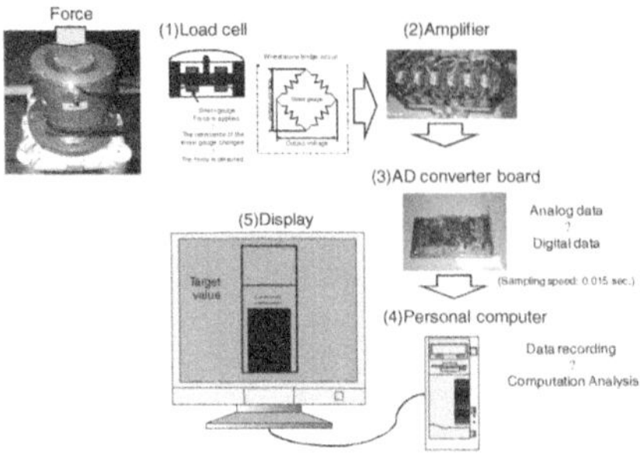

Fig. 2. Experimental setup

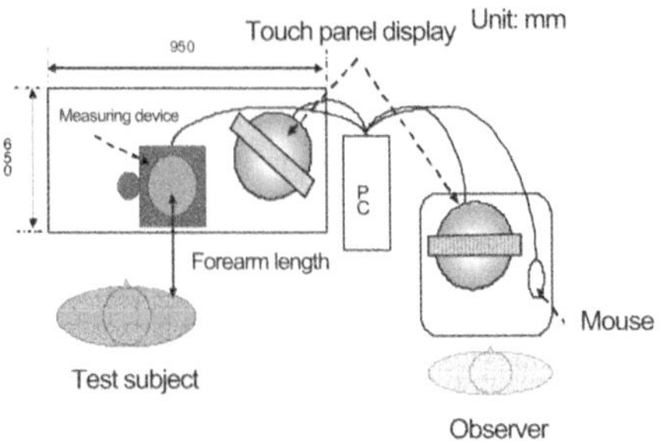

Fig. 3. Layout

The layout of the experiment in this study is shown in Figure 3. The force measurement device is placed at a position in front of the test subject at a length obtained by adding the forearm length to the acromial point (uppermost point of the shoulder) of his/her better arm. The observer operates the personal computer behind the test subject to the right, where the observer cannot be seen. The working position is set in front of the test subject, at a point determined by adding the forearm length of the subject to the acromial point (so that the working position remains constant). By setting and fixing the work position in this way, the working posture of the test subject will also stay somewhat fixed. More postural parameters, such as the degree to which the subject stretches the muscles of his/her back or open his/her arm, are less precisely controlled. For general control, the subject is instructed to relax and take a natural posture. This seemed a suitable approach, as the best posture for easy application of force may differ from person to person.

2.3 Characteristic Value and Evaluation Index of the Force Waveform

The difference between the maximum applied force at each test and the target value is recognized as the control error, and the control error during the test is used as the evaluation index for the training difficulty.

2.4 Internal Gauge

Given that human aesthesia can always somehow be categorized [6], there is assumed to be a force gauge internal for every test subject (the "internal gauge"). And if an internal gauge does exist to each test subject, it is very likely that the internal gauge affects the training result, and analysis of internal gauge during training will be necessary. For this reason, we researched the internal gauges of the test subjects in this study.

For measurement of a psychological amount (such as an internal gauge), we decided to use reaction words for judgment, following the earlier example from the study by Campbell and others [7]. Specifically, we decided to use three reaction words: "weak," "normal," and "strong." In an earlier study on weight lifting stimulation [8], modifiers (e.g., "very") were added to reaction words (e.g., "light"), resulting in additional categories such as "very light" or "very heavy." In designing the present study, we speculated whether too many extra categories would make it difficult for the test subject to select between adjacent categories. To keep the things simpler for our subjects, we had them express their internal gauges as described above, in the categories of "weak," "normal" and "strong."

Thus, our research on the internal gauges proceeded in the following steps: 1) using the experimental setup described above, the test subjects were randomly subjected to 23 levels of force, set in increments of 5N within a range of 0~120N; 2) the subjects selected "weak," "normal," or "strong" on the touch panel after each force experience; 3) the test results were graphed (left side of Figure 4). When a test subject gave a conflicting response (e.g., by selecting "normal" for 70N but "strong" for 60N), the range in which the conflict appeared was regarded as the boundary of the internal gauge. No measurements were taken for ranges of less than 10N. Forces of 10N or less were assumed to be "weak" internal gauges, because the weight of a hand alone represents 5~10N, and thus corresponds to a state of zero force applied. Likewise, measurements of ranges of greater than 120N also went unmeasured. All 20 test subjects answered "strong" for 120N or higher in the preliminary test, hence 120N or higher was assumed to be "strong."

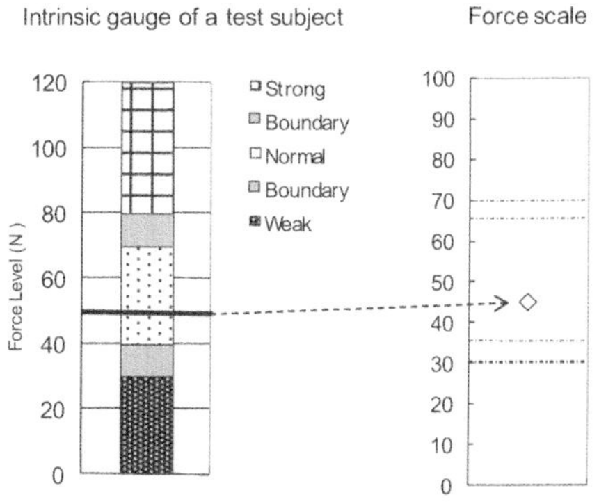

Fig. 4. Internal gauge and force scale

To clearly describe where a certain force *f* is relatively positioned in the internal gauge of a test subject, the internal gauge was expressed by a value in the range 0~100, and a force scale of S(*f*) was made according to the steps described below.

1. To express the internal gauge by the numbers of 0~100, the value 30, that is, approximately one third of 100, is allocated to "weak," "normal," and "strong," and the remaining 10 is divided into two and allocated to two gray zones, one between "weak" and "normal" and one between "normal" and "strong." As a result, the force scale range where the test subject feels "weak" is 0~30, the gray zone force scale between "weak" and "normal" is 30~35, the force scale range for "normal" is 35~65, the gray zone force scale between "normal" and "strong" is 65~70, and the force scale range for "strong" is 70~100.
2. We also surveyed the subjects to determine the category ("weak," "normal" or "strong") into which *f* falls for the test subject, then computed the force scale S(*f*)

by formula (1), assuming that the minimum value of force within the applicable range is *F*min, the maximum value of force within the applicable range is *F*max, the maximum value of the force scale within the applicable range is *S*max, and the minimum value of the force scale within the applicable range is *S*min.

$$S(f) = \frac{f - F_{min}}{F_{max} - F_{min}} \times (S_{max} - S_{min}) + S_{min} \quad \text{Formula (1)}$$

Figure 6 explains the procedure used to obtain the force scale *S*(50) for *f*=50N, when a certain test subject perceives 40~70N as "normal." Because *f* is a force the test subject perceives as "normal," the range of interest in this case is "normal." The minimum value that this test subject feels "normal" is 40N and the maximum value is 70N. Therefore, *F*min=40N and *F*max=70N. Furthermore, because the force scale perceived as "normal" is 35~65, *S*min=35 and *S*max=65. Therefore, we can obtain the result as shown in formula (2) below. Furthermore, the unit of force scale is S, and is described as 45S in the example above.

$$S(50) = \frac{50 - 40}{70 - 40} \times (65 - 35) + 35 = 45\text{S} \quad \text{Formula (2)}$$

3 Experiment on Force and Training Difficulty

3.1 Purpose of the Experiment

The purpose of the experiment was set as investigating the relationship between the size of the target and the training difficulty level by assuming a training to acquire technique to control force in manual assembly work, establishing multiple target values, and measuring the training difficulty level at each target value.

3.2 Experimental Plan

Eight target values were set up for the experiment: 30, 40, 50, 60, 70, 80, 100, and 120N. The targets were set at intervals of 10N between 30 to 80N, and at intervals of 20N between 80 to 120N, to adjust for muscle fatigue of the subjects. The test subjects were 10 male students, aged from 22 to 24. The experiment was conducted in the following steps in an environmentally controlled room maintained at a temperature of 20~25℃ and humidity of 50~55%.

Experiment procedure

1. The order of training for the above 8 target values was determined randomly.
2. The feedback bar was shown on the display, and the subjects memorized the target values in 10 consecutive training sessions.
3. The feedback bar was not shown on the display, and the subject was asked to keep the force at the target value level without the guidance of the feedback. Five consecutive tests were conducted.
4. The subject took a 30 minute break.
5. The target values were changed according to the order determined in 1 above, and steps 2 - 4 were repeated until the experiment was completed for all of the target values.

6. When the training and testing were completed for all of the target values, the investigation for the internal gauge was conducted. This was done after the experiment to prevent the internal gauge from biasing the experimental results. This was considered a risk, as a prior investigation of the internal gauge may have induced the subjects to form the internal gauge intentionally during the subsequent experiment.

3.3 Results of the Experiment and Observation

Figure 5 shows the average value and standard deviation of the training difficulty for all the test subjects, by target value (horizontal axis, target value; vertical axis, training difficulty). Judging from the figure, we can conclude as follows: there is little difference or variation in the training difficulty in the 30~70N range; the training difficulty rises as the target value increases in the range of 100N or higher; the variation also tends to rise at 120N.

Figure 6 shows the average value and standard deviation of the training difficulty by target value for each test subject (horizontal axis, target value; vertical axis, training difficulty). In looking at the relationship between the force and training difficulty of each test subject, we find two types of subjects: those such as h, who encounter lower training difficulty than the other test subjects for all of the target values; and those such as a, c, d, f and J, who encounter large training difficulty for certain target values. Based on this result, we can see that the training difficulty for each test subject tends to go up as the average target value increases (Figure 5) for all test subjects, while there should be some difference from one individual to another. We also note, significantly, that each test subject scores abnormal points when encountering high training difficulty in a discontinuous manner in the target value range of 50~70N (dotted line in Figure 6).

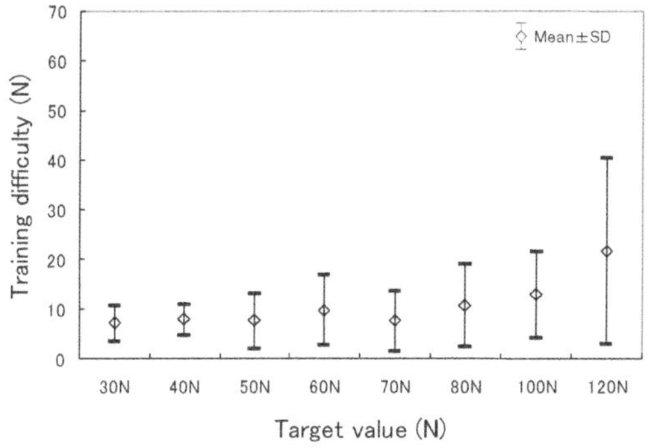

Fig. 5. Relationship between the target value and training difficulty

Weber's law and Steven's power law can explain the rise in the training difficulty to a high level as the target value rises, but they cannot explain why abnormal points appear in the 50~70N range. Therefore, we show the force scale of abnormal points in Figure 7 (horizontal axis: test subject, vertical axis: force scale) to clarify the relationship between the internal gauge of each test subject and the abnormal points. Each point in the figure shows the eight target values, and the white circles show the

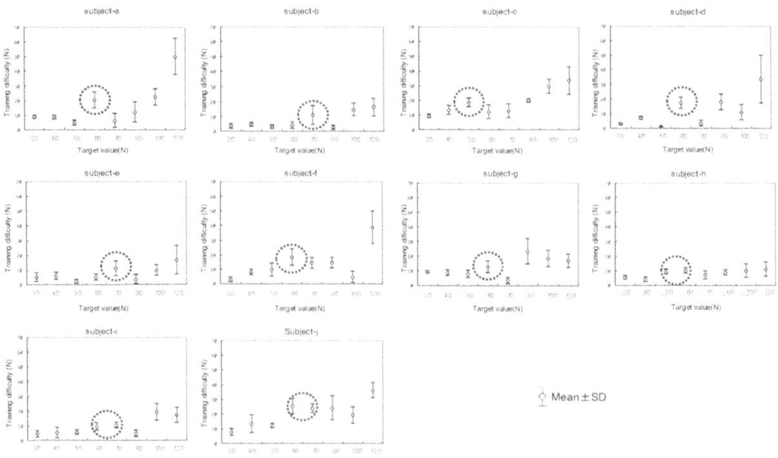

Fig. 6. Training difficulty for each test subject

abnormal points. As the figure demonstrates, the abnormal points of each test subject appear within the force range of 35~65S (a range the subjects perceived as "normal").

In comparing Figures 6 and 7, we find the following. The training difficulty was small in the range of 30S or smaller (a range the subjects perceived as "weak"), and the difficulty change little even when the target value changed. The training difficulty also tended to rise as the target value rose in the range of 70S or larger (a range the subjects perceived as "strong"). Based on this result, the internal gauge of the test subject can be assumed to affect the training difficulty.

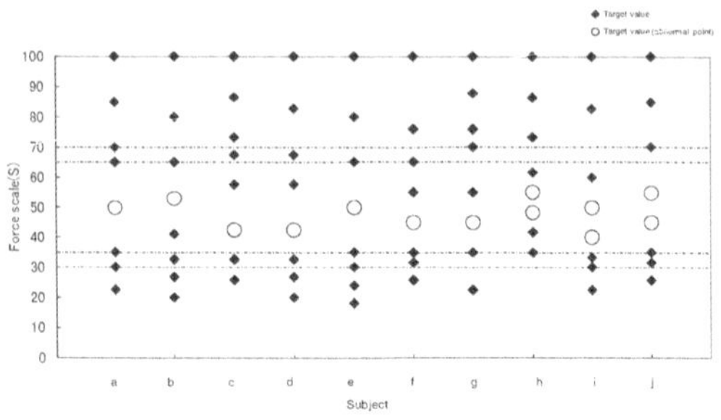

Fig. 7. Force scale for each target value

3.4 Characteristic of the Behavior of Abnormal Point

To analyze the actual forces applied by test subjects during the tests, Figure 8 graphs values obtained by converting the maximum abnormal points onto a force scale (horizontal axis, test subjects; vertical axis, force). Given that the maximum abnormal points are distributed around 68S, as the graph shows, we can assume that a so-called "centering trend" is taking effect [9]. When a neutral value is set as a reference point for making a judgment, a centering trend is at work when the values greater than the neutral point are underestimated and the values smaller than the neutral point are overestimated. If, however, we set 68S as a neutral point in our experiment here, the test subjects tend to apply a force larger than the target value in the test even for the target values greater than this neutral point, without any evident bias back towards 68S. Thus, we can consider this a phenomenon different from the centering trend. Based on this fact, there is a trend that the target value is overestimated for the target value of high training difficulty level, while the subject restrains the applied force more for target values of 35~65S, as the standard of the internal gauge of 68S for forces larger than the target value gives the subject the impression that the force isn't "strong." As a consequence, the force can be assumed to be assimilated into the force around 68S, the threshold between the force perceived as "normal" and the force perceived as "strong."

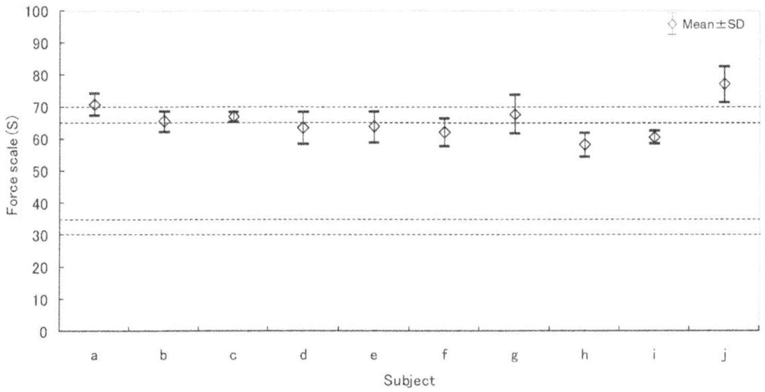

Fig. 8. Maximum value at abnormal point (Force scale)

Based on the above analysis of the behavior of force, we see that it is important, when designing the training program, to have the trainees recognize their own tendency to overestimate the target values of high training difficulty. It also seems necessary that the training program add the adjustment capability at an early point, starting from lower forces which are easier to learn, such as 32S or 68S, and to have the subjects learn the size of the target value relative to the other values.

4 Conclusions

In this study, a training to learn the technique to control force is assumed, and experiments were conducted with an objective to clarify the impact of the target value size to the training difficulty level.

Based on Weber's law or Stevens' power law, we assumed that the training difficulty would continuously increase as the applied force for training increased. As it turned out, however, the training difficulty was low at target values of 32S or lower, stayed almost the same even as the target value changed, rose to high levels at the target level of 50S, dipped at the target level of 68S, and rose to high levels again at the target values higher than 68S. Based on this result, we concluded that a target value with high training difficulty for a test subject could be clarified by measuring an internal gauge of the test subject and computing a force scale.

Once the relationship between the force and training difficulty is clarified, the number of training session needs to be increased or a method of feedback needs to be devised if the training session is conducted for a force-control technique at a target value with high training difficulty. One way to improve work efficiency, in the design stage, is to eliminate any maneuver which requires a force of high training difficulty for many workers.

In the future we plan to develop a training method which is efficient for target values of high training difficulty.

References

1. Shida, K., Matsumoto, T., Kanazawa, T.: The development of training system for self-threading screw tightening operation. Japanese Journal of Ergonomics 37(1), 1–10 (2001)
2. Fechner, G.T.: Elemente der Psychophysik, Leipzig (1860)
3. Stevens, S.S.: On the Psychophysical Law. Psychol. Rev. 64, 158–181 (1957)
4. Toriizuka, T.: Characteristic and Application of Skillwork on Industrial Plant Maintenance, Keio Univ. (2005)
5. Matsumoto, T., Tashiro, H., Kanazawa, T.: Proposal of a set of new indices for evaluating the quality of Screwdriving Operation and a measurement device. Journal of the Society of Plant Engineers Japan 4(4), 12–20 (1993)
6. Bravo, L., Mayzner, M.S.: Assimilation and Contrast Effects of Anchoring Stimuli on Judgments, A partial replication of the Sherif, Taub and Hovland study. J. Psychol. 52, 333–334 (1961)
7. Compbell, D.T., Lewis, N.A., Hunt, W.A.: Context Effects with Judgmental Language that is Absolute Extensive, and extra-experimentally anchored. J. Exp. Psychol. 55, 220–228 (1958)
8. Helson, H.: Adaptation-Level Theory. Harper and Row, New York (1964)
9. Sherif, M., Taub, D., Hovland, C.I.: Assimilation and Contrast Effects of Anchoring Stimuli on Judgments. J. Exp. Psychol. 55, 150–155 (1958)

A Tangible Mixed Reality Interface for the AMI Automated Meeting Assistant

Jochen Ehnes

Centre for Speech Technology Research, University of Edinburgh,
10 Crichton Street, Edinburgh, UK
jehnes@inf.ed.ac.uk

Abstract. In this paper we describe our approach to support ongoing meetings with an automated meeting assistant. We propose an alternative user interface for the AMIDA Content Linking Device. In order for the system to be less distractive and more collaborative than the original laptop screen based one, we developed a system that projects documents onto the table tops right in front of the meeting participants. This way they appear as if they were printed on paper, lying in front of the participants. We describe our setup as well as the user interface we built to handle and share these documents.

Keywords: meeting assistants, meeting processing, mixed reality (MR), projected user interface, tangible user interface (TUI).

1 Introduction

Started in January 2004, the European AMI (Augmented Multi-party Interaction) integrated project, as well as its follow on project AMIDA (Augmented Multi-party Interaction with Distance Access), has been about building systems to enhance the way meetings are run and documented. AMI research revolves around instrumented meeting rooms which enable the collection, annotation, structuring, and browsing of multimodal meeting recordings. So far, meeting participants can use AMIs JFerret browser [1] to go through previous meetings and get themselves up to date. While the possibility to look through recordings of previous meetings and being able to search for important sections by keywords is a very helpful tool, it requires direct action by the user. Furthermore, as the users have to interact with the system on a personal computer, which draws their attention to it and distracts from a conversation, the browser is more useful in preparation for a meeting than during the meeting itself.

The aim of this work is to support ongoing meetings with an automated meeting assistant based on the AMIDA Content Linking Device [2]. The basic idea is that the assistant monitors the ongoing meeting. As it gets the transcript from the speech recognition system, it tries to determine the topic of the conversation, searches for relevant documents from previous meetings and displays them. While the results of this content linking process were displayed on a laptop in the original setup, we then aimed at making the system less distracting and more supportive

M.J. Smith and G. Salvendy (Eds.): Human Interface, Part I, HCII 2009, LNCS 5617, pp. 485–494, 2009.

of collaboration. We built a system that projects documents on table tops in front of meeting participants, resembling pieces of paper. By presenting information of relevance to the current discussion in this way instead of personal computers at each participants place, we expect the ongoing meeting to be disrupted less as the participants do not have to interact with their computers. Also, a projection on the tabletop supports the communication within the group as all participants can see the projected documents and move them around easily, just as traditional documents on paper. As a result, the participants can discuss these documents in a more natural way. In order to provide enough space for several users, we use multiple projection systems, one for each user.

Recently back-projection systems with camera based multi touch screens (such as Microsofts surface computers) have become quite popular. However, they require significant volume below the surface, which would become a disadvantage for large display areas, as meeting participants would not be able to sit at the table comfortably. Furthermore we were aiming for a technology that can provide the additional functionality without requiring big changes to the existing meeting environment, as we already had a lot of equipment installed in our instrumented meeting room. For those reasons we decided to go for a system projecting from above onto the existing tables.

In order to make the interaction with the system as direct as possible, we aimed to make the projected objects graspable. Therefor we track physical objects carrying fiducial markers that serve as grabbing devices. In order to move around the projected objects, a physical document grabbing device is associated with them. As long as this connection exists, the virtual document follows the grabber. And it does so not only on one users space, but across multiple projection units. Also, this connection continues after long pauses of visibility, even between different runs of the projection system. This way important documents can be kept in reach by keeping them connected to grabbers. In a similar way keyboards can be connected to documents.

The meeting assistant brings up new documents in regular intervals, often several documents at the same time. Consequently it is not enough to make the documents appear at a default location (e.g. the center of the tabletop). We implemented a way to spread them out automatically. Of course the space on the table is not unlimited, so we scale down documents and rearrange them after a while before they get removed eventually.

In addition to augmenting the table, the system incorporates the whiteboard as well. However, as earlier meeting recordings showed, people usually have an inhibition to get up and go to the whiteboard to write something onto it. In fact, a special task had to be designed to make them make use of it. In order to reduce this inhibition, we developed an alternative way to enable participants to interact with content on the whiteboard directly from their place and move content between their space and the shared whiteboard space easily. We considered the whiteboard as a form of shared space. By making this shared space available at each projection unit in form of a second layer on top of the normal display area, the content of the whiteboard can be manipulated from every participants place.

In the following sections we will describe the concepts and details of the interaction with our system, as well as an overview of its implementation. We will also present some results from our prototype setup for two users spaces and a whiteboard.

2 Previous and Related Work

This work builds on the AMIDA Content Linking Device [2]. The system consists of a Document Bank Creator (DBC) that gathers documents that are of potential interest for an upcoming meeting, a Document Indexer (DI) that creates an index over the document bank created by the DBC, a Query Aggregator and a User Interface, all connected via a central Hub. During a meeting, the Query Aggregator performs document searches at regular time intervals using words and terms from the automatic speech recognition system. It produces a list of document names, ordered by relevance, based on the search results, as well as on a persistence model, assuming that documents that come up during several searches are likely to be more relevant than others that do not. The User interface finally displays the results to the user(s). For this work we replaced the flash based user interface with a mixed reality interface, projecting virtual documents onto the tabletop in front of the meeting participants.

The idea to use the table top as an interface to computers is not new. The first system of that kind known to the authors was DigitalDesk [3,4]. Its main intention was to bring together electronic- and paper-documents.

In [5,6] a similar setup consisting of video projector and camera, I/O-Bulb as the authors call it, mounted above the table was used to create applications that are manipulated using physical objects. Applications include the simulation of holography setups using physical placeholders for optical elements such as lasers, mirrors etc. or the simulation of fluids flowing around objects standing on the tabletop. An obvious advantage of this kind of user interface is their collaborative nature as several users can manipulate different physical objects on the tabletop at the same time, instead of being restricted to a single mouse in a conventional computer setup.

While being able to see what everybody else sees is a very important factor for collaboration, it sometimes is necessary to have some private space as well in order to work out an idea before presenting it to the whole group. In [7] the authors presented a system that supports the discussion of virtual prototypes by a group of designers/engineers sitting around a projection table. However, the difference to other 3D viewers or the applications running on the I/O-Bulbs was that the content does not occupy the whole screen space. Instead the virtual prototype would be visible on a virtual piece of paper. Apparently conventional plots of CAD drawings were still used frequently during design review meetings, as it was so easy to sketch on them with a pencil to point out flaws or suggest improvements. Furthermore, one could just work out an idea on a piece of paper before presenting it to other meeting participants. In order to make computer systems more usable for such applications, the above mentioned prototype was developed. As the 3D models were displayed on the virtual pieces of paper, they were visible to everyone and they could be moved around using a tracked puck, so that they could be brought closer and rotated to a single person to allow for a more personal use. By grabbing two points, one with a puck and one with a pen, the virtual paper could be scaled, similarly to the two finger pinch scaling gesture known from the iPhone. Using tracked pens, participants could draw lines on the objects to annotate them. Furthermore, the system allowed to connect each piece of paper to one of several tracked pairs of shutter glasses to get a full three dimensional impression of the object. But as the stereo view certainly hindered others looking at the object, it could easily be switched off again by putting the glasses down during discussions. While we do not display 3D objects in our content linking system, we use the concept of having virtual pieces of paper that can be moved around using physical devices such as the puck. The Shared Design Space [8], a system consisting of four projectors for an interactive surface on a tabletop and a projector to create an interactive wall, is of interest as it not only use video cameras to track objects for interaction. Anoto Pens, digital pens that can track a pattern of tiny dots on the surface they are writing on, are used to control the system as well as to draw onto the virtual documents (images). As we aimed for a simple interface to view existing documents, we don't provide such a feature at the moment.

3 Setting

As our instrumented meeting room plays an important role within the AMI project, the meeting assistant had to be integrated into this environment. It also meant that the new system should require as little changes to the meeting room as possible. Back projection systems, such as Microsoft's Surface Computers became quite popular recently, as they allow to track fingers and objects touching the display surface. However, they obviously require transparent tabletops as screens, or they come as complete units which would require significant changes to the existing room. Furthermore, as they require sufficient room below the surface for the projection as well as the video tracking of touches, they would not allow meeting participants to put their feet below the table and sit comfortably. For those reasons we chose to build an on projection onto the existing tabletops, even if that does not allow to track fingers touching the surface directly. As the system has to accomodate several meeting participants, we decided to use several projection units, one for each user.

4 User Interface

In order to make the interaction with the system as direct as possible, we aimed to make the projected objects graspable. We decided against hand tracking as it is difficult to distinguish between gestures meant to manipulate documents and gesturing during discussions. This is especially true as the current setup does not allow to detect if the user's hands touch the desk top. Instead we track physical objects that serve as interaction devices using the AR-ToolKit+ tracking library.

4.1 Document Handling

In order to move projected documents around, a physical object (paper grabber) is associated with them. As long as this connection exists, the virtual document follows the grabber. The grabber objects consist of a piece of cardboard containing three markers, one of them elevated on a box (see figures 1 and 1(a)).

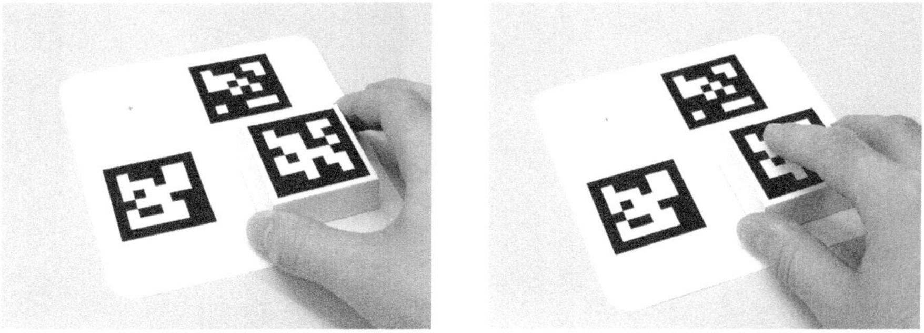

(a) Not grabbing (b) Grabbing

Fig. 1. The grabbing device and its switch marker

The elevated marker has the functionality of a switch. By blocking the visibility of this marker using a finger for example, one tells the system to grab the document below the device. If grabbed only on the sides, so that the marker on top is fully visible, the marker is disconnected and can be moved freely. Once the marker is placed on a virtual piece of paper, users can grab the document by holding the box like a mouse and thereby covering the top marker. Because the switch marker is on the box that user's grab, they do not have to think consciously about covering the marker or not. They just have to remember the two ways of holding the grabber device; On the sides to lift the grabber from the paper or with the hand on top of it, pressing it onto the paper they want to move. Once grabbed, the document stays connected to the grabber until it is released again, i.e. the top marker gets recognized again. This may be on the same or another user's projection system.

4.2 Keyboard Forwarding

Instead of providing virtual, projected keyboards as it is usually done with touch screen interfaces, we chose to use standard wireless keyboards. In order to allow keyboard based input, a Keyboard identified by the two markers attached to it (see figure 3(b)), can be placed on a displayed document. This allows to route keyboard events to the display applications that create the graphical representation of the documents. It replaces the physical connection (which keyboard is connected to which projection unit) with a virtual connection between keyboards and documents.

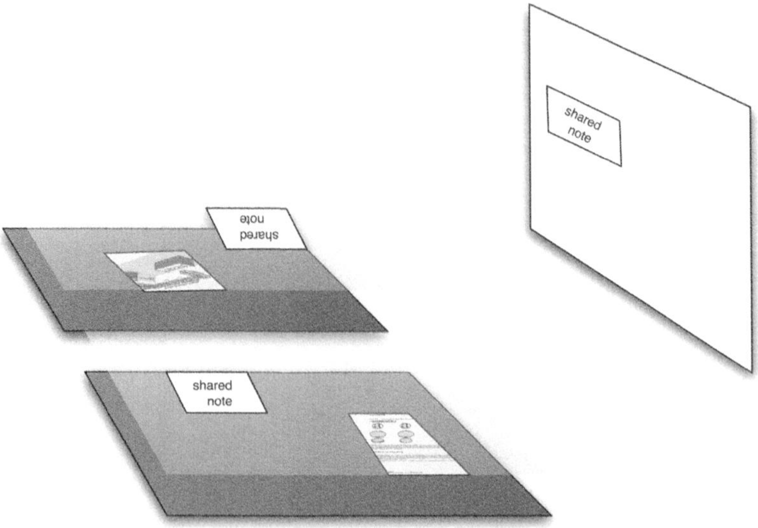

Fig. 2. The shared space as a semitransparent layer on top of the normal space. It can be (de)activated by covering a switch marker.

4.3 Sharing

In addition to augmenting the table, we wanted the system to incorporate the whiteboard as well. This way, participants are able to interact with content on the whiteboard directly from their place and move content between their space and the shared whiteboard space easily. While Hyperdraging as described in [9] would allow participants to do that in principle, it relies on a laptop with a conventional interaction device such as a touchpad. Using hyperdraging therefore would work against our goal to let the computer disappear.

We believe it is better to 'bring the shared screen to the participant' on the press of a button, or in our case when a marker is covered by the user. We therefore implemented a shared semitransparent layer (see figure 2) on top of the normal projection area which can be activated and deactivated by covering a marker placed on the projection area for that purpose. The presentation screen is the only exception here, as it does not have a private layer. It always displays the shared layer. Documents can be moved between the private and shared layers by grabbing them on one layer before switching to the other one. Once on the shared layer, all state changes such as position, orientation or which page of a multi page document is shown are forwarded immediately to other systems displaying the shared layer.

4.4 Auto-arrangement and Auto-iconizing

The Content Linking Device brings up new documents in regular intervals. In fact it often finds several documents to be displayed at the same time. In this situation it is obviously not enough to make the documents appear at a fixed location (eg. the center of the tabletop). Of course the space on the table is not unlimited, so a method had to be developed to prioritize documents and remove less relevant ones gradually. We implemented a system to arrange and iconize documents automatically. It behaves as follows: New documents, are appended to an array of automatically arranged documents. If the number of elements in this array grows above a given limit (two documents in our case), the first one is removed and appended to the array of iconized documents. Additionally, a timer is set for each document added to the array of automatically arranged documents. Once the timer fires, the document gets iconized as well. This gives meeting participants the chance to have a glimpse at them and makes sure that documents that do not appear to be relevant to users are removed after 30 seconds. If the number of elements in the array of iconized documents grows above its limit (ten documents in our case), the first one, which has not been looked at for the longest, is removed and closed. Whenever documents are added to or removed from these arrays, all managed documents are sent a new goal position and scale factor according to the array they are in and their position within that array. The first auto arranged document is displayed on the left side. The second (and latest) one is positioned next to it in the center of the projection, leaving the right side for documents the user places there to read. Their scale factor is 1.0 so they are displayed in full size. The automatically iconized documents on the other hand are scaled down to 0.3 times their size and arranged along the front edge of the table with the oldest one being displayed on the left and newer ones being added to the right. When documents are sent to new positions or receive new scale factors, they dont change to these values immediately. Instead they animate towards these values over a given duration (1.5 seconds seemed best). This way it becomes obvious when the ranking changes and it is easy to follow what is going on. This is very useful when a document that is already open is deemed to be relevant by the query aggregator again, as one can see it move form its previous (possibly iconized) position to the position of the newest document (center). If the user places a paper grabber or keyboard connector on top of a virtual paper, it prevents the paper from being affected by the auto arrangement/iconizing system. If placed on an iconized paper, the paper is also scaled up to full size again. Now the user may move the document to where it can be read conveniently without interference of the auto arrangement system. Once the user removes the paper handling device and/or keyboard connector from the projected document, the system will take responsibility for the document again and iconize it after 30 seconds to clean up the tabletop. For the case that the document the query aggregator determined as relevant is already displayed as the latest document or controlled by the user, a visual ping has been implemented. If pinged, a document visually vibrates for a short period of time. It is animated by scaling it slightly up and down from its original size using a sine function. The amplitude of this vibration is scaled down to zero within 1.5 seconds to fade out the effect smoothly.

5 Our Setup

As mentioned before, we designed our system to consist of one projection unit per meeting participant. For our first prototype we decided to support two participants, as that would allow us to demonstrate the distributed nature of our system with a minimal budget. Each of the two projection units consisted of a projector (Optoma EP709), a camera (ImagingSource DFK 21BF04) and a computer (Mac Mini). Figure 3(a) shows the two projection units mounted around the projector for the shared screen. We used a MacBook Pro (not in the picture) to generate the content for this projector. A weakness of our first prototype,

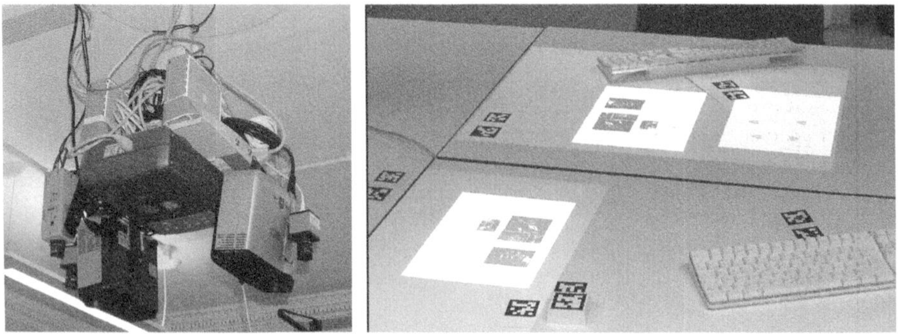

(a) Setup of our first prototype (b) Projection by our second prototype, devices

Fig. 3. Projection units and projected user interface

the resolution of its projectors (1024x768), became obvious quite soon. While this resolution was ok for documents containing little text in large fonts, such a meeting agenda or Power Point slides, it was not sufficient to display regular text documents as a virtual sheet of A4 paper. In order to address the resolution problem we replaced one of our Optoma EP709 projectors with an Optoma HD800X, a projector capable of projecting full HD video (1920x1080). In order to get an even better resolution from this projector, we used it in portrait mode, effectively augmenting only half of the users' table space. As documents are usually printed in portrait format, this enabled us to make better use of the projected pixels. While the EP709 projector would give us 768 pixels in height, the HD800X in portrait mode gives us 1920 pixels for the same height, an increase in resolution by a factor of 2.5 per dimension. Figure 3(b) shows the the projection areas for comparison, the larger area created by the lower resolution projector on the opposite side of the table and the smaller area originating from the HD projector on the front side.

6 Software Architecture

In order to keep the number of projection systems scalable, we divided our projection system into two parts. "Smart Projector", an application running on all projection units and a central "Projection Manager". This display system is connected to the Content Linking Device via a third application, the "Hub Interface". These applications, as well as the other components of the Content Linking Device, may all run on different machines as they communicate via network. In our test setup however, we ran all these components on the laptop that rendered the contents of the shared screen. Only the two instances of Smart Projector augmenting the users' places, which had to perform video tracking as well, ran on separate machines, the two Mac minis.

7 Results

We developed a scalable projection system to be used in meeting environments. The way it is set up allows for easy installation in existing environments. After all, the camera, projector and computer could be integrated into a single unit mounted above the tables. We implemented software components that allow for easy management and coordination of the projection units as well as a user interface based on tracked interaction devices. We demonstrated that it is easily possible to move documents around on one, as well as between different projection units or between private and shared spaces. Figure 3(b) shows both projection units displaying a document on the shared space, while the unit on the oposite side shows a (slightly dimmer) local document as well. Furthermore, the system is able to connect to the central Hub of the AMI/AMIDA project. This way it can be used to display documents the content linking device deems relevant for the ongoing discussion. Additional functionality to manage the displayed content automatically was implemented to cope with the stream of new documents being introduced by the content linking device.

8 Future Work

In order to be able to evaluate the usefulness of the system, we will have to support more users with it, as the typical AMI/AMIDA meetings have four participants. While we believe that the higher resolution, which we achieved for one user in our second prototype, is necessary for every participant, we are not sure that augmenting half of a user's space is sufficient. In some situations a participant may use a laptop and other objects as well, so only half of one's space may be available anyway. In other situations however it may be important to have a glimpse at many documents at once, as they all could be important. In that case we would have to use two HD projectors per user. That however could lead to serious heat and noise issues in the meeting room.

Acknowledgements

This work is supported by the European IST Programme Project FP6-0033812 (AMIDA) as well as the Marie Curie Intra European Fellowship (IEF) FP7-221125 (NIPUI). This paper only reflects the authors' views and funding agencies are not liable for any use that may be made of the information contained herein.

References

1. Fapso, M., Schwarz, P., Szöke, I., Smrz, P., Schwarz, M., Cernocký, J., Karafiát, M., Burget, L.: Search engine for information retrieval from speech records. In: Proceedings of the Third International Seminar on Computer Treatment of Slavic and East European Languages, pp. 100–101 (2006)
2. Popescu-Belis, A., Boertjes, E., Kilgour, J., Poller, P., Castronovo, S., Wilson, T., Jaimes, A., Carletta, J.: The AMIDA automatic content linking device: Just-in-time document retrieval in meetings. In: Popescu-Belis, A., Stiefelhagen, R. (eds.) MLMI 2008. LNCS, vol. 5237, pp. 272–283. Springer, Heidelberg (2008)
3. Wellner, P.: The digitaldesk calculator: Tangible manipulation on a desk top display. In: Proc. ACM SIGGRAPH Symposium on User Interface Software and Technology, pp. 107–115 (1991)
4. Wellner, P.: Interacting with paper on the DigitalDesk. Communications of the ACM 36(7), 86–97 (1993)
5. Underkoffler, J., Ishii, H.: Illuminating light: An optical design tool with a luminous-tangible interface. In: CHI, pp. 542–549 (1998)
6. Underkoffler, J., Ullmer, B., Ishii, H.: Emancipated pixels: real-world graphics in the luminous room. In: Rockwood, A. (ed.) Proceedings of the 26th annual conference on Computer graphics and interactive techniques, pp. 385–392. ACM Press/Addison-Wesley Publishing Co. (1999)
7. Ehnes, J., Knöpfle, C., Unbescheiden, M.: The pen and paper paradigm supporting multiple users on the virtual table. In: Proceedings of the Virtual Reality 2001 Conference (VR 2001), p. 157. IEEE Computer Society, Los Alamitos (2001)
8. Haller, M., Brandl, P., Leithinger, D., Leitner, J., Seifried, T., Billinghurst, M.: Shared design space: Sketching ideas using digital pens and a large augmented table-top setup. In: Pan, Z., Cheok, A.D., Haller, M., Lau, R.W.H., Saito, H., Liang, R. (eds.) ICAT 2006. LNCS, vol. 4282, pp. 185–196. Springer, Heidelberg (2006)
9. Rekimoto, J., Saitoh, M.: Augmented surfaces: A spatially continuous workspace for hybrid computing environments. In: Proceedings of CHI 1999 (1999)

Brain Activities Supporting Finger Operations, Analyzed by Neuro-NIRS

Miki Fuchigami[1], Akira Okada[2], and Hiroshi Tamura[2]

[1] Osaka City University, Sugimotocho, Osaka City, Japan
[2] Institute for HUMAN INTERFACE, C1803 Portpia plaza, 6-14 Minatojima nakamachi, Chuoku, Kobe City, Japan 650-0046
tamura@mobilergo.com

Abstract. Brain activities are analyzed for the button selection task using Neuro-NIRS, which is analysis method of quick component of NIRS. Quick component of NIRS is numerically quantifies by short time average of absolute value. The average is represented by, maximum and density. Since button selection task is short time event, it seems the instantaneous activities, like maximum is playing major roles in finger selection to operate button.

Keywords: neural activities, NIRS, quick components, finger selection, button operation.

1 Introduction

Working brain needs more oxygen supply. Thus hemodynamic studies expect the areas supplied with rich oxygen are tentatively working. But this is only an expectation, but no evidence. We have to find evidences related to parts of the brain actually functioning. Hemodynamic studies are interested in slow components of NIRS, especially oxyHb. People engaged in hemodynamics, however have realized the existence of quick signal while recording NIRS. So far the signal is regarded as noise, and skill of NIRS measurement is to find the location where the noise can be minimized.

So called noise is so large to be regarded as measurement noise, it must be substantial physiological signal, which we named quick component of NIRS. Attempts have been made to analyze the signal and to make the model of the signal source [5]. Evidence are convincing that the quick component is result of neural activities under recording and successful results are promising in relation to daily human activities, like hand up/down, finger sign, and text entry, etc., to which hemodynamic studies are powerless.

Hemodynamic and neural functions are both significant aspects of brain studies. Cerebral infarction might cause hemodynamic problems, which might result in neural dysfunctions. Rehabilitation is the process of recovering hemodynamics and neural function. Neuro-NIRS is a new significant aspect of NIRS studies.

M.J. Smith and G. Salvendy (Eds.): Human Interface, Part I, HCII 2009, LNCS 5617, pp. 495–504, 2009.

2 Quick Components

2.1 NIRS Coordinates

Fig. 1 is showing Oxy, Deoxy, and Total Hb (hemoglobin) recording of 6 channels from NIRS. The upper trace corresponds to oxy, the middle to total, and the lower to deoxy Hb. The channel number is indicated at the middle of the left side of each chart. Looking to the channel 26, oxy and deoxy Hb contains big high frequency components, which we call quick components, the total Hb contains less. The traces of 29 and 32 contain less quick components, in both oxy and deoxy, as well as total Hb.

Dynamics of Oxy and deoxy Hb can be represented by total and difference Hb as below:

$$\text{Total Hb:} \quad \text{Total=Oxy +Deoxy} \tag{1}$$

$$\text{Difference Hb:} \quad \text{O\_D= Oxy –Deoxy} \tag{2}$$

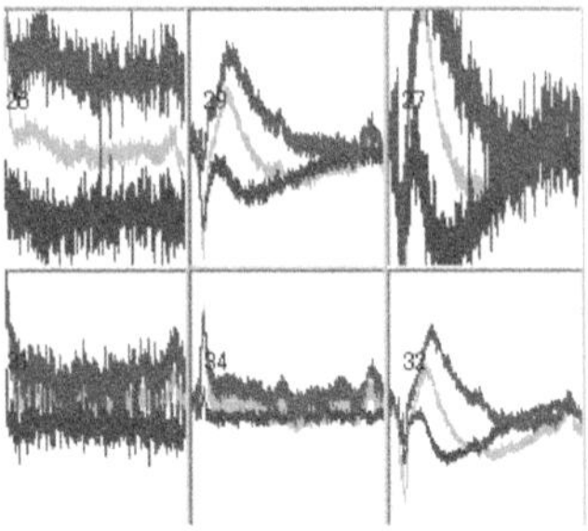

Fig. 1. Raw traces of NIRS

Fig. 2 is showing traces of above variables. The quick components of Oxy Hb and Deoxy Hb are in opposite phases, thus by the addition of quick components, they are almost compensated each other, while by subtraction, the quick component is almost doubled in amplitude.

Applying moving average of 2 second, quick and slow components of Total and O_D can be derived.

$$\text{Slow total Hb:} \quad \text{slow total=Average (Total)} \tag{3}$$

$$\text{Slow difference Hb:} \quad \text{slow O\_D= Average (O\_D)} \tag{4}$$

$$\text{Quick total Hb:} \quad \text{Quick T= Total} - \text{Average(Total)} \tag{5}$$

$$\text{Quick difference Hb:} \quad \text{Quick O\_D = O\_D} - \text{Average(O\_D)} \tag{6}$$

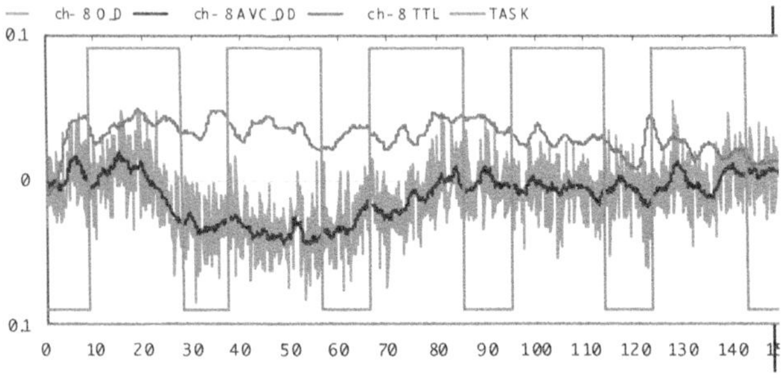

Fig. 2. Total Hb(red) and difference Hb(green) and the slow O_D (blue) among the middle of difference Hb (O_D) trace. Square trace indicates on and off of task.

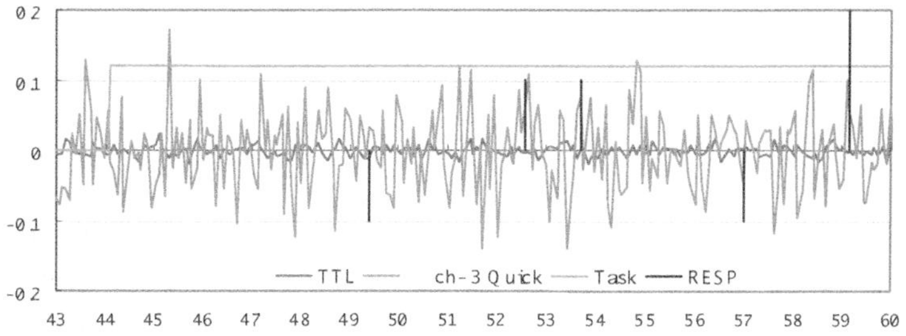

Fig. 3. Traces of Quick O_D and Quick total, sampled at every 0.1 sec

Thus the analysis of quick components of NIRS can be concentrated to the quick O_D. The quick components derived from eq. (5) and (6) are shown in Fig. 3. The magnitude of quick total is about 1/10 of quick O_D.

3 Button Selection Task

3.1 Experimental Systems

NIRS (Near Infra-Red Spectroscopy) facility was a product of Shimadzu co. (OMM3000/12) which is installed in Future ITC Research Center, KARC (Kansai Advanced ICT Research Center, Akashi, Kobe, Japan). The authors were honored by Joint Research Treaty with the center to use the NIRS facility. The NIRS had 12-16 light transmitter and receiver pairs and was capable of processing up to 54 channels of brain activities depending on transmitter and receiver arraignment. The arraignment we adopted in measurement is shown in Fig. 4. By this arraignment, 48 channel brain activities are collected at every 0.1 second interval. Beside NIRS data, the machine is capable of 8 external analogue signals, which are used to record start, stop, button operations and task completion signals together with NIRS data.

Experimental task design and the analysis methods are developed by authors. At this moment, the subjects participated in the measurement are graduate students, volunteered with informed consent in accordance with the Institutional Review Board of Kobe Women's University. Analysis for larger population should be reported elsewhere.

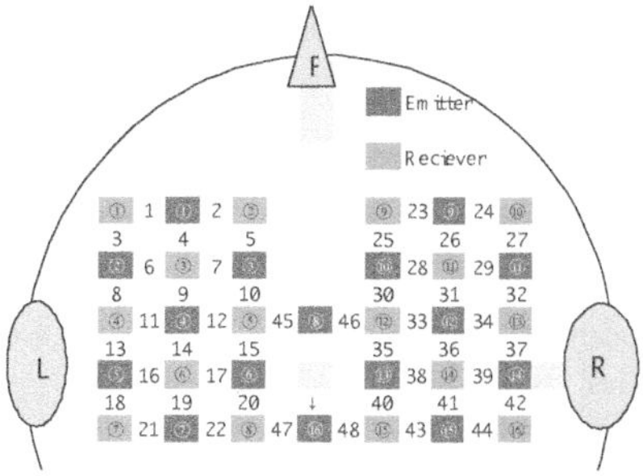

Fig. 4. Transmitter/receiver arraignment 48 channel data are collected at 0.1 sec. interval

3.2 Finger Selection Task (Button Selection)

Subject rest her hand (left or right) on the button box. Subject operated a button in accordance to the instructions generated by computer and presented either on visual display or by computer reproduced voice. Instructions were to choose a button either one of right, left, up or down. The subject fingers are allocated at home position, in which middle finger stay in the location of center. Fig. 5 shows the case, the box is operated by left hand from left side. In case of up or down instruction, index or ring finger might be selected to operate a red button. In case of right or left, middle finger has to change its location by changing the bending angle. The change in necessary action might be related to different brain activities to be analyzed in this paper.

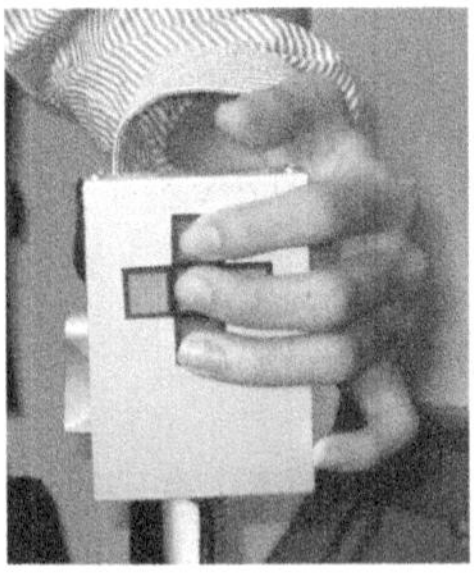

Fig. 5. Button selection by left hand (side entry)

Two methods of visual presentation are used in case of visual display, i.e. by letter (上, 下, 左, 右) and by pictogram (↑, ↓, ←, →). An instruction to operate is presented once by 3 second in average; the order of instruction is randomized. The instruction is presented on display for 1.1 second. The operator completed the button push tasks almost within 1.1 second. The average response time was in between 0.4 - 0.8 second depending on the instruction. The responses were all correct. Analysis of this paper will be limited within the left hand, side entry data (20 trials).

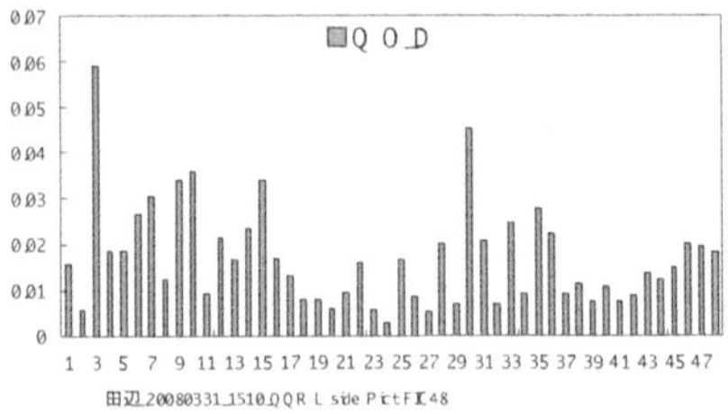

Fig. 6. Quick component distribution around the channels. Ordinate scale in mMol/l·cm

The essential bases of quick component analysis will be explained by applying the method to the button selection data. Total analysis and comparison of all the data are beyond the scope of this paper.

3.3 Analysis of Quick Components

Unified quantitative formulation of quick component is necessary, for the comparison of quick components, derived from various channels at various tasks. An appropriate function applicable to noise like signal as quick components is Avedev (Microsoft Excel function), which is average of absolute difference of data and their mean value. In our analysis, slow component is subtracted from O_D signal, thus the mean value of quick component is assumed to be zero. In short, avedev is time average of absolute value of quick components.

In the previous papers, long term (>30second) distributions of quick components around different channels and their changes related to tasks are reported [4]. A distribution proper to the case of this paper is shown in Fig. 6, where pictogram is used for instruction display, and operator used left hand at side entry position.

Avedev is stable in value, when averaging interval is taken 10 seconds or longer. It change considerably as averaging interval is shorten, reflecting change of brain activities under the recording. In this paper, a method to evaluate short term change in brain activities is analyzed by introducing two parameters, M (magnitude) and D (density), related to Q.(avedev of quick component). The relation of these parameters is formulated as below.

$$Q = M * D \tag{7}$$

Here

$$M = \max(\max(\text{member}), -\min(\text{member})) \tag{8}$$

member: data within averaging interval. Denoting
N: the number of data in the interval, inequality below is derived.

$$0 < 1/N < D < 1 \tag{9}$$

Inequality (9) states that the density remains within the bound of 0 and 1. Further studies are concentrated to the change of M and D, in succession of averaging cycles.

4 Response Analyses

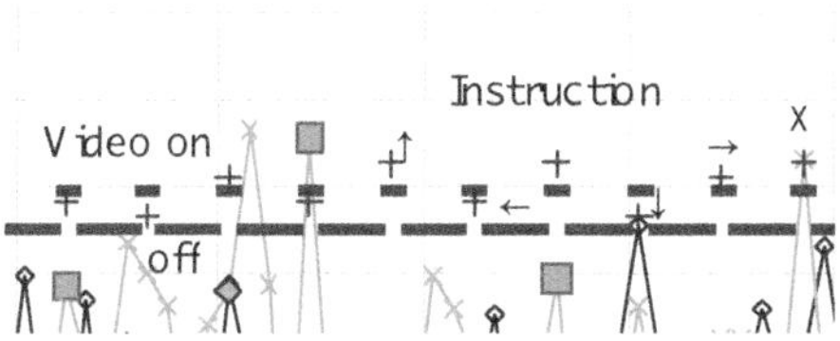

Fig. 7a. Video on/off (- up/down), button instruction (↑,↓,←,→) and the button responses (+)

Quick component is analyzed below taking the averaging interval 1.1 second. Some time traces of specific channels are shown in Fig. 7 b) - e). Notations for video on/off and button selection are explained in Fig. 7 a). Instruction display is on when - symbol is in higher position, and while instruction is off, the - symbol is in lower position.

Result of button selection is shown by distance from display interval line (-). Instruction is shown by pictogram (↑ , ↓ , ←, →) adjacent to corresponding + symbols in Fig. 7a). Symbol (+) touching just above video on sign is correct response to right (→) instruction, while one touching just below is corresponding to (←) instruction. Symbol (+) above and below apart from video on sign is corresponding to up(↑) and down (↓) instruction.

4.1 Visual Analysis

Fig 7b) illustrates -min. and 0.1D (density) of each averaging interval in succession, together with display and response detail while performing button selection task. Blue trace illustrates the –min. Readers might realize many peaks of blue trace are synchronized to the video and the response sign. The peaks synchronized to response signs are illustrated by filled symbol of enlarged size. Time location of synchronization is denoted by letter N.

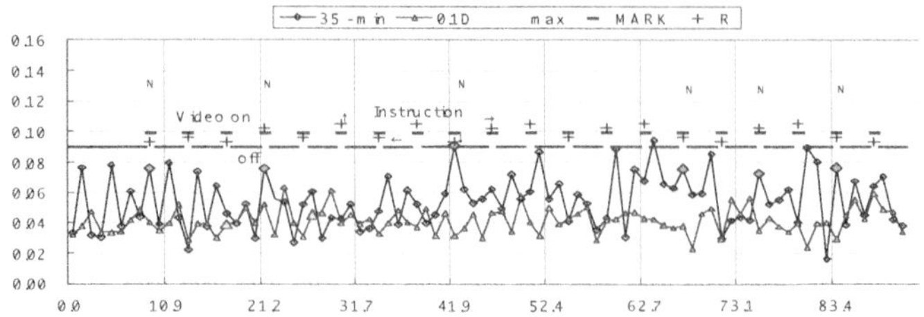

Fig. 7b. Response synchronization of –min. wave form to various instructions, denoted by "N"

In general synchronizations of peaks and responses occur frequently in response to various instructions, but not consistently to a definite instruction. But ch.35 is a specific channel, which shows synchronization to one specific instruction.

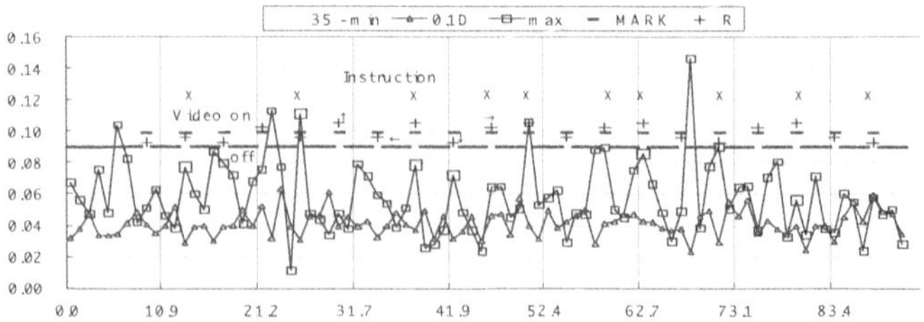

Fig. 7c. Response synchronization of max. wave form to various instructions, denoted by "X"

Fig. 7c). is the similar trace as Fig.7b). The only difference is replacement of –min. by max. waveform. Synchronization also happens with max. Synchronization points are indicated by X marks, and square notations of enlarged size. Total 10 X marks are indicated in the figure. Five of them are in responses to up instructions. All the up instructions are responded by synchronization of max. waveform. This is what we mean by consistency. Certain channel of the brain seems to respond consistently to certain instruction response combination.

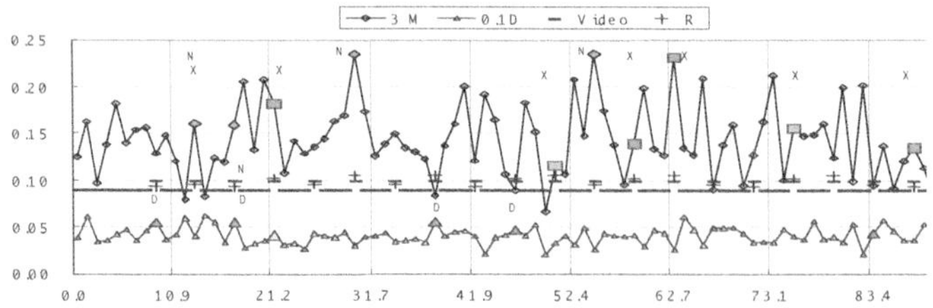

Fig. 7d. Response synchronization of magnitude (max. –min) and density (D) wave forms of ch. 3 to various instructions, denoted by "N, X and D". Double synchronizations (N+X) and (N+D) are identified in time zone between 10.9 and 21.2.

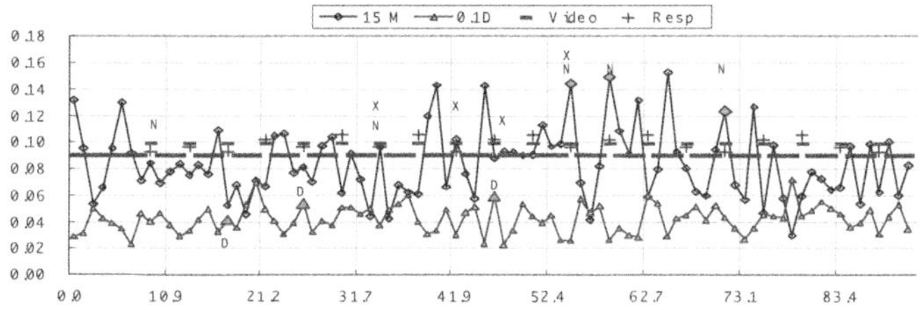

Fig. 7e. Response synchronization of magnitude (max. –min) and density (D) wave forms of ch. 15 to various instructions, denoted by "N, X and D". Double synchronizations, two (N+X) and one (N+D), are identified in time zone between 31.7 and 62.7.

Without saying, channels with large magnitude of quick components might show synchronization. It is true. Two major channels, ch. 3, and 15 are selected out from channels of large magnitude of quick components and the synchronization points are shown by colored notation on magnitude trace in Fig.7d) and e). Synchronization happens not always with one waveform. When max. and –min waveforms are close to each other temporarily, double synchronization happens. Double synchronization cannot be indicated by a mark in the graph. Thus synchronization are denoted by text "N" and "X", beside notation mark. In many cases, marks often overlaps with other marks, double indications are useful. Double synchronizations also happen between density waveform and max. or -min. Some cases of double synchronization between N and X, and between X and D are shown in Fig. 7d) and in Fig.7e).

Prominent synchronizations are remarkable but consistency is not found in Fig. 7d) and e).

4.2 Computer Analysis

For the further minute studies, visual analysis of every one peak is powerless.

The computer algorithm, shown below, is applied to1920 data points, which are 48 channels*20 trials*2 wave forms (max,-min).

The results are shown in Table 1 and 2. Table 1 is concerned with 7 channels with big magnitude, and Table 2 is concerned with 18 channels with some consistency. Other 23 channels not belonging to either groups are to be mentioned elsewhere.

The consistency is graded in three levels, full (>80%), fair (80 ~ 60%) and low (<60%). In Table 1, channel 35 showed full consistency to up (↑) instruction. Channel 2 to right (→) and 9 down(↓) showed fair consistency. There were no candidate channels for consistency in N synchronization.

5 Consistency Channels

In the Table 2 are collected synchronization distributions of consistent group. First five line from the above are N synchronization for up instruction(↑). Full consistency channels are not found in these lines. In the second 5 lines from the above, X synchronization for up instruction(↑) shows full consistency in 11 and 35 channels. Fair consistency is also found in 22 and 39 channels.

Table 1. Channels with large quick components

X/N	Instr	TIME	Channel 3	7	9	10	15	30	35
N	↑	29.5	23			5			
N	↑	37.8				14		11	
N	↑	50.2		7					
N	↑	62.7			9			9	
N	↑	79.3		10					
X	↑	29.5							5
X	↑	37.8			7				8
X	↑	50.2	12						11
X	↑	62.7	23			8			9
X	↑	79.3							6
N	→	21.2							8
N	→	46.1				12			
N	→	58.5		5			15		
N	→	75.1		6					7
X	→	21.2	18					17	
X	→	46.1			11		9		
X	→	58.5	14	9					
X	→	75.1	12			11			
N	↓	8.7				9	8		8
N	↓	17	14					11	
N	↓	41.9							9
N	↓	70.9		8			12		
N	↓	87.6		12		16			
X	↓	8.7			10				
X	↓	17		7	6				
X	↓	41.9					10		7
X	↓	70.9			6	10			9
X	↓	87.6	13	8	10	11			6
N	←	12.9	10					10	
N	←	25.3				12			
N	←	33.6					8	12	
N	←	54.4	23	6			10	13	
N	←	66.8		7					8
N	←	83.4		10					8
X	←	12.9	16	7				8	8
X	←	25.3			8	8			11
X	←	33.6					10		
X	←	54.4					14	12	
X	←	66.8						11	
X	←	83.4		7		9			

Algorithm for synchronization detection

m = max or -min of current interval
m_E : m of previous interval
m_N : m of succeeding interval

m = IF(and(m*g> m_E, m*g> m_N), m*100," ")
g= 0.9: peak margin
In order to avoid decimal point, m's in the table are shown after multiplied by 100.

In the third 4 lines from the above, N synchronization for right (→) instructions are shown. Full consistency channels are not found, but fair consistency is found in channel 20. In the next 4 lines from the above, X synchronization for right (→) instructions are shown. Full consistency is found in channel 2, and fair consistency is found in channel 3 and 31. In the 5 five lines by the middle of the Table2, N synchronization for down(↓) instructions are shown. Full consistency is found in channel 33 and fair consistency is found in channel 27. Next 5 lines below are X synchronization for down(↓) instructions are shown. No full synchronization but two fair synchronization channels are found in channel 4 and 9. In the bottom of Table 2, N and X synchronization for left (←) instruction are shown. No full consistency channels but fair consistency is found in some channels. One question is why no full consistency for left (←) instruction. It may be related to the experimental set. This experiment used the left hand, side entry task as explained in chapter 3. In this set up, to touch left and right button middle finger of left hand has to be extended or constricted. These actions will be more complex compared to up and down action which is simple touching to a button. The effects of different finger selection and actions can be examined by changing hand rest position, and by comparing results of left and right hands.

Table 2. Channel group related to response consistency

X/N	Instr	TIME	CHANNEL 1	2	3	4	9	11	20	21	22	27	31	33	34	35	38	39	40	46
N	↑	29.5	2		23							1		5						
N	↑	37.8				4		3										2		
N	↑	50.2		2					2	3									4	6
N	↑	62.7		1			9		2						2				4	5
N	↑	79.3	3			5									3					
X	↑	29.5						3		4	5	1	4		2	5		2		6
X	↑	37.8	6				7	3								8		2		
X	↑	50.2		1	12	4		2		2	6					11		2		5
X	↑	62.7			23			2			3	1				9				
X	↑	79.3	3			5		2			4		6			6	2	2	3	4
N	→	21.2	5						2						3	8		2	3	
N	→	46.1	4			7		2									3			
N	→	58.5				4			2		5		7	7			3			
N	→	75.1		1					1		4				2	7		2		6
X	→	21.2		1	18	5		3	2	3		1	4					3		
X	→	46.1		1			11		2		5						3			
X	→	58.5		1	14			2		3			8	7						
X	→	75.1		1	12								4						4	
N	↓	8.7	4								4	1	5	8		8		2	3	
N	↓	17		1	14			2		3		2		6	2		3			
N	↓	41.9							2			1		7		9	3	2	3	
N	↓	70.9						2			4			4						6
N	↓	87.6				7				3		1		7					5	
X	↓	8.7	3			5	10			2		1		6			3			6
X	↓	17				6	6			3		1	5	6	2					6
X	↓	41.9				7								6		7				
X	↓	70.9	5	2			6		2							9			3	
X	↓	87.6		1	13	5	10					2			3	6	4			6
N	←	12.9	3		10	6		3			3							2		
N	←	25.3								3		1		5						3
N	←	33.6	4	2		4					4		5		2				3	7
N	←	54.4	4		23	7								8	2				2	
N	←	66.8	4						2	3				6	2	8	3		3	6
N	←	83.4						2	1	3	4	1	5			8	3		3	4
X	←	12.9			16	4								7	2	8				
X	←	25.3					8			2					3	11				
X	←	33.6																	2	6
X	←	54.4				6					4				2					6
X	←	66.8				5			2	2				5	2			2	3	
X	←	83.4						2				2		6				1		

X full consistancy
X. fair consistancy
N full consistancy
N fair consistancy

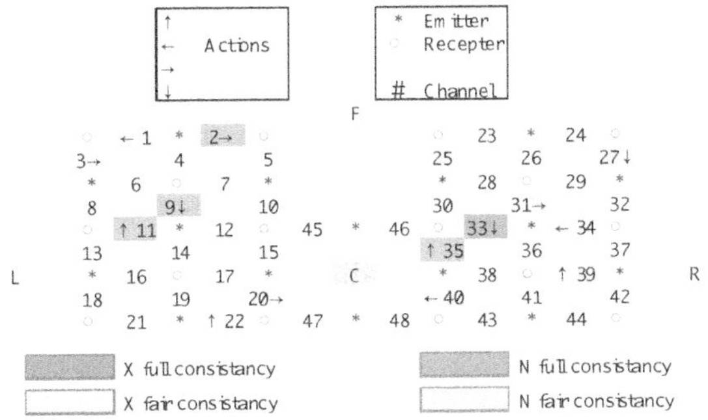

Fig. 8. Allocation of full and fair consistency channels for left hand side entry operation. Instruction and action direction are indicated by arrows.

In conclusion of synchronization analysis, consistency allocation for the left hand side entry operation is illustrated in Fig. 8.

Table 1 is channel group with active quick components and effecting to almost all activities but not always specifically related to certain tasks. Table 2 is the channel group which are not always active in quick component but certain consistency to finger selection is confirm by this analysis. Some more channels are supposed to be included to Table 2 group based on further analysis related to left/right hand and top/side entry variations. Still more channels are expected to enter the group based on instruction presentation variations. At the moment 23 more channels remain out of either group.

Further analysis is done with the density related to finger selection. The NIRS analysis[7] of continuous events so far suggests that the channels actually working are less variable, so that standard deviations are small, while channels out of work show large deviation. Similar trend are observed also in finger selections task.

6 Conclusion

Button operation by finger is still most essential activities supporting human computer interactions, to which not much is known about the brain functions.

Quick component analysis is successfully applied to finger selection of button operation task.

Magnitude and density analysis will extend the scope of brain function studies to real time activities.

References

[1] Tamura, H., Omori, M., Choui, M.: NIRS trajectories in Oxy-Deoxy Hb plane and the trajectory map to understand brain activities related to human interface. In: Smith, M.J., Salvendy, G. (eds.) HCII 2007. LNCS, vol. 4557, pp. 994–1003. Springer, Heidelberg (2007)

[2] Fuchigami, M., Okada, A., Tamura, H., Omori, M.: A Comparative Study of Brain Activities Engaged in Interface Operations by Means of NIRS Trajectory Map. In: Smith, M.J., Salvendy, G. (eds.) HCII 2007. LNCS, vol. 4557, pp. 830–839. Springer, Heidelberg (2007)

[3] Tamura, H., Fuchigami, M., Okada, A., Omori, M., Hasegawa, S., Fujikake, K., Choui, M.: On Physiological Roles of Quick Components in Near Infrared Spectroscopy. In: Proc. Mobile Interactions 2008, pp. 97–102. Society of Mobile Interactions (2008) (in Japanese)

[4] Tamura, H., Fuchigami, M., Okada, A.: On Physiological Roles of Quick Components in Near Infrared Spectroscopy. In: Proc. Mobile Interactions 2008, pp. 97–102. Society of Mobile Interactions (2008)

[5] Tamura, H., Fuchigami, M., Okada, A.: Analysis of Text Entry behavior by NIRS Quick Components. In: Proc. Mobile Interactions 2009, pp. 109–114. Society of Mobile Interactions (2009)

[6] Fuchigami, M., Okada, A., Tamura, H.: Analysis of NIRS Quick Components in direction-selective button operations. In: Proc. Mobile Interactions 2009, pp. 115–120. Society of Mobile Interactions (2009)

[7] Tamura, H., Fuchigami, M., Okada, A.: Neuro-NIRS: Analysis of Neural Activities using NIRS. In: HCII 2009 (2009)

Effects of Practice with Foot- and Hand-Operated Secondary Input Devices on Performance of a Word-Processing Task

Fredrick P. Garcia and Kim-Phuong L. Vu

Center for Usability in Design and Accessibility
Department of Psychology, California State University, Long Beach
1250 N. Bellflower Blvd. Long Beach, CA 90840, USA
garcia.fp@gmail.com, kvu8@csulb.edu

Abstract. This study compared the performance of users operating a foot controlled input device (foot mouse) with that of users operating a hand controlled input device (hand trackball). Four different tasks that required (1) direct manipulation of on-screen objects and menu command activation, (2) keyboard entry and direct manipulation, (3) keyboard entry and command activation, and (4) keyboard entry, direct manipulation, and command activation were assessed. Performance on each task was measured both before and after participants practiced computing tasks with one of the devices for 750 trials. For all 4 tasks, practice improved performance with the foot mouse but not the hand trackball. However, overall performance was better with the hand trackball.

Keywords: input device, foot input, hand input, practice, word processing.

1 Introduction

Graphical User Interfaces (GUIs) allow for a more efficient interaction with the computer environment than the command prompt interfaces that preceded them, but they still have costs associated with their use. One such cost relates to the fact that to be used to their greatest potential, GUIs require that a separate input device (e.g., a mouse) be operated in addition to the keyboard. Research has shown that a consequence of using the hand to control the secondary input device is increased musculoskeletal injuries [4], or at the very least, exposure to awkward angles identified as risk factors for musculoskeletal injuries [1, 3]. Also, a significant amount of time and accuracy are lost when the resources required by the secondary input device overlap with those that are already put to use from typing with the keyboard. These performance decrements come from physically switching back and forth between devices [2] and mentally preparing to execute the physical actions after attention has been switched from one task to the other [6]. However, there is a possibility that this lost time could be recovered if there were a way to prevent those resources required by the devices from overlapping. In attempting to prevent resource overlap, speech recognition software has been developed, but it is not advanced enough to handle keyboard responsibilities [9]. Furthermore, voice input cannot be used to perform all the

M.J. Smith and G. Salvendy (Eds.): Human Interface, Part I, HCII 2009, LNCS 5617, pp. 505–514, 2009.

activities that can be performed with a hand controlled input device (e.g., cursor positioning, dragging and dropping items, etc.) or take full advantage of the GUI interface (e.g., simple recognition of icons rather than recall of key combinations or voice commands). However, a foot controlled input device can perform these activities.

Previous research evaluating the appropriateness of using a foot controlled input device for working with computers has shown that people are consistently slower using their foot to control an input device than using their hand [7, 8, 10]. However, past research has failed to examine the role of practice in creating an efficient human-computer system involving a foot device. In past studies, while devices with which participants had worked before were operated with their hands, they were provided novel input devices to operate with their feet and allowed little practice. Another issue that may have led to slower foot input times in previous studies is the omission of foot specific input devices that are commercially available. Commercially available foot specific input devices would be expected to help users work as efficiently as possible. Previous studies have also limited the types of tasks used to test performance with foot controlled input devices. There have been no studies designed with tasks that take advantage of the unique ability of the foot controlled input device to be used in parallel with the keyboard, even though there are many multi-step computer tasks that require using both the keyboard and some other input device. Though simple tasks favor the hand operated input device, complex tasks may allow better overall performance with the foot operated input device.

The purpose of this study was to examine the effect of practice with a commercially available foot-specific secondary input device (a foot mouse) and a hand-controlled secondary input device that was not overpracticed like the hand mouse (a hand trackball) on word processing task performance. Typical word processing activities fall into three categories: textual input (typing), direct manipulation of on-screen objects (e.g., dragging and dropping objects or selecting text), and command activation (e.g., choosing to "Print" from a command menu) [5], so completion time was measured for 4 tasks requiring varying amounts of these activities. Participants used either the foot mouse or hand trackball with the keyboard (when needed) to complete the tasks before and after 750 trials of practice with one of the secondary input devices.

2 Method

2.1 Participants

Sixteen participants were recruited from the campus of California State University, Long Beach. Participants received \$5.00 for each of the 10 sessions of the experiment they completed, except for sessions 1 and 10 for which they received \$10.00 (\$5.00 for each of 2 parts), and a \$20.00 bonus for completing the entire experiment. Participants ranged from 21-32 years of age (M = 25 years). All participants had normal vision, were right handed and footed, and had full use of their hands and feet.

Participants reported being relatively experienced with computers (M = 1.27, with 1 being very familiar and 5 not at all familiar for this and subsequent questions), spending an average of 34 minutes a day using a Windows based computer, and being very familiar with the Microsoft Word word-processing program (M = 1.27). Participants also reported that they were very familiar with using the hand mouse (M = 1.27), not very familiar with using the hand trackball (M = 3.64), and not at all familiar with using a foot input device to interact with the computer (M = 4.93).

2.2 Apparatus

All parts of the experiment were programmed in Visual Basic (Microsoft Corporation). The program provided the workspace for the practice and word processing tasks and recorded task completion times. Both the practice trials and the word processing performance tests were run on a Pentium 4 based PC using a 19" color monitor. Text input was handled by a standard Dell QWERTY keyboard, while for command activation and direct manipulation activities the Logitec Trackman Wheel trackball was used as the hand-controlled input device and the NoHands Mouse by Hunter Digital was the foot-controlled input device.

2.3 Materials

The materials and task domain used to evaluate word-processing task performance were modified from four tasks developed for a previous experiment on word processing performance using multiple input devices [5]. For the first task, participants were given an unformatted document and told to reformat it to a predefined style (a preformatted version of the document served as the model and remained visible during the task, see Fig. 1) using six text formatting tools that were selected from a drop-down menu. This task was completed using only direct manipulation and commands executed with the secondary input device (i.e., keyboard was not used).

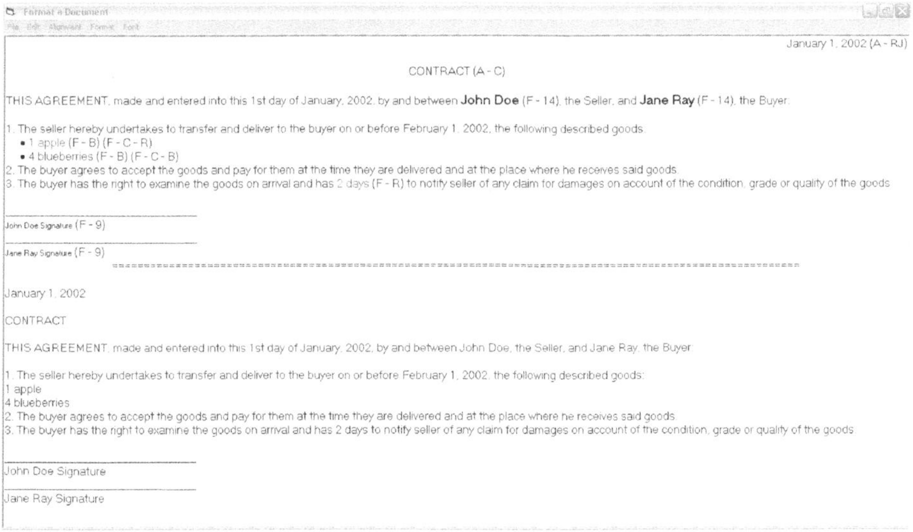

Fig. 1. Screenshot of Task 1 workspace

The second task required participants to type a short scientific formula (1) containing subscripted and superscripted text, and Greek symbols. To complete this task there was a minimum ratio of 1.48 required keystrokes to 1 required formatting command (approximately the same amount of keystrokes and commands were needed). Participants were asked to type the formula from left to right as they activated the necessary commands with the secondary input device. This task was completed using the keyboard and commands executed with the secondary input device.

$$d^2(r_v^{\ 1})/dt^2=-\mu^1(r_v^{\ 1}/r_v^{\ 3})+\mu_m(-d_{mv}\ d_{mv}^{\ \ 3}-r_m/r_m^{\ 3})\ . \tag{1}$$

The third task involved participants building a table consisting of the first 12 letters of the Greek alphabet and their English alphabet equivalents, using the copy and paste keyboard accelerator commands (Ctrl-C and Ctrl-V, respectively) while navigating the document by scrolling up and down. Scrolling was necessary because the list of letters and the table were not on the same page. To enter each letter in the table participants were asked to: 1) select the letter with the secondary input device, 2) copy it, 3) memorize the English alphabet equivalent corresponding to the letter, 4) scroll down to the table with the secondary input device, 5) paste the letter in the table, and 6) type in the English alphabet equivalent. If the participant did not remember the English equivalent, they needed to scroll back up to read it as many times as was necessary to correctly type it in the table. To create a keyboard to secondary input device dynamic that was similar to the previous task, there was a minimum ratio of 1 required keystroke to 1 required direct manipulation with the secondary input device.

The last word processing task required participants to type a short paragraph containing subscripted, superscripted, italicized, underlined, bolded text, and Greek symbols. All format changes were created using command activation (e.g., if a word was to be bolded, the Bold command was first selected from the Format drop down menu and then the word was typed). Greek symbols were created by selecting them from the Symbols command menu. Participants were required to type the text from left to right and activate the menu commands when needed. This task had a minimum keystroke to formatting command ratio of 16.31 to 1. However, this was the upper bound of the minimum keystroke to secondary input device activity ratio as the task also required that participants used the secondary input device to drag the scroll bar (direct manipulation) to compare what they were copying to what they were typing as the paragraphs were on separate pages. If participants wanted to see every line as they typed it, the minimum keystroke to secondary input device activity ratio was 10.02 to 1. This task incorporated all three word processing components: typing, command activation, and direct manipulation. The paragraph to be typed was:

> The teacher had the students turn to *Section 2* in their <u>Modern Mathematics</u>[1] textbook, which had replaced their previous text: <u>Old Math</u>[2]. The math lesson was on variables together with exponents, like x^5. The teacher was going to review the **Pythagorean Theorem**[3], where $a^2 + b^2 = c^2$, as an example. She also wanted to talk about finding the **area of a circle** using the formula π**r**^{2}, where **r** stands for the radius of the circle. After Math, she was going to give a Chemistry lesson. She was going to begin by discussing H_2O, which she was *sure* everyone would know about.

2.4 Procedure

Upon arrival, each participant was asked to read and sign a consent form. Participants then filled out a demographic questionnaire and were reminded about the multiple session commitment required for the experiment. They were then asked to sit in front of a computer monitor at a comfortable viewing distance. Participants worked through each of the 4 word processing tasks using either the foot- or hand- operated secondary input device first (half of the participants used the device they had been assigned to practice on first, and half used the device they would not be practicing first).

Before beginning the 4 tasks with an input device, participants completed a block of practice trials to familiarize themselves with the device. This was the same activity that they later practiced for 750 trials in the practice sessions. Each trial consisted of the same 3 steps: 1) find and click the "1" button randomly located in the top 1/4 of the workspace, 2) scroll down to find and highlight the string of characters randomly located in the bottom 1/4 of the workspace, 3) scroll up to find and press button "2" randomly located in the top 1/4 of the workspace.

After participants performed 30 practice trials of this type they worked through the 4 word processing tasks described earlier (taking a 3-5 minute break between tasks). This completed part 1 of test session 1. After a 15 minute break, participants returned for part 2 of test session 1. During part 2, participants performed 30 practice trials with the other device and then worked through all 4 tasks with it as well. Participants always worked through the tasks in the following order: task 1, 2, 3 and then 4. Each task was explained to the participant just before they began. Participants took a break between tasks. Each part of test session 1 took less than an hour.

After initial testing with the 4 word processing tasks on the first day, participants returned for 8 practice sessions with the secondary input device with which they were assigned to receive practice. Which device each participant was assigned to practice with was determined randomly, but was counterbalanced across participants. Each of the practice sessions were no more than 1 hour long and contained between 50 and 150 practice trials of the sort previously described. More practice trials were able to be included in later practice sessions because performance improved with practice.

When participants finished all practice sessions, they returned for a final test session with the 4 word processing tasks. Participants used whatever input device they used first during the first test session. Again, each task was explained to the participant just before they began and any questions were answered. Participants took a 3 - 5 minute break between tasks and a 15 minute break when secondary input devices were switched. After completing all tasks using both the foot and hand, participants answered a post-experiment questionnaire and were then verbally debriefed as to the purpose of the experiment. The final testing session took 1 hour.

3 Results and Discussion

Separate 2 (practiced device: foot mouse or hand trackball) x 2 (operated device: foot mouse or hand trackball) x 2 (test session) mixed ANOVAs were performed on task completion times for each of the 4 word processing tasks. Practiced device is the between-subject factor, and operated device and test session are within-subject factors.

3.1 Task 1

There was a main effect for operated device such that participants were faster with the hand trackball (M = 121 s) than the foot mouse (M = 246 s), $F(1, 14) = 45.69$, $p < .05$. There was also a main effect of test session, where participants were faster at test session 2 (M = 160 s) than at test session 1 (M = 207 s), $F(1, 14) = 11.49$, $p < .05$. These main effects were qualified by a significant practiced device x operated device x test session 3-way interaction $F(1, 14) = 6.18$, $p < .05$, see Fig. 2.

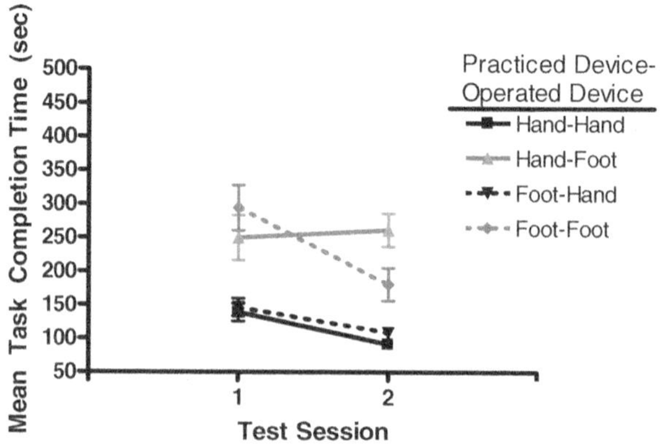

Fig. 2. Mean Task 1 Completion Time as a function of Practice Device, Operated Device, and Test Session

When this interaction was examined further, a significant interaction of practiced device x test session was found to be present when the foot mouse was the operated device, $p < .05$, but not when the hand trackball was the operated device, $F < 1.0$. When participants operated the foot mouse and had practiced with the foot mouse, their completion times improved from test session 1 to test session 2, $p < .05$. However, when they operated the foot mouse and had practiced with the hand trackball, completion times were not significantly different between test sessions, $p = .77$ (see Fig. 2). In sum, practice effects were evident with the foot operated secondary input device but not the hand operated one.

3.2 Task 2

There was a main effect for operated device such that participants were faster with the hand trackball (M = 151 s) than the foot mouse (M = 225 s), $F(1, 14) = 35.54$, $p < .05$. There was also a main effect of test session where participants were faster at test session 2 (M = 159 s) than at test session 1 (M = 217 s), $F(1, 14) = 50.20$, $p < .05$. These main effects were qualified by a significant practiced device x operated device x test session 3-way interaction, $F(1, 14) = 8.32$, $p < .05$, see Fig. 3.

When this interaction was examined further, a significant interaction of practiced device x test session was found to be present when the foot mouse was the operated device, $p < .05$, but not when the hand trackball was the operated device, $F < 1.0$. When participants operated the foot mouse and had practiced with either the foot mouse or hand trackball, their completion times improved from test session 1 to test session 2, *ps* < .05. However, greater improvement was observed between test session 1 and 2 when participants practiced with the foot mouse than when they practiced with the hand trackball, $p < .05$. Again, these results show that practice improved performance on the foot mouse but not the hand trackball.

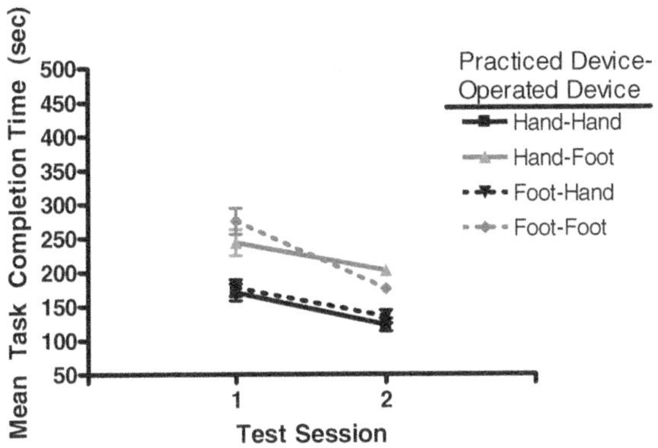

Fig. 3. Mean Task 2 Completion Time as a function of Practice Device, Operated Device, and Test Session

3.3 Task 3

The main effects for operated device and practice session were significant, $Fs(1, 14) > 26.57$, $ps < .05$. Participants were faster with the hand trackball (M = 204 s) than the foot mouse (M = 424 s), and at test session 2 (M = 272 s) than at session 1 (M = 356 s). These effects were qualified by a significant practiced device x operated device x test session interaction, $F(1, 14) = 16.18$, $p < .05$, see Fig. 4.

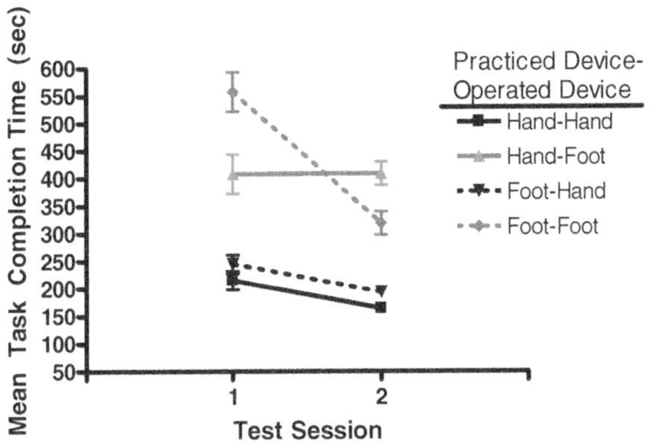

Fig. 4. Mean Task 3 Completion Time as a function of Practice Device, Operated Device, and Test Session

When this interaction was examined further, a significant interaction of practiced device x test session was found to be present when the foot mouse was the operated device, $p < .05$, but not when the hand trackball was the operated device, $F < 1.0$. When participants operated the foot mouse and had practiced with the foot mouse, their completion times improved from test session 1 to session 2, $p < .05$. However, when they operated the foot mouse and had practiced with the hand trackball, completion times were not significantly different between test sessions, $p = .984$. Again, practice improved performance on the foot mouse but not the hand trackball.

3.4 Task 4

There was a main effect for operated device such that participants were faster with the hand trackball ($M = 494$ s) than the foot mouse ($M = 728$ s), $F(1, 14) = 20.12$, $p < .05$. There was also a main effect of test session where participants were faster at test session 2 ($M = 530$ s) than at test session 1 ($M = 693$ s), $F(1, 14) = 39.78$, $p < .05$. These main effects were qualified by a significant practiced device x operated device x test session 3-way interaction, $F(1, 14) = 5.06$, $p < .05$, see Fig. 5.

When this interaction was examined further, a significant interaction of practiced device x test session was found to be present when the foot mouse was the operated device, $p < .05$, but not when the hand trackball was the operated device $F < 1.0$. When participants operated the foot mouse and had practiced with the foot mouse, their completion times improved from test session 1 to test session 2, $p < .05$. However, when they operated the foot mouse and had practiced with the hand trackball, completion times were not significantly different between test sessions, $p = .345$. Thus, similar to Tasks 1-3, practice improved performance on the foot mouse but not the hand trackball.

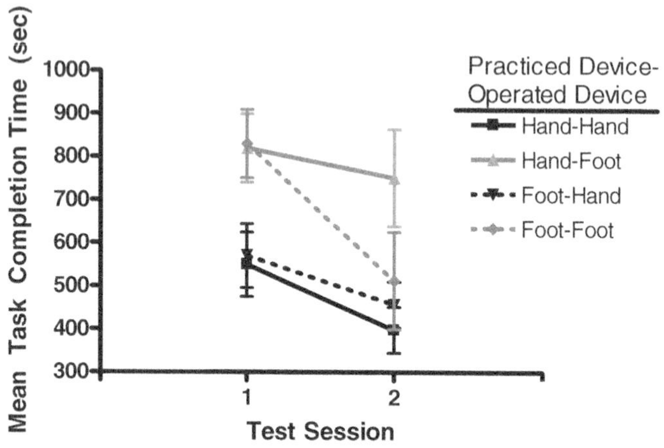

Fig. 5. Mean Task 4 Completion Time as a function of Practice Device, Operated Device, and Test Session

4 Conclusion

For all tasks, there was a significant main effect of operated device where participants were faster with the hand trackball than the foot mouse, and a significant main effect of test session such that participants were faster at test session 2 than test session 1. Also, for all tasks, these main effects were qualified by a 3-way interaction of practiced device x operated device x test session. Further analysis of these interactions showed that practice improved performance with the foot mouse but not with the hand trackball.

That practice did not improve performance when the hand trackball was the operated device suggests a ceiling effect for performance with the hand trackball. That is, participants may have already been so skilled with using their hand for activities similar to operating the trackball that 750 trials of practice could not further their hand trackball skills enough to show effects over the improvement that came from task repetition (i.e., improvement on each task from test session 1 to 2 regardless of device operated or practiced). A ceiling effect on performance with the trackball is somewhat surprising, as participants (on average) reported they were somewhat unfamiliar with the hand trackball. Moreover, finding a ceiling effect on a hand-controlled device, which was rated as being somewhat unfamiliar by users, calls into question all past research that compared performance with hand-controlled secondary input devices with foot-controlled secondary input devices.

Participants may be so skilled at using their hand to perform fine motor activities similar to operating a computer input device that using an unpracticed hand-operated input device may allow for a level of performance close to what could be expected if it were practiced. Even beyond general fine motor skills, it is possible that skills specific to using the hand to operate an indirect pointing device (where movement of an input device in the horizontal plane must be mapped to the vertical plane of the display and some gain is applied to the device where its velocity and acceleration are augmented in the display) to interact with a computer have been learned by the general population through repetitive hand mouse use. This previous experience with an indirect pointing device using the hand and not the foot could also contribute to unpracticed hand trackball performance resembling practiced performance and add to performance differences between the foot mouse and hand trackball.

The finding that practice significantly improves performance with a foot mouse, but not with a hand trackball, suggests that the foot mouse should be practiced many times before it is used in an experiment so that a more accurate determination can be made about the level of performance that can be expected with the device. It would be useful to look at the effect of different amounts of practice and whether further practice can continue to improve performance. It would also be beneficial to examine the effect of practicing the foot-controlled secondary input device with the keyboard rather than practicing it on component tasks separate from the keyboard.

One limitation of the current experiment concerns the foot-controlled secondary input device. During the experiment, the commercially available foot mouse progressively lost sensitivity until it could no longer be used. When a replacement was obtained it was more sensitive than the previous device had ever been. It would be expected that a commercially available foot controlled input device that is at least as reliable and consistent as commercially available hand controlled input devices would only increase performance improvements due to practice with the foot device.

References

1. Burgess-Limerick, R., Shemmell, J., Scadden, R., Plooy, A.: Wrist Posture During Computer Pointing Device Use. Clinical Biomechanics 14, 280–286 (1999)
2. Card, S.K., Moran, T.P., Newell, A.: The Keystroke-Level Model for User Performance Time with Interactive Systems. Communications of the ACM 23(7), 396–410 (1980)
3. Fagarasanu, M., Kumar, S.: Carpal Tunnel Syndrome Due to Keyboarding and Mouse Tasks: A Review. International Journal of Industrial Ergonomics 31, 119–136 (2003)
4. Jensen, C., Borg, V., Finsen, L., Hansen, K., Juul-Kristensen, B., Christensen, H.: Job Demands, Muscle Activity and Musculoskeletal Symptoms in Relation to Work with the Computer Mouse. Scandinavian Journal of Work and Environmental Health 24, 418–424 (1998)
5. Karl, L., Pettey, M., Shneiderman, B.: Speech-Activated Versus Mouse-Activated Commands for Word Processing Applications: An Empirical Evaluation. Technical Report No. CAR-TR-630. University of Maryland Center for Automation Research, College Park, MD (1992)
6. Monsell, S.: Task Switching. TRENDS in Cognitive Sciences 7(3), 134–140 (2003)
7. Pakkanen, T., Raisamo, R.: Appropriateness of Foot Interaction for Non-accurate Spatial Tasks. In: CHI 2004 Extended Abstracts on Human Factors in Computing Systems, pp. 1123–1126. ACM Press, New York (2004)
8. Pearson, G., Weiser, M.: Exploratory Evaluation of a Planar Foot-Operated Cursor-Positioning Device. In: O'Hare, J.J. (ed.) Proceedings of the SIGCHI Conference on Human Factors in Computing Systems - CHI 1988, pp. 13–18. ACM Press, New York (1988)
9. Schneiderman, B.: The Limits of Speech Recognition. Communications of the ACM 43(9), 63–65 (2000)
10. Springer, J., Siebes, C.: Position Controlled Input Device for Handicapped: Experimental Studies with a Footmouse. International Journal of Industrial Ergonomics 17, 135–152 (1996)

Manipulation with Fingers in a 3-D Physical Space

Yuki Kanai, Makoto Oka, and Hirohiko Mori

Musashi institute of Technology, 28-1, Tamadutsumi 1-chome,
Setagaya-ku, Tokyo, 158-8557, Japan
g0781709@sc.musashi-tech.ac.jp, oka@ergo.ie.musashi-tech.ac.jp,
mori@si.musashi-tech.ac.jp

Abstract. Recently, input devices in a three dimensional physical space become popular. However these devices get used only with easy interaction and do not get used for complex interaction yet. In this study, proposes a character recognition technique drawn in a three dimensional physical space considering wearable computing environments. The input device that we developed is worn to a finger. The sensor used the acceleration sensor. The recognition rate is verified through the experiment. As a result of doing various experiments, the recognition rate was 65%.

Keywords: Manipulation, 3-D physical space, OLCR, Japanese.

1 Introduction

Recently, input devices to detect motion in a three dimensional physical space, such as game controllers, mice, become popular. These devices, however, can detect only simple operations such as big motion, cursor movements, and do not get used for complex interaction yet. We premised the system that is able to write memos anytime, anywhere and easily. This premised system is able to easily write the text. We thought about the input device for this system.

As the related works, there is the input device named Airpen [1]. This device is able to draw on the virtual canvas through the HMD in a three dimensional physical space. However, this device needs the button manipulation for drawing the character and the picture. Therefore, it is equal to carrying about the mouse.

This paper, considering wearable computing environments, proposes a character recognition technique of a drawn character in a three dimensional physical space with the device worn to a finger.

There are some problems to solve in recognizing a character written in a 3-D space comparing to the case of 2-D planes. First, to recognize a character written in three dimensional physical space, it is necessary to detect the description plane, because the finger swings in all directions. Second, though the detection of each strokes an important role in ordinary OLCR systems in a two dimensional plane, it is difficult to divide a one seamless line in to some strokes in case of three dimensional space.

In this study, three-axis acceleration sensor is used as an input device and it is used to detect a finger motion in a 3-D space dealing in recognizing the motion of the tip of a finger, the 3-axis value of the acceleration obtained with the device is converted into

M.J. Smith and G. Salvendy (Eds.): Human Interface, Part I, HCII 2009, LNCS 5617, pp. 515–523, 2009.

the two dimensional coordinates. In this way, we can obtain relative position of the tip of a finger on two dimensional. This conversion is done about the description plane above mentioned considerately. The stroke of a character written by the tip of a finger is able to be obtained by the converted values. The computer software of OLCR learns a character written the stroke without pause. Finally, the character is converted to the computer text.

2 System Architecture

Fig. 1 shows the system image we developed. The acceleration sensor mounted on the device can detect acceleration from -2G to +2G. The data is converted into 10bit data by A/D Converter of Microcomputer. This converted data is transmitted to PC by using USB. The interval in about ten milliseconds is put, and it saves the data in PC.

The preserved data is processed with PC. Then, the moved distance of the tip of finger is converted into the pixel data. This means the movement for two directions, x-axis and y-axis.

CellWriter [2] is used as OLCR software in this study.

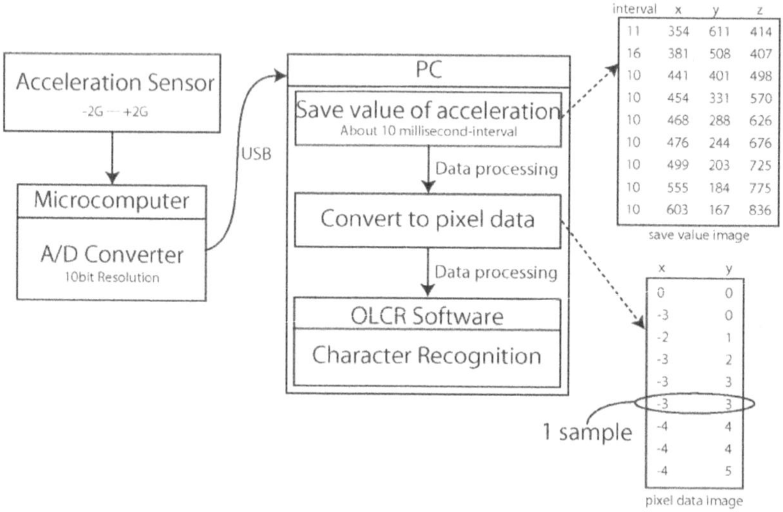

Fig. 1. System architecture

The obtained value from the acceleration sensor is possible to calculate two values such as static acceleration and dynamic acceleration. The value of static acceleration is gravity that hangs in the accelerometer. The tilt of the finger is decided by using this value. The value of dynamic acceleration is acceleration caused by motion of the sensor. The detection position of fingers in a three dimensional physical space is decided by using this value.

To evaluate my system, I used Japanese characters of Hiragana and Katakana respectively by 46 characters. Through the experiment, we obtained 11 sets of these characters each from ten subjects. In these data, 7 set are used for learning of OLCR and 4 sets are done for the test (Fig. 2 shows a learning method image).

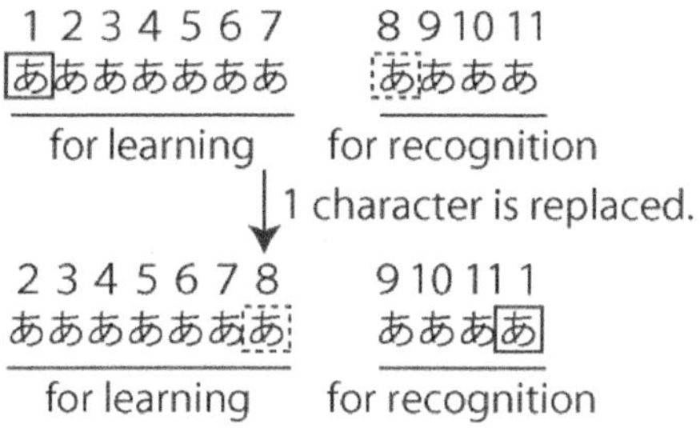

Fig. 2. Learning method

3 Exploratory experiment

3.1 Method

We conducted the experiment which was aimed to investigate whether it is possible to recognize the value in obtained state. The description plane was detected as shown in Fig. 3.

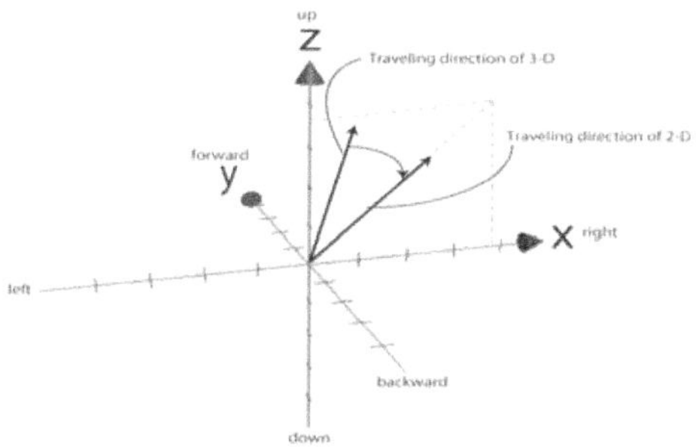

Fig. 3. The description plane for exploratory experiment

Moreover, it is stored in the frame when converting the pixel data (Fig 4.). The frame is 40 pixels in width, and 50 pixels in height.

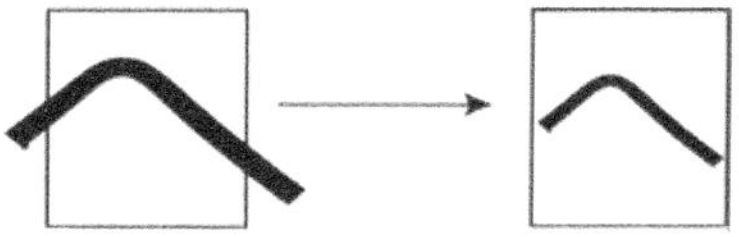

Fig. 4. Regularization image. This image shows that it is stored in the frame. This character is Japanese 'he'.

3.2 Results and Consideration

The result of the recognition rate in this experiment was 53% and it was worse than we expected.

We considered that there were two big problems. First, the same character was different the number of the obtained value from acceleration sensor. Therefore, we were thought that was necessary to equate the same as much as possible. Second, we thought that we had to verify about the detection of description plane.

4 Experiment 1

4.1 Method

We conducted the experiment which was aimed to investigate whether the recognition rate improves after the acceleration data is corrected by the following methods. We also conducted the experiment which was aimed to investigate whether the recognition rate improves by changing the detection method of description plane.

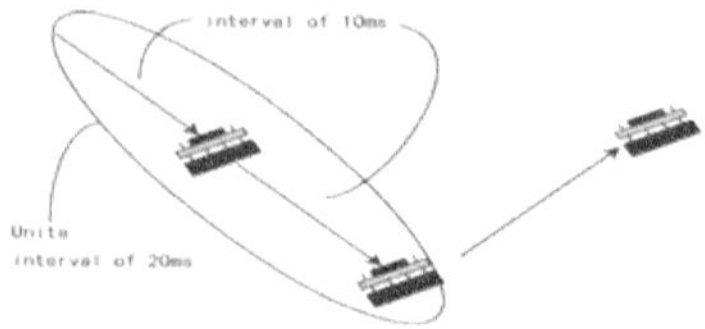

Fig. 5. Unite Image

Here we propose two correction methods. First, the obtained value from acceleration sensor unites if it is the same traveling direction (Fig. 5). The method of deciding the same direction is as follows. Hereafter, this is called 1st method. It calculates an angle from the value of the obtained acceleration (Fig. 6). And, the difference of the angle is calculated in order of the time series. If the difference does not exceed the threshold, it is decided the same direction. In the value of the threshold, θ is 0.50, ω is 0.01 and γ is 0.50.0.

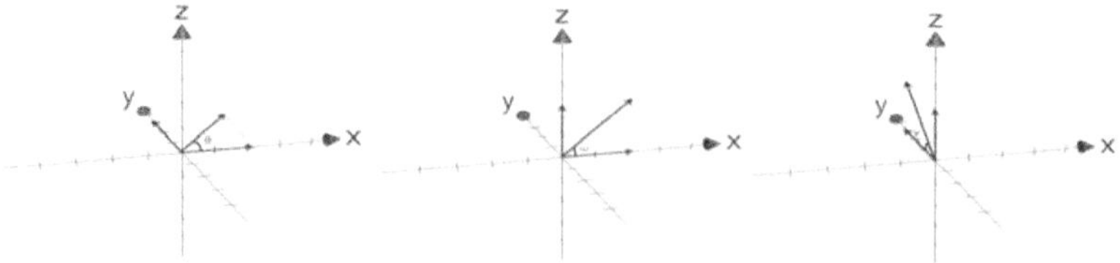

Fig. 6. The angle for deciding the same direction

There seems to be a new problem happens in this method (Fig. 7). The group is made when the inclination of finger of the finger is the same. Then, the first method is applied in the group. We thought that we were able to solve a previous problem by this method. Hereafter, this is called 2nd method. Additionally, the acceleration data after the above-mentioned method is applied is converted into pixel data. The movement of x and y has the part where the pixel data both becomes 0. In that case, the part is deleted.

The description plane such as the x-y plane, x-z plane, and y-z plane in addition to method of exploratory experiment is verified. The x-y plane is only the value of x-axis and y-axis is used. The other is also similar.

4.2 Results and Consideration

Table 1 indicates the recognition rate. To our regret, the recognition rate fell by correcting it.

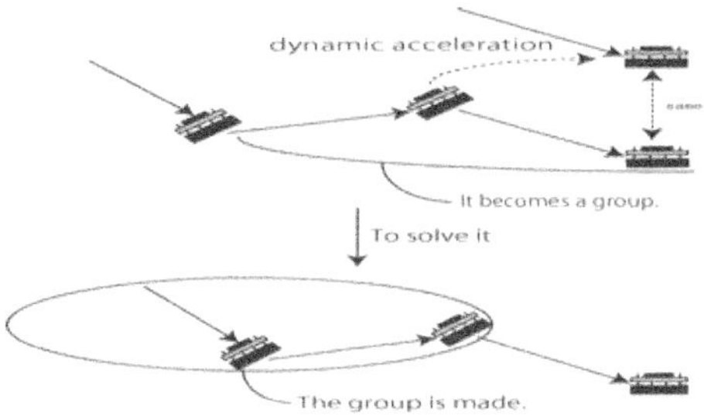

Fig. 7. The problems of 1st method and the solutions

As the tendency to the description plane, the recognition rate of x-y plane is the highest.

Table 1. Experiment 1 result.Unit:%.

	xy	xyz	xz	yz
no correct	64	54	61	42
1st method	35	30	32	25
2nd method	52	44	48	36

We paid attention to drawing of the character. Fig. 8 shows the same character that the same subject wrote. The right character looks smaller than the left. When regularizing it for the pixel data conversion, width and height were converted into the same ratio this time. We thought that the verification was necessary for another ratio.

Fig. 8. Drawing Japanese characters. This is Japanese 'ki'.

5 Experiment 2

5.1 Method

We conducted an experiment which was aimed to investigate whether the recognition rate improves after the pixel data conversion as shown in Fig. 9. No correcting and the 2nd method proposed in the experiment 1 were used here.

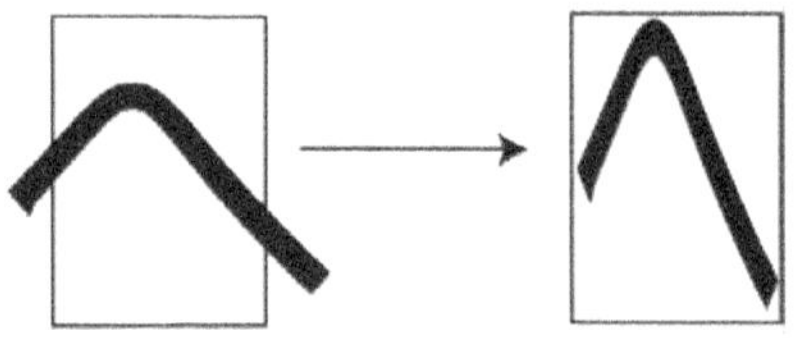

Fig. 9. New regularization Image

5.2 Results and Consideration

Table 2 indicates the recognition rate. To our regret, there isn't so much of a difference between result of experiment 1 and result of experiment2. As the tendency to the description plane, the recognition rate of x-y plane is the highest.

Table 2. Experiment 2 result.Unit:%.

	xy	xyz	xz	yz
no correct	64	49	58	40
2nd method	53	42	47	34

In this experiment, the method shown in Fig.9 was applied to obtained data for all characters to convert to the pixel data. However, this method seems to be not necessarily suitable for all the character data. In the case of one subject, the recognition rate of Japanese characters of katakana 'ni' worsened from 90.9% to 47.7 compared with experiment 1, although the recognition rate of Japanese characters of katakana 'ke' improved from 38.6% to 72.7%. This result implies that the pixel data conversion method proposed in Experiment1 is suitable for some characters and the method proposed in this section is so for the other characters and therefore, we should switch the pixel data conversion method proposed in experiment 1 to the method proposed experiment 2 depending on characters.

And, the result of the 2nd method is bad. To improve the rates, it is necessary to adjust the threshold value and to verify it

6 Experiment 3

6.1 Method

We conducted the experiment which was aimed to investigate whether the recognition rate improves by adjusting the threshold value of 2nd method proposed in Experiment 1.

Three kinds of threshold value were set as shown in the table 3. The description plane is the x-y plane, the x-z plane and the x-y-z plane proposes with exploratory experiment.

Table 3. The threshold for 2nd method. Unit:radian.

	θ	ω	γ
No.1	0.50	0.01	0.01
No.2	0.01	0.01	0.01
No.3	0.50	0.50	0.50

6.2 Results and Consideration

Table 4 indicates the recognition rate. The 2nd method worsened the recognition rate.

Table 4. Experiment 3 result.Unit:%.

	xy	xyz	xz
No.1	55	49	50
No.2	59	49	52
No.3	36	29	31

This result must be caused by destroying the features of the characters.

7 Experiment 4

7.1 Method

We conducted the experiment which was aimed to investigate whether the ratio of each character is changed at the pixel data conversion and the recognition rate improves (Fig. 10). Also, we conducted the experiment which was aimed to investigate whether the stored frame is a square at the pixel data conversion and the recognition rate improves.

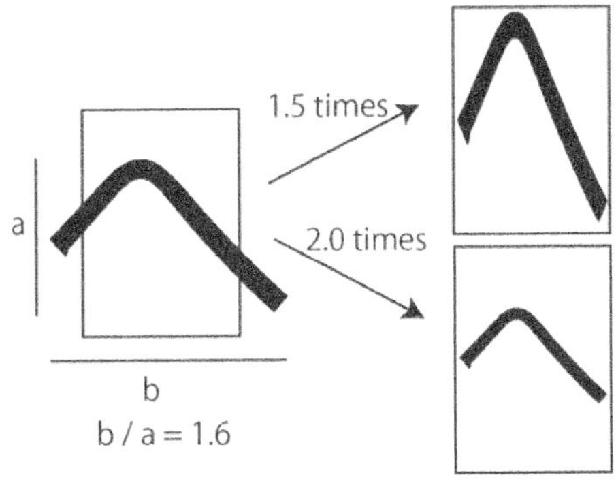

Fig. 10. Experiment 4 method image

When there was a difference the threshold value or more width and height, the regularization method is changed. 1.50 times and 2.00 times are verified at a rectangular frame. A square frame is assumed that width and height are 40 pixels. 1.0 times, 1.25 times, 1.50 times and 2.00 times are verified at a square frame. And, the description plane is the x-y plane and the x-z plane.

7.2 Results

Table 5 indicates the recognition rate.

Table 5. Experiment 4 results. Unit:%.

	1.00	1.25	1.50	2.00
xy 40×50	-	-	64	64
xy 40×40	65	65	65	64
xz 40×50	-	-	59	60
xz 40×40	60	60	60	61

To our regret, the results were not improved so much comparing to the results of experiment 3.

8 Consideration

Among the experiments, it was observed that the system misjudges a character as other ones which have the similar letterforms. It is thought that there is a limitation to recognize without dividing a whole locus into the stroke without pause.

Using the method in this section, correction to the value obtained from the acceleration sensor, the recognition rate worsened. This must be happened that the feature of the character was destroyed by the correction because the recognition rate of the character that didn't have the difference of the drawing after correction changed, too. This is because the feature of the other characters was changed by correction and the characters depended on high-ranking candidate. Fig. 8 is one of the examples and the recognition rate of experiment 1 worsened from 47.7% without correction to 31.8% with 2nd method.

Through the experiments, we found that the recognition rate was the best when the x-y plane was set as the description plane, though the subjects thought that they wrote a character on the x-z plane. At this time, we do not identity this reason, but the results shows that the features of the characters are represented on the x-y plane in the case of the hand-written character in 3-D space. To clarify the reason and to examine the motion of z axis must improve the recognition rate.

We also found that it had to be adjusted to put a square frame at the pixel conversion, because, the recognition rate was the best, although the improvement was not seen as a result of the experiment 4.

9 Conclusion

In this paper, we describe some method to recognize characters which is drawn in the 3-D space. Though the result of the recognition rate was 65% which is not as high as we expected, we found that the mapping a locus drawn in the 3-D space into the x-y plane must cause good results despite the user believes that he/she draw a character in the x-z plane. We also found that some value of the acceleration should not be corrected because of the error of the device. Additionally, it should be stored in the square frame when converting the pixel data.

Three kinds are thought as a method of approaching the future. The first is an approach method of judging the pause of the stroke and extracting a necessary part that composes character. The second is an approach method of developing OLCR software for a 3-D physical space. The third is approach method of guessing the stroke of a 2-D plane from the stroke of a3-D physical space. We are thinking which method is suitable.

References

1. Yamamoto, Y., Shio, I.: Airpen: Information sharing by virtual drawing on real space, IPSJ (1999)
2. CellWriter, http://risujin.org/cellwriter/

Design of Wearable Interface Considering Touch Communications

Mariko Kato and Naoki Saiwaki

Guraduate school of Humanities and Science, Nara Women's University, Kitauoya-cho, Nara-shi, Nara. 630-8334, Japan
{ebm.katou,saiwaki}@cc.nara-wu.ac.jp

Abstract. Wearable computing has the image of the high-level information processing which uses the small-sized computer. However, a motion of the body is barred by the complexity of wiring or apparatus and the function of assistance of operation which is the function of original dress is disregarded in many cases. Moreover, it tends to be kept at a distance that externals look like the cyborg by a general person. On the other hand, the interface that uses the movement of the body recently is actively researched. Then, the prototype system of a good design and fit was developed for the college woman and evaluated in this research. The system has the touch interfaces that pay attention to touch communications and can do the music sharing and the voice conversation by two-way, and the switch can be done by touch.

Keywords: Touch Communication, Wearable Interface.

1 Introduction

Wearable computing tend to be researched and developed focused on functionality to be able to carry advanced information devices such as a computer, a display, input devices and so on. But on the other hand, the original functions of clothes such as comfortability or supporting of the movement were considered less serious. For example, the motion of the body was limited by the complexity of wiring or attaching devices. Moreover, daily fashion characteristic is tending to be neglected. The appearance to install the wearable equipment is like the strange cyborg. It is not so matured to unite refined design and advanced function yet. In addition, it is a problem that the degree of freedom of the arrangement of the wearable equipments like various interfaces is extremely low. The research that uses electroconductive clothes instead of wiring is paid attention to improve these problems recently [2].

But, the input or operational device which reflects the movements of a body, like a remote controller of Nintendo Wii, is paid attention in resent years. These interfaces enable intuitive and intelligible system operation. From the viewpoint of interface research, it is important that development of the easy operation interface for wearable system based on the idea that why and how we can or should wear these devices. It is thought that not only the miniaturization of device of personal computer but also the development of the interface with the special function that can be realized because the

M.J. Smith and G. Salvendy (Eds.): Human Interface, Part I, HCII 2009, LNCS 5617, pp. 524–533, 2009.

equipment is installed in the body is important from the viewpoint of the interface research in the wearable computing.

Then, in this study, we designed the control interface for wearable system focusing a body touch communication [1]. Prototype systems were developed as two clothes and evaluated. These have good and comfortable design for Japanese women's university student, and can switch music sharing or voice conversation between 2 persons automatically with touch pose.

To construct the prototype system based on this idea, we studied from three angles.

1. Consideration of the touch communications
2. Hearing survey of fashion sense of general women's university student for easily accepting and smarter installing arrangement of various devices
3. Evaluation of electronic conductive textile (e-textile) as new material for constructing the interface

It is expected that the e-textile solves the problems of complexity of the wiring and improves appearances and comfortability as clothes.

2 Touch Communication

Wearable computing treats a machine very close. From the viewpoint of user interface it is important to think to operate a machine more intuitively. Basically, the present input interface of wearable system is similar to usual PC's such as how to press a key of the specific function. But it is possible to construct better and easy-to-use interface for the users if our purpose and situation of using such system are limited. By using wearable interface installed in clothes and extracting the meaning of user's action such as gesture, attitude and touch by person and person in communication, more effective interaction of wearable system will be realized. There are some researches that the feature of the gesture and the operation is analyzed as the vital signal and the image information for utilizing more intimate interface [3, 4]. In this research, we focus on development of the interface of body touch communication between 2 persons.

The human knows to live in this world by sense of touch first. Then, the human grows up with abundant haptic stimuli from infant to adults, and learns human relation and fundamental confidence for the neighboring world [5]. The expectations, the excitement feeling and the feeling of fullness that the tactile stimuli bring cannot be replaced by other senses. The custom of the body contact, for instance, frequency, the place, and the meaning are different according to the culture. However, the hand, the shoulder, the forehead, the head, the neck, and the forearm are the contacted places in many cases [6].

Then, we picked up and thought about the following often cases.

1) You can share your intention or feeling by the continuance of touch like handshake.
2) You can show your intention or feeling by the changing position or frequency or strength of touch like hitting the shoulder or high touch.

In case 1, the touch is very similar to information sharing. So, we designed an example of music sharing function by the conductive clothes. If conductive fabric can be used for making clothes instead of a conduction line, the lead of a portable music player becomes unnecessary. In case 2, such actions mean the switch the information or the mode. For example, the person when speaking, touching the shoulder, pulls note. We thought about effective use when two people get on a motorcycle. When a person who rides back touches the waist of driver they can share music, next when he touches the shoulder of driver they can talk through a microphone. We developed the prototype wearable interface which can switch music sharing and conversation by touch place.

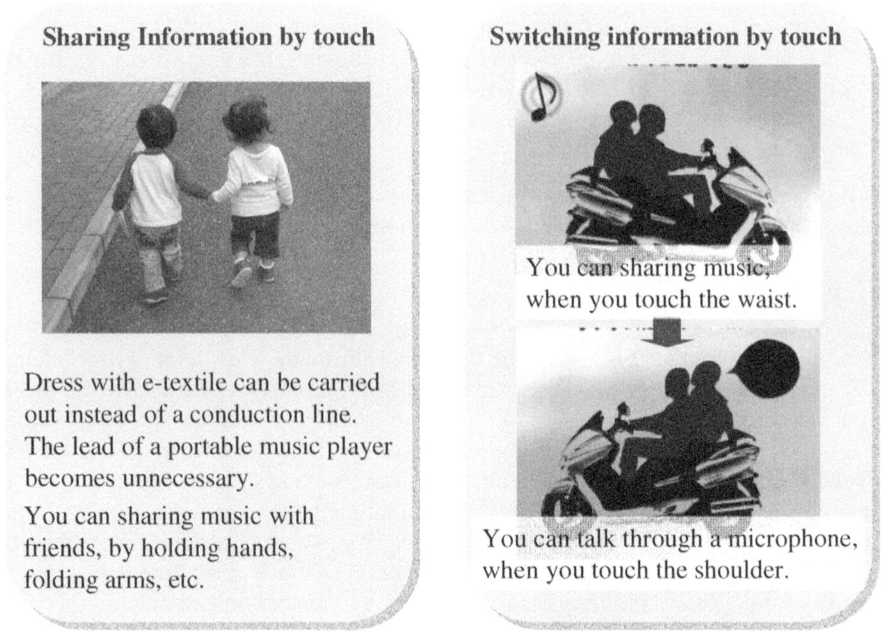

Fig. 1. The example of Touch communication

3 Fundamental Examination for Touch Interface

3.1 Appearance Attitude Survey about Wearable Interface Position

The improvement of the fashionability is one of the problems of the wearable computing for the wide acceptance. This time, if a wearable equipment of how much size was arranged at which position on clothes, it was investigated whether general college women may put on the clothes.

A round sticker was passed to ten testers (college woman in her twenties). The sticker was considered to be a wearable equipment, and pasted on the surface of women's M size jacket. The number of the stickers and places and lab hour were not limited.

Fig. 2. Place where two people or more pasted sticker

The place where two people or more put the sticker on Fig.2 is shown with red frames. The most popular place is the neck and is the basic place where decoration such as brooches is put. Most people like the design which does not give strange feeling as clothes which installed wearable equipment. From this survey, red frame places of Figure 2 were decided as the equipment installation position of our wearable system. These fixing places were the results of corresponding to the body area where the touch communications were done, and supporting the interface construction possibility by the touch communications. Moreover, it is also possible to put out the equipment to the surface of clothes if accepted as a design.

3.2 Electrical Specification of e-Textiles

The cloth made with the fiber with conductivity is called electronic conductive textile (e-textile). Those can because instead of a conduction line or a circuit board. Therefore, the applicability to wearable computing of these is easily expected. However, because the electrical specification of the electroconductive cloth is various depending on the material and the manufacturing method of the cloth, it is necessary to select according to the purpose.

Therefore we measured the electrical resistance and the frequency response of the e-textiles. Three kinds of e-textiles which were sold as cloth for an electromagnetic wave shield were used as an experiment sample. We experimented on three kinds of e-textiles as 1, 2 and 3 respectively (Table. 1).

3.2.1 Measurement of the Electric Resistance for Alternating (AC) Signal Current

First we examined the electrical resistance that was the most basic characteristic. We cut e-textiles into 2 rectangles of 2 cm * 60 cm and 30 cm * 60 cm respectively as the samples. The size of the cutting e-textile (30cm * 60cm) was almost the half of body of a garment in consideration of making the female clothes with the electroconductive

Table 1. Specification of three kinds of e-textiles

1 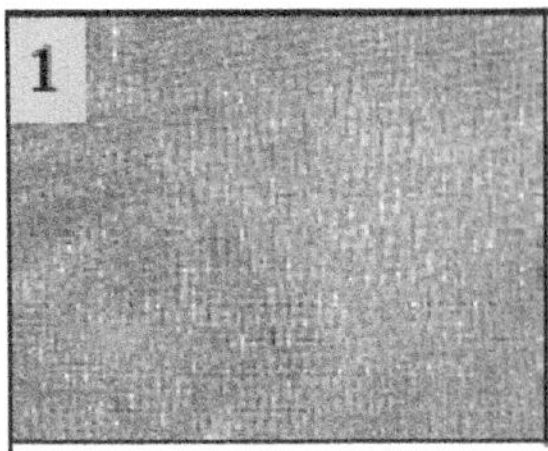	2 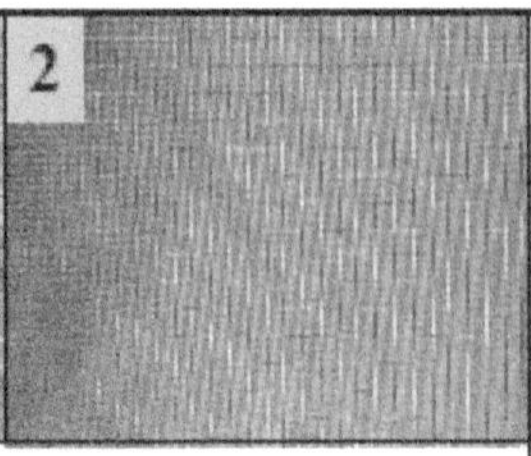	3 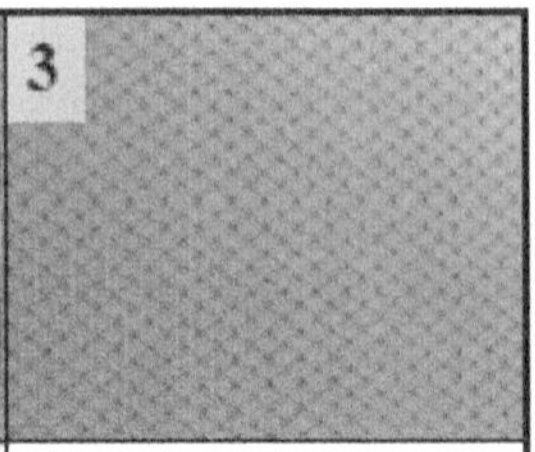
Base material: nylon knitted PA6-20f1 denier Silver coating (purity < 99.9%) Heat resistance: -30~90 degrees		Base material: 87% polyester knit 13% carbon fiber
Weight: 35g/m^2±5%	Weight: 75g/m^2±10%	Weight: 130g/m^2 ± 10%

cloth in this experiment. We were aimed at checking the difference of electrical characteristic with the wide shape and the narrow shape that was close to the usual electric conduction line. The resistance between the vicinity of length was measured (three times for narrow sample, five times for wide samples), and the mean value was calculate respectively.

As for the result, electric resistance of the wider is one digit smaller than the narrow. Because e-textiles are weaved by the electroconductive string, the route where electricity flows becomes the net which has many irregularity, and very complex. But generally, resistance decreases gradually as width is expanded though even details were not analyzed this time. Therefore, in usually use, it can be considered as resistance of the parallel connection. Therefore, it is clear that e-textile doesn't have any problem even if it connected direct or alternative current power source.

3.2.2 Frequency Response

Then, the frequency response to the alternative signal was examined as follows. It is because the phase and the gain change greatly if the alternative signal of the high frequency is input to the electroconductive material like the net, and there is a possibility of negatively affecting the signal. We prepared 12 samples in total by three kinds of e-textiles by 2 shapes and set by 2 patterns in the following. There are two methods when e-textile is used instead of the wire. One is the parallel array and the other is piling them up. In both case, two e-textiles are not to touch each other.

1) Two e-textiles were separated in 5cm and put them in parallel. This corresponded to set e-textiles to one side of dress.

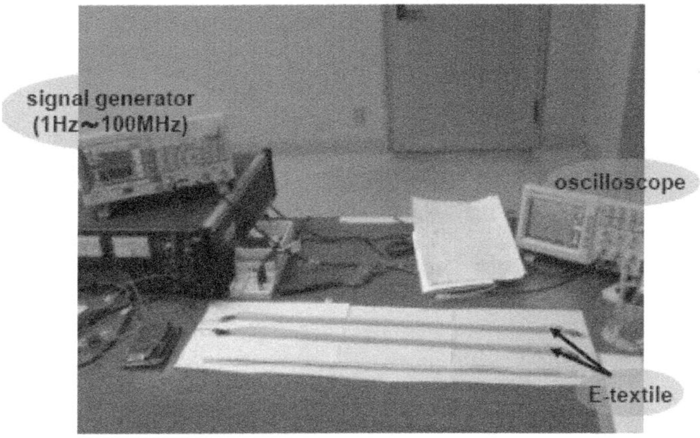

Fig. 3. Frequency response experiment

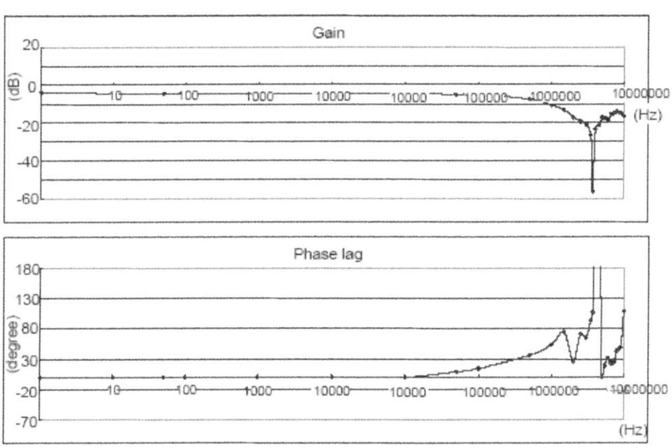

Fig. 4. Gain and phase of signal in each frequency of sample 2

2) Two e-textiles sandwiched the non-conductive textile. This corresponded to set e-textiles to both sides of dress like TextileNet system [1].

Two cutting e-textiles used by 3.2.2 were prepared respectively, and the signal of the frequency from 1Hz to 100MHz was input to the e-textile by a signal generator. And, the gain and the phase were calculated from the pair of input/output signals Fig.3).

The result of the gain and the phase are shown in Fig.4 as an example. This sample is e-textile 1, narrow and parallel type. These results are below;

1) Any condition passed the signal through 1MHz without trouble. It seems that it is possible to use it for the bandwidth of VHF on practical use without trouble.
2) The specific frequency to which the signal extremely attenuates exists though the level is different according to arrangement and shape of e-textiles.
3) When the cloth was piled, the level of the attenuation of the signal was larger than that of the case arranged in parallel.

First of all, it is possible to use e-textile as wire if the gain and the phase are stable. If the gain and the phase greatly change, the energy of the signal is lost as heat or as the electromagnetic radiation discharged from the e-textile like the antenna. However, the discharge of heat and the electromagnetic radiation were not seen while experimenting. It seems that using such frequency should be avoided anyway. Finally, it becomes a capacitor when the e-textiles are piled up. And there is worry that influences the signal according to the frequency. When the e-textiles are sewn up to clothes, it is necessary to note it. Anyway, it was basically confirmed that electric characteristic of it was no problem in the signaling property by our preliminary experiment.

4 Construction and Evaluation Experiment of the Prototype System

Based on the examination described by Chapters 2, 3, we made prototype wearable interface system which can switch music sharing to conversation by touch.

4.1 Construction of the Prototype System

We made two cloths including e-textiles and equipments. Fig.5 and Fig.6 show that whole of the prototype system.

Fig. 5. The prototype system whole

In rider's jacket of gray on the left in the Fig.5 and Fig.6, e-textile is sewn in inside the arm, side and the shoulder, and everything has been connected electrically. In shirt of pink on the right in the Fig.5 and Fig.6, e-textile is sewn in inside the arm, the cuffs and the shoulder, and it is connected electrically from the cuff to the shoulder and divided by a left body and a right body.

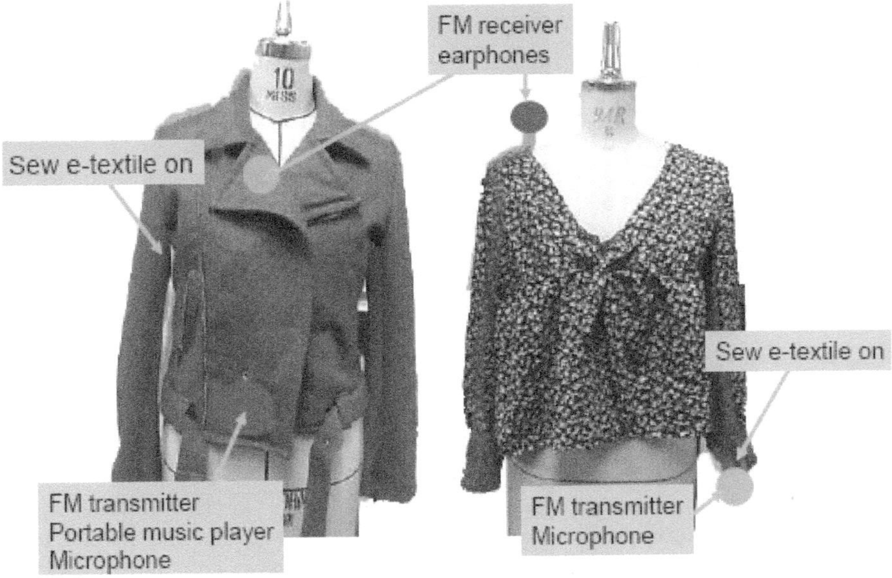

Fig. 6. The prototype system whole

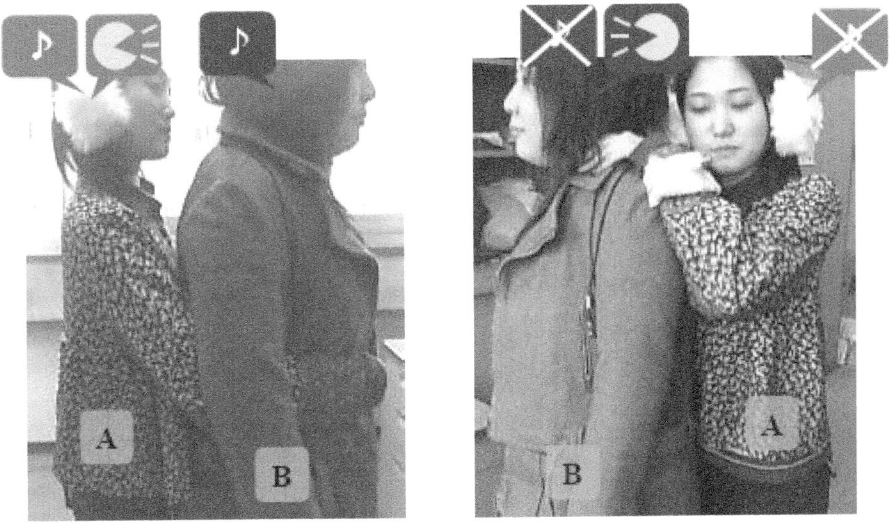

Fig. 7. Example of touch communications by prototype system

It designed like clothes marketed for the college woman so as not to consider user's wearing a special, wearable device to the body. The portable music player and FM transmitter were put in the pocket of a left rider's jacket. Music is transmitted through the e-textile. The user who put on the rider's jacket can listen to music with FM receiver installed on the neck. On the other hand, the user who put on a right floral print shirt can share the same music by touching the rider's jacket. Because both clothes are connected by the e-textile, and the signal from the rider's jacket is received from the cuff and transmitted to the shoulder. And, FM receiver and the earphone enter a pink earmuffs and music is played. It is a situation riding the motorcycle by two people while listening to music.

On the other hand, the microphone and FM transmitter are buried under the white scrunchy applied to the left hand of the floral print shirt, and the user can be switched to the conversation mode by touching the shoulder of the rider's jacket. It is operation to touch other's shoulder to take the other's attention. The microphone is installed in the helmet of the rider's jacket user also, and rider's voice is transmitted to the floral print shirt. Fig.7 shows the example of such a touch communications.

4.2 Evaluation Experiment of the Prototype System

We did the evaluation experiment that used prototype system. Four general women's university students as subjects used the system in pairs. And we ask them for feedback as to the system freely.

There was an opinion that it wanted to use this system to share feelings and the topic like not only the motorcycle travel but also the haunted house and the movie theater and so on, in the place where it was difficult to speak. The following answers were in the good point that the subjects had pointed out.

1. The evaluation on the interface side: Operation that moves the hand to the shoulder when speaking connects operation with the meaning knowing by intuition and is comprehensible.
2. The evaluation as a wearable device: Neither the size nor the wiring for the equipment are not anxious.
3. The evaluation of externals: the design is natural and is lovely. Using the e-textiles as an accent is cool.

Therefore, effectiveness in the touch communications interface that had been constructed this time was shown. There was also negative opinion that the operation with touch is hard to understand, and operating the traditional flip switch is easier to use. There was other negative opinion about unstableness of the touch operation it self and the bad of the sound quality.

5 Conclusion and Future Work

In this study, we studied from three angles.
1. Wearable interface based on touch communication.
2. Smart and refined design for arrangement of various devices.
3. Experiment on electrical resistance and a frequency response of the e-textile.

In this research, the interface which could switch the function by the touch communication were proposed and developed.

Other various possibilities are thought by the touch interface. The interface of the new kind of thinking also has the problem that the user is not accustomed to use it. However, every user can recognize the effectiveness of the touch communication interface in short time and be accustomed.

On the other hand, there is a limit in our system that uses an analog circuit like this time for the accuracy of touch and the improvement of the function and tone quality. The digitalization of the system is scheduled to be examined in the future.

References

1. Kato, M., Terada, T., Akita, J., Toda, M., Saiwaki, N.: Design of Wearable Interface considering Touch Communications. In: Human Interface Society of Japan symposium 2008, pp. 879–880 (2008)
2. Akita, J., Shimmura, T., Murakami, T., Toda, M.: Flexible Network System for Wearable Computing. Transactions of Information Processing Society of Japan 47(12), 1–10 (2006)
3. Fukumoto, M.: Whisper: A Wristwatch Style Wearable Handset. In: Proc. of CHI 1999, pp. 112–119 (1999)
4. Kume, Y., Shimada, M.: A Wearable Input Device Using Wristwatch-like Camera and Reflection Marker. The Journal of The Institute of Image information and Television Engineers 60(2), 249–253 (2006)
5. Iwamura, Y.: Touch; Igaku-Shoin (2001)
6. Nomura, M.: Miburi to shigusa no jinruigaku [Anthropology of a body language and gesture]; Chuokoron-Shinsha (1996)

f-MRI Study of Brain Activation in Tactile Feeling

Yuka Koda[1], Maki Taniguchi[1], Yukiyasu Kamitani[2], and Naoki Saiwaki[3]

[1] Graduate School of Humanities and Sciences, Nara Women's University
[2] ATR Computational Neuroscience Laboratories
[3] Faculty of Human Life and Environment Nara Women's University
acid51613@gmail.com

Abstract. This research examined a tactile objective evaluation using ICPF (Ionic Conducting Polymer gel Film) actuators and real materials paying attention to the brain activation. The ICPF actuator is a display that presets artificial tactile feelings like fabrics, which we presented in HCI International 2005 [1]. The procedure of this evaluation method is as follows. 1: applying four kind of artificial tactile stimuli to subjects in f-MRI chamber, 2: measuring of the brain activity by f-MRI, 3: analyzing the difference between the brain activities evoked by different stimuli, revealed by decoding method, 4: examining an effectiveness of the result. In addition, we prepared four real materials tactile stimuli for comparison. Experimental results showed that a human brain could classify significantly both real materials and artificial tactile stimuli and that we could confirm them by decoding method with f-MRI. This fact derived the idea that measuring brain activity was one example of tactile feeling objective evaluation method. To verify the idea, we carried out a test with stimuli that narrowed the tactile gap. This experiment showed an insignificant result from its decoding value. But, the decoding results of all subjects exceeded a *chance level. The chance level* is chance of success in this paper. These experiment results didn't deny the brain activity measurement by f-MRI as a technique for evaluating the sense of touch and the effectiveness of the decoding method. Therefore, it is thought significant researching continuously in the further.

1 Introduction

There are many different points between the internet shopping and the purchase in a shop. In one of them, we can't pick up the item that we want until buying it in online shopping. The person who shops online cannot confirm the difference of touch. This difference is the demerit of online shopping.

You may be enough even if I cannot touch the article before the purchase. For example, a visual information in the case of books, or a audio information like a audition in the case of CDs. It may be enough that the confirmations by the audition for CD and by the screen for the books and the photographs ever if it isn't possible to take them beforehand. But, it is natural that we want to touch in case of wearing items like clothes before buying them. For example, we usually want to confirm, and try on the feeling of texture and quality of liked clothes. The artificial tactile feeling technology is effective to solve these problems.

M.J. Smith and G. Salvendy (Eds.): Human Interface, Part I, HCII 2009, LNCS 5617, pp. 534–542, 2009.

Actually, the concern for the artificial tactile display has risen for the application to the VR technology and the achievement of the nursing-care robot. Therefore, a lot of researchers are working on the research of the artificial tactile feeling display and the tactile sensing. However, the practical use of the artificial tactile feeling display technology is still for. One of the reasons is that the relations between the synthesized parameters of the artificial tactile and the impression generated by artificial tactile stimuli are indistinct. For example, a color is determined by value of RGB. There is a clear definition between the established value and the color. But, the relativity between the synthesized parameters and the generated stimuli are not clear in case of the artificial tactile feeling display. The display which we reported in HCI 2005 is similar. A clear model how tactile feeling controlled has not been established yet, though it was shown to be able to present complex tactile feelings by the psychology experiment with our device. That is, it is necessary to find the method of objectively evaluating the similarity between the artificial tactile stimuli and the real one for the quality improvement of the artificial tactile display.

Then, because tactile information was processed by the brain, we thought that it was an effective technique to evaluate the similarity of tactile stimulus paying attention to the brain activity. This time, we used f-MRI to obtain the minute brain activity generated by the tactile stimulus. The results are reported as follows.

2 Structure of Human Skin and Concepts of Selective Stimulus

2.1 The Elements of Tactile Sense

Human skin has a number of sense receptors of elementary sensations such as touch, pressure, vibration, pain, temperature, joint, muscle, etc. In dexterous manipulation, complex sensations such as weight, shape, tactile feel, and total impression include important information in addition to the elementary sensations. Such complex sensations are created by the fusion of information sensed by various receptors. In this study, we focus on mechanoreceptors in human skin. Figure1 shows the structure and location of mechanoreceptors in human skin.

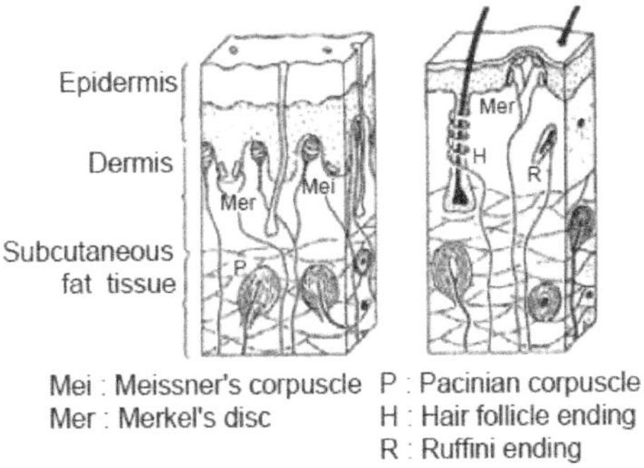

Fig. 1. Structure and location mechanoreceptors [2]

There are four types of mechanoreceptors embedded in human fingers, FA I (Meissner's corpuscle), SA I (Merkel corpuscle), FA II (Pacinian corpuscle), and SA II (Ruffini endings). It is known that each receptor has response characteristics for mechanical stimulus and causes subjective sensation corresponding to its responsive deformation. For example, SA I detects static deformations of skin and produces static pressure sensation, and FA I detects the velocity of the deformation and produces the sense of fluttering vibration. Tactile impression is an integrated sensation of these elementary sensations. To present tactile feel arbitrarily, stimuli applied to these receptors should be controlled selectively. Figure 2 shows the concepts of the selective stimulus method.

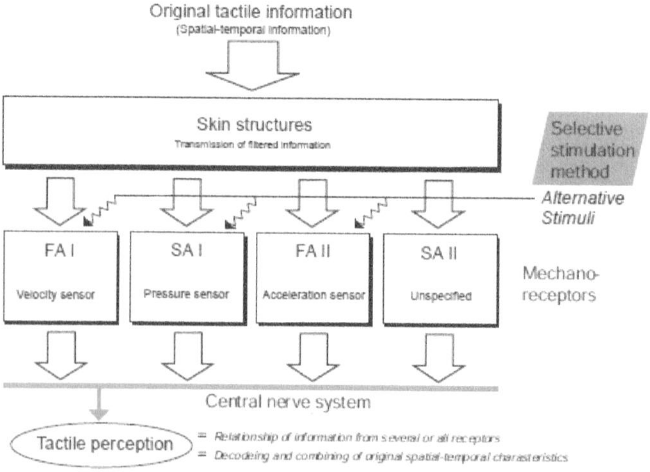

Fig. 2. Concept of selective stimulus method

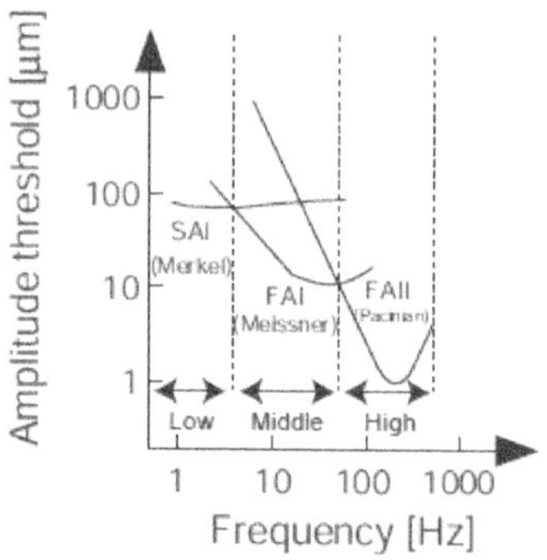

Fig. 3. Thresholds of tactile receptors for vibratory stimulus and selective stimulus ranges

To stimulate each receptor selectively, we focus on the frequency response characteristics of tactile receptors. Figure 3 illustrates the human detection threshold of vibratory stimuli, which represents sensibility of each receptor for the frequency variation. The smaller amplitude threshold means higher sensibility. This figure shows that there are three frequency ranges in which the most sensitive receptor changes. That is, in the low frequency range illustrated in the figure, SA I(Merkel) is most sensitive relatively, and the best becomes FAI in the middle range and FA II in the high range, respectively. This suggests that the selective stimulus can be realized utilizing these frequency characteristics, and arbitral tactile feels can be produced by synthesizing several frequency components.

2.2 ICPF Actuator

ICPF is a new material discovered in 1992 [3]. It is a composite of PFS (PerFluoro-Sulfouic acid; Nafion) membrane and thin gold surface layers. It makes bending motion in water or under wet condition by internal ionic motion when an electric field is applied between the surfaces. This actuator has remarkable features as follows [3]. 1: softness (Young's modulus E = 2.2 × 10 s Pa), 2:low driving voltage (1.0-2.0 V), 3:high speed response (> 200 Hz), 4:miniaturizability (mm order), 5:motion in water or under little wet condition. The ICPF actuator could generate various difference tactile feelings by adjusting spectrum frequencies and output intensity of adding signal. Figure 4 shows the structure of the ICPF actuator.

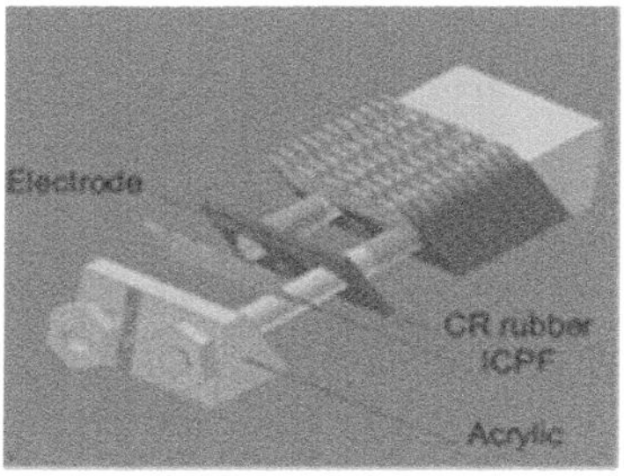

Fig. 4. ICPF actuator

3 Brain Activity Analysis at the Time of the Tactile Feel Recognition by f-MRI

3.1 In the Case of Real Materials Stimulus

(1) Experimental conditions

The experimenters made subjects touch real materials in the f-MRI chamber and measured their brain activity. Real materials for using in this experiment were four kinds of fabrics - towel, fleece, boa, leather (Figure5). These four, two pieces for each, eight pieces in total were prepared and it made on the bunch and it passed it to the subjects as shown in Figure 6. We didn't direct touch speed to subjects, but directed them to touch at a steady speed.

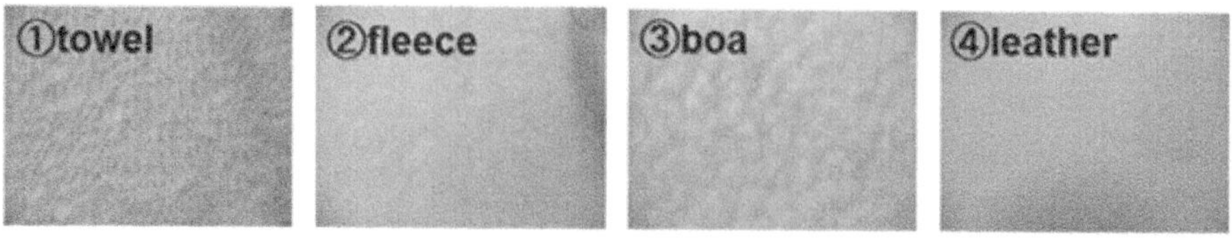

Fig. 5. Real materials

Fig. 6. Bunch of four real materials

Figure 7 shows this experimental model. The block design that repeated task(16sec.) and rest(8sec.) eight times was assumed to be 1Session. And 10Session repeated in the total. The order of the sample changed at random in each session. We controlled this experimental condition by directing subjects to close their eyes and to be in a quiet state in the f-MRI chamber. Additionally, we directed them to move only their arm and hand during their task. These instructions were given from outside of the f-MRI chamber. The signs of task start and stop to subjects were real voice recorded as "Start" and "Stop" beforehand. The f-MRI that we used for this experiment was Siemens AG's product (3.5T MAGNETOM Trio).

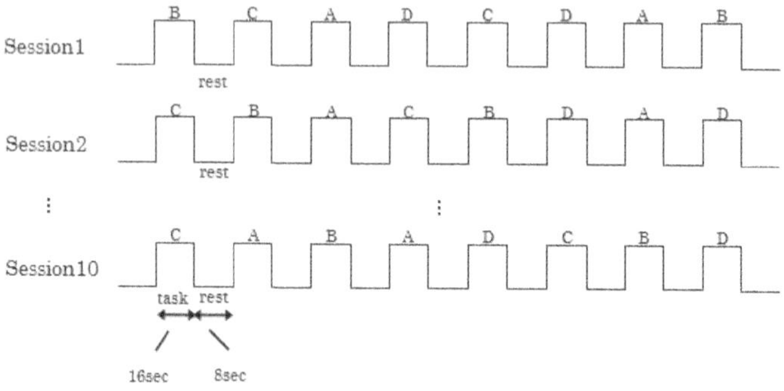

Fig. 7. Experimental model

(2) Subjects

Subjects were four healthy adults from twenties to forties. (one male and three females) All the experiments were executed under informed consent.

(3) Method of analysis
The analysis method of this experiment was *the decoding approach. the decoding approach* which was reported in "Nature Neuroscience" by Kamitani at ATR in May, 2005 [4]. This approach classifies brain activity pattern based on signal strength of each voxel of activated area estimated by f-MRI. The reliability of pattern classification is calculated as a chance level. It can be said that the brain activities can classify if the chance level reaches the significant value.

3.2 In the Case of Artificial Tactile Stimulus

(1) Experimental conditions
The artificial tactile stimuli were given to the subjects by the ICPF actuator in the f-MRI chamber and measured their brain activity. The actuator was set on subject's first forefinger joint of right hand. The virtual stimuli had been programmed beforehand, and these were given automatically to subjects.

Virtual stimuli by the ICPF actuator are simple harmonic stimuli. Table1 shows the virtual stimuli of this experiment. They were four stimuli which had different frequency and output intensity. Many subjects could feel the difference among them, so we adopted them. They were all sine waves. The experiment conditions except stimuli were similar to 3.1.

Table 1. Four kinds of artificial stimuli generated by ICPF actuator

		frequency	
		10Hz	150Hz
output intensity	1.5V	A	B
	2.5V	C	D

(2) Subjects
Subjects were ten healthy adults from twenties to forties. (two males and eight females) All the experiments were executed under informed consent.
(3)Methods of analysis
Methods of analysis and f-MRI is same as 3.1.

3.3 Result

The brain activities of all subjects were most obvious on motor area and somatosensory area of left hemisphere. Therefore, we selected these voxels as an analysis pattern by the decoding approach. In the case of real materials, measured results of three subjects were analyzed, because one subjects' measuring data was lack. Three subjects' decoding data were 0.3, 0.56 and 0.56. These values exceeded the chance level (The chance level of this experiment is 0.25). In contrast, Figure8 shows decoding results of artificial tactile stimuli as a sample. Though accuracy was uneven by the subject and stimulus, human brain seemed to be able to recognize tactile differences of both real materials and artificial tactile stimuli. That result showed that analysis of brain activity measurement with tactile stimuli by decoding approach had possibilities of one of the index for objective evaluation of tactile feeling.

Consequently, we experimented the limit of this method by the experiment that gave a extremely minute difference in the next stage.

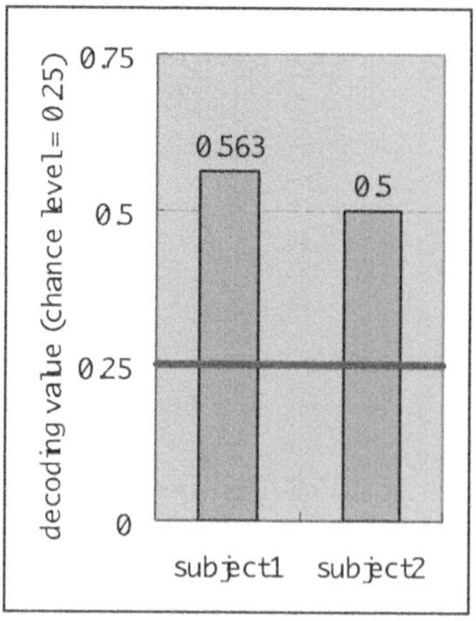

Table 2. Decoding results in the case of artificial tactile stimuli

4 Experiment on Minimal Sensitivity of Tactile Recognition That Uses Nano Fiber

This experimental model was same as 3.1. Meanwhile, we prepared eight test samples, and let subjects touch each sample once in 1session. This experimental circumstance was basically same as 3.1 also, but just in case, we let subjects close their eyes and wear an eye mask.

(1)Experiment condition
We made subjects touch nano-fibers with their right hand finger in the f-MRI chamber and measured their brain activity. Nano-fibers used in this experiment were eight kinds. They are different only as thickness of string –7000nm, 4100nm, 2100nm, 1600nm, 700nm, about 400nm, about 300nm and about 200nm. Figure8 shows these fibers. These fibers replaced order at random and were passed by the subject.

Fig. 8. Nano-fibers(7000nm,4100nm…about 400nm,about 300nm,about 200nm)

(2)Subjects
Subjects were seven healthy adults from twenties to forties. (one male and six females) All the experiments were executed under informed consent.
(3)Result
Only one subject's brain showed activation area in right-hemisphere. We analyzed measurement results of six except the one by decoding approach. Table3 shows this decoding approach result.

Table 3. Decoding results

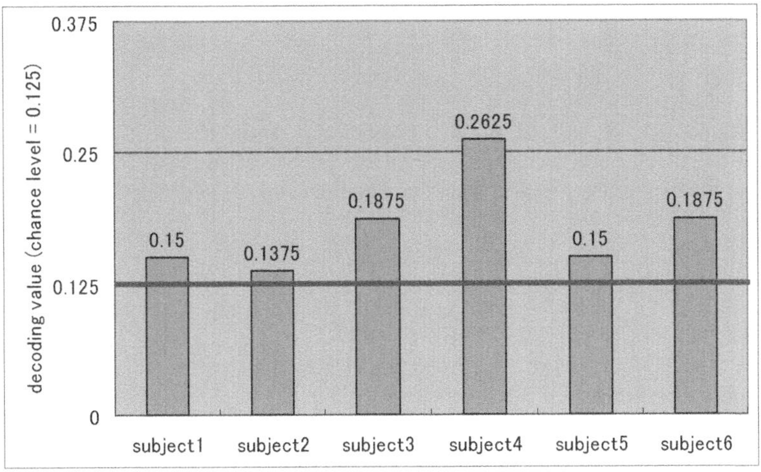

Reliable evidence was not obtained that human brain could classify the minute tactile difference between nano-fibers in this result. However, on the other hand, the decoding values of all subjects were over the chance level. It is shown that the possibility to be able to classify by the brain activity even in the tactile stimuli of minute difference remains though there might be a problem in this experimental methodology. It is necessary to experiment with a string that is fatter a little more than what used to experiment this time to verify this result of contradiction.

5 Conclusions

In this research, it was investigated whether the brain activity pattern classification by f-MRI was possible or not, when a different tactile stimuli were given by touching with real clothes, artificial tactile display and nano-fibers. The decoding method that we had developed was used for the pattern identification. That result showed that analysis of brain activity measurement with tactile stimuli by decoding approach had possibilities of one of the index for objective evaluation of tactile feeling. Consequently, we experimented the limit of this method by the experiment that gave a extremely minute difference by nano-fiber..

The following findings were obtained from our experiments. Stimuli which are able to distinguish the difference easily by our sense, like both real materials and artificial tactile stimuli, can be classified significantly. The brain activity pattern did not seem to be able to be classified in the case of very minute tactile feeling like nano-fiber.

In the future, I want to investigate the limit that can detect the difference of the sense of tactile feeling with brain activities. And I want to investigate relativity with the activation pattern between tactile, auditory and visual area also.

References

[1] Yoshida, A., Taniguchi, M., Saiwaki, N., Tadokoro, S.: Artificial Tactile Feel Display for Textile Fabrics. Human Computer Interaction (2005)

[2] Hoshimiya, N.: Bio information engineering, pp. 193–228. Morikita Publ. (1986)

[3] Tadokoro, S., Yamagami, S., et al.: Multi-DOF device for soft micromanipulation consisting of soft gel actuator elements. In: Proc. IEEE ICRA, pp. 2177–2182 (1999)

[4] Kamitani, Y., Tong, F.: Decoding the visual and subjective contents of the human brain. Nature Neuroscience 8, 679–685 (2005)

[5] Iwamura, Y.: Touch. Igaku-Shoin Ltd. (2001)

[6] Saiwaki, N., Yoshida, A., Tadokoro, S., Taniguchi, M., Chiba, I., Nagato, C.: Basic Research on Virtual Tactile Feelings Creation and Recognition. Information Processing Society of Japan Kansai Branch, 113–116 (2004)

[7] Taniguchi, M., Yoshida, A., Saiwaki, N., Tadokoro, S.: Analysis of Human Brain Activity in Tactile Feeling. In: FAN Symposium 2005 in Kyoto (2005)

[8] Saiwaki, N.: Creation of Artificial Tactile Feeling and Analysis of Human Brain Activity in Tactile Feeling. Information Processing Society of Japan Kansai Branch (2006)

[9] Uchida, H., Kamitani, Y., Taniguchi, M., Saiwaki, N.: f-MRI Image-Based Decoding Artificial Tactile Feeling. In: Mechanism of Brain and Mind The 7th Summer Workshop (2006)

[10] Saiwaki, N., Koda, Y., Taniguchi, M., Uchida, H., Konyo, M., Tadokoro, S.: Basic Research on Artificia Tactile Feeling Display and f-MRI Image-Based Decoding. Human Interface Society (2007)

[11] Koda, Y., Taniguchi, M., Saiwaki, N.: Evaluation of Fabric Texture by f-MRI. Japan Society of Home Economics Kansai Chapter (2008)

[12] Naoki, S., Maki, T., Hajime, U., Yukiyasu, K.: Basic Research on Brain Activity under Virtual Tactile Feeling Stimulus. The Japan Research Association for Textile End-Uses 47, 12, 25–33(2006)

Using 3D Touch Interaction for a Multimodal Zoomable User Interface

Florian Laquai, Markus Ablassmeier, Tony Poitschke, and Gerhard Rigoll

Institute for Human-Machine Communication
Technische Universität München
Theresienstrasse 90, 80333 Munich, Germany
{laquai,ablassmeier,poitschke,rigoll}@tum.de

Abstract. Touchscreens are becoming the preferred input device in a growing number of applications. They are interesting devices which are more and more introduced into the automotive domain. Current implementations impose problems like precise pointing or high visual attention and therefore the capabilities of projected capacitive touchscreens are investigated. Unlike traditional sensing techniques like resistive or optical measurements, projected capacitive sensors register the presence of a hand or finger even before the actual touch. This enables such devices to be used not only for 2D input but for interaction using the position and distance of the users finger relatively to the screen. The additional distance information is then applied to control a Zoomable User Interface. In this study touch and speech interaction is being applied to allow fuzzy input and lower visual attention demand than conventional Touchscreens. A demonstrator was developed using the mentioned input strategies for the automotive domain and finally evaluated in a user study.

1 Introduction

As displays become larger and resolutions are increasing, touch interaction is the input modality of choice in a growing number of devices. While such screens have been around in PDAs for several years and have proven to be capable of everyday use, the real breakthrough was achieved by portable music players, Tablet PCs or industrial applications. The traditional technique for touch recognition is resistive sensing which has the nature of wearing out after extensive use and therefore reducing recognition performance. A new way of achieving touch sensitivity is capacitive touch sensing which can be divided into "projected" and "contact" sensing. Both work with the fact, that the human body can store an electrical charge and thus can be used in a measurement process. The technology used here is projected capacitive sensing which means that the human body can be thougt of as one plate of a plate capacitor which means that no actual touch is needed. The sole presence of a finger can therefore be detected by a sensor array and the distance to it can be measured.

Given this technology, the finger distance information can be used to control a Zoomable User Interface, see [1], [2]. Such interfaces use zooming and magnification effects to manage large amounts of displayable data and usually give graphical objects

M.J. Smith and G. Salvendy (Eds.): Human Interface, Part I, HCII 2009, LNCS 5617, pp. 543–552, 2009.

a geographically defined place in the menu structure. This interface class can roughly be split into two subclasses, where zooming can be applied to the overall interface or just to portions of it. In this application the latter version is used. The amount of data displayed by an object is defined by its zoom level which can be interpreted as a selective information reduction or increase. With the finger distance controlling the zoom level and the 2D finger position controlling an invisible cursor such an interface can be navigated very intuitively and efficiently. As a result users can point to an object to increase the zoom level and preview the information behind it.

Speech input is an additional modality which can be used in combination with finger position to support interaction with small graphical objects. This way minimal visual distraction from the main task (e.g. driving a car) can be achieved while maintaining the possibilities of direct manipulation. The employed technology is being described in the first part which covers the touchscreen, the finger detection algorithm and the speech recognition engine. After that the demonstrator for the Zoomable User Interface and multimodal input is presented and finally the results of a user study are being discussed.

2 Technology

This first section gives an overview regarding the technical components. As a key component the touchscreen will be described in the following part.

2.1 Projected Capacitive Touchscreens

Unlike most traditional touch sensing technologies, Projected Capacitive sensors can detect the presence of a finger or a hand even without an actual touch. This is achieved by using conductive sensor pads made of copper or ITO (Indium Tin Oxide). In this case the human body acts as one plate of a plate capacitor and the sensors as the other plate. Depending on the finger distance, the capacity changes following the basic plate capacitor equation

$$C = \varepsilon_0 \varepsilon_r \cdot \frac{A}{d}$$

where C is the Capacity, ε the dielectric coefficient, A the surface area and d the plate distance. Thus the capacity of a sensor pad is related to the finger distance and can be used as an measurement variable. A capacity measurement for a sensor pad can be accomplished using a relaxation oscillator as seen in Figure 1 on the right. The output value of the timer component is directly related to the measured capacity and consequently to the finger distance. The device from Elo Touch Systems used for this application is built of two layers of glass and 16 sensor strips for x- and y-axis located between these layers.

All 32 sensors are connected to a single controller which sends out the raw data on a serial port. The data string is encoded as follows:

[X0][X1]...[X15][Y0][Y1]...[Y15][NULL]

A visualization of such a data set can be seen in Figure 2. From this raw data output the static offset first has to be measured and then subtracted from every following

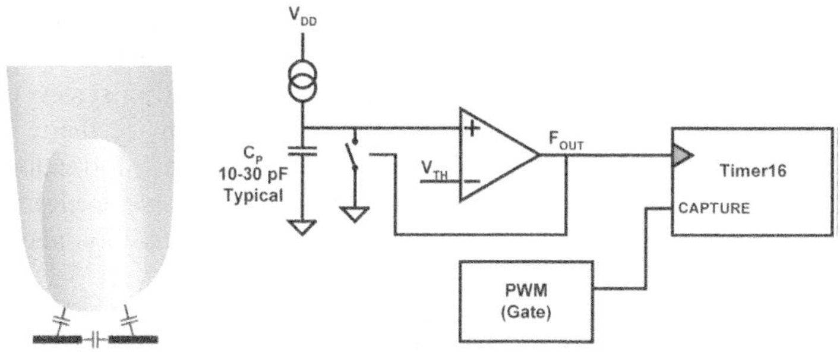

Fig. 1. Projected Capacitive sensing with a relaxation oscillator (Cypress Semiconductor)

sample. In the following step the values for each axis are smoothed using a sliding average filter with a window length of 2 samples in both directions to minimize introduced errors. After that a temporal averaging is performed over the actual and last data set. These measures significantly improve noise robustness while maintaining a quick sensor response.

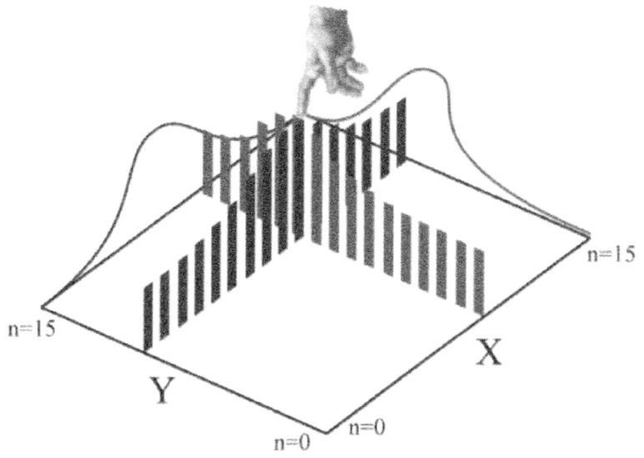

Fig. 2. Visualisation of sensor data output

2.2 Finger Detection Algorithm

After preprocessing the data a detection algorithm is being applied. For the first tests two detection algorithms were implemented:

- Cross Correlation
- Relative Maximum Search

For the Cross Correlation a set of 300 pre-calculated reference samples is used. These samples are created using a Gaussian which was fitted into a typical finger shape. The shape is then moved from one screen edge to the other simulating all possible finger positions. During operation the algorithm computes the correlation coefficient ρ_i for every reference sample using the following equation:

$$\rho_i = \sum_{n=0}^{15} sensor_n \cdot sample_{i,n}$$

Where

- i: index of reference sample
- $sensor$: data vector of one axis
- $sample$: data vector of the reference sample

In the next step the absolute maximum of all coefficients for every axis is determined and a quadratic interpolation around the maximum is performed to exceed the spatial resolution of 16 sensor strips per axis. The resulting interpolated maximum then corresponds to the finger / hand position.

The "Relative Maximum Search" uses the fact, that the user's index finger is the first distinct relative maximum when searched from the top left corner. This can be assumed because only one hand coming from a very predictable direction can be used in a automotive setup. Thus the absolute maximum is searched as a reference and then the index finger is assumed at the first relative maximum that exceeds a fixed percentage of the absolute one. The next step is like in the above algorithm a quadratic interpolation using the points around the finger position.

For the finger distance calculation the difference between the absolute extremums averaged for both axis is calculated ($\Delta MaxMin_{x/y}$). The following equation shows this relationship, whereas the constants 60 and 10 have been determined empirically and z is the finger distance.

$$z = 60 - 10 \cdot \sqrt{\frac{1}{2} \cdot (\Delta MaxMin_x + \Delta MaxMin_y)}$$

It was found that the usable working space of the sensor is approximately 0 - 4 cm above the screen distributed over the whole sensor area.

During the first expert evaluations the Relative Maximum Search algorithm proved to be more robust and precise. The Cross Correlation was too prone to false detections for various touching techniques.

Whilst still better than Cross Correlation, the Relative Maximum Search still did not show the necessary accuracy when actually touching the screen surface. Therefore a combination was used with the manufacturer driver handling clicks and dragging while the described algorithms (correlation and maximum search) handle operation when the finger is distant to the screen.

Problems arise when large parts of the palm are closer to the screen than the index finger due to signal ambiguities (see [7]). This can only be solved by applying more sophisticated detection algorithms.

2.3 Speech Recognition

The commercial engine "ScanSoft VoCon 3200" was used for speech recognition. It provides the possibility of a constantly open microphone, which was favored in this application. The vocabulary to be recognized can be defined using a BNF (Backus-Naur-Form). For every recognition iteration the engine returns a list of possible words which is sorted by the corresponding confidence value. This is an index for the estimated recognition quality and can be used as a criterion for rejecting or accepting a recognition result.

3 Application

Both input modalities "Touch" and "Speech" are fed into the demonstrator application depicted in Figure 3. The screenshot shows the main menu with the user's finger pointing at a "Contacts" menu and selecting a list entry. The application includes zooming and previewing content as well as multimodal input, window management and text input using zoom techniques.

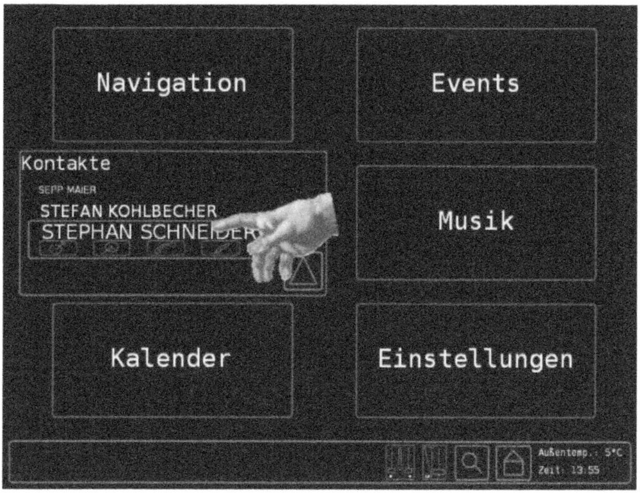

Fig. 3. Screenshot of main menu with one zoomed menu

In the following sections the different features are discussed in detail.

3.1 Zooming

As mentioned in the introduction, partial zooming is being used in this implementation. The desired item to be zoomed is selected by pointing towards it in the sensor's working space. If the finger hovers over a graphical object, the object can change its size and state depending on the finger distance. The concepts shown here divide the working distance into four zones:

- finger not detectable
- finger is distant
- finger is close
- touch

The zones and the corresponding states of a list item are depicted in Figure 4. This way the user can literally „dive" into parts of the user interface and preview or even select previewed items. All interactions are designed in a way that pointing to an object and zooming it never causes data to be manipulated. This can only be done with an actual touch that naturally gives haptic feedback.

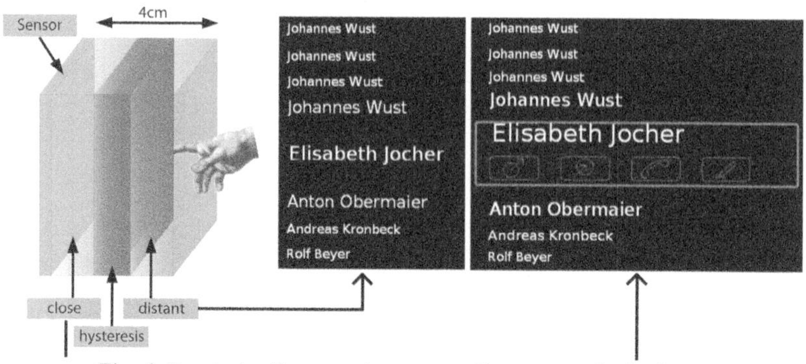

Fig. 4. Proximity Zones and corresponding states of a list item

Figure 3 shows the possibility to select the "Contacts" menu by pointing to it in the "distant" sensor area. This causes the menu to be increased in size and a list of contacts is shown.

3.2 Zoomable Lists

The list uses a kind of "Fisheye Zooming" where the item under the user's finger is the largest while the others are shifted aside. As a result the item can be touched more easily while maintaining the amount of displayed data.

By moving the finger closer to the screen into the "close" sensor area a frame is being displayed around the item and buttons are attached that represent the most common functions for this item class. These can be "Call contact", "Use as route destination" or "Edit contact" for the contact list or "Play song" and "Add to playlist" for a song list.

In consequence it is possible to

- preview and search a list entry
- select it and choose an action
- execute the action

with just one finger movement and a final touch of the screen. Scrolling in the list is achieved by hovering the finger over the top or bottom end of the list as can be seen in Figure 5. The more the finger reaches the list end, the faster the content is scrolled.

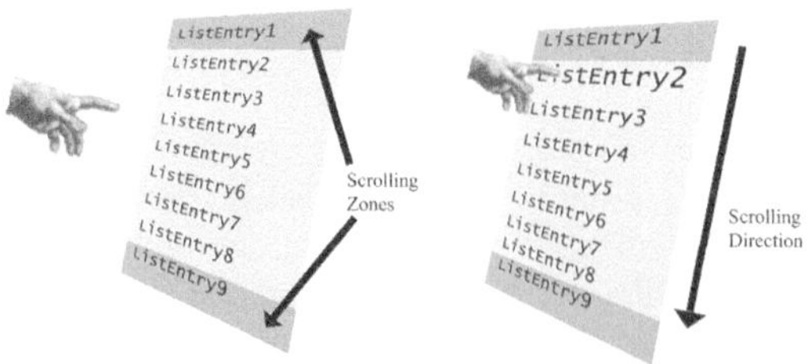

Fig. 5. Fisheye Zooming for lists and scrolling zones

3.3 Text Input

The distance information can also be used to enhance existing or create new text input methods. As a first example an onscreen keyboard was developed. It uses zooming effects combined with distance related key displacement. This means that the zoom level of each key in the area around the user's finger is indirectly proportional to the 3D-finger distance. To avoid overlapping, the adjacent keys are being moved away from the projected finger position, see Figure 6 left. The combination of the two effects makes it possible to minimize space requirements of the virtual keyboard which is of special interest on small screens. Consequently the user still has a good overview of the actual context while typing in opposite to a full screen version as it is often used in portable devices. The maximum zoom factor for the keys is 2.5 which equals 6.25 times the initial key area. This improves both readability and error rate significantly compared to a static keyboard of equal size. The keyboard is activated by touching a text input area and placed with the arrow on the bottom left corner pointing to the selected input line.

The second keyboard implementation (Figure 6 right) arranges the keys in a square raster which is divided into 3x3 sections, see [5]. Seven of these sections contain 3x3 keys and the remaining sections are used as "Space" and "Backspace" key. This results in a total of 65 usable keys. Again the required space is to be minimized and therefore the key size is chosen so that a finger tip covers about 4 keys. Here the finger proximity causes underlying section of 9 keys to be zoomed over the whole keyboard area (Figure 6 right shows the zooming status). As described for the menu states, the detectable finger distance is divided into two sections corresponding with the two zoom states and a hysteresis for increased stability in the transition area.

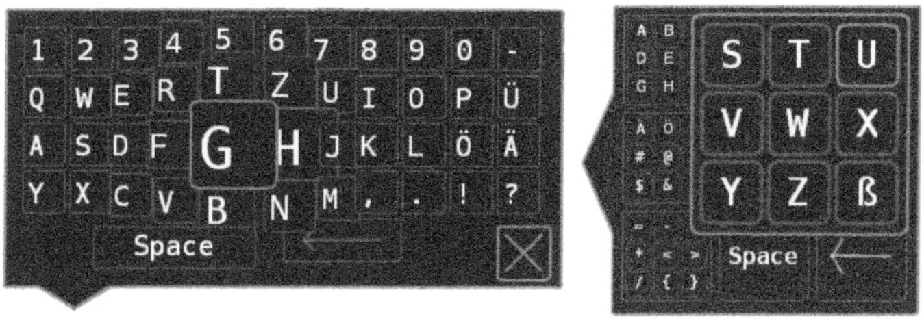

Fig. 6. Fisheye QWERTZ keyboard and Matrix keyboard

3.4 Combining Modalities

In this application two kinds of multimodal interaction have been implemented. Both parallel / independent and sequential interactions (see [3], [4]) are used for different commands. Main menu items like „Music" or „Navigation" can be activated at any time and System state. In that special case speech input is even more powerful than touch input. An example for mode-independent and equally available manipulation is shown in Figure 7 for a volume slider element. It can either be pointed at for zooming and dragged or be changed by saying „Volume Up" or „Volume down".

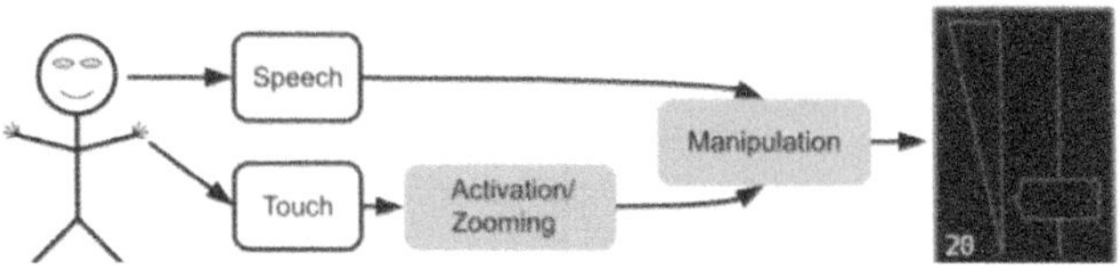

Fig. 7. Independent use of modalities for a volume slider

By combining finger position information with speech input, the demand for visual attention can be reduced while maintaining input possibilities. Figure 7 shows an application where list items are selected by pointing at them and saying the desired action. So the user does not need to touch a certain small area but only points roughly towards it.

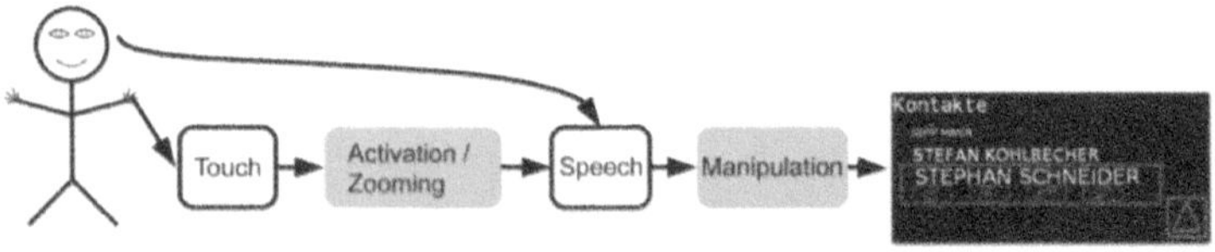

Fig. 8. Selecting list entries by touch and issueing commands by speech

The same actions that are available in the surrounding box in the „Close" state can then be triggered by speech input.

4 User Study

The system evaluation was conducted in the driving simulator of the Institute for Human-Machine Communication. It consists of a real car cockpit with the Touchscreen installed in the center console and a projection screen in front of the car for the environment visualisation. After a short system introduction several tasks had to be performed by the 20 test persons while driving in a highway scenario without other cars. The time for the completion of each subtask was measured and after that the participants filled out a questionnaire about their subjective impression.

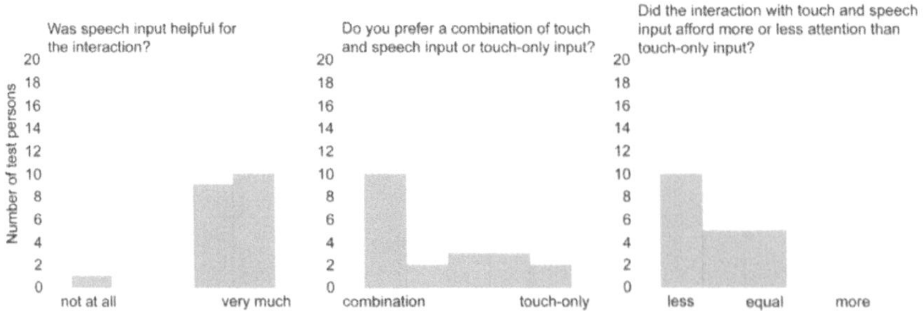

Fig. 9. Comparison of multimodal and touch-only interaction

First the findings regarding multimodal input shall be discussed. Generally multimodal input was considered superior to touch-only input and less stressful, see Figure 9.

Next, the Zoomable User Interface structure had to be rated by the participants. Again the overall opinion was very positive for the intuitiveness and usefulness of such a menu system.

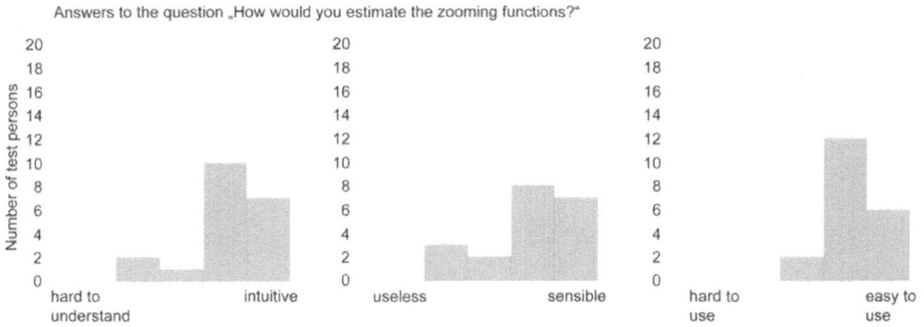

Fig. 10. User estimations for zooming functions

Finally the average times for text input were measured and compared against the BMW iDrive speller. This systems uses a rotary controller for character selection and word estimation. The average time needed for the city name „Schrobenhausen" with the iDrive system was 0:36 min, for the Fisheye keyboard 0:56 min and the Matrix keyboard 1:06 min. The main reason for the much longer input times of the touch input variants was identified to be the accuracy and jittering of the developed touch driver software.

5 Conclusion

The presented system shows the potential of capacitive Touchscreens with proximity detection. A Zoomable User Interface was developed to provide a proof of concept and test environment for user studies. The main findings are a high acceptance of such Graphical User Interfaces and the combined touch and speech interaction.

The main weakness at this point is the accuracy of the touch driver software which sometimes missed a touch event or jittery cursor position. This could be improved by applying more sophisticated finger detection algorithms like neural networks or classical image recognition algorithms.

References

1. Good, L., Stefik, M., Bederson, B.: A Comparison of Zoomable User Interfaces and Folders For Grouping Visual Objects. HCIL-2004-33, CSTR-4641, UMIACS-TR-2004-83 (November 2004)
2. Bederson, B.B., Meyer, J.: Implementing a Zooming User Interface: Experience Building Pad++. J. Software: Practice and Experience (1998)

3. Oviatt, S.: Handbook of Human-Computer Interaction. In: Jacko, J., Sears, A. (eds.). Lawrence Erlbaum, New Jersey (2002)
4. Seifert, K.: Evaluation multimodaler Computer-Systeme in frühen Entwicklungsphasen. Dissertation, Fakultät V - Verkehrs- und Maschinensysteme, TU Berlin (2002)
5. Ingmarsson, M., Dinka, D., Zhai, S.: TNT – A Numeric Keypad Based Text Input Method. In: CHI 2004, Vienna, Austria, April 24-29 (2004)
6. Hassenzahl, M., Burmester, M., Beu, A.: Engineering Joy. IEEE Software 1(2), 70–76 (2001)
7. Westerman, W.: Hand Tracking, Finger Identification and Chordic Manipulation on a Multi-Touch Surface, Dissertation, University of Delaware (1999)

A Framework for Fairness Guaranteed Multi-streaming Transport Protocol

Wonhyuk Lee[1,4], Hyuncheol Kim[2], Doowon Seo[3], and Jinwook Chung[4]

[1] Korea Institutes of Science and Technology Information
livezone@kisti.re.kr
[2] Dept. of computer science, Namseoul University
hckim@nsu.ac.kr
[3] Dept. of computer information, Daeduk college
dwseo@ddc.ac.kr
[4] Department of electrical and Computer Engineering, Sungkyunkwan University
{whlee,jwchung}@songgang.skku.ac.kr

Abstract. High-speed service and bulk transmission were available by the development of the network, and communication technologies. Moreover, voice-oriented communication service is undergoing a change by various services including data and video services. This paper presents a revised, improved congestion control scheme such as SCTP-SCF that includes SCTP multi-streaming to solve performance problems at high speed and evaluates the method's performance. The performance of the proposed scheme is evaluated using an OPNET simulator. Proposed SCTP-SCF overcomes existing problems in the traditional SCTP when a multi-streaming feature is used.

Keywords: NGcN, SCTP, Fairness.

1 Introduction

Just a few years ago, communication networks were divided into point-to-point public telecommunication networks and packet-based communication networks. Recently, to satisfy various customer demands, interest has been increasing in new types of services that can accommodate various kinds of traffic, such as voice, data, and video, even on so-called "specialized networks."

A Next Generation convergence Network (NGcN) is a network that allows for the accommodation of voice, high-speed Internet, frame relay, and other functions on a single packet-based network infrastructure that combines operation management and control. High-speed service and bulk transmission were available by the development of the network, and communication technologies. Moreover, voice-oriented communication service is undergoing a change by various services including data and video services.

Stream Control Transmission Protocol (SCTP) is a general-purpose transport protocol for IP network data communications that resembles Transmission Control Protocol (TCP) and User Datagram Protocol (UDP). The Internet Engineering Task

M.J. Smith and G. Salvendy (Eds.): Human Interface, Part I, HCII 2009, LNCS 5617, pp. 553–562, 2009.

Force (IETF) published it as RFC2960 [1]. The main purpose of SCTP is providing reliable end-to-end message transportation services over IP-based networks. The network community has paid a lot of attention, regarding both research and development, to SCTP because of its beneficial features such as multi-streaming and multi-homing in many institutes and universities [2][3][4][5].

A vital aspect of any transport protocol is congestion control. SCTP congestion control acts similarly to TCP congestion control. SCTP congestion control is derived from TCP's well established rate-adaptive window-based congestion control scheme, thus guaranteeing that SCTP will lower its sending rate during periods of network congestion and stop congestion collapses in shared networks. SCTP finds lost, reordered, duplicate, or corrupt packets and gives reliable transmission through retransmitting lost or corrupt packets. Since standard SCTP congestion control operates using the same basic functions as TCP, the same kinds of problems can occur in high speed communications.

Multi-streaming makes data transmission through multiple streams possible on an association. Even if one stream fails, only a portion of the data is lost and needs to be resent. If message loss occurs, only the stream by which it was delivered is affected. Thus, SCTP is able to prevent Head-Of-Line (HOL) blocking [6].

This paper presents a revised, improved congestion control scheme such as SCTP-SCF that includes SCTP multi-streaming to solve performance problems at high speed and evaluates the method's performance. The performance of the proposed scheme is evaluated using an OPNET simulator.

The remainder of this paper is organized as follows. Section 2 examines some previous studies that try to enhance the congestion control technique of the SCTP and summarizes the problems that it entails. Section 3 describes the characteristics of the proposed congestion control mechanism and it primarily focuses on the congestion control scheme along with a possible solution. Section 4 compares the existing congestion control scheme with the proposed congestion control scheme through simulation and numerical analysis in order to analyze and contrast their quantified performances. Section 5 concludes this paper by summarizing some key points made throughout and assessing the representation of analyzed results.

2 Related Works

2.1 TCP Connection and SCTP Association

In this subsection, we discuss applications that have to exchange various types of data between hosts. Sending multiple kinds of data in-parallel between endpoints has depended on four kinds of approaches as shown in Fig. 1. In all four cases, host A wants to transmit three kinds of data (known as data 1 through data 3) to host B.

In the first case of Fig. 1, host A starts three TCP connections with host B, with one connection for each type of data. Although this method logically separates the data based on its type, multiple connections overcome TCP-friendly congestion control by letting an application obtain an unfair part of accessible bandwidth to the detriment of other data flows in the network.

In the second case of Fig. 1, host A multiplexes and de-multiplexes the three kinds of data on just one connection. Applications that employ this method retain TCP-friendly congestion control, but this method makes it more complex for the application programmer because now the application itself has to manage the difficult work of efficiently and fairly multiplexing the data transmission.

The third case is to have a multimedia application on host A use UDP to send data to host B. This method is very similar to the second method, but application programmers also have to provide their own trustworthy service because of UDP's untrustworthy, connectionless service.

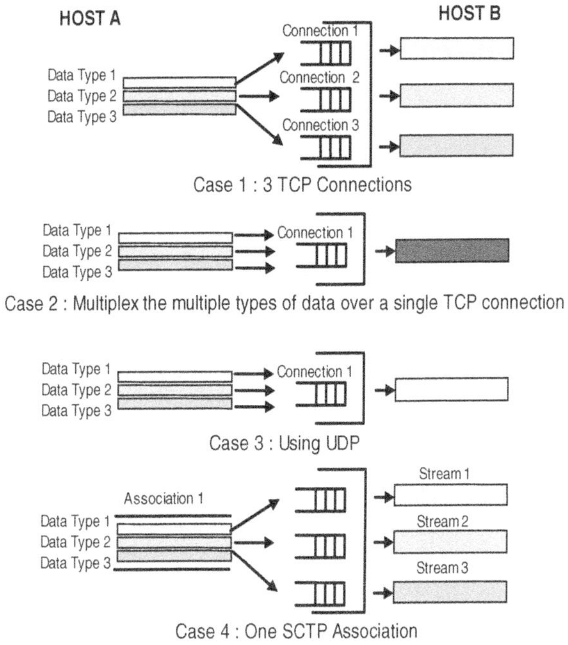

Fig. 1. Multi-streaming in SCTP

The last case of Fig. 1 illustrates an example situation in which hosts A and B have set up a multi-streamed association. In this instance, host A would like to send three different kinds of data. As a result, during association setup, host A asks for three streams from host B (numbered as streams 0 to 2). Host B has only one kind of data to transmit to host A and so requires and keeps one stream to host A [7].

2.2 SCTP Congestion Control

SCTP employs the same congestion control mechanism as TCP Reno. Its basis is that TCP-friendliness is a critical aspect when we consider the usage of SCTP, since TCP is the most commonly used transport-layer protocol in the Internet today. Table 1 compares SCTP congestion control with TCP congestion control.

Table 1. TCP congestion control vs. SCTP congestion control

	SCTP	TCP
Congestion window	Increased by the number of acknowledged bytes	Increased by the number of ACK segments received
SACK Usage	Mandatory	Optional
Congestion Avoidance	*cwnd* can only be increased when the full *cwnd* is utilized	TCP does not have SCTP's restriction
Fast Retransmit Algorithm	Based on SACK gap reports, similar to that of TCP SACK. Begins after four DupACKs	Begins after three DupACKs are received
Fast Recovery	Does not have an explicitly fast recovery phase. SCTP achieves fast recovery automatically with SACK usage	
Initial *cwnd*	2 * MTU	One MTU
if *ssthresh==cwnd*	Slow start	Slow start or congestion avoidance

Congestion control is one of SCTP's basic functions. For some applications, it may be probable that sufficient resources will be allocated to SCTP traffic to assure prompt arrival of time-critical data. Thus, it seems improbable that transmissions will meet extreme congestion conditions during normal operations. An endpoint needs to use the slow start and congestion avoidance algorithms to adjust the number of data being put into the network. SCTP's generic congestion control is used regarding the association, not a single stream.

3 SCTP-SCF

3.1 Motivation

If we make a typical request to an HTTP server to download a web page when surfing the Internet, normally we will receive several different files, containing text, graphics, or sound. In Fig. 2, the left side represents the source, which sends three small files upon the request of the client, and the right side is the destination. In Fig. 2, we see what happens when the three files are sent using a single TCP connection: as the first datagram is lost, even though the second and third files arrived entirely at the client they cannot be delivered to the upper user.

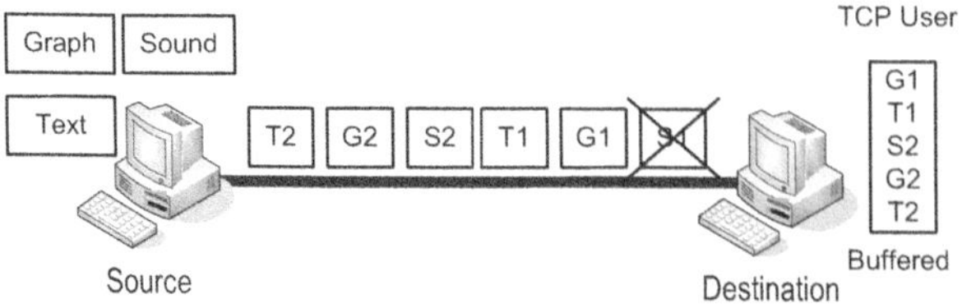

Fig. 2. A single TCP Connection

Usually, the client opens several different TCP connections, one independent connection per file, and closes each one once the file is completely transferred. However, we still suffer from the delay involved in opening and closing the TCP connections, and what is worse, we are wasting resources by having several TCP connections open at the same time between the same two endpoints, as servers have a limitation in the number of open TCP connections they can have at the same time. Apart from saving resources and avoiding delay by establishing these TCP connections, all the streams belong to the same association [8].

Using multi-streaming, data transmitted with a single SCTP association may be divided into multiple streams. The delivery sequence of these streams is independently handled. Within each stream, it is also possible to independently set the ordered or unordered delivery. Packet loss in one stream has no effect on the remaining streams, which assists them in avoiding the well-known TCP HOL blocking problem. When using the multi-streaming feature, the in-sequence delivery of the transmitted data may only be guaranteed within each stream, not the whole association. At the same time, a transport is performed within a single association, so all of the streams are under a common flow and congestion control mechanism [9].

Since congestion and flow control are respectively done for each path and association, a stream congestion control is performed per application. Thus, when single loss occurs for a particular stream, SCTP suffers *cwnd* reductions per path and halves the whole stream values each time. Consequently, the current specification of SCTP's congestion control does not ensure precise congestion control for each stream when there is a single loss or multiple losses within a certain stream [10].

3.2 Standard SCTP Module

The standard SCTP module consists of association management module, global congestion control, and queue management modules. Stream queue management: The endpoint must manage the operation and resources for every stream. For outbound streams, the endpoint will have to assign stream sequence numbers for every outgoing message. For inbound streams, the endpoint will have to monitor the continuity of stream sequence numbers of the arriving messages and undertake message reordering when needed.

Global congestion controller: The end point must check whether a congestion problem exists in the network. If it finds that network congestion is happening, the endpoint will have to adjust its sending operation to help mitigate the network congestion.

3.3 SCTP-SC

Proposed SCTP using Separated Congestion controller (SCTP-SC) is an SCTP using separated congestion control. Since congestion and flow control are respectively done for each path and association, a stream congestion control is performed per application. Thus, when a single loss occurs for a particular stream, SCTP suffers *cwnd* reductions per path and halves the whole stream values each time. SCTP-SC will settle this problem.

The SCTP-SC module is made up of association management module, per stream congestion controller, queue management module, and windows update scheme. The

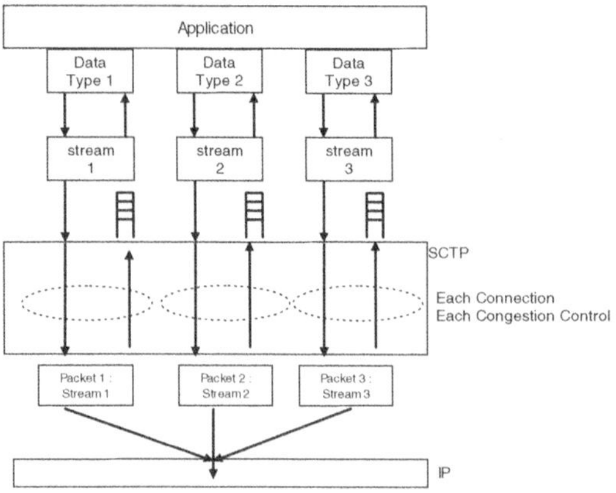

Fig. 3. The layered operation of SCTP-SC's

per stream congestion controller and windows update scheme operate on each stream. Fig. 3 shows the layered operation of SCTP-SC.

3.4 SCTP-SCF

In this section, we propose an algorithm to improve the SCTP fairness based on the fairness guarantee scheme and the window adaptation scheme.

SCTP using Separated Congestion controller with Fairness guarantee module (SCTP-SCF) use separated congestion control with a fairness guarantee module. SCTP-SC increases the total throughput but makes a new problem that generates unfairness. SCTP-SCF controls the congestion window based on its own information and information from another stream. SCTP-SCF is more efficient than the original SCTP. By applying a separated congestion control module, SCTP-SCF created and can be used for various applications. SCTP-SCF overcomes existing problems in the standard SCTP when the multi-streaming feature is used.

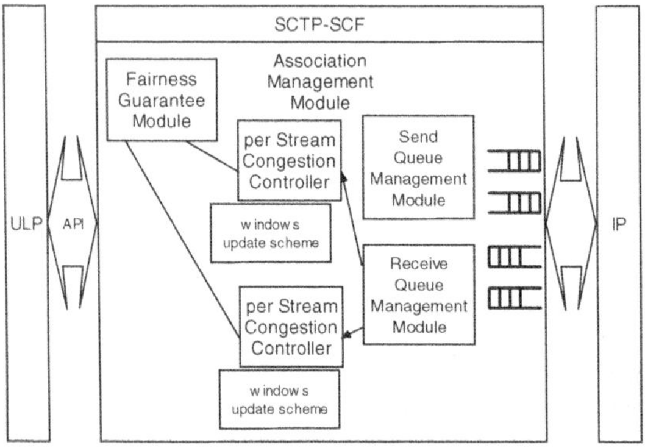

Fig. 4. SCTP-SCF's functional modules

Unfairness supports unfair service to applications. When the status of fairness is not very good, a situation can be generated where several packets are unable to transfer accidently. At such a time, the average throughput of SCTP is degraded.

Fig. 4 shows the SCTP-SCF module. The SCTP-SC module is made up of an association management module, per stream congestion controller, queue management module, windows update scheme, and fairness guarantee module. The fairness guarantee module uses congestion control for aggregates where flows have the same end-to-end path.

The architecture's core is the fairness guarantee module, which coordinates path information to guarantee correct congestion behavior and permits applications to readily adapt to network congestion. To react more effectively with high BDP networks, the window adjustment algorithm needs to be fairer than it has been. It is worthwhile to attempt a fairer approach depending on the information the estimated extra data gives. Layered operation of SCTP-SCF is same as that of SCTP-SC.

We introduce four zones, OverLoadZone (OLZ), LoadZone(LZ), Base Zone(BZ), and Idle Zone(IZ), as shown in Fig. 5.

```
if ( (α + RanF) < Diff )
   OLZ
else if ((α < Diff) && ( Diff < (α + RanF)))
   LZ
else if (((α - RanF) < Diff) && ( Diff < α ))
   BZ
else if ( Diff < (α - RanF))
   IZ
◦Expected = cwnd / BaseRTT
◦Actual =cwnd / RTT
◦ Diff = Expected – Actual
◦ RanF: Range Factor of α
```

Fig. 5. Classification of zone

where α is extra buffer counts that are occupied by the connection in the network.

The OLZ is a zone where overload traffic is repeatedly in a collision status. At OLZ, SCTP-SCF can increase the congestion window size based on its own cwnd, cwnds from other streams, and a compensation factor such as equation (1).

$$Cwnd[n] = Cwnd[n] + \left(\mu \times \frac{Cwnd[n]}{\sum_{i=0}^{n} Cwnd[i]} \right) \quad (1)$$

where n is the number of stream, μ is compensation factor. If the effect of μ is rapid, compensation factor μ is exchanged to μ/n. Equation (1) is changed to equation (2).

$$Cwnd[n'] = Cwnd[n'] + \left((\mu / n) \times \frac{Cwnd[n']}{\sum_{i=0}^{n} Cwnd[i]} \right) \tag{2}$$

At IZ, cwnd could be decreased based on information from other streams as in equation (3) and equation (4). The LZ is a zone where overload traffic is sometimes in a collision status.

$$Cwnd[n] = Cwnd[n] - \left(\mu \times \frac{Cwnd[n]}{\sum_{i=0}^{n} Cwnd[i]} \right) \tag{3}$$

$$Cwnd[n'] = Cwnd[n'] - \left((\mu / n) \times \frac{Cwnd[n']}{\sum_{i=0}^{n} Cwnd[i]} \right) \tag{4}$$

4 Performance Evaluation

4.1 Modeling

This Chapter describes the SCTP module developed for the network simulator OPNET. OPNET is a discrete event simulator developed for network research. Our SCTP implementation is based on RFC 2960, CINT's SCTP module, and the University of Delaware's SCTP module. The latter module is in wide use and provides a strong baseline for SCTP functionality.

OPNET is implemented in C with a modular framework for extensibility. Since we are dealing with a simulation environment, we abstracted some functions of SCTP from all the detailed functions. We also simplify other procedures such as association establishment and association shutdown. The target network configuration of our simulation model is shown in Fig. 6.

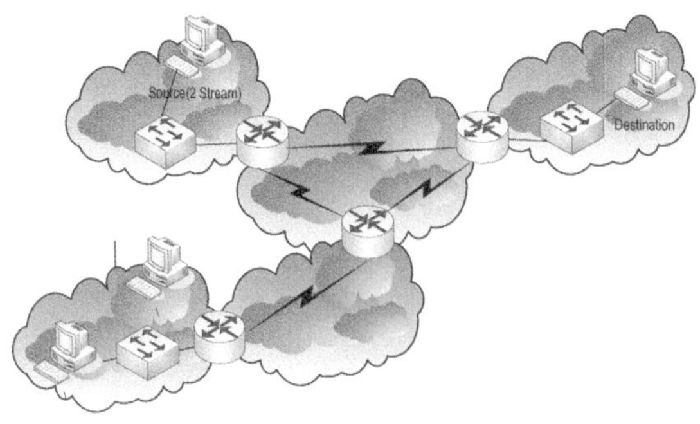

Fig. 6. Logical representation of a target network

4.2 Simulation Results

We constructed two scenarios. The first scenario is for SCTP-SC, and the second scenario is for SCTP-SCF at the *OLZ*. In each scenario, we used the SCTP stream0 process model and SCTP stream1 process model in order to use a multi-streaming function. Traffic losses were generated only on stream 0.

Jain's fairness index function is used to justify the fairness of TCP schemes. Jain fairness index function is used to justify the fairness of SCTP schemes.

From Fig. 7, it can be seen that SCTP-SCF performs better than SCTP-SC. SCTP-SCF outperforms SCTP-SC because SCTP-SCF finds the peer stream's measurement and increases *cwnd*. Measurement data of a packet loss, a timeout and *cwnd* are considered.

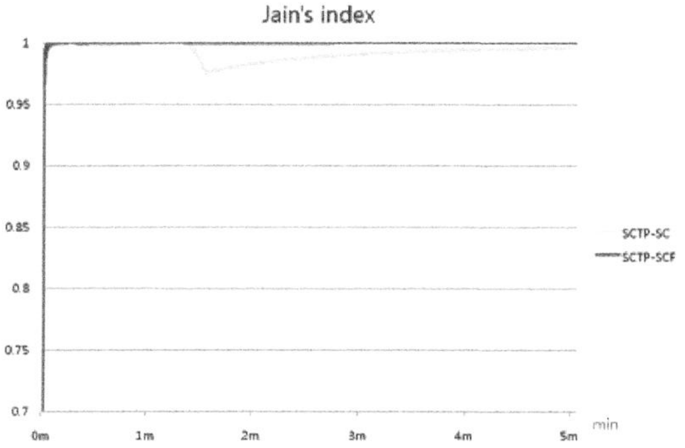

Fig. 7. Jain's fairness index

5 Conclusion

Congestion and flow control at SCTP are executed once for each path and association, respectively. Contrarily, congestion control is executed for each application. Therefore, when a certain stream has a single loss, SCTP undergoes reductions in *cwnd* size for each path, and the whole-stream values are halved in value every time. Consequently, the current specification of SCTP's congestion control does not make certain that congestion control is accurate for every stream when a single loss or multiple losses occur within a particular stream.

Multi-streaming with only one congestion control cannot settle the HOL blocking problem. To settle the problem clearly, a separate congestion control on each stream is needed. However, a new problem will be generated in that fairness will not be guaranteed. It seems to be a worthwhile subject to settle the fairness problem. As it will be a great help in upgrading the total throughput over networks.

We investigated an important aspect of fairness adjustment between streams of SCTP, the underlying congestion control mechanism. To improve SCTP's performance and to guarantee fairness in a new approach that refers to another stream's information such as *cwnd* and RTT, we introduce a fairness guarantee module. Proposed SCTP-SCF overcomes existing problems in the traditional SCTP when multi-streaming feature is used. The network simulator OPNET was used to run the performance evaluation. Through simulation, our algorithms show an improvement over standard SCTP.

References

1. Stewart, R., Xie, Q., Morneault, K., Sharp, C., Schwarzbauer, H., Taylor, T., Rytina, I., Kalla, M., Zhang, L., Paxson, V.: Stream Control Transmission Protocol. RFC 2960 (October 2000)
2. Park, E.-M.: A study of SCTP Streaming Scheme for Effective Bandwidth Aggregation. Thesis for the Degree of Master, Korea University (June 2005)
3. Jung, K.-M.: An Efficient Video Transmission Scheme Using Predictive SCTP. Thesis for the Degree of Master, Korea University (December 2005)
4. Ye, G.: Congestion Control for SCTP in Wireless Ad-Hoc Networks. Ph.D. dissertation, the City University of New York (2004)
5. Yi, Z.: Analytical Modeling of SCTP and Load Balancing SCTP. Ph.D. dissertation, the City University of New York (2005)
6. Kanpurwala, T.Y.: SCTP multi-streaming: Study of transmission of partially reliable and reliable data using dynamic stream priorities, Thesis for the Degree of Master, Oklahoma State University (2006)
7. Heinz II, G.J., Amer, P.D.: Priorities In Stream Transmission Control Protocol (SCTP) Multistreaming. In: SCI 2004, Orlando, FL (2004)
8. Rodríguez, I.A.: Stream Control Transmission Protocol The design of a new reliable transport protocol for IP networks. Helsinki University of Technology (Feburary 2002)
9. Dainotti, A., Loreto, S., Pescap, A., Ventre, G.: SCTP performance evaluation over heterogeneous networks. In: Published online 3 in Wiley InterScience, April 2007, pp. 1207–1218 (2007)
10. Choi, E.-H.: A SCTP Traffic Control Scheduler for Effective Multi-streaming Management. Thesis for the Degree of Master, Korea University (December 2005)

A Study on Computing Resource Partition for Increasing Efficiency of Grid System

Changsun Lim[1], Seongjin Ahn[2], and Jinwook Chung[1]

[1] Dept. of Computer Engineering, Sungkyunkwan Univ, Korea
{cslim,jwchung}@songgang.skku.ac.kr
[2] Dept. of Computer Education, Sungkyunkwan Univ, Korea
sjahn@songgang.skku.ac.kr

Abstract. This study deals with resolving the indefinite wait forced on normal resource reservations caused by the presence of priorities between normal resource reservations and advanced resource reservations. The algorithm we propose modifies the algorithm of the PLUS system that uses the GT4 toolkit and suggests improving flexibility by partitioning the normal reservation sector from the advanced reservation sector. With this partition, we are able to offer a more guaranteed use of normal resources. After we have discovered the specifications regarding the ratio of normal jobs to advanced reservations and guaranteeing advanced reservations, further studies will need to find ways to promote the efficiency of resource use.

Keywords: Advanced Reservation, GridARS, GlobusARS, Grid, Globus Toolkit.

1 Introduction

Resource reservation systems provide a way to efficiently utilize resources in a grid environment. A large project has to wait a long time before it can leverage all the resources that it needs when it chooses to request for timed resource use instead without resource reservation. Since grids are intended to create a virtual supercomputing environment for large research projects, they must be guaranteed the use of bulky amounts of resources. Therefore, we can guarantee the allocation of resources to large projects through resource reservation under the grid environment. PLUS[10] is a meta-scheduler designed to provide a grid environment through mutual interaction with GridARS[10]. PLUS can run independently and it supports the AR function. The AR function receives and executes tasks with either normal reservations or advanced reservations where it prioritizes advanced reservations over normal reservations. However, this prioritization scheme diminishes the stability and reliability of the overall system because it does not take into account the stability of normal tasks.

This paper is organized as follows. In Session 2, we study grid environment, resource reservation, and the fundamentals of PLUS and its problems. Session 3 shows how resource allocation can solve the resource reservation problem of PLUS. Session 4 concludes the resource allocation methodology proposed in this study.

M.J. Smith and G. Salvendy (Eds.): Human Interface, Part I, HCII 2009, LNCS 5617, pp. 563–569, 2009.

2 Related Studies

2.1 Grid Environment

Grids are represented by lattice structures of networks ranging from as small as private networks to as large as the internet. The ultimate purpose is to form a single supercomputing resource by allowing resources across connected computers to be shared throughout. The volume of tasks is currently growing exponentially. Computing resource requests are increasing in various fields ranging from studies on atomic structures to astronautic simulations. These jobs require a tremendous amount of time when executed on single computers. This kind of time can be effectively reduced with supercomputing resources by providing resources tens to millions of times greater than those offered in single computers.

There are ongoing efforts to develop grid middleware for building grid environments. Some instances include Globus, glite, and UNICORE where the most well-known is Globus Toolkit of Globus. GridARS dealt with in this study is also a GT4-based system interlinked with PLUS. [5]

2.2 Resource Reservation Technology

Resource reservation technology is one of the important resource sharing technologies in grid environments. Resource reservation guarantees stability in supplying resources to tasks that require stability and projects that use bulks of resources. However, the reservation module must be implemented in a standard grid structure in order to carry out reservation tasks for multiple computer resources where this module must be implemented in the lowest resource node or in the job scheduler at the top of the cluster. The latter is preferred because implementing the module in the lowest resource node diminishes the efficiency and cost-effectiveness.

Meta schedulers that support resource reservation systems are under constant development even today with completed developments including meta schedulers tapped with resource reservation modules such as CSF, Gridars, and GARA. Most resource reservation methods use algorithms that prioritize resources reservations over normal jobs.

2.3 PluS

PLUS is a meta scheduler that aids the operation of the resource reservation system in hand with GridARS. GridARS requests for jobs reserved with PLUS in the upper node of PLUS where PLUS receives the requests and sends job information to the batch scheduler of each node. In PluS, jobs residing in the reservation queue have higher priority than the jobs in the local queue. This way, a preempted request from the reservation queue when a job with normal reservation is using a computing resource will force the already-running job back in the local queue to wait. Here, when there are insufficient computing resources with heavy job requests, jobs with normal requests will be placed on hold indefinitely. This can also cause a turnaround in the priorities when a job with a lower priority arriving at a later time is completed before jobs with higher priorities [10].

2.4 Execution Flow

In the previous version of plus, normal reservations requested locally and advanced reservations requested by the super scheduler shared all computing resources. This means there are cases when a normal reservation under execution is forced to give up its execution to a newly arrived advanced reservation. Tasks preempted in this manner are placed in the local queue that they were originally in and wait until a usable resource becomes available.

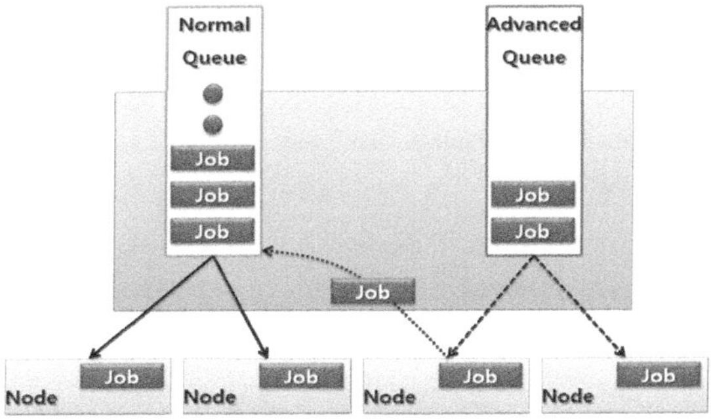

Fig. 1. Example of preemption flow of computing resource on PluS

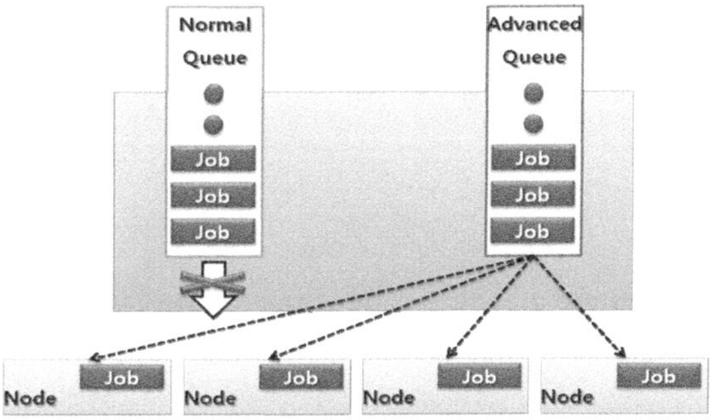

Fig. 2. Starvation by advanced reservation

In this case, however, preempting jobs in execution in order to handle jobs with advanced reservations as shown in [Fig. 2] will cause jobs in the local queue to wait indefinitely as they continue to pile up.

3 Guaranteeing Normal Reservations by Resource Sector Dispersion

3.1 Resource Sector Dispersion

The most imminent issue among those mentioned earlier is to prevent normal jobs from waiting indefinitely. With the existing method, normal jobs have no other option but to wait until all the prioritized advanced reservations are taken care of. Therefore, in this chapter, we introduce a scheduling method in which resources are divided into those that are exclusive to advanced reservations and those that are exclusive to normal reservations.

In this scheme, as shown in [Fig. 3], we separate computing resources that can be exclusively utilized by preempted reservations from resources utilized by only normal reservations. According [Fig. 3], the three nodes on the left represent resources that can be shared across the local queue and the reservation queue. These resources can be preempted through advanced reservation. The two nodes located on the right indicate nodes that jobs in the local queue can leverage exclusively. Thus, this mechanism guarantees that normal reservations won't have to wait indefinitely but will run at some time. As mentioned earlier, there are two types of nodes that provide computing resources where the first type is nodes that both queues can use and the other is nodes that only the local queue can use.

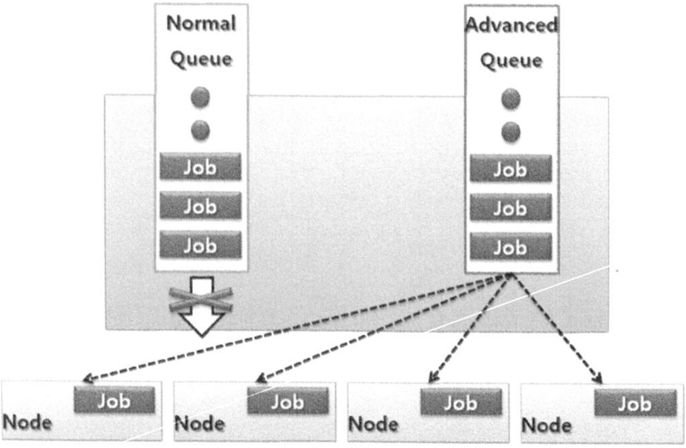

Fig. 3. Scheduling with distributed computing resources

In the proposed system, advanced reservations can only reserve resources that are usable by nodes that they have access to. Meanwhile, normally reserved jobs in the local queue can leverage assigned nodes to complete their instructions without having to wait for resources to become available even if multiple advanced reservation requests arrive and preempt the resources at the same time because a separate block of resources has been allocated to use for the local queue only.

3.2 Simulation

Although the proposed system can resolve the aforementioned problems, there still exist outstanding issues such as how much computing resources to allocate as shared resources and how much of those to allocate as resources exclusive to the local queue. Over-allocating resources for the local queue will impose a limit on jobs with advanced reservation. On the other hand, under-allocating resources for the local queue will prolong the time resources in the local queue will have to wait, thereby violating fairness. Therefore, in this chapter, we measured the change in performance with changing a ratio of resources shared by the local queue and the reservation queue to resources exclusive to the local queue through experiments and observation.

The experiment used a program written in C language where we defined a unit of job as an arbitrary time slot with basic unit time of 0.5 sec. We also assumed that jobs waiting in the local queue of the reservation queue are of identical types, or types that use computing resources. Here, the random number function is used to generate numbers ranging 1~5 sec, which are used for the expected running time for each job. In this experiment, we measured how performance shifts according to the ratio of amount of resources shared by the local queue and the reservation queue to amount of resources used only by the local queue by modifying the ratio of normal reservations to advanced reservations.

Table 1. Reservations efficient on divide resource

Kind of reservation \ Shard resources		5	10	15	20	25	30	35	40
Normal reservation	Advanced reservation								
300	100	182	167	163	160	159	158	157	273
300	200	212	182	173	168	165	163	161	291
300	300	242	197	183	175	171	168	165	299
300	400	272	212	193	183	177	173	170	306
300	500	303	227	203	190	183	177	174	314

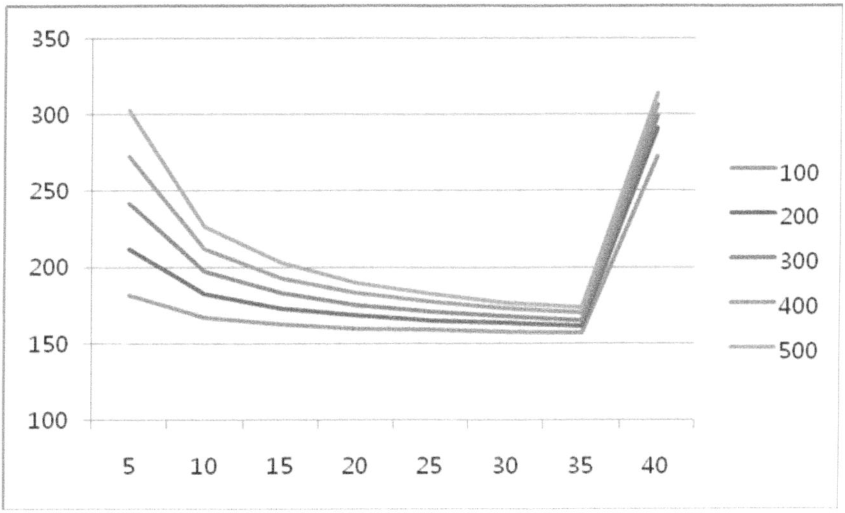

Fig. 4. Reservations efficient on divide resource

[Table 1] shows average values obtained through 1000 simulations when assuming there are 40 computing resources. Taking a closer look at the values, we can see that the ratio of jobs reserved normally to jobs scheduled in advance doesn't cast a significant impact on the performance. Also, the best performance came when the proportion of resources shared by the local queue and the reservation queue was 90% of the entire resources.

4 Conclusion

In this paper, we proposed a grid computing resources reservation system that distinguishes advanced reservations from normal reservations. In this system, the advanced reservation function is handled by the central scheduler and the normal reservation function is managed by the local scheduler. Computing resources, which are shared by the central scheduler and the local scheduler, are monitored by the local scheduler. This way, an advanced reservation can preempt a normal reservation when both of them request a resource by tagging a higher priority to the advanced reservation.

Thus, by experimenting and observing, we measured changes in performance according to the ratio of resources shared by the local queue and the reservation queue to resources used only by the local queue. Experiment results show that the ratio of jobs scheduled by normal reservation to jobs scheduled by advanced reservation does not pose any significant impact on performance. Also, we witnessed best performance when the proportion of resources shared by the local queue and the reservation queue is 90% of the entire pool.

However, due to the limited job execution times, limited number of jobs, and limited computing resources, more experiments will need to follow-up in order to improve the reliability of the results that we obtained. Studies on specific algorithms and execution orders will also follow in the future.

References

1. Bode, B., Halstead, D.H., Kendall, R., Lei, Z.: The Portable Batch Scheduler and the Maul Scheduler on Linux Clusters. In: Proceedings of the 4th Annual Linux Showcase and Conference, Atlanta, Georgia (October 2000)
2. IBM Load Leveler, `http://www.ibm.com/products/`
3. Portable Batch System, `http://www.openpbs.org/`
4. Maui Scheduler Open Cluster Software, `http://mauischeduler.sourceforge.net/`
5. Globus Toolkit, `http://www.globus.org/`
6. Litzkow, M.J., Livny, M.: Experience With The Condor Distributed Batch System. In: Proceedings of the IEEE Workshop on Experimental Distributed Systems, October 1990, pp. 97–101 (1990)
7. Ferris, M.C., Munson, T.S.: Modeling languages and Condor: metacomputing for optimization. Mathematical Programming 88(3), 487–505 (2000)
8. Romberg, M.: The UNICORE Grid infrastructure. Scientific Programming 10(2), 149–157 (2002)
9. Dietmar, W.: Erwin UNICORE - a Grid computing environment

10. Nakada, H., Takefusa, A., Ookubo, K., Kudoh, T., Tanaka, Y., Sekiguchi, S.: An Advance Reservation based computation resource manager for glocal scheduling. In: Proceedings of gca 2007 (2007)
11. Open Grid Forum, http://www.ggf.org/
12. Maui, http://www.clusterresources.com/
13. Globus alliance, http://www.globus.org/
14. Farooq, U., Majumdar, S., Parsons, E.W.: Efficiently Sche-duling Advance Reservations in Grids, http://www.sce.carleton.ca/~ufarooq/ufarooq_ar_sched.pdf
15. Yi, D.-H., Kim, J.: Shared Resource Management Scheme in Advance and Immediate Reservations for Effective Resource Allocation. Korea Institute of Communication Sciences (July 2004)
16. Gupta, A.: Advance reservation in real-time communication services. In: Proc. of the IEEE 22nd Annual Conference on Local Computer Net-works (1997)
17. Foster, I., Kesselman, C., Lee, C., Lindell, R., Nahrstedt, K., Roy, A.: A Dis-tributed Resource Management Architecture that Supports Advance Reservations and Co-Allocation. In: The Proceedings of the 7th International Workshop on Quality of Service, London, UK (May 1999)
18. Smith, W., Foster, I., Taylor, V.: Scheduling with Advanced Reservations. In: The Proceedings of the IEEE/ACM 14th International Parallel and Distributed Processing Symposium, Cancun, Mexico (May 2000)

Richbiff: E-Mail Message Notification with Richer Clues

Mitsuru Minakuchi and Hisashi Miyamori

Kyoto Sangyo University, Faculty of Computer Science and Engineering
Motoyama, Kamigamo, Kita-ku, Kyoto 603-8555, Japan
{mmina,miya}@cse.kyoto-su.ac.jp

Abstract. Traditional e-mail notification applications, which notify the user of newly arrived messages, are frequently insufficient because users daily receive a vast number of messages. Hence, we propose an e-mail notification application that peripherally displays summaries of messages in the form of kinetic typography. In addition, it infers the possibility of a message being read to control notification levels and conceals parts of words to avoid exposing excessive personal information. We have qualitatively confirmed the effects of the application through trial use; it can strongly attract users' attention and motivate the user to check significant messages.

Keywords: notification, kinetic typography, peripheral display, ambient display.

1 Introduction

As e-mail has become popular, people receive a vast number of e-mail messages daily and can find it annoying to check them. Simple notification applications or devices that inform the user of newly arrived messages are adequate when he or she receives only a limited number of messages. However, in many cases, they are almost no use because the user may be notified very often of unimportant messages. The `biff` command displays headers and the first few lines of messages, but is highly interruptive because the user has to read the lines.

In contrast, users sometimes desire to find newly arrived e-mail messages as fast as possible. For example, some senders expect quick replies. Also, users may benefit from quickly grasping information, such as invitations for limited-time offers. However, it is too exhausting for users to read all messages.

We propose an application, Richbiff, that peripherally displays information about e-mail messages in attractive forms with a limited but sufficient amount of information for users to judge whether they should read them or not.

2 Approach

We adopted four features to fulfill the requirements for a notification tool.

2.1 Use of Kinetic Typography

Because e-mail messages consist of textual information, displaying text itself is straightforward. However, the user may easily miss new text when the display is full

M.J. Smith and G. Salvendy (Eds.): Human Interface, Part I, HCII 2009, LNCS 5617, pp. 570–578, 2009.

of text. Notification tools should actively attract the user's attention. We thus adopted kinetic typography [1]. Kinetic typography can be noticed even if it is displayed in the peripheral visual field, though it is less legible. It is also useful to represent attributes of information with motion patterns, which can help the user to estimate its significance or urgency, and to judge whether it is worth reading carefully [2]. In addition, it is aesthetically pleasing and enjoyable. This feature is suitable for ambient displays that peripherally present various kinds of information.

2.2 Adequate Amount of Information

Notification tools should provide an adequate amount of clues to support instant decisions on whether to read a message. A trade-off exists between the number of clues and understandability; too many clues take more time for the user to read, while too few are insufficient for decision-making.

The sender, recipients, subject, message body and attachment files are constituent elements of an e-mail message. Richbiff displays the sender, the subject, and keywords in the message body because they are considered to be essential parts.

2.3 Control of Notification Levels

The probability of an e-mail message being read can be estimated with some degree of accuracy. Richbiff estimates it from reading histories and determines a motion pattern according to an estimated possibility, i.e., vigorous motion if it is likely to be read and quieter motion if it is not likely to be read.

2.4 Information Censoring

If notification is performed in a public space, filtering functions are required for privacy. We assume that concerned persons still can understand partially corrupted or incomplete information because they can complement it with their knowledge. For example, people who are proficient in English can understand the scrambled sentence shown in Fig. 1[1], though it may be difficult for beginners to understand. Similarly, we expect that concealment of some characters in words makes it difficult for outsiders to understand. This method also helps prevent embarrassing words from excessive exposure in a public space.

> Aoccdrnig to a rscheearch at Cmabrigde Uinervtisy, it deosn't mttaer in waht oredr the ltteers in a wrod are, the olny iprmoetnt tihng is taht the frist and lsat ltteer be at the rghit pclae. The rset can be a toatl mses and you can sitll raed it wouthit porbelm. Tihs is bcuseae the huamn mnid deos not raed ervey lteter by istlef, but the wrod as a wlohe.

Fig. 1. An example of scrambled text. The order of letters except the first and the last one in each word are randomly shuffled.

---

[1] The source is unknown. Matt Davis reported a discussion on the description in his Web page http://www.mrc-cbu.cam.ac.uk/~mattd/Cmabrigde/, which is currently inaccessible. It can be found at some Internet archive sites, such as Wayback Machines and Google's cache (retrieved on Feb. 1, 2009).

3 Implementation

Fig. 2 shows the architecture of Richbiff. The mail user agent communicates with a mail server and checks new e-mail messages. The text processing module applies morphological analysis to messages to extract words and determine their parts of speech. The keyword extraction module calculates tf/idf values of the words in the messages and selects significant words. The reading-possibility estimation module infers whether the user will read a new message using a Bayesian filter that is trained with senders, words in subjects and message bodies, and reading histories.

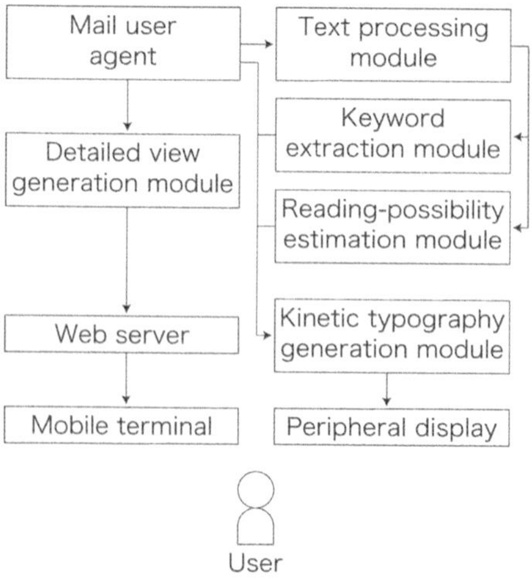

Fig. 2. Architecture of Richbiff

The kinetic typography generation module composes moving texts for each e-mail message showing the sender, the subject, and keywords by applying them to an animation template [3]. It displays e-mail addresses without the real names of the senders to conceal private information from other people. It divides the subject into segments and displays them with a scrolling motion from the bottom to the top of the screen. The keywords are displayed with zooming pop-up motion; final scaling factors are

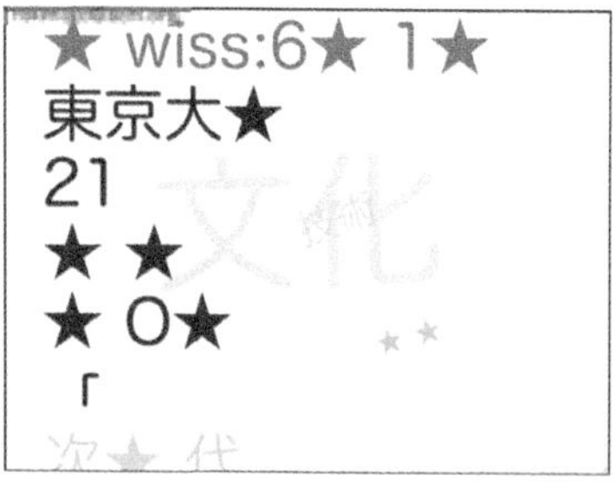

Fig. 3. Screenshot of Richbiff. The sender's address at the left top is hidden to protect personal information.

determined according to their tf/idf values. It displays messages with high reading probability in active motion. Letters in each word are randomly replaced with symbols (the current implementation uses asterisks, "*") according to a preset censoring rate. Fig. 3 shows a screenshot of Richbiff. Fig.4 shows an example sequence of scrolling and pop-up motion.

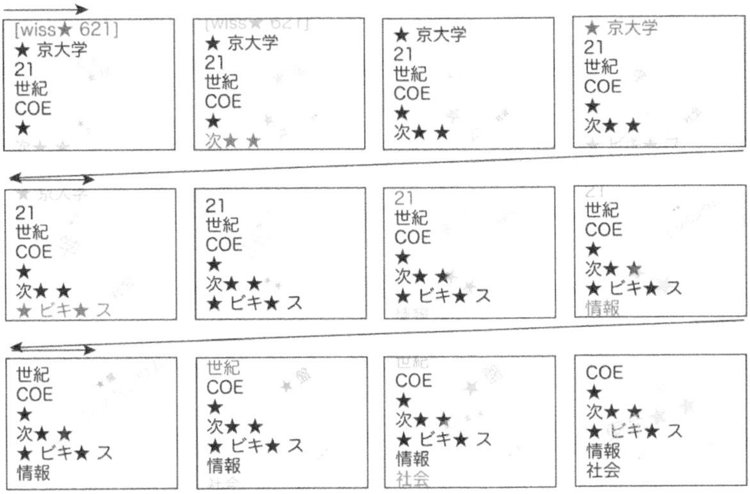

Fig. 4. Example sequence of scrolling and pop-up motion (five fps)

Richbiff plays the generated animation of text using a customized version of the kinetic typography engine [1]. At the same time, it generates detailed views of the messages on a local Web server. The user can read the entire content of the current message with a Web browser on a mobile terminal (Fig. 5).

Fig. 5. Actual operation example. The picture on the left shows a peripheral display of Richbiff. On the right is a Web browser showing a current e-mail message reported by the peripheral display.

4 Discussion

We have qualitatively confirmed the following effects through trial use of Richbiff, though we have not quantitatively evaluated them.

4.1 Effects of Kinetic Typography

Kinetic typography was able to strongly attract the user's attention.

We have not compared the degree to which it interrupts users' work with that of other notification media, for example, sound and static texts. According to our subjective views and audience comments at demonstrations, kinetic typography interrupted the user at the same level as, or a little bit more weakly, than an alarm sound. However, it could provide enough information to show digests of e-mail messages. In contrast, audio requires a certain period of time to provide such digests.

The user could notice its motion even if it was displayed in the user's peripheral visual field because its brightness varied. This achieved an effect similar to that of ambient displays, which convey information to the user with abstract changes, such as lighting patterns. Kinetic typography was superior in notification to static text, which the user could easily miss when it was displayed in his or her peripheral visual field.

Continuous display of kinetic typography sometimes caused environmental disorder, that is, it conveyed visual restlessness. However, it will be acceptable to the user that the system uses vigorous motion only for messages likely to be read. In addition, the user may become accustomed to the restlessness with long-term use.

4.2 Adequacy of Amount of Information

The amount of information provided by Richbiff was adequate for understanding outlines of e-mail messages in a short time. Some newly arriving e-mail messages of high importance and/or urgency are related to the user's ongoing tasks or discussion. Thus, outlines were enough to trigger the user to check such significant messages, of which the contents were predictable with a few clues.

We inspected the effects of information about the sender, the subject, and keywords by comparing displays with and without them:

Displaying the sender was useful for judging whether the message was worth reading immediately, though it did not contribute as much to the understandability of outlines. For example, users paid great attention to e-mail messages from their friends and colleagues and often neglected e-mail newsletters. The current implementation displays only bodies of e-mail addresses without the senders' real names. This is enough for recipients to understand who the sender is. The user was sometimes anxious that strangers might figure out at least the sender's affiliation when it was displayed in a public space. Abbreviation of domain parts, for example `mmina@CKAJ` instead of `mmina@cse.kyoto-su.ac.jp`, will be effective in such a case.

Subjects provided essential information, because it was difficult to guess contents of messages in many cases when they were hidden. However, they sometimes required too much user's attention to read them.

Keywords helped users to understand outlines of e-mail messages of which the subjects were inadequate or vague. The pop-up display pattern was able to give impressions of messages without requiring active reading. Simultaneous use of popped-up keywords and scrolling subjects was not confusing because they were displayed in different areas and orientations.

4.3 Effects of Control of Notification Levels

It was effective to select display patterns according to notification levels, i.e., the system displayed messages of high reading probability with active motion. As discussed in section 4.1, kinetic typography with great variation in visual expression tends to attract the user's attention too much. This mechanism of notification level control reduced such unnecessary intrusion on users' on-going tasks.

We have not quantitatively evaluated the accuracy, precision, and recall. According to our subjective view, it was nearly as useful as spam filters, which use the same technique of Bayesian filtering.

4.4 Effects of Information Censoring

It was possible to understand content with certain degrees of concealment of letters. Fig. 6 shows examples of displays at various censor rates.

Our empirical observation suggested that most messages at low censor rates (under 0.25) were understandable with little difficulty, and that those at high censor rates (over 0.5) were difficult to understand even if the messages were highly relevant to the user. Mid range censor rates (between 0.25 and 0.5) could make messages moderately obscure: the user could understand such messages if he or she knew the topic but could only vaguely understand them if he or she had no context for them. According to these observations, we expect that the effect is applicable to outsiders who do not

Fig. 6. Examples of display at various censor rates

share the recipient's background, though we have not fully confirmed this. Quantitative evaluations, including differences in understandability for concerned persons and others would be priority in our future work.

We implemented the random concealment algorithm that randomly replaces letters in words. However, it sometimes caused spotty concealment when the censor rate was in the middle range. It is not preferable because important words may be concealed too well to understand, and sensitive information may be inadequately concealed. It would be useful to conceal at least one letter in a word. A black-list of words would also be useful to mandatorily censor specific words.

4.5 Other Discussion

- Animated display had sometimes progressed only part of the way when the user noticed it. In these cases, the user may not guess the content of messages because of missing an important part. Thus, we implemented rewind and fast-forward functions for both the peripheral display and the Web page, which enabled the user to navigate to the previous and next messages. We used a keypad for the peripheral display, but it was not suitable for some situations, for example, when the user did not have the keypad at hand. Other multimodal operations, such as voice and gesture, would be useful.
- Richbiff might also be useful for reminding the user about e-mail messages he or she has not read or forgot to reply to. We are planning to improve it to report such missed e-mail messages. However, handling all received messages is not effective. We should introduce inference techniques of significance to report only possibly important messages.
- Although we designed Richbiff for a single person's use, it can be extended to multi-user environments. This could make good use of peripheral displays in shared places, such as small offices and living rooms. In that case, Richbiff needs to display messages in such a way that users can distinguish their messages from others. For example, it can change the background color, the display position of each element, or the animation pattern according to the recipient.

5 Related Work

According to a taxonomy of ambient information systems by Pousman and Stasko [4], a few existing studies were classified into medium or somewhat high levels of information capacity and notification level, the levels for which Richbiff aimed. We considered that one reason is a lack of a proper display form. Also, this kind of notification tool was excluded from a definition of ambient information systems. In addition, they defined high-throughput textual displays that have low aesthetic emphasis. However, Richbiff achieved aesthetic representation of richer information. Hence, kinetic typography can make a new kind of ambient display that has medium information capacity, medium notification level, high representational fidelity, and high aesthetic emphasis.

Vogel and Balakrishnan proposed interactive public ambient displays that supported the transition from implicit to explicit interaction with both public and personal information [5]. They represent messages in an abstract form of color bars when the user is far from the display. When the user approaches the display, it shows content of messages with small letters in order to prevent others from reading it. Though this approach is a useful way to handle personal information in a public space, it requires sensors to detect the distance between the user and the display. Moreover, the color bars may be too abstract to trigger implicit interaction.

Many automatic text summarization techniques have been proposed. They can improve Richbiff by allowing it to show better summaries of e-mail messages to reduce the need to access the entire content with a Web browser.

As for information concealment, Nakamura and Tanaka proposed an information-filtering system that detects and hides entities, for example, results of sport games, referring to the user's recording schedule of TV programs [6]. This approach, that is, targeting parts of information by referring to external information, is useful for hiding appropriate parts of e-mail messages.

6 Conclusion

We developed an email notification application named Richbiff that peripherally displays information about e-mail messages. Richbiff has the following four features: use of kinetic typography, providing an adequate amount of information, control of notification levels, and information censoring. Our trial use proved their effectiveness. Kinetic typography could strongly attract the user's attention and could realize aesthetically appealing and enjoyable peripheral textual displays. The amount of information that Richbiff provided was adequate to motivate the user to check the full text of significant e-mail messages. The motion pattern selection was useful to reduce unnecessary intrusion on user's on-going tasks. Empirical observation suggested that the concealment of letters with the proper range of randomness worked well.

In future work, we plan to quantitatively evaluate the effects, including noticeability and intrusiveness of kinetic typography and understandability of partially hidden words.

Acknowledgements. This work was partially supported by MEXT Grants-in-Aid for Scientific Research, 20500120, and 20300042, and Kyoto Sangyo University Research Grants. The current implementation utilizes Lee et al.'s kinetic typography engine [1], the word vector tool [7], the Classifier4J [8], and the Sen morphological analyzer [9].

References

1. Lee, J.C., Forlizzi, J., Hudson, S.E.: The Kinetic Typography Engine: an Extensible System for Animating Expressive Text. In: Proceedings of the 15th Annual ACM Symposium on User Interface Software and Technology (UIST 2002), pp. 81–90. ACM Press, New York (2002)
2. Minakuchi, M., Kidawara, Y.: Kinetic typography for ambient displays. In: Proceedings of the 2nd International Conference on Ubiquitous Information Management and Communication (ICUIMC 2008), pp. 54–57. ACM Press, New York (2008)

3. Minakuchi, M., Tanaka, K.: Automatic Kinetic Typography Composer. In: Proceedings of the 2005 ACM SIGCHI International Conference on Advances in Computer Entertainment Technology (ACE 2005). ACM Press, New York (2005)
4. Pousman, Z., Stasko, J.: A Taxonomy of Ambient Information Systems: Four Patterns of Design. In: Proceedings of the Working Conference on Advanced Visual Interfaces (AVI 2006), pp. 67–74. ACM Press, New York (2006)
5. Vogel, D., Balakrishnan, R.: Interactive Public Ambient Displays: Transitioning from Implicit to Explicit, Public to Personal, Interaction with Multiple Users. In: Proceedings of the 17th Annual ACM Symposium on User Interface Software and Technology (UIST 2004), pp. 137–146. ACM Press, New York (2004)
6. Nakamura, S., Tanaka, K.: Temporal Filtering System to Reduce the Risk of Spoiling a User's Enjoyment. In: Proceedings of the 12th International Conference on Intelligent User Interfaces (IUI 2007), pp. 345–348. ACM Press, New York (2007)
7. The Word Vector Tools, http://nemoz.org/joomla/content/view/43/83/lang,en/ (retrieved on Feburary 18, 2009)
8. The Classifier 4J, http://classifier4j.sourceforge.net/ (retrieved on Feburary 18, 2009)
9. Sen, https://sen.dev.java.net/ (retrieved on Feburary 18, 2009)

An Approach for the Design of Secure Communication in Embedded Systems

Pavel Ocenasek

Brno University of Technology, FIT, Bozetechova 2, 612 66 Brno, Czech Republic
ocenaspa@fit.vutbr.cz

Abstract. This paper deals with the novel evolutionary approach used for the automatic design of security protocols that also employs the novel grammatical generation of protocols according to the design and security requirements. The first part outlines the introduces security protocols Then, in the main part of this paper, we propose the new algorithm based on natural evolution that is utilized for the automated design of secure communication and security protocols. These protocols may be used for authentication, key distribution and/or for providing other security functions and properties.

Keywords: Security protocol, design, evolutionary algorithm, genetic algorithm, secure communication.

1 Introduction

The increasing popularity of distributed computing and applications like internet banking and electronic commerce has created both tremendous risks and opportunities. Many of risks stem from security breaches, which can be ruinously expensive. One of the cornerstones of security is the use of security (cryptographic) protocols in which information is exchanged in a way intended to provide security guarantees.

Security protocols are becoming widely used and many new protocols are being proposed. Since security protocols are notoriously difficult to design, computer assistance in the design process is desirable.

In this paper we describe the goals of security protocols – user authentication and/or key distribution and we propose an evolutionary technique that may be useful in designing new security protocols.

2 Security Protocol

A security protocol is a recipe that describes the operations in which the subjects should achieve some security goals. Protocols are often described using informal notation, for example as a sequence of instructions explaining the actions taken by the subjects.

Each step describes an event $A \rightarrow B$: X, which states that A exchanges the message X with B. Messages consist of atoms, like subject names and nonces (randomly

M.J. Smith and G. Salvendy (Eds.): Human Interface, Part I, HCII 2009, LNCS 5617, pp. 579–586, 2009.

generated strings), and can be composed. Moreover, messages may be encrypted using keys of subjects. Because security protocols may contain certain flaws, finding such attacks is the purpose of formal validation using various approaches [1] [2] [5].

2.1 Introduction

A protocol is a recipe that describes how subjects should act to achieve some goal. Protocols are often described using informal notation, for example as a sequence of instructions explaining the actions taken by the subjects.

Each step describes an event $A \rightarrow B$: X, which states that A exchanges the message X with B. Messages consists of atoms, like subject names and nonces (randomly generated strings), and are composed by tupling. Moreover, messages may be encrypted using keys of subjects.

However, describing the protocol components, cryptographic properties and requirements in details is beyond the scope of this abstract. For more information please refer e.g. to [6].

2.2 Needham – Schroeder Protocol

Probably the best known security protocol used for authentication is the **Needham-Schroeder** protocol (in this example we use the public-key version of this protocol):

1. $A \rightarrow S$: A, B
2. $S \rightarrow A$: $\{K_B, B\}_{K_S^{-1}}$
3. $A \rightarrow B$: $\{N_A, A\}_{K_b}$
4. $B \rightarrow S$: B, A
5. $S \rightarrow B$: $\{K_A, A\}_{K_S^{-1}}$
6. $B \rightarrow A$: $\{N_A, N_B\}_{K_a}$
7. $A \rightarrow B$: $\{N_B\}_{K_b}$

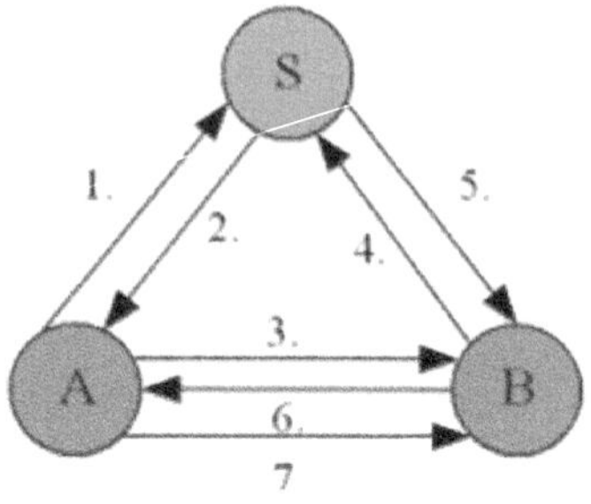

Fig. 1. Needham-Schroeder Protocol

This protocol can be considered as the interleaving of two logically disjoint protocols: messages 1, 2, 4 and 5 are concerned with obtaining public keys, whereas messages 3, 6 and 7 are concerned with the authentication of A and B.

The protocol was found unsecured by Lowe and new fixed version was published. The proposed attack allows an intruder to impersonate another agent.

3 Motivation

There are five primary goals of the research:

1. Modification of standard evolutionary techniques for the use in the field of security protocol design. Designing new protocols is a completely different problem, other

than those previously solved by these techniques. Protocols are programs with instructions that are evolved and verified for security.

2. Optimization of fitness function calculation that is used to evaluate the protocol. Calculating the fitness function is a crucial point in the evolutionary process. The value will be computed from the results produced by external security verification tools. In principle, the fitness corresponds to the percentage of met security presumptions, e.g. the protocol length, satisfactions of initial requirements (sets of knowledge and belief) and security properties.

3. Optimization of the generation of elementary instructions and message components used for representing security protocols. The number of elementary instructions and components is another crucial point. High number of invalid instructions increases the size of search space. Lower number of instructions and components increases the design performance.

4. Analysis of the heuristic strategies that can be used in the design process. The evolution process itself is not powerful enough to efficiently explore the space of possible protocols. Additional heuristics might be needed to accelerate the evolution process, e.g. the heuristic crossover operator, heuristics for fitness computation, etc.

5. Design and implementation of an application, evaluation of the performance of proposed approach and the influence of proposed heuristic techniques.

This research aimed to address the shortcomings of existing approaches, namely to minimize the requirement of human-intervention. So the general goal is to make the protocol design process highly automated.

4 Protocol Design

As mentioned above, the protocol is a set of rules and conventions that define the communication framework between two or more subjects. These rules may be elementary instructions that consist of operations such as sending a message, encrypting/decrypting a message with a secret key. This chapter proposes a novel algorithm for automated design of communication and security protocols.

4.1 Protocol Encoding

The very important task was to choose the right and efficient form for protocol representation. Security protocols are encoded into chromosomes as strings of elementary instructions. Each chromosome (individual) in population represents one random protocol.

The chromosomes, unlike in standard genetic algorithms, may be of variable length. This is due to the optimization of protocol instructions.

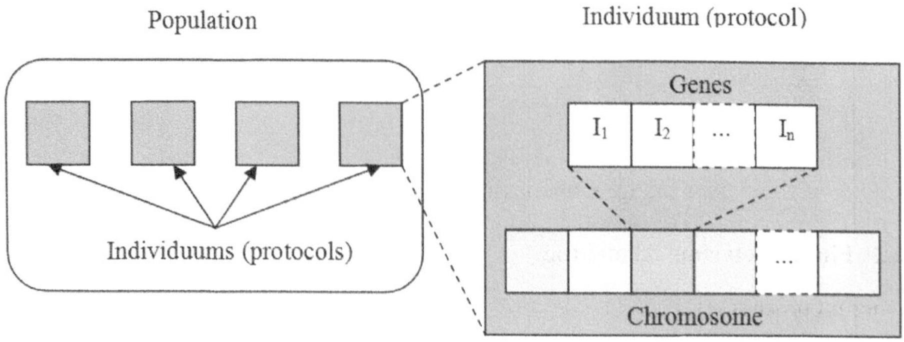

Fig. 2. Structure of Population. Protocol (individuum) is encoded as a sequence of instructions (I_1, …, I_n).

4.2 Algorithm

While the protocol runs, each instruction affects the corresponding sets of knowledge and belief, described using BAN logic.

From the evolutionary-optimization point of view, a chromosome of variable length represents the sequence of protocol instructions. There could be generated as many random chromosomes as required for initial population. Each chromosome represents a different protocol and its fitness is computed by simulating its run, according to the changed sets of knowledge and belief.

Security goals

Before we start the design of new security protocol, we need to state the initial presumptions and security goals to reach. We should specify what should or shouldn't contain sets of knowledge and belief for each involved subject, desired protocol length, forbidden knowledge for each involved subject, etc.

Some of the initial requirements are set in the generator, e.g. which message is secret (implicitly keys are considered as secret) and cannot be sent unencrypted via unsecured channel, etc.

Some of the security goals can be specified within the grammar, e.g. format of protocol instructions, number of recursive encryptions and some specific security properties.

The main algorithm (see figure below) derives from the genetic algorithm and describes the whole flow we use to design security protocols. The algorithm is described in detail in the following paragraphs.

Initial Population

Randomly generated protocols (instruction sequences) are encoded into the chromosomes. In this step the fitness of all individuals is calculated.

Definition: The context free grammar (CFG) for generating common security protocols is a 4-tuple defined as $G = (N, T, P, S)$ where N is a finite set of non-terminals, T is a finite set of terminals, P is a finite set of production rules and S is an element of N the distinguished sharing non-terminal. Elements of P are of the form $N \rightarrow (T \rightarrow N)^*$.

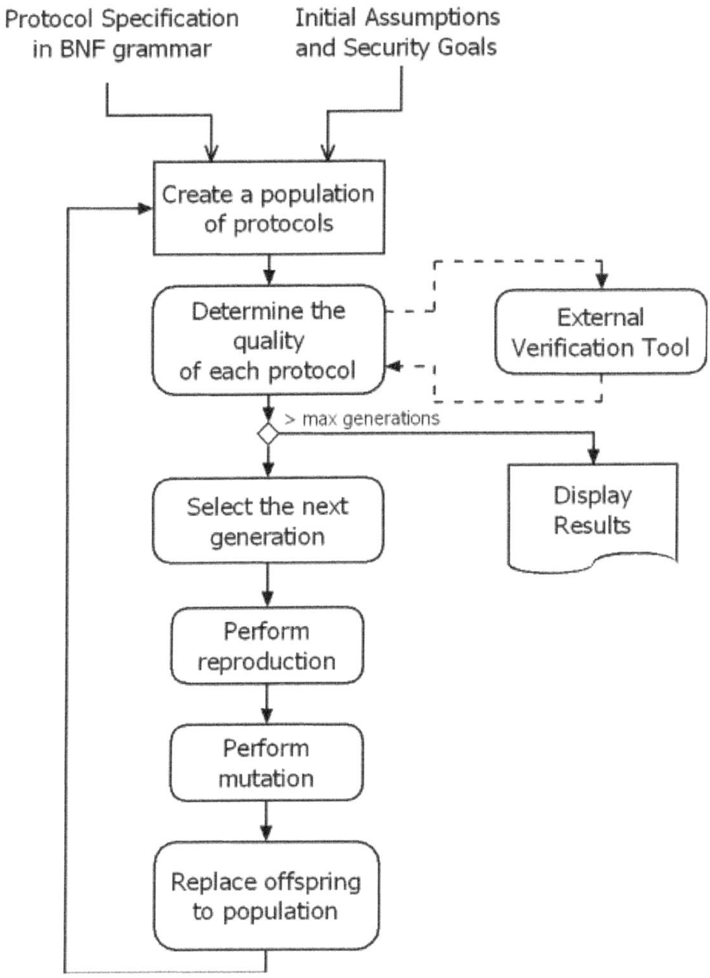

Fig. 3. Evolutionary Algorithm for Security Protocol Design

For the grammar producing the protocol messages, the sets are defined as follows:

```
N = {start, message, next, crypt, genkey, nonce, sub}

T = {send, :, ,, {, }, key, N, N-1, a, b, s}

S = <start>

P :

          start → sub send sub : message
```

```
message → sub next
        | nonce next
        | crypt next
        | genkey next
next → , message | ε
crypt → { message } genkey
genkey → key sub sub | key sub
nonce → N sub | N-1 sub
sub → a | b | s
```

The intelligent generator creates a random population of protocols using the BNF form (Backus Naur Form) of context free grammar. The generated protocols are valid, so they are executable. The generator considers the security of channels and properly maintains the sets of knowledge for each subject: the subject is able to send only messages stored in his set of knowledge.

In this research we use the presented grammar in our examples for creating common security protocols.

Evaluation

In this step, the evaluation of each protocol is performed. The fitness value depends highly on the presumptions and satisfaction in each state of the protocol run. The employment of some additional verification tools for finding certain flaws might be helpful. To calculate the fitness of solutions - generated security protocols - the instruction sequences must be traced to simulate their outcomes. In security protocols, we use various measures to quantify the quality of a generated protocol. The higher is the fitness, the less security flaws exist and the more initial presumptions are satisfied.

The fitness is computed (implicitly) from the following parameters:

- required / received knowledge,
- required / received belief,
- protocol length,
- duplicate instructions,
- forbidden instructions,
- invalid instructions

Selection

Like in standard genetic algorithms, the individuals with the best fitness are stochastically chosen to be parents for mating. For the proposed approach we have tested various selection operators for protocols. The best seemed to be the roulette wheel selection.

Crossover

The choice of the right crossover operator and locations in chromosomes is very important. It may highly affect the chances for generating a chromosome with better fitness. The basic idea for mating is that two chromosomes may cross at selected states (instructions) if both have corresponding sets of knowledge and belief. This means that we have to prove that after crossing the instruction strings, the rest of protocol makes sense for both individuals.

Replacement with offspring

The produced individuals are replaced to the new population and the evolution process starts over again from the step "Choosing parents".

The whole design is finished when some individuals (with best fitness) satisfy the initial presumptions or the maximum number of generation is reached.

Finishing the evolution

The whole design is automated and evolves according to the algorithm specified above. The result is the chromosome with the best fitness (that mostly satisfies security requirements), which can be interpreted as a sequence of instructions in the security protocol. The final optimization is performed among the resulting solutions. This means generally that the duplicate instructions (instruction that perform the similar operations) are removed.

Although we may obtain during the evolution solution with the minimum protocol length (as the length parameter is considered in the fitness calculation too), the minimization would be performed during a very long process of evolution and thus we could obtain the protocols not clearly minimized. This is the main reason for performing the final optimization separately.

In practice, it is valuable to consider as the result the top-n protocols (e.g. $n = 10$) and verify them using some external verification tools. This is due to the principles of evolutionary approach, where the solution with the best fitness isn't always the best solution in the real-world conditions. Although the solution may have the best calculated fitness, it might contain several security flaws.

4.3 Summary

As we can see, the process of protocol design can be highly automated. The use of this approach is desirable especially in the design of complex protocols where e.g. many subjects are involved.

The crossover and mutation operators have to be modified for generating simple security protocols as well as the process protocols evaluation process. The usability for more complex protocols requires addition of heuristic strategies (e.g. implemented intelligent instruction injection). Although the security requirements are specified at the beginning of the automatic generation, the use of additional verification techniques might be useful to check other security properties and the protocols correctness.

5 Conclusions

There are many areas of practical applications of security protocols. These protocols are used in all the communication between subjects, where the authentication process or key establishment is needed. The security protocols are widely used in the computer network communication (e.g. SSL protocol) where they serve for establishing the encrypted communication between server and client (commonly used in the area of electronic banking – e.g. client and bank). The other usability is in the area of embedded systems or electronic commerce.

Acknowledgement. The research has been supported by the Czech Ministry of Education in frame of the Research Intention MSM 0021630528: Security-Oriented Research in Information Technology, MSM 0021630503 MIKROSYN: New Trends in Microelectronic Systems and Nanotechnologies, and by the Grant Agency of the Czech Republic through the grant GACR 102/08/1429: Safety and security of networked embedded system applications.

References

1. Abadi, M., Needham, R.: Prudent Engineering Practice for Cryptographic Protocols. In: Proceedings of the 1994 IEEE Symposium on Security and Privacy, pp. 122–136. IEEE Computer Security Press, Los Alamitos (1994)
2. Abadi, M., Tuttle, N.: A Semantic for a Logic of Authentication. In: Proceedings of the ACM Symposium on Principles of Distributed Computing, pp. 201–216 (1991)
3. Agray, N., van der Hoek, W., de Vink, E.: On BAN Logics for Industrial Security protocols. In: CCEMAS, p. 8 (2001)
4. Burrows, M., Abadi, M., Needham, R.: A Logic of Authentication. ACM Transactions on Computer Systems 8(1), 18–36 (1990)
5. Gritzalis, S.: Security protocols over open networks and distributed systems: Formal methods for their Analysis, Design and Verification. University of Aegean, Greece, Computer Communications 22(8), 695–707 (1999)
6. Ma, L., Tsai, J.: Formal Verification Techniques for Computer Communication Security Protocols. In: Handbook of Software Engineering and Knowledge Engineering, p. 23. World Scientific Publishing Company, Singapore (2000)
7. Očenášek, P., Trchalík, R.: Modal Logics Used for Authentication Protocols Analysis: Survey and Comparison. In: Proceedings of the 7th International Carpathian Control Conference, Ostrava, CZ, VŠB-Technical University of Ostrava, pp. 401–404 (2006)
8. Očenášek, P., Očenášek, J.: Designing Secure Communications Using Evolutionary Approach. In: Genetic and Evolutionary Computation Conference GECCO 2006, Seattle, WA, US. ACM, New York (2006)
9. Očenášek, P.: Towards Selected Problems in the Security Protocol Design and Verification. In: 1st Doctoral Workshop on Mathematical and Engineering Methods in Computer Science, MEMICS, Znojmo, CZ (2005)
10. Shmatikov, V., Stern, U.: Efficient Finite-State Analysis for Large Security Protocols. In: Proceedings of the 11th IEEE Computer Security Foundation Workshop, p. 10. IEEE Computer Society Press, Los Alamitos (1998)
11. Alves-Foss, J., Soule, T.: A weakest precondition calculus for analysis of cryptographic protocols. In: DIMACS Workshop on Design and Formal Verification of Crypto Protocols (1997)

Towards Security Issues in ZigBee Architecture

Pavel Ocenasek

Brno University of Technology, FIT, Bozetechova 2, 612 66 Brno, Czech Republic
ocenaspa@fit.vutbr.cz

Abstract. This paper discusses selected security issues of the Zigbee/802.15.4 standard. The first part gives an introduction to the 802.15.4 standard and describes the security of the two lowest layers (MAC, PHY). The second part focuses on ZigBee architecture and its security functions. The last part outlines security problems in the specified standards, the network vulnerabilities and tries to draw new research topics in the area of sensor networks security.

Keywords: Sensor Networks, ZigBee, Security, Attack, Security Protocol.

1 Introduction

A wireless sensor network [3] [4] is a network composed of a large numbers of sensors that (1) are psysically small, (2) communicate wirelessly among each other, and (3) are deployed without prior knowledge of the network topology. Due to the limitation of their physical size, the sensors tend to have storage space, energy suuply, and communication bandwidth so limited that every possible means of reducting the usage of resources is aggressively sought.

2 Sensor Networks

The 802.15.4 specification [3] for sensor networks defines the first two layers of standard OSI model:

1. Physical (PHY) layer
2. Medium access control (MAC) layer.

In this chapter we focus on security services provided by this standard, especially the second (link, MAC) layer.

The MAC layer security protocol provides four basic security services: access control, message integrity, message confidentiality, and reply protection.

Access control and message integrity. Access control means the link layer protocol should prevent unauthorized parties from participating in the network. Legitimate nodes should be able to detect messages from unauthorized nodes and reject them. Also, a secure network should provide message integrity protection: if an adversary modies a message from an authorized sender while the message is in transit, the receiver should be able to detect this tampering. Including a message authentication code (MAC) with eachpacket provides message authentication and integrity. A MAC

M.J. Smith and G. Salvendy (Eds.): Human Interface, Part I, HCII 2009, LNCS 5617, pp. 587–593, 2009.

can be viewed as a cryptographically secure checksum of a message. Computing it requires authorized senders and receivers to share a secret cryptographic key, and this key is part of the input to the computation. The sender computes the MAC over the packet with the secret key and includes the MAC with the packet. A receiver sharing the same secret key recomputes the MAC and compares it with the MAC in the packet. The receiver accepts the packet if they are equal, and rejects it otherwise. Message authentication codes must be hard to forge without the secret key. Consequently, if an adversary alters a valid message or injects a bogus message, she will not be able to compute the corresponding MAC, and authorized receivers will reject these forged messages.

Confidentiality. Condentiality means keeping information secret from unauthorized parties. It is typically achieved with encryption. Preferably, an encryption scheme should not only prevent message recovery, but also prevent adversaries from learning even partial information about the messages that have been encrypted.

One implication of semantic security is that encrypting the same plaintext two times should give two different ciphertexts. If the encryption process is identical for two invocations on the same message, then semantic security is clearly violated: the resulting ciphertexts are identical. A common technique for achieving semantic security is to use a unique nonce for each invocation of the encryption algorithm. A nonce can be thought of as a side input to the encryption algorithm. The main purpose of a nonce [6] is to add variation to the encryption process when there is little variation in the set of messages. Since the receiver must use the nonce to decrypt messages, the security of most encryption schemes do not rely on nonces being secret. Nonces are typically sent in the clear and are included in the same packet with the encrypted data.

Replay Protection. An adversary that eavesdrops on a legitimate message sent between two authorized nodes and replays it at some later time engages in a replay attack. Since the message originated from an authorized sender it will have a valid MAC, so the receiver will accept it again.

Replay protection prevents these types of attacks. The sender typically assigns a monotonically increasing sequence number to each packet and the receiver rejects packets with smaller sequence numbers than it has already seen.

3 ZigBee Architecture

The layers defined by the 802.15.4 standard:
1. Physical (PHY) layer
2. Medium access control (MAC) layer.

The Zigbee standard is specified in [1] and defines upper layers, such as:
3. Network (NWK) Layer
4. Application Layer (APL)

The architecture of all the defined ZigBee layers is presented in Figure 1.

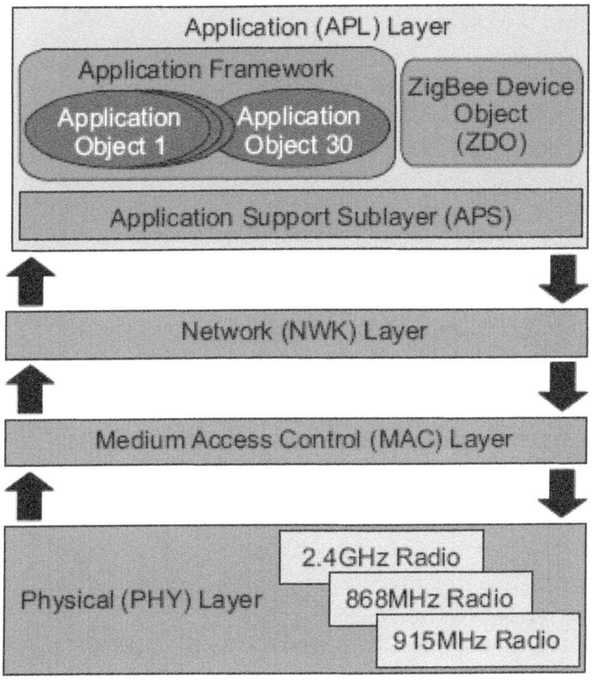

Fig. 1. Zigbee architecture

3.1 Security

The ZigBee standard and its security specification deals with some basic security protections.

- ZigBee provides Freshness
- ZigBee provides Message Integrity
- ZigBee provides Authentication
- ZigBee provides Encryption

ZigBee could add headers to the data frames at the MAC, NWK, and APS layers (see Figure 2).

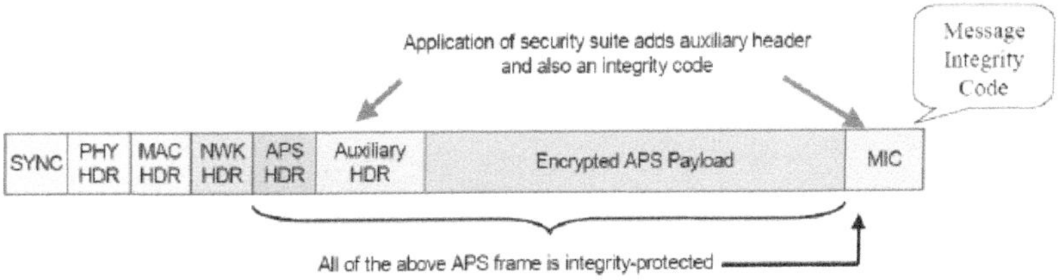

Fig. 2. Zigbee frames with Security

The ZigBee standard deals with two operation modes:

- Residential mode (RM)
- Commercial mode (CM)

Residential mode: The secure wireless network (see Figure 3) that can be installed and mantained by a homeowner with no knowledge of security. The security is transparent during the setup process and the network must still provide best security possible.

The homeowners take no active role in mantaining security of network.

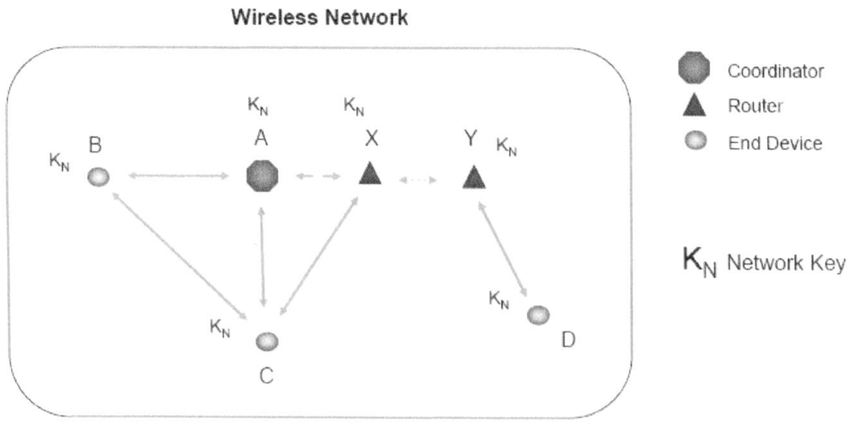

Fig. 3. Residential Mode

In this mode, the minimum number of keys/storage is used for low cost. Residential Trust center only needs to store network key. This cause the storage minimization, also the low capability device can act as trust center and trust center can be easily replaced with another device without homeowner intervention.

In this mode, all the devices require network key and frame counters. The network provides only network level authentication, integrity and encryption protection, so it is vulnerable to insider attacks.

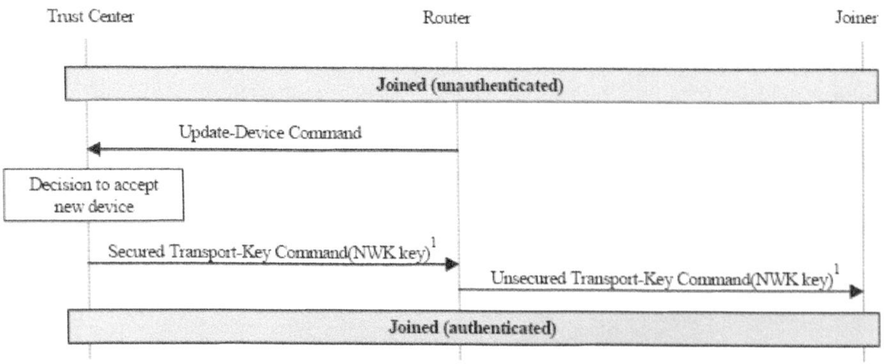

Fig. 4. Example of RM authentication procedure

Commercial mode: Wireless network (see Figure 5) is controlling mission critical applications (alarm, production monitoring and control, ...). A wireless network is actively monitored and mantained (sheduled key updates, controlled adition of new devices, revocation of discarded devices).

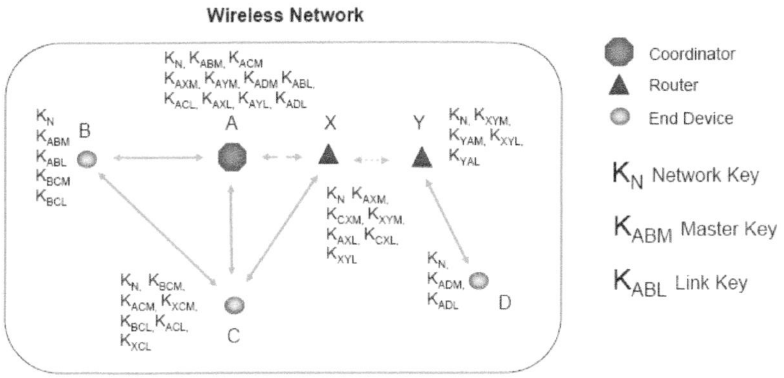

Fig. 5. Commercial Mode

In this mode, the Trust center should only admit new devices when manually enabled. This prevents unauthorized devices from joining the network. The Trust center should also update network key for legitimate the devices periodically. Network key should only be used by the network layer. This prevents attackers from using network key to control devices.

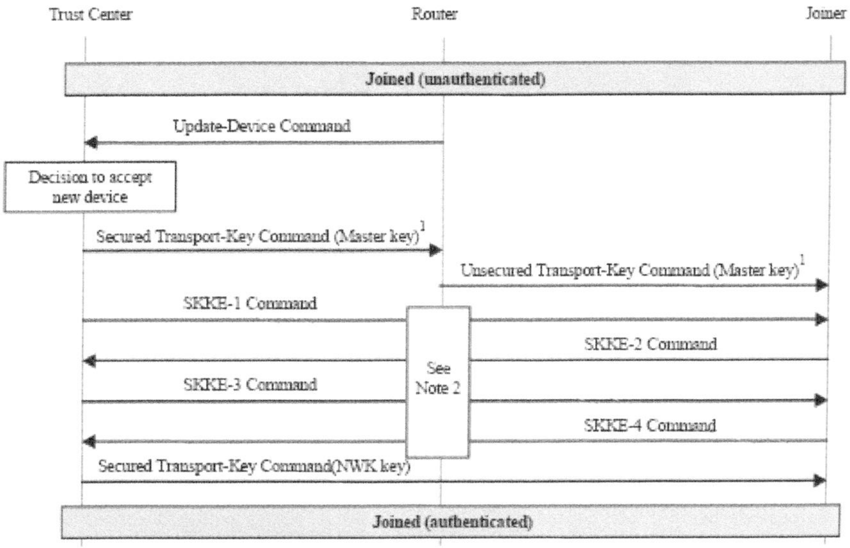

Fig. 6. Example of CM authentication procedure

3.2 Security Problems

The security issues of ZigBee/802.15.4 standard [1] [2] [5] could be generally divided into the following cathegories:

IV Management Problems
- Same Key in Multiple ACL-Entries
- There are 255 different ACL entries used to store different keys and their associated nonce. However, there is a bug if the same key is used in two different ACL entries.
- Power Interruptions Causes Loss of ACL State
- The node will emerge with a cleared ACL table when power is restored.

Insufficient Integrity Protection
- Unauthenticated Encryption Modes
- Denial-of-service Attacks on AES/CTR
- Acknowledgment Packets: No Integrity

Key Management Problems
- No Support for Group Keying
- Network Shared Keying Incompatible with Replay Protection
- Pairwise Keying Inadequately Supported

The source of the difficulties with the key management problems is partly a result of confusing the role of a nonce [6] and a replay counter. The nonce sent in outgoing packets serves two purposes: it provides a non-repeating value that protects confidentiality; and, it provides a monotonically increasing counter that prevents replay attacks. To protect confidentiality, the sender must ensure that it never uses the same nonce twice for the same key.

4 Recommendation

When designing/implementing the standard for communication in sensor networks, we recommend following the robustness principles:

- be very clear about the security goals and assumptions,
- be clear about the purpose of an encryption (secrecy, authenticity, etc.): do not assume that its use is synonymous with security
- be careful that your protocol does not make some unexamined assumption about the properties of the underlying cryptographic algorithm,
- be sure to distinguish different protocol runs from each other,
- do not assume that a message you receive has only a particular form, even if you can check this,
- if timestamps are used as freshness guarantees by reference to absolute time, then the difference between local clocks at various machines must be less than the

allowable age of message deemed to be valid; furthermore, the time maintenance mechanism everywhere becomes part of the Trusted Computing Base,

- where the identity of a subject is essential to the meaning of a message, it should be mentioned explicitly in the message,
- sign before encrypting; if a signature is affixed to encrypted data, then one cannot assume that the signer has any knowledge of the data; a third party certainly cannot assume that the signature is authentic, so non-repudiation is lost.

5 Conclusions

The new ZigBee standard is progressive wireless communication standard, targeting automation and control applications. It provides reliable data transmission at low rates, with very low power consumption at very low-end device cost.

This paper described the security principles used in the ZigBee/802.15.4 standards and pointed out the actual security weaknesses of this architecture.

Acknowledgement

The research has been supported by the Czech Ministry of Education in frame of the Research Intention MSM 0021630528: Security-Oriented Research in Information Technology, MSM 0021630503 MIKROSYN: New Trends in Microelectronic Systems and Nanotechnologies, and by the Grant Agency of the Czech Republic through the grant GACR 102/08/1429: Safety and security of networked embedded system applications.

References

1. ZigBee Alliance: ZigBee Specification v 1.0. ZigBee Alliance Board of Directors (2004), `http://www.zigbee.org/`
2. ZigBee Alliance: ZigBee Security Specification Overview (2005), `http://www.zigbee.org/`
3. Wireless medium access control and physical layer specications for low-rate wireless personal area networks. IEEE Standard, 802.15.4-2003 (2003) ISBN 0-7381-3677-5
4. Law, Y.W., Doumen, J., Hartel, P.: Survey and benchmarks of block ciphers for wireless sensor networks. ACM Transaction on Sensor Networks 2(1), 65–93 (2006)
5. Sastry, N., Wagner, D.: Security considerations for IEEE 802.15.4 networks. In: Proceedings of the 2004 ACM workshop on Wireless security WiSe, pp. 32–42 (2004)
6. Ocenasek, P.: Security Protocols: Analysis and Design. In: Proceedings of the Junior Scientists Conference 2006, Vienna, AT, TUV, pp. 51–52 (2006)
7. Očenášek, P., Trchalík, R.: Reasoning About Security Protocols in the ZigBee Standard. In: Proceedings of the 8th International Carpathian Control Conference, Košice, SK, TU v Košiciach (2007)
8. Švéda, M., Trchalík, R., Očenášek, P.: Design of Networked Embedded Systems: An Approach for Safety and Security. In: Preprints of IFAC Workshop on Programmable Devices and Embedded Systems PDeS, Ostrava, CZ, IFAC, pp. 131–136 (2009)

A Study on Color Conversion for Color Deficient People to Identify Color

Makoto Oka[1], Naoki Ozawa[2], Hirohiko Mori[1], and Akito Sakurai[2]

[1] Musashi Institute of Technology, 1-28-1, Tamadutumi, Setagaya-ku Tokyo
[2] Keio University, 3-14-1 Hiyoshi, Kohoku-ku,Yokohama, Japan
oka@ergo.ie.musashi-tech.ac.jp, o_naoki@ae.keio.ac.jp,
mori@si.musashi-tech.ac.jp, sakurai@ae.keio.ac.jp

Abstract. The sensitivity of the M-cones and L-cones of an anomalous trichromat is lower than that of normal trichromats. By intensifying the light to the cones with lower sensitivity, the output ratio from the cone approached the level of a normal trichromat, and it was presumed that achieving close to original color recognition was a possibility. In order to increase luminous intensity, in the LMS space, the boundary was set as the plane surface representing pseudoisochromatic color, and the distance from that plane was expanded. When making large changes in light intensity near the surface representing pseudoisochromatic color, improvements in color differentiation and degree of color identification were observed. We plan to implement this method in mobile devices and to evaluate the experimentation results.

1 Introduction

In recent years, large amounts of information containing color information are being transmitted. In many cases, this color information contains important messages or brings attention to the importance of information. The color-blind are unable to receive color information accurately. 5% of Japanese males (around 3 million), and 8% of Caucasian males, have what is called "red-green color blindness," in which color differentiation becomes difficult for color ranges containing red or green [3].

What are the options to address the issues arising from color-blindness? One is for those transmitting information to take action in consideration of those who are color blind. The other is for the color-blind to take action on the receiving end. This paper deals with the latter option. Methods to increase color differentiation for anomalous trichromats will be considered, as well as observation of changes in perception of images occurring as a result, and methods to decrease the difficulty of color identification.

2 Background

Color recognition, the process of how color-blindness develops, and the simulation method for color-blind image perception will be explained.

M.J. Smith and G. Salvendy (Eds.): Human Interface, Part I, HCII 2009, LNCS 5617, pp. 594–600, 2009.

2.1 Color Recognition

Color is not a quality of matter, but a sensation created by the brain based on the wavelength and intensity information received by the eye. Light in the range of 360nm - 830nm can be recognized as color. Light with varying wavelengths are recognized as different colors. Light at 540nm is recognized as the color green, 580nm yellow, and 660nm as red. When light at 540nm and at 660nm is mixed, the light is recognized as yellow. Our recognition is not distinguishing differences in physical properties of light [2] [4].

Photoreceptor cells are divided into rods and cones. Depending on the intensity of available light, rods mainly function in relatively low light environments, and cones function mainly in brighter environments. Cones are divided into S (Short) cones, M (Middle) cones, and L (Long) cones. The wavelength in which each type of cone responds is different and depends on the quality of the visual pigment inside the cone. S-cone (blue cone) has blue visual pigment (absorption maximum wavelength 419nm), M-cone (green cone) has green visual pigment (absorption maximum wavelength 531nm), and L-cone (red cone) has red visual pigment (absorption maximum wavelength 558nm). Each cone responds through the visual pigments depending on the wavelength element of the light entering the eye.

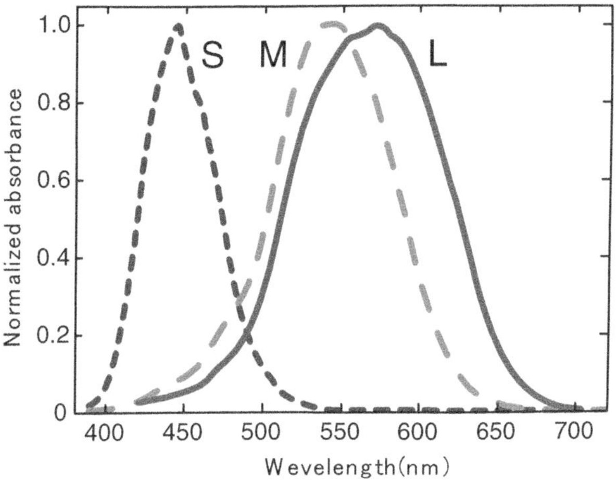

Fig. 1. Normalized responsively spectra of human cone cells, S, M, and L types

2.2 The Process of Color-Blindness Development

When mutation occurs in the visual pigment genes, visual pigments cease to exist or the qualities of visual pigments change radically. This results in dichromacy or anomalous trichromacy.

Mutation of red visual pigment is called protonomaly; mutation of green visual pigment deutronomaly, and mutation of blue visual pigment is called tritanomaly. Out of Japanese males, 1.5% has protonomaly, 3.5% has deutronomaly, and 0.001% has tritanomaly.

The red and green visual pigments have overlapping absorption spectra. If there is a problem in either, color differentiation becomes difficult in the red through green range, and thus is called "red-green color-blindness." In third color-blindness, the color difference in the yellow through blue wavelength range becomes difficult to recognize and thus is called "blue-yellow color-blindness."

2.3 Color-Blindness Simulation

The method used on normal trichromats to simulate color-blindness is explained [1] [5].

The RGB value of each pixel in an image is converted to the amount of LMS stimulation received by the 3 types of cones. After compensation, the RGB value is re-calculated to create the compensated image.

Equation 1 is color stimuli Q.

$$Q = (L_Q, M_Q, S_Q) \tag{1}$$

Color stimuli Q are determined by the amount of each wavelength included in the original light. The original light spectra distribution is expressed as $\varphi_q(\lambda)$, the spectral sensitivity function of each cone as $\bar{l}(\lambda)$, $\bar{m}(\lambda)$, $\bar{s}(\lambda)$, k is a constant determined to complete Eq. 2. Color stimuli Q are expressed in Eq. 3.

$$L_Q - M_Q = 1 \tag{2}$$

$$\begin{aligned} L_Q &= k\int \varphi_Q(\lambda)\bar{l}(\lambda)d\lambda \\ M_Q &= k\int \varphi_Q(\lambda)\bar{m}(\lambda)d\lambda \\ S_Q &= k\int \varphi_Q(\lambda)\bar{s}(\lambda)d\lambda \end{aligned} \tag{3}$$

P_R, P_G and P_B are the colors produced when RGB values are maximized on a monitor. The $\varphi_R(\lambda)$, $\varphi_G(\lambda)$ and $\varphi_B(\lambda)$ are measured at that time, and L_i, M_i and S_i $(i = R, G, B)$ are found with Eq. 3. Table 1 shows the results.

Table 1. LMS tristimulus values for the red, green, and blue primaries

	$i = R$	$i = G$	$i = B$
L_i	0.1992	0.4112	0.0742
M_i	0.0353	0.2226	0.0574
S_i	0.0185	0.1231	1.3550

Q, the color displayed on the monitor is expressed in Eq. 4.

$$Q = (R_Q P_R + G_Q P_G + B_Q P_B) \tag{4}$$

R_Q, G_Q and B_Q are a number greater than 0 and less than 1.

The method of converting RGB values to stimulation values LMS are shown below. The original color V is Eq. 5.

$$V = \begin{pmatrix} R_Q \\ G_Q \\ B_Q \end{pmatrix} \tag{5}$$

Eq. 6 is the conversion matrix T using the values in table 1.

$$T = \begin{pmatrix} L_R & L_G & L_B \\ M_R & M_G & M_B \\ S_R & S_G & S_B \end{pmatrix} \tag{6}$$

Q is expressed in Eq. 7.

$$Q = TV \tag{7}$$

Q is adjusted according to the color-blind user as Q'. Using Eq. 8, V' corresponding to the color stimuli Q' recognized by the color-blind user is found. V' is used to create the compensated image.

$$V' = T^{-1}Q' \tag{8}$$

3 Proposal

Anomalous trichromacy is between normal trichromacy and dichromacy. Anomalous trichromats have lower M and L-cone sensitivity compared to normal trichromats. Only the light wavelengths responding to M or L-cone from the original image are intensified. By intensifying the light wavelengths where sensitivity was low, the reaction ratio of the cone approaches that of normal trichromats, and possibly causes color recognition similar to that of normal trichromats.

Using the method stated in the color-blindness simulation, the RGB of each pixel is converted to LMS and intensified in the axial L and M direction. The boundary was set as the plane surface representing pseudoisochromatic color, and the coordinates greater in comparison were emphasized more, while coordinates smaller in comparison were emphasized less. The overall distance between each coordinate and the plane surface representing pseudoisochromatic color is made larger. By doing so, it is presumed that the color recognition of the anomalous trichromat can be changed from a state close to dichromacy, to that of a normal trichromat.

3.1 The Emphasis Range

The method used for color-blindness simulation is used for the given color stimuli Q, the point on the planar surface representing pseudoisochromatic color is found, and this value is set as 0. If intensifying in the M axis direction, L and S are fixed while M is changed. If intensifying in the L axis direction, M and S are fixed while L is changed. If in the M axis direction, the largest M value is set as 1 in the reproducible monitor range, and if in the L axis direction, the largest L value is set as 1. Normalization is conducted so that the plane surface representing pseudoisochromatic color becomes 0, and the largest reproducible value is 1. Normalization is also conducted so that the smallest reproducible value becomes 1.

The plane surface representing pseudoisochromatic color spreads beyond the reproducible monitor range. It is possible that when intensifying in the M axis direction from a given value, the result is beyond the reproducible range. When this happens, given that the plane surface representing pseudoisochromatic color is 0, there are 2 possibilities. Either, both the largest possible value and the smallest value of the reproducible range are both lined up greater than 0, or, conversely, they are both lined up less than 0. In this situation, the one farther from the plane surface representing pseudoisochromatic color is set as 1. The distance between the ignored point and the plane surface representing pseudoisochromatic color is an irreproducible range, and thus it is impossible for that value to be obtained from the original image.

3.2 Calculation Method of Color Stimuli Q'

LMS is converted to find Q'. The 3 conversion methods are shown below.

(a) The original large-small relationship is maintained in the axis direction.
(b) Changes are not made at the 0 point.
(c) Containment within the monitor reproducible range.

In this paper, the functions shown below Eq. 9 are used to find the emphasis range.

$$f(x) = \begin{cases} (1+20a)x & (0 \le x \le 0.05) \\ x+a & (0.05 \le x \le 0.95) \\ (x+20a)/(1+20a) & (0.95 \le x \le 1 \end{cases} \tag{9}$$

Eq. 9 is used to find Q' from Q.

4 Experiment

Color blind subjects were used for the test. The test subjects were shown the 1 original image and converted image, and color recognition was tested.

4.1 Experiment

Differentiation and impression was tested. The test subjects were shown a photo, image converted from an original image. The following were tested for the test subjects.

- Ease of color differentiation
- Whether the converted image retains the impression of the original image

(a) Original Image

(b) Converted Image

Fig. 2. The original image and the color converted image

4.2 Results of Experiment

- Dichromats did not feel any difference between the converted image and the original image.
- Anomalous trichromats felt an improvement in ease of color differentiation for converted image.

5 Discussion

The effectiveness of the proposed method will be discussed based on the results.

5.1 Difference in Effectiveness, by Type of Color Blindness

The proposed method was intended for anomalous trichromats and the possibility of there being no positive effects on dichromats was expected. Based on the results from experiment 3, the method is ineffective for dichromacy.

In the survey answers from dichromats, some answers suggest that they recognized the changes between the converted image and the original image. Further study is needed to verify these claims.

From the results of experiment, the method had positive effects for anomalous trichromats. However, further study is needed regarding the conversion methods.

5.2 Evaluation of the Conversion Method

By using the proposed method, anomalous trichromats were able to feel a difference in the color of the original image and the converted image.

Three test subjects felt an improvement in both ease of color differentiation and color identification for converted image.

6 Application for Mobile Devices

Recently, mobile devices are often equipped with a digital camera and color display. We are implementing technology that would allow the color blind user to take a picture of the image he would like to see, convert the image, and view it on the display of his mobile device. We plan to evaluate the color conversion application implemented on a mobile device.

7 Conclusion

The cone of the anomalous trichromat has low sensitivity; however, by intensifying the color stimuli, there was a response, like a normal trichromat cone. As a result, improvements were observed in ease of color differentiation and level of color differentiation.

8 Future Issues

There are also large differences in the cones of individuals. Especially in anomalous trichromacy, there is great variation, and this can possibly result in minute variations of recognition. Therefore, genetic variance needs to be considered. A system where the color blind can choose parameters that meet their needs, or a system that adapts to the needs of the color blind is needed.

References

1. Brettel, H.: Computerized simulation of color appearance for dichromats. Optical Society of America 14(10), 2647–2655 (1997)
2. Stockman, A., MacLeod, D.I.A., Johnson, N.E.: Spectral sensitivities of the human cones. Optical Society of America 10(12), 2491–2521 (1993)
3. Okabe, M., Ito, K.: Color Universal Design (CUD) - How to make figures and presentations that are friendly to Colorblind people, `http://jfly.iam.u-tokyo.ac.jp/color/index.html`
4. Colour and Vision Research Labs, Colour & Vision database, `http://cvrl.ioo.ucl.ac.uk/`
5. Vischeck, Vischeck, `http://vischeck.com/vischeck/vischeckImage.php`

Widgets for Faceted Browsing

Jan Polowinski

Technische Universität Dresden, Nöthnitzerstr. 46, 01187 Dresden, Germany
jan.polowinski@tu-dresden.de

Abstract. Faceted Browsing, also known as Faceted Search, recently became a popular interface paradigm and is used as a visual selection mechanism to browse data collections. It is particularly suitable for structured, but heterogeneous data with explicit semantics. As the amount of this data will grow in respect to a rising Semantic Web, it is worth looking at the current state of Faceted Browsers. In this paper we extract characteristics of several browsers from various domains. We then propose the use of advanced widgets for Faceted Browsers and how to select appropriate widgets.

Keywords: Faceted Browsing, Faceted Search, Dynamic Taxonomies, User Interface, Semantic Web, Facet Widgets.

1 Introduction

A Faceted Browser can be used to explore structured data sources such as media collections with annotated metadata, data stored as part of a Semantic Web [3] or data from Semantic Desktops [25]. Faceted Browsing can also be seen as a means to construct complex queries without writing them by hand [8]. In contrast to visual query languages [7], Faceted Browsing does not return the constructed query explicitly. Instead it performs an incremental refinement of a set of results by selecting values of the data's facets that are turned into restrictions on the data set. The choices for creating new restrictions are adjusted, so that, by construction, only queries leading to a non-empty result set can be created. The paradigm is based on the librarian principle of Faceted Classification [21] which is difficult to implement in the physical world, but easy to do in a virtual world. Therefore it became popular the last years although often being implemented in a relaxed version (Sect. 2). In advantage to a fix predetermined taxonomic order [24] multiple navigations path exist and allow the user to browse task-oriented. Another point is that the user does not need to know what exactly she is looking for and recalling the exact terms for searching is not required. Thereby Faceted Classification is well suited for a process of exploration. Recently many Faceted Browsers have been implemented that are based on the idea of Faceted Classification, some of them following the early and famous Flamenco [9] browser. Additionally pragmatic facet-like systems have wide spread in the field of e-commerce. Now it is time to compare those approaches in order to find possibilities for further improvement of these browsers.

The rest of the paper is structured as follows: In Sect. 2 we will first describe commonalities of Faceted Browsers. In Sect. 3 we will then extract characteristic

M.J. Smith and G. Salvendy (Eds.): Human Interface, Part I, HCII 2009, LNCS 5617, pp. 601–610, 2009.

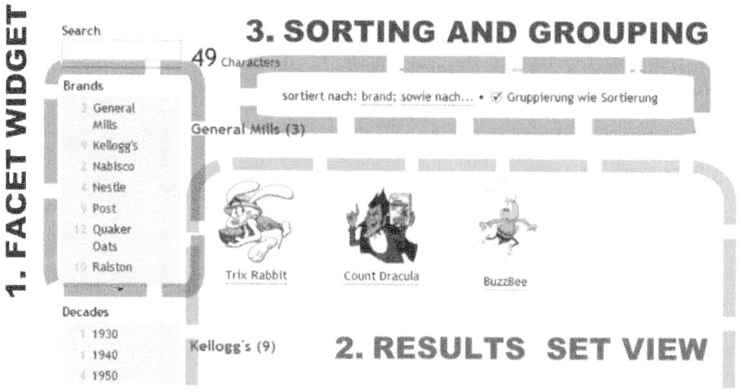

Fig. 1. GUI Elements of Faceted Browsers

features of existing Faceted Browsers, divided into the fields of filtering and displaying results. In Sect. 4 we will propose a reflection of the facets structure to achieve better widgets.

2 Commonalities

In the context of Faceted Browsing we distinguish between *facets*, *facet values*, *restrictions* and *categories*. Every property of the artifacts in focus is seen as a facet in a Faceted Browser. For all items in the system this property may have different facet values. Priss mentions that according to Ranganathans original definition of Faceted Classification [21] an item can only have one value per facet. This is often not the case for many facet-like systems in computer science [22]. Additionally the properties are not even required to be fully orthogonal to be called a facet [13].

Putting a restriction to a facet *A* can be seen as a logical expression, reducing the set of items to those that have special facet values for facet *A*. Finally, the term *category* is used in the following sections for a special facet that is usually hierarchically structured by a *subClassOf* relationship and serves as a hierarchical backbone that already classifies a domain in one possible way.

GUI Elements. Faceted Browsers offer both, a visualization of the selection mechanism and a means to visualize the data itself. Therefore the user interface of Faceted Browsers (Fig. 1) consists of: 1) widgets for each facet to select the facets values, 2) a view of the currently selected result set, 3) means to sort and possibly group the results. Additionally many browsers visualize the applied restrictions, i.e., the constructed query, as "breadcrumbs".

Instant Update of Facets and Results. Common to the majority of all Faceted Browsers is the fact that the facet values the user has to choose from are instantly restricted to the possibilities which do not lead to an empty result set. This guarantees that a search always returns at least one instance. While most browsers completely remove facet values that lead to an empty result set, another approach is to only gray out those values. This allows the user to get an impression of items she might be able to indirectly navigate to [15].

3 Characteristics

We extracted a set of characteristics to compare Faceted Browsers. These can be divided into general characteristics referring to a browser as a whole (Sect. 3.1), characteristics regarding the filtering process with the facet widgets (Sect. 3.2) and the presentation of the results (Sect. 3.3).

3.1 General Characteristics

Domain Independence (without Configuration). Faceted Browsers can be divided into the group of domain-independent general approaches and domain-specific ones, which are usually existing applications that have been extended with Faceted Browsing capabilities.

Facet Set Context Depending. Especially if the domain is very broad and many classes are included, it is beneficial to show only a subset of facets. Very specific ones, applying to only a few items, can be shown only after certain other conditions are met. This might be, for instance, when a certain class in a hierarchically structured class-facet is selected. To give a concrete example: Amazon [2] displays the facet *resolution* only if the class *camera* was chosen, not right from the beginning. Foundations for selecting a facet set automatically based on facet ranking and metrics are discussed by Oren et. al. [20]. Hearst describes the selection of important facets based on query logs and click logs [13]. Context-dependency cannot only be applied to facets but also to facet values (e.g., eBay [5] does this) and can not only regard the selection of a subset to be displayed, but also the order. When a subset of possibilities is presented it is a visual convention to insert a link labeled *more...* [12].

Arbitrary Restriction Order. Restrictions can usually be added (zoom-in) and cleared again (zoom-out) in any order for all facets. This allows for exploring the data set on multiple paths, depending on the users intention. Other navigation modes such as Shift and Pivot are described in detail in [8].

Facet Set Configurable by the User. This characteristic regards the flexibility of the user interface in terms of changing the set of facets and changing their order. Letting the user choose a set of appropriate facets and put those facets into a convenient order adds to the personalization of the navigation path.

3.2 Facet Widgets

Result Cardinalities as Feedback. The vast majority of Faceted Browsers shows the number of items with a certain value for every value of each facet in the current result set. Consequently, this is also the number of items that will remain in the result set if this value is turned into a new restriction. This way, the cardinality can be seen as a preview and warning that anticipates the next potential restriction step. Stefaner and Müller [28] use color in addition to numbers to visualize the cardinality.

Value Range Selection. Choosing a single value is often undesirable. For some facets, such as time or price, the user often wants to define a range of values between a (often vague) maximum and a minimum. Three different approaches can be found for the selection of ranges: The simplest one is the possibility to select a range by clicking on a start and an end value while pressing a key such as *Ctrl* or *Shift*, or by selecting multiple discrete values with the mouse. However, this approach is not suitable for continuous values. A second possibility is the predefinition of ranges, e.g., price from *10,00 - 20,00 EUR*; *21,00 - 30,00 EUR*. Finally, the most elegant way is the selection of special widgets using sliders, for instance.

Selection of Continuous Values. The ability to define arbitrary values for continuous facets can be realized by text input fields, selection wheels or sliders. Since entering one continuous value alone will most likely not hit any item, the input of continuous values will usually be combined with the definition of ranges. While sliders and input fields in particular are a convenient way to edit restrictions, they make it harder to give feedback on remaining possible facet values.

Multiple Selection. (MS) An alternative to the selection of ranges is the possibility of selecting multiple values of a facet. However, this has its limitations for facets with a large number of values and is only possible for discrete values. MS is possible, for example, in option lists, lists of checkboxes and could also be implemented for Tree Maps [17]. Sacco [24] stresses the importance of building disjunctions of multiple facet values to allow the user to form custom groups spontaneously without needing to select the whole super class[1]. MS is an important feature to allow the construction of complex queries with the help of Faceted Browsing. However, please note that by MS alone, the expressiveness of the constructed queries is still limited to conjunctions of disjunctions such like *(facetA=ValueX OR facetA=ValueY) AND (facetB=ValueZ)*[2]. Since Faceted Classification in computer science usually allows the assignment of multiple values per facet, it would be useful to allow building queries that can represent conjunctions of values within one facet, for instance: "All countries being part of Europe and Asia at the same time." An intuitive graphical representation for this definition could be challenging however. Related is the problem of including objects with no value assigned for a given facet. EBay, has introduced an additional label *not specified* that can be additionally selected to give users the possibility of keeping those objects in the result set.

Free Input. In contrast to the selection of predefined values, the free input of discrete, nominal values (by their string identifier) requires recall. Therefore free input is better suited to the selection of continuous (i.e., alphanumerical) values or ranges. However, advanced auto-completion functions could be a considerable solution to tackle the problem of recall.

Integration of Free String Search. Most browsers allow the combination of faceted search with a search based on string comparison. This can be realized in several flavors. First we can distinguish whether the items (the content) or the facet values (the

[1] Selecting a super class requires that a hierarchical widget is employed (Sect. 4).

[2] Parallax [15] (submitted) is a very recent approach that allows for construction of more complex queries, i.e., related sets of instances.

metadata) are filtered[3]. Here, a second criterion is the scope of search. There are browsers offering only a single search field for all facets together and there are others using one search field for each facet. In the letter case only the facet values that are left for further restriction of this facet are compared to the string.

Integration of Standard Classification. In contrast to an explicit query, all browsers require the use of some predefined static classification, because otherwise, starting with the top concept *Thing*, the set of facets becomes too large for the user interface to be efficiently used. However, multiple variants for this exist: i) The trivial case is when a single class is chosen by the author and all facets apply to this class (not considered in Table 1). This principle is used in domain specific browsers for media collections, for instance, where all objects are media-items. ii) A list of important (domain dependent) classes aka *Startingpoints*, *Primaries*[4] or *Focal Points* [23] is offered as a list or tree, this is an approach some generic browsers take. We can further distinguish when the user selects the class. Either j) she does this on one or more separate screens in the beginning of the exploration process (e.g., e-commerce solutions), or jj) the fixed classification is seen as one facet among others.

3.3 Presentation of the Results

Display Paradigm. The way the result set is represented varies between the browsers and often multiple views are offered. The most common paradigm is a list representation. However, others can be found such as table, map, timeline, grid, thumbnail view.

Sorting of Results Configurable. While nearly all browsers allow for sorting the result set by some facet, not all of them allow the user, or at least the admin, to define the facet for sorting. Another criterion is the possibility to sort by multiple facets with different priority. E.g., it can be filtered first by country then by company. (This is something that most sortable table implementations do not make explicit.)

Grouping of Results. The (sorted) result set can often be additionally grouped into items with the same value for a given facet. When icons are grouped by multiple facets this results in a tree that can either be displayed explicitly or, for instance, by nested boxes. The facet for grouping can be chosen by the user in only a few browsers. Whenever sorting, grouping and a modification of the facet set were configurable this could be done separately for sorting, grouping and filtering.

Search for Similar Resources. When a specific item is displayed in the result view, some browsers allow for searching for similar items directly. This can be done by linking other items with the same values for most of the facets (such as the product recommendations in e-commerce applications). Similar manual navigation modes are referred to as (extensional and conceptual) shift navigation in [8].

4 Advanced Widgets for Faceted Browsing

For filtering, suitable widgets to choose the facet values have to be offered. We propose a selection of different widgets based on the characteristics of the facet to be

[3] There are cases where we have no annotated main content, such as knowledge bases.

[4] http://www.w3.org/2005/04/fresnel-info/manual/#primary

filtered. In contrast, most of the current Faceted Browsers use the same widgets for all facet types and rarely has been paid attention to the fact that the facets have a structure besides alphabetical order, which can helpfully be reflected by the widget. The standard widget for selecting facet values is a list where the single values can be selected. However, for some data types and purposes other widgets are much more useful. E.g., when supporting the selection of value ranges some (but few) Faceted Browsers offer specific widgets such as sliders or two free input fields for a minimum and a maximum value. Specific widgets could either be chosen automatically, depending on the facets data type, or be defined manually for selected facets.

In the field of statistics, values are categorized into nominal, ordinal and quantitative data representing different scales of information on the relationship between the values. Depending on the category to which the facet values belong to, different widgets are appropriate to reflect this information.

Nominal Data. If there is no relationship between the facet values, except for the fact that they are all different and can be distinguished by their name, the data is (only) nominal and theoretically could be represented as an unordered set of choices. A trivial sorting however can be performed using the alphabetical order of the labels.

Ordinal Data. If there is an order relation between the values, the data can be referred to as ordinal. The topology build by the order relation might be a sequence, a tree or a DAG. As an example, for a facet that has a tree structure, take the facet *takesPlaceIn* defined as an RDF[5] property with a domain *Event*. It could have the class *GeographicalRegion* as a range. For the moment, assume that geographical regions are related to each other via an exclusive *partOf* relationship. Other *partOf* relationships can be found between body parts or composed components. The *subClassOf* relation which is typically used within taxonomies but also others such as *knows* can build a tree structure. The relationship between the facet values that is reflected by the widget does not necessarily have to be the facet relationship itself. A facet widget can be structured by any other relation that fulfils the condition of not leaving the facets range. Consequently, the range and the domain of a potential alternative structuring relation have to be the same and have to equal the facets range (Fig. 2b). The special case of a widget that is structured by the facet itself is shown in Fig. 2a. Sequences can be displayed by slider and option list widgets. Facets that have a hierarchical topology can be displayed by the classical directory tree as already seen in Fig. 2a but could also be displayed by a Tree Map widget as shown in Fig. 2d. There are a few other existing widgets specialized to hierarchical facets: FacetZoom [4] is a novel zooming (facet) widget specialized to the visualization of tree structures and lends itself to be integrated with Faceted Browsers, too. There is a variety of hierarchical facets that could be efficiently displayed with the widget, including *partOf* relationships (e.g., for time spans, geographic regions or body parts). Also, in the code repository of Exhibit [6], some specific widgets such as hierarchical facet widgets (and also sliders for range selection) can be found which have, however, not been released so far. Another distinction in the context of tree navigation is whether the selection of a certain level can be performed with one click, because the whole

---

[5]The examples in this section use the Resource Description Framework (RDF, http://www.w3.org/RDF/) as a data representation format.

hierarchy is explicitly shown in an expanded state, or if the selection is performed by an incremental refinement, level by level. With an explicit representation of the facets hierarchy the user can easily restrict a facet on several levels of concreteness. mSpace [17] uses the second approach when it displays two facet widgets, one for genre and another for subgenre. DAGs can be displayed, either as a graph, or by transformation to a tree by copying values.

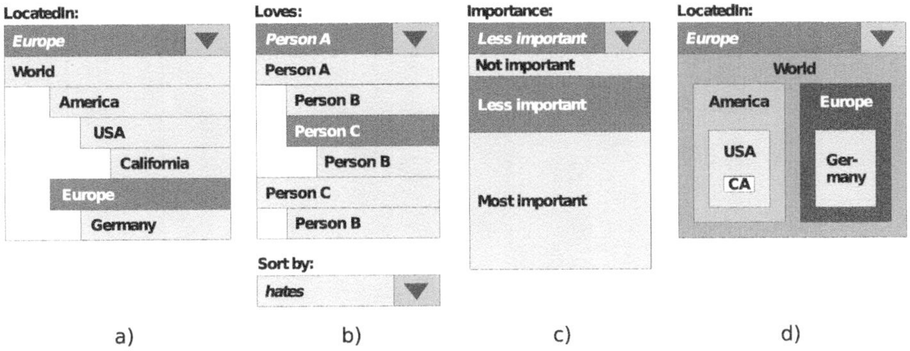

Fig. 2. Selection Widgets for Tree Facets (a), Tree Facets Structured by a Differing Relationship (b), Quantitative Data (c) and Tree Map Facet Widget (d)

Although hierarchical facet navigation already existed in Flamenco [12] and some widgets exist, we argue that widgets depending on the facets structure have not been consequently implemented in many later approaches. Often only typical taxonomical *subClassOf* relations are in the focus (for example in the context of dynamic taxonomies). Data from the semantic web offers explicit semantics and various relations. In [1] this knowledge is exploited for additional navigation modes. But it can also be used to construct advanced widgets representing this structure which might be flat (ordered or unordered) and also a hierarchy or a DAG.

Quantitative Data. provides information on the ratio of values or at least on the ratio of differences between pairs of values (depending on the fact if *Null* was chosen in a meaningful way, or not). This quantitative information can be reflected by the area that is consumed by the entry for the value in the list, for example (Fig. 2c). A Tree Map is able to provide quantitative information by the proportions of areas and also a Pie Chart could be used as a widget. The Option List Widget can also be extended to represent quantitative information by using different row heights for each value to reflect the ratio to other values (Fig. 2d). Stefaner and Müller [28] have developed a widget that they call Elastic Lists. Here the widget already reflects the proportional value of one facet value in respect to all other values in the same facet. E.g., it does not simply draw equal rows for each facet value, but maps cardinality to the rows height. Likewise brightness stands for characteristicness and unusualness of one facet value among all others. Quantitative data always provides sequence, so it is always possible to handle only the ordinal data aspect. All numeral simple data types such as the XML-Schema data types *xsd:float* and *xsd:integer* offer quantitative data.

Table 1. Comparison of some Existing Faceted Browsing Applications

	Domain	Environment	Filtering: Arbitrary Restriction Order	Filtering: Result Cardinalities Shown	Filtering: Multiple Selection	Filtering: Range Selection	Filtering: Standard Classification Integrat	Filtering: String Search Integrated	Filtering: Facet Values Explained	Filtering: Special Widgets	Display of Results: Sorting of Results Configurable	Display of Results: Grouping of Results	Display of Results: Search for Similar Items	Display of Results: Display Paradigm	Others: Facet Set Context Depending	Others: Facet Set Configurable by User
iTunes [22]	Audio File Collections	Desktop	no	no	yes	yes	no	yes	no	no	yes	yes[3]	no	List, Table	no	no
foobar2000 + foo_facets [23]	Audio File Collections	Desktop	yes	no	yes	yes	yes	yes[10]	no	no[16]	yes[1]	yes[1]	no	List, Table	no	yes
Google Base [24]	e-Commerce	Web	yes	no	yes	yes[7]	yes	yes	no	yes[5]	no[3]	no	no	List, Grid	yes	no
amazon [26]	e-Commerce	Web	yes	yes	no	no	yes	yes	no	no[15]	no[3]	no	yes	Grid, List	yes	no
eBay [27]	e-Commerce	Web	yes	yes	yes	yes[7]	yes	yes	no	yes[13]	no[3]	no	no	List, Grid	yes	yes
SEEK 1.0.1 [28]	e-Mail	Desktop	yes	yes	yes[14a]	yes	no[9]	yes[11]	no	no[14a]	yes	no	no	Tree[18], Table, List	no	no
Facet Map [9]	Domain Independent with Configuration	Desktop	yes	yes	no	no	no	yes	no	no		yes[2]	no	TreeMap	yes	
Flamenco [12]	Domain Independent with Configuration	Web	yes	yes	no	no	no	yes	no	no[15]	yes[14]	yes[1]	yes	Grid	no	no
Honeycomb [25]	Domain Independent with Configuration	Web	yes	no	yes	yes	no	yes	no	yes[12]	no[3]	yes[1,3]	no	TreeMap	no	no
mSpace [14]	Domain Independent with Configuration	Web	no	no	yes	no	no	yes	yes	no	no[3]	no	no	List	no	yes
Exhibit [11]	Domain Independent	Web	yes	yes	yes	yes[7]	no	yes[10]	no	no[14]	y.[17,1]	yes[19]	no	Grid, List, Table, Map, TimeLine	no	no
Longwell [13]	Domain Independent	Web	yes	yes	yes[6]	no	no[8]	yes[11]	no	no	yes[1,3]	no	no	List	no	no

General	1) configurable by user, 2) configurable by admin, 3) fixed facets only, 4) prices only 5) zip only
Multiple Selection	6) but in a second step
Range Selection	7) fixed
Standard Classification	8) preview, only primaries, 9) missing subfolder integration
String Search	10) all facets searched , 11) each single facet searchable
Special Widgets	12) sliders, 13) sliders for price spans with very high values only, 14) more advanced widgets are in the source repository, but not yet published, 14a) some facets have multiple selection disabled
Structure Explicit	15) categories as tree, 16) multi-columns represent some facet structure (e.g. date | month)
Sorting	17) sorting by multiple facets
Grouping	18) displays tree relationship of exchange of letters on demand, 19) configuration follows sorting

5 Conclusion

In this paper we identified characteristics of existing Faceted Browsing applications. We conclude that, although Faceted Browsing becomes more and more frequently integrated into user interfaces for several domains and although some domain independent solutions exist, the browsers support different useful features and could benefit from adopting each other's principles. While there is an increasing number of different views for result presentation such as maps, timelines and tables, this does not apply equally to facet widgets. Especially the generic approaches are often limited to simple unified facet widgets that could better be replaced by facet-depending widgets. A basis for creating such specific widgets for the case of data with explicit semantics has been introduced. Based on the extracted characteristics, a detailed tabular comparison of existing Faceted Browsers, including both generic approaches and Faceted Browsing extensions of existing domain specific applications, has been done (Table 1). However, the details cannot be published here due to space limitations and will be available at http://www.polowinski.de/work/facetedbrowsing.

References

1. Allard, P., Ferré, S.: Dynamic Taxonomies for the Semantic Web, Database and Expert Systems Application. In: DEXA 2008 (2008)
2. Amazon, http://www.amazon.com
3. Berners-Lee, T., Hendler, J., Lassila, O.: The Semantic Web. Scientific American 284(5), 28–37 (2001)
4. Dachselt, R., Frisch, M., Weiland, M.: FacetZoom: A Continuous Multi-Scale Widget for Navigating Hierarchical Metadata. ACM Press, Florence (2008)
5. eBay, http://www.ebay.com
6. Exhibit, http://simile.mit.edu/wiki/Exhibit
7. Fadhil, A., Haarslev, V.: GLOO: A Graphical Query Language for OWL Ontologies. OWL: Experience, Directions, pp. 215–260 (2006)
8. Ferré, S.: Agile Browsing of a Document Collection with Dynamic Taxonomies, Database and Expert Systems Application. In: DEXA 2008, pp. 377–381 (2008)
9. Flamenco, http://flamenco.berkeley.edu/
10. foobar2000 + foo facets, http://foobar2000.audiohq.de/foo_facets/
11. Google Base, http://base.google.com
12. Hearst, M.: Design Recommendations for Hierarchical Faceted Search Interfaces. In: ACM SIGIR Workshop on Faceted Search (2006)
13. Hearst, M.: UIs for Faceted Navigation: Recent Advances and Remaining Open Problems. In: HCIR 2008 Second Workshop on Human-Computer Interaction and Information Retrieval (2008)
14. Honeycomb, http://www.hivegroup.com/
15. Huynh, D.F.: Parallax and Companion: Set-based Browsing for the Data Web. In: WWW 2009 (2009) (submitted, to be published)
16. iTunes, http://www.apple.com/de/itunes/
17. Johnson, B., Shneiderman, B.: TreeMaps: A space-filling approach to the visualization of hierarchical information structures. In: Proc. Visualization 1991, pp. 284–291 (1991)
18. Longwell, http://simile.mit.edu/wiki/Longwell
19. mSpace, http://demo.mspace.fm
20. Oren, E., Delbru, R., Decker, S.: Extending faceted navigation for RDF data. In: Cruz, I., Decker, S., Allemang, D., Preist, C., Schwabe, D., Mika, P., Uschold, M., Aroyo, L.M. (eds.) ISWC 2006. LNCS, vol. 4273, pp. 559–572. Springer, Heidelberg (2006)

21. Priss, U.: Faceted Information Representation. Working with Conceptual Structures. In: Contributions to Proceedings of the 8th International Conference on Conceptual Structures (ICCS 2000), pp. 84–94. Shaker Verlag, Achene (2000)
22. Priss, U.: Facet-like Structures in Computer Science. Axiomathes 18, 243–255 (2008)
23. Rutledge, L., Hardman, L.: Making RDF presentable: integrated global and local semantic Web browsing. In: Proceedings of the 14th international conference on World Wide Web, pp. 199–206 (2005)
24. Sacco, G.: Some Research Results in Dynamic Taxonomy and Faceted Search Systems. In: SIGIR 2006 Workshop on Faceted Search (2006)
25. Sauermann, L., Bernardi, A., Dengel, A.: Overview and outlook on the semantic desktop. In: Proceedings of the 1st Workshop on the Semantic Desktop at the ISWC 2005 Conference (2005)
26. Seek, http://simile.mit.edu/wiki/Seek
27. Smith, G., Czerwinski, M., Meyers, B., Robbins, D., Robertson, G., Tan, D.S.: FacetMap: A Scalable Search and Browse Visualization. IEEE Transactions on Visualization and Computer Graphics 12(5), 797–804 (2006)
28. Stefaner, M., Müller, B.: Elastic lists for facet browsers. Database and Expert Systems Applications. In: 18th International Conference on DEXA 2007, pp. 217–221 (2007)

Applications of Visible Light Path Laser Projector

Nobuchika Sakata, Shu Okamoto, and Shogo Nishida

Division of Systems Science and Applied Informatics Graduate School of Engineering Science, Osaka University, Machikaneyama-cho 1-3, Toyonaka city, Osaka 560-8531 Japan
{sakata,okamoto,nishida}@nishilab.sys.es.osaka-u.ac.jp

Abstract. We proposed and developed a method to visualize the light path of a laser with jetting mists along light axis of a laser pointer. The estimation accuracy for the position of a laser spot occluded behind an object was improved when the light path of the laser was visualized. In this paper, we propose visible light path laser projector (VLLP) for remote collaboration. VLLP is equipped with a laser projector and a mist generator. With the laser projector, VLLP can instruct with not only a laser spot but also simple line drawings. Furthermore, we propose several VLLP application prototypes.

Keywords: Remote collaboration, Visible Light Path Laser, VLLP, Laser projector.

1 Introduction

Work conducted by a local worker under the instructions of a remote instructor is called remote collaboration. Using a telecommunication terminal, the remote instructor and the local worker transmit and receive sounds and videos to accomplish their work since they can not share voices and views directly. On the other hand, a worker and an instructor sometimes communicate regarding objects and places in real work spaces in local collaborative works [1-3]. To conduct such communication smoothly, a support system sends the remote instructor's instructions including the place of the work to the local worker. In some research, a teleoperated laser pointer or laser projector is adopted as a means of sending such instruction [4-8] This method enables a remote collaboration system to be realized with a compact device. The teleoperated laser pointer has been adopted as a tool to point at a target of a work for compact systems [6,7]. However, although the instructor can watch the laser spot, this method can be difficult for workers observing the laser spot when it is occluded behind the object being pointed at or other obstacles. Research has shown that laser-pointing enables smooth remote collaboration for simple tasks, such as selecting and specifying real-world targets. In addition, a laser spot not only provides awareness information about a remote instructor's point of view, but also links a real-world position and virtual information to a local worker. Therefore, the remote collaboration cannot be conducted smoothly when the laser spot providing important information is hidden from the local worker's sight. The proposed method thus allows for smooth remote collaboration by realizing 3D directional expression to enable a worker to

M.J. Smith and G. Salvendy (Eds.): Human Interface, Part I, HCII 2009, LNCS 5617, pp. 611–618, 2009.

estimate the position of a laser spot from the visualized light path, even if the laser spot is occluded.

Originally, this research aims to support remote collaboration, including instructions identifying real objects with a compact device. In previous research, we proposed a new method for pointing to objects for remote collaboration support devices and provides an assessment of the method. The method used 3D directional expression, which is similar to using a pointing stick, by visualizing the light path of a laser pointer. In user test we conducted, we found that with this method, a worker can be expected to estimate the position of an occluded laser spot from the visualized light path. In this paper, applying knowledge of visible light path laser pointer, we propose visible light path laser projector (VLLP) for remote collaboration. Furthermore, using 3D directional expression of VLLP, we propose not only applications of remote collaboration, but also several VLLP application prototypes.

2 Related Works

The teleoperated laser pointer is adopted in some research as a pointing tool for remote collaboration. Telepointer [4] is a wearable device mounted on the chest. Telepointer is equipped with a fixed camera and a teleoperated laser pointer which enables remote collaboration targeting objects in real-world. A Wearable Active Camera with a Laser pointer (WACL) [7] is a hands, head, and eye-free wearable device mounted on a worker's shoulder. WACL is comprised of a camera and a laser pointer on a biaxial camera platform. Sakata et al. conducted an experiment to compare settings when a laser spot is projected and not projected on targets [9]. The results showed that the laser spot played an important role, named the visual link, linking information shown on a chest worn display and the real-world position the information regarded. GestureCam [5] is not a wearable device but a device placed in a work space. GestureCam has a camera mounting a laser pointer on it. An instructor can pan and tilt the laser pointer on the camera to point at real-world objects. CTerm [6] is also a device placed in a work space. CTerm is equipped with a camera and a laser pointer which are remotely controlled.

Some studies have realized systems to convey gestural expression to remote sites. These systems can express 3D direction. GestureMan [10] is a system equipped with not only a teleoperated laser pointer but also a robot head and a robot arm. The robot head and the robot arm trace the motion of the remote instructor. Therefore, GestureMan is assumed to be able to convey the target of instruction even if the laser spot is occluded. However, the robot arm and the robot head prevent the system from being reduced in size for compactness because the robot arm and the robot head require a certain size to express gestures and orientation. Telecommunicator T2 [11] is a wearable communication system with gesture expressions. Telecommunicator T2 has a teleoperated camera and a compact arm. Using this system, a remote person can communicate a wearer with gestures of the arm. However, the system does not aim to support remote collaboration with instructions regarding real-world objects. Moreover, the size of the arm is limited to enable mounting on the shoulder. Therefore, pointing with Telecommunicator T2 is not as accurate as using a laser pointer. Real World Video Avatar [12] is a telecommunication system which can convey human gestures. The system shows images at all angles by spinning a large display, on which a filter enhancing the directivity is attached, with the images and the angle of the display synchronized. Although Real World Video Avatar has rich expression, the device is large in scale.

As described above, remote collaboration systems with rich gestural expressions are large-scale, although they are assumed to be able to convey instructions regarding real-world objects without relying upon a laser spot. Thus, there is a trade-off between rich 3D directional expression and compactness. There has been no research aimed at realizing rich 3D directional expression with a compact device.

3 Visible Light Path Laser Projector (VLLP)

Named the Visible Light path Laser Projector (VLLP), for remote collaboration is being constructed. VLLP is equipped with a laser projector and a mist generator as shown in Fig. 1. With the laser projector, VLLP can instruct with not only a laser spot but also simple line drawings. We describe visible light path as main function of VLLP in section 3.1.

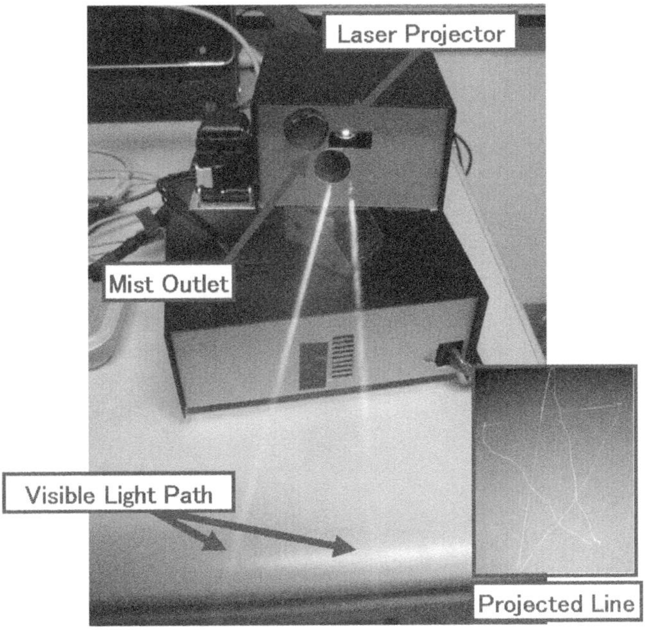

Fig. 1. Visible light path laser projector

3.1 Visible Light Path Laser Pointer

The light path of a laser is visualized as a bright line when the laser is scattered by particles. Scattering a high-power laser with dust is a possible method for visualizing the light path of a laser. However, a high-power laser is not suitable for remote collaboration since it can harm a worker. Although scattering a low-power laser by filling up the workspace with mist is harmless, time is required to fill up the workspace with an artificial mist even if a large-size device is applied. Therefore, we suggests a harmless method which requires neither a large-scale device, nor a long wait time. The proposed method scatters a laser with an artificially generated mist jetted in the direction of the laser with a fan. This allows a reduction in the necessary amount of mist and reduces the required size of the device compared with filling up the workspace with mist. Moreover, this method has low danger since the method can visualize the light path of the laser even at low power. An analogous method was suggested by Shimizu and Takaguchi, who proposed shooting a "light arrow" by shooting smoke from an air cannon along a laser [13]. Their method has merit in that less smoke is required and the smoke can reach farther than the method proposed here. However, the light path is difficult to watch since it is intermittent. In addition, an air cannon is not suitable for mounting on a compact device. For these reasons the method proposed here uses not an air cannon but a fan.

A visible light path laser pointer was developed experimentally by applying the suggested method and is shown in Fig. 2. The device consists of a mist generator and a laser pointer. A mist generated by the mist generator is transferred to a hose by a fan, and then jets from the hose end. A laser pointer is mounted on the hose end along the direction of jetted mists, so the laser goes through the mist. The mist generator is equipped with two ultrasound atomizer HM-2412 (Honda Electronics Co., Ltd.). One HM-2412 can atomize approximately 250 ml/h of water. The size of a mist particle is approximately 3 μm. The size of one circuit board is 45 × 70 mm2, and the size of one oscillator for the mist generator is 42 × 30 × 13 mm3. However, there are many

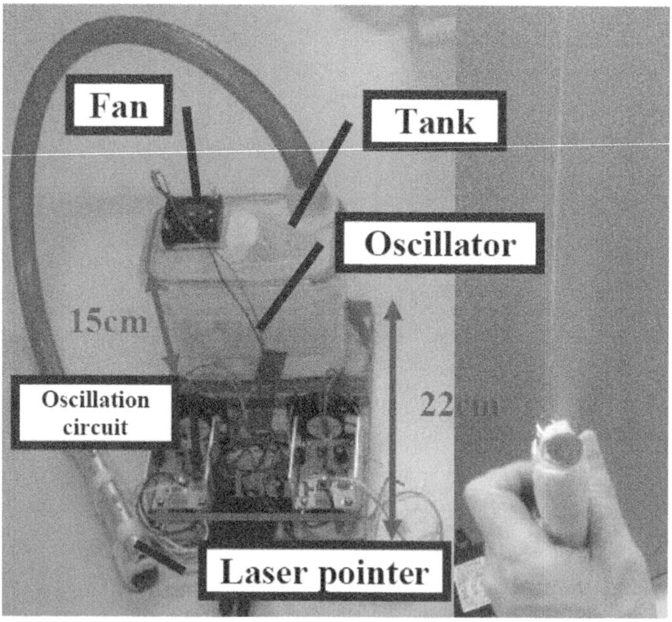

Fig. 2. Laser pointer with jetting mist along laser light path

products smaller than HM-2412 but provide the same function. The laser pointer is a commercially available product with an on/off push switch. The laser is green with a wavelength of 532 nm, maximum output of less than 1 mW, and a beam width of 5 mm. The length from the switch to the end of the laser pointer is 46 mm, and the length from the switch to the end of the hose is 55 mm.

The light path of the visible light path laser pointer with atomized water could be observed under room lights of 460 lx illuminance. The length of the visible part of the path was approximately 1 m. The length was slightly short compared to a pointing stick, probably caused by the ready vaporization of the water mist. To prevent vaporization, fog liquid (FLG1, Stage Evolution), which is a liquid for smoke machines, was atomized. When fog liquid diluted to 20% with water was atomized, the length of the visible part was extended to approximately 3 m since the fog liquid vaporized less easily than water. However, the amount of mist decreased because FLG1 was not atomized by HM-2412. As a result, the brightness of the light path decreased compared with the case of water.

Though the light path was bright enough to be easily observed from the front of the laser pointer, the light path was less easy to see from the side. In addition, when the direction of the laser pointer was changed, the visible light path took a short time until the mist followed the movement. The delay was approximately 0.5.1 s when water was used. Moreover, in windy circumstances such as outdoors, the mist was blown away, shortening the visible light path.

When a mist is jetted onto a hand, glass, or plastic at a distance of 5 cm for several minutes, these objects became wet. On the other hand, jetting at a distance of 40 cm caused little wetting. Therefore, the visible light path laser pointer should not cause any trouble even if in a room with machinery or computers unless mists are jetted onto them directly for a long time. Proper installing of the visible light path laser pointer should prevent moisture from collecting.

The function of the visible light path is similar to a pointing stick. However, a pointing stick cannot point at targets with shorter distances than the length of the pointing stick because the stick may make contact with the target. Moreover, a pointing stick has the danger of contact with a worker or object since a worker may have to work near a device mounted with a pointing stick. Therefore, a pointing stick is not suitable as a device supporting remote collaboration. In contrast, the visible light path laser pointer can point at targets with shorter distances than the length of the visible light path. In addition, the visible light path laser pointer contains no danger of contact with a worker or objects. Therefore, the visible light path laser pointer is suitable for mounting on a compact device supporting remote collaboration.

We conducted user test to confirm effectiveness of visible light path laser pointer. We assume the situation that when an instructor facing a worker tries to point to an object between the worker and the instructor with a laser pointer, the laser spot is likely to be out of worker's line of vision, as shown on the left side of Fig. 3. If the worker can estimate the position of the laser spot from the visualized light path even when the worker cannot see the laser spot, as shown on the right of Fig. 3, the remote collaboration targeting real-world objects can be conducted smoothly. After small user test, the proposed method improves the accuracy when estimating the position of an occluded laser spot. In addition, the proposed method was effective in terms of accuracy when distinguishing an object being pointed at from other objects, even when the laser spot is occluded.

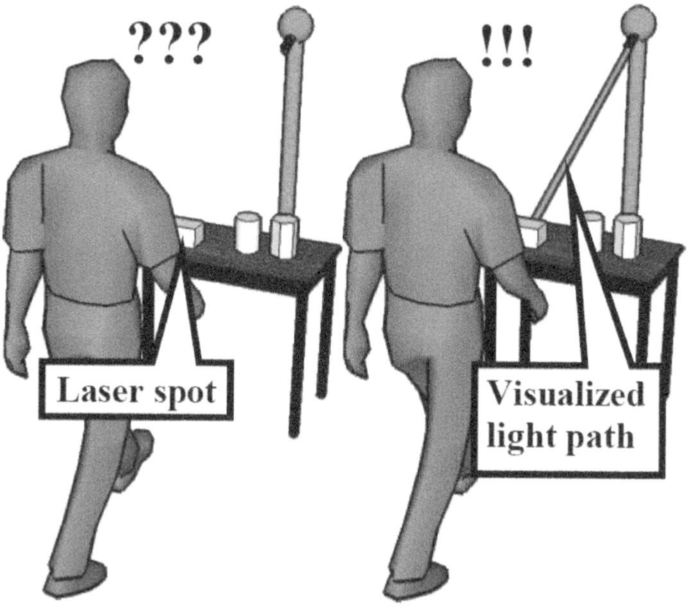

Fig. 3. Laser spot is occluded (left) and light path is visualized (right)

4 Applications of VLLP

In this section, we describe two applications of VLLP.

4.1 Remote Collaboration Terminal for Visual Light Path Laser Projector

Basically, we suppose to apply VLLP to remote collaboration terminal. We install a microphone, a camera and a speaker to VLLP. We intend to attach a fish-eye camera for observing remote work field widely. With the laser projector, VLLP can instruct with not only a laser spot but also simple line drawings. We assume that combining feature of visible light path and line drawing function with a laser projector, another significant advantage of VLLP is boosted. That is multiple instructions. Teleoperated laser pointer including WACL, telepointer and Gesture Laser can instruct only one point at the same time. On the other hands, teleoperated laser projector can instruct multiple points at the same time. However, laser projector, which is even vector scan type, can not watch the moving laser spot while scanning. For example, remote instructor point three buttons which are spread spatially in work space. And then, remote instructor says "Could you press the buttons?". However, field worker can not find two or three spread button immediately because the instructed button might be existed behind the field worker. In case of applying VLLP, observing visible light path from laser projector, the field worker can sense a direction and place where laser spots exist on. Finally, field worker can detect instructed places easily. We believe this is one big feature of VLLP in remote collaboration.

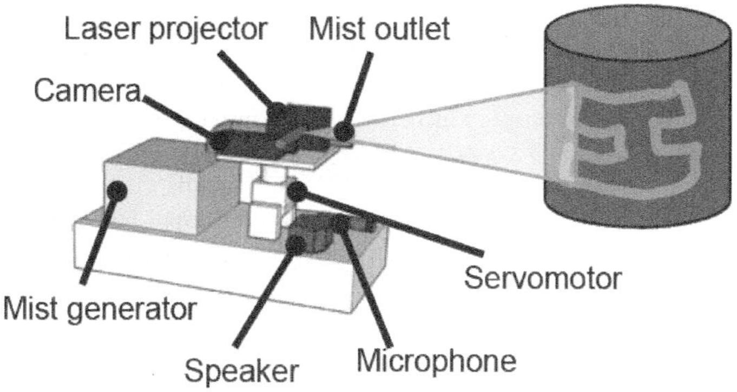

Fig. 4. Visible light path laser projector for remote collaboration

4.2 Guide Robot Installed VLLLP

We suppose to use VLLP as one part of guide robot[14][15]. In those researches, various robot body parts such like body size of robot, arms, neck and head take an important role for guide in terms of gestural expressions. Due to this, the size of robot should be kept to generate gestural expressions, consequentially. On the other hands, a risk of clash and spatial storage are increased. For these reasons, we mount VLLP to robot as a part for increasing those gestural expressions (Fig. 5). Using visible light pass as kind of pointing stick, even small robot represents large gestural expressions.

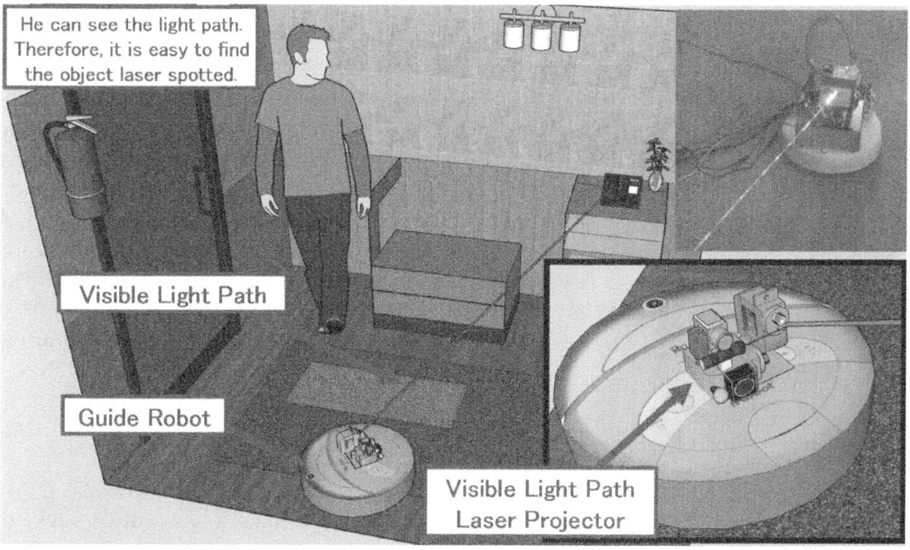

Fig. 5. Visible light path laser projector

5 Conclusion

In this paper, we propose and describe VLLP is equipped with a laser projector and a mist generator. With the laser projector, VLLP can instruct with not only a laser spot but also simple line drawings. Furthermore, we propose two application targets, "Remote collaboration" and "Guide robot". In case of "Remote collaboration", we suggest possibility of multiple instructions. Also, in case of "Guide robot", VLLP can increase gestural expressions instead of neck, head, arms and body. In the future, we develop mobile type VLLP and proposed applications and evaluate features of VLLP.

References

1. Kuzuoka, H.: Spatial workspace collaboration: a Sharedview video support system for remote collaboration capability. In: CHI 1992, pp. 533–540 (1992)
2. Fussell, S.R., Setlock, L.D., Kraut, R.E.: Effects of head-mounted and scene-oriented video systems on remote collaboration on physical tasks. In: Proc. CHI 2003, pp. 513–520 (2003)
3. Kraut, R.E., Miler, M.D., Siegal, J.: Collaboration in performance of physical tasks: Effects on outcomes and communication. In: CSCW 1996, pp. 57–66 (1996)
4. Mann, S.: Telepointer: Hands-free completely self-contained wearable visual augmented reality without headwear and without any infrastructural reliance. In: ISWC 2000, pp. 177–178 (2000)
5. Kuzuoka, H., Kousuge, T., Tanaka, M.: GestureCam: a video communication system for sympathetic remote collaboration. In: CSCW 1994, pp. 35–43 (1994)
6. Mikawa, M., Matsumoto, M.: Smooth and easy telecommunication using CTerm. In: IEEE SMC 1999, pp. 732–737 (1999)
7. Sakata, N., Kurata, T., Kato, T., Kourogi, M., Kuzuoka, H.: WACL: Supporting telecommunications using wearable active camera with laser pointer. In: ISWC 2003, NY, USA, pp. 53–56 (2003)
8. Palmer, D., Adcock, M., Smith, J., et al.: Annotating with light for remote guidance. In: 2007 conference of CHISIG of Australia on computer-human interaction, pp. 103–110 (2007)
9. Sakata, N., Kurata, T., Kuzuoka, H.: Visual assist with a laser pointer and wearable display for remote collaboration. In: 2nd international conference on Collaboration Technologies (CollabTech. 2006), Tsukuba, Japan, pp. 66–71 (2006)
10. Kuzuoka, H., Furusawa, Y., Kobayashi, N., Yamazaki, K.: Effect on Displaying a Remote Operator's Face on a Media Robot. In: ICCAS 2007, pp. 758–761 (2007)
11. Tsumaki, Y., Fujita, Y., Kasai, A., Sato, C., Nenchev, D.N., Uchiyama, M.: Telecommunicator: A novel robot system for human communications. In: IEEE ROMAN 2002, pp. 35–40 (2002)
12. Tanikawa, T., Suzuki, Y., Hirota, K., Hirose, M.: Real world video avatar: real-time and real-size transmission and presentation of human figure. In: 2005 international conference on Augmented tele-existance, pp. 112–118 (2005)
13. Shimizu, N., Takiguchi, T.: Rokujurossundou. In: VRSJ 8th Annual Conference, p. 289 (2003) (in Japanese)
14. Burgard, A.B., Cremers, D., Fox, D., Hähnel, G., Lakemeyer, D., Schulz, W., Steiner, W., Thrun, S.: Experiences with an interactive museum tour-guide robot. Artificial Intelligence 114(1-2), 3–55 (1999)
15. Kuzuoka, H., Kosaka, J., Yamazaki, K., Yamazaki, A., Suga, Y.: Dual Ecologies of Robot as Communication Media: thoughts on Coordinating Orientations and Projectability. In: CHI 2004, pp. 183–190 (2004)

Virtual Convex Polygon Based Hole Boundary Detection and Time Delay Based Hole Detour Scheme in WSNs

Inyoung Shin, Ngoc Duy Pham, and Hyunseung Choo*

School of Information and Communication Engineering
Sungkyunkwan University, Korea
choo@skku.edu

Abstract. In wireless sensor networks, an important issue often faced in geographic routing is the "local minimum phenomenon." To mitigate the local minimum issue, when the routing process becomes stuck at hole boundary nodes, the existing perimeter routing tends to route data packets along the boundaries of the holes. However, this may enlarge the hole, causing the "first hole diffusion" problem. On the other hand, the existing hole detour scheme based on the virtual ellipse forwards data packets to outside the virtual ellipse. This may generate other holes around the existing hole - the "second hole diffusion" problem. Therefore, we propose a novel virtual convex polygon based hole boundary detection and time delay based hole detour scheme. The proposed scheme solves first and second hole diffusion problems. Comprehensive simulation results show that the proposed scheme provides approximately 22% and 16% improvements in terms of the packet delivery ratio and the network lifetime, respectively.

Keywords: Wireless Sensor Networks (WSNs), Geographic Routing, Hole Problem.

1 Introduction

Geographic routing protocol [1], efficient and scalable strategy, can minimize the hops from the source to the destination by forwarding the data packet to the 1-hop neighbor which is closest to the destination. However, the geographic routing fails if there is no neighbor that is closer to the destination than the current node, this is well known as the local minimum problem [2]. To mitigate the local minimum issue, the several schemes have been proposed [1-11]. These existing schemes, however, still have the first and second hole diffusion problems.

In this paper, we propose a novel hole boundary detection method based on the virtual convex polygon to solve the first hole diffusion problem. It can reduce the overhead incurred by distribution of hole information of all nodes inside the virtual polygon since the number of nodes inside a convex polygon is smaller than other polygons. In addition, unnecessary establishment of the detour routing path is eliminated. We also present energy efficient hole detour scheme based on time delay

* Corresponding author.

M.J. Smith and G. Salvendy (Eds.): Human Interface, Part I, HCII 2009, LNCS 5617, pp. 619–627, 2009.

mechanism. When the data packet reaches the boundary node of virtual convex polygon, it is forwarded to outside of the convex polygon along the dynamic path founded by time delay mechanism. This reduces the energy consumption and data congestions on the hole boundary nodes. Moreover, the hole detour method using time delay evenly distributes the load among peripheral nodes around the virtual convex polygon. Therefore, the second hole diffusion problem could be relieved.

The reminder of this paper is organized as follows: in Section 2, we briefly discuss related work. Section 3 presents our hole boundary detection and detour schemes. The performance of our proposed scheme is evaluated in Section 4. Finally, we conclude with the main findings and contributions of our research in the last section.

2 Related Work

Geographic greedy forwarding [1] is the most promising routing scheme in wireless ad-hoc sensor networks (WASNs). In such a scheme, it is assumed that each node knows its own location and the location of its 1-hop neighbors, source knows the location of the destination and encapsulates the destination location in each data packet, a node sends data packets to 1-hop neighbor closest to the destination. However, geographic forwarding suffers from the so called local minimum phenomenon. Specifically, a packet gets stuck at a node whose 1-hop neighbors are all further away from the destination. In the existing perimeter routing scheme [1][2], the data packets tend to be routed along the boundaries of holes to solve local minimum phenomenon. However, hole diffusion problem may arise due to the energy exhaustion of the hole boundary nodes. In addition, data collisions may occur in the hole boundary nodes if multiple data streams are bypassing a hole simultaneously.

Hole Detour scheme based on Ellipse (HDE) [3][4] detects the hole using virtual ellipse. The node which firstly detects a hole (the initiator) sends out a Hole Boundary Detection (HBD) packet along the boundary of the hole by the well-known right hand rule. This process repeats until the HBD packet has traveled around the hole and eventually been received by the initiator. The initiator gets the location information of all boundary nodes of the hole from receiving HBD and then calculates the ellipse which can cover the hole exactly. Then initiator distributes an Ellipse Distribution (ED) packet which includes all information about the ellipse, and sends out the ED packet to all nodes inside the ellipse. The mission of the defined the ellipse is to prevent data packets from entering the ellipse. When a node locating on the boundary of the ellipse receives the data packet, HDE forwards the data packet to outside of virtual ellipse by geographic forwarding mechanism. HDE solves the first hole diffusion problem. However, HDE increases overhead incurred by distribution of hole information of all nodes inside the virtual ellipse and generate unnecessary hole detour since the ellipse does not cover in the exactly close proximity to the hole. HDE also generates the second hole diffusion problem if the source and destination are same.

3 Proposed Scheme

3.1 Virtual Convex Polygon Based Hole Boundary Detection

In this section, we introduce the method of the virtual convex polygon based hole boundary detection. The node that firstly detects a hole sends a HBD packet including its location information along the boundary of the hole by the right hand rule [2]. The node receiving the HBD packet refers to the table of movement patterns as in Table 1 to determine if its upstream node is a convex or concave node. In a polygon drawn by connecting the nodes forming the boundary of the hole, a convex node is a vertex bulging outward further than its neighboring nodes. The use of a movement pattern may reduce the size of the packet that traverses the boundary nodes to form the information of a convex polygon. The possible movement patterns of the convex and concave nodes that may take place while traversing hole boundary nodes, in accordance with the right hand rule, are shown in Table 1. Blank spaces in Table 1 may contain any of +, –, and 0.

Table 1. The movement patterns of the convex and concave nodes

Pattern of Concave node	Pattern of Convex node	Pattern of common node
(+,) → (, +)	(+,) → (, –)	**No change of pattern** **Ex.** (+, +) → (+, +)
(, –) → (+, –)	(, –) → (–,)	
(, +) → (–,)	(, +) → (+,)	
(–,) → (, –)	(–,) → (, +)	

Convex nodes and concave nodes may be determined using the movement pattern as follows: Assume that the HBD packet is transferred through nodes N_1, N_2 and N_3, in that order, as shown illustrated in Fig. 1. Given the node location information (x, y), the transfer of the packet from N_1 to N_2 would result in the location information changing from (10, 20) to (12, 24). As both x and y values increased, the movement in this case may be summarized as (+, +). Using the same method, the movement from N_2 to N_3 would be (+, –). As shown in the drawing, N_2 is a convex node that is bulging outward than N_1 and N_3. Therefore, a given node is determined to be a convex node if the movement from its upstream node and that to its downstream node are (+, +) and (+, –), respectively.

Each node judges its upstream node by making reference to the movement pattern, and if the upstream node is a convex node, it states in the HBD packet that the upstream node is convex, and then adds the location information of the current node to the packet before handing it to its downstream node. The location information of convex nodes is stored while the packet traverses boundary nodes. The location information of non-convex nodes is deleted from the HBD packet when it becomes no longer needed. As the location information of any node is required to judge the node as well as its immediate upstream and downstream nodes, the information of any non-convex node is communicated no further than a boundary node following the downstream node. This process continues until the HBD packet has traveled around the hole and eventually been received by the initiator. The initiator then uses the Graham scan algorithm [12] on the list of the convex nodes extracted. This process results in a complete convex polygon.

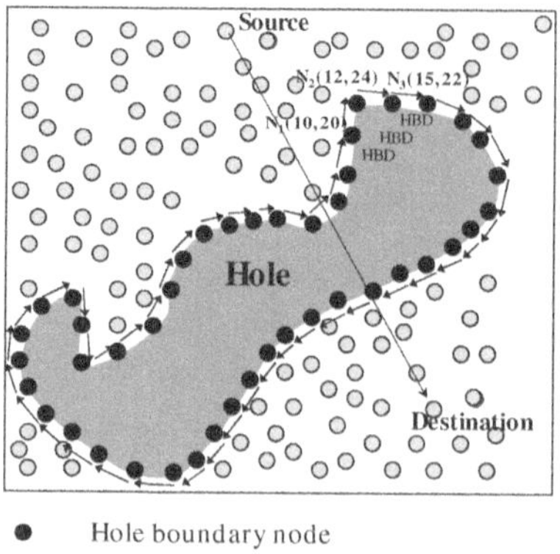

Fig. 1. Tracing the hole information

The initiator assembles a Convex Polygon Distribution (CPD) packet for use the information of the recognized convex polygon to search for detour points. The CPD packet contains the information of the convex polygon and transfers it to all nodes found to form the polygon. Each node receiving the CPD packet uses the convex polygon information to determine if it belongs to the polygon, and obtains the maximum and minimum x and y values to store information O (minimum x, maximum y), P (maximum x, maximum y), Q (minimum x, minimum y) and R (maximum x, minimum y) which are the detour points, as shown in Fig. 2.

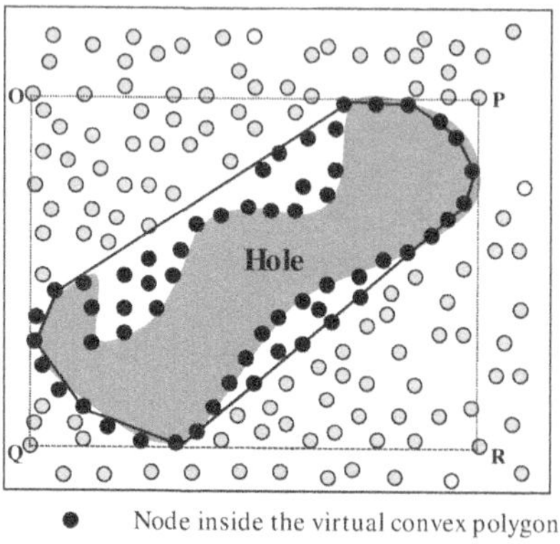

Fig. 2. Selecting 4-detour points

3.2 Time Delay Based Hole Detour Scheme in WSNs

In this section, we depict the data transmission process from source to destination. When a data packet is transferred, a hole is detoured in a following manner: Inside each data packet is a node flag that determines the mode of transfer. The data packet is transferred in the same manner as in Greedy Perimeter Stateless Routing (GPSR) [1] until it reaches a node containing hole information, where the transfer mode of the packet is a basic mode. If a data packet has reached a node containing hole information, appropriate detour points to avoid the hole are determined. If the calculation of detour points result in the conclusion that no detouring is necessary, the transfer mode of the packet remains the basic mode. If detour points may be calculated, the transfer mode of the packet is switched to the detour mode. The detour mode is a transfer mode indicating that the dynamic transfer route must be determined with energy and distance taken into account until the packet nears the calculated detour points.

The detour points are determined using the intersection relationship between (i) segment $\overline{SD}$ connecting the source (S) and destination (D), and (ii) segments $\overline{OP}, \overline{OP}, \overline{OQ}$, and $\overline{PR}$ connecting O, P, Q and R as in Table 2. Detour points are established only when $\overline{SD}$ intersects two segments. If there are two possible detour routes, a route offering the shortest distance to D is selected. After the selection of detour points, the transfer mode of the packet is switched to the detour mode, and the next node is found by using the time delay mechanism while the packet goes through the detour points.

Table 2. The selected detour points by intersection relationship

The intersection of the segment $\overline{SD}$	Detour Points
$\overline{OP}$, $\overline{QR}$	(O, Q) or (P, R)
$\overline{OQ}$, $\overline{PR}$	(O, P) or (Q, R)
$\overline{OP}$, $\overline{PR}$	P
$\overline{OQ}$, $\overline{QR}$	Q
$\overline{OQ}$, $\overline{OP}$	O
$\overline{PR}$, $\overline{QR}$	R

The Time Delay value (TD) is calculated by (1).

$$TD = \begin{cases} \omega \times N_e + (1-\omega) \times N_d, & if \ \forall_{N_e} \rangle 0 \\ \infty, & if \ \exists_{N_e} = 1 \end{cases} \tag{1}$$

where, N_e denotes the residual energy level using logarithmic properties; N_d denotes the distance between the current node and the destination. Each attribute can be adjusted by weight factor, $\omega \in [0,1]$. A large ω gives more weight to the node's residual energy than to the distance.

$$N_e = \min\{1, -\log_{10}\frac{E_{residual}}{E_{initial}}\}$$
$$N_d = \min\{1, \frac{(D_{destination} - D_{shortest})}{(D_{longest} - D_{shortest})}\} \quad (2)$$

where, $E_{residual}$ and $E_{initial}$ are defined as the residual energy and initial energy, respectively; $D_{destination}$ is the distance between the current node and the destination; $D_{longest}$ is the longest distance between the neighbor and the destination; and $D_{shortest}$ is the shortest distance between the neighbor and the destination. When the data packet reaches the boundary node of the convex polygon, it sets the data packet to detour mode, and adds the detour location information to the data packet. This node broadcasts Request Time Delay (RTD) packet to its neighbors. The nodes receiving the RTD packet sets a timer after calculating TD. The timer value is proportion to TD. TD increases when the residual energy decreases and distance to the destination increases. The neighbor whose timer expires at first sends Clear of Time Delay (CTD) packet to the node sending RTD packet. The nodes sending RTD packets forward data packet to the node firstly sending CTD packet. This process repeats until the packet reaches the node closest to the detour point. When the node closest to the detour location receives the data packet, it resets the data packet to basic mode, and then transmits the data packet to destination directly by the existing geographic routing. The time delay mechanism could guarantee more uniform energy consumption among peripheral nodes around virtual convex polygon including boundary nodes of the hole, so this can relieve the second hole diffusion problem.

4 Performance Evaluation

We have implemented the proposed scheme, GPSR, and HDE using a simulator built in JAVA to evaluate their performance. In the simulations, nodes with a transmission radius of 40m are deployed to cover an interest area of 500m x 500m. The residual energy below 2.5J is randomly assigned to each node. We manually set one hole in the center of the network. The main parameters of our simulation are listed in Table 3.

Table 3. Simulation parameters

Initial energy	2.5J
Data packet size	500Bytes
Control packet size	15Bytes
Energy consumption model E_{tx}	α_{11}, $\alpha_2 r^2$
Energy consumption model E_{rx}	α_{12}
α_{11}, α_{12}	80nJ/bit
α_2	1pJ/bit/m^2

Our simulation environment uses the following energy model [13]: $E_{tx}=\alpha_{11}+\alpha_2 r^2$, $E_{rx}=\alpha_{12}$ where E_{tx} and E_{rx} denote the energy consumed to transmit and receive a bit over a distance r, respectively. α_{11} is the energy/bit consumed by the transmitter electronics. α_2 is the energy dissipated in the transmit op-amp and α_{12} is the energy/bit consumed by the receiver electronics.

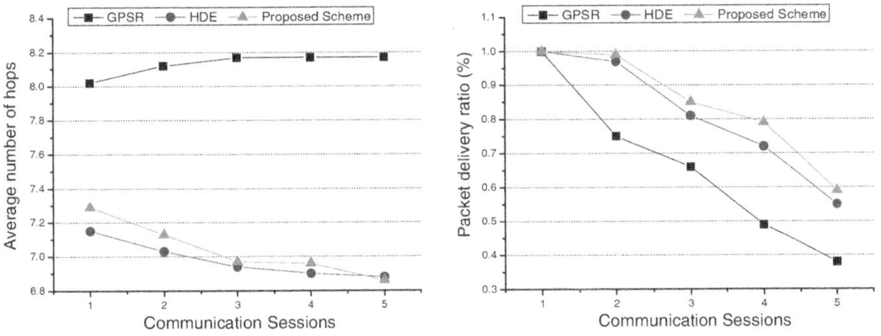

(a) (b)

Fig. 3. Average number of hops and Packet delivery ratio

We compare the average number of hops of routing path, packet delivery ratio, maximum number of transmitted messages, and average residual energy of boundary nodes with GPSR, and HDE. Fig. 3(a) shows the average number of hops of routing path. Comparing with HDE and the proposed scheme, GPSR need more hops because they forward the data packet by right hand rule regardless of the location of destination. Fig. 3(b) shows the delivery ratio with different number of communication sessions when 500 packets are forwarded. GPSR forwards data packets along hole boundaries. Therefore, with the number of communication session increasing, multiple communication sessions may need to bypass a hole simultaneously. So the probability that data collisions occur in the nodes around the hole increases with an increasing number of communication sessions. However, the

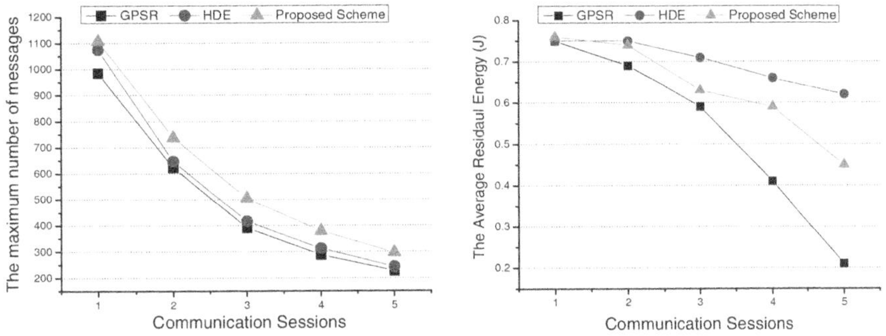

(a) (b)

Fig. 4. The maximum number of message and the average residual energy

data packets in HDE and the proposed scheme are redirected once they encounter a hole, the detour location is different with different sources or destinations. HDE uses same path when source and destination are same. On the other hand, the proposed scheme changes the path from first boundary node to detour node using time delay mechanism although source and destination are same. Thus, the performance of the proposed scheme is slightly better than that of HDE.

Fig. 4(a) shows the proposed scheme attains approximately 16% improvement in terms of the maximum number of messages before first node on the path depletes its batteries. The maximum number of messages is the performance metric to show the network lifetime indirectly. Fig. 4(b) shows the average residual energy of the nodes around the holes after 500 packets are forwarded. The average residual energy of GPSR is drastically decreased with an increasing number of communication sessions because data packets are forwarded along the boundary of the hole. Thus the energy consumption of nodes on boundary holes is quite high in GPSR. The average residual energy of HDE is always higher than that of other schemes because it does not use boundary nodes. However, there is high probability to make other holes beside the existing hole because the boundary nodes of the virtual ellipse tend to be depleted quickly. The performance of the proposed scheme is higher than that of GPSR and lower than that of HDE. This result means that the proposed scheme evenly distributes the load among peripheral nodes around the virtual convex polygon.

5 Conclusion

In this paper, we propose the virtual convex polygon based hole boundary detection and hole detour scheme using time delay in WSNs. The virtual convex polygon based hole boundary detection solves the first hole diffusion problem. In addition, hole detour method using time delay evenly distribute the load among peripheral nodes around virtual convex polygon, so the second hole diffusion problem can be relieved. The network lifetime of the proposed scheme could be improved by reducing speed of the hole diffusion. This also results in the increasing delivery ratio. In our future work, we will extend our approach for selection of detour points. Because the proposed scheme only considers 4-detour points for detouring the hole, this is the cause of earlier exhaustion of their energy.

Acknowledgments. This research was supported by MKE, Korea under ITRC IITA-2008-(C1090-0801-0046).

References

1. Karp, B., Kung, H.T.: GPSR: Greedy perimeter stateless routing for wireless networks. In: Proceedings of ACM/IEEE International Conference on Mobile Computing and Networking, August 2000, pp. 243–254 (2000)
2. Fang, Q., Gao, J., Guibas, L.J.: Locating and bypassing routing holes in sensor networks. In: Proceedings of IEEE INFOCOM, March 2004, pp. 2458–2468 (2004)

3. Yu, F., Lee, E., Choi, Y., Park, S., Lee, D., Tian, Y., Kim, S.H.: A Modeling for Hole Problem in Wireless Sensor Networks. In: Proceedings of International Wireless Communications and Mobile Computing Conference, August 2007, pp. 370–375 (2007)
4. Tian, Y., Yu, F., Choi, Y., Park, S., Lee, E., Jin, M., Kim, S.H.: Energy-Efficient Data Dissemination Protocol for Detouring Routing Holes in Wireless Sensor Networks. In: Proceedings of IEEE International Conference on Communications, May 2008, pp. 2322–2326 (2008)
5. Yu, F., Park, S., Tian, Y., Jin, M., Kim, S.H.: Efficient Hole Detour Scheme for Geographic Routing in Wireless Sensor Networks. In: Proceedings of IEEE Vehicular Technology Conference, May 2008, pp. 153–157 (2008)
6. Yu, F., Choi, Y., Park, S., Lee, E., Tian, Y., Kim, S.H.: Virtual Circle Based Geometric Modeling of Holes for Geographic Routing. In: Proceedings of IEEE Wireless Communications and Networking Conference, March 2008, pp. 2426–2431 (2008)
7. Yu, F., Choi, Y., Park, S., Lee, E., Kim, S.H.: An Edge Nodes Energy Efficient Hole Modeling in Wireless Sensor Networks. In: Proceedings of IEEE Global Telecommunications Conference, November 2007, pp. 4724–4728 (2007)
8. Jiang, Z., Ma, J., Lou, W., Wu, J.: An Information Model for Geographic Greedy Forwarding in Wireless Ad-Hoc Sensor Networks. In: Proceedings of IEEE INFOCOM, April 2008, pp. 825–833 (2008)
9. Olariu, S., Stojmenovic, I.: Design guidelines for maximizing lifetime and avoiding energy holes in sensor networks with uniform distribution and uniform reporting. In: Proceedings of IEEE INFOCOM, April 2006, pp. 1–12 (2006)
10. Wu, X., Chen, G., Das, S.: On the energy hole problem of nonuniform node distribution in wireless sensor networks. In: Proceedings of IEEE International Conference on Mobile Ad Hoc and Sensor Systems, October 2006, pp. 180–187 (2006)
11. Chen, S., Fan, G., Cui, J.: Avoid "void" in geographic routing for data aggregation in sensor networks. International Journal of Ad Hoc and Ubiquitous Computing 1(4), 169–178 (2006)
12. Graham, R.L.: An Efficient Algorithm for Determining the Convex Hull of a Finite Planar Set. Information Processing Letters 1(4), 132–133 (1972)
13. Bhardwaj, M., Garnett, T., Chandrakasan, A.P.: Upper bounds on the lifetime of sensor networks. In: Proceedings of IEEE International Conference on Communications, June 2001, pp. 785–790 (2001)

iFeel_IM! Emotion Enhancing Garment for Communication in Affect Sensitive Instant Messenger

Dzmitry Tsetserukou[1], Alena Neviarouskaya[2], Helmut Prendinger[3], Naoki Kawakami[1], Mitsuru Ishizuka[2], and Susumu Tachi[1]

[1] Department of Information Physics and Computing, University of Tokyo, Japan
dima_teterukov@ipc.i.u-tokyo.ac.jp,
{kawakami,tachi}@star.t.u-tokyo.ac.jp
[2] Department of Information and Communication Engineering, University of Tokyo, Japan
lena@mi.ci.i.u-tokyo.ac.jp, ishizuka@i.u-tokyo.ac.jp
[3] National Institute of Informatics, Japan
helmut@nii.ac.jp

Abstract. In the paper we are proposing a conceptually novel approach to reinforcing (intensifying) own feelings and reproducing (simulating) the emotions felt by the partner during online communication through specially designed system, iFeel_IM!. The core component, Affect Analysis Model, automatically sense nine emotions from text. The detected emotion is stimulated by innovative haptic devices (HaptiHeart, HaptiHug, HaptiTickler, HaptiCooler, and HaptiWarmer) integrated into iFeel_IM! The implemented system can considerably enhance emotionally immersive experience of real-time messaging.

Keywords: Affective haptics, affective user interface, wearable devices.

1 Introduction

"All emotions use the body as their theater..."
Antonio Damasio [1]

Pleasure, euphoria, sadness, fear, anger – these and other emotions color our lives, enrich our experiences, and imbue our actions with passion and character. Research on emotion in both psychology and neuroscience indicates that emotions also play a pivotal role in social interaction, decision-making, perception, attention, memory, learning, etc. Affective Computing [2], which aims to realize emotionally intelligent human-computer interaction, is a well-established research area that deals with investigation of the role of emotions in technology and poses a question: "How can we make computer to be aware of emotions?" While conducting our research in an emerging frontier of Affective Haptics, we are attempting to answer another question: "How to elicit (or enhance) the user's affective state in human-human mediated communication by haptic technologies?"

Besides emotions conveyed through text, researchers developed an additional modality for communicating emotions in Instant Messenger (IM) through tactile interfaces with vibration patterns [3,4,5,6]. However, in the proposed methods users

M.J. Smith and G. Salvendy (Eds.): Human Interface, Part I, HCII 2009, LNCS 5617, pp. 628–637, 2009.

have to memorize the vibration or pin matrix patterns and cognitively interpret the communicated emotional state. Demodulation of haptically coded emotion is not natural for human-human communication, and direct evocation of emotion cannot be achieved in such system.

Driven by the motivation to enhance social interactivity and emotionally immersive experience of real-time messaging, we are proposing a conceptually novel approach to reinforcing (intensifying) own feelings and reproducing (simulating) the emotions felt by the partner through specially designed system, iFeel_IM!. The philosophy behind the iFeel_IM! (**i**ntelligent system for **Feel**ing enhancement powered by affect sensitive **I**nstant **M**essenger) is "***I feel*** [therefore] ***I am!***". The emotion evoked by physical stimulation might imbue our communication with passion and increase the emotional intimacy, ability to be close, loving, and vulnerable. We argue that interpersonal relationship and the ability to express empathy grow strongly when people become emotionally closer through disclosing thoughts, feelings and emotions for the sake of understanding.

In this work, we focus on implementation of innovative system and devices for generation of physical stimulation aimed to convey the emotion experienced during online conversations. We attempt to influence on human emotions by physiological changes, physical stimulation, and social touch.

1.1 Physiological Changes Evoke Human Emotions

Our emotional states include two components: physical sensations (heart rate, respiration, etc.) and conscious feelings (joy, anger, etc.). Emotional states are mediated by a family of peripheral, autonomic, endocrine, and skeletomotor responses. When frightened, we not only feel afraid but also experience increased heart rate and respiration, dryness of the mouth, tense muscles, and sweaty palms [7].

According to James-Lange theory [8], the conscious experience of emotion occurs after the cortex receives signals about changes in physiological state. Researchers argued that feelings are preceded by certain physiological changes. Thus, when we see a venomous snake, we feel afraid because our cortex has received signals about our racing heart, knocking knees, etc. In [9], Prinz revived the James-Lange theory, supported it with reference to recent empirical research, and thoroughly criticized cognitive theories of emotions for downplaying the role of the body.

Among a variety of physiological changes accompanying the experience of emotion (heart rate, temperature, respiration, pulse, blood pressure, muscle action potentials), for our research we selected those (heart rate and temperature) that can be affected externally in a safe manner. The ability of false heart rate feedback to change our emotional state was reported in [10]. Briese [11] found a tight relationship between emotions and body temperature.

To our knowledge, the only work concentrating on the emotion induction through physical means is described in [12]. The interactive chair changes emotional state of the subject watching the movie with rich emotional content. However, user study did not show any significant effect of the chair on intensity of experienced emotion. The reported reasons are: (1) weakness of the technical solutions (in many cases users did not perceive any physical influence, and complained on interruption effect on the movie watching); (2) user-study procedure was not well elaborated.

1.2 Mediated Social Touch and Haptic Communication of Emotions

In a real world, whenever one person interacts with another, both observe, perceive and interpret each other's emotional expressions communicated through a variety of signals. Valuable information is also transferred by non-verbal communication (e.g. social touch). It is well known that touching is one of the most powerful means for establishing and maintaining social contact. The fact that two people are willing to touch implies an element of trust [13]. Expressive potential of touch is the ability to convey and elicit strong emotions. Although current on-line interaction mainly relies on senses of vision and hearing, there is a substantial need in mediated social touch. Comprehensive review of research on social touch is presented in [14].

Among many forms of physical contact, the most emotionally charged one is hug. It conveys warmth, love, and affiliation. The Hug Shirt allows people who are missing each other to send physical sensation of the hug over distance [15]. User can wear this shirt, embedded with actuators and sensors, in everyday life. The communication is performed through cell phone facilitated with Bluetooth and Java software. DiSalvo et al. [16] devised "The Hug" interface. When person desires to communicate hug, he/she squeezes the pillow that results in vibration and temperature changes in the partner's device.

Even though such devices can potentially increase interactivity, they have such drawbacks as: (1) the functionality is highly limited and is not extendable (only hug pattern can be transmitted); (2) inconvenient design (users found them difficult to use in real life); (3) inability to resemble natural hug sensation and, hence, elicit strong affective experience (only slight pressure is generated by vibration actuators); (4) lack of visual representation of the partner.

We believe that it is possible to highly increase the emotional immersion through the iFeel_IM! system featuring automatic affect recognition from text, visualization, and physical stimulation feedback.

2 iFeel_IM! Architecture

Nowadays, communication through IM and chat is very popular. However, during on-line communication people are concentrating on textual information and are only slightly affected emotionally. Conventional mediated systems usually (1) support only simple textual cues like emoticons; (2) lack visual emotional signals such as facial expressions and gestures; (3) support only manual control of expressiveness of graphical representations of users (avatars); (4) completely ignore such important channel of social communication as sense of touch.

In the iFeel_IM! system, great importance is placed on the automatic sensing of emotions conveyed through textual messages, visual reflection of the detected emotions and communicative behaviour through a 2D cartoon-like avatars, enhancement of user's affective state, and reproduction of feeling of social touch (e.g., hug) by means of haptic stimulation in a real world. The architecture of the iFeel_IM! system is presented in Fig. 1.

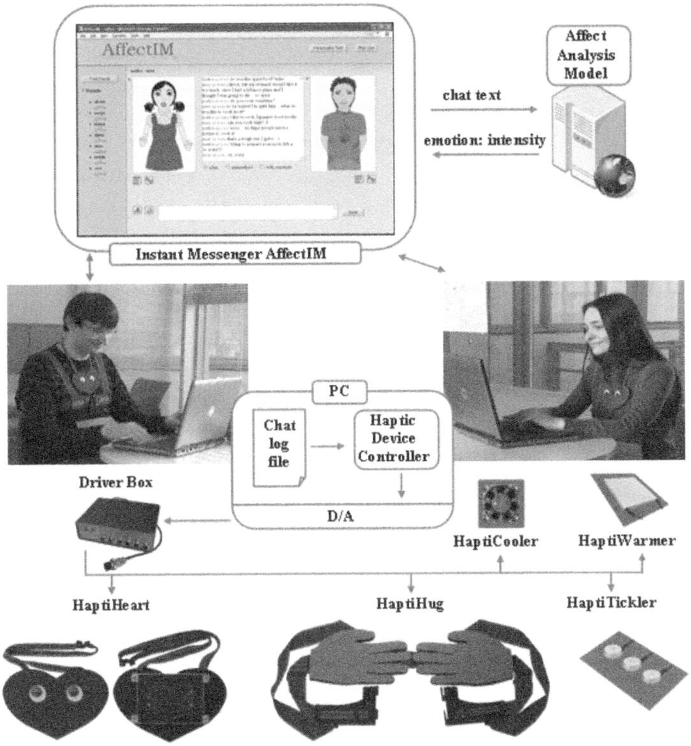

Fig. 1. Architecture of iFeel_IM! system

In order to communicate through iFeel_IM! system, users have to wear innovative haptic devices (HaptiHeart, HaptiHug, HaptiTickler, HaptiCooler, and HaptiWarmer) developed by us.

As a media for communication, we developed AffectIM [17] instant messenger. AffectIM receives the typed utterance, sends it to the Affect Analysis Model [18] responsible for recognition of emotions conveyed through text and located on the server, gets the result (dominant emotion and intensity for each sentence), and visually reflects the sensed affective state through avatar facial expression and gestures.

The results from the Affect Analysis Model are stored along with chat messages in a file on local computer of each user. Haptic Device Controller analyses these data in a real time and generates control signals for Digital/Analog converter (D/A), which then feeds Driver Box for haptic devices with control cues. Based on the transmitted signal, the corresponding haptic device (HaptiHeart, HaptiHug, HaptiTickler, HaptiCooler, or HaptiWarmer) worn by user is activated.

3 AffectIM: Intelligent Affect Sensitive Instant Messenger

AffectIM, supported by the Affect Analysis Model, was developed as a web-based application running in the Internet browser.

For affect categorization, we have decided to use the subset of emotional states defined by Izard [19]: 'anger', 'disgust', 'fear', 'guilt', 'interest', 'joy', 'sadness', 'shame', and 'surprise'. While constructing our Affect Analysis Model we took into account crucial aspects of informal online conversation such as its specific style and evolving language. The affect sensing algorithm consists of five main stages: (1) symbolic cue analysis; (2) syntactical structure analysis; (3) word-level analysis; (4) phrase-level analysis; and (5) sentence-level analysis. Analyzing each sentence in sequential stages, our method is capable of processing sentences of different complexity, including simple, compound, complex (with complement and relative clauses), and complex-compound sentences (details are given in [18]). The working flow of the Affect Analysis Model is presented in Fig. 2.

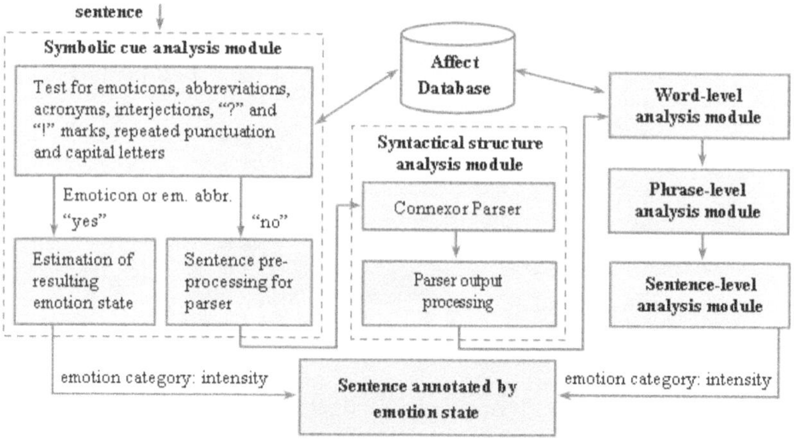

Fig. 2. Working flow of the Affect Analysis Model

An empirical evaluation of the Affect Analysis Model algorithm showed promising results regarding its capability to accurately classify affective information in text from an existing corpus of informal online communication. In a study based on blog entries, the system result agreed with at least two out of three human annotators in 70% of the cases.

The example of conversation in AffectIM (with emotion annotations from Affect Analysis Model) is given below:

A: *Hi, good to see you there!* [JOY]
B: *Hi, it is great that AffectIM eliminates spacial distance and connects us.* [SURPRISE]
A: *Right {}* [HUG]
B: *Oh, I really feel the warmth of your hug!!*
A: *Yes, it's so exciting, as if we met in a real world.* [INTEREST]
B: *How is your study at the University of Tokyo going on?*
A: *I have successfully graduated last month.* [JOY]
B: *Congratulations, I am very proud of you!* [JOY]
A: *My mind is now filled with pleasant memories about student life, but at the same time I regret this period, full of fun, flew away so fast.* [GUILT]

B: *You certainly deserve rest.*
A: *Former students of our Lab have invited me to the trip to Ski resort.*
B: *I remember the first time I tried to balance on a snowboard was definitely a very thrilling experience! Have you ever been descending a steep slope on a snowboard?* [FEAR]
A: *I always feel angry at myself for not having enough courage for snowboarding. Shamefully, I prefer safe skiing.* [ANGER, SHAME]
B: *Sure, extreme sport needs thorough training. My first attempt resulted in a broken leg. But now my experience and protective gear allow me to enjoy snowboarding.* [SADNESS, JOY]
A: *For me, risky snowboarding might be safe only in a virtual world.* :) [JOY]

The twenty-user study conducted on AffectIM [17] showed that the IM system with automatic emotion recognition function was successful at conveying users' emotional states, thus enriching expressivity and social interactivity of online communications.

4 Affective Haptic Devices

In order to support the affective communication, we implemented several novel haptic gadgets embedded in iFeel_IM!. They make up three groups. First one is intended for emotion elicitation implicitly (HaptiHeart, HaptiCooler, HaptiWarmer), second type evokes affect in a direct way (HaptiTickler), and third one uses sense of social touch (HaptiHug) for mood influence. All these devices produce a sense of touch including kinesthetic and coetaneous channels. Kinesthetic stimulations, produced by forces exerted on body, are sensed by mechanoreceptors in the tendons and muscles [7]. This channel is highly involved in sensing the stimulus produced by HaptiHug device. On the other hand, mechanoreceptors in the skin layers are responsible for cutaneous stimulation perception. Different types of tactile corpuscles allow us sensing thermal property of the object (HaptiCooler, HaptiWarmer), pressure (HaptiHeart, HaptiTickler), vibration frequency (HaptiTickler), and stimuli location (localization of stimulating device position enables association with particular physical contact).

4.1 HaptiHug: The Way of Realistic Hugging over Distance

When people are hugging, they generate pressure on the back of each other by hands, and on the chest area simultaneously. The key feature of the developed HaptiHug is that it physically reproduces the hug pattern similar to that of human-human interaction. The hands for a HaptiHug are sketched from a real human and made from soft material. The important point is that the hands were designed in such a way that the user feels as if the friend's hands actually contact him. Couple of oppositely rotating motors are incorporated into the holder placed on the user's chest area. Soft Hands, aligned horizontally, contact back of the human. Shoulder strips, supporting motor holder and Soft Hands, allow aligning vertical position of the device on human torso. Once 'hug' command is received, couple of motors tense the belt, pressing thus hands and breast part of HaptiHug in direction of human body (Fig. 3).

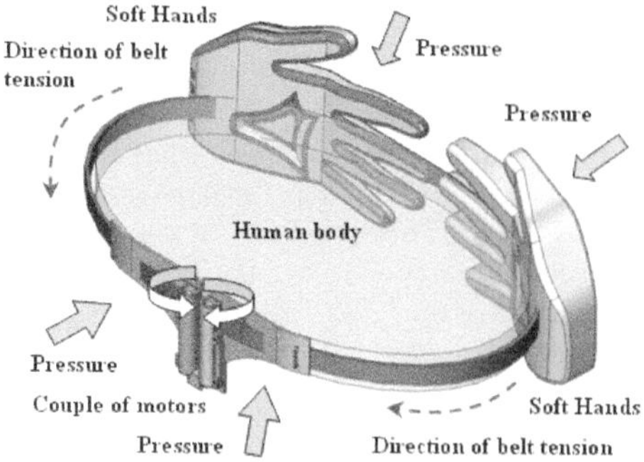

Fig. 3. Structure of wearable HaptiHug device

The duration and intensity of the hug is controlled by the software according to the detected emoticon or keyword. The Driver Box regulates the current magnitude in DC motor from 0 mA to 500 mA. For presentation of plain hug level (e.g., '(>^_^)>', '{}', '<h>'), big hug level (e.g., '>:D<', '{{}}'), and great big hug level (e.g., 'gbh', '{{{}}}'), the pressure of 200 N/m with duration of 2 sec, the pressure of 300 N/m with duration of 3 sec, and the pressure of 450 N/m with duration of 4 sec, was applied on the user's back and chest, respectively.

The technical specification of the device is as follows. DC motors RE 10 1.5 W in combination with planetary gearhead GP 10 A (gear ratio of 64:1) generate stall torque of 192.6 mNm. The pressure produced on the human chest by tensed belt equals 450 N/m at the most. Motor holder was manufactured on 3D printer Dimension DS 768 from ABS plastic material. Special slots were provided to make the pass for pressing belt.

Preliminary, Soft Hands were printed on 3D printer from ABS with thickness of 1 mm. However, plate flexibility was not enough to achieve touch sensation similar to human hand. Therefore, we decided to make hands from compliant rubber-sponge material. The contour profile of Soft Hand is sketched from the male human and has front-face area of 155.6 cm^2. Two identical pieces of Soft Hand of 5 mm thickness were sandwiched by narrow belt slots and connected by plastic screws. Such structure provides enough flexibility to tightly fit to the human back surface while being pressed by belt. Moreover, belt can loosely move inside the Soft Hands during tension period. The structure of HaptiHug is presented in Fig. 4.

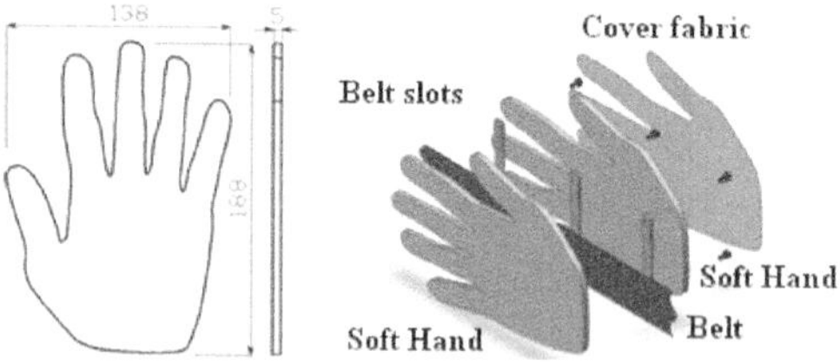

Fig. 4. Structure of Soft Hands

HaptiHug device has lightweight tri-glide bucklers and side release fastener integrated into the motor holder to facilitate easy adjustment of the belt size to her/his body sizes and detaching it in a natural and rapid manner.

4.2 HaptiHeart, HaptiCooler, HaptiWarmer, HaptiTickler. Or How We Can Enhance and Influence on Our Emotions by Haptics

Each emotion is characterized by a specific pattern of physiological changes. Fear features cold, sweaty hands, tensed muscles, increased heart rate, joy – warm hands, increased heart rate. We selected four distinct emotions with strong physical features: 'anger', 'fear', 'sadness', and 'joy'.

There is no doubt that feelings are intuitively connected with the heart, and our lexicon confirms this. The research on interplay between heart rate and emotions revealed that different emotions are associated with distinct patterns of heart rate variations [20]. We developed heart imitator HaptiHeart to produce special heartbeat patterns according to emotion to be conveyed or elicited (sad associated with slightly intense heartbeat, anger with quick and violent heartbeat, fear with intense heart rate).

We take advantage of the fact that our heart naturally synchronizes with the heart of a person we hold or hug. Thus, the heart rate of a user is influenced by haptic perception of the beat rate of the HaptiHeart. Furthermore, false heart beat feedback can directly be interpreted as real heart beat, so it can change the emotional perception.

The HaptiHeart consists of two modules: flat speaker FPS 0304 and speaker holder. The flat speaker sizes (66.5 x 107 x 8 mm) and rated input power of 10 W allowed us to design powerful and relatively compact HaptiHeart device producing realistic heartbeating sensation with high fidelity. The 3D model is presented in Fig. 5.

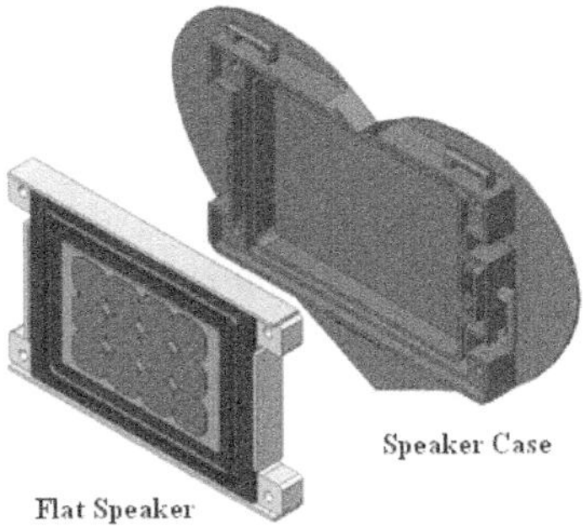

Fig. 5. HaptiHeart layout

We recorded the heartbeat patterns for the cases of experiencing fear, anger, and sadness. The pre-recorded sound signal with low frequency generates the pressure on the human breast through vibration of the speaker surface.

HaptiTickler is responsible for joy emotion evocation. The idea behind this device is to reproduce effect of *"Butterflies in the stomach"* (fluttery or tickling feeling in the stomach felt by people experiencing love) by means of circular arrays of vibration motors attached to user abdomen area.

HaptiWarmer is intended for simulation of warmth on the human skin to evoke either aggression or pleasant feeling. It was proved that uncomfortably hot temperatures arouse anger feelings [21]. On the other hand, temperatures within the comfortable region can elicit positive emotions. Based on this evidence, we change the temperature of Peltier element inside HaptiWarmer to be in the limits of 25-36 ℃ in the case of joy and temperature range of 40-50 ℃ for influencing on anger state. With motivation to boost the fear, we designed HaptiCooler that produces "*Coldness on back*" effect simulated by the cold airflow along with cold side of Peltier element.

5 Conclusion and Future Research

In a nutshell, while developing the iFeel_IM! system, we attempted to bridge the gap between mediated and face-to-face communications by enabling and enriching the spectrum of senses such as vision and touch along with cognition and inner personal state. In the paper we described the architecture of iFeel_IM! and development of novel haptic devices, such as HaptiHeart, HaptiHug, HaptiTickler, HaptiCooler, and HaptiWarmer. These devices were designed with particular emphasis on natural and realistic representation of the physical stimuli, modular expandability, and ergonomic human-friendly design. User can perceive the intensive emotions during online communication, use desirable type of stimuli, comfortably wear and easily detach devices from torso. We are planning to conduct the user study experiments to estimate the performance of the developed system.

On-line communicating users are usually invisible for each other (non-verbal channel is deactivated), and can dissemble actual emotions. This can lead to confusion during text messaging. The iFeel_IM! system can make communication emotionally truthful, so that users will tend to express emotions they actually perceive. It should be noted, that our system can not only enhance the emotional state of the user but also generate emotional feedback (empathy) in communicating person. In future research, we will investigate the topic "how can we make the partner to feel our own emotions". We believe that iFeel_IM! will encourage users to get in "touch" with their emotions and to make social contact with other people in on-line communication richer and more enjoyable.

Computer gaming is very interesting application area for affective haptic devices. The success of Wii has demonstrated that better graphics alone (PS3, Xbox360) is not the key to the game success. Its popularity is mainly due to high involvement of the user into the game through interactivity. By using kinaesthetic and tactile senses with Wii controller, players experience greater physiological arousal. We are planning to design two types of interactive affective game with emotional feedback to create fantastic and emotionally charged playing experiences: (1) player goal is to change emotional state of the virtual characters by performing social touch actions (hugging, patting on the back, handshaking, etc.); (2) during on-line communication players attempt to convey own emotions to partner.

Acknowledgments. The research is supported in part by a Japan Society for the Promotion of Science (JSPS) Postdoctoral Fellowship for Foreign Scholars.

References

1. Damasio, A.: The Feeling of What Happens: Body, Emotion and the Making of Consciousness. Vintage, London (2000)
2. Picard, R.W.: Affective Computing. MIT Press, Cambridge (2000)
3. MacLean, K., Enriquez, M.: Perceptual Design of Haptic Icons. In: EuroHaptics, pp. 351–363 (2003)
4. Shin, H., Lee, J., Park, J., Kim, Y., Oh, H., Lee, T.: A Tactile Emotional Interface for Instant Messenger Chart. In: 12th International Conference on Human-Computer Interaction, pp. 166–175. Springer, Heidelberg (2007)
5. Rovers, A.F., Van Essen, H.A.: HIM: A Framework for Haptic Instant Messaging. In: 26th SIGCHI Conference on Human Factors in Computing Systems, pp. 1313–1316. ACM, New York (2004)
6. Mathew, D.: vSmileys: Imagine Emotions through Vibration Patterns. In: Alternative Access: Feeling and Games (2005)
7. Kandel, E.R., Schwartz, J.H., Jessel, T.M.: Principles of Neural Science. McGraw-Hill, New-York (2000)
8. James, W.: What is an Emotion? Mind 9, 188–205 (1884)
9. Prinz, J.J.: Gut Reactions: A Perceptual Theory of Emotion. Oxford University Press, Oxford (2004)
10. Decaria, M.D., Proctor, S., Malloy, T.E.: The Effect of False Heart Rate Feedback on Self Reports of Anxiety and on Actual Heart Rate. Behavior research & Therapy 12, 251–253 (1974)
11. Briese, E.: Emotional Hyperthermia and Performance in Humans. Physiology & Behavior 58(3), 615–618 (1995)
12. Kok, R., Broekens, J.: Physical Emotion Induction and Its Use in Entertainment: Lessons Learned. In: 1st IFIP Entertainment Computing Symposium, pp. 33–48. Springer, Heidelberg (2008)
13. Collier, G.: Emotional Expression. Lawrence Erlbaum Associates Inc., New Jersey (1985)
14. Haans, A., Jsselsteijn, W.I.: Mediated Social Touch: A Review of Current Research and Future Directions. Virtual Reality 9, 149–159 (2006)
15. Hug Shirt. CuteCircuit Company, `http://www.cutecircuit.com`
16. DiSalvo, C., Gemperle, F., Forlizzi, J., Montgomery, E.: The Hug: An Exploration of Robotic Form for Intimate Communication. In: 12th IEEE International Workshop on Robot and Human Interactive Communication, pp. 403–408. IEEE Press, New-York (2003)
17. Neviarouskaya, A., Prendinger, H., Ishizuka, M.: User Study of AffectIM, an Emotionally Intelligent Instant Messaging System. In: Prendinger, H., Lester, J.C., Ishizuka, M. (eds.) IVA 2008. LNCS, vol. 5208, pp. 29–36. Springer, Heidelberg (2008)
18. Neviarouskaya, A., Prendinger, H., Ishizuka, M.: Textual Affect Sensing for Sociable and Expressive Online Communication. In: Paiva, A.C.R., Prada, R., Picard, R.W. (eds.) ACII 2007. LNCS, vol. 4738, pp. 220–231. Springer, Heidelberg (2007)
19. Izard, C.E.: Human Emotions. Plenum Press, New York (1977)
20. Anttonen, J., Surakka, V.: Emotions and Heart Rate while Sitting on a Chair. In: 27th SIGCHI Conference on Human Factors in Computing Systems, pp. 491–499. ACM, New York (2005)
21. Anderson, C.A.: Temperature and Aggression: Ubiquitous Effects of Heat on Occurrence of Human Violence. Psychological Bulletin 106(1), 74–96 (1989)

An Interactive System Based on Semantic Graphs

Johann Vandromme, Samuel Degrande, Patricia Plénacoste, and Christophe Chaillou

LIFL, Université de Lille 1, Cité scientifique
Bâtiment IRCICA, 50 Avenue Halley, Haute Borne, 59650 Villeneuve d'Ascq, France
{johann.vandromme,samuel.degrande,patricia.plenacoste, christophe.chaillou}@lifl.fr

Abstract. This paper introduces an original method towards post-WIMP desktops. Our main contribution involves a new way of organizing desktops with respect of the Instrumental Interaction model and relies on semantic graphs. A semantic graph expresses semantic dependencies of documents or components like a scene graph expresses geometrical dependency. This structure allows users to visualize the desktop given a chosen criterion. We also propose polymorphic tools that enable Direct Manipulation: tools directly modify the semantic visual factor leading to a semantic modification of object. A prototype is proposed to illustrate this concept.

Keywords: software design, interfaces, graphs, semantic, desktop, tools.

1 Introduction

In WIMP systems, application functionalities are represented by buttons grouped in menus, sub-menus and tool boxes. Given the increasing number of these functionalities, graphical interfaces grow in complexity. Some rules and habits exist to design menus and interfaces but they are limited to common operations such as Open File, Save File, Print, etc. For specific tools, interaction modalities and their location in the interface may vary from one application to another (e.g. layer manipulation in 2D image editing software).

The second element that induces such complexity is the lack of interoperability between applications. To edit a specific type of file, one needs to launch a specialized application, then export the file in a generic format. The generic format can eventually be imported into another application. This export/import operation tends to be replaced by drag-and-drop between applications but is not generalized at the moment.

Current desktop managers and tools tend to provide features that are document-centered. The file-tree is about to disappear for indexation systems based on Meta data (WinFS[1], spotLight[2], beagle[3], google desktop[4]). The documents are linked in

[1] http://msdn.microsoft.com/en-us/magazine/cc164028.aspx
[2] http://support.apple.com/kb/HT2531?viewlocale=en_US
[3] http://beagle-project.org/
[4] http://desktop.google.com/features.html

M.J. Smith and G. Salvendy (Eds.): Human Interface, Part I, HCII 2009, LNCS 5617, pp. 638–647, 2009.

several ways, using some semantic criterion. This increases the document research power that is not limited to document name, type, size or modification date.

Our first aim is to extend this principle to the documents content. We would like to bring uniformity in the desktop description with a generic structure for documents, components and system data.

Our second aim is to create a unique tool box for the whole system: each tool can act on every type of document. The main interest is to reduce the cognitive charge of the user by reusing polymorphic tools. In current systems, an application (tools and opened documents) are embedded into a single window. Given this unique toolbox, it will induce a window less system: the desktop is made of generic tools and a set of documents.

The third aim is to bind a visual parameter to a given semantic aspect. This allows a quick overview of documents sharing a common property and to extend direct manipulation by using graphical tools to modify semantic values.

2 State of Art

This section recaps previous works which introduce attempts to replace WIMP systems with direct manipulation. We will first present theory about instrumental interaction, then the software toolkits that implement this concept.

2.1 Direct Manipulation and Instrumental Interaction

In Shneiderman's direct manipulation definition [1], the user manipulates documents rather than files. This principle was first introduced by Xerox Star in 1981[2] and respects the three main principles of:

- Continuous representation of the objects and actions of interest: graphically modifying an object rather than using a dialog box;
- Physical actions or presses of labeled buttons instead of complex syntax: encourage graphical interfaces to command lines;
- Rapid incremental reversible operations whose effect on the object of interest is immediately visible.

These principles stimulate discovery and interface exploration and reduce the learning time by proposing intuitive interactions [3].

As an extension, Michel Beaudouin Lafon proposes the Instrumental Interaction model [4] in order to simplify current interfaces and to allow more flexibility for interaction development. The principle is to replace application-based interfaces by document-centered ones: Objects of the domain (i.e. documents) are made of a set of attributes that can be modified via interaction instruments: instruments are defined as attributes modifiers. As a complement for his model, he advises developers of interactive applications to make interaction as first class object and to separate the interaction code from the tool executive code [5]. These principles allow to create high level and post-wimp interactions, such as bi-manual ones.

Implementing high level interactions is difficult in current WIMP system. These 2D interfaces are built upon DOM trees [6] (e.g. XUL[5]). DOM trees are an equivalent of 2D scene graphs where each node is a wimp component (button, text box and layer) and expresses the geometrical dependence for a node to its parent. In such a structure, each component embeds its own interaction code as events. Therefore a high level interaction, or an interaction that needs a combination of different components (such as drag & drop), is split inside the DOM tree and becomes very hard to design or maintain; Designing interaction as first class object avoids the code of a high level interaction to be split among the nodes of a scene graph.

Beaudouin Lafon defined several principles for his model [7]:

- Reification is the process by which concepts are turned into objects – as an example: an interaction instrument is a reification of a command and instruments are objects that can be operated by Meta instruments.
- Reuse: an interaction or a set of interactions can be reused as pattern. For instance, the redo command can replace a set of interactions or the copy-paste allows reusing the result of previous user commands.
- Polymorphism is the property that enables a single command to be applicable on objects of different types. One can create polymorphic tool several ways: for current file format systems with multiple existing formats, it is necessary to create a tool with a code for each type of document. For a new type of document a new tool code has to be developed. The second possibility requires designing a common structure for each type of document. Thus, a tool can act on any type of document by modifying the common structure.

Instrumental interaction is a pertinent model oriented on direct manipulation. This is a formal model with several guidelines but without associated specification. Michel Beaudouin Lafon has tested it on a specific application dedicated to colored Petri nets [8]. This application includes a full set of modern interactions such as floating palettes, tool glasses and marking menus.

2.2 Related Toolkits

To achieve a system based on instrumental interaction model, toolkits will have to respect the separation of interaction with representation and introduce a generic structure for every type of documents for tool polymorphism.

Toolkits based on Instrumental Interaction rely on scene graph for representation, which is a classical design for 3D and now 2D graphical toolkits (like SVG[6] , or [9] [10]). This common model defines the relationship container/contained. However, programming interactions in such a graph is equivalent to current WIMP interaction definition.

To separate interaction from display, Huot & Dumas propose to design two different graphs: the DOM tree that deals with display and an interaction tree for the object's behavior [11]. In MaggLite based application, interaction graphs rely on ICON (Input Configurator) model [12]: the interaction graph is composed of devices. A

[5] http://developer.mozilla.org/en/XUL_Reference
[6] http://www.w3.org/TR/SVG

device is a black-box that can receive input values and can generate output values. The graph begin state is a system device (input peripheral), the internal states are a set of Library Devices (transform an input signal to get a modified one) and end states are Application devices (control the application objects). At the moment, this toolkit has been used to design classical windows-based applications and there is no proposition about managing a whole desktop system.

The DPI model (Document / presentation / interaction) [13] is an implementation model based on Michel Beaudouin Lafon's Instrumental Interaction. This generic model manipulates different types of document and can design a desktop system. The components of this model are documents (containing the domain data), Presentation (the representation of the document's data) and instruments (for interaction). The separation between presentation and documents allows multiple presentations coexistence.

An advantage of the DPI model is to provide a document structure to design polymorphic tools: Objects are defined as a set of attributes but unlike the Instrumental Interaction model, instruments don't modify directly attributes of documents: they create actions (tools are action producers) on components (action consumers). This system allows polymorphism: when an instrument tries to modify a document, it compares produced action (what the tool can do) to consumable action of the document (what is modifiable). If the actions are compatible, the object is modified. The DPI model is an important step in modeling a workspace centered on documents and instruments.

2.3 Limits

The previous systems present several drawbacks:

- Firstly tools are defined as attributes modifiers: they operate only on attributes and thus the structure of the geometric tree cannot be modified. The problem is, sometimes one needs to modify the scene graph (for instance, for a drag & drop operation). To solve this problem, one can define a specific tool that modifies the tree structure.
- Secondly these toolkits are dedicated to document edition. Their tools can only modify components, but cannot be used to manage documents.
- Finally interactions are based on graphical representations of documents. This principle limits the range of actions to graphical modifications. For instance, changing a geometrical aspect of a component using direct manipulation is possible but changing an attribute that expresses the relevance of a component often requires a dialog box.

3 Proposal

Our will is to extend the principles of instrumental interaction to manage a full desktop system. Our proposition must deal with components, documents or activities. Activities are several types of documents or components that share common semantic properties; to add semantic information, one classical possibility is to add attributes in each node in the DOM tree, but this principle doesn't deal with hierarchic semantic links. We propose to extend above toolkits capabilities by structuring the whole desktop using semantic multi-graph structures.

3.1 Structuring Desktop with Semantic Graphs

We have chosen to use several tree structures called semantic graphs. These structures will allow to group objects having hierarchical semantic properties (Fig 1).

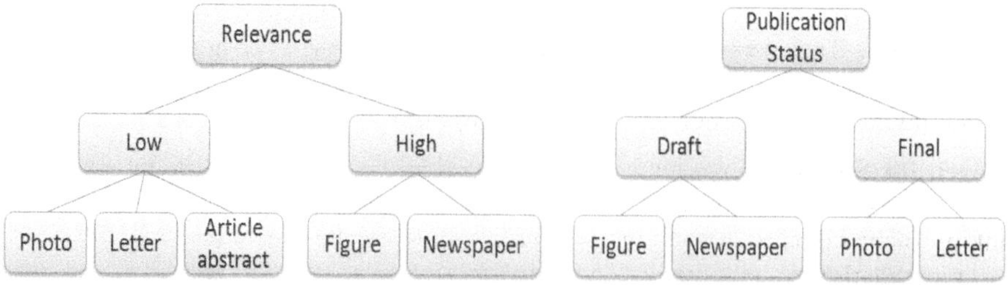

Fig. 1. Two examples of semantic graphs used in the desktop. On the left, the relevance graph groups documents depending on their relevance value: the figure and newspaper are highly relevant, the photo, letter and article abstract are not. On the right, the publication graph groups elements in two categories: draft or final. We can notice that the same objects can be present in different graphs.

A semantic graph expresses semantic dependencies of documents or components like a scene graph expresses geometrical dependency. It is defined this way: semantic properties are nodes and objects of the desktop are leafs. The objects of the desktop are documents or components (parts of documents). This definition allows to add granularity or precision in the semantic definition of an object by adding intermediate nodes in the graph. For instance, one can refine the relevance category by adding intermediate nodes that express the deadline time for each document.

According to that definition, a given object can be present in multiple graphs, given that an object can have different semantic properties.

The whole desktop is composed of semantic graphs:

- Some of them represent document structure. For example, the document's DOM-tree defines how components are assembled.
- Some are used by the system. As an example, document write-protection or the objects backups are stored as graph.
- Others can be created by the user to manage his/her activities: he/she can define his/ her own semantic values in order to group and retrieve objects the way he/she chooses.

3.2 Polymorphic Tools

Since the whole system is structured the same way, this leads to a harmonization of the desktop description and allows the tool polymorphism. The same tools can be used to modify a component (modifying an image), to manage the documents (positioning the image in the documents) or to manage the desktop (moving documents in the desktop): so, in our system, tools can modify objects' attributes (as in Instrumental Interaction) but also semantic links between objects.

These generic tools can be gathered in a unique tool box. So, an application defined as a container that gathers manageable documents and associated tools is no longer valid. It can be replaced to create a document centered system. Our desktop is made of a set of documents, a tool box and a set of graphs. Using the same tool box in a whole system allows transfer of knowledge when facing a new type of document: the user is familiar to his/her toolbox and knows the capabilities of the tools. Our tools respect interaction modality based on direct manipulation. In current applications, the user chooses the document to edit by opening it in an application: the choice of tools is defined by the application. In our system, one selects a tool and then applies it on a document. This modality is close to the real world where a user can take the same pen (thanks to polymorphism) to annotate a post-it, then to correct student's copies or to color a drawing.

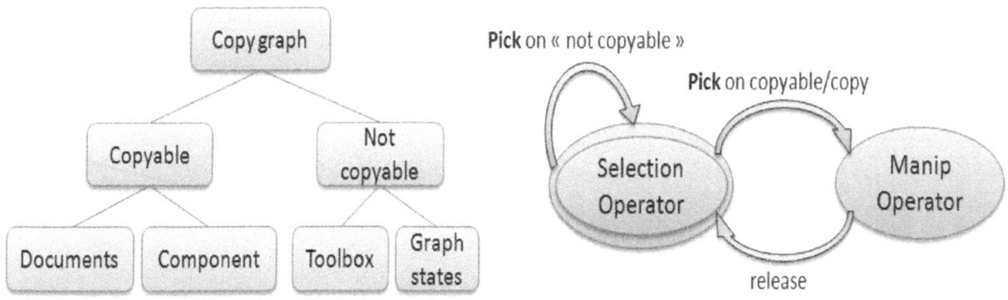

Fig. 2. An automaton that deals with the copy capability of an object: Combination of automaton rule and semantic graph won't allow copy of toolbox and graph state objects. The transition that loops on selection operator is defined via a rule.

Such a proposal (where a tool can act the same way with documents, components or system) is too permissive. First, some special actions must not be permitted in some cases. Secondly, some specific tools (tools that can be applied only on some types of documents) exist, so a mechanism is necessary to create exceptions. Our Tools are designed as interaction automata [14]: basic operators can be linked together to create a high level tool. A system of rules can modify the default behavior of automaton. Given a specific criterion, a rule can prevent a state change or can perform a specific action. As an example, there can be objects in the desktop system that cannot be copied (only one instance of the object is allowed). These objects can be defined as leaves for semantic parent "not copy-able". A rule will avoid copy operation for all objects that share that property when a displacement operation is made (see Fig 2).

This system of rules will be used to apply specific behavior for one tool on one type of document or to manage the mechanism that deal with specific tools.

3.3 Direct Manipulation and Semantic Graphs

Our method allows direct manipulation extension. By associating a visual parameter to a semantic property, one can visualize a semantic aspect of the desktop. This system allows the user to display his documents from a chosen semantic point of view (Fig 3).

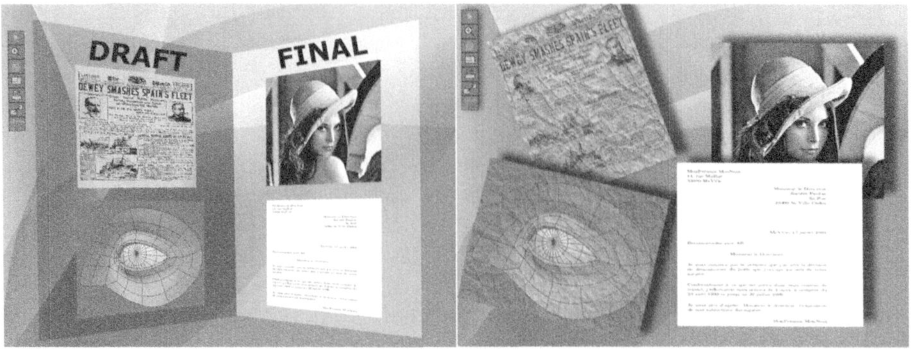

Fig. 3. A-B. Two different examples of semantic visualization of the desktop with the "publication status" semantic criteria: on the left, we use a labeled container for each value (draft or final). On the right, a crumple and rotation effect is applied to draft documents.

Fig. 4. The upper left desktop is visualized with "relevance" semantic value as scale visual factor. The upper right picture is displayed with blur effect as completion factor. The lower picture is an example of semantic composition of the two previous semantic values.

As a consequence, having a graphical representation of a semantic graph allows the use of generic graphical tools to change a semantic value of an object. This can enhance the use of direct manipulation and reduces the use of dialog-boxes by providing a way to use generic visual tools to modify semantic attributes. However, a "bijection" is required between visual effect and semantic value for the visual modifications to be translated back to semantic values. In the example of the publication graph (Fig 3.A), moving a document from Draft to Final will modify its status.

A mechanism is necessary to select which semantic values will be set and how it will be visualized. It must be able to compose with different semantic properties to get a view that mixes several semantic values (Fig 4). This mechanism is independent of our semantic graph model, it depends only on the implementation of the model.

3.4 Predict User's Objectives

In specific cases, the system can help predict user's objective. Let us consider an object displacement from a container to another. A system solely based on the scene graph cannot determine whether the object is simply translated or a modification of the scene graph is required. The first case induces the modification of the position attribute and the latter case induces that the user wants to perform a drag-and-drop operation (the object hierarchy must change and induces to unlink the object from its parent to link it to the new container). A classical scene graph only provides the information that a shape has moved from one position to another. In our system, the system can obtain several semantic aspects of a component to determine what to do.

With a drag-and-drop operation in Fig 3.A, our solution can provide additional information to the system that a document has moved from a container "draft" to a container "final" which implies that the parent relationship must change.

4 Demonstration Desktop

To illustrate our proposition, we have developed a demonstration desktop based on semantic graphs. This desktop is document centered: it is made of a set of documents and a set of generic tools gathered in a unique toolbox. Several generic tools (which can act on different semantic aspects using direct manipulation) are proposed (relevance, publication, geometry...).

There are several possible strategies to manage specific tools. One can choose to make every tool available at any time. But this choice can enhance complexity of the tool box. Another possibility is to attach specific tools to specific documents, but this method is close to actual windowing systems that link documents to their tools via application context. We have chosen to manage with specific tools this way: specific tools of all displayed documents are added to the tool bar. This allows the user to have every tool for documents gathered and available and to keep the interaction modality used in real world: "choose the tool, then apply the tool". This is coupled with visual clues when a specific tool is activated to inform the user whether the tool can be applied on documents or not.

Our demonstration is based on OpenGL, but is still a 2D-desktop. Indeed, the 3D engine is used to display 3D effects, like overlapping elements using different depths cues. Another advantage of using a modern 3D engine is the possible use of extended visual effects. These effects can be obtained and modified in real time using Shaders. Our aim is to use these different visual effects such as lights, shades or blur to express semantic aspects of documents or elements.

For future works, we would like to integrate a Meta mode view. In this mode, the user can choose which graphs will be displayed from a list of semantic properties. The selection of several graphs will display every element that shares these properties. To apply a graphical effect to a semantic property, one can apply a visual tool directly on a selected node of a graph representation. The system will link this modification to the visual representation of every element that shares that property (Fig 5).

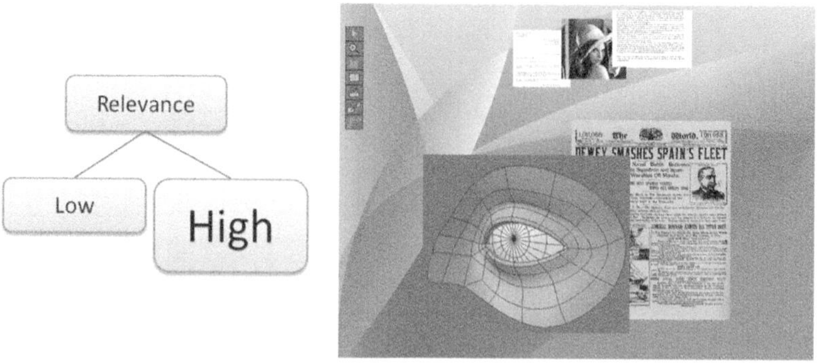

Fig. 5. In Meta mode, modifying the representation of the graph can modify associated objects: rescaling the "High" node in relevant graph (with the generic scale tool) rescales the associated objects on the desktop.

5 Conclusion

We propose a method for a desktop based on instrumental interaction with the use of semantic graphs. Semantic graph gathers documents or components that share common semantic values. Displaying a semantic graph allows to get different points of view of the desktop and to extend direct manipulation by using graphical tools to change a semantic value. The semantic graph structure is common to every component of the desktop system, thus enabling polymorphic tools creation. Every tool is gathered in a unique toolbox and a system of rules is set up to manage specific tools. A demonstration desktop is introduced as an implementation of our proposition. Several features are shown to prove our system capabilities. Future works will involve the creation of a Meta mode and evaluations to validate our approach.

References

1. Shneiderman, B.: Direct manipulation for comprehensible, predictable and controllable user interfaces. In: Moore, J., Edmonds, E., Puerta, A. (eds.) Proceedings of the 2nd international Conference on intelligent User interfaces, IUI 1997, Orlando, Florida, United States, January 6-9 (1997)
2. Johnson, J., Roberts, T.L., Verplank, W., Smith, D.C., Irby, C.H., Beard, M., Mackey, K.: The Xerox Star: A Retrospective. Computer 22(9), 11–26, 28–29 (1989)

3. Blanch, R.: PhD Thesis: Architecture logicielle et outils pour les interfaces hommes-machines graphiques avancées, http://en.scientificcommons.org/17563015
4. Beaudouin-Lafon, M.: Instrumental interaction: an interaction model for designing post-WIMP user interfaces. In: Proceedings of the SIGCHI Conference on Human Factors in Computing Systems, CHI 2000, The Hague, The Netherlands, April 1-6 (2000)
5. Beaudouin-Lafon, M.: Designing interaction, not Interfaces. In: Proceedings of the working conference on Advanced visual interfaces, Gallipoli, Italy (2004)
6. Document Object Model (DOM) Level 2 Core Specification - W3C Recommendation November 13 (2000), http://www.w3.org/DOM/
7. Beaudouin-Lafon, M., Mackay, W.E.: Reification, polymorphism and reuse: three principles for designing visual interfaces. In: Proceedings of the Working Conference on Advanced Visual interfaces, AVI 2000, Palermo, Italy (2000)
8. Beaudouin-Lafon, M., et al.: CPN/Tools: A Post-WIMP Interface for Editing and Simulating Coloured Petri Nets. In: Colom, J.-M., Koutny, M. (eds.) ICATPN 2001. LNCS, vol. 2075, pp. 71–80. Springer, Heidelberg (2001)
9. Bederson, B., Meyer, J., Good, L., Jazz, L.: An extensible Zoomable User Interface Graphics Toolkit in Java. In: Proceedings of IUST 2000: 13th ACM Symposium on User Interface and Software Technology, ACM/SIGCHI, San Diego, CA, November 2000, pp. 171–180 (2000)
10. Lecolinet, E.: A molecular architecture for creating advanced interfaces. In: CHI Letters, pp. 135–144. ACM Press, New York (2003)
11. Huot, S., Dumas, C., Dragicevic, P., Fekete, J.D., Hégron, G.: The MaggLite post-WIMP toolkit: draw it, connect it and run it. In: Proceedings of the 17th Annual ACM Symposium on User interface Software and Technology, UIST 2004, Santa Fe, NM, USA, October 24-27 (2004)
12. Dragicevic, P., Fekete, J.D.: Input Device Selection and Interaction Configuration with ICON. In: Proceedings of IHM-HCI 2001, People and Computers XV – Interaction without Frontiers, Lille, France, September 2001, pp. 543–548. Springer, Heidelberg (2001)
13. Beaudoux, O., Beaudouin Lafon, M.: OpenDPI: a toolkit for developing Document-Centered environments. In: Actes Neuvièmes Journées sur l'Interaction Homme-Machine, IHM 1997, Poitiers, septembre 1997, Cépaduès-Editions (1997)
14. Martin, N., Degrande, S., Chaillou, C.: Adaptation dynamique de modalités d'interaction 3D par règles de filtrage. In: Journées de l'Association Français de Réalité Virtuelle (AFRV) (2006)

Evaluation for Adjustment Method of Vehicle's Location by Recognizing Crosswalks

Yoshihisa Yamaguchi[1], Takashi Nakagawa[1], Hirokazu Kato[2], and Shogo Nishida[3]

[1] Advanced Technology R&D Center, Mitsubishi Electric Corporation,
8-1-1, Tsukaguchi-honnmachi, Amagasaki, Hyogo, 661-8661, Japan
{Yamaguchi.Yoshihisa@dp,
Nakagawa.Takashi@dy}.MitsubishiElectric.co.jp
[2] Graduate School of Information Science, Nara Institute of Science and Technology
kato@is.naist.jp
[3] Graduate School of Engineering Science, Osaka University,
1-3, Machikaneyama, Toyonaka, Osaka, 560-8531, Japan
nishida@sys.es.osaka-u.ac.jp

Abstract. We have developed AR-Navi, an in-vehicle navigation system that captures real-time video via an in-vehicle camera and overlays guidance on a video display. In this study, we conducted a driving experiment to evaluate the location-adjustment-method of AR-Navi. We also discuss the registration accuracy of the route guidance that AR-Navi gives at intersections.

Keywords: Augmented reality, In-vehicle navigation systems, Route guidance.

1 Introduction

In recent years, over 32 million in-vehicle navigation systems have been produced in Japan [1], and have become popular. Their basic functions, such as map indication and route planning, perform sufficiently well, so manufacturers have been mainly improving their user interface design. For example, newer functions are being provided, such as using 3D computer graphics (CG) and mixing aerial photos and existing maps to show cityscapes.

To improve the user interface design, some researchers have developed in-vehicle navigation systems that use video-based augmented reality technology [2-4]. The system has an in-vehicle camera that captures real-time video, and the system then overlays guidance on to the video. Because the video is almost the same as what the driver is seeing, the system provides easily understandable guidance. However, the system has some problems: if registration gaps between real-time video and CG occur driver may be confused. Sawano [2] describes extracting geometry information of roads and drawing CG roads on to real-time video. Hu [3] describes extracting future points of roads and adjusting the location between the guidance, which is drawings of the roads, and real-time video. These works detail the high-precision alignment between real-time video and CG with image analysis. The image analysis technology used in these works, however, also has some problems. First, the technology needs

M.J. Smith and G. Salvendy (Eds.): Human Interface, Part I, HCII 2009, LNCS 5617, pp. 648–656, 2009.

much time complexity. Second, it is not robust enough for conditions such as rain, night, backlight, and so on, so it is not used in current in-vehicle navigation systems. Therefore, we have developed a registration adjustment method base on recognizing crosswalks and have tried to develop a commercializable system using this method, which needs only a camera and can adapt to various environmental conditions. We call the system AR-Navi.

In this paper, we explain the adjustment method used for AR-Navi. Then we detail the evaluation of the method by driving on public roads and discuss the registration accuracy of the route guidance that AR-Navi shows at intersections.

2 Registration Adjustment Based on Recognizing Crosswalks

2.1 Location Gap

When AR-Navi overlays information on to the video, it adjusts positions between the information and the video on the basis of the location and direction of the driver's vehicle. If the location and direction have errors, gaps occur in the overlaid image.

Almost all current in-vehicle navigation systems calculate their own location and direction on the basis of two steps [5]. At the first step, the systems estimate their location and direction using a combination of Global Positioning System (GPS) data and autonomous navigation data such as that from gyros and speed sensors. At the second step, they use a map matching method, which compares the result of the first step with road shapes. Furthermore, the systems correct the vehicle location if the driving greatly changes (such as at turns) or there are differences between the driving and road shapes. Therefore, immediately after a vehicle turns, the location and direction changes highly accurately. However, huge errors sometimes happen if a vehicle goes straight for a long time without enough GPS signal. We have divided the gaps between location and direction into four categories and consider solutions as follows.

- Back-and-forth gap: As I have indicated before, the longer a vehicle goes straight, the more back-and-forth gap widens. Because a huge back-and-forth gap makes drivers confuse a wrong intersection with the right one, it is necessary to narrow this gap.
- Right-and-left gap: Because of using map matching, the calculated location rarely leaves the road. Therefore, the right-and-left gap does not exceed road width. If it does not depart from the road, AR-Navi does not give wrong guidance, so this gap is not a practical issue.
- Rotation gap: Current in-vehicle navigation systems estimate their direction on the basis of gyros. The rotation gap widens easily because gyros integrate the change in angle. Moreover, the measurement is delayed. It misaligns guidance and video when the vehicle direction changes greatly, such as when making a U-turn. Thus, it is necessary to narrow the rotation gap.
- Pitch-angle gap: If a vehicle goes along at a constant speed, the vehicle is parallel to the ground, but while accelerating or braking the vehicle inclines. The pitch-angle gap, however, is not a practical problem because it is only a temporary phenomenon.

Back-and-forth and rotation gaps must be adjusted when in-vehicle systems overlay CG information on to the video. Because we use an image-filtering method to remove images that have large rotation gaps [4], we only have to deal with back-and-forth gaps. To resolve this problem, we propose registration adjustment based on recognizing crosswalks.

2.2 Proposed Method

To decrease back-and-forth gaps, we propose adjusting the location at intersections, which are found using image analysis, corresponding with intersections on the maps. It takes the following conditions to use the algorithm in current in-vehicle navigation systems.

- Little time complexity: For the algorithm to work in current in-vehicle navigation systems, time complexity needs to be small.
- High robustness: The algorithm has to be robust to work in different weather and levels of brightness.

Considering the above conditions, this method finds crosswalks by analyzing only one line that is set on a captured image and located a certain distance ahead of the driver's vehicle. The line is called the scan line.

Then it calculates the back-and-forth gap by comparing the found crosswalk with the corresponding crosswalk on the map, because it is assumed that there is an intersection between a pair of crosswalks.

This method has the following merits.

- It needs little time complexity because only the image of a narrow area, the scan line, is analyzed.
- If the scan line is set near enough, it is rarely interfered with by other vehicles while driving, so image analysis is stable.
- If environmental conditions are unfavorable, such as rain and darkness, the scan line is lit by the car's headlights.

This method works as follows.

1. It judges whether a zebra pattern exists or not.
2. It stores the position where vehicle goes into and leaves out of the zebra pattern.
3. It judges whether the zebra pattern is a crosswalk or not.
4. It recognizes a pair of crosswalks.
5. It calculates the distance between the two crosswalks.
6. It calculates midway between a location going into and location coming out of a crosswalk, if the distance is not too short or too long.
7. It adjusts the back-and-forth gap by difference between the calculated midway and the corresponding crosswalks on the map.

Judging whether a Crosswalk Exists or not. To find crosswalks, a crosswalk empathies filter is applied to the brightness of each pixel on the scan line. Fig. 1 shows the crosswalk empathies filter. It is easy to obtain high-accuracy binarized results because the filter turns the raw brightness into the brightness shown in Fig. 2.

Then this method judges whether a crosswalk exists or not on the scan line by checking the number of black and white runs. If the lengths of each run are almost same as the widths of the white lines of a crosswalk in the image and the black and white runs alternate, this method judges it to be a crosswalk.

-1	0	0	0	0	2	0	0	0	0	-1

Fig. 1. Crosswalk empathies filter

Fig. 2. Result of applying of Crosswalk empathies filter

Judging whether a Pair of Crosswalks is Appropriate or not. As was discussed in section 2.2, to find a crosswalk, two crosswalks must be found. Even if two crosswalks are found, however, there is a possibility of image analysis failing. For this reason, this method does not adjust location if the width of a crosswalk or the distance between the two crosswalks is too short or too long.

3 Evaluation of Proposed method

3.1 Conditions

We installed AR-Navi based on our in-vehicle navigation system and a lens, the specifications of which were 320 * 240 pixels, 6 mm focal length, and 47-degree angle of view. We drove in urban areas using AR-Navi and measured the back-and-forth gaps.

When registration adjustments not caused by the proposed method occurred, the effect of the proposed method was not measured. We set the following conditions:

- No turning: After a vehicle turns, its location is adjusted by the map matching method. Therefore, we needed to drive straight and not turn.
- No using GPS: General in-vehicle navigation systems adjust the vehicle's location if there is huge difference between the location based on the GPS signal and one recognized by the systems. Therefore, we needed to remove the GPS.

We selected a road that had four lanes and ran from north to south for about 4.6 kilometers after considering the above reasons. We drove along the road with AR-Navi and counted the number of times that the proposed method adjusted registration, and we measured the adjustment length to examine the effectiveness of the proposed method. During this driving experiment, AR-Navi was not connected to a GPS receiver.

3.2 Discussion

Adjustment lengths are shown in Table 1. The proposed method was carried out nine times and lengths adjusted an average of 6.1 meters in the direction traveled. Fig. 3 shows the relationship between the adjustment length and distance from the adjusted point to the previous adjusted point. (The first adjusted point is not plotted in Fig. 3 because it has no previous point.) The dashed line of Fig. 3 shows straight-line approximation. The correlation coefficient between the vertical axis and the horizontal

Table 1. Adjustment lengths of the proposed method

Length [m]
10.6
3.2
9.4
5.4
4.9
2.6
3.7
8.1
6.8

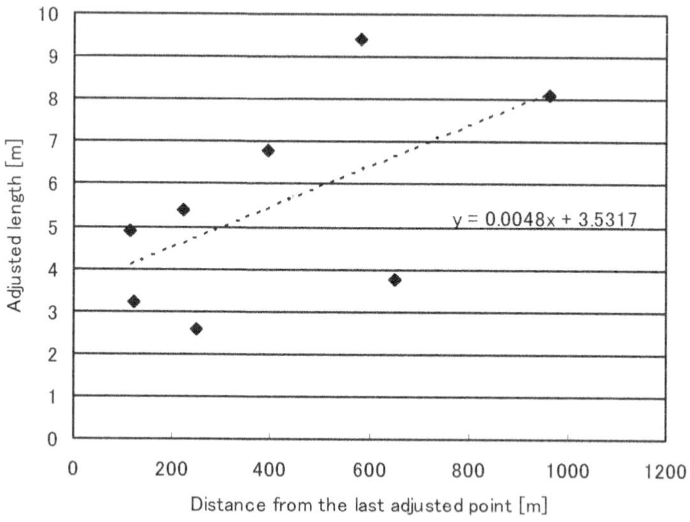

Fig. 3. Relationship between adjustment length and distance to the last adjusted point

axis is 0.59. This shows that the adjustment length decreased along with distance from the last adjustment point. Thus, we presume that the back-and-forth gap disappears immediately after the registration adjustment occurs but the gap widens with driving.

Fig. 3 shows the vehicle's location is adjusted by several meters even when a vehicle drives only 100 meters after registration adjustment. Therefore, the adjustment length includes the following gaps:

1. Random gap
2. Constant gap
3. Growing gap

Random gaps (1) are caused by the frame rate of AR-Navi and the accuracy of the maps. First, AR-Navi uses a 10-frame-rate video because the navigation system has limited update currency. AR-Navi sometimes could not recognize crosswalks correctly. For example, a vehicle runs one meter per frame when it runs at 36 km/h. Second, maps used in conventional in-vehicle navigation systems are not perfectly accurate; because there are errors in the maps, random gaps occur.

Constant gaps (2) are caused by time lags in carrying out the proposed method. Registration adjustment is sometimes not carried out immediately because many processes are running and some processes, such as drawing maps, have higher priority. Because a vehicle runs while its location is being adjusted, differences come about between the adjusted length and actual necessary length, that is, the next adjustment length increases. According to the y- intercept of dashed line in Fig. 3, the constant gap is 3.5 meters. Because this gap can be removed by modifying programs, we discuss the following without reference to constant gaps.

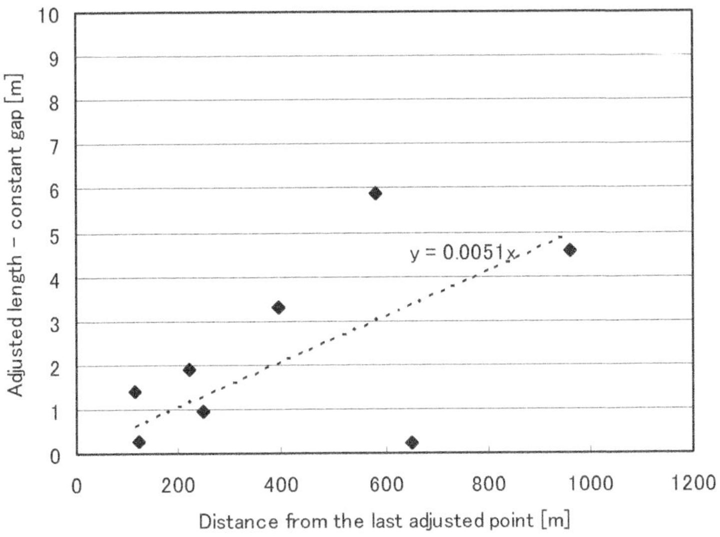

Fig. 4. Relationship between growing gap and distance to the last adjusted point

Growing gaps (3) are caused by autonomous navigation. To clear the growing gap, Fig. 4 shows the relationship between the adjustment length minus the constant gap (3.5 meters) and distance from the current adjusted point to the previous adjusted point. The average size of a growing gap is 2.3 meters and the maximum growing gap's size is less than 6 meters. The dashed line in Fig. 4 shows straight-line approximation and represents the gap-growing model. To examine the model, we drove on the same roads as the above experiment without using the proposed method and measured the back-and-forth gap between the actual and estimated locations using an in-vehicle navigation system. The measured gap was 20 meters after driving 4.6 kilometers. On the other hand, the theoretical gap is 23.5 meters according to the dashed line. Because these two gaps are quite different, the model is correct. The back-and-forth gap of AR-Navi increases by 0.51 meters every 100 meters.

In the next section, we discuss the accuracy of route guidance of AR-Navi on the basis of commercialization.

4 Discussion about Accuracy of Route Guidance

If route guidance information is overlaid on an intersection area in a real-time video, the driver recognizes the intersection where he or she will turn (hereafter the target intersection). The conventional in-vehicle navigation systems do not display roads that are less than 5.5 meters wide except when vehicles are close to their destinations. Therefore, we discuss whether AR-Navi can overlay information on 5.5-meter wide intersections. This is the most taxing of conditions because intersections are normally wider than 5.5 meters. To overlay information on a 5.5-meter intersection, we calculated the probability that the proposed method limited the back-and-forth gap to 5.5 meters. In the following discussion, only growing gaps are considered because random gaps do not occur on average and constant gaps can be removed.

We model the probability that the proposed method recognizes a pair of crosswalks and adjusts the vehicle's location accordingly. Because where crosswalks exist is random natural phenomenon, the probability follows Poisson distribution. Poisson distribution indicates that if the expected number of occurrences in a certain interval is λ, the probability that they are exactly k occurrences is equal to formula (1) (e is the base of the natural logarithm).

$$P_{(N=k)} = e^{-\lambda}\lambda^{k}/k! \tag{1}$$

The limit to which growing gaps do not exceed 2.75 meters is 539.2 meters. Therefore, we used Poisson distribution with a 539.2-meter interval and calculated the probability that the vehicle's location is adjusted at least once per an interval. λ, that is the expected number of times registration adjustment occurs in 539.2 meters, is calculated as formula (2), because registration adjustment operates once every 412.6 meters on average according to the experiment.

$$\lambda = 539.2/412.6 \approx 1.31 \tag{2}$$

Also, the probability that registration adjustment does not operate is equal to formula (3).

$$P_{(N=0)} = 0.270\cdots \approx 0.27 \tag{3}$$

In other words, the probability that back-and-forth gaps do not exceed 2.75 meters and AR-Navi correctly overlays information on an intersection is 73 percent by using the proposed method even in the most demanding of conditions.

We have evaluated in the most difficult conditions, which were based on the narrowest roads. Nevertheless, considering average roads, it is possible that AR-Navi overlay information on an intersection if back-and-forth gaps are greater than 2.75 meters. Moreover, because AR-Navi can estimate back-and-forth gaps, AR-Navi can switch information displayed in accordance with back-and-forth gaps and the intersections' sizes. For example, conventional CG guidance is displayed if the estimated gap exceeds the size of the target intersection. Therefore, even if a back-and-forth gap is big, AR-Navi can still display correct route guidance.

In addition, considering general situations, there is less probability that the gap exceeds 2.75 meters at the target intersection because of following reasons.

- Frequency of turning
- Guidance timing

First, it is unusual that a vehicle would go straight on local roads without GPS signal for long distances as we did in the experiment. The vehicle's location is frequently adjusted in general situations because turnings trigger registration adjustment. Furthermore, if there is a big gap between the vehicle's location in the real world and that on its in-vehicle navigation system, the system adjusts the location via the GPS signal. Therefore, back-and-forth gaps occur less than in the experimental results.

Second, guidance is not always displayed if the vehicle's location is not adjusted for long distances and the gap becomes bigger. If AR-Navi recognizes a pair of crosswalks before reaching the target intersection, AR-Navi adjusts the vehicle's location and the gap disappears.

For the above reasons, the probability that AR-Navi cannot overlay CG information on the target intersection is less than 27 percent in general driving. Therefore, AR-Navi gives exact guidance practically thanks to the proposed method.

5 Conclusion

In this paper, we conducted a driving experiment to evaluate the registration adjustment based on recognizing crosswalks and showed that it occurs once every 412.6 meters and adjusts 2.3 meters on average. In addition, we discussed the accuracy of route guidance when using the proposed method and showed AR-Navi can display exact and easy-to-understand guidance. An evaluation of AR-Navi's usability should be a future work.

References

1. Japanese Ministry of Land, Infrastructure, Transport and Tourism, Road Bureau ITS Homepage, http://www.mlit.go.jp/road/ITS/j-html/vics_pdf/navi_vics.pdf
2. Sawano, H., Okada, M.: A Study on an Augmented Reality-based Rendering Method for Car Navigation Utilizing Real-time Video. In: FCV 2004 - Tenth Korea-Japan Joint Workshop on Frontiers of Computer Vision, Fukuoka Japan (February 2004)
3. Hu, Z., Uchida, K.: Solution of Camera Registration Problem via 3D-2D Parameterized Model Matching for On-Road Navigation. International Journal of Image and Graphics 4(1), 3–20 (2004)
4. Yamaguchi, Y., Nakagawa, T., Akaho, K., Honda, M., Kato, H., Nishida, S.: AR-Navi: an in-vehicle navigation system using video-based augmented reality technology. In: Proceedings of the 12th International Conference on Human-Computer Interaction, pp. 1139–1147 (2007)
5. Quddus, M.A., Ochieng, W.Y., Zhao, L., Noland, R.B.: A General Map Matching Algorithm for Transport Telematics Applications. The GPS Solutions Journal 7(3), 157–167 (2003)

Part V

User-Centered Design and User Experience in Information Systems

Improving the User Experience of Our Own UX Ideas

Joel Eden

Infragistics, Inc., User Experience, 50 Millstone Road
Princeton, NJ 08520, USA
jeden@infragistics.com

Abstract. Systems thinking is a perspective that focuses on looking at interdependent relationships between entities in an organization, making visible otherwise hard to see leverage points. Evaluation is an inherent part of designly thinking, even though many practicing UX professionals view evaluation as something that follows "design." These two views are discussed related to understanding how to get more traction for our UX ideas in the existing design processes within our organizations.

Keywords: User Experience, Systems Thinking, Evaluation as Design.

1 Introduction

What can we as user experience professionals do to better understand and become part of the process of moving ideas to products? If everyone involved in the process of creating great user experiences (desirable, useful, and usable products and services) is working with the best intentions, why is it always so difficult to move forward from the space of possibility (i.e. the ideas) to the final product? You have video of users, quotes from users, market research, statistically valid experimental data, and focus groups all showing clearly (you think) what could and should be changed to make a product improve, but why do some ideas end up in the product and some don't.

Whether working in small or complex organizations, and whether working on small to complex problems, I have found that issues with getting traction for ideas has at least as much to do with understanding organizational and social structure on a holistic level as it does with the specific product design idea, usability issue, or whatever it is we are focused on. In addition, contrary to what we preach outward, the way user experience professionals present ideas may not always adhere to even our own standards.

Two interrelated ideas are explored in this paper. First, a systems thinking view may allow us to better discover, understand, and communicate our ideas in a way that gives them increased traction in the product development process. Systems thinking involves viewing the interdependent relationships between people and artifacts in a situation. Taking a systems thinking perspective can make visible otherwise hard to see leverage points, where changes to the system may have the most impact. In this paper, systems thinking will be explored with respect to how it can enable UX professionals to understand the leverage points in the system made up of their own stakeholders (e.g. how to best integrate UX findings).

M.J. Smith and G. Salvendy (Eds.): Human Interface, Part I, HCII 2009, LNCS 5617, pp. 659–664, 2009.

Second, the way we think about design in general will impact how we think about the role of evaluation. This is turn will impact how we and others in our organization try to place us in the product development process. Looking at evaluation and design as inseparable, or even seeing evaluation as design can give a different perspective on how to better integrate UX ideas into the overall product design process.

Are we practicing what we preach? Are we using current user experience guidelines, patterns, and designerly thinking to generate our own ideas? Using the two ideas above (systems thinking and the role of evaluation in design), examples of current UX practices and how we typically present UX findings will be explored in order to understand how we might ourselves provide a better user experience for those who use the information we provide (i.e. more useful, usable, and desirable findings and deliverables for stakeholders).

2 Applying Systems Thinking

2.1 Introducing Systems Thinking

Systems thinking is a perspective that focuses on looking at interrelated, interdependent relationships between entities in an organization, as opposed to a focus on the individual entities [4]; moreover this view can be contrasted with a view of the organization as a set of linear relationships. Rather than viewing the organization as a static collection of entities, systems thinking leads you to try and view the organization as an active, dynamic process where many of the attributes of the overall system are a result of interactions between constituent entities; these systemic attributes can be said to be attributes of the system as a whole rather than being attributable to any given individual entity.

Taking a systems view of your organization gives you a better chance at seeing the leverage points that you can use to your advantage in getting more traction for your ideas. These types of systems can be thought of as a combination of interrelated feedback loops, loops that both excite and suppress information flows and their relevance to stakeholders. What matters to UX professionals is the fact that depending on where in this overall system you interject your ideas, they may be amplified or suppressed. Understanding where the leverage points are in your organization will enable you to have a better chance at impacting product design; e.g., your goal is to amplify your UX thinking throughout the organizational ecosystem, rather than having your ideas going effectively unnoticed.

Those who study systems thinking have identified many "archetypes," which are patterns of interrelated loops that seem to show up across many types of systems. One such archetype relevant to us is called "fixes that fail." Fixes that fail describes situations when short term fixes to a problem result in unintended consequences that show up later, ultimately resulting in reproducing or making worse the initial problem that the short term fixes were supposed to solve. Now let's explore how this fixes that fail archetype can show up in our UX work.

2.2 UX Fixes That Fail

We find lots of types of potential issues, ranging in severity and impact, but for many reasons we usually end up focusing on "low hanging fruit" issues. As soon as you finish usability testing, or an expert review, you probably get a request for the low hanging fruit so that the team can get started on fixing these issues. A problem with this is that in many cases these low hanging fruit issues are really just symptoms of larger, more systemic issues. Rather than a fruit picking metaphor, it may be more appropriate to use a lawn care metaphor. Mowing your lawn every week to get rid of weeds will not get rid of weeds; it actually just chops them up and spreads them around causing more widespread weed growth. You need to use some sort of product or method to actually deal with the underlying problem leading to the systemic growth of weeds. Similarly, if you only deal with user experience issues at a superficial level, i.e. the low hanging fruit issues you can fix quickly, you are likely never going to get deep enough into the real systemic issue to stop similar issues from recurring (just like the weeds that will continue to grow back every week).

Let's use the fixes that fail archetype to look at this fairly typical manner in which results from user experience evaluation are integrated back into the design process. Different projects work within their own budgets, and therefore results from usability testing, expert reviews, and other forms of design critiques are usually focused specifically on the goals of the given project (and likely on the specific questions that lead into the evaluation activity). Potential issues discovered and the related recommended fixes will likely focus on near term issues that seem more important to the current design project, potentially ignoring the relationship between these issues and underlying systemic problems.

The major problem with all of this is that quick fixes for a given current project probably don't even impact the underlying systemic issues that are plaguing all projects; e.g. a company may simply have a weak understanding of what customers need or want, so their offerings are inherently not seen as useful, even if usable. In most cases, the project managers cannot afford to takes steps that save time and money in the long run; this is a common paradox in that there never seems to be funding for the long term underlying issues that keep causing the same seemingly smaller issues to reoccur (which is what funding usually focuses on). Once the word gets out that later on in a project timeline, employing expert reviews or usability testing finds a lot of issues, the UX staff will be inundated with work related to symptoms (i.e. the unintended consequence of creating more busywork unrelated to the real underlying issues). Therefore all of these quick fixes that UX professionals can offer in many cases may actually hurt us by not ever letting us get to the more systemic causes.

What you can do

One way to deal with the fixes that fail issues described above is to simply try to make your solutions transcend the current issue or project. Focus on communicating issues, recommendations, and results across projects and organizational groups. If you create and use a common library of design patterns, you can refer to these patterns and try to get others to use them as well. This will help to work towards a common language around design and UX, across all of your organization.

Make it easier for issues to be noticed and dealt with further upstream in the design process. If you can let go of needing to get credit for UX ideas, it may become easier to get smaller pieces of your ideas injected throughout the organization. Get others to spread the message, so it comes from people within each group that already have credibility. For example, if you can get projects managers and developers seeing value in just some of the ideas you have, get them to bring them up at their meetings; not because you ask them to, but because you've shown them the value in the ideas. Therefore you should care more about the ideas getting propagated than ownership of the ideas. This is a good example of how to use systems thinking to understand that currents of information flow already exist in your organization, and rather than swimming upstream (i.e. you trying to shove information at people as the UX "outsider"), you should let the information flow through these existing currents (i.e. get others to sell the value of the UX ideas for you). You need to find and take advantage of these existing leverage points in your organization.

3 Evaluation as Design

The relationship between evaluation and design

One way to view the role of evaluation in an overall design process is that before you can even talk about evaluation you need to have a design to evaluate; i.e. the designers work on finishing "the design" while people with titles like "human factors engineer" or "usability engineer" wait in the wings for their turn (the evaluation). Taking this perspective of design and evaluation, we normally think of usability testing as the main evaluation activity that is carried out. Unfortunately, too many user practitioners and the companies they work for adhere to this thinking. This idea of conceptually splitting design and evaluation misses some of the most important aspects of good design thinking, or as some call it, the designerly way of thinking. The designerly way of thinking is all about viewing design as the consideration of many alternatives; as Schon [3] has phrased it, design is about having "a conversation with the situation," meaning that the design process is best carried out in a manner that allows the designer to get feedback about design ideas while in the moment of design. Rather than viewing design as preceding evaluation, or even being separate from evaluation, design inherently includes evaluation; we can in some ways view evaluation as design.

"...she takes Northover's comments as a criticism of her drawing, yet it is clear that she sees drawing not as thought-experimenting but as a way of presenting ideas. Northover seems to be saying 'You are not really designing at all. You are simply having 'ideas' and putting them down on paper. The moves you make have consequences that are testable, but you must draw to scale and in section in order to test them. The whole process of designing is lost to you because you will not do these things.'" [3], p. 130.

The above quote related to the inherent relationship between evaluation and idea creation in the design process from Donald Schon [3] may make interaction designers jealous of architects; imagine if all you had to do to get an estimate of the usability of a set of alternative designs was to "draw to scale and in section to test them" a priori.

Schon's concept of reflection-in-action places emphasis on the opportunity for designers to see the consequence of each design move.

Christopher Alexander argues that it is easier to recognize misfit between a designed artifact and its context of use than it is to recognize fit [1]. Henry Petroski shares this same view when he proposes that irritation, not necessity, is the mother of invention [2]. Relating Schon and Alexander, reflection-in-action can be fostered by providing designers with pseudo-immediate feedback regarding the consequence (misfit) of their design assumptions on the context of use. As the time between design choices and feedback increases, the opportunity for reflection-in-action fades away towards reflection-on-action, and it is a slippery slope beyond this towards our current state of waiting weeks, months (i.e., too late to be useful) to get results of usability testing in order to understand the consequence of design choices.

What you can do

If you want your UX ideas (e.g., design ideas, evaluation results) to be treated as first class citizens of the product design process, you need to start treating them that way. Pay attention to the language you use when discussing UX topics; if you talk a lot about the value of evaluation or testing, people may assume you mean these ideas are best applied later in the process. If what you are trying to get across is the inherent relationship between creative and evaluative activities in the design process, then talk about that. Early user research evaluates our understanding of what people need to accomplish; you are evaluating ideas rather than wireframes. Consideration of alternative design ideas is evaluation, whether it be during a hallway conversation with a co-worker, or with a participant during a formal usability test, or in front of whiteboard with a few co-workers.

Remember that it's all about communication between designers, clients, and all stakeholders throughout your organization. Leverage the ideas on systems thinking discussed above to inject your evaluation ideas into the design process using the leverage points that already exist in your organization.

Create multiple versions of your deliverables that can be used for different leverage points. Beware of showing deliverables in levels of fidelity that do not match the expectations that you already know are held by specific stakeholders. For example, if you know certain project managers consider high fidelity wireframes or prototypes as a sign that a concept has been finalized, instead create an alternative version that is lower fidelity. Don't fall prey to the urge to show them how good you are with making it look high fidelity if in the end you will get less of what you want (the ability to have more time to explore design alternatives). No matter how much explaining you do that the ideas represented in the high fidelity wireframe are really still up for debate, and that you want to explore many more alternatives, the inherent behavior of these types of project managers, partly due to their place within the product development responsibility chain, will be to treat the design ideas as further along. Remember from the discussion on systems thinking that many of the behaviors that each of us exhibits will be partly driven by our interrelated, interdependent placement within the overall organizational system; i.e. project managers have pressures and goals that may work against you if you choose to show higher fidelity deliverables, that ultimately betray your goal of getting feedback and more time to iterate ideas.

4 Conclusions

Two interrelated ideas were explored in this paper. I discussed the value of taking a systems thinking perspective as well as viewing the inherent role of evaluation in design thinking and the design process. These two views work together to help us understand how to get more traction for our UX ideas in the existing design processes within our organizations. Systems thinking involves viewing the interdependent relationships between people and artifacts in a situation, making visible otherwise hard to see leverage points. This can enable UX professionals to understand the leverage points in the system made up of their own stakeholders, and therefore understand how to best integrate UX thinking into existing information and political currents throughout their organization.

Looking at evaluation and design as inseparable, and even seeing evaluation itself as a form of design thinking provides perspective on how to better integrate UX ideas into the overall product design process.

This all adds up to the question, are we practicing what we preach? Are we using current user experience guidelines, patterns, and designerly thinking to generate and communicate our own ideas? Using the two ideas discussed (systems thinking and the role of evaluation in design), we can hopefully provide a better user experience for those who use the information we provide (i.e. more useful, usable, and desirable findings and deliverables for stakeholders).

References

1. Alexander, C.: The Timeless Way of Building. Oxford University Press, Oxford (1979)
2. Petroski, H.: Success Through Failure: The Paradox of Design. Princeton University Press, Princeton (2006)
3. Schon, D.: Educating the Reflective Practitioner. Jossey-Bass (1990)
4. Senge, P.: The Fifth Discipline: The Art and Practice of the Learning Organization. Currency (2006)

Construction of the Data Save System of the Cusp Surface Analysis Using Web Application

Yasufumi Kume and Jee Seob Seo

Department of Mechanical Engineering, Kinki University,
3-4-1 Kowakae, Higashiosaka, Osaka 577-8502, Japan
kume@mech.kindai.ac.jp

Abstract. This paper describes cusp surface analysis system constructed to examine the effectiveness of the program on a web server. According to operating on the web server of cusp surface analysis, client can easily use a cusp surface analysis program without a preparing for an analysis program or the system environment etc. And the paper also examine characteristic of the human visual perception using 315 kinds of ambiguous figures. As the results of the experiments, the cusp catastrophe phenomenon occurs as kinds of the ambiguous figures and the levels of detail figures in human visual perception.

Keywords: cusp surface analysis, web server, kinds of ambiguous figures, levels of detail figures, visual perception.

1 Introduction

The catastrophe theory is applied to many fields such as behavioral science, psychology, physics, biology etc. French mathematician Rene Thom classifies the elementary catastrophe of seven types in catastrophe theory [1]. The cusp model with two control factors is frequently applied to discontinuous phenomenon. Furthermore, Loren Cobb proposed the cusp surface analysis. A cusp surface is a statistical response surface model, based on the cusp model of catastrophe theory [2]. Conventional cusp surface analysis performed with a stand-alone computer. But, in this paper, cusp surface analysis system is constructed to examine the effectiveness of the program by web server. According to operating on web server of cusp surface analysis, client can easily use a cusp surface analysis program without a preparing for an analysis program or the system environment etc.

This paper examines the relationship of kinds of ambiguous figures, levels of detail figures and human visual perception using 315 kinds of ambiguous figures. Fisher introduced ambiguous man/girl figure, and Attneave made 8 figures by embedding the figure in the sequence, and Tim Poston and Ian Stewart made 32 figures. 315 kinds of ambiguous figures by interpolation of figure of Tim Poston and Ian Stewart figure are made by Atsuo Murata etc [3]. An array of 315 ambiguous man/girl figures has 21 steps of column from a man's face to girl's shape, and has 15 row levels of detail. The experiment data apply to cusp surface analysis.

M.J. Smith and G. Salvendy (Eds.): Human Interface, Part I, HCII 2009, LNCS 5617, pp. 665–674, 2009.

2 Construction of Cusp Surface Analysis System

Fig.1. above shows interchange information between Web application server and the client. Web browser indicates the web application to client on Fig.1. A system developed cusp surface analysis program is deployed on web server. The cusp surface analysis program is opened and performed by the URL like a web page. When registered URL on web server is appointed, a cusp surface analysis program is corresponding to it and carrying out as HTML in a web browser via a web server.

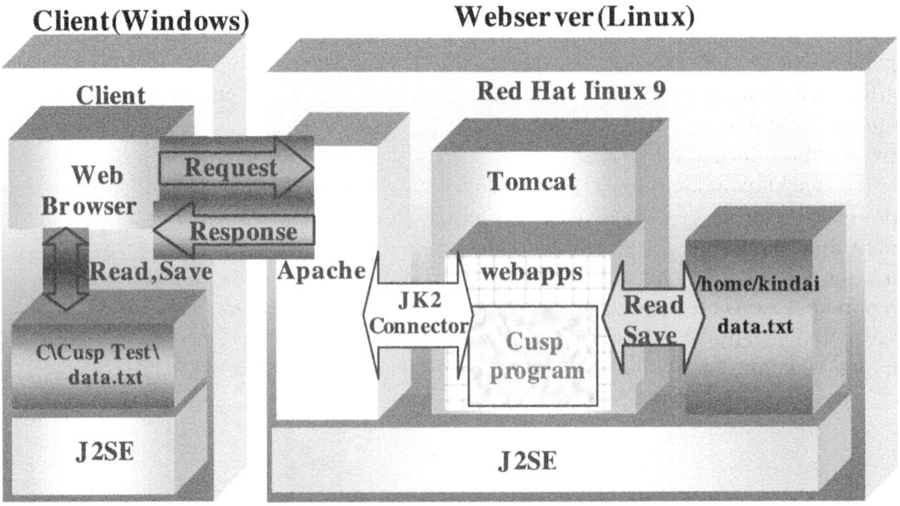

Fig. 1. Structure of client and Web server

The cusp surface analysis system is a system performing calculation in the cusp surface analysis on a web server from cooperation of Linux, Apache, Tomcat and Java. Apache of web server software is installed in the web server. In addition, Tomcat of the web server software is carried to treat Java servlet and Java server pages(JSP). Tomcat is done plug-in into Apache. Tomcat independently plays a role of the web server. But the reason of the plug-in is the improvement of the processing performance and the problem of security etc. If the client has web browser and Java, client can use the cusp surface analysis program on the web server. The giving and receiving of the web server is performed in protocol called the HTTP (Hyper Text Transfer Protocol). The client requests "the cusp surface analysis program" as HTTP, and the web server responses the service of "the cusp surface analysis program" as HTTP. Web browser on the client provides service for the client as HTML(Hyper Text Markup Language). Web browser plays a role to change HTTP gotten from web server into HTML. The cusp surface analysis system is available that the client can save data to both web server and local directory. And saved data from both web server and local directory are also able to open. Internet Explorer (I.E) is started to start cusp surface analysis program, and following URL is input: http://163.51.55.136/cusp2008/cusp.html.

3 Limitation in Applet

The cusp surface analysis program is built with Java applet, and Java isolate applet in a frame called Send Box in order not for applet to damage in every new performance. Thus, applet has basic limitations as following [4];

1. An applet cannot ordinarily read or write files on the host that is executing it.
2. An applet cannot make network connections except to the host that it came from.
3. An applet cannot start any program on the host that is executing it.
4. An applet cannot read certain system properties.

Cusp surface analysis program has loading and saving function of data. But client can not use the functions.

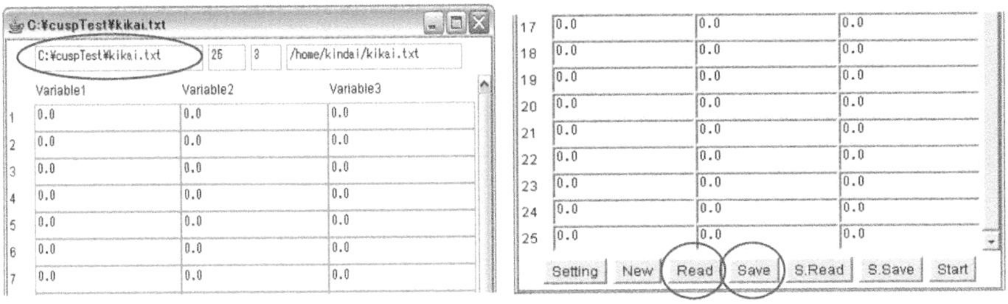

Fig. 2. Reading function of Data

Because reading and writing the file of performing Host is prohibited according to 1) limitation above. However, the Policy File [5] (.java.policy) makes it possible in spite of applet limitation. That is, the set up of Policy file is necessary for opening saved data to program. Clients have to set up the Policy file and then 'Read', 'Save' buttons would be available.

4 Experimental of Ambiguous Figures

This experiment is to examines that a characteristic of the human visual perception. The result of experiment shows the relationship among 21 kinds ambiguous figures which changes from man's face to woman's shape, 15 levels detail figures, and visual perception. 315 kinds of ambiguous figures are composed of 'kinds of ambiguous figures' and 'levels of detail figures'. 315 kinds of ambiguous figures are used to examine the visual perception by 'kinds of ambiguous figures' and 'levels of detail figures'. Figures size used in this experiment is 2.2×4.0cm. In the kinds of ambiguous figures, man's face is A and woman's shape is U. Among 15 levels of detail figure, level① is most detail figure and most indistinct figure is level⑮. This experiment ①, ③, ⑤, ⑦, ⑨, ⑪ level of detail figures are only used. ⑫-⑮ level of detail figures are barely recognized so that ①-⑪ level of detail figures are selected for experiment. In this experiment, A subject of 20's man is shown at random 126 figures and 3 times. Visual perception change points are set as following: subject answers Man's face - (0.0), Yes and no - (0.5), and subject answers woman's shape - (1.0).

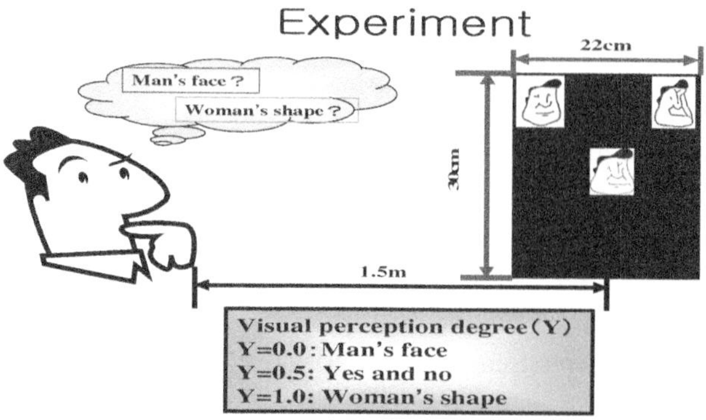

Fig. 3. Experimental method

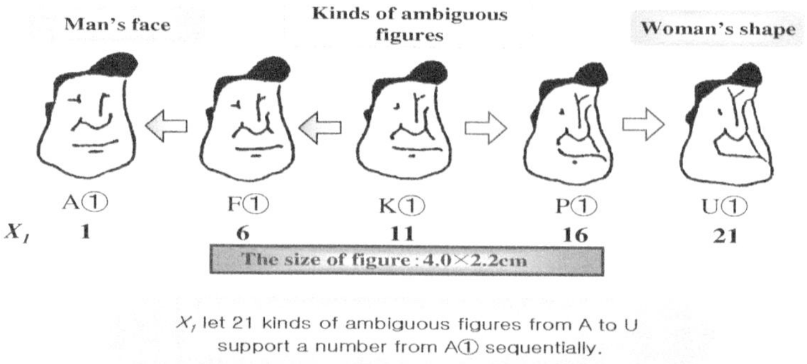

Fig. 4. In the case of level ① of detail figures, 21 kinds of ambiguous figures

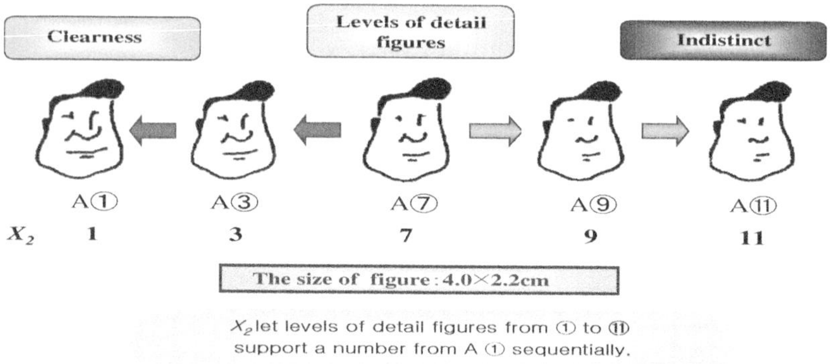

Fig. 5. In the case of the kind A of ambiguous figures, 11 levels of detail figures

Two figures past on the upper side of a black board so that the subject can judge easily. Two figures are the highest probability to be seen by man's face and woman's shape.

A subject is shown 1.5 meter far from figure. Since back ground colors may influence subject's judgment, 22×30cm black board is pasted on the back of each figure. This getting data are used for a cusp surface analysis.

Table 1. The results of 21 kinds ambiguous figures, 15 levels detail figures, and visual perception (Case1~Case378) Experimental data (126×3=378)

Case		X_1	X_2	Y
1	1	20	9	1
2	2	7	9	0.5
3	3	10	11	1
4	4	21	11	1
5	5	15	3	1
6	6	19	9	0.5
7	7	10	5	1
8	8	10	1	1
9	9	16	3	0.5
10	10	20	11	0.5
11	11	13	3	1
12	12	18	1	1
		⋮		
119	119	8	3	0
120	120	16	11	0.5
121	121	6	7	0.5
122	122	7	5	0.5
123	123	3	3	0
124	124	13	5	1
125	125	9	1	0
126	126	1	11	0

Case		X_1	X_2	Y
127	1	12	1	1
128	2	16	11	0.5
129	3	19	1	1
130	4	20	9	0.5
131	5	4	1	0
132	6	20	1	1
133	7	8	11	0.5
134	8	15	7	0.5
135	9	12	3	1
136	10	11	11	0.5
137	11	20	7	0.5
138	12	11	7	0.5
		⋮		
245	119	11	3	1
246	120	4	11	0.5
247	121	7	9	0
248	122	18	11	0.5
249	123	9	11	0
250	124	2	3	0
251	125	21	9	0.5
252	126	15	11	0.5

Case		X_1	X_2	Y
253	1	5	5	0
254	2	21	3	1
255	3	5	1	0
256	4	15	3	1
257	5	20	7	0.5
258	6	21	5	1
259	7	7	9	0.5
260	8	15	11	0.5
261	9	10	7	0.5
262	10	21	7	1
263	11	17	1	1
264	12	10	11	0.5
		⋮		
371	119	18	5	1
372	120	12	11	0.5
373	121	12	1	1
374	122	21	11	0.5
375	123	9	1	0.5
376	124	16	11	0.5
377	125	9	3	0.5
378	126	17	5	1

X_1: Kinds of the ambiguous figures, X_2: Levels of detail figures, Y : Visual perception

5 Application of Cusp Surface Analysis System to Experimental Result

Linear regression model's parameter is presumed.

The linear regression model's parameter is as follows:

$$A(\underline{X})=0,\ B(\underline{X})=B_0,\ C(\underline{X})=C_1 X_1+\ldots+C_v X_v,\ D=0\ . \tag{1}$$

Linear regression model's parameter is substituted for the expression.

$$0=B_0[Y-(C_1X_1+\ldots+C_vX_v)]\ . \tag{2}$$

Linear model's parameter is estimated by using the least square method.

The correlation coefficient (γ) is obtained from the data of Table1 by the next expression. (S_{xy} : covariance S_x, S_y : sum of squares)

$$\gamma=\frac{S_{xy}}{\sqrt{S_x}\cdot\sqrt{S_y}}$$

$$= \frac{\sum_{i=1}^{n}(x_i - \bar{x})\cdot(y_i - \bar{y})}{\sqrt{\sum_{i=1}^{n}(x_i - \bar{x})^2 \times \sum_{i=1}^{n}(y_i - \bar{y})^2}} \quad (3)$$

Table 2. Sum, Mean, and Sum of squares of the experimental data

	X_1	X_2	Y
Sum $\left(\sum x_i\right)$	4172	2268	181
Mean $\left(\bar{x} = \frac{\sum x_i}{n}\right)$	11.037	6	0.479
Sum of squares $\left(S_x = \sum (x_i - \bar{x})^2\right)$	13803.48	4410	58.331

Table 3. Covariance and Correlation coefficient of the experimental data

	$X_1 \cdot X_2$	$X_1 \cdot Y$	$X_2 \cdot Y$
Covariance $\left(S_{xy} = \sum (x_i - \bar{x})\cdot(y_i - \bar{y})\right)$	-14	671.796	-131
Correlation coefficient $\left(\gamma = \frac{S_{xy}}{\sqrt{S_x}\cdot\sqrt{S_y}}\right)$	-0.00179	0.74868	-0.25829

As a result, the following correlation matrix was obtained.

Table 4. Correlation matrix of the experimental data

	X_1	X_2	Y
X_1	1.000	-0.00179	0.74868
X_2	-0.00179	1.000	-0.25829
Y	0.74868	-0.25829	1.000

The inverse matrix of the correlation matrix is as follows.

Table 5. Inverse matrix of table 4

2.499	-0.513	-2.003
-0.513	1.177	0.688
-2.003	0.688	2.677

Linear r^2 value is obtained that is squared multiple correlation coefficient r.

When assume the element of the inverse matrix of multiple correlation coefficient r to be r^{22} .

$$r^2 = 1 - \frac{1}{r^{22}} = 0.626 \tag{4}$$

Log likelihood $L1$ of the linear regression model can be obtained from linear r^2 value. (N : Number of data)

$$\begin{aligned} L_1 &= -\frac{N}{2} \times \left(1.0 + \ln\left(2\pi \times \left(1 - r^2\right)\right)\right) \\ &= -\frac{378}{2} \times (1.0 + \ln(2\pi \times (1 - 0.626))) \\ &= -350.48 \end{aligned} \tag{5}$$

This is the initial values used by the maximum likelihood method when it begins its iterative search for the best fitting coefficients for the catastrophe model.

$$\begin{aligned} L_2 &= \ln \prod_{i=1}^{N} f[Y|X] \\ &= \ln \prod_{i=1}^{N} \exp\left[\psi + A(\underline{X}) \times Z + \frac{1}{2} B(\underline{X}) \times Z^2 - \frac{1}{4} D \times Z^4\right] \\ &= \sum_{i=1}^{N} \ln \exp\left[\psi + A(\underline{X}) \times Z + \frac{1}{2} B(\underline{X}) \times Z^2 - \frac{1}{4} D \times Z^4\right] \\ &= \sum_{i=1}^{N} \left[\psi + A(\underline{X}) \times Z + \frac{1}{2} B(\underline{X}) \times Z^2 - \frac{1}{4} D \times Z^4\right] \end{aligned} \tag{6}$$

The maximum value of log likelihood L_2 can be obtained by the Newton-Raphson method, it becomes L_2=-235.983. When log likelihood L_2 convergence with Newton-Raphson method, a standard partial regression coefficient is shown in Table 6.

Table 6. Standard partial regression coefficient

Var	A_v	B_v	C_v	D
Const	0.026	1.941	0.047	3.572
1	1.883	-0.983	0.03	
2	-0.56	-1.728	-0.088	

$$0 = A(\underline{X}) + B(\underline{X})[Y - C(\underline{X})] - 3.572[Y - C(\underline{X})]^3 \tag{7}$$

$$\begin{aligned} A(\underline{X}) &= 0.026 - 1.883X_1 - 0.56X_2 \\ B(\underline{X}) &= 1.941 - 0.983X_1 - 1.728X_2 \\ C(\underline{X}) &= 0.047 + 0.03X_1 - 0.088X_2 \end{aligned} \tag{8}$$

X_1: Kinds of the ambiguous figures, X_2: Levels of detail figures, Y : Visual perception.

When the level of detail is ① (X2=1) , human visual perception changes according to kinds of ambiguous figures from Man's face (X1=1) to woman's shape (X1=21). Thus, jump phenomenon is appeared. However, increasing the level of detail makes jump phenomenon decreasing and finally it disappears when level is ⑪ (X2=11) . Especially, on Fig. 6, visual perception dramatically changes

A→B→C→D from man's face to woman's shape. Thus, when the level of detail is ① (X2=1) , we can see visual perception is suddenly changed at near 11 of ambiguous figure. In other words, the human visual perception makes a change near to the figure of Fisher. As levels of detail figures are not clear, the subject recognize that the inside line becomes faint and makes the point. The change of the visual perception regards the point as man's face. And that time as the unclear degree increase, the subject doesn't know the figure which outside shape is changed. The cusp catastrophe phenomenon occurs as kinds of ambiguous figures and levels of detail figures in human visual perception.

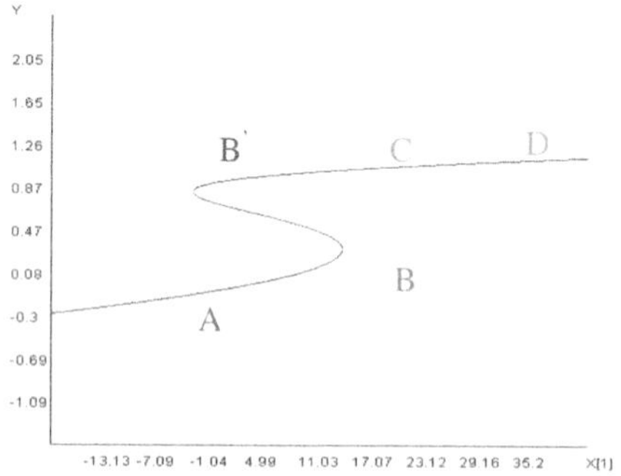

Fig. 6. Visual perception by kinds of ambiguous figures in the case of level① of detail figures

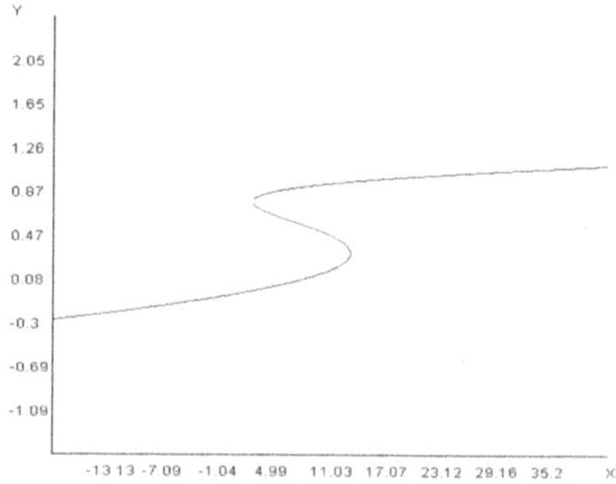

Fig. 7. Visual perception by kinds of ambiguous figures in the case of level③ of detail figures

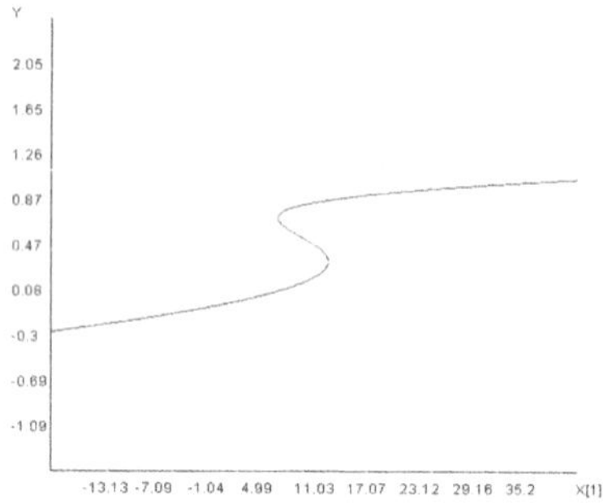

Fig. 8. Visual perception by kinds of ambiguous figures in the case of level⑤ of detail figures

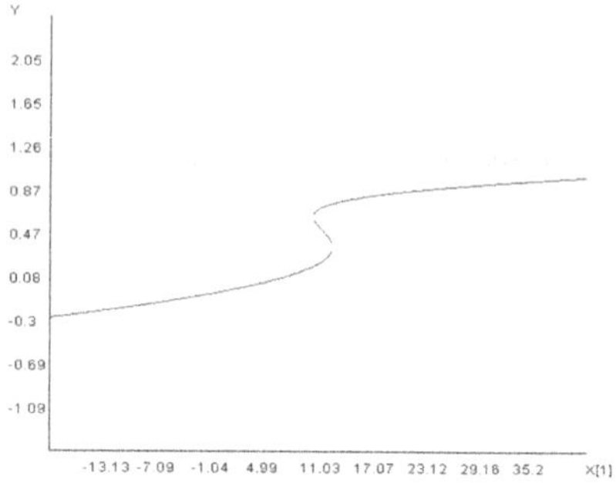

Fig. 9. Visual perception by kinds of ambiguous figures in the case of level⑦ of detail figures

Fig. 12. shows that the visual perception expresses the 3D graph according to kinds of ambiguous figures and levels of detail figures.

Fig. 13, 14 show the probability density function that the variety of ambiguous figures at the levels of detail is ①(X2=1). The figure of equal probability is Fisher's figure of K. But the experiment test result shows that the ambiguous figure to have equal probability exist between H and I in this case. Fig. 15 shows the probability density function 3D graph according to the kinds of ambiguous figures and the visual perception.

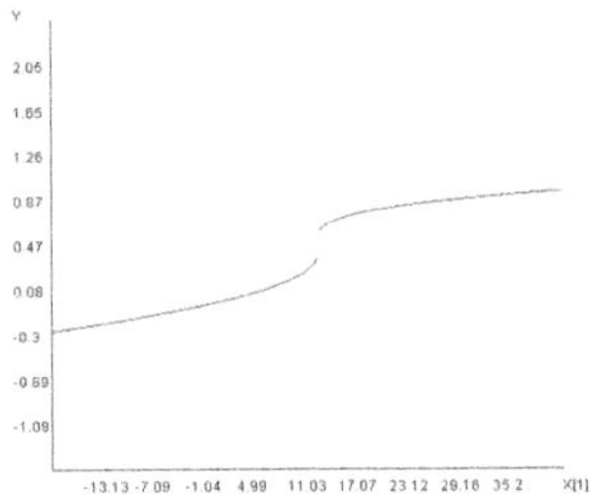

Fig. 10. Visual perception by kinds of ambiguous figures in the case of level⑨ of detail figures

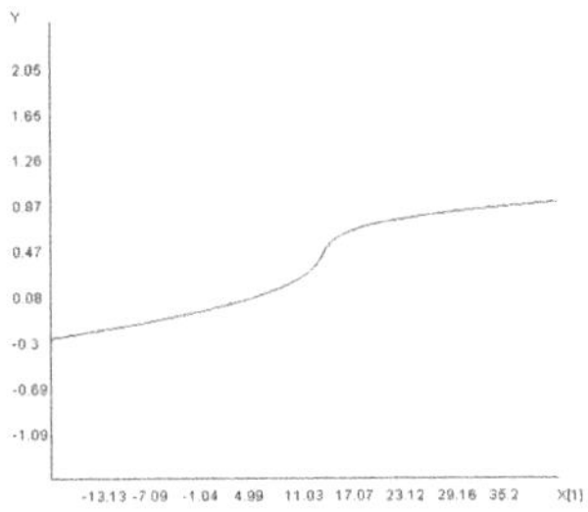

Fig. 11. Visual perception by kinds of ambiguous figures in the case of level⑪ of detail figures

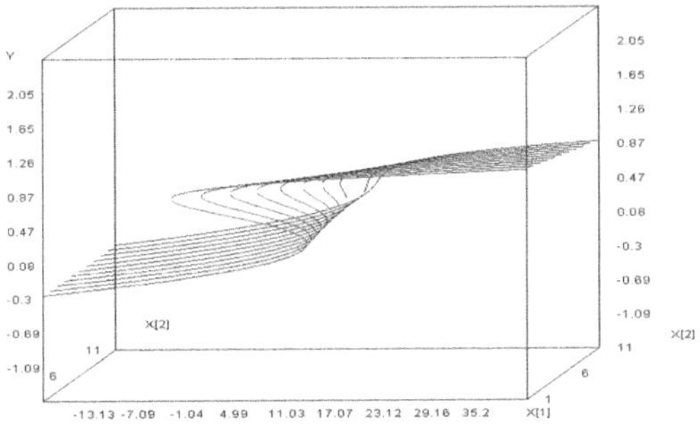

Fig. 12. Effect of Kinds and Detail in Ambiguous Figures to Visual Perception

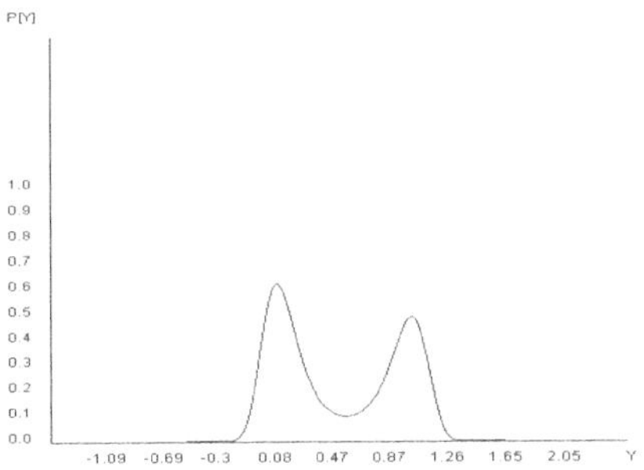

Fig. 13. Probability density function of ambiguous figure H in level① of detail figures

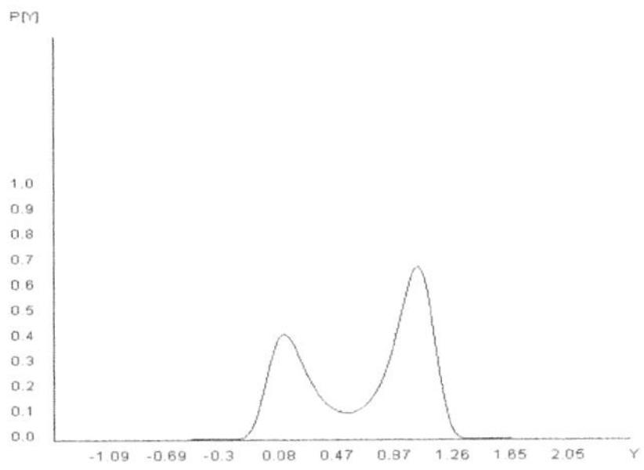

Fig. 14. Probability density function of ambiguous figure I in level ①of detail figures

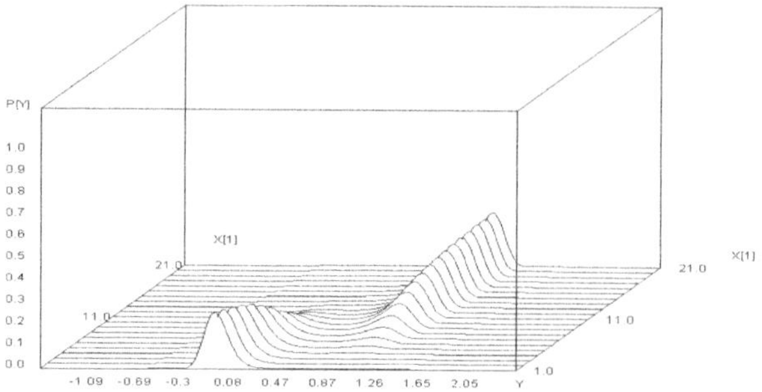

Fig. 15. 3D Probability Density Function

6 Conclusions

The results can be summarized in this case as follows.

1. As the cusp surface analysis can be opened to the public on the web server, the clients input URL he or she can perform the cusp surface analysis.
2. It is possible that the cusp surface analysis data's loading and saving in local directory and web server directory.
3. As the results of the experiments, the cusp catastrophe phenomenon occurs as kinds of the ambiguous figures and the levels of detail figures in human visual perception.

References

1. Thom, R.: Structural stability and morphogenesis. Benjamin-Addison Wesley, New York (1975)
2. Cobb, L.: An Introduction to Cusp Surface Analysis (1998), `http://www.aetheling.com/models/cusp/Intro.htm` (retrieved on January 28, 2009)
3. Murata, A., Kume, Y., Hashimoto, F.: Cusp Catastrophe Phenomenon in Visual Organic Functions. Japan Industrial Management Association 35(3), 150–155 (1984) (in Japanese)
4. Security Restrictions, `http://java.sun.com/docs/books/tutorial/deployment/applet/security_practical.html` (retrieved on January 28, 2009)
5. Java Security Evolution and Concepts, Applet Security, `http://java.sun.com/developer/technicalArticles/Security/applets/index.html` (retrieved on January 28, 2009)

Data Pattern for Allocating User Experience Meta-Data to User Experience Research Data

Li Li, Hong Ji, Xuejiao Chen, and Xiaowei Yuan

Beijing ISAR User Interface Design,
237-101, Nanhuxiyuan Wangjing, Chaoyang District, 100102, Beijing, China
li_li@isaruid.com

Abstract. The vision of user experience is making life of users of products as convenient as possible, especially during the interaction with a productor a service. An important aspect of perceived convenience is the user experience of a product. The visual design and especially the interaction design has a major influence on this perception. In order to achieve the vision, user experience experts apply different types of tasks. One type of task is to analyze how users carry out tasks and what user's needs or problems are. Another type of task is to design user experience solutions and other typical task type deals with carrying out usability evaluations in order to find problems in using software application. In the course of user experience activities, many data are being collected. Many of the collected data relates to certain activities of users. For the user experience area there exist just a few tools, which support typical tasks in different ways. None of the tools supports linking results of user experience work to user experience meta-data. Why is it a problem? The current tools do not support an access to user experience project data with generic search and filter criteria like "industry", "application area", "use case" etc. This makes the access to user experience research data difficult and the comparison of user experience project data between different projects inefficient. In general, results of different user experience projects are difficult to reuse. The core idea of data pattern for allocating user experience Meta-Data to User Experience research data is to associate user experience project data with user experience meta-data. The data pattern considers associating user experience project data with user experience meta-data partially automatically and partially manually by the user. The key idea is that we want to reuse project data like the project sponsor, the application area, the industry, use cases etc. as user experience meta-data and assign them to user experience research data. The benefits of the data pattern are: Reusing results of user experience research projects. Making access to available results more efficient. Direct comparison of available results is supported and more efficient. UX Office is the typical instance for application.

Keywords: User Experience, Project Data, Research Data, Meta-Data.

1 Background of Data Pattern

1.1 The Field of Data Pattern

The vision of user experience is making life of users of products (e.g. software products, home appliances, consumer products etc.) as convenient as possible, especially

M.J. Smith and G. Salvendy (Eds.): Human Interface, Part I, HCII 2009, LNCS 5617, pp. 675–684, 2009.

during the interaction with a product (e.g. mobile phone, PC) or a service (e.g. online book shop). An important aspect of perceived convenience is the user experience of a product. The visual design and especially the interaction design has a major influence on this perception.

In order to achieve the vision, user experience experts apply different types of tasks. One type of task is to analyze how users carry out tasks and what user's needs or problems are (use context analysis). Another type of task is to design user experience solutions (design) and other typical task type deals with carrying out usability evaluations (e.g. summative or formative usability tests) in order to find problems in using software application (usability evaluation). In the course of user experience activities or the typical user experience tasks (e.g. Analysis, Concept, Evaluation, Prototyping, Implementation), many data are being collected. Many of the collected data relates to certain activities of users. For example: A typical question is which problems to users have in sending a picture message to somebody via a mobile phone. This activity could be in the focus of the use context analysis (in order to understand the problems), in the focus of the design (in order to design a user experience solution) and of the evaluation (in order to identify problems of the user while s/he is using the solution).

1.2 Information of Related Field

For the user experience area there exist just a few tools, which support typical tasks in different ways. There are tools, which support the use context analysis, e.g. card sorting tools. There are typical tools, which support design activities, like rapid prototyping tools, picture editing tools, vector graphic tools. And there are also tools, which support usability evaluation activities, like tools supporting usability tests.

But all of these tools can support some user experience activity only. These tools likely belong to the different companies, which can bring a big matter. we find that it's too difficult to connect the information or result of a activity to another activity. For example: when we use some tools to help us to complete the usability tests, we hope the result of the test can be directly used in design work and guided to fix the design. In another word, It should not be a report which covered by all words, and instead of it, we prefer to get the visualized result which can be used in the design tools.

Through the experience of ISAR, user experience data can be divided into two parts, Project Data and Research Data. Project Data means any kind of data which describe characteristics of the user experience project. It's just a data type not an instance. For examples, project name, project objective, project success criteria, project approach, user experience activity, user experience method, project tasks, project deliverables, project resources, project risks, project status, project flag, target user group, use case, scenario, and so on. Research Data is any kind of data which is a result of a user experience activity. In general, it's also a data type. For example: requirements, storyboards, strengths, weaknesses, goals, goal trees, tasks, storyboards, process requirements, product requirements, service requirements, completion time, completion information, usability flaws, usability flaw cluster, usability severity, effort to fix, flaw classes, recorded sounds of user experience activities, results of usability analysis, and so on.

1.3 Related Problems

So, what is the problem for us? That is none of the mentioned tools supports linking results of user experience work (herein called user experience research data) to user experience meta-data. And why is it a problem? The current tools do not support an access to user experience project data with generic search and filter criteria like "industry", "application area", "use case" etc. We can't abstract the useful data from the data of project. And it's difficult to setup a connection between the project data and the research data by ourselves. Even we don't know which data should be saved in project library in order that we can use it in the next similar project or product. This makes the access to user experience research data difficult and the comparison of user experience project data between different projects inefficient. In general, results of different user experience projects are difficult to reuse. And even the same project, the results arise from different activities are difficult to connect automatic.

2 Summary of the Data Pattern

Through the experience of ISAR, user experience data can be divided into two parts, Project Data and Research Data. Project Data means any kind of data which describe characteristics of the user experience project. It's just a data type not an instance. For examples, project name, project objective, project success criteria, project approach, user experience activity, user experience method, project tasks, project deliverables, project resources, project risks, project status, project flag, target user group, use case, scenario, and so on. Research Data is any kind of data which is a result of a user experience activity. In general, it's also a data type. For example: requirements, storyboards, strengths, weaknesses, goals, goal trees, tasks, storyboards, process requirements, product requirements, service requirements, completion time, completion information, usability flaws, usability flaw cluster, usability severity, effort to fix, flaw classes, recorded sounds of user experience activities, results of usability analysis, and so on.

The core idea of the user experience data pattern is to associate user experience project data with user experience meta-data. Including, user experience meta-data is a type of data which is assigned to user experience Research Data. The Data pattern considers associating user experience project data with user experience meta-data partially automatically and partially manually by the user. The key idea is that we want to reuse user experience project data as user experience meta-data and assign them to user experience research data.

The benefits of the data pattern are:

- Reusing results of user experience research projects
- Making access to available results more efficient
- Direct comparison of available results is supported and more efficient

3 Detailed Description of the Data Pattern

3.1 Core Idea

User experience Meta-Data stem usually from user experience Project Data. They can be extended with user defined user experience Meta-Data. An instance of user experience Meta-Data is a single value of a user experience Meta-Data, like the use case, target user group.

User experience data pattern covers two parts: The first part includes how user experience Meta-Data are identified and assigned to user experience Research Data. The second part contains how user experience Meta-Data are used in order to realize the benefits of the user experience Meta-Data.

This data pattern allows deriving user experience Meta-Data automatically from available user experience Project Data. The data pattern includes also assigning user experience Meta-Data to user experience Research Data automatically and supports the user in assigning user experience Meta-Data to user experience Research Data manually.

It is assumed that user experience Project Data are already stored in the system and some user experience Research Data are already available in the system. The data pattern can be applied if the user enters new user experience Project Data and new user experience Research Data later to the system.

The user experience Meta-Data are selected from user experience Project Data which can be assigned to selected user experience Research Data. Part of the data pattern can manage the allocation between selected user experience Project Data and user experience Research Data. This allocation is a set of data which establishes the user experience Meta-Data. And the allocation can be changed by the user.

3.2 Allocation Rules

A core part of the user experience data pattern is how to assigns user experience Project Data to user experience Research Data. In order to do so, the mechanism requires allocation rules which defines which user experience Research Data and which user experience Project Data can be allocated. It considers an allocation conditions which needs to be fulfilled. If an allocation condition is fulfilled, a related allocation rule is being selected. According to project experience for user experience from ISAR, some user experience Project Data can be directly used in Research Data, and some should associate other data. For example, it is assumed that information instances of user experience Project Data instances of user experience Research Data are available. The allocation rules determine which instances of user experience project data are associated with which instances of user experience research data.

Examples of allocation rules are:

- Conditions: Usability flaw available (user experience Research Data), test scenario available (user experience Project Data); Allocation rule: Allocate test scenario (user experience Project Data) to usability flaw (user experience Research Data)
- Conditions: Problem available (user experience Research Data), target user group available (user experience Project Data); Allocation rule: Allocate target user group (user experience Project Data) to problem (user experience Research Data)
- Conditions: Storyboard available (user experience Research Data), use case available (user experience Project Data); Allocation rule: Allocate storyboard (user experience Project Data) to use case (user experience Research Data)

3.3 Allocation Mechanism

Another core part of the user experience data pattern is an allocation mechanism which allocates instances of user experience Project data to instances of user experience Research data, based on the allocation rules. After an instance of the user experience research data is available, the allocation mechanism allocates instances of user experience Project Data by carrying out the respective allocation rule as defined.

Example of the allocation mechanisms are: If the user enters a usability flaw in the context of a usability test (instance of a user experience Research Data) for a preselected scenario (instance of a user experience Project Data), the allocation mechanism reads the allocation rule.

3.4 Allocation Results

Another core part of the data pattern is to store the results of the allocation mechanism and make them available as user experience Meta-Data. This is specifically a connection between instances of user experience Project Data and user experience Research Data.

There are different ways to implement the allocation results. One way is to store it in a central storage where a reference from a user experience Product Data instance is connected to a reference to a user experience Research Data instance and vice versa (centralized solution). The instances can be grouped in user experience Product Data and user experience Research Data each. This allows an efficient access to the data.

Another way is to store the user experience Meta-Data decentralized. Each instance of a user experience Project Data is connected with at least one or more placeholder for references which can be connected to user experience Research Data (and vice versa) (decentralized solution).

An extension of the data pattern is that the user can edit the allocation results directly or indirectly and therefore change it. Changing the allocation results does not necessarily mean that the user changes the related user experience Project Data instances or the user experience Research Data instances, but the user experience Meta-Data which reflects the connection between user experience Research Data instance(s) and user experience Project Data instance(s).

3.5 Selection Mechanism (via User Experience Research Data)

The user experience data pattern includes also a selection mechanism by selecting user experience Research Data in order to realize the benefits. The required input for the selection mechanism is typically user experience Research Data, e.g. usability flaws, flaw cluster etc. Typically, the user needs to select which user experience Research Data he/she is interested in before initiating the selection mechanism. Another option is not to restrict the selection of user experience Research Data and let the mechanism select all available data. The mechanism uses the allocation results in order to execute the selection and collect the results.

Example of selection requests are:

- What are usability flaws (user experience Research Data) for a certain use case (user experience Project Data)?

- What are usability flaws (user experience Research Data) for a certain research object, e.g. a mobile phone (user experience Project Data)?
- What are the differences of usability flaws (user experience Research Data) between different research objects (e.g. between mobile phone A and mobile phone B)
- What are the differences of usability flaws (user experience Research Data) between different countries (user experience Project Data) for a certain research objects?
- What is the usability industry benchmark (user experience Research Data) of a certain research object (e.g. a mobile phone)?
- What are most relevant problems (user experience Research Data) for a certain target user group (user experience Project Data) for a certain domain?
- What are the most critical usability flaws (user experience Research Results) for a certain research object (user experience Project Data) and what are the related (user experience Project Data)?

3.6 Selection Mechanism (via User Experience Meta-Data)

The user experience data pattern includes also a selection mechanism for the instances of user experience Meta-Data in order to realize the benefits. The required input for the selection mechanism are instances of user experience Meta-Data, e.g. use cases, domains, applications, research objects, target user groups etc. The results are typically user experience Research Data. Typically, the user needs to select which user experience Research Data he/she is interested in before initiating the selection mechanism. Another option is not to restrict the selection and let the mechanism select all available information. The mechanism uses the allocation results in order to execute the selection and collect the results.

Example of user experience Meta-Data are:

- For a certain use case (instance of user experience Meta-Data): Select all usability flaws, all completion time, all participants subjective ratings, all videos with usability flaws (all user experience Research Data)
- For a certain research object (instance of user experience Meta-Data): Select all usability flaws, all critical usability flaws, average participants subjective ratings, all videos with critical usability flaws (all user experience Research Data)
- For a application (instance of user experience Meta-Data): Research object with best usability results, (all user experience Research Data)
- For a domain (instance of user experience Meta-Data): Critical usability flaws, usability flaw clusters, use cases with least usability flaws (all user experience Research Data)

The connection between the instance of project data and the instance of research data is the Meta-data. The project data and research data can build some relation via allocation rules and selection Mechanism. Generally, the concept of Meta-data can be used in database setup in computer field. Now we use this data pattern in user experience data. UX Office is the typical instance for application.

4 Application

UX Office system is developed by Beijing ISAR User Interface Design L.td and German partners together, used to conduct the Usability Evaluation and the user experience management.

The system was composed by two main modules: the usability evaluation and project management, and four assisted modules: client management, data management, resource management and administration.

In a professional usability evaluation process-oriented, guide and assist all types of personnel to test user experience, data analysis and report writing of different kinds of products, the main part include:

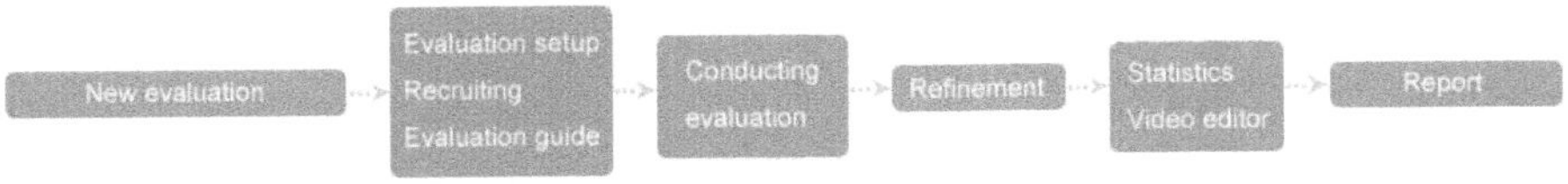

Fig. 1.

The data pattern of the meta-data has been used in the concept and the function of this system.

When a new evaluation starts, the preparation should be done first, it concludes Evaluation setup, Recruiting and evaluation guide. The Evaluation setup contents evaluation type, evaluation product, evaluation objective, use case, user group, location, evaluation units, evaluation teams, facility and experiential design. The function of Evaluation setup is to define the related data of the preparation for pre-test, and the Recruiting aims to recruit participant which meets the definition of evaluation unit. And the Evaluation guide includes Scenario, Questionnaires and Checklist. The function here is to guide the user experience experts to establish the test process.

When the preparation is finished, the test could be conducted, and in this part, the data of the test could be recorded. If the data is not recorded perfectly during the conducting process,(the time that a recorder can use is very limited during the conduct evaluation process) the user experience experts can use the Refinement to refine the recorded data.

When the data of the test has been collected well, the system supplies statistics and video editor to analysis the collected data. The statistics contain the analysis of flaw, questionnaires and scenarios. The video editor can make a high light video from the whole video data for a specific flaw or a usability problem.

When the data analysis has been finished, the user experience experts move to the report part, the system supply the CIF report template, and the test editing function to finish a report.

The data in the whole process is related to each other, now we can take the evaluation unit as an example. The data in the evaluation setup content both the project data and the research data, the evaluation unit contents such data types: evaluation product, location, user group, the number of participant in the evaluation unit and the start date and the end date of the test. The data type is defined under the guidance of this data pattern.

The data of the Evaluation setup, the data of the Recruiting and the data of the Evaluation guide are associated to each other. In this system, the data defined in one part can be used in other parts which are related with the originally part.

For example, the information of the user group, evaluation product and the location affect the data and the definition of Evaluation unit. And when the user experience expert has already defined the user group, evaluation product and the location, the information of these parts can be used automatically when the Evaluation Unit need to be done. And when an evaluation has been established by such information, the recruiting has to collect the participant under the profile in the user group, which is defined when a user group is established.

Also the data in preparation (content evaluation setup, recruiting and evaluation guide) is the basic of the work when conducing evaluation, when a test is conducting, the data during the test should be recorded by some rules and some information. In this system, the observer can record the test data by the guidance of the Checklist, Scenarios and the Questionnaires. For example, the test content is according to the scenario which is define in the evaluation guide before, the usability problem which would be found in the test has much to do with the use case, when a usability flaw is recorded, it has belong to one scenario, and the flaw has many other property, for it related to the user group, the use context and so on. And the whole test is conducted by the arrangement of the recruiting and evaluation unit.

When the data is collected already, the analysis of the data should be done. When doing statistics, the data can be analyzed in user group, location, evaluation product, scenarios, questionnaires, evaluation units and participant. For example, question A has been define in the Questionnaires part of the evaluation guide, and the answer of each participant has been recorded during the conducing test process, then the data of this question A can be statistics by different method for different need. The data can be analyzed the answer between different user group in the different location, or the different evaluation unit in each scenario, and so on.

When the data has been analyzed, and the result has been created, the user experience experts would write the report of the usability evaluation. We supply the CIF template as a standard structure, all the data before in any part, which is useful for the report, can be used here automatically or manually. For example, when the content of the report template is about the product, the information of the evaluation product can

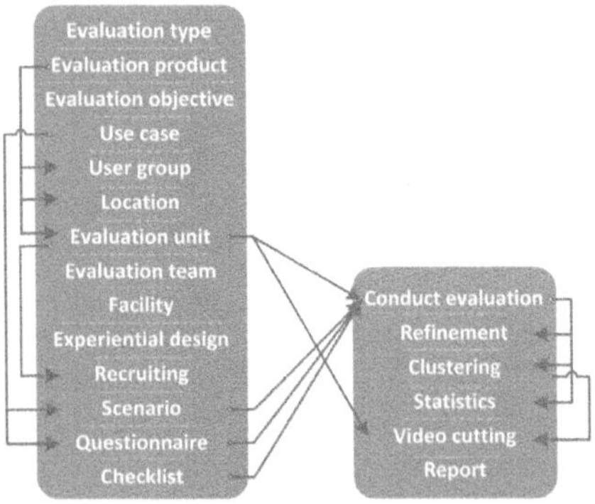

Fig. 2.

be used here automatically, and when the data is not in the default content, the data can be transferred from anywhere in the system. If a usability flaw has to be show with 2 related analysis results, the results can be transferred from the analysis part easily, but not to input them manually.

Fig. 3.

From the above description, the conclusion can be seen that the data pattern is used in the whole system, and the whole relationship can be seen in the following figure.

The examples above are all about the way that to store and use the Meta-Data decentralized. And another way to store the Meta-Data is to store all of them in storage. In this system, the Meta-Data is stored in the Library Module by the different type, the structure can be managed and edited by the administrator. The data in the library can be used in any new related project, and it also can be compared when the data is accumulated.

By using the data pattern, the system, you can get these benefits as following:

1) Provide professional processes and methods of usability evaluation
2) Effective coordination of the work of user experience team
3) Support multiple people observation and recording of the usability test at the same time.
4) Capture the real-time video data, observe and record the test data in the progress of usability test
5) Analyze record data, create data chart and table
6) Classify the video clips and create video groups Collection
7) To create report speedily, and export to MS Word document

5 Conclusion

The user experience pattern is same as thought of the database setup. The data structure of user experience Office is guided by this user experience data pattern. It can support the users to store, reuse of data and data evaluation (such as statistical forms, questionnaires, and other information) rapidly, also user can define data and templates, greatly increased the efficiency. The user experience Project Data can be used in user Research Data, and as the connection, the user experience Meta-Data is the core of this system. The data pattern makes the system more automatically and more intelligent, and it help us to do the user experience project more efficiently.

References

1. Dumas, J.S., Redish, J.C.: A Practical Guide to Usability Testing. Intellect, Ltd., UK (1999)
2. Cooper, A., Reimann, R., Cronin, D.: About Face 3.0. Wiley, Chichester (2007)
3. Buxton, B.: Sketching User Experiences. Morgan Kaufmann, San Francisco (2007)
4. Nielsen, J.: Usability engineering. Academic Press, London (1993)

Transcending Human-Centered Design by Service Sciences

Hiroyuki Miki[1], Naotsune Hosono[2], and Sakae Yamamoto[3]

[1] Oki Electric Ind. Co., Ltd., System Platform Center
1-16-8 Chuou, Warabi-shi, Saitama-ken 335-8510, Japan
hmiki@cf.netyou.jp
[2] Oki Consulting Solutions Co., Ltd.
hosono903@oki.com
[3] Tokyo Univ. of Science
sakae@ms.kagu.tus.ac.jp

Abstract. Human-Centered Design (HCD), which emphasizes the user's point of view, has brought many good results to date. For instance, with detailed analysis of user's context of use, developers of products and services can make them easier to use. Despite such good results, some limitations of HCD have also been pointed out. For example, since most methods of HCD are qualitative, they require exhaustive work every time with small improvements in efficiency. Although many qualitative quick methods have been proposed and tried to solve the problem, their results have not brought big differences. On the other hand, in the area of Service Sciences (SS), quantitative methods that explicate details of human activities based on a large number of data have brought some good results in producing new added-values and higher efficiency. Methods in SS emphasize the users' point of view as with those in HCD, and have come to be associated with innovation theories. Considering these current situations, this research emphasizes the users' viewpoint and combines HCD, SS, and innovation theories to come up with new methods that lead to new added-values and higher efficiency. As an initial effort of the research, this paper first clarifies relations between HCD, SS, and innovation theories, and then discusses issues in transcending HCD.

Keywords: Human-Centered Design, Service Sciences, Innovation Theories, Optimal Design Loop, Persona, Field Studies (Ethnography).

1 Introduction

Recently, situations surrounding business organizations have been changing rapidly and extensively. Due to the wide availability of broadband Internet in many countries, international division of labor, namely offshoring, has become quite common, and global competition is growing intense [1]. As a result, business organizations must seek to produce added-values and higher efficiency to survive the competition.

At the same time, big changes can be seen in modern technologies such as IT equipment, consumer electronics and services. Not only have their prices dropped

M.J. Smith and G. Salvendy (Eds.): Human Interface, Part I, HCII 2009, LNCS 5617, pp. 685–692, 2009.

rapidly as they become commodity, but they come to be avoided more when the designs are technology-driven and do not meet user needs [2].

To summarize, business organizations are required to consider the user's viewpoint, namely context of use, and at the same time pursue competitive added-values and higher efficiency for survival.

2 Current Human-Centered Design

The development considering user's viewpoint with their way of use is called Human Centered design (HCD), and it has been applied in many developments [3-6].

HCD is a development method based on design activities as shown in Figure 1. First, it requires observing the situation in which users actually try to use the equipment or service (1) and then clarifying the requirements for the design (2) followed by development (3). After development, evaluation is performed preferably accompanied by the users (4).

HCD is a cyclic activity similar to PDCA (Plan, Do, Check, Action) used in the process of production or quality administration. However, HCD is quite different from PDCA in that HCD emphasizes user standpoint by observing them use products or makes them participate in activities. HCD not only covers electronic products and IT equipments but also services.

The effects of HCD are;

- Resolution of usability issues.
- Creation of new products and services through the understanding of actual usage. This eventually leads to products or services with higher degree of user satisfaction.

HCD brought an innovation in the sense that it focuses not on technology but on user manners. However, accepting too much user needs can have negative side effects like excess functionality or lack of consistency [2].

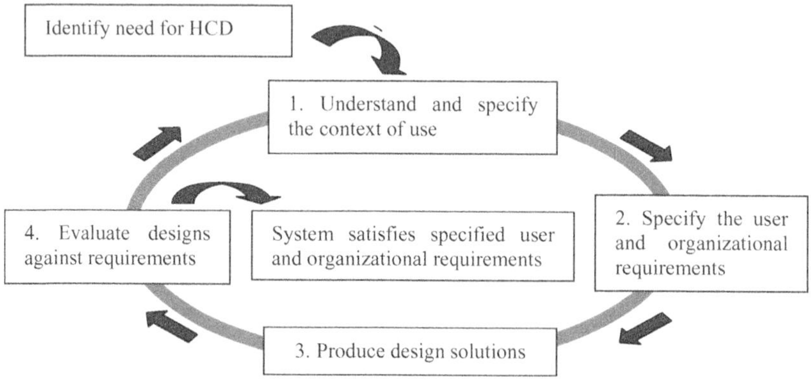

Fig. 1. Human-Centered Design activities [3]

Furthermore, the limitation in the efficiency of HCD has been pointed out due to the enormous labor required for its qualitative method such as field studies (Ethnography).

3 Optimal Design Loop in Service Sciences

Recently, an optimal design loop is gaining acceptance for its efficiency and effectiveness in the field of Service Sciences (SS) [7]. The loop is similar to HCD based on user's viewpoint and cyclic loop, and the method is much more quantitative and promises to solve many issues of HCD. SS and the optimal design loop are discussed in sequence below.

Tertiary industries are represented by immaterial services that bring benefits to its customers. There have been special emphases on the productivity improvement of these services in recent times [8, 9]. What was once dependent on experience or intuition, SS pursues approaches based on science and engineering [10].

SS is defined by the following formula. Service productivity is added-value divided by labor input.

$$\text{Service productivity} = \text{Added-value} / \text{Labor input} \tag{1}$$

Following the formula in (1), the improvement of service productivity requires improvement in the denominator, the numerator or both conditions. SS is expected as a way to bring significant progress by improving both the denominator and the numerator. Namely, service productivity will rise through less labor input by way of process improvement and cost cutting as well as higher added-values such as customer satisfactions.

A method called optimal design loop is in practice for a general approach to improve productivity (Figure 2) [7]. This design loop is rather general in design and consists of processes with data observation and accumulation, analysis, design, and application. It is not so much emphatic on users as HCD.

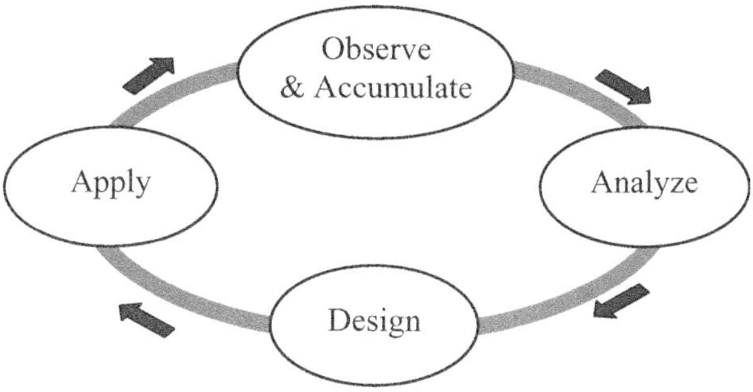

Fig. 2. Optimal Design Loop in Service Sciences (This is created based on Motomura [7]. The emphasized steps will be different depending on the types of objects and innovations explained in Figure 3.)

The advantage of the loop is that once the prediction model is built using the data relation analyzed from observation and accumulation, prediction of the entire data will be possible even if it is later given only partial observation data. By repeating the loop in such a process, precision enhancement and expanded prediction area are eventually expected. An example is a scale with a body fat monitor found on the market today. Information about body weights, electric conductivities and body fat percentages from hundreds of people are observed and collected in a database. Then the analyzed data formulate the relations of body information. The manufacture designs a scale with a body fat monitor using the formula. The formula is applied when the scale is used to calculate body fat percentages, the skeletal muscle rate, and the basal metabolism based on the weight and the electric conductivities. At the same time, the observation data uploads to accumulate by means of the Internet. Then the analyzed data contribute to the next generation product with enhanced precision and expanded functions.

When accumulation and analysis of quantitative mass data clarify actual human activities, and prediction precision is improved by rotating the cyclic loop, creation of new added-value and efficiency improvement, which was a HCD weak point, can be expected.

4 Consideration of Market Lifecycle

The purpose of working on SS is the significant improvement of service productivity. The approach with such a purpose is generally called "Service-related Innovation" or simply "Service Innovation" [7, 10].

Geoffrey A. Moore, who proposed the "Chasm theory" on technical innovation, explains four types of innovation in a market lifecycle: Product & Service Leadership type, Customer Intimacy type, Operational Excellence type and Category Renewal type [11]. This innovation framework not only covers products but also services.

The four types of innovation proposed by Moore are shown in Figure 3. The four rotating arrows are Optimal Design Loops in Figure 2. * indicates the improvement of the numerator (added-value) in formula (1) whereas - indicates the denominator (labor input) improvement.

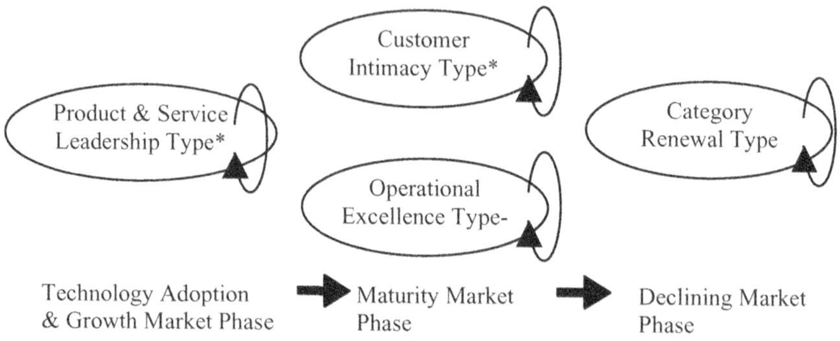

Fig. 3. Four types of innovation according to market lifecycle (This is created based on Moore [11])

In Figure 3, "Product & Service Leadership" type innovation is at the beginning stage or introduction and growing phase of the market. The innovation at this phase requires mainly technical excellence and performance. There is less importance on customer satisfaction or marketing since it is positioned before the mature market phase where fundamental features of user requirements are fulfilled. The improvement of service productivity at this phase primarily focuses on its performance and added-value and requires a good deal of R&D investment.

"Customer Intimacy" type innovation is positioned after the period of market introduction or growth. It is a mature time where the performances required by users are primarily satisfied and less attention is paid to technical features. The innovation at this time is to concentrate on making the service even the slightest more attractive than the competitors [2].

"Operational Excellence" type innovation is to make cost-cuts by optimizing the operations of service providers. The previous two innovations are improvements of the numerator in formula (1), whereas this is of the denominator. At the time of maturity market, maintaining customer intimacy and improving operational excellence lead to improvements of both the numerator and the denominator in formula (1). Then significant service productivity improvement can be expected.

Finally, "Category Renewal" type innovation is located at the end of lifecycle where future value creation is no longer possible. Since improvement of service productivity cannot be expected, the innovation at this phase is to identify problems coming from important customers and connect to "Product Leadership" type innovation of the next generation.

Optimal design loop of Figure 2 can be applied to each of the four types of innovation in Figure 3. For example, in the case of the scale with body fat monitor mentioned before, the original scale without the body fat monitor was improved functionally taking user research and feedback into account in "Customer Intimacy" type innovation of "Mature Market" phase. Cost reduction effort was also conducted in "Operational Excellence" type innovation. In "Declining Market" phase, "Category Renewal" type innovation was conducted, which sought unknown fundamental issue of important customers. It was the body fat monitor in the example. Once such issue was found, "Product & Service Leadership" type innovation was started investing R&D expense to produce a new innovative product, namely the scale with body fat monitor.

As mentioned above, the major issues of the service productivity improvement are quite different depending on the lifecycle status of the market.

5 Discussion: Transcending HCD

So far, relations between HCD, SS, and innovation theories were clarified to some extent. Following the arguments above, future issues of HCD are discussed here; the first two are related to the optimal design loop and the next two are related to the category maturity life cycle.

1) Participation of users:

In HCD, participation of users in the four activities of Figure 1 is considered quite valuable. Now that the notion has been well accepted and executed, it is becoming harmful rather than useful. For example, Donald A. Norman claimed that too much attention not to user activities but to the users would lead to too much listening resulting in confusion as mentioned in the second chapter [2]. In addition, innovation theorist Clayton M. Christensen points out that successful companies in "Maturity Market" phase tend to listen to the users too much and add too much functionality to their products or services, resulting in loss of their market. It is because of assault by new comers that provides "disruptive" products or services with low-price and limited functionality that are sufficient for the previous non-users [12].

In the optimal design loop, which applies to four types of innovation, the users become continual providers of valuable quantitative data that are indispensable for the prediction model. The users can save time attending interviews and user tests which is required in HCD. Furthermore, accumulating data according to predetermined parameters of the prediction model can prevent excessive user participation.

A new data acquisition and utilization method is required for HCD to save precious user time and to obtain more real-time and continual data.

2) Prediction model based on quantitative data:
In HCD, "conceptual model" which represents abstract structures and functionality of a product is often created [2]. Although a good conceptual model eases user understanding of a product, it requires many researches on the context of use to clarify important points for the understandings. Refining the conceptual model also requires almost the same amount of effort.

On the other hand, prediction model of the optimal design loop will be more accurate over time with more quantitative data from the users. It is often the case that the data is acquired almost automatically with advanced sensor technologies.

Since sensor technologies are so advanced to continually obtain various real-time data of user activities, such prediction model based on quantitative data should be used in HCD as well for more detailed tracking and understanding of user situations.

3) "Total Solution" which solves user problems completely:
Moore claimed that in "Technology Adoption & Growth Market" phase it is crucial to provide "Total Solution" for a niche market and repeat it until an enough size of the market is acquired [11].

For this purpose, HCD based on qualitative methods has two limitations: reusability of results of a niche market and situation-dependent solutions. In regard to reusability of results, since solutions derived from user's real goal and the context of use analyses are so specific to the situation that the solutions can not be applied to other niche markets. Same efforts are required for other niche markets. Meanwhile in the prediction model stated above, if relations between a setting, preferences, and activities are learned for instance, relations between preferences and activities will be reused to some extent in other settings as well. In this sense, the prediction model is superior to HCD in reusability of results resulting in better efficiency.

With respect to situation-dependent solutions, while solutions in HCD are usually static as stated above, the prediction model can usually simulate many combinations based on real-time data. Hence, the prediction model is superior to HCD in producing solutions in response to real-time situation changes.

Since HCD tends to concentrate on highest usability and highest user experience, it usually provides a partial solution rather than a "Total Solution". Although the prediction model does not always provide a "Total Solution", HCD is required to see such broader issues.

4) Transcending elaboration-focused HCD:
In most HCD projects, elaboration of products or services is focused on taking the context of use and user's real goal into account since they were created with little care of such issues and do not meet user needs. Meanwhile, for innovation in SS where even definition of market is difficult, research efforts such as generating ideas before building a specific user profile ("Persona") and clarifying hidden logics are requisite rather than elaboration.

Although there were some research-oriented HCD efforts in Universal Design area, where common-sense guess does not work at all [13], more situation-oriented models which cover a huge number of situational variations are required. For this purpose, HCD specialists such as ergonomists need to cooperate more with researchers in a quantitative modeling area such as Bayesian Network in Artificial Intelligence and in the innovation theory area.

6 Concluding Remarks

In general, HCD has been applied firstly to usability improvement, and then to User Experience and Universal Design with the framework enhanced.

To cope with SS, HCD needs further enhancements. Cooperating with experts from different fields, promising services must be identified and their trial results must be shared to create new methods which transcend HCD.

References

1. Friedman, T.L.: The World Is Flat: A Brief History of the Twenty-first Century. Farrar Straus & Giroux (2005)
2. Norman, D.A.: Invisible Computer: Why good products can fail, the personal computer is so complex and information appliances are the solution. MIT Press, Cambridge (1998)
3. ISO: ISO 13407: Human-centred design processes for interactive systems. ISO (1999)
4. Kurosu, M., Horibe, Y., Hirasawa, N., Miki, H.: Understanding ISO1307. Ohmsha, Tokyo (2001) (in Japanese)
5. Miki, H., Hosono, N.: IT Universal Design, Maruzen, Tokyo (2005) (in Japanese)
6. Miki, H., Akatsu, H., Hosono, N.: Universal Design Models of Development Lifecycle Process. In: Proc. of 2nd Int. Conf. for Universal Design in Kyoto 2006, Int. Assoc. for Universal Design (2006)
7. Motomura, Y., Nishida, Y., Mochimaru, M., Akamatsu, M., Naito, K., Hashida, K.: Large Scale Data Observation, Modeling, Service Design and Applying Loop for Service Innovation. Journal of the Japanese Society for Artificial Intelligence 23(6), 736–742 (2008) (in Japanese)
8. Council on Competitiveness: Innovate America: National Innovation Initiative Summit and Report (2005), `http://www.compete.org/publications/detail/202/innovate-america/`

9. OECD: Innovation and Knowledge-Intensive Service Activities, OECD 2006 (2006)
10. Maglio, P.P., Srinivasan, S., Kreuren, J.T., Spohrer, J.: Service systems, service scientists, SSME, and innovation. CACM 49(7), 81–85 (2006)
11. Moore, G.A.: Dealing with Darwin: How great companies innovate at every phase of their evolution. Portfolio (2006)
12. Christensen, C.M.: The Innovator's Dilemma: The revolutionary book that will change the way you do business. Harvard Business School Press (1997)
13. Akatsu, H., Miki, H., Hosono, N.: Design Principles Based on Cognitive Aging. In: Jacko, J.A. (ed.) HCI 2007. LNCS, vol. 4550, pp. 3–10. Springer, Heidelberg (2007)

Effects of a Mnemonic Technique on Subsequent Recall of Assigned and Self-generated Passwords

Deborah L. Nelson and Kim-Phuong L. Vu

Department of Psychology, California State University, Long Beach
1250 N. Bellflower Blvd., Long Beach, CA 90840, USA
deborah.l.nelson@gmail.com, kvu8@csulb.edu

Abstract. Participants were trained on how to use a mnemonic strategy for memorizing assigned passwords or for generating new passwords. Memory for these passwords was examined at short and long recall delays. There was a significant interaction between type of password and recall delay for both the amount of time and number of attempts needed for participants to accurately recall their passwords. Participants trained in how to use the mnemonic technique to generate their own passwords were able to recall them more quickly and accurately than participants who were trained in how to use the mnemonic technique to memorize their assigned passwords. The impact of self-generated passwords on memory was discussed as well as the relative value of the mnemonic training strategy. Areas of future research were identified that may lead to the development of mnemonic training strategies to better enable users to recall their passwords.

Keywords: memory training programs, mnemonics, retrieval strategies, passwords, proactive password checking.

1 Introduction

It is well-documented that people have a hard time remembering their passwords, and current trends indicate that the number of password-protected accounts managed by a person will only continue to grow [1]. Because of the difficulty people have recalling the passwords they have created, people tend to choose passwords that are simple and relatively easy to remember [2] and often share the same password across multiple accounts [1]. During the past few years, many alternatives to alphanumeric password methodologies; such as biometrics, keystroke rate, graphical patterns, and voice recognition have been touted as providing better security and/or authentication procedures [3], [4], [5], [6], [7]. However, due to the high cost and limited availability of many of these alternative methods, alphanumeric passwords are likely to remain the primary method of user-authentication for the foreseeable future.

People have done little to improve password security on their own. Passwords analyzed from a UNIX time-sharing system in 1979 [8] were shown to be an average length of 4 characters and were primarily comprised of all lower case letters, digits, or a combination of the two. Thirty years later, the passwords of half a million Microsoft

M.J. Smith and G. Salvendy (Eds.): Human Interface, Part I, HCII 2009, LNCS 5617, pp. 693–701, 2009.

users were found to have changed little [1]. These passwords were, on average, 6 characters in length and were comprised mostly of single case letters, digits, or a combination of the two. It is painfully obvious that users need to be educated in methods for creating more secure and memorable passwords.

People are generally responsible for creating their own passwords and have been found to do an inadequate job of creating passwords that are both secure and memorable [2], [9], [10]. The poor memory people have for secure passwords is called the security/memory tradeoff [11], meaning that the more secure a password, the harder it is for a user to remember. Secure passwords are usually difficult to remember because they resemble random combinations of letters, digits, and special characters. Previous research has indicated that training in the use of first-letter and passphrase mnemonic techniques can improve recall of secure passwords [9], [10], [12].

The purpose of this study was to compare the memorability of secure passwords that were assigned to users versus self-generated passwords created using the same mnemonic technique. Half of the participants were trained on how to use a mnemonic technique to memorize "secure" (likely to be crack-resistant) passwords that had been generated using a phrase-based mnemonic technique [10]. The other half was trained on how to use the same phrase-based mnemonic technique to generate a password that met several proactive password checking (PPC) restrictions. Although passwords generated with this technique are not guaranteed to be secure, we were more interested in the impact of self-generated passwords on memory.

2 Method

2.1 Participants

Twenty participants were recruited from flyers posted around the Psychology Building at California State University, Long Beach. All participants were unpaid volunteers.

2.2 Design

This study used a 2 (Type of password: assigned or generated) by 2 (recall delay: 10 minutes or 1 week delay) mixed design. Type of password was a between-subjects variable, and recall delay was a within-subjects variable. The dependent variables included length of time for password recollection and number of trials for correct input of password. Password forgetting rates were also measured.

2.3 Apparatus

This study was conducted on a personal computer equipped with a 17" LCD monitor. Participants sat in an adjustable chair placed before the computer workstation that was located in a quiet room. The program used to run this study was written in Visual Basic 6 and consisted of a typical graphical user-interface password input box in which participants were asked to separately enter the passwords for each of 5 different accounts (see Fig. 1). Participants used a standard QWERTY keyboard to enter their passwords.

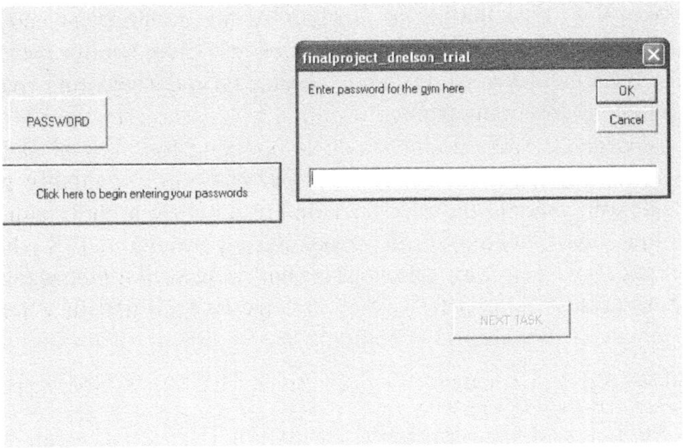

Fig. 1. Interface for password entry used in the present study

2.4 Materials

Participants were provided training on how to use a phrase-based mnemonic technique to memorize assigned passwords or to generate new ones. The assigned passwords were taken directly from Vu et al.'s [10] study. These passwords were selected because they were secure, that is unable to be cracked by the LC5 program within 12 hours, see Table 1 for samples of assigned and self-generated passwords.

Table 1. Some samples of the assigned passwords versus generated passwords

Assigned Passwords	Generated Passwords
eye8egg$4B	1M@cD@ddy
!H8work!	Iwork4$$$
Go2cla$$	Luv2Read!
eyeH8^u!	<32Yahoo!
Ih8rhino$	M3y3*Space*
ILike^00	1Ne^^orld

Mnemonic Training for Participants Assigned Passwords. Participants were provided training on how to use a mnemonic technique to memorize and later recall their assigned passwords. Participants were informed that past researchers had observed that some passwords look like random strings of letters and numbers, but are in fact organized into meaningful patterns. The participants were told that, when inspected closely, the passwords were really mnemonic sentences, phrases, or words. They were provided with an example of how the password "EyeD8tedM@" could be remembered as the sentence "I dated Matt". Participants were asked to note how the sentence was transformed into the password using the phrase-based mnemonic technique: "I" to "Eye" because Matt has pretty blue eyes; "dated" to "D8ed" because of the phonetic similarity between the "words"; "Matt" to "M@" because of phonetic similarity and that the special character, @, was needed to satisfy a PPC requirement. Additionally, capitalization of the first letter of each of the "words" contained in the password was used to cue changes in case. A second example providing similar instructions for the meaningful replacement and inclusion of digits, special characters, and case change was also provided to the participants.

Mnemonic Training for Participants who Generated Passwords. Participants who generated their own passwords were provided training on how to use the same mnemonic technique described above. However, these participants were asked to use the technique to generate their own passwords. They were walked through two examples, step-by-step, detailing the process of how to modify each part of a sentence, phrase, or word to create a meaningful mnemonic password, as in the assigned password condition.

Post-study Survey and Questionnaire. Following the second recall session, participants were asked to complete a demographic survey and post-recall questionnaire. Participants were asked to provide standard demographic information and as well as information regarding their computer experience and password security knowledge, and their password generation and management practices.

2.4 Procedure

Each participant was tested individually over two sessions, each lasting approximately 30 minutes. Participants were provided informed consent forms to read and sign upon arrival to the experiment. They were informed that they were participating in a study examining the memorability of passwords. They were told that all passwords used during the study would be recorded, and that if they were asked to generate any passwords to please not use any passwords that they personally knew to be in existence.

Following the receipt of a participant-signed consent form, the experimenter escorted the participant into the test room. Depending on the condition to which the participant was assigned, mnemonic instructions written on an 8.5 by 11 inch sheet of paper were placed on the table in front of the participant, in the space between the participant and the keyboard. Participants were given written and verbal task instructions. A list of PPC restrictions was then placed before the participant, and explained verbally. The list of restrictions was left on the table within view of the participant during the entire course of the experiment.

After reading and verbally reviewing the mnemonic instructions, password examples, PPC restrictions, and answering any participant questions, the participants were instructed to turn on the monitor and follow the instructions written on the screen. The accounts were the same for each of the two groups, a bank account, a book store account, an email account, a social networking account, and a computer account.

Once the participants had successfully learned or generated their passwords, they were asked to complete a series of simple math tasks to their completion or until 10 minutes had passed, whichever came first. Following a brief break while the computer was reset, the participants were brought back into the test room and asked once again to sit in the chair placed before the same computer. When ready, the participant was asked to begin the recall phase of the first session.

Participants were again provided written instructions on the screen and were asked to separately enter the passwords for each of the same 5 accounts for which they had earlier learned or generated their passwords. Each account was presented in random order and appeared a total of four separate times. For each account appearance, participants were provided with 10 opportunities to input the correct password for the account. If the correct password was not entered by the tenth attempt, the program informed the participant that the maximum number of allotted trials had been reached and the participant was moved along to the next account. Once participants had the opportunity to enter all of their passwords, they were presented with a screen that announced that they had successfully completed session 1 of the experiment. Prior to leaving the first session, participants were asked not write down any of their passwords used during the study, not discuss any part of the experiment with anyone else, and were provided with a reminder slip for the following week's scheduled session (session 2).

Upon arriving for the second session the following week, participants were greeted and brought into the same room and seated before the same computer used during the first session. Once again, they were asked to separately enter the passwords for each of the same 5 accounts for which they had entered passwords during the prior session. Once participants had the opportunity to enter all of their passwords, they were presented with a screen that announced that they had successfully completed the experiment.

Prior to leaving, participants were asked to complete the post-recall questionnaire and demographic survey. Upon completion of the paperwork, participants were provided with a debriefing form, asked if they had any questions, and thanked for their time.

3 Results and Discussion

The current study compared performance of participants who were assigned "secure" passwords that were generated using a phrase-based mnemonic technique with participants who were trained on how to use the technique to generate passwords. Results indicated that there was no significant difference in the average amount of time needed for participants to either learn ($M = 72.5$ s) or generate ($M = 79$ s) their passwords. However, participants were more variable in the time spent learning the assigned passwords (range 24 s to 284 s) than generating them (range 54 s to 132 s). Additionally, the results indicated that there was no significant difference in the number of attempts needed for participants to either learn (i.e., enter their password correctly without having access to it, $M = 1.23$, $SD = 0.42$) or generate ($M = 1.30$, $SD = 0.33$) their own passwords.

3.1 Password Recall Time

Mean recall time (in seconds) was calculated for each participant as a function of type of password and recall delay. There was a marginal effect of type of password for the amount of time needed for participants to recall their passwords accurately, $F(1,18) = 3.42$, $p = .080$. Participants who generated their passwords tended to take less time to recall the passwords (M = 43 s) than did those who were assigned passwords (M = 66 s). There was a significant main effect of recall delay, $F(1,18) = 23.35$, $p < .001$. Participants took longer to recall their passwords following the week-long delay (M = 79 s) than following the 10-minute delay (M = 31 s).

More important, there was a significant interaction between of type of password and recall delay, $F(1,18) = 9.50$, $p < .01$, see Fig. 2. Participants' recall times were similar for both of type of passwords following a short 10-minute delay (M = 34 s for assigned passwords and M = 28 s generated passwords). However, following the longer delay of one week, participants who received instruction on how to use mnemonic techniques to recall their assigned passwords took significantly longer (M = 105 s) than did those who received mnemonic instructions on how to generate their passwords (M = 52 s).

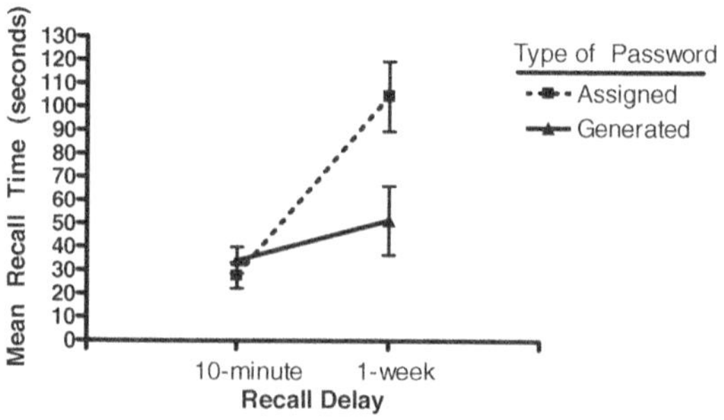

Fig. 2. Mean recall time (in seconds) as a function of recall delay and training method

3.2 Password Recall Attempts

There was a significant main effect of type of password for the number of attempts, $F(1,18) = 7.54$, $p < .015$. Participants who generated their passwords needed fewer attempts to accurately recall their passwords (M = 1.73) than participants who were assigned their passwords (M = 2.39). There was also a main effect of recall delay on the number of attempts, $F(1,18) = 8.78$, $p = .01$. Participants required more attempts to recall their passwords following a one-week delay (M = 2.50) than following a 10-minute delay (M = 1.62). Unlike with recall time, though, these main effects were not qualified by a significant interaction between type of password and recall delay, $F(1,18) = 2.86$, $p > .10$.

3.3 Forgetting

There was significant main effect of forgetting between the type of passwords, $F(1,19) = 4.99$, $p = <.04$. Participants who received instruction on how to use the mnemonic technique to generate their passwords forgot significantly fewer (M = 0.60) passwords than participants who received instruction on how to use the mnemonic technique to recall their assigned passwords (M = 1.8), see Fig. 3.

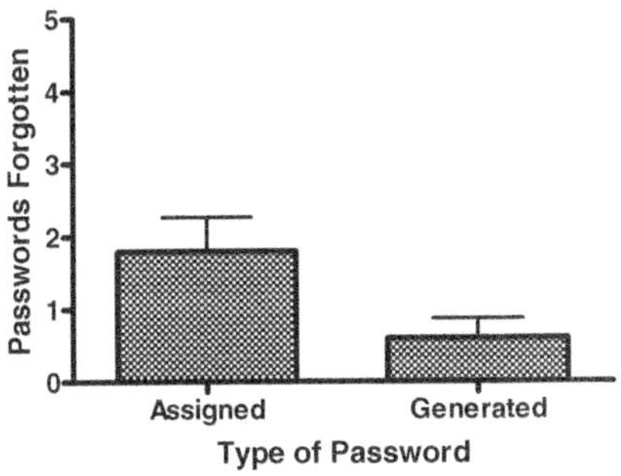

Fig. 3. Mean number of passwords forgotten as a function of the type of password

3.4 Post-recall and Demographic Questionnaires

Post-recall questionnaire and demographic data indicated that participants had over 12 years of computer experience. Most participants (85%) reported having average, or above average, computer knowledge. However, 75% reported having little knowledge of computer security. Participants reported managing an average of 15 password-protected accounts, with about 6 unique passwords that were shared across accounts. Within their own passwords, all participants reported that they include numbers, 65% incorporate both upper and lowercase letters, and only 28% include special characters. Over two-thirds of the participants reported that more than one of their passwords could easily be personally associated with them, and slightly less than two-thirds reported that an average of two of their passwords were a name followed by a number. All participants reported a belief that the requirement of a passwords increases security; however, only 63% of the participants indicated that password policies were somewhat important to them.

4 Discussion

Participants trained to use the phrased-based mnemonic technique to generate passwords were able to recall more of those passwords, and do so more quickly, than those assigned secure passwords and trained to use the same mnemonic technique to memorize the passwords. Memory-based explanations are evaluated in regard to these findings.

The use of mnemonic strategies has been shown to enhance memory by imposing organization and structure to information that needs to be recalled [13]. In the present study participants were trained to use a passphrase text-based mnemonic strategy to either generate or memorize an assigned password (i.e., transforming the phrase "Later I ate tofu" into the password "L8erEye8+ofu"). For both groups of participants, the structure of the sentence should aid memory for the passwords. However, because both groups were trained with the same mnemonic technique, the technique itself cannot account for group differences. The group differences are likely due then to how the technique was used. For the assigned condition, participants were more variable in the time spent using this technique to learn the passwords. It may be the case that more time needs to be spent using the technique to learn the password in order for it to be as effective as using the technique to generate passwords. The time spent learning the assigned password was correlated with performance variables following the one-week recall delay. All correlations were negative (recall time: $r = -.25$, $p = .45$; number of attempts: $r = -.45$, $p = .196$; forgetting: $r = -.42$, $p = .22$), but not significant. Thus, there was a trend indicating that more study time for the assigned passwords might benefit subsequent recall. The issue of whether the mnemonic technique may effectively be used to enable users to better recall assigned passwords following longer study periods should be further evaluated, preferably with a larger test group (more than 10 participants).

In addition, research has indicated that memory for items can be enhanced when participants are required to generate the items that they are attempting to recall, as opposed to simply reading them from a list [14]. Thus, it is likely that the benefit for the self-generated password group is due to the generation effect and to users tending to generate password phrases that are relevant to themselves. The resulting password would then have meaning pertinent to the user, which would increase the likelihood that the user would be more likely to recall the password at a later time. Rogers, Kuiper, and Kirker [15] asked participants to determine which words on a provided list described them personally. They found that participants showed the highest levels of recall for the words they felt were self-descriptive.

This study supports the practice that users should be allowed to generate passwords and not have to memorize computer-assigned passwords. Even after training users on how to relate assigned passwords to a sentence structure, these passwords were more likely to be forgotten than self-generated passwords. Mnemonic techniques used to generate passwords should allow users to impose a structure that can be meaningfully tied to the password. Without this, it becomes near impossible for the placement of digits, special character, and case changes to be accurately recalled over time.

Acknowledgments. This research was supported in part by a CSULB Psychology Department 22B Assigned Time Award to K.-P. L. Vu.

References

1. Florêncio, D., Herley, C.: A Large Scale Study of Web Password Habits. In: Proceedings of the Sixteenth International World Wide Web Conference, WWW 2007, Banff, BC, pp. 657–666 (2007)
2. Adams, A., Sasse, M.A.: Users Are Not the Enemy. Communications of the ACM 42, 40–46 (1999)
3. DeAngeli, A., Coventry, L., Johnson, G., Renaud, K.: Is a Picture Really Worth a Thousand Words? Exploring the Feasibility of Graphical Authentication Systems. Intl. J. of Human-Computer Studies 63, 128–152 (2005)
4. Ives, B., Walsh, K.R., Schneider, H.: The Domino Effect of Password Reuse. Communications of the ACM 47, 75–78 (2004)
5. Stubblefield, A., Simon, D.R.: Inkblot Authentication. Technical Report: Microsoft Corporation (2007)

6. Tullis, T.S., Tedesco, D.P.: Using Personal Photos as Pictorial Passwords. In: Proceedings of ACM CHI 2005 Conference on Human Factors in Computing Systems, pp. 1841–1844. ACM Press, New York (2005)
7. Wiedenbeck, S., Waters, J., Birget, J.-C., Brodskiy, A., Memon, N.: PassPoints: Design & Longitudinal Evaluation of a Graphical Password System. Intl. J. of Human-Computer Studies 63, 102–127 (2005)
8. Morris, R., Thompson, K.: Password Security: A Case History. Communications of the ACM 22, 594–597 (1979)
9. Vu, K.-P.L., Bhargav, A., Proctor, R.W.: Imposing Password Restrictions for Multiple Accounts: Impact on Generation and Recall of Passwords. In: Proceedings of the 47th annual meeting of the Human Factors and Ergonomics Society, HFES, California, pp. 1331–1335 (2003)
10. Vu, K.-P.L., Proctor, R.W., Bhargav-Spantzel, A., Tai, B.-L., Cook, J., Schultz, E.E.: Improving Password Security and Memorability to Protect Personal and Organizational Information. Intl. J. of Human-Computer Studies 65, 744–757 (2007)
11. Proctor, R.W., Lien, M.-C., Vu, K.-P.L., Schultz, E.E., Salvendy, G.: Improving Computer Security for Authentication of Users: Influence of Proactive Password Restrictions. J. of Behavior Research Methods, Instruments, & Computers 34, 163–169 (2002)
12. Yan, J., Blackwell, A., Anderson, R., Grant, A.: Password Memorability and Security: Empirical Results. IEEE Security & Privacy 2, 25–31 (2004)
13. Bellezza, F.S.: Mnemonic Devices: Classification, Characteristics, and Criteria. Review of Educational Research 51, 247–275 (1981)
14. Slamecka, N.J., Graf, P.: The Generation Effect: Delineation of a Phenomenon. J. of Experimental Psychology: Human Learning & Memory 4, 592–604 (1978)
15. Rogers, T.B., Kuiper, N.A., Kirker, W.S.: Self-reference and the Encoding of Personal Information. J. of Personality and Social Psychology 35, 677–688 (1977)

Control of Personal Tempo to Support Individual Action and Cognition

Naoya Nobutani and Yoshio Nakatani

Ritsumeikan University
1-1, Noji-Higashi 1, Kusatsu, Shiga 525-0058, Japan
cc011044@is.ritsumei.ac.jp, nakatani@is.ritsumei.ac.jp

Abstract. We live in our own rhythms in daily lives. Human life is closely related to rhythm as the words "talk rhythm," "getting into the rhythm" and the "Biorhythm" show. The rhythm is important for us to live in order and cheerfully. Each person has his/her own action rhythm called the "Personal Tempo." It is the basic rhythm observed in person's voluntary actions, such as tapping on a table. Psychology has been studying this phenomenon for long years. By measuring the personal tempo, person's psychological condition can be inferred in all situations because personal tempo changes within a certain width according to person's psychological condition, while it basically shows homeostasis. Most studies have focused on relation between personal tempo and physiological measures, such as cardiac rate, but not on control of cognitions and actions by controlling personal tempo. In this paper, a control method of individual action by controlling the personal tempo is proposed. Drum beat pattern is used to control the personal tempo. We show the effectiveness of this method by applying this to cooking, especially chopping by a knife.

Keywords: personal tempo, rhythm, action control, sound rhythm.

1 Introduction

The times when music is carried every times. The spread of the portable music player is who makes i-pod of the Apple Co in recent years.

In previous works, due to rhythm sound causes synchronous reaction, listening rhythm sound cause physical rhythm, but there are no psychology studies about individual rhythm or characteristic and there has been no study that tried to materialize rhythm supporting.

This paper discusses rhythm sound based personal tempo is valuable for physical rhythm. We investigate how effect does the sound when monotonous action by experiments. We discuss the future of supporting method by sound rhythm based personal tempo from investigate result in the seventh chapter.

M.J. Smith and G. Salvendy (Eds.): Human Interface, Part I, HCII 2009, LNCS 5617, pp. 702–708, 2009.

2 Related Research

2.1 About a Personal Tempo

A personal tempo is a rhythm feeling that exists inside the individual, and the principle that has been researched in physiology from of old. There is a personal tempo between 68bpm (bit per minute) - 158bpm. Moreover, the speed of the tempo is admitted to change by the interaction with the mental status, and when the changing tempo is removed, too the homeostasis that returns to former state with the time passage is admitted to be had [2]. Personal tempo is reported to be no origin from the physiology function, peculiar to the individual, and change by feelings and the environment [3], but details of them are not clear.

The personal tempo is individual and different from others. As for the relation between human and the rhythm are deep as words such as "Rhythm of life" and "Rhythm of the conversation" show from our experience. It is thought that it has some influences on the action so that it is suitable from there to rhythm, not suitable or differences of others' tempo speed are admitted. In communications with others, there is something that the relation between speakers and communications of the smoothness degree is examined as the research theme of having the personal tempo [4]. As for communications with the others having similar rhythm feeling, the thesis that the tune tendency is seen at the early stage of communications was reported.

2.2 Relation between Locomotor Rhythm and Sound Rhythm

As for related research and related psychology experiment concerning the sound rhythm, especially, many research papers from musical aspect are announced. The early research on the locomotor rhythm and the sound rhythm is the research of “Effect of drawing in” that appears when the walking operation is done while listening to music. This paper reported that we cause a synchronous reaction easily as for the rhythm, the sound rhythm causes the locomotor rhythm, and the locomotor rhythm causes the sound rhythm[5]. Moreover, many researches of the music recommendation system have reported in recent years from to being able the easy treatment of music by data and becoming of individual music data are hugeness.

From such the background, many products and research about relation between the music recommendation system and the movement are reported.

2.3 About a Personal Tempo and the Locomotor Rhythm

We feel some rhythms in living, and act as synchronizing with the rhythm. As the example, it is seen a lot in monotonous constant operation such as the walking operation and the pedaring of bicycle, and their rhythms of the action are matched to the rhythm of listening music. Relation between the action that have constant rhythm and the personal tempo is admitted by a lot of related researches. Moreover, when it is made to think and to come round by something to worry about, our finger and foot might be moved in unconsciousness by a constant rhythm. This is the example of the natural embodiment feeling the rhythms. The relation between rhythm perception and a personal tempo can be expected from these while thinking.

2.4 Support of Monotonous Locomotor Rhythm

In this research, we reports on the verification results that the sound rhythm based on a personal tempo gives what influence and drawing in to individual feelings and the locomotor rhythm in a monotonous behavior pattern. The system that supports our action and cognition by personal tempo based on personal tempo is useful, if there is considerable validity in the results.

In the verification experiment, we use the drum rhythm based on a personal tempo and focus attention on cutting the material in cooking as action of locomotor rhythm, we consider what influence the action synchronized in the rhythm based on a personal tempo. Figure1 shows the rhythm support model. The early research reports the utility of the rhythm support of monotonous locomotor rhythm such as the walking operation and pedaring of the bicycle, and we develop it and focus attention on a little complex work pattern of cutting rhythm basic operation in this research. Cutting is work that the rhythms exists, and the good effect can be expected.

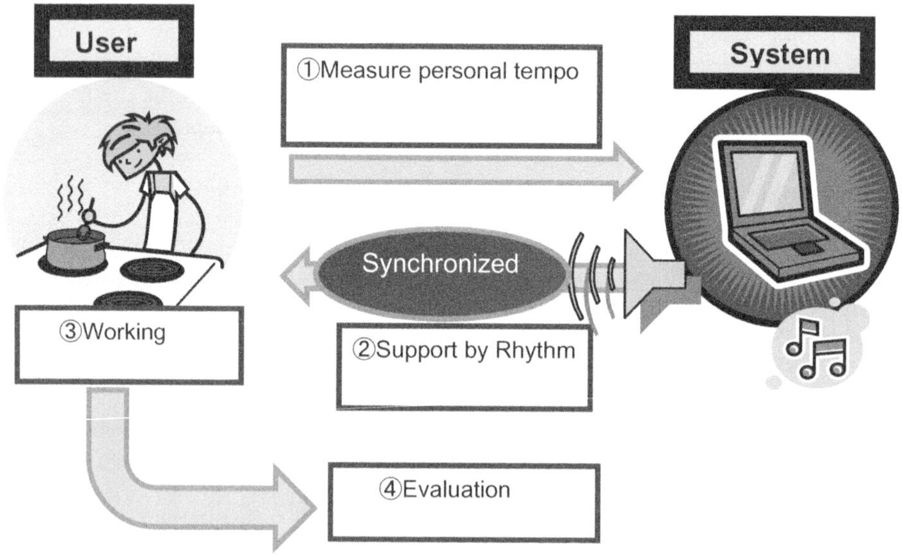

Fig. 1. A Model Which Supports Rhythm in Daily Life

3 Outline of Systems

The third chapter presents the experimental system that uses it for the verification and describes which the sound rhythms to select in a prior experiment.

3.1 Support by Drum Rhythm Pattern

In this research, it paid attention only to the drum rhythm as the sound rhythm for the rhythm support. We need to specify the sound rhythm to support it because the music is composed by various sounds and musical instruments at the same time. The drum rhythm is bases, and supports the basis of the rhythm of the music. Various patterns exist in the drum rhythm, and it is composed by the combination and repetition.

We use 5 kinds of typical examples and 7 kinds Basic examples of each musical genre of the drum rhythm pattern in this research.

Typical examples are;

Metronome (2beat) / 4beat / 8bea t/ 16beat / Shuffle Beat

Basic examples of rhythm of each musical genre are;

Rock / Samba / House / Techno / Funk / Ska / EuroBeat

Total 12kinds

12 all kinds of patterns were made with MIDI sound source software LM-7 by using Cubase (by the Steinberg).

3.2 Details of the Experimental Systems

This system measures user's personal tempo immediately before working, because the personal tempo is not always constant. It shows in Figure 2 as an example of the system screen.

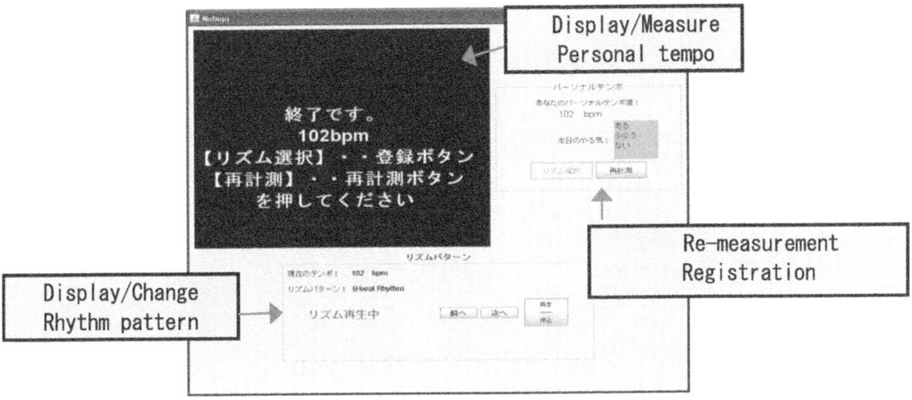

Fig. 2. A Screen Image of System

It is measured by keyboarding the Space key in user's comfortable rhythm according to the instruction of the system screen. In this research, we define the keyboarding tempo is the personal tempo, BPM (bit per minutes) signify the speed. The speed and the rhythm pattern of the rhythm can be changed and reproduction/be stopped on the system screen.

4 Experiments

In this research, it experiments on the rhythm support as the behavior pattern that the singsong is admitted for the work cut while cooking. First of all, we verify which rhythm pattern worked easily in the work that each testee cut in experiment1. Next, we verify what change was seen by changing the speed of the sound rhythm of the rhythm pattern in experiment 2. The rhythm pattern is result of experiment1.

The Testees were ten university students of 20-24 years old, and they are independently verified it respectively.

4.1 Experiment 1

The measurement of the personal tempo is input by using an experimental system immediately before work begins. It was to measure after it had been taught to testees, "Please strike key at a pleasant speed when it walked not too fast but not too late" in the measurement of the personal tempo. The measurement was finished, the sound rhythm started, the testees started cutting the material. The flowing rhythm flowed sequentially the one made at the same speed as a personal tempo, reproduced one by one by about two minutes one rhythm pattern, and made it answer the evaluation item by oral. Is the evaluation item natural compared with each rhythm pattern. Do you feel that it is fast unnatural. The sensibility was evaluated by the SD (Semantic Differential) method of seven stages and six items in a comparative content of feeling were done.

4.2 Experiment 2

One rhythm pattern is chosen by the rhythm pattern evaluation of each testee is the highest in experiment1,

1. speed that makes the personal tempo +10bpm fast
2. speed that makes the personal tempo +20bpm fast

In two rhythm support, it is made to answer the evaluation by oral just like experiment 1, and we consider what change is seen between result of experiment1 and above two cases in the sensibility evaluation.

5 Consideration

5.1 Verification Result

In the paper, we propose the rhythm support based on a personal tempo that existed inside the individual, and we do the verification experiment by the drum rhythm system based on user's personal tempo. It is because the early research that was not considering a rhythm feeling different depending on the individual. Our report proves the personal tempo is within 68bpm-158bpm (Refer to Table 1).

Table 1. Personal Tempos of Subjects

Testee	A	B	C	D	E	F	G	H	I	J	Average
Personal tempo(bpm)	100	68	69	78	128	80	95	100	117	80	91.5

The tendency to which the preference was not so shown because suitable for the rhythm of the cut work and not suitable became important in the evaluation when be working, and it was selected by a pattern like the *Samba* rhythm and the *House* rhythm. It is because their rhythm is not too monotonous but not too complex. This is thought that the evaluation was high from the point corrected that it is possible to synchronize again when the gap is caused in the work rhythm and the heard sound rhythm easily. Therefore, it can be said that the pattern that is not liked monotonous sound rhythm even if it is a monotonous work rhythm but a little complex composition that can be corrected will be liked.

Next, the tendency that the rhythm support evaluation when being working by presenting the sound rhythm that hastened +10bpm from a personal tempo becomes higher was shown in the same rhythm pattern since the evaluations of experiment 1 and experiment 2 were compared. This is because a personal tempo is drawn in to the rhythm by fast support, and the rhythm of the cut work is promoted more of +10bpm because the act of cutting the material is done from a personal tempo a little by a fast rhythm. Moreover, we pay attention to the fall of the evaluation in the support of +20bpm fast rhythm. There were a lot of opinions that it was not easy to work because the sound rhythm was too fast as for +20bpm though it became easy to work in +10bpm.

We targeted the cutting rhythm to support of the work in the research. The utility of support by the sound rhythm based on a personal tempo was confirmed in the locomotor rhythm that the work of the cut work was drawn in to the sound rhythm in the pattern that the testees felt worked easily from the result of the questionnaire. There were a lot of opinions of having felt support and the constant rhythm existed.

5.2 Problem

The problem in this research has the measuring method of a personal tempo. It is necessary to examine the method of putting besides the rhythm expression by the hand because the measurement of a personal tempo is thought expressing targeting timing and the rhythm of key stroke is not only the finger but also expresses the rhythm by the foot and the body. In the measuring method, it was necessary to investigate user's psychological condition more in detail, and to arrange the condition like the same place and time zone.

The personal tempo changes on each occasion. It is thought that it is load for the users to measure it in every case in the action of daily life, and they have to measure it when we take the system within daily life. It is necessary to make the system automatically measure it from the rhythm expression of the user, and to reduce the user's load.

Moreover, the user wants to extract it psychological condition automatically in the future. It is thought that the psychological condition is predictable to measure all testee's information with compress strength of keyboard when the personal tempo is expressed, the ventricular rate, and the psychology rating equipment of the stress meter. We want to verify continuously in the future, and to increase the number of case samples.

6 Future Work

The experiment verification in the effect of drawing in of the sound rhythm by an experimental system based on a personal tempo was done in this text.

It is thought that this research has a good influence on the working user when it is reluctant compared with work and work that monotonous operation continues excluding the operation targeted by this research. It relates to the car stereo based on the verification result of this text, and the support of the driver who drives a motorcar will be examined in the future. The driver's personal tempo and psychological condition are recommended, and the system takes into consideration, and recommends the best music for each occasion and the sound rhythm that takes the place of it from the information automatically by the car stereo. As the support target, it wants to take various information and support that not only the support of work of driving of controlling the attention rousing and irritation but also support, the road situation, and the weather of the place that communications in the car are related to the driver of consideration into consideration, and to propose the supporting method based on a personal tempo.

References

1. Taniguchi, T.: The Sound Changes Music in the Mind, Kitaoji-shobo (2000) (in Japanese)
2. Suginohara, M., Matsuda, S., Taira, S.: Fundamental Study on Personal Tempo (7). Hiroshima Shudo University Thesis Collection 23(2), 120 (1982) (in Japanese)
3. TAkenaka, M., Okai, S., Ohara, Y., Inoue, K.: Research on Physiological Response by Rhythm Listening of Tempo Based on Heart Beat. Clinical Educational Psychology Research 31(1), 43–45 (2006) (in Japanese)
4. Oishi, S., Oda, M.: Influence that Difference of the Personal Tempo between Speakers Exerts on Smooth Communications. Corporation Institute of Electronics, Information and Communication Engineers 105(536), 31–36 (2006) (in Japanese)
5. Nagashima, Y.: Effect of Drawing in that Musical Beat Causes for Perception of Image Beat 3(1), 108–109 (2004) (in Japanese)

The Effects of Practice and Speed Stress with Different Stimulus-Response Mappings

Kim-Phuong L. Vu, Audrey Rabas, and Richard Roberson

California State University, Long Beach,
1250 North Bellflower Blvd, Long Beach, California
kvu8@csulb.edu, audreyelise@hotmail.com,
rroberson69@hotmail.com

Abstract. Stimulus-response (S-R) compatibility refers to better performance with compatible over incompatible S-R mappings. We investigated how a stressor in the form of a time constraint influences performance in choice-reaction tasks with S-R mappings that varied in degree of compatibility. A 600-ms response deadline did increase participants' stress levels as indicated in subject workload reports. Furthermore, the time constraint decreased reaction time and increased error rate more for incompatible (mirror-opposite, mixed and random) mappings compared to compatible mappings. Participants who learned to respond with the incompatible mappings reverted to the more natural, corresponding responses when stressed. However, the effect of the time constraint was reduced when the incompatible mapping was systematic compared to when it was random. Thus, there are benefits of applying systematic rules when designing products for a user population.

Keywords: Stimulus-response compatibility, display-control compatibility, stress, practice effects.

1 Introduction

In everyday life, we are bombarded with stimuli. Whether we choose to respond to a stimulus and how we respond to it depends on natural response tendencies and past experiences. Imagine, for example, walking towards a door for the very first time and seeing a handle on it. Would you pull or push on the handle to open the door? Due to the response affordance of handles, people have developed a response tendency to pull on handles rather than to push on them. However, some doors are designed in a manner that requires a user to push on the handle to open the door rather than to pull on it. Can users interact with this type of door successfully? The answer is "yes" even when the push response does not match users' response tendencies. With practice, people are able to learn that a particular type of door handle requires a push and not a pull response. However, what would happen if a person had to quickly exit a building with handled doors that require push responses? Would the person follow his/her natural response tendencies and pull on the door handle or would s/he follow the learned response to push the handled door? Being able to determine the ways in

M.J. Smith and G. Salvendy (Eds.): Human Interface, Part I, HCII 2009, LNCS 5617, pp. 709–717, 2009.

which humans respond to situations under both time constraints and no time constraints can help improve the ability to design systems and interfaces that maximize user's performance and efficiency (low reaction time, RT; low error rate, ER). The aim of the current study is to examine human response tendencies in the laboratory in choice-reaction tasks when speed stress, in terms of a time constraint, is applied compared to when no time constraint is imposed.

Stimulus-response (S-R) compatibility is a topic that has been studied since Human Factors emerged as a field [1]. The S-R compatibility effect refers to the fact that performance is better when (a) the stimulus configuration matches the response configuration than when it does not (set-level compatibility), and (b) when individual stimuli are mapped to their corresponding responses than when they are not (element-level compatibility [2]. Fitts and Seeger [1] demonstrated set-level compatibility by examining performance when three different display configurations were paired with three response configurations. Stimulus set A consisted of eight lights arranged in a circle and the response set A used a response apparatus that consisted of a stylus that could be moved along eight paths of a circle (see Fig. 1). Response sets B and C were arranged in a square or as two lines, one horizontal and one vertical. RT and ER were lower when stimulus set A was mapped to response set A than to response sets B and C because the configurations corresponded. In other words, performance was better when the stimulus set closely matched the response set than when it did not.

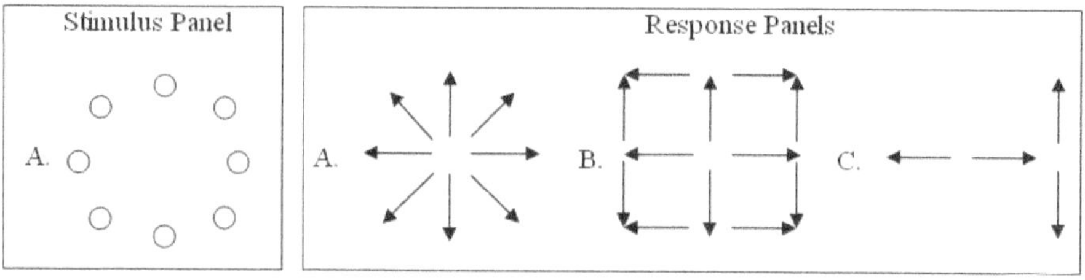

Fig. 1. Illustration of Stimulus Panel A and Response Panels A-C used by Fitt's and Seeger (1953)

Fitts and Deininger [3] illustrated element-level compatibility effects using stimulus set A and response set A. The mapping of stimulus locations to response locations was corresponding, mirror-opposite, or random (see Fig. 2). For the corresponding mapping, the illumination of the top light required the stylus to be moved to the top position of the response apparatus. For the mirror-opposite mapping, the illumination of a light to the right of a vertical midline required the stylus to be moved to the left position of the response apparatus, opposite of what a corresponding response would be. For the random mapping, there was no systematic relationship between the stimulus and response locations, and participants had to memorize the specific S-R pairings. RT was lowest for the corresponding mapping, intermediate for the mirror-opposite mapping, and highest for the random mapping. Thus, there was maximal benefit for maintaining spatial compatibility, but performance also benefited from a systematic S-R rule (e.g., respond with the mirror opposite location to the stimulus location).

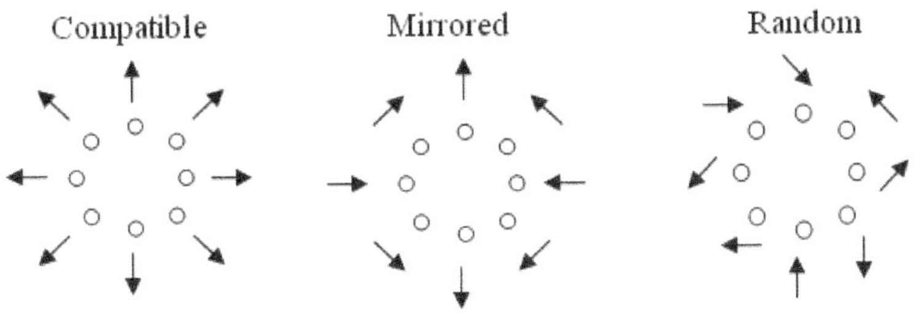

Fig. 2. Illustration of element-level mappings used by Fitts and Deininger [3]. The circles represent stimulus locations, and arrows indicate the direction of the response to be made to the stimulus.

Kornblum, Hasbroucq, and Osman [4] contended that element- and set-level compatibility work together to influence performance. To obtain the shortest RT and lowest ER, both set- and element-level compatibility should be high. Thus, any interface design should take into account both set- and element-level compatibility effects. Morin and Grant [5] illustrated the degree of element-level compatibility by using five different mappings based on a correlation of direct spatial mapping, where $r = 1$ (completely corresponding) represented a direct mapping of a stimulus to a response, and matched participants natural response tendencies. The other four mappings ranged from correlations of $r = .86$ (non-corresponding) to $r = .00$ (completely random). Performance was the best when the mapping was completely corresponding; and was proportional to the correlational level (lower correlation = lower performance) when S-R mappings deviated from the completely corresponding mapping.

Speed stress is one factor that can affect how much time people have to respond to a stimulus event and the errors that people make. Castaneda and Lipsitt [6] presented subjects with a stimulus set of eight lights horizontally aligned and a response set of 8 switches located directly below the eight lights. Half of the lights had corresponding S-R mappings, where the switch below the light turned the light off and the other half had non-corresponding S-R mappings; the switch to turn the light off was adjacent to the lit light, either to the left or to the right. All participants were shown the same stimulus sequence and were placed into either a stressed or non-stressed group. Stress was induced in the form of a 1-s time constraint, where participants had to flip the correct switch within one second. The non-stressed group did not have a time limit for responding to the stimulus.

ERs were calculated for both the corresponding and non-corresponding mappings. Fewer errors were committed with the corresponding mappings than with the non-corresponding mappings. In addition, fewer errors were committed for the corresponding stressed group than for the corresponding non-stressed group. However, the pattern for the non-corresponding group was reversed, with more errors committed in the non-corresponding stress group than in non-corresponding non-stressed group. Based on these findings, Castaneda and Lipsitt [6] concluded that speed stress facilitated performance when the mapping was simple and matched participants' natural response tendencies. However, when the mapping was non-corresponding and complex, the stress interfered with performance. Although, this experiment showed that speed stress could influence compatible and incompatible mappings in different ways, it did not examine performance with S-R mappings that varied in degree of compatibility.

The purpose of the current study was to investigate participants' performance with learned S-R mappings that vary in degree of compatibility, under conditions with and without speed stress. As in Castaneda and Lipsitt's [6] study, the stressor employed was a time constraint. In addition, we investigated the type of errors made within each mapping condition, to determine if they were random or systematic. A random error is an errant key strike, and a systematic error is selecting the key that spatially corresponds to the stimulus location, since it is the most natural mapping. We hypothesized that participants in the compatible mapping condition would yield the shortest RT and be most resistant to the effect of speed stress. For the incompatible mappings, performance should be better and more resistant to speed stress when the S-R assignments are more systematic than when they are not.

2 Method

2.1 Participants

Forty-eight students (18 males; 30 females; *M* age = 19.5 yrs, *sd* = 2.15 yrs) participated in this study for credits toward a Psychology course requirement. All participants reported having normal or corrected-to-normal vision.

2.2 Apparatus and Stimuli

The stimuli were presented on a personal computer using the Micro Experimental Laboratory (MEL) program. The responses were keypresses on a traditional QWERTY keyboard. This study employed a four-choice reaction task with four different S-R mappings (see Fig. 3). For the compatible mapping, a spatially corresponding mapping was employed, where each stimulus was mapped to its corresponding response. For the mirror-opposite mapping each stimulus was mapped to it mirror-opposite response. For the mixed mapping condition the left pair of stimuli was mapped to their corresponding responses and the right pair to mirror-opposite responses. For the random mapping, there was no rule that guided overall response selection, and participants had to remember each of the four S-R assignments.

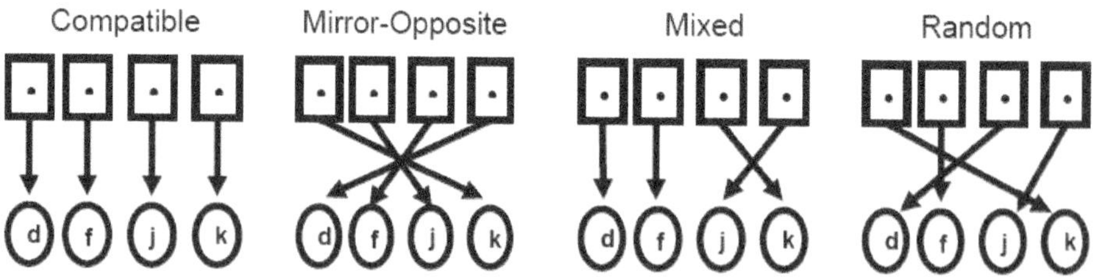

Fig. 3. Illustration of the Four Mapping Conditions Used in the Present Study

Past research has shown that a time constraint decreases performance [7]. The time constraint will induce speed stress because the participants know that they have a short amount of time to respond to the stimulus. To gauge the amount of stress perceived, participants completed a subjective workload rating scale, the NASA Task Load Index (TLX) [8] after each block. At the end of study participants also filled out stress and personality questionnaires. These questionnaires were filled out for exploratory purposes that will not be discussed in the present paper.

2.3 Procedure

The study took place in a laboratory in the Psychology building at California State University Long Beach. Participants were tested individually in a single session lasting approximately two hours. Each participant read and signed a consent form before the experiment began, and were then seated in a room in front of a computer screen.

Participants in all conditions were given written and verbal instructions about the task, assigned mapping, and hand placement on the keyboard. The stimulus was a white circle presented on a black background that appeared in one of four white boxes aligned horizontally. Participants were told that the stimulus would appear in one box and to make the response assigned to the particular stimulus based on their mapping condition. A standard QWERTY keyboard was used for responding, in which the middle and index finger on the left hand were placed on the D and F keys, respectively, and the middle and index finger on the right hand were placed on the K and J keys, respectively.

Once the participant read the instructions, he/she pressed the space bar to begin the first phase of the experiment, which consisted of 1,600 trials. In this learning phase, participants had two seconds to respond to the stimulus. This allowed them enough time to respond without feeling any speed stress. If a response was not made in two seconds or a wrong choice was selected, an error tone sounded, and the next trial was presented. The intertrial interval was set at 1 s. After the first 800 trials, a message informed the participant to see the experimenter to fill out a questionnaire (the NASA TLX) prior to taking a break. After the 5-min break, participants returned to the room to complete the remaining 800 learning trials and to fill out another NASA TLX scale.

After the learning phase was completed, participants took a 10-minute break prior to performing the experimental phase. The experimental phase employed the same mapping as the training phase, but was divided into three parts: 300 trials with the 2-sec response time deadline, 600 trials with the 600-ms deadline, followed by 300 more trials with the 2-sec deadline. The NASA TLX was given at the end of each part along with five-minute breaks. After the experimental phase concluded, participants filled out a demographics questionnaire, a stress inventory, and a personality inventory.

3 Results

3.1 NASA TLX

As a manipulation check, a repeated measure ANOVA was performed on the composite NASA TLX scores. NASA-TLX scores range from 0 (no workload) to 100 (high workload). There was an effect of stress, $F(2,98) = 71.43$, $p < .001$. Workload ratings in the first experimental block ($M = 44$) increased when the 600-ms deadline was imposed ($M = 68$) and returned to pre-stress levels when it was removed ($M = 45$).

3.2 Practice Session

RT less than 200 ms and trials where participants did not respond within the given time frame (2 s or 600 ms) were excluded (less than 2% of all trials). Mean correct RT and ER for each participant were submitted to a 4 (mapping condition: corresponding, mirror-opposite, random, mixed) x 16 (practice blocks of 100 trials) mixed ANOVA, with mapping condition as the between-subjects variable.

Reaction time. The main effects of practice block, $F(15,660) = 13.14$, $p < .001$, and mapping condition $F(3,44) = 19.38$, $p < .001$ were significant. However, these effects were modified by a 2-way interaction between practice block and mapping condition, $F(45,660) = 2.19$, $p < .001$, see Fig. 4. RT decreased across the learning phase, and there was a benefit for compatible over incompatible mappings. However, RT decreased more for the incompatible mapping conditions than for the compatible mapping.

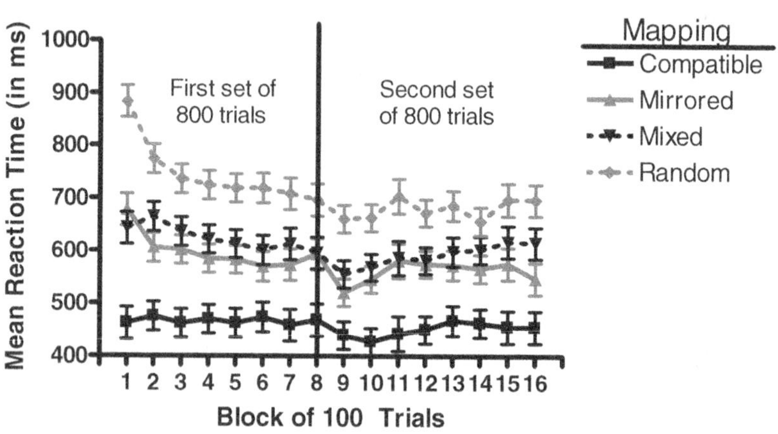

Fig. 4. Mean RT (in ms) for the Four Mapping Conditions as a Function of Practice

Error rate. An error was counted if participants did not respond within the given time frame (2 seconds for the no stress or 600 ms for the stress condition) or if they selected an incorrect response. As with RT, the main effects of block, $F(15,660) = 14.13$, $p < .001$, and mapping condition, $F(3,44) = 12.09$, $p < .001$, were significant. These effects were modified by a 2-way interaction of the two variables, $F(45,660) = 4.17$, $p < .001$. Similar to the RT analysis, the decrease in ER with practice was larger for the incompatible (Mean Difference, *MD*, = 12.59% for random, 5.89% for mirror-opposite and 4.6% for mixed) mapping conditions than for the compatible (*MD* = .26%) one.

3.3 Experimental Session

A 2 (stress, no stress) X 4 (mapping condition: corresponding, mirror-opposite, random and mixed) X 6 (blocks of 100 trials) mixed ANOVA was performed on RT and ER.

Reaction time. The main effect of stress, $F(1,220) = 75.43$, $p < .001$, showed that RT was shorter when the response deadline was applied than when it was not. The main effect of trial block, $F(15,660) = 13.14$, $p < .001$, showed a practice effect. The main effect of mapping, $F(3,44) = 19.38$, $p < .001$, showed that RT was shortest for the compatible mapping, intermediate for the mirror-opposite and mixed mappings, and longest for the random mapping.

Mapping condition entered into two-way interactions with stress, $F(3,220) = 5.53$, $p < .05$, and trial block, $F(15,220) = 7.74$, $p = .001$, and a three-way interaction with stress and block, $F(15,220) = 2.61$, $p = .001$ (see Fig. 5). In general, with no stress, RT decreased more for incompatible mapping conditions than for the compatible mapping condition. When the stress was applied; however, RT was relatively constant across blocks.

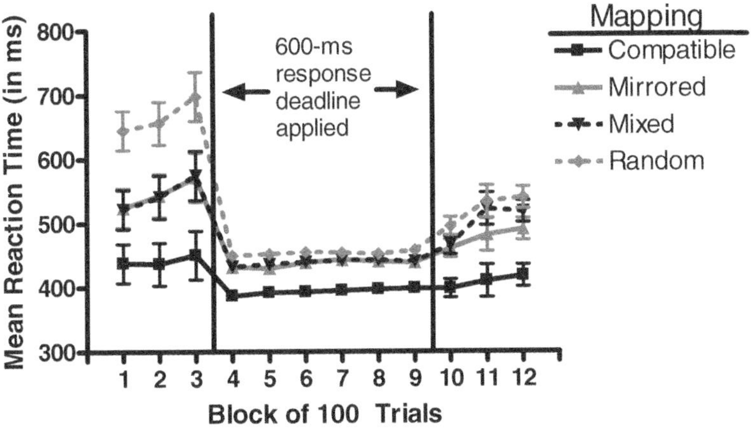

Fig. 5. Mean RT (in ms) for the Four Mapping Conditions as a Function Trial Block and Stress

Errors. Only trials in which participants selected the incorrect response were counted in the error analysis, see Fig. 6. The main effects of stress, $F(1,220) = 58.79$, $p < .001$, block $F(5,220) = 15.99$, $p < .001$, and mapping condition, $F(3,44) = 9.85$, $p < .001$, were significant. The only other significant effect was the interaction of stress and mapping condition, $F(3,220) = 9.91$, $p < .001$. ER between the stress and the no stress blocks increased by the smallest amount for the compatible mapping, by an intermediate amount for the mirror-opposite and mixed mappings, and by the largest amount for the random mapping. That is, when the stress was applied, the ER increased more for the incompatible than for the compatible mappings, especially for the random mapping.

Table 1. Types of Errors (%) Made in No Stress, and Stress Blocks by Mapping Condition

	No Stress blocks		Stress blocks	
Condition	Corresponding	Other (ind*)	Corresponding	Other (ind*)
Corresponding	N/A	47 (15.7)	N/A	53 (17.7)
Mirrored	18	23 (11.5)	29	30 (15)
Random	25	16 (8)	37	22 (11)
Mixed	23	25 (12.5)	42	10 (5)

*Percent of time that one of the possible alternative "other" responses would be selected

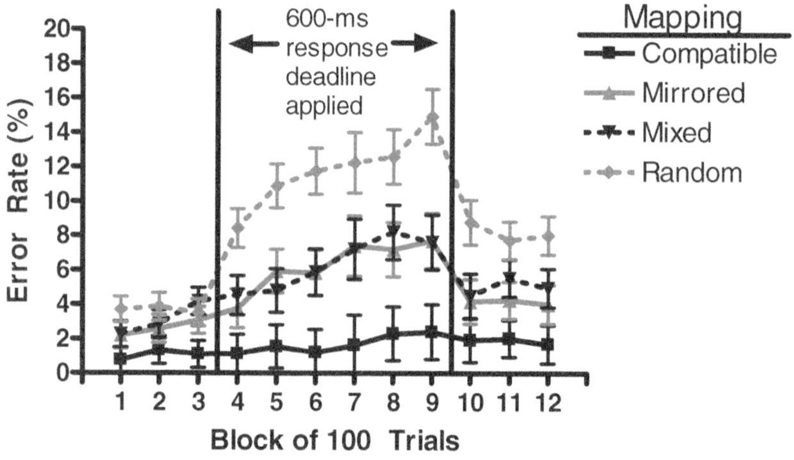

Fig. 6. Mean error rate for the Four Mapping Conditions as a Function Trial Block and Stress

To investigate the types of errors participants made, the percentage of errors involving a corresponding response versus other errors was calculated (see Table 1). In the no stress blocks, participant made corresponding errors more often than other errors. The same pattern was seen in the stress blocks, but the effects were enhanced.

4 Discussion

Previous research has investigated the S-R compatibility effect with mappings that varied in degree of compatibility [2, 5] and with time-constraint stressors [7], but few, if any, have combined both approaches as in the present study. Our results showed that with practice, RT for the incompatible mapping conditions decreased more than for the compatible one. This indicates a larger room for improvement for incompatible mappings [see also 2]. In addition, when a time-constraint stressor was added, the RT for the incompatible mappings decreased to the level of the compatible one. However, the ER data from the stress blocks shows a different story, where the ER increases instead of decreasing. Thus, in the stress blocks, there appears to be a tradeoff between speed and accuracy for the incompatible mapping conditions, but not for the compatible one.

For all incompatible mapping conditions, participants tended to revert back to responding with the compatible response under stress. However, the time-constraint stressor exerts differential impact on performance for the three incompatible mapping conditions. The mirror-opposite and mixed mappings yielded lower RT and ER, and were more resistant to stress, than the random mapping. Moreover, performance may be more impacted by the systematic nature of the S-R mappings rather than by the number of rules employed. Performance for the mirrored and mixed conditions was similar, even though the former conforms to one rule and the latter to two rules.

In general, display-control interfaces should be designed in a manner that is compatible with users' natural response tendencies. However, if a design cannot match users' natural response tendencies, due to design limitation or space constraints, then a systematic mapping rule should govern the mapping of display to control elements in the design. Moreover, when an interface is designed to be used under conditions where rapid responding is required and has a non-corresponding mapping, the interface should be tested under speeded conditions using different display-control configuration.

Acknowledgments. This research was supported in part by a CSULB Provost's Summer Undergraduate Research Award to R. Roberson and a CSULB Psychology Department 22B Assigned Time Award to K.-P. L. Vu.

References

1. Fitts, P.M., Seeger, C.M.: S-R compatibility: Spatial characteristics of stimulus and response codes. J. of Exp. Psychol. 46, 199–210 (1953)
2. Proctor, R.W., Vu, K.L.: Stimulus-response compatibility principles: Data, theory, and application. CRC Press, Boca Raton (2006)
3. Fitts, P.M., Deininger, R.L.: S-R compatibility: Correspondence among paired elements within stimulus and response codes. J. of Exp. Psychol. 48, 483–492 (1954)
4. Kornblum, S., Hasbroucq, T., Osman, A.: Dimensional overlap: Cognitive basis for stimulus-response compatibility–A model and taxonomy. Psychol. Rev. 97, 253–270 (1990)
5. Morin, R.E., Grant, D.: Learning and performance on a key-pressing task as a function of the degree of spatial stimulus-response correspondence. J. of Exp. Psychol. 49, 39–47 (1955)
6. Castaneda, A., Lipsitt, L.P.: Relation of stress and differential position habits to performance in motor learning. J. of Exp. Psychol. 57, 25–30 (1959)
7. Castaneda, A.: Effects of stress on complex learning and performance. J. Exp. Psychol. 52, 9–12 (1956)
8. Hart, S.G., Staveland, L.E.: Development of NASA-TLX (Task Load Index): Results of empirical and theoretical research. In: Hancock, P.A., Meshkati, N. (eds.) (eds), pp. 139–183. North-Holland, Amsterdam (1988)

Application of Population Stereotypes to Computerized Tasks

Jeff Wiebe[1] and Kim-Phuong L. Vu[2]

[1] DISH Network Services, L.L.C.
9601 S. Meridian Blvd., Englewood, CO 80112
[2] Department of Psychology, California State University, Long Beach
1250 N. Bellflower Blvd. Long Beach, CA 90840, USA
{jwiebe_66@yahoo.com, kvu8}@csulb.edu

Abstract. Control-display configurations are the primary means by which people interact with machines and equipment. Regardless of the level of complexity, these configurations should be designed in accordance with the preferred responses of the user population to achieve a successful interaction. These responses are known as population stereotypes. This study examined whether mouse-click response tendencies adhere to Warrick's, scale-side, and clockwise-to-increase principles. More stereotype consistent responses were obtained for dial-turn questions than for mouse-click questions, implying that the movement of the control may be an important factor when predicting users' responses based on these three principles. These findings indicate that care must be taken when applying population stereotypes to programs that use mouse-click responses.

Keywords: human factors, population stereotypes, Warrick's principle, scale-side principle, clockwise-to-increase principle, control-display configurations.

1 Introduction

Control-display configurations are the primary means by which people interact with machines and equipment. Regardless of the level of complexity, these configurations should be designed in accordance with the preferred responses of the user population to achieve a successful interaction. These preferred responses are known as population stereotypes [1, 2]. Control-display configurations designed with established population stereotypes taken into consideration have been found to be advantageous in that they yield higher performance, more consistent performance, and increased resistance to stress and fatigue [3]. In addition to these benefits, confusion on the part of the user is decreased, creating a product that is more intuitive to use [3].

Incorporating population stereotypes into design has proven to be important in real world applications. Loveless [4] gives the example of a case in which a hydraulic press was ruined as the result of an incorrect decision made by the operator. An emergency situation occurred, and the operator moved the lever up to move the ram up because people tend to associate upward movement of a control with upward

M.J. Smith and G. Salvendy (Eds.): Human Interface, Part I, HCII 2009, LNCS 5617, pp. 718–725, 2009.

movement of the corresponding device. However, with this device, upward movement of the control resulted in downward movement of the ram. As a result, the heavy hydraulic press was wrecked. The reason for this error was that the operator was dealing with a control-display configuration that was not compatible with natural control-display relations. Accidents like this could be avoided by using more compatible control-display mappings that adhere to population stereotypes.

Though there are many suggested stereotypes in the field of human-machine interface design, there are several specific stereotypes that, when applicable, have become accepted as prevalent and critical to design requirements [3]. These stereotypes include Warrick's principle, the clockwise-to-increase/anything principle, and the scale-side principle [5]. Warrick's principle states that when the control is located adjacent to the display, the pointer or indicator on the display will move in the same direction as the side of the control that is closest to it (see Fig. 1a). Thus, this principle is not applicable when the control is at the end of the display, or when then display passes through the center of the control.

The clockwise-to-increase principle follows the logic of circular movement in which the clockwise motion of a control will always elicit an increase in a display (see Fig. 1b). This principle does not, however, apply to situations in which the function of the control is to release a material thought to be under pressure or force. For example, increasing water flow is accomplished with a counterclockwise turn of the tap. The dial on a gas cooker is also turned counter-clockwise to increase the flow [5]. The scale-side principle indicates that an operator will assume that a display indicator will move in the same direction as the side of the control dial that is on the same side as the scale on the display (see Fig. 1c).

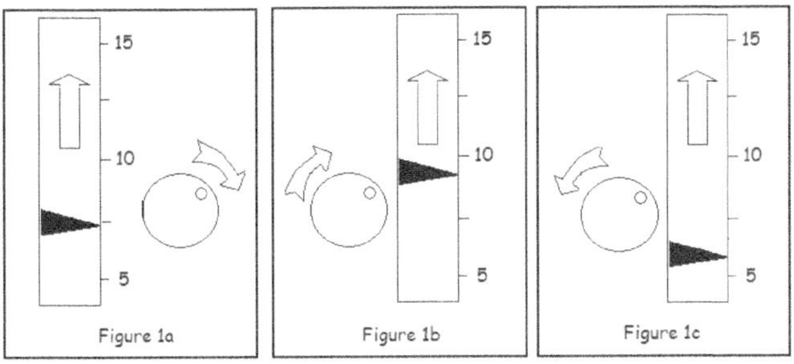

Fig. 1a (left). Warrick's principle predicts a clockwise movement to increase the display setting.
Fig. 1b (middle). The clockwise-to-increase principle predicts a clockwise movement.
Fig. 1c (right). The scale-side principle predicts a counter-clockwise movement.

The ideal display-control arrangement would be one in which all three principles are in agreement. This would provide the least ambiguous setup for users to operate [7]. The dangers of designing a control-display relationship utilizing only one principle are identified in a study by Petropoulos and Brebner [6], who found that, when applicable, Warrick's principle was almost always confirmed. However, the influence of Warrick's principle was significantly weakened when it conflicted with the scale-side principle (see, e.g., Fig. 1b). As implied by these results, it is best to design control-display relationships with agreement among several established principles. When conflict amongst the principles occurs, certain principles have been found to override the others [3]. This finding implies that Warrick's principle should take precedence in the design of control-display arrangements. However, because Warrick's principle relies on the rotary control movement along the side of the indicator, it may not play as big a role for representations on computer displays that do not require dial rotations such as depictions of dials that require mouse-click responses on the dial.

Computerized representations of control-display configurations exist in many types of software, anywhere from online games to aviation simulation training software. To date, though, there have been no studies of which we are aware that examined response tendencies for computer mouse clicks as opposed to clockwise and counter-clockwise dial rotations. For computerized tasks, users are required to make judgments regarding the manipulation of a rotary control to accomplish a given increase or decrease in the corresponding display. This must be done with a click of a computer mouse rather than a rotation of a control. Thus, with the lack of a movement of the control dial to map onto the display movement, it is not clear whether Warricks', the clockwise-to-increase, and the scale-side principles will affect responding in the same way as when operating a control dial. Yet, population stereotypes for mouse-click responses must be established if one wants to assume positive transfer from computerized training to real life control-display configurations.

2 Method

2.1 Participants

Participants consisted of 106 students recruited from Introductory Psychology courses at California State University Long Beach. Approximately 53% of the participants indicated that they use a computer multiple times every day, 25% use the computer about once a day, 13% use the computer 3-5 days per week , 5% use the computer 2-3 days per week, and 2% indicated that they do not use computers frequently at all. Because the study is examining population stereotypes with mouse click responses, participants were asked if they had experience with flight simulation software, where rotary dials are often depicted. 61% of participants indicated having no prior experience with flight simulation software, 23% almost no experience, 9% occasional experience, 2% frequent experience and 2% very frequent experience.

2.2 Stimuli

Based on previous research [3], the stimuli were selected from a set consisting of combinations of 6 control locations, 2 pointer types (directional or neutral), 2 directions of scale (increase or decrease), 2 scale sides (left or right), and 2 directions of movement instruction (increase and decrease). These various factors yielded 128 possible configurations. Because the purpose of the study was to examine whether stereotype agreement exert similar influence on mouse-click responses as dial-turn responses, and not the influence of the 5 aforementioned factors on responding, only configurations for which all stereotypes are applicable were selected. The four conditions of stereotype agreement of interest are as follows: All three principles are in agreement (Category 1), Warrick's and scale-side principles are in agreement (Category 2), Warrick's and clockwise-to-increase principles are in agreement (Category 3), and scale-side and clockwise-to-increase principles are in agreement (Category 4).

Of the 128 original stimuli, there are 48 configurations that fit into Category two, 32 configurations that fit into Category four, and 24 configurations that fit into Categories one and three. To equate the number of stimuli in each condition, only 24 configurations from each category were used. In conditions two and four, there were more than 24 applicable configurations, so 24 configurations were randomly chosen to represent each of those conditions. For each category of stereotype agreement, there were 12 questions asking for dial-turn responses and 12 questions for mouse-click responses. Within each response type, 6 questions were devoted to increasing the display setting and 6 to decreasing the display settings. The control location, pointer type, direction of scale, and scale side were not controlled but selected randomly from the initial set to provide different visual cues for the different display-control configurations.

Each of the 96 control-display configurations were presented along with a question asking users to indicate which response manipulation would they perform to increase (or decrease) the display setting. The questions were presented with multiple-choice response options below the display-control configuration. For the mouse-click questions, participants were instructed to indicate which of the six designated locations on the control must be clicked in order to increase (or decrease) the display. For the dial-turn questions, participants were instructed to indicate which direction the dial should be turned in order to increase (or decrease) the display.

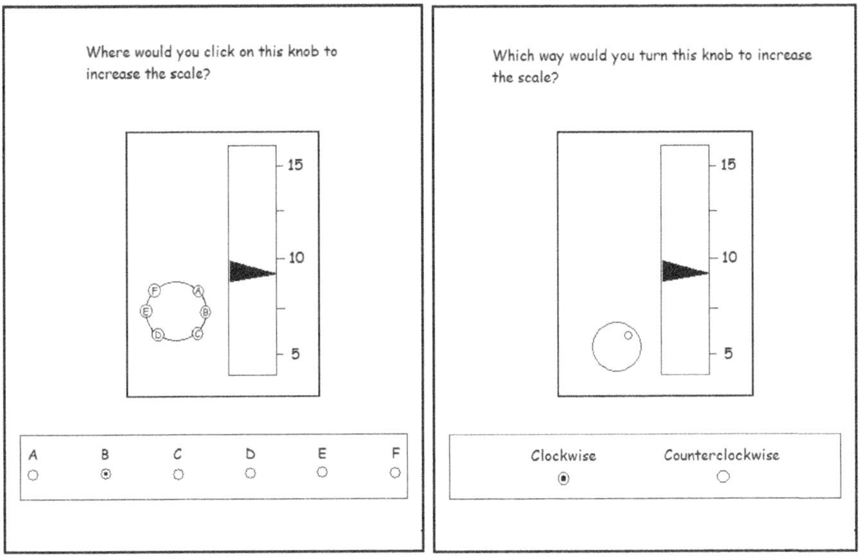

Fig. 2. A sample mouse-click survey question (left panel) and dial-turn question (right panel)

2.3 Apparatus

The experiment was conducted using the online survey program www.surveyconsole.com. The hyperlink to the survey was revealed after the participant signed up for the experiment. Participants were instructed to complete the survey as an individual effort. Images and text were presented to the participant on the screen (see Fig. 2 for samples) and their responses were recorded and downloaded to a database.

2.4 Procedure

Participants accessed the survey through the California State University Long Beach PSY 100 subject pool's Experimetrix program and received additional instructions regarding how to complete it. The instructions indicated that participants should address all control-display configurations as though they were within a computer software application.

Participants were instructed to sit comfortably in front of a computer and to use the mouse with their dominant hand for the experiment. First, the participant completed a short demographic survey providing information such as handedness, gender, and average daily computer usage. In the survey that followed, there were two types of questions. For the mouse-click questions, instructions indicated that there were six possible responses for each question. These six possible responses were represented visually with the letters A, B, C, D, E, and F on the control dial. These letters represented where the participant would click on the dial to make a response. For the dial-turn questions, instructions indicated that there were two possible responses consisting of either clockwise or counter-clockwise. The participant was instructed to select the response they thought would provide the desired increase or decrease in the scale. Half of the participants received all questions relating to dial-turn instructions before receiving questions relating to mouse-click instructions, and the remaining half received all questions relating to mouse-click instructions before receiving questions relating to dial-turn instructions.

There was no enforced time limit for each question or for the completion of the study. Participants were instructed to indicate only one response per item, and that the response should not be changed once selected.

3 Results

3.1 Coding

The questions requesting dial-turns have responses consisting of either clockwise or counter-clockwise. The questions requesting mouse-clicks consisted of six options: A, B, C, D, E, or F, see Fig. 2. These responses were coded into two categories: A, B, and C for clockwise rotation and D, E, and F for counterclockwise rotations. With this coding established, the relationships between the stereotypes and influence of the stereotypes were directly compared for mouse-clicks as opposed to dial turns.

3.2 Analysis

The percentages of subject responses consistent with what was expected based on stereotype agreement were submitted to a 2 (Representation: mouse-click or dial-turn) x 4 [Category: 1 (all three stereotypes agree), 2 (Warrick's and scale-side agree), 3 (Warrick's and clockwise-increase agree), or 4 (scale-side and clockwise-increase agree)] ANOVA, with all factors being within-subjects (see Table 1 for means).

Table 1. Mean percentages of responses consistent with what was predicted by the stereotypes

Stereotypes in Agreement	Mouse-Click questions Stereotype Consistent	Mouse-Click questions Not Stereotype Consistent	Dial-turn questions Stereotype Consistent	Dial-turn questions Not Stereotype Consistent
W, SS, C-I (Category 1)	56.15%	43.85%	63.39%	36.61%
W, SS (Category 2)	42.21%	57.79%	43.85%	56.15%
W, C-I (Category 3)	58.41%	41.59%	59.27%	40.73%
SS, C-I (Category 4)	53.12%	46.88%	58.41%	41.59%

There was a main effect of Representation, $F(1, 106) = 5.75$, $MSE = 526$, $p < .02$. Participants responded as predicted more often to dial-turn questions ($M = 56.23$) than to the mouse-click questions ($M = 52.47$). There was also a main effect of Category, $F(3, 318) = 20.65$, $MSE = 621$, $p < .001$. Bonferonni pairwise analyses revealed that participants responded as predicted more often to questions in Category 1 ($M = 59.77$) than to Category 2 ($M = 43.03$) questions, $p < .001$. Moreover, predicted responses for Category 2 questions were less than Category 3 ($M = 58.84$), $p < .001$, and Category 4 ($M = 55.76$), $p < .001$, responses. No other pairwise comparisons were significant. The interaction between representation and Category was not significant, $F < 1.0$.

4 Discussion

For computerized tasks, users are required to make judgments regarding the manipulation of a rotary control to accomplish a given increase or decrease in the corresponding display with a click of a computer mouse rather than a rotation of a dial. Thus, with the lack of movement of the control dial to map onto the display movement, it is not clear whether Warrick's, the clockwise-to-increase, and the scale-side principles will affect responding in the same way as when operating a control dial. This study examined whether mouse-click and dial-turn response tendencies adhere to these principles for rotary dial control representations.

It was hypothesized that, if participants make a connection between clicking on the left half of a control dial with a counter-clockwise response, and clicking on the right half of a control dial with a clockwise response, that all principles would be predictive of the clockwise versus counterclockwise responses. However, if the movement of the control is necessary in order to predict users' responses, then it was expected that the population stereotypes would have no effect or a reduced effect for predicting mouse-click responses. Overall, results indicated that the movement of the control is an important factor when predicting users' responses based on Warrick's and the scale-side principles, as more stereotype-consistent responses were obtained for dial-turn questions than for mouse-click questions when these stereotypes were applicable. For mouse-click responses, more stereotype-consistent responses occurred when the clockwise-to-increase principle was in effect.

Participants responded in agreement with predicted direction of motion most often when all three stereotypes were in agreement (Category 1 responses) and least often when only Warrick's and the scale-side principles were in agreement (Category 2 responses). Contrary to prior studies, though, Warrick's principle was not stronger than the clockwise-to-increase principle. In other words, Warrick's principle did not dominate when it was in conflict with the clockwise-to-increase principle. The dominance of the clockwise-to-increase principle in the present study over Warrick's and side-scale principles may be related to the participants recruited in the present study being Introductory Psychology students. Hoffman [8] conducted a study examining the presence of population stereotypes for psychology students and engineering students. Differences were found between the two groups. Engineers were more likely to respond in accordance with Warrick's principle, whereas psychology students were more strongly affected by the clockwise-to-right (also known as clockwise to increase) principle. These differences may be due to an increase in mechanical knowledge amongst engineering students.

4.1 Limitations

The participants for this study were recruited from the Psychology 100 subject pool. As noted, there has been evidence that different preferences can be obtained with Psychology and Engineering students [8], so the findings may not generalize to other populations. Moreover, stereotypes can be specific to a variety of sub-populations, such as profession or ethnic culture. Specifying the population is an important step in the design process because there can variations in control-display responses and preferences across cultures [7, 9].

References

1. Courtney, A.J.: The effect of scale0side, indicator type, and control plane on direction-of-turn stereotypes for Hong Kong Chinese subjects. Ergon. 37, 865–877 (1994)
2. Bergum, B.O., Bergum, J.E.: Population Stereotypes: An attempt to measure and define. In: Proceedings of the 25th Annual Meeting of the Human Factors Society, pp. 662–665. Human Factors Society, Santa Monica (1981)

3. Chan, A.S., Shum, V.Y., Law, H.W., Hui, I.K.: Precise effects of control position, indicator type and scale side on human performance. International J. of Adv. Manufacturing Tech. 22, 380–386 (2003)
4. Loveless, N.E.: Direction of motion stereotypes, a review. Ergon. 5, 357–383 (1962)
5. Courtney, A.J.: Hong Kong Chinese direction-of-control stereotypes. Ergon. 37, 417–426 (1994)
6. Petropoulos, H., Brebner, J.: Stereotypes for direction-of-movement of rotary controls associated with linear displays: the effects of scale presence and position, of pointer direction, and distances between the control and the display. Ergon. 24, 43–151 (1981)
7. Wong, C.K., Lyman, J.: American and Japanese control-display stereotypes: possible implications for design of space station systems. In: Proceedings of the 32nd Annual Meeting of the Human Factors Society, pp. 30–34. Human Factors Society, Santa Monica (1988)
8. Hoffman, E.R.: Strength of component principles determining direction of turn stereotypes—linear displays with rotary controls. Ergon. 40, 199–222 (1997)
9. Courtney, A.J.: Control-Display Stereotypes for Multicultural User Systems. IEEE Transactions on Systems, Man, and Cybernetics 22, 681–687 (1992)

Designing Transportation Services Based on HCD

Kiko Yamada-Kawai[1], Naotake Hirasawa[1], Shinya Ogata[1], and Shou Ohtsu[2]

[1] User Experience Research Division
[1,2] CBC at Otaru University of Commerece
3-5-21 Midori, Otaru 047-8501, Japan
{kawai,ogata_shinya,hirasawa,ohtsu}@res.otaru-uc.ac.jp

Abstract. A "Social System" enabled by ICT, such as Intelligent Transportation System affects society itself as well as users' behavior. When designing these systems, an approach of having users or citizens in the center of the system is most crucial to derive planned use of the system. In this study, examples of social systems in the fields of transportation and e-government are first reviewed to reveal the history and issues, and then defined the definition of a "social system." A case example is made on ITS in regard with obtaining higher quality in use by employing Human-centered Design. The necessity of broadening conventional HCD methodologies is further discussed.

Keywords: Human-centered Design, social system, Intelligent Transportation System, ITS, ISO13407, e-government, policy making process.

1 Introduction

Human-centered Design or HCD approach with process models and various methodologies is already well implemented in the fields of product and interactive system design. An international standard on HCD, ISO 13407[1], has been effective and it requires to iterate four steps throughout the system development processes. The four steps denotes to 1) specify the context of use, 2) specify user and organizational requirements, 3) produce design solutions, and 4) evaluate designs against user requirements. These four steps are well used for designing user interface for embedded system development[2], but ISO 13407 does not limit its coverage only to product and UI design. In fact it only states "interactive systems" and leaves room for any kind of "system" that has interaction with users.

When looking at some public infrastructures, the government and related bodies have started working on how to place the users at the center of consideration, in response to the criticism on wasting public resources in constructing roads. This move is motivated to ensure accountability for public works and effectiveness of the system. In the case of road administration, which is operated by national and local governments and recently-privatized highway companies, started annually publishing performance and achievement reports to state and check against their annual goals set by projects[3]. In the case of Intelligent Transportation System or ITS which is operated by the government and highway companies with innovative technologies introduced by automobile and ICT industries, they are trying to shift their motivation from

M.J. Smith and G. Salvendy (Eds.): Human Interface, Part I, HCII 2009, LNCS 5617, pp. 726–735, 2009.

"utilizing innovation by ITS"[4][5] to enabling "human-oriented ITS"[6]. In the case of e-government in Japan, like the case in U.K., the human-centered approach started from the criticism of relatively small number of use compared to the large amount of budget allocated [7][8].

Administrators of those systems seek for measures to meet user needs and improve effectiveness for allocating resources while the requirement for accountability and limitation for the resources are also taken into consideration, but in Japan's case, well established Human-centered design approach is not yet fully utilized.

2 Examples towards User-Centered Policy Making

2.1 e-Government

In the case of U.K., the problem of delay in passport issuance occurred in 1999 when the authorities launched a new computer system to issue more secure passport, resulting in a significant period of delay in processing[9]. National Audit Office analyzed that it was partly caused by lack of communication between staff in operation and the public. Later, Identity & Passport Service (IPS) started issuing post-implementation assessment reports on its five key projects[10]. In the report, IPS declares the cancellation of developing the second-generation online application of passports (EPA2) projects "on the grounds that it was no longer viable due to rising costs and too short a period to recover the investment."

The Government of Japan has been working on e-Japan and u-Japan initiatives[11]. Policies include the development and installation of an e-government system for online applications. In FY2007, while approximately 13,000 out of 14,000 total applications, or about 92%, were able to be processed online; however, merely 20.5% of the actual applications were done by online, which shows some increase from 11.3% in FY2005[12]. However, online passport service has been suspended by the ministry since February 2007[13] because of the extremely small number of users due to the complicated procedures required for the application.

Responding to the inefficient e-Government in Japan, the Cabinet Office inaugurated a committee consisting of two subcommittees to establish usability and security guidelines for developing e-government system[14]. This committee aims to publish two guidelines that shall be followed by e-Government system developers and the clients, in this case the government.

2.2 Road Administrations

The Government of Japan has started reorganizing their activities in terms of "user-oriented" or "user-centered." In the case of infrastructure of road administration, in 2003 the government implemented publishing performance/achievement annual reports by admitting that their management policies had been mostly targeted at meeting the growing demand as well as post-WW2 recovery[3]. The annual reports aim to realize the process of reflecting users' needs to road administration. Executing and disclosing the cost benefit analysis in the project planning process are required since 2003 as well. A web survey on road user's satisfaction has been conducted every year since 2002[15]. Each survey drew approximately 20,000 respondents with a questionnaire of 5-point rating scale on users' satisfaction on roads, such as "most used roads", "highways the respondent used", "traffic flow and congestions on most used roads", and so on. It is the first consecutive survey on road users' satisfaction, however, an in-depth investigation on quality in use or the structure of users' satisfaction was not conducted.

As is the case of the e-government, roads are built and managed mostly by the government and are subject to the criticism of "cost-efficiency." Questions are raised if the government(s) failed to properly forecast demand for the road systems.

Demand forecasting framework has been established in the field of transportation engineering. Author's earlier work was in regard with improving accuracy of micro-economic choice models to describe transportation behavior[16]. Yai et.[17] estimated the effectiveness of ways to describe demand with cases by analyzing the structure of citizens attitudes to demand forecast. Not only these works, numerous models have been introduced, however the demand forecasting framework is closely linked with the policy making process by the government[18] leaving little room to introduce new process or framework.

2.3 Intelligent Transportation System

ITS or Intelligent Transport System is a system that manage transportation behavior of users and roads with innovative ICT. It started as a national project in 1995 in cooperation with ministries, industries and academia[19]. It is illustrated as implementation and penetration of ETC (Electronic Toll Collecting system) terminals, car navigation and VICS (Vehicle Information and Communication System).

A concept of ITS was often discussed in early age of its implementation. But the discussion was limited to the scheme of the system that involved hardware and software, but not users in 1999. Ogawa[20] pointed out the necessity of drawing and sharing the overall picture of the ITS from the perspective of developing a large computer system, employing the failed example of integrating banking systems at the merger of the newly created Mizuho Financial Group, Inc.

The development of ITS was first focused on the implementation of infrastructures, and then moved to the IT utilization of technologies including the dissemination of in-vehicle units[21]. However, the development of ITS in Japan has been affected by legacy systems of each ministry and agency involved, namely the three systems of "VICS: Vehicle Information and Communication System," "ETC: Electronic Toll Collection System" and "AHC: Advanced Cruise-assist Highway System."

Yamada[22] described the three systems, VICS, ETC and AHS in relation with the common platform "SMART WAY." In his description, ITS is composed of the infrastructure "SMARTWAY," automobiles "Smart Cars" and telecommunication tools "Smart Gateway." A figure[23] is referred to describe the structure of ITS, highlighting the SMARTWAY as a common platform for those three systems. The platform consists of three layers: hardware, software and data base.

ITS is a social system with various technologies incorporated. The system has been considered as a bundle of technologies and the services that utilize these technologies. A cross-organizational party, ITS Japan has conducted a survey to draw future scenarios of the implementation of ITS services[24]. The scenarios, written by experts of the technologies, are based on the progress of development of each technology.

Needless to say that ITS is a large system growing further with innovative technologies; there are always users or "human" involved in the system, making the perspective of users inevitable. As such system, ITS has been expected to provide more various services than ETC and VICS that are widely well known by general users. The importance of user-oriented approach to the development of ITS has been repeatedly argued, however very few research can be found.

3 Developing a Social System

3.1 Defining a Social System

Defining a social system aims to apply HCD process as it is applied to UI design and computer system design. Moreover it broadens the application of HCD from merely the UI of software to laws and regulations that control the society, which does not require computer software.

In this context, the feature of a social system can be described as a system:

- which has interaction feature with users,
- with which laws, rules and regulations are closely involved,
- of which stakeholders include public body,
- which provides services to meet needs by society, and
- which consists of two or more sub-systems.

In this definition, social system includes e-government, various transportation systems, and more over laws, rules and regulations provided by public body.

3.2 ITS as a Social System

To follow the definition above, Intelligent Transportation System can be described as:

- users of ITS are citizen drivers,
- relevant legal provisions on traffic, wireless communication, highway involve in ITS,
- four ministries as well as private sector involve in ITS, and
- ITS directly affects citizens' transportation behavior.

In Japan, the most provided services of ITS are ETC and VICS. ETC has been operated only for collecting tolls on expressways, while VICS provides vehicles with real-time traffic information. ETC terminals have been prevailed to be currently 27,000,000 vehicles in total equipped with the terminal, or roughly 50% of registered vehicles. And 5.4 million vehicles/day or 76.4% of all the vehicles that go through the toll gates are ETC-equipped. The government is now extending its service fields from the toll gates of expressways to towns by revising relevant laws that prohibited private sector from installing and using ETC system.

Cases expected by the government[19] are as follows:

- Harbor: to simplify the boarding procedures,
- Shopping malls: to provide shopping information,
- Parking lots: to collect charge,
- Gas stations: to collect gas charge,

- Taxi bay: to allocate cabs, and
- Drive-through restaurants: to collect food and drink charge.

3.3 Development of ITS

These cases mentioned above require transmitting more information to the data center and service providers.

In the case of ITS in Japan, so-called "system architecture" (Figure 1) was published in 1996. This system architecture consists of nine areas of projects; each is composed of "user services" in hierarchical manner. In addition, requirements of each project or the system of ITS are defined as 172 "user services." Here a system concept is described from a bundle of projects, in other words, the projects are not planned under the system concept of ITS.

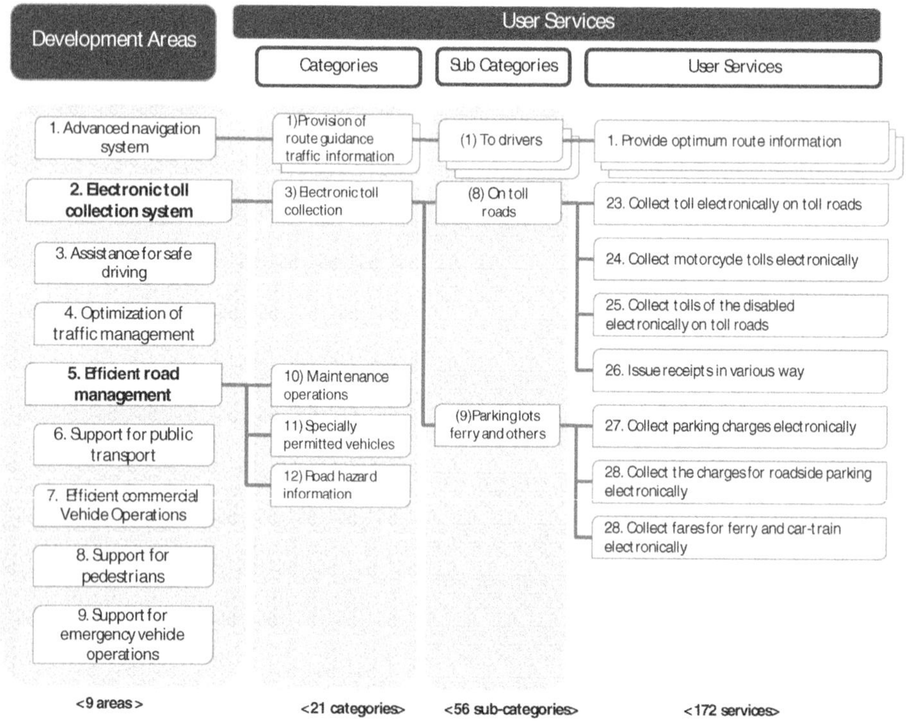

Fig. 1. User services determined in the "System Architecture of ITS"[19] consists of nine areas with 21 categories, 56 sub-categories and 172 user services. In the case of ETC, it consists of seven services, two sub-categories and one category.

Figure 3 denotes a simplified diagram of the development process of a system. To realize a particular system, concept of the system shall be first drawn to enable breaking down to the projects or subsystems. For each project, designing concept, planning, then implementation shall be executed before operation. The control over the system is thus the system concept that is drawn prior to projects configuration.

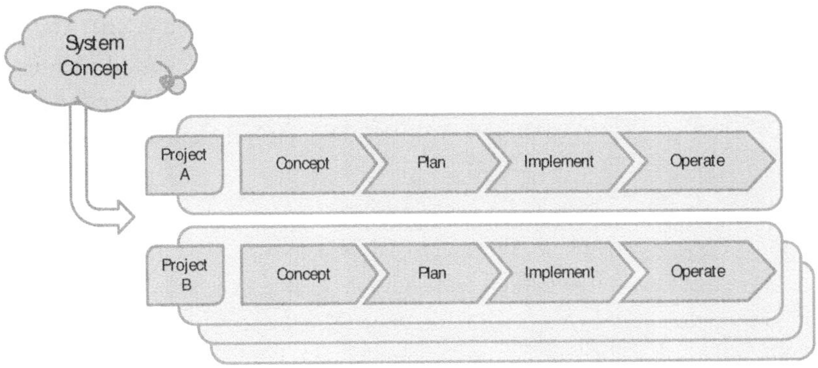

Fig. 2. Development process of a system starts from drawing a system concept. Each project is planned to realize the system concept employing processes starting from planning concept to operation.

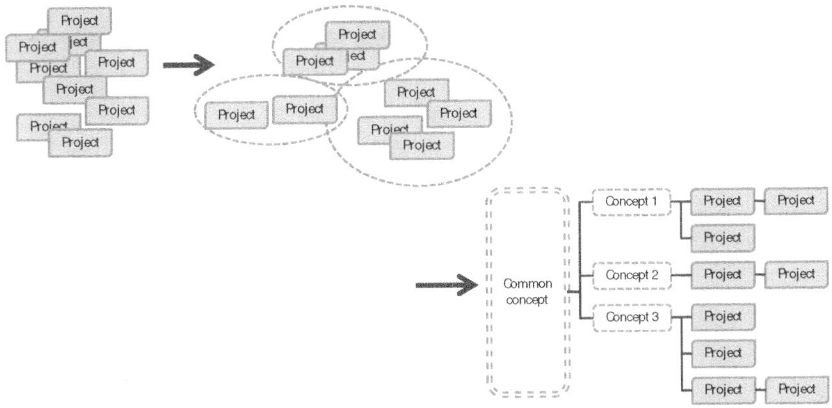

Fig. 3. Actual planning process of a social system may not follow the system development process. Concept of a system is drawn after organizing smaller projects that are already in progress in the administration body such as government.

Social systems such as ITS and e-government in Japan are planned in manners project-dependent (Figure 2) or technical seeds-oriented. This is often the case in developing a comprehensive plan by local governments. The first step taken by the department responsible is to ask all other departments to submit information of relevant projects they are handling. Then projects are organized into categories and concepts are developed to make a framework (Figure 3).

4 HCD in Social Systems

Considering the situation described in the previous chapter, questions arise how HCD approach is able to contribute with the framework of HCD process diagram in ISO13407 (Figure 4). They include:

- What scopes shall be insisted for social systems?
- How can the methodologies be applied?
- What and how the solutions can be evaluated?

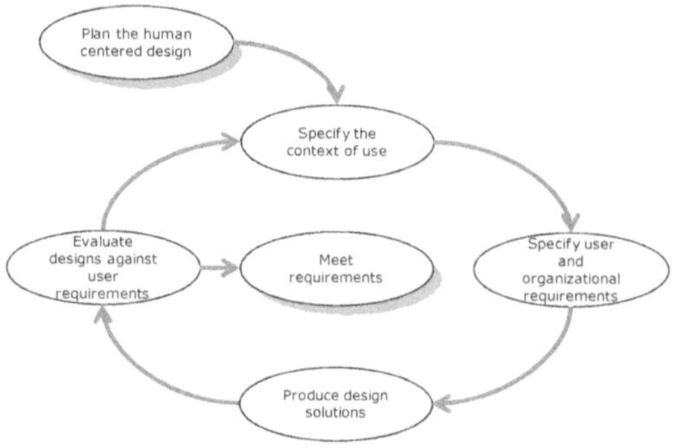

Fig. 4. Simple diagram for Human-centered design stated in ISO 13407[1]. This diagram shall be operated in every process of system development.

Scopes. Examples of social systems in Japan have been targeting at utilizing element technologies or "technical-centered approach." On the other hand, HCD approach focuses on services and experiences provided to users. When implementing HCD approach to social systems, it is indispensable to cope with the two approaches or the two scopes (Figure 5). Adding to the issues on scopes, size of the users shall be in consideration. The users of a social system are expected to vary in a wide range in the population. Most social systems have more target users than that of a product. And as the system provides public services, and thus the system shall not reject any users, which make the target user and user requirements more difficult.

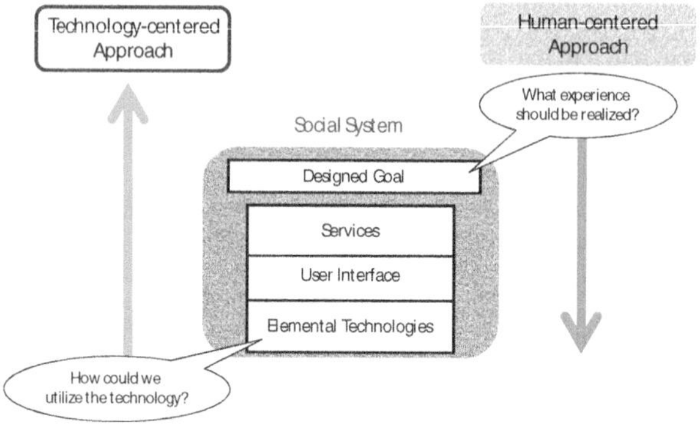

Fig. 5. Approaches to develop a social system can be described as "Technology-centered approach" and "Human-centered approach." Scope of "Technology-centered approach" targets on elemental technologies while HCD on the services provided to users.

HCD methodologies. The conventional methodologies that are often used in HCD process such as persona, scenario and prototype need to be explained in the context of developing social systems. For instance the appropriate number of personas and scenarios will be asked when employed to a social system and to form a prototype as well. Relatively larger size and wider range of user population seemed to be affecting these issues.

Evaluation. The objective of integrating HCD into social system is to have the system more understood and shared among the public as well as researchers of technologies and service operators. As a social system that influences the society, sharing future "context of use" and "quality of use" among stakeholders help improve the system to be more effective and affective. Considerations of the procedures stated in the laws and regulations are crucial.

Evaluation after the implementation of a system, for example, road, bridge, railway, railway station, and transportation policy, is often discussed[25]. However, these discussions are often conducted after or during the period of implementation, and concentrate on a subject such as if the prediction methods and estimates of demand by the project were proper and acceptable, which were expected to affect the society at the planning stage[26].

Incorporating conventional planning and implementing processes into context of V-model (Figure 6) would be essential for further discussions.

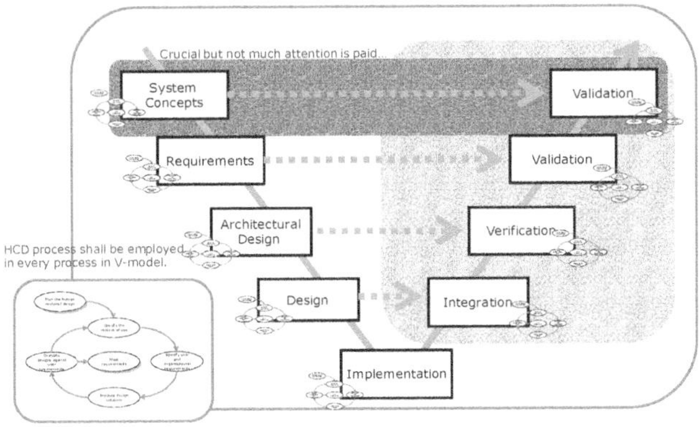

Fig. 6. System development process is described in V-model where system concepts are the basis for validation, requirements for validation and so on. When HCD approach is employed, HCD diagram shall be iterated in every process of V-Model.

5 Conclusions

In this research, the needs for Human-centered design in the field of social systems are explained, focusing on e-government, road administration and ITS in Japan. Then ITS was employed as an example of a social system for HCD to be implemented. In analyzing ITS as a social system, lack of design and examination of the system concepts were pointed out. In terms of a system before physically implementing a system (or a part of a system) derives start operating the system without evaluation.

In other words, conventional HCD processes do not fully cover the area of designing and implementing large social systems such as ITS. Adding to this is that in-depth understanding of policy making procedures usually authorized by laws and regulations, or by government is inevitable to actually adopt the procedure.

Extension of our research focuses on defining the specific difference between product developing processes and social system developing processes, along with extending HCD methodologies for a system with larger user population.

References

1. International Organization for Standardization: ISO 13407 - Human-centered design processes for interactive systems (1999)
2. Ogata, S., Hirasawa, N., Urokohara, H., Yamada-Kawai, K.: A UI Design Methodology Integrated into Embedded System Development Processes, working paper (2009) (in Japanese)
3. Road Bureau, Ministry of Land, Infrastructure and Transport of Japan: Performance Management of Road Administration in Japan, http://www.mlit.go.jp/road/management-e/index.html
4. Road Bureau, Ministry of Land, Infrastructure and Transport of Japan (eds.): ITS Handbook 2001-2002, Highway Industry Development Organization (2001) (in Japanese)
5. ITS Japan: ITS Annual Report 2008, ITS in Japan – For Promoting the Second Stage in Collaborative among Industry, Government and Academia (2008) (in Japanese)
6. The 7th Symposium on ITS (in Japanese), http://www.its-jp.org/topics/topics066.htm
7. The UK National Audit Office: The United Kingdom Passport Agency: The Passport Delays of Summer (1999), http://www.nao.org.uk/pn/9899812.htm
8. The Ministry of Foreign Affairs: Information on On-line Application (in Japanese), http://www.mofa.go.jp/mofaj/procedure/mado/
9. The UK National Audit Office: The United Kingdom Passport Agency: The Passport Delays of Summer (1999), http://www.nao.org.uk/pn/9899812.htm
10. The UK Identity & Passport Service: Report on Key Projects Implemented in 2007 (2008)
11. Cabinet Office of Japan: IT Strategic Headquarters, http://www.kantei.go.jp/foreign/policy/it/index_e.html
12. Ministry of Internal Affairs and Communications of Japan: Development of Online Application Procedures for FY 2007 (2008) (in Japanese)
13. The Ministry of Foreign Affairs: Information on On-line Application (in Japanese), http://www.mofa.go.jp/mofaj/procedure/mado/
14. The Cabinet Secretariat of Japan: Announcement of Inaugurating Committee for e-Government Guidelines (in Japanese), http://www.kantei.go.jp/jp/singi/it2/guide/konkyo.html
15. Road Bureau, Ministry of Land, Infrastructure and Transport of Japan: Users' Satisfaction Survey (in Japanese), http://www.mlit.go.jp/road/ir/ir-user/ir-user.html
16. Morikawa, T., Ben-Akiva, M., Yamada, K.: Forecasting Intercity Rail Ridership Using Revealed Preference and Stated Preference Data, Transportation Research Record, No. 1328, Transportation Research Board, pp.30–35 (1991)
17. Yai, T., Fukuda, D., Nehashi, K.: An Attitudinal Analysis of Citizens' Credibility Gap for Travel Demand Forecasting. Journal of JSCE 62(1), 131–144 (2006) (in Japanese)

18. Road Bureau, Ministry of Land, Infrastructure and Transport of Japan: Traffic Demand Forecasting (in Japanese), http://www.mlit.go.jp/road/ir/ir-data/ir-data.html
19. ITS Handbook 2007-2008, Highway Industry Development Organization, Tokyo (2008)
20. Ogawa, T.: Overall Picture of ITS - System Architecture of ITS, Cover Story - Does ITS Solve the Issues in Transportation? ITS Related Technology Development, Research and Business in Civil Engineering Society. Journal of JSCE 86(11), 7–11 (2002) (in Japanese)
21. Cabinet Secretariat of Japan: Annual Report on ITS for 2008, ITS Japan, pp.17–19 (2008) (in Japanese)
22. Yamada, H.: SMARTWAY and Civil Engineering, Cover Story - Does ITS Solve the Issues in Transportation? ITS Related Technology Development, Research and Business in Civil Engineering Society. Journal of JSCE 86(11), 14–15 (2002) (in Japanese)
23. ITS Handbook 2001-2002, Highway Industry Development Organization, Tokyo (2001)
24. ITS Japan, the Mechanical Social Systems Foundation: Research Report on Developing Scenarios of Introducing ITS Services (2006), http://www.its-jp.org/topics/topics042.htm
25. Special Issue; Current Status and Future Direction on Road Project Evaluation, Traffic Engineering 43(1), 1–55 (2008) (in Japanese)
26. Engineering Affairs Division, Minister's Secretariat, Ministry of Land, Infrastructure and Transport: The Public Project Appraisal at Ministry of Land, Infrastructure and Transport, Government of Japan, Traffic Engineering 43(1), 26–32 (2008) (in Japanese)

Author Index

Abe, Seimei 459
Ablassmeier, Markus 543
Abu Hassan, Lailatul Faizah 375
Ahn, Seongjin 563
Allen, Robert B. 333
Ando, Takahisa 339
Anggreeni, Irene 217
Anse, Michiko 77
Aoki, Yuuki 447
Aoyama, Kouji 181
Arent, Michael 104
Arsic, Marjan 22
Ashley, Jeremy 3
Avilés-López, Edgardo 466

Beresniewicz, John 243
Bezuayehu, Lulit 190
Brangier, Eric 345
Brombacher, Aarnout 227

Chaillou, Christophe 638
Chen, Cheih-Ying 263
Chen, Hong-Sheng 13
Chen, Wei-Lin 355
Chen, Xuejiao 675
Choo, Hyunseung 619
Chung, Jinwook 113, 553, 563
Claros Gómez, Ivan Dario 365
Cobos, Ruth 365
Courage, Catherine 233

Davcev, Danco 22
de Bruin, Renate 227
Degrande, Samuel 638
Demong, Rochin 375
Dinet, Jérôme 345
Doyo, Daisuke 77, 475
Duman, Hakan 384

Eden, Joel 659
Egi, Hironori 31
Ehnes, Jochen 485
Eilrich, Laurent 345

Federoff, Melissa 233
Franklin, Christopher 154
Fuchigami, Miki 495
Funaki, Kentaro 94

Garcia, Fredrick P. 505
García-Macías, J. Antonio 466
Gerken, Peter 154
Ghanea-Hercock, Robert 384
Goldberg, Joseph H. 243
Goode, A. Walkyria 253

Hashimoto, Hideki 404
Healing, Alex 384
Helfman, Jonathan I. 243
Henderson, Sarah 394
Hijikata, Yoshinori 404
Hirasawa, Naotake 726
Hofmann, Paul 104
Horváth, Ádám 40
Hosono, Naotsune 49, 685
Houri, Shotaro 31

Ichino, Junko 59
Ilievski, Dalibor 22
Inoue, Tsuyoshi 94
Isahara, Hitoshi 59
Ishibashi, Masashi 412
Ishii, Gaku 69
Ishizuka, Mitsuru 628
Iwakura, Tomoya 339

Ji, Hong 675

Kamitani, Yukiyasu 534
Kanai, Yuki 515
Kaneko, Takumi 163
Kangyal, András 40
Kato, Hirokazu 648
Kato, Mariko 524
Kato, Yukari 31
Kawakami, Naoki 628
Kawashimo, Takashi 77
Kawata, Masaaki 419
Kim, Hyuncheol 553
Kim, Min-Jeong 87
Koda, Yuka 534
Kojima, Hiroyuki 94

Kondo, Toshiyuki 459
Koo, Jahwan 113
Kumar, Janaki 104
Kumari, S. Anitha 437
Kume, Yasufumi 665

Laquai, Florian 543
Laufer, László 40
Lee, ChiHoon 113
Lee, Doohyung 113
Lee, Wonhyuk 553
Lee, Ying-Jye 263
Leuchter, Sandro 271
Li, Li 675
Lim, Changsun 563
Lin, Chiuhsiang Joe 123
Lin, Shi-Bin 123
Lin, Shiau-Feng 123
Ling, Chen 429
Lu, Yuan 227

Matsumoto, Toshiyuki 131
Miki, Hiroyuki 49, 685
Minakuchi, Mitsuru 570
Mistrzyk, Tomasz 140
Miyamori, Hisashi 570
Miyoshi, Tetsuya 279
Moreno-Llorena, Jaime 365
Mori, Hirohiko 515, 594
Mühlenberg, Dirk 271

Nakagawa, Masaki 31
Nakagawa, Masao 279
Nakagawa, Takashi 648
Nakatani, Yoshio 323, 412, 702
Nakayasu, Hidetoshi 279
Nalluru, Sumanth 333
Nelson, Deborah L. 693
Neviarouskaya, Alena 628
Ngugi, Benjamin 171
Nishida, Shogo 404, 611, 648
Nobutani, Naoya 702
Nozawa, Takayuki 459

Ocenasek, Pavel 149, 579, 587
Ogata, Shinya 726
Ogawa, Katsuhiko 419
Ohtsu, Shou 726
Oka, Makoto 515, 594
Okada, Akira 495
Okamoto, Seishi 339
Okamoto, Shu 611
Ozawa, Naoki 594

Park, Jonghun 87
Pham, Ngoc Duy 619
Plénacoste, Patricia 638
Poitschke, Tony 543
Polowinski, Jan 601
Prasad, E.V. 437
Prendinger, Helmut 628

Rabas, Audrey 709
Redenius, Alexander 140
Regli, Susan Harkness 154
Rice, Sean 190
Rigoll, Gerhard 543
Roberson, Richard 709
Rosenberg, Dan 104
Rottermann, Andrea 287

Saiwaki, Naoki 524, 534
Sakata, Nobuchika 611
Sakoda, Masayuki 447
Sakuma, Masatake 69
Sakurai, Akito 594
Samoylov, Alexei 154
Sanghavi, Mansi 171
Sato, Noriaki 77
Schlegel, Thomas 296
Seo, Doowon 553
Seo, Jee Seob 665
Seta, Kazuhisa 447
Shi-jie, Wang 205
Shida, Keisuke 131
Shiga, Satoko 339
Shin, Inyoung 619
Shirai, Hideki 131
Soderston, Candace 306
Song, Guangfeng 429
Srinivas, Mettu 437
Srinivasan, Ananth 394
Stibler, Kathleen 154
Supreethi, K. Pujari 437
Suzuki, Michio 49
Sveda, Miroslav 149

Tabe, Tsutomu 77, 163, 197
Tachi, Susumu 628
Taguchi, Yumiko 163, 197
Takeuchi, Kazuhiro 59

Tamura, Hiroshi 495
Taniguchi, Maki 534
Tarasewich, Peter 171
Teng, Wei-Guang 355
Terada, Tatsuya 31
Tomita, Yukata 49
Tremoulet, Patrice 154
Tsetserukou, Dzmitry 628
Tsuji, Hiroshi 447
Tsukahara, Wataru 31
Tuan Besar, Tuan Badrol Hisham 375

Ugai, Takanori 181

van der Voort, Mascha 217
Vandromme, Johann 638
Vaughan, Misha W. 3
Venkatacharya, Patañjali 190
Vu, Kim-Phuong L. 287, 505, 693, 709, 718

Wada, Yoshihisa 447
Wichansky, Anna M. 316
Wiebe, Jeff 718
Wu, Fong-Gong 263

Xinmin, Zhang 205

Yamada-Kawai, Kiko 726
Yamaguchi, Yoshihisa 648
Yamamoto, Sakae 685
Yang, Chih-Wei 123
Yatsuzuka, Daisuke 197
Ying-chun, Wang 205
Yonezawa, Yuri 323
Yong-xian, Liu 205
Yuan, Xiaowei 675

Zulkifli, Zulhaimi 375

GPSR Compliance

The European Union's (EU) General Product Safety Regulation (GPSR) is a set of rules that requires consumer products to be safe and our obligations to ensure this.

If you have any concerns about our products, you can contact us on ProductSafety@springernature.com

In case Publisher is established outside the EU, the EU authorized representative is:

Springer Nature Customer Service Center GmbH
Europaplatz 3
69115 Heidelberg, Germany

Batch number: 10844766

Printed by Printforce, the Netherlands